NETWORK PROTECTION & AUTOMATION GUIDE

Protective Relays, Measurement & Control

GRID

we are shaping the future ALSTOM

Network Protection & Automation Guide

NETWORK PROTECTION & AUTOMATION GUIDE, EDITION MAY 2011

Previously called Protective Relays Application Guide

First Edition	June 1966
Reprinted	January 1967
	August 1968
	November 1970
	September 1971
	February 1973
	January 1974
Second Edition	March 1975
Reprinted	November 1977
	December 1979
	November 1982
	October 1983
	October 1985
Third Edition	June 1987
Reprinted	September 1990
	March 1995

Network Protection & Automation Guide

First Edition	July 2002

ISBN: 978-0-9568678-0-3

Published by Alstom Grid
Alstom Grid Worldwide Contact Centre
www.alstom.com/grid/contactcentre

Tel: +44(0) 1785 250 070
www.alstom.com/grid

CONTENTS

Chapter 1
Introduction

Since 1966, the Network Protection and Automation Guide (formerly the Protective Relays Application Guide) has been the definitive reference textbook for protection engineers and technicians. For 2011, Alstom has capitalised on its pool of experts at the St Leonards Centre of Excellence in Stafford UK to launch a new edition.

New chapters treat topics such as system integrity protection and remedial action schemes, phasor measurements and wide area schemes. The digital substation, including IEC 61850, Ethernet station bus, GOOSE, process bus, and precision time synchronising is also detailed. Advancements in protection and control application engineering have assisted the authors in exploring and integrating the new techniques and philosophies in this edition, whilst retaining vendor-independence – as we continue to deliver the genuine, impartial, reference textbook.

This book is a précis of the Application and Protection of Power Systems (APPS) training course, an intensive programme, which Alstom (and its predecessor companies at Stafford) has been running for over 50 years. This course, by the ingenuity and dedication of the trainers, is vibrant and evolving. As APPS progresses, the Network Protection and Automation Guide advances too, whilst never losing sight of the key basic principles and concepts. Beginners and experts alike will each feel satisfied in their search for relaying, measurement, communication and control knowledge.

In the list opposite, we name a mix of new authors for this edition, and key historical figures at Stafford who have contributed significantly to the advancement of APPS and NPAG, and hence the quality and integrity of our book. We sincerely hope that this book assists your navigation through a challenging and rewarding career in electrical power engineering. Protection and control has long been termed an art, rather than a precise science - this book offers a mix of both.

We acknowledge and thank Alstom colleagues in the wider Alstom Grid and Alstom Power organisations for photographs used within this book.

Michael Bamber
Michael Bergstrom
Andrew Darby
Susan Darby
Graham Elliott
Peter Harding
Graeme Lloyd
Alan Marshall
Allen Millard
Andrew Myatt
Philip Newman
Anthony Perks
Steve Pickering
Stephen Potts
Simon Richards
Jack Royle
Peter Rush
Brendan Smith
Mark Stockton
Paul Wilkinson
Alan Wixon
John Wright

Chapter 2
Fundamentals of Protection Practice

2.1 INTRODUCTION

The purpose of an electrical power system is to generate and supply electrical energy to consumers. The system should be designed to deliver this energy both reliably and economically. Frequent or prolonged power outages result in severe disruption to the normal routine of modern society, which is demanding ever-increasing reliability and security of supply. As the requirements of reliability and economy are largely opposed, power system design is inevitably a compromise.

A power system comprises many diverse items of equipment. Figure 2.1 illustrates the complexity of a typical power station Figure 2.2 shows a hypothetical power system.

Figure 2.1: Modern power station

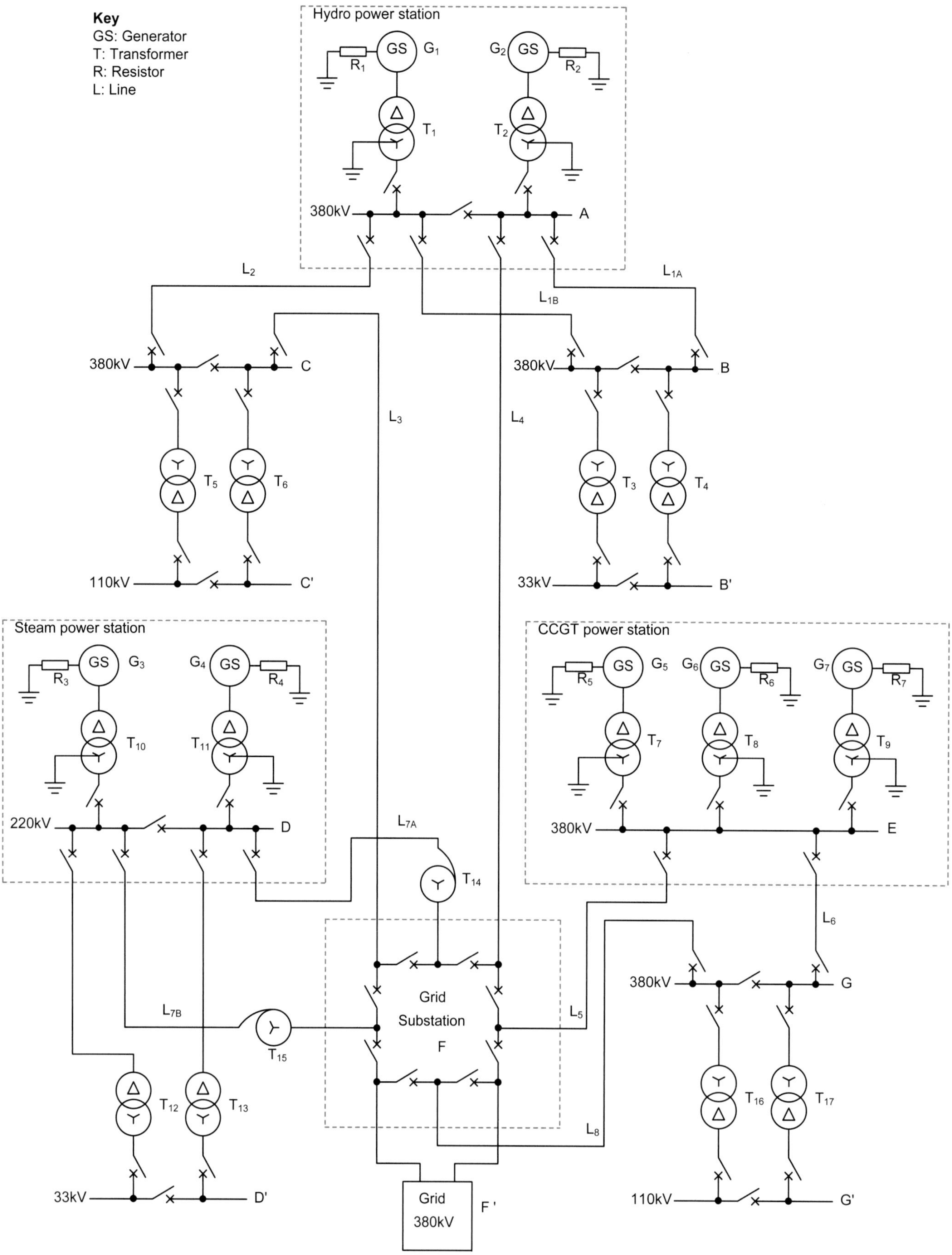

Figure 2.2: Example power system

Figure 2.3: Onset of an overhead line fault

Many items of equipment are very expensive, and so the complete power system represents a very large capital investment. To maximise the return on this outlay, the system must be utilised as much as possible within the applicable constraints of security and reliability of supply. More fundamental, however, is that the power system should operate in a safe manner at all times. No matter how well designed, faults will always occur on a power system, and these faults may represent a risk to life and/or property. Figure 2.3 shows the onset of a fault on an overhead line. The destructive power of a fault arc carrying a high current is very large; it can burn through copper conductors or weld together core laminations in a transformer or machine in a very short time – some tens or hundreds of milliseconds. Even away from the fault arc itself, heavy fault currents can cause damage to plant if they continue for more than a few seconds. The provision of adequate protection to detect and disconnect elements of the power system in the event of fault is therefore an integral part of power system design. Only by doing this can the objectives of the power system be met and the investment protected. Figure 2.4 provides an illustration of the consequences of failure to provide adequate protection. This shows the importance of protection systems within the electrical power system and of the responsibility vested in the Protection Engineer.

Figure 2.4: Possible consequence of inadequate protection

2.2 PROTECTION EQUIPMENT

The definitions that follow are generally used in relation to power system protection:

- Protection System: a complete arrangement of protection equipment and other devices required to achieve a specified function based on a protection principle (IEC 60255-20)
- Protection Equipment: a collection of protection devices (relays, fuses, etc.). Excluded are devices such as Current Transformers (CTs), Circuit Breakers (CBs) and contactors
- Protection Scheme: a collection of protection equipment providing a defined function and including all equipment required to make the scheme work (i.e. relays, CTs, CBs, batteries, etc.)

In order to fulfil the requirements of protection with the optimum speed for the many different configurations, operating conditions and construction features of power systems, it has been necessary to develop many types of relay that respond to various functions of the power system quantities. For example, simple observation of the fault current magnitude may be sufficient in some cases but measurement of power or impedance may be necessary in others. Relays frequently measure complex functions of the system quantities, which may only be readily expressible by mathematical or graphical means.

Relays may be classified according to the technology used:

- electromechanical
- static
- digital
- numerical

The different types have varying capabilities, according to the limitations of the technology used. They are described in more detail in Chapter 7.

In many cases, it is not feasible to protect against all hazards with a relay that responds to a single power system quantity. An arrangement using several quantities may be required. In this case, either several relays, each responding to a single quantity, or, more commonly, a single relay containing several elements, each responding independently to a different quantity may be used.

The terminology used in describing protection systems and relays is provided in Appendix A. Different symbols for describing relay functions in diagrams of protection schemes are used, the three most common methods (IEC, IEEE/ANSI and IEC61850) are provided in Appendix B.

2.3 ZONES OF PROTECTION

To limit the extent of the power system that is disconnected when a fault occurs, protection is arranged in zones. The principle is shown in Figure 2.5. Ideally, the zones of protection should overlap, so that no part of the power system is left unprotected. This is shown in Figure 2.6(a), the circuit breaker being included in both zones.

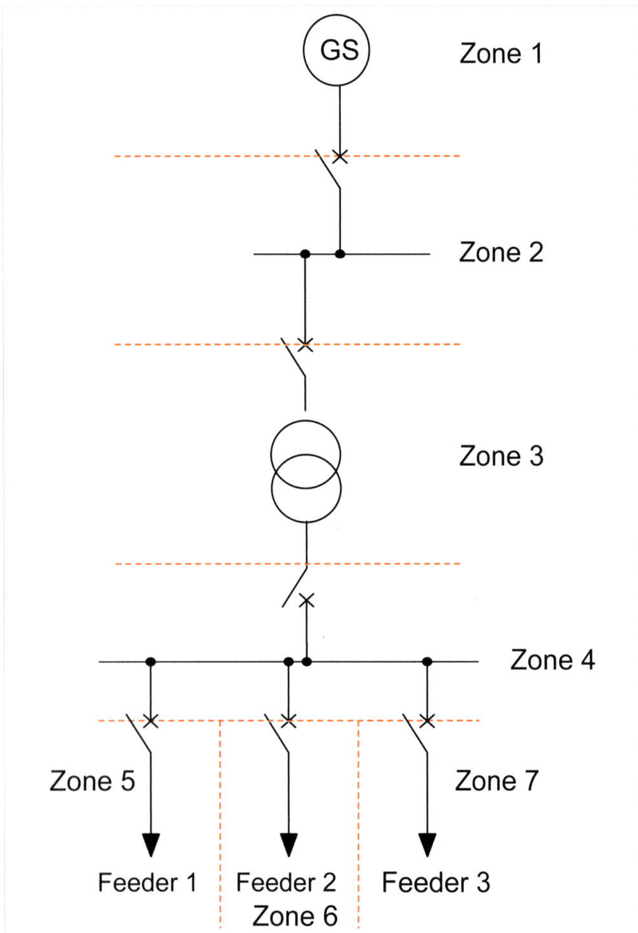

Figure 2.5: Division of power systems into protection zones

For practical physical and economic reasons, this ideal is not always achieved, accommodation for current transformers being in some cases available only on one side of the circuit breakers, as shown in Figure 2.6(b). In this example, the section between the current transformers and the circuit breaker A is not completely protected against faults. A fault at F would cause the busbar protection to operate and open the circuit breaker but the fault may continue to be fed through the feeder. If the feeder protection is of the type that responds only to faults within its own zone (see section 2.5.2), it would not operate, since the fault is outside its zone. This problem is dealt with by intertripping or some form of zone extension, to ensure that the remote end of the feeder is also tripped. These methods are explained extensively in chapters 11 and 12.

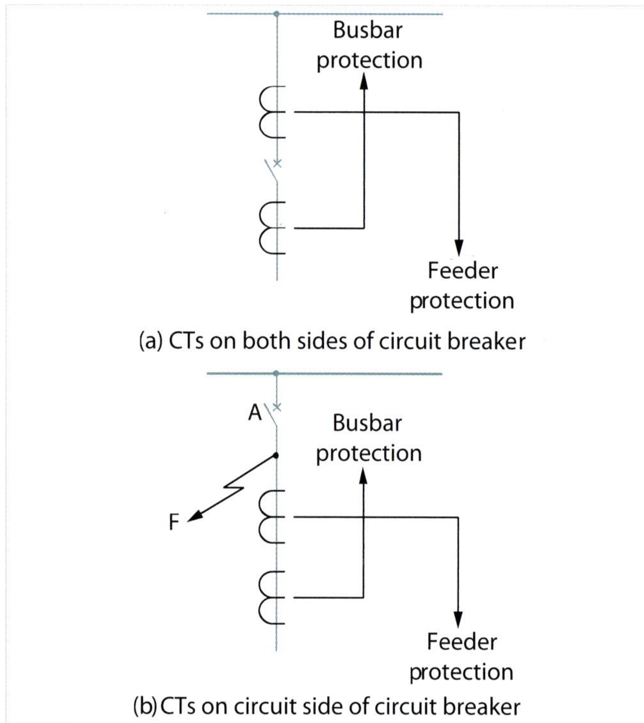

Figure 2.6: CT locations

The point of connection of the protection with the power system usually defines the zone and corresponds to the location of the current transformers. Unit type protection results in the boundary being a clearly defined closed loop. Figure 2.7 shows a typical arrangement of overlapping zones.

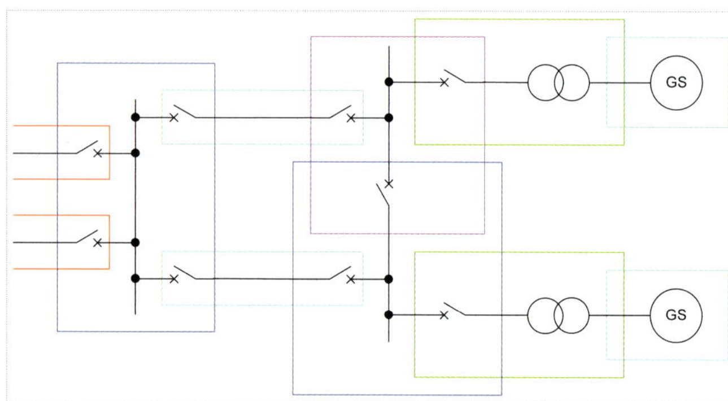

Figure 2.7: Overlapping zones of protection systems

Alternatively, the zone may be unrestricted; the start will be defined but the extent (or 'reach') will depend on measurement of the system quantities and will therefore be subject to variation, owing to changes in system conditions and measurement errors.

2.4 RELIABILITY

The need for a high degree of reliability has already been discussed briefly. Reliability is dependent on the following factors:

- incorrect design/settings
- incorrect installation/testing
- deterioration in service

2.4.1 Design

The design of a protection scheme is of paramount importance. This is to ensure that the system will operate under all required conditions, and refrain from operating when so required. This includes being restrained from operating for faults external to the zone being protected, where necessary. Due consideration must be given to the nature, frequency and duration of faults likely to be experienced, all relevant parameters of the power system and the type of protection equipment used. Of course, the design of the protection equipment used in the scheme is just as important. No amount of effort at this stage can make up for the use of badly designed protection equipment.

2.4.2 Settings

It is essential to ensure that settings are chosen for protection relays and systems which take into account the parameters of the primary system, including fault and load levels, and dynamic performance requirements, etc. The characteristics of power systems change with time, due to changes in loads, location, type and amount of generation, etc. Therefore, setting values of relays may need to be checked at suitable intervals to ensure that they are still appropriate. Otherwise, unwanted operation or failure to operate when required may occur.

2.4.3 Installation

The need for correct installation of protection systems is obvious, but the complexity of the interconnections of many systems and their relationship to the remainder of the system may make checking the installation difficult. Site testing is therefore necessary. Since it will be difficult to reproduce all fault conditions correctly, these tests must be directed towards proving the installation itself. At the installation stage, the tests should prove the correctness of the connections, relay settings, and freedom from damage of the equipment. No attempt should be made to 'type test' the equipment or to establish complex aspects of its technical performance.

2.4.4 Testing

Testing should cover all aspects of the protection scheme, reproducing operational and environmental conditions as closely as possible. Type testing of protection equipment to recognised standards is carried out during design and production and this fulfils many of these requirements, but it will still be necessary to test the complete protection scheme (relays, current transformers and other ancillary items). The tests must realistically simulate fault conditions.

2.4.5 Deterioration in Service

Subsequent to installation, deterioration of equipment will take place and may eventually interfere with correct functioning. For example: contacts may become rough or burnt due to frequent operation, or tarnished due to atmospheric contamination, coils and other circuits may become open-circuited, electronic components and auxiliary devices may fail, and mechanical parts may seize up.

The time between operations of protection relays may be years rather than days. During this period, defects may have developed unnoticed until revealed by the failure of the protection to respond to a power system fault. For this reason, relays should be periodically tested in order to check they are functioning correctly.

Testing should preferably be carried out without disturbing permanent connections. This can be achieved by the provision of test blocks or switches.

The quality of testing personnel is an essential feature when assessing reliability and considering means for improvement. Staff must be technically competent and adequately trained, as well as self-disciplined to proceed in a systematic manner to achieve final acceptance.

Important circuits that are especially vulnerable can be provided with continuous electrical supervision; such arrangements are commonly applied to circuit breaker trip circuits and to pilot circuits. Modern digital and numerical relays usually incorporate self-testing/diagnostic facilities to assist in the detection of failures. With these types of relay, it may be possible to arrange for such failures to be automatically reported by communications link to a remote operations centre, so that appropriate action may be taken to ensure continued safe operation of that part of the power system and arrangements made for investigation and correction of the fault.

2.4.6 Protection Performance

Protection system performance is frequently assessed statistically. For this purpose each system fault is classed as an incident and only those that are cleared by the tripping of the correct circuit breakers are classed as 'correct'. The percentage of correct clearances can then be determined.

This principle of assessment gives an accurate evaluation of the protection of the system as a whole, but it is severe in its judgement of relay performance. Many relays are called into operation for each system fault, and all must behave correctly for a correct clearance to be recorded.

Complete reliability is unlikely ever to be achieved by further improvements in construction. If the level of reliability achieved by a single device is not acceptable, improvement can be achieved through redundancy, e.g. duplication of equipment. Two complete, independent, main protection systems are provided, and arranged so that either by itself can carry out the required function. If the probability of each equipment failing is x/unit, the resultant probability of both equipments failing simultaneously, allowing for redundancy, is x^2. Where x is small the resultant risk (x^2) may be negligible.

Where multiple protection systems are used, the tripping signal can be provided in a number of different ways. The two most common methods are:

- all protection systems must operate for a tripping operation to occur (e.g. 'two-out-of-two' arrangement)
- only one protection system need operate to cause a trip (e.g. 'one-out-of two' arrangement)

The former method guards against false tripping due to maloperation of a protection system. The latter method guards against failure of one of the protection systems to operate, due to a fault. Occasionally, three main protection systems are provided, configure in a 'two-out-of three' tripping arrangement, to provide both reliability of tripping, and security against unwanted tripping.

It has long been the practice to apply duplicate protection systems to busbars, both being required to operate to complete a tripping operation. Loss of a busbar may cause widespread loss of supply, which is clearly undesirable. In other cases, important circuits are provided with duplicate main protection systems, either being able to trip independently. On critical circuits, use may also be made of a digital fault simulator to model the relevant section of the power system and check the performance of the relays used.

2.5 SELECTIVITY

When a fault occurs, the protection scheme is required to trip only those circuit breakers whose operation is required to isolate the fault. This property of selective tripping is also called 'discrimination' and is achieved by two general methods.

2.5.1 Time Grading

Protection systems in successive zones are arranged to operate in times that are graded through the sequence of protection devices so that only those relevant to the faulty zone complete the tripping function. The others make incomplete operations and then reset. The speed of response will often depend on the severity of the fault, and will generally be slower than for a unit system.

2.5.2 Unit Systems

It is possible to design protection systems that respond only to fault conditions occurring within a clearly defined zone. This type of protection system is known as 'unit protection'. Certain types of unit protection are known by specific names, e.g. restricted earth fault and differential protection. Unit protection can be applied throughout a power system and, since it does not involve time grading, it is relatively fast in operation. The speed of response is substantially independent of fault severity.

Unit protection usually involves comparison of quantities at the boundaries of the protected zone as defined by the locations of the current transformers. This comparison may be achieved by direct hard-wired connections or may be achieved via a communications link. However certain protection systems derive their 'restricted' property from the configuration of the power system and may be classed as unit protection, e.g. earth fault protection applied to the high voltage delta winding of a power transformer. Whichever method is used, it must be kept in mind that selectivity is not merely a matter of relay design. It also depends on the correct co-ordination of current transformers and relays with a suitable choice of relay settings, taking into account the possible range of such variables as fault currents, maximum load current, system impedances and other related factors, where appropriate.

2.6 STABILITY

The term 'stability' is usually associated with unit protection schemes and refers to the ability of the protection system to remain unaffected by conditions external to the protected zone, for example through-load current and faults external to the protected zone.

2.7 SPEED

The function of protection systems is to isolate faults on the power system as rapidly as possible. One of the main objectives is to safeguard continuity of supply by removing each disturbance before it leads to widespread loss of synchronism and consequent collapse of the power system.

As the loading on a power system increases, the phase shift between voltages at different busbars on the system also increases, and therefore so does the probability that synchronism will be lost when the system is disturbed by a fault. The shorter the time a fault is allowed to remain in the system, the greater can be the loading of the system. Figure 2.8 shows typical relations between system loading and fault clearance times for various types of fault. It will be noted that phase faults have a more marked effect on the stability of the system than a simple earth fault and therefore require faster clearance.

System stability is not, however, the only consideration. Rapid operation of protection ensures minimisation of the equipment damage caused by the fault. The damaging energy liberated during a fault is proportional to the time that the fault is present, thus it is important that the protection operate as quickly as possible. Speed of operation must be weighed against economy, however. Distribution circuits, which do not normally require a fast fault clearance, are usually protected by time-graded systems. On the other hand, generating plant and EHV systems require protection systems of the highest attainable speed and reliability, therefore unit systems are normal practice.

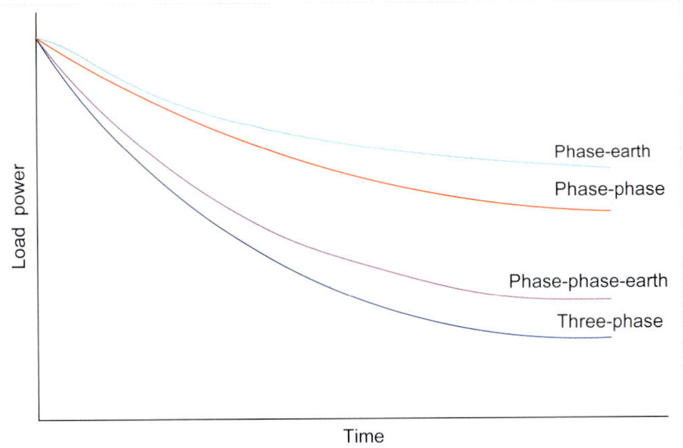

Figure 2.8: Typical power/time relationship for various fault types

2.8 SENSITIVITY

Sensitivity is a term frequently used when referring to the minimum operating level (current, voltage, power etc.) of relays or complete protection schemes. Relays or protection schemes are said to be sensitive if their primary operating parameters are low.

With older electromechanical relays, sensitivity was considered in terms of the measuring movement and was measured in terms of its volt-ampere consumption to cause operation. With modern digital and numerical relays the achievable sensitivity is seldom limited by the device design but by its application and associated current and voltage transformer parameters.

2.9 PRIMARY AND BACK-UP PROTECTION

The reliability of a power system has been discussed earlier, including the use of more than one primary (or 'main') protection system operating in parallel. In the event of failure or non-availability of the primary protection some other means of ensuring that the fault is isolated must be provided. These secondary systems are referred to as 'back-up protection schemes'.

Back-up protection may be considered as either being 'local' or 'remote'. Local back-up protection is achieved by protection that detects an un-cleared primary system fault at its own location, which then trips its own circuit breakers; e.g. time graded overcurrent relays. Remote back-up protection is provided by protection that detects an un-cleared primary system fault at a remote location and then issues a trip command to the relevant relay; e.g. the second or third zones of a distance relay. In both cases the main and back-up protection systems detect a fault simultaneously, operation of the back-up protection being delayed to ensure that the primary protection clears the fault if possible. Normally being unit protection, operation of the primary protection will be fast and will result in the minimum amount of the power system being disconnected. Operation of the back-up protection will be, of necessity, slower and will result in a greater proportion of the primary system being lost.

The extent and type of back-up protection applied will naturally be related to the failure risks and relative economic importance of the system. For distribution systems where fault clearance times are not critical, time delayed remote back-up protection may be adequate. For EHV systems, where system stability is at risk unless a fault is cleared quickly, multiple primary protection systems, operating in parallel and possibly of different types (e.g. distance and unit protection), will be used to ensure fast and reliable tripping. Back-up overcurrent protection may then optionally be applied to ensure that two separate protection systems are available during maintenance of one of the primary protection systems.

Back-up protection systems should, ideally, be completely separate from the primary systems. For example, a circuit protected by a current differential relay may also have time-graded overcurrent and earth fault relays added to provide circuit breaker tripping in the event of failure of the main primary unit protection. Ideally, to maintain complete redundancy, all system components would be duplicated. This ideal is rarely attained in practice. The following compromises are typical:

- Separate current transformers or duplicated secondary cores are often provided. This practice is becoming less common at distribution voltage levels if digital or numerical relays are used, because the extremely low input burden of these relay types allows relays to share a single CT
- Voltage transformers are not duplicated because of cost and space considerations. Each protection relay supply is separately protected (fuse or MCB) and continuously supervised to ensure security of the VT output. An alarm is given on failure of the supply and where appropriate, unwanted operation of the protection is prevented
- Trip power supplies to the two protection types should be separately protected (fuse or MCB). Duplication of

tripping batteries and of circuit breaker trip coils may be provided. Trip circuits should be continuously supervised.

- It is desirable that the main and back-up protections (or duplicate main protections) should operate on different principles, so that unusual events that may cause failure of the one will be less likely to affect the other

Digital and numerical relays may incorporate suitable back-up protection functions (e.g. a distance relay may also incorporate time-delayed overcurrent protection elements as well). A reduction in the hardware required to provide back-up protection is obtained, but at the risk that a common relay element failure (e.g. the power supply) will result in simultaneous loss of both main and back-up protection. The acceptability of this situation must be evaluated on a case-by-case basis.

2.10 RELAY OUTPUT DEVICES

In order to perform their intended function, relays must be fitted with some means of providing the various output signals required. Contacts of various types usually fulfil this function.

2.10.1 Contact Systems

Relays may be fitted with a variety of contact systems for providing electrical outputs for tripping and remote indication purposes. The most common types encountered are as follows:

- Self-reset: The contacts remain in the operated condition only while the controlling quantity is applied, returning to their original condition when it is removed
- Hand or electrical reset: These contacts remain in the operated condition after the controlling quantity has been removed.

The majority of protection relay elements have self-reset contact systems, which, if so desired, can be modified to provide hand reset output contacts by the use of auxiliary elements. Hand or electrically reset relays are used when it is necessary to maintain a signal or lockout condition. Contacts are shown on diagrams in the position corresponding to the un-operated or de-energised condition, regardless of the continuous service condition of the equipment. For example, an undervoltage relay, which is continually energised in normal circumstances, would still be shown in the de-energised condition.

A 'make' contact is one that is normally open, but closes on energisation. A 'break' contact is one that is normally closed, but opens on energisation. Examples of these conventions and variations are shown in Figure 2.9.

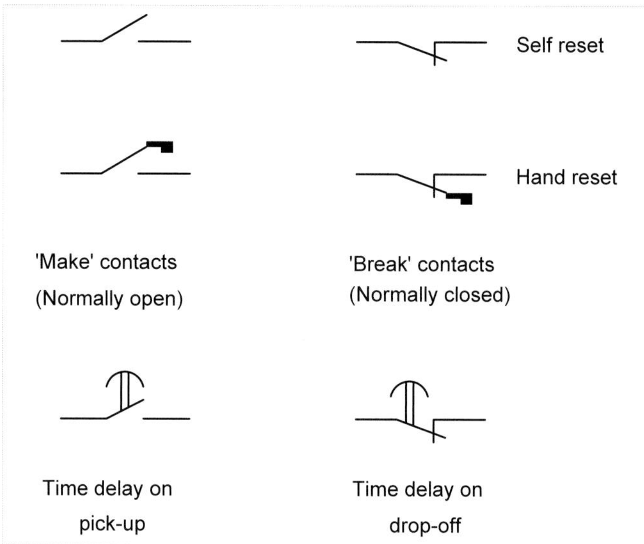

Figure 2.9: Contact types

A 'changeover' contact generally has three terminals; a common, a make output, and a break output. The user connects to the common and other appropriate terminal for the logic sense required.

A protection relay is usually required to trip a circuit breaker, the tripping mechanism of which may be a solenoid with a plunger acting directly on the mechanism latch or an electrically operated valve. The power required by the trip coil of the circuit breaker may range from up to 50 W for a small 'distribution' circuit breaker, to 3 kW for a large, EHV circuit breaker.

The relay may energise the tripping coil directly, or through the agency of another multi-contact auxiliary relay, depending on the required tripping power.

The basic trip circuit is simple, being made up of a hand-trip control switch and the contacts of the protection relays in parallel to energise the trip coil from a battery, through a normally open auxiliary switch operated by the circuit breaker. This auxiliary switch is needed to open the trip circuit when the circuit breaker opens since the protection relay contacts will usually be quite incapable of performing the interrupting duty. The auxiliary switch will be adjusted to close as early as possible in the closing stroke, to make the protection effective in case the breaker is being closed on to a fault.

Where multiple output contacts or contacts with appreciable current-carrying capacity are required, interposing contactor type elements will normally be used.

Modern numerical devices may offer static contacts as an ordering option. Semiconductor devices such as IGBT transistors may be used instead of, or in parallel with, conventional relay output contacts to boost:

- The speed of the 'make' (typically 100μs time to make is achieved)

- Interrupting duty (allowing the contacts to break trip coil current.

In general, static, digital and numerical relays have discrete measuring and tripping circuits, or modules. The functioning of the measuring modules is independent of operation of the tripping modules. Such a relay is equivalent to a sensitive electromechanical relay with a tripping contactor, so that the number or rating of outputs has no more significance than the fact that they have been provided.

For larger switchgear installations the tripping power requirement of each circuit breaker is considerable, and further, two or more breakers may have to be tripped by one protection system. There may also be remote signalling requirements, interlocking with other functions (for example auto-reclosing arrangements), and other control functions to be performed. These various operations may then be carried out by multi-contact tripping relays, which are energised by the protection relays and provide the necessary number of adequately rated output contacts.

2.10.2 Operation Indicators

Protection systems are invariably provided with indicating devices, called 'flags', or 'targets', as a guide for operations personnel. Not every relay will have one, as indicators are arranged to operate only if a trip operation is initiated. Indicators, with very few exceptions, are bi-stable devices, and may be either mechanical or electrical. A mechanical indicator consists of a small shutter that is released by the protection relay movement to expose the indicator pattern.

Electrical indicators may be simple attracted armature elements, where operation of the armature releases a shutter to expose an indicator as above, or indicator lights (usually light emitting diodes). For the latter, some kind of memory circuit is provided to ensure that the indicator remains lit after the initiating event has passed.

The introduction of numerical relays has greatly increased the number of LED indicators (including tri-state LEDs) to enhance the indicative information available to the operator. In addition, LCD text or graphical displays, which mimic the electrical system provide more in-depth information to the operator.

2.11 TRIPPING CIRCUITS

There are three main circuits in use for circuit breaker tripping:

- series sealing
- shunt reinforcing
- shunt reinforcement with sealing

These are illustrated in Figure 2.10.

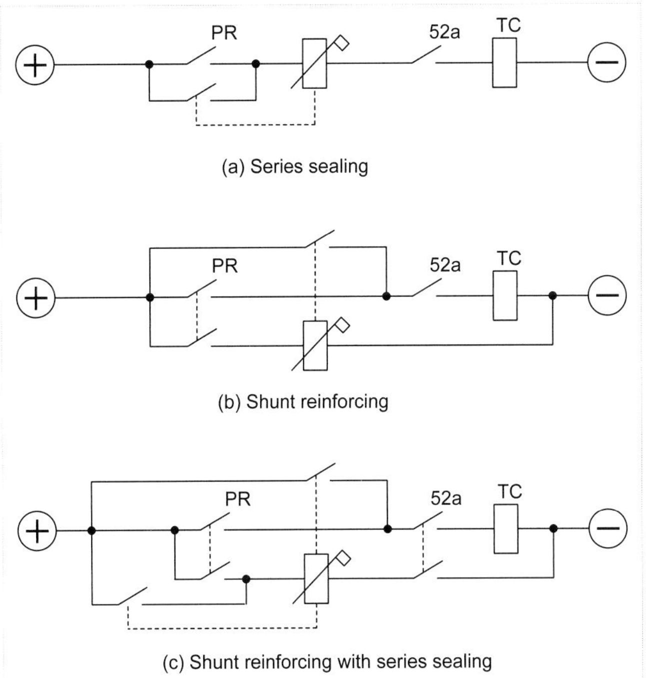

Figure 2.10: Typical relay tripping circuits

For electromechanical relays, electrically operated indicators, actuated after the main contacts have closed, avoid imposing an additional friction load on the measuring element, which would be a serious handicap for certain types. Care must be taken with directly operated indicators to line up their operation with the closure of the main contacts. The indicator must have operated by the time the contacts make, but must not have done so more than marginally earlier. This is to stop indication occurring when the tripping operation has not been completed.

With modern digital and numerical relays, the use of various alternative methods of providing trip circuit functions is largely obsolete. Auxiliary miniature contactors are provided within the relay to provide output contact functions and the operation of these contactors is independent of the measuring system, as mentioned previously. The making current of the relay output contacts and the need to avoid these contacts breaking the trip coil current largely dictates circuit breaker trip coil arrangements. Comments on the various means of providing tripping arrangements are, however, included below as a historical reference applicable to earlier electromechanical relay designs.

2.11.1 Series sealing

The coil of the series contactor carries the trip current initiated by the protection relay, and the contactor closes a contact in parallel with the protection relay contact. This closure relieves the protection relay contact of further duty and keeps the tripping circuit securely closed, even if chatter occurs at the main contact. The total tripping time is not affected, and the indicator does not operate until current is actually flowing through the trip coil.

The main disadvantage of this method is that such series elements must have their coils matched with the trip circuit with which they are associated.

The coil of these contacts must be of low impedance, with about 5% of the trip supply voltage being dropped across them.

When used in association with high-speed trip relays, which usually interrupt their own coil current, the auxiliary elements must be fast enough to operate and release the flag before their coil current is cut off. This may pose a problem in design if a variable number of auxiliary elements (for different phases and so on) may be required to operate in parallel to energise a common tripping relay.

2.11.2 Shunt reinforcing

Here the sensitive contacts are arranged to trip the circuit breaker and simultaneously to energise the auxiliary unit, which then reinforces the contact that is energising the trip coil.

Two contacts are required on the protection relay, since it is not permissible to energise the trip coil and the reinforcing contactor in parallel. If this were done, and more than one protection relay were connected to trip the same circuit breaker, all the auxiliary relays would be energised in parallel for each relay operation and the indication would be confused.

The duplicate main contacts are frequently provided as a three-point arrangement to reduce the number of contact fingers.

2.11.3 Shunt reinforcement with sealing

This is a development of the shunt reinforcing circuit to make it applicable to situations where there is a possibility of contact bounce for any reason.

Using the shunt reinforcing system under these circumstances would result in chattering on the auxiliary unit, and the possible burning out of the contacts, not only of the sensitive element but also of the auxiliary unit. The chattering would end only when the circuit breaker had finally tripped. The effect of contact bounce is countered by means of a further contact on the auxiliary unit connected as a retaining contact.

This means that provision must be made for releasing the sealing circuit when tripping is complete; this is a disadvantage, because it is sometimes inconvenient to find a suitable contact to use for this purpose.

2.12 TRIP CIRCUIT SUPERVISION

The trip circuit includes the protection relay and other components, such as fuses, links, relay contacts, auxiliary switch contacts, etc., and in some cases through a considerable amount of circuit wiring with intermediate terminal boards. These interconnections, coupled with the importance of the circuit, result in a requirement in many cases to monitor the integrity of the circuit. This is known as trip circuit supervision. The simplest arrangement contains a healthy trip lamp or LED, as shown in Figure 2.11(a).

The resistance in series with the lamp prevents the breaker being tripped by an internal short circuit caused by failure of the lamp. This provides supervision while the circuit breaker is closed; a simple extension gives pre-closing supervision.

Figure 2.11(b) shows how, the addition of a normally closed auxiliary switch and a resistance unit can provide supervision while the breaker is both open and closed.

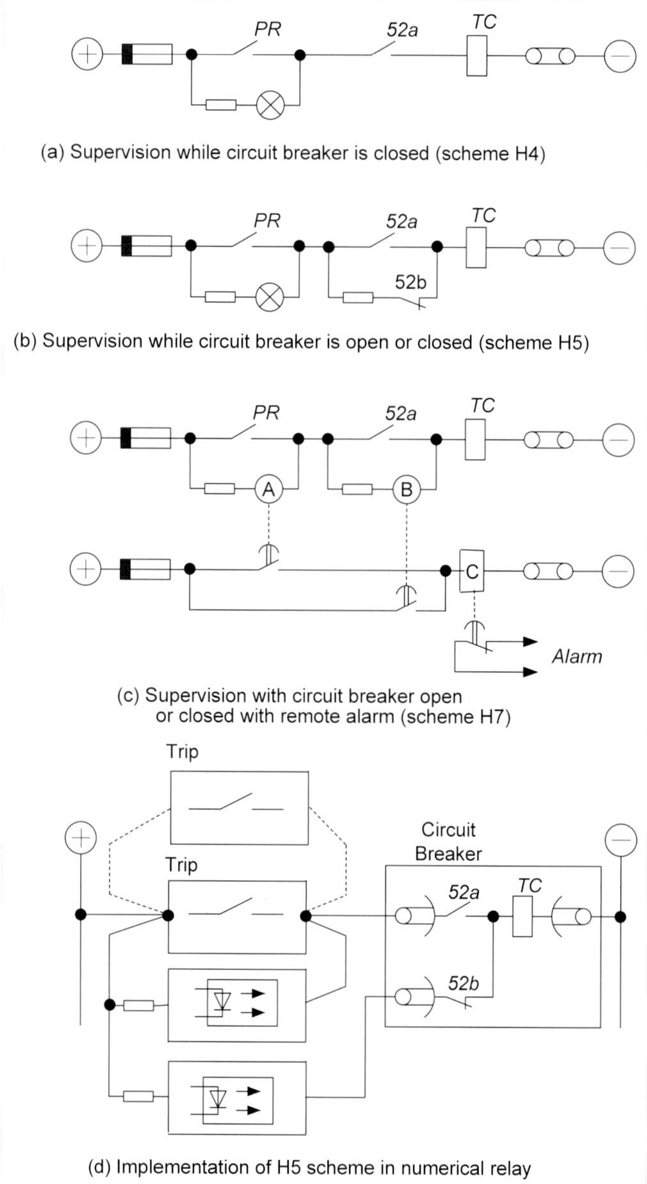

Figure 2.11: Trip circuit supervision circuit

In either case, the addition of a normally open push-button contact in series with the lamp will make the supervision indication available only when required.

Schemes using a lamp to indicate continuity are suitable for locally controlled installations, but when control is exercised from a distance it is necessary to use a relay system. Figure

2.11(c) illustrates such a scheme, which is applicable wherever a remote signal is required.

With the circuit healthy either or both of relays A and B are operated and energise relay C. Both A and B must reset to allow C to drop-off. Relays A, B and C are time delayed to prevent spurious alarms during tripping or closing operations. The resistors are mounted separately from the relays and their values are chosen such that if any one component is inadvertently short-circuited, tripping will not take place.

The alarm supply should be independent of the tripping supply so that indication will be obtained in case of failure of the tripping supply.

The above schemes are commonly known as the H4, H5 and H7 schemes, arising from the diagram references of the utility specification in which they originally appeared. Figure 2.11(d) shows implementation of scheme H5 using the facilities of a modern numerical relay. Remote indication is achieved through use of programmable logic and additional auxiliary outputs available in the protection relay.

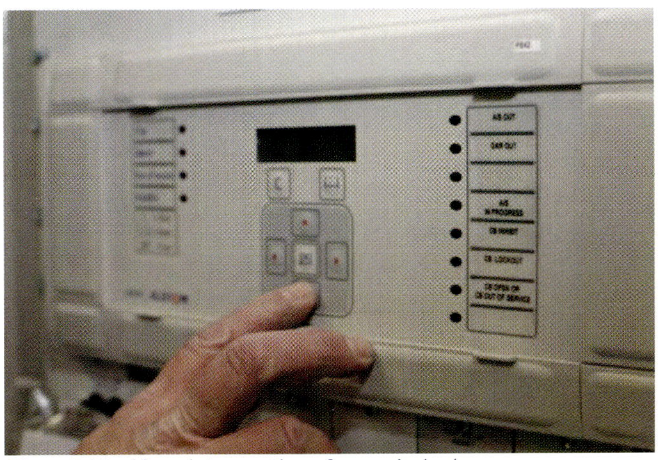

Figure 2.12: Menu interrogation of numerical relays

Chapter 3
Fundamental Theory

3.1 INTRODUCTION

The Protection Engineer is concerned with limiting the effects of disturbances in a power system. These disturbances, if allowed to persist, may damage plant and interrupt the supply of electric energy. They are described as faults (short and open circuits) or power swings, and result from natural hazards (for instance lightning), plant failure or human error.

To facilitate rapid removal of a disturbance from a power system, the system is divided into 'protection zones'. Protection relays monitor the system quantities (current and voltage) appearing in these zones. If a fault occurs inside a zone, the relays operate to isolate the zone from the remainder of the power system.

The operating characteristic of a protection relay depends on the energising quantities fed to it such as current or voltage, or various combinations of these two quantities, and on the manner in which the relay is designed to respond to this information. For example, a directional relay characteristic would be obtained by designing the relay to compare the phase angle between voltage and current at the relaying point. An impedance-measuring characteristic, on the other hand, would be obtained by designing the relay to divide voltage by current. Many other more complex relay characteristics may be obtained by supplying various combinations of current and voltage to the relay. Relays may also be designed to respond to other system quantities such as frequency and power.

In order to apply protection relays, it is usually necessary to know the limiting values of current and voltage, and their relative phase displacement at the relay location for various types of short circuit and their position in the system. This normally requires some system analysis for faults occurring at various points in the system.

The main components that make up a power system are generating sources, transmission and distribution networks, and loads. Many transmission and distribution circuits radiate from key points in the system and these circuits are controlled by circuit breakers. For the purpose of analysis, the power system is treated as a network of circuit elements contained in branches radiating from nodes to form closed loops or meshes. The system variables are current and voltage, and in steady state analysis, they are regarded as time varying quantities at a single and constant frequency. The network parameters are impedance and admittance; these are assumed to be linear, bilateral (independent of current direction) and constant for a constant frequency.

3.2 VECTOR ALGEBRA

A vector represents a quantity in both magnitude and direction. In Figure 3.1 the vector OP has a magnitude $|Z|$ at an angle $|\theta|$ with the reference axis OX:

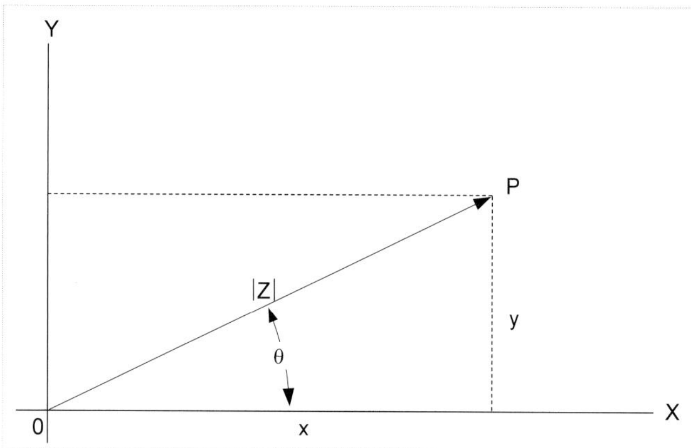

Figure 3.1: Vector OP

The quantity may be resolved into two components at right angles to each other, in this case x and y. The magnitude or scalar value of vector Z is known as the modulus $|Z|$, whilst the angle θ is the argument and is written as arg $\overline{Z}$. The conventional method of expressing a vector $\overline{Z}$ is to write $|Z|\angle\theta$. This form completely specifies a vector for graphical representation or conversion into other forms.

It is useful to express vectors algebraically. In Figure 3.1, the vector Z is the resultant of adding x in the x-direction and y in the y direction. This may be written as:

$$\overline{Z} = x + jy$$

Equation 3.1

where the operator j indicates that the component y is perpendicular to component x. The axis OC is the 'real' axis, and the vertical axis OY is called the 'imaginary' axis.

If a quantity is considered positive in one direction, and its direction is reversed, it becomes a negative quantity. Hence if the value +1 has its direction reversed (shifted by 180°), it becomes -1.

The operator j rotates a vector anti-clockwise through 90°. If a vector is made to rotate anti-clockwise through 180°, then the operator j has performed its function twice, and since the vector has reversed its sense, then:

$$j^2 = -1 \text{ giving } j = \sqrt{-1}$$

The representation of a vector quantity algebraically in terms of its rectangular co-ordinates is called a 'complex quantity'. Therefore, $x + jy$ is a complex quantity and is the rectangular form of the vector $|Z|\angle\theta$ where:

$$|Z| = \sqrt{(x^2 + y^2)}$$

$$\theta = \tan^{-1} \frac{y}{x}$$

$$x = |Z|\cos\theta$$

$$y = |Z|\sin\theta$$

Equation 3.2

From Equations 3.1 and 3.2:

$$\overline{Z} = |Z|(\cos\theta + j\sin\theta)$$

Equation 3.3

and since $\cos\theta$ and $\sin\theta$ may be expressed in exponential form by the identities:

$$\sin\theta = \frac{e^{j\theta} - e^{-j\theta}}{2j}$$

$$\cos\theta = \frac{e^{j\theta} + e^{-j\theta}}{2j}$$

By expanding and simplifying this equation, it follows that:

$$\overline{Z} = |Z|e^{j\theta}$$

Equation 3.4

A vector may therefore be represented both trigonometrically and exponentially.

3.3 MANIPULATION OF COMPLEX QUANTITIES

In the above section, we have shown that complex quantities may be represented in any of the four co-ordinate systems given below:

- Polar $Z\angle\theta$
- Rectangular $x+jy$
- Trigonometric $|Z|(\cos\theta+j\sin\theta)$
- Exponential $|Z|e^{j\theta}$

The modulus $|Z|$ and the argument θ are together known as 'polar co-ordinates', and x and y are described as 'cartesian co-ordinates'. Conversion between co-ordinate systems is easily achieved. As the operator j obeys the ordinary laws of algebra, complex quantities in rectangular form can be manipulated algebraically, as can be seen by the following:

$$\overline{Z_1} + \overline{Z_2} = (x_1 + x_2) + j(y_1 + y_2)$$

Equation 3.5

$$\overline{Z_1} - \overline{Z_2} = (x_1 - x_2) + j(y_1 - y_2)$$

Equation 3.6

$$\overline{Z_1}\,\overline{Z_2} = |Z_1||Z_2|\angle\theta_1 + \theta_2$$

$$\frac{\overline{Z_1}}{\overline{Z_2}} = \frac{|Z_1|}{|Z_2|}\angle\theta_1 - \theta_2$$

Equation 3.7

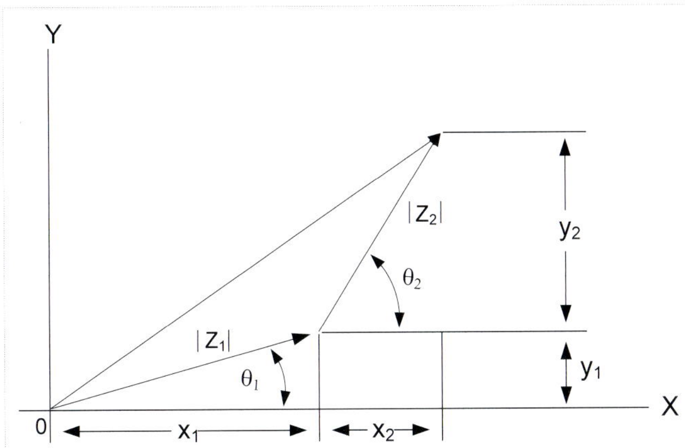

Figure 3.2: Addition of vectors

3.3.1 Complex Variables

In the diagrams shown in Figure 3.1 and Figure 3.2, we have shown that complex variables are represented on a simple chart, where the y-axis is perpendicular to the x-axis displaced by 90°. The argument, or angle of incidence with respect to the x-axis is also known as the phase. So a quantity lying along the y-axis is 90° out of phase with a quantity lying along the x-axis. Because we are rotating in an anti-clockwise direction, the quantity y is then leading the quantity x by 90°.

If we take a simple sinusoidal waveform of frequency f, where one cycle of the waveform (360°) takes T seconds (1/f) we can see that the phase angle can be represented by the angular velocity multiplied by the time taken to reach that angle. At this point, we should move away from using degrees to measure angles and move over to radians. There are 2π radians in one cycle so:

- 360° = 2π radians
- 270° = 3π/2 radians
- 180° = π radians
- 90° = π/2 radians

Thus

$$|Z|\angle\theta = |Z|(\cos\theta + j\sin\theta) = |Z|(\cos\omega t + j\sin\omega t)$$

where θ is the angle moved in time t, of a quantity moving at ω radians per second.

Some complex quantities vary with time. When manipulating such variables in differential equations it is useful to express the complex quantity in exponential form.

3.3.2 The 'a' Operator

We have seen that the mathematical operator j rotates a quantity anti-clockwise through 90°. Another useful operator is one which moves a quantity anti-clockwise through 120°, commonly represented by the symbol 'a'.

Using De Moivre's theorem, the nth root of unity is given by solving the expression.

$$1^{1/n} = (\cos 2\pi m + j\sin 2\pi m)^{1/n}$$

where m is any integer. Hence:

$$1^{1/n} = \cos\frac{2\pi m}{n} + j\sin\frac{2\pi m}{n}$$

where m has values 1, 2, 3, ... (n - 1)

From the above expression 'j' is found to be the 4th root and 'a' the 3rd root of unity, as they have four and three distinct values respectively. Below are some useful functions of the 'a' operator.

$$a = -\frac{1}{2} + j\frac{\sqrt{3}}{2} = e^{j\frac{2\pi}{3}}$$

$$a^2 = -\frac{1}{2} - j\frac{\sqrt{3}}{2} = e^{j\frac{4\pi}{3}}$$

$$1 = 1 + j0 = e^{j0}$$

$$1 + a + a^2 = 0$$

$$1 - a = j\sqrt{3}a^2$$

$$1 - a^2 = -j\sqrt{3}a$$

$$a - a^2 = j\sqrt{3}$$

$$j = \frac{a - a^2}{\sqrt{3}}$$

3.4 CIRCUIT QUANTITIES AND CONVENTIONS

Circuit analysis may be described as the study of the response of a circuit to an imposed condition, for example a short circuit, where the circuit variables are current and voltage. We know that current flow results from the application of a driving voltage, but there is complete duality between the variables and either may be regarded as the cause of the other. Just as the current flowing through the primary winding of transformer is as a result of the voltage applied across the primary terminals, the voltage appearing at the secondary terminals of the same transformer is as a result of current flowing through the secondary winding. Likewise, the current flowing through a resistor is caused by a voltage applied to either side of the resistor. But we can just as well say that the voltage developed across the resistor is as a result of the current flowing through it.

It is possible to represent any circuit with five circuit elements:

- Voltage source
- Current source
- Resistance
- Capacitance
- Inductance

When a circuit exists, there is an interchange of energy between these elements. A circuit may be described as being made up of 'sources' and 'sinks' for energy. For example, voltage and current sources are energy sources, resistors are energy sinks, whereas capacitors and inductors (in their pure form) are neither sinks nor sources, but are energy stores. They merely borrow energy from the circuit then give it back.

The elements of a circuit are connected together to form a network having nodes (terminals or junctions) and branches (series groups of elements) that form closed loops (meshes).

In steady state a.c. circuit theory, the ability of a circuit to impede a current flow resulting from a given driving voltage is called the *impedance* (Z) of the circuit. The impedance parameter has an inverse equivalent (1/Z), known as admittance (Y). The impedance of a circuit is made up its *resistance* (R) from resistors and its *reactance* (X) from inductors and capacitors. Likewise the admittance of a circuit comprises the *conductance* (G) from resistors and *susceptance* (B) from inductors and capacitors.

Impedance

If a steady state dc voltage is applied to a circuit, a current will flow, which depends only on the resistance of the circuit according to ohms law V=IR. The circuit's reactive components will not play a part in the long term. However if a changing voltage source is applied, the subsequent flow in current depends not only on the resistance of the circuit, but also the reactance of the circuit, according to the equation:

$$V = IZ$$

where Z is the circuit impedance consisting of the resistive part R and the reactive part X:

Consider the following circuit:

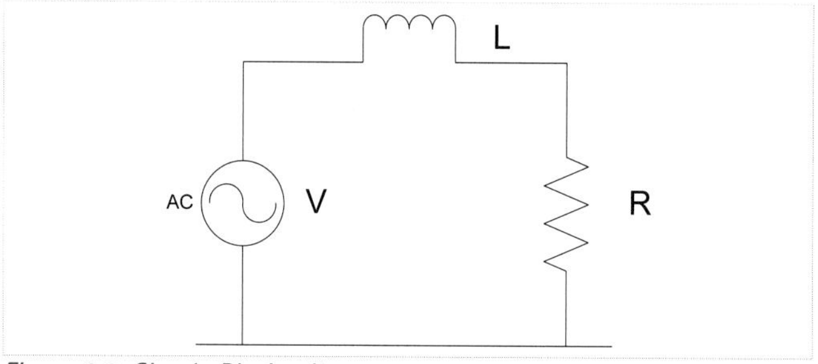

Figure 3.3: Simple RL circuit

When the voltage is changing, the inductive component L inhibits the subsequent change of current. So in addition to the resistance, the circuit offers *reactance* to the changing voltage according to the equation:

$$V_L = L\frac{di}{dt}$$

where V_L is the instantaneous voltage across the inductor

The equation that defines the voltage of the circuit is thus:

$$V = iR + L\frac{di}{dt}$$

It can be seen that in this circuit, the higher the frequency the higher the impedance.

As a series inductance offers impedance to alternating current flow, a series capacitance will offer admittance. Consider the following circuit:

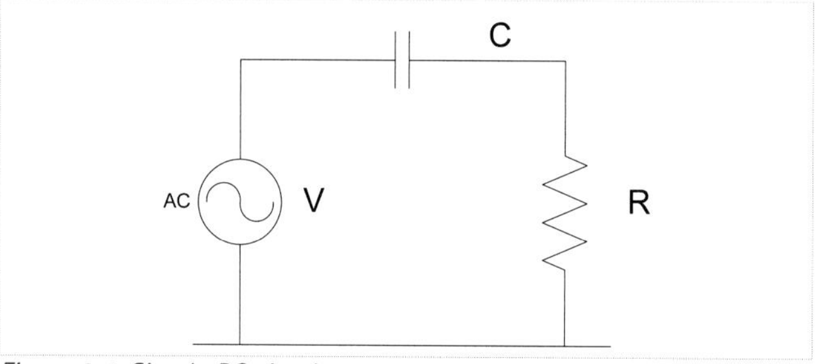

Figure 3.4: Simple RC circuit

When the current is changing, the series capacitance C inhibits the voltage build-up on the capacitor. The reactance of the series capacitor is given by:

$$V_C = \frac{1}{C}\int idt$$

where V_C is the instantaneous voltage across the capacitor

In this circuit, the complete voltage equation is as follows:

$$V = iR + \frac{1}{C}\int idt$$

It can be seen that in this circuit, the lower the frequency the higher the impedance.

If the voltage waveform applied to an inductor is

$$V_{(t)} = V_m \, sin(\omega t)$$

where $V_{(t)}$ is the voltage as a function of time, V_m is the maximum voltage, ω is the angular velocity and t is the time, then:

$$V_m \, sin(\omega t) = L\frac{di}{dt}$$

therefore

$$\frac{di}{dt} = \frac{V_m}{L} sin(\omega t)$$

and

$$I = -\frac{V_m}{\omega L} cos(\omega t)$$

The reactance X is defined as the voltage across the reactive component divided by the current flowing through the reactive component, therefore

$$X = \frac{V_{(t)}}{I_{(t)}} = V_m \, sin(\omega t) \Big/ -\frac{V_m \, cos(\omega t)}{\omega L}$$

therefore

$$X = \omega L$$

Likewise, it can be shown that the reactance of a capacitor is:

$$X = -\frac{1}{\omega C}$$

Phase Angle

It has been explained that in an inductor, the current lags the voltage. When one considers a sinusoidal waveform, the current lags the voltage by 90° (This assumes a pure inductor with zero resistive component). Likewise in a pure capacitor, the current leads the voltage by 90°.

As the reactive components introduce a 90° phase shift between the current and the voltage, the waveforms can be represented by the impedance by a complex number, such that:

$$Z = R + jX$$

where Z is the overall impedance, R is the resistive (or real) component and X is the reactive (or imaginary) component.

The modulus of the impedance is:

$$|Z| = \sqrt{R^2 + X^2}$$

and the angle is:

$$\angle Z = tan^{-1}\frac{X}{R}$$

The impedance of a resistor in series with a capacitor in series with an inductor is:

$$Z = R + j\omega L + \frac{1}{j\omega C} = R + j\left(\omega L - \frac{1}{\omega C}\right)$$

3.4.1 Circuit Variables

AC current and voltage are (in the ideal case) sinusoidal functions of time, varying at a single and constant frequency. They can be regarded as rotating vectors.

For example, the instantaneous value, e of a voltage varying sinusoidally with time is:

$$e = E_m \, sin(\omega t + \delta)$$

Equation 3.8

where:

E_m = the maximum amplitude of the waveform

ω = the angular velocity, measured in radians per second

δ = the phase of the vector at time $t = 0$

At $t=0$, the actual value of the voltage is $E_m sin\delta$. So if E_m is regarded as the modulus of a vector, whose argument is δ, then $E_m sin\delta$ is the imaginary component of the vector $|E_m|\angle\delta$. Figure 3.5 illustrates this quantity as a vector and as a sinusoidal function of time.

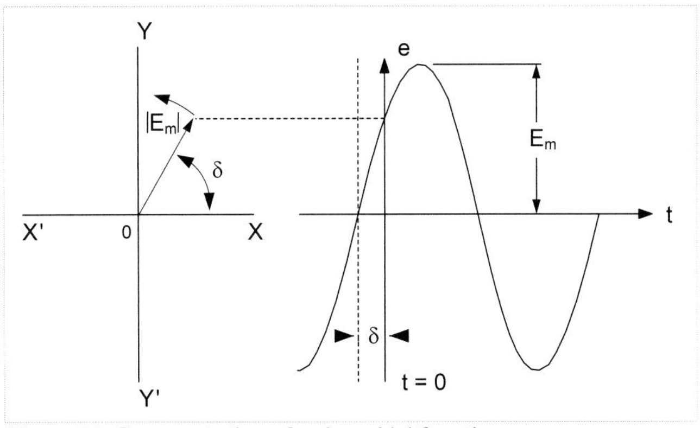

Figure 3.5: Representation of a sinusoidal function

The current resulting from applying a voltage to a circuit depends upon the circuit impedance. If the voltage is a sinusoidal function at a given frequency and the impedance is constant the current will also vary harmonically at the same frequency, so it can be shown on the same vector diagram as the voltage vector, and is given by the equation

$$i = \frac{|E_m|}{|Z|} sin(\omega t + \delta - \phi)$$

Equation 3.9

where:

$$|Z| = \sqrt{R^2 + X^2}$$

$$X = \left(\omega L - \frac{1}{\omega C} \right)$$

$$\phi = tan^{-1} \frac{X}{R}$$

Equation 3.10

From Equations 3.9 and 3.10 it can be seen that the angular displacement ϕ between the current and voltage vectors and the current magnitude $|I_m|$ is dependent upon the impedance $\overline{Z}$. In complex form the impedance may be written $\overline{Z} = R + jX$. The 'real component', R, is the circuit resistance, and the 'imaginary component', X, is the circuit reactance. When the circuit reactance is inductive (that is, $\omega L > 1/\omega C$), the current 'lags' the voltage by an angle ϕ, and when it is capacitive (that is, $1/\omega C > \omega L$) it 'leads' the voltage by an angle ϕ.

Root Mean Square

Sinusoidally varying quantities are described by their 'effective' or 'root mean square' (r.m.s.) values; these are usually written using the relevant symbol without a suffix.

Thus:

$$|I| = \frac{|I_m|}{\sqrt{2}}$$

and

$$|E| = \frac{|E_m|}{\sqrt{2}}$$

Equation 3.11

The 'root mean square' value is that value which has the same heating effect as a direct current quantity of that value in the same circuit, and this definition applies to non-sinusoidal as well as sinusoidal quantities.

3.4.2 Sign Conventions

In describing the electrical state of a circuit, it is often necessary to refer to the 'potential difference' existing between two points in the circuit. Since wherever such a potential difference exists, current will flow and energy will either be transferred or absorbed, it is obviously necessary to define a potential difference in more exact terms. For this reason, the terms voltage rise and voltage drop are used to define more accurately the nature of the potential difference.

Voltage rise is a rise in potential measured in the direction of current flow between two points in a circuit. Voltage drop is the converse. A circuit element with a voltage rise across it acts as a source of energy. A circuit element with a voltage drop across it acts as a sink of energy. Voltage sources are usually active circuit elements, while sinks are usually passive circuit elements. The positive direction of energy flow is from sources to sinks.

Kirchhoff's first law states that the sum of the driving voltages must equal the sum of the passive voltages in a closed loop. This is illustrated by the fundamental equation of an electric circuit:

$$e = iR + L\frac{di}{dt} + \frac{1}{C}\int idt$$

Equation 3.12

where the terms on the left hand side of the equation are voltage drops across the circuit elements. Expressed in steady state terms Equation 3.12 may be written:

$$\sum \overline{E} = \sum \overline{IZ}$$

Equation 3.13

and this is known as the equated-voltage equation [3.1].

It is the equation most usually adopted in electrical network calculations, since it equates the driving voltages, which are known, to the passive voltages, which are functions of the currents to be calculated.

In describing circuits and drawing vector diagrams, for formal analysis or calculations, it is necessary to adopt a notation which defines the positive direction of assumed current flow, and establishes the direction in which positive voltage drops and increases act. Two methods are available; one, the double suffix method, is used for symbolic analysis, the other, the single suffix or diagrammatic method, is used for numerical calculations.

In the double suffix method the positive direction of current flow is assumed to be from node 'a' to node 'b' and the current is designated $\overline{I}_{ab}$. With the diagrammatic method, an arrow indicates the direction of current flow.

The voltage rises are positive when acting in the direction of current flow. It can be seen from Figure 3.6 that $\overline{E}_1$ and $\overline{E}_{an}$ are positive voltage rises and $\overline{E}_2$ and $\overline{E}_{bn}$ are negative voltage rises. In the diagrammatic method their direction of action is simply indicated by an arrow, whereas in the double suffix method, $\overline{E}_{an}$ and $\overline{E}_{bn}$ indicate that there is a potential rise in directions *na* and *nb*.

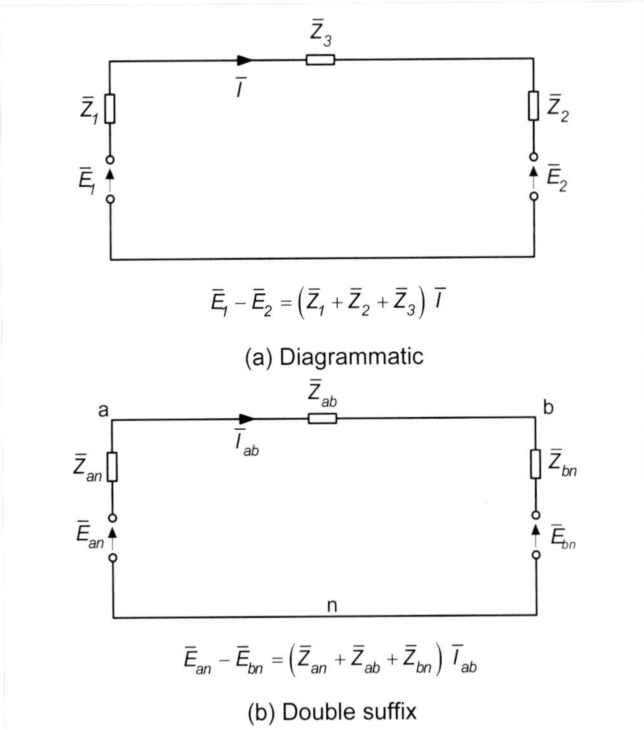

Figure 3.6: Methods of representing a circuit

Voltage drops are also positive when acting in the direction of current flow. From Figure 3.6(a) it can be seen that $\overline{Z}_1 + \overline{Z}_2 + \overline{Z}_3$ is the total voltage drop in the loop in the direction of current flow, and must equate to the total voltage rise $\overline{E}_1 - \overline{E}_2$. In Figure 3.6(b) the voltage drop between nodes *a* and *b* designated V_{ab} indicates that point *b* is at a lower potential than *a*, and is positive when current flows from *a* to *b*. Conversely V_{ba} is a negative voltage drop.

Symbolically:

$$\overline{V}_{ab} = \overline{V}_{an} - \overline{V}_{bn}$$

$$\overline{V}_{ba} = \overline{V}_{bn} - \overline{V}_{an}$$

(where n is a common reference point)

Equation 3.14

3.4.3 Power

The product of the potential difference across and the current through a branch of a circuit is a measure of the rate at which energy is exchanged between that branch and the remainder of the circuit. If the potential difference is a positive voltage drop the branch is passive and absorbs energy. Conversely, if the potential difference is a positive voltage rise the branch is active and supplies energy.

The rate at which energy is exchanged is known as power, and by convention, the power is positive when energy is being absorbed and negative when being supplied. With a.c. circuits the power alternates, so, to obtain a rate at which energy is supplied or absorbed it is necessary to take the average power over one whole cycle. If

$e = E_m sin(\omega t + \delta)$ and $i = I_m sin(\omega t + \delta - \phi)$, then the power equation is:

$$p = ei = P[1 - cos\,2(\omega t + \delta)] + Q\,sin\,2(\omega t + \delta)$$

Equation 3.15

where:

$$P = |E||I|\cos\phi$$

and

$$Q = |E||I|\sin\phi$$

From Equation 3.15 it can be seen that the quantity P varies from *0* to $2P$ and quantity Q varies from $-Q$ to $+Q$ in one cycle, and that the waveform is of twice the periodic frequency of the current voltage waveform.

The average value of the power exchanged in one cycle is a constant, equal to quantity *P*, and as this quantity is the product of the voltage and the component of current which is 'in phase' with the voltage it is known as the 'real' or 'active' power.

The average value of quantity *Q* is zero when taken over a cycle, suggesting that energy is stored in one half-cycle and returned to the circuit in the remaining half-cycle. *Q* is the product of voltage and the quadrature component of current, and is known as 'reactive power'.

As *P* and *Q* are constants specifying the power exchange in a given circuit, and are products of the current and voltage vectors, then if S is the product EI it follows that:

$$S = P + jQ$$

Equation 3.16

The quantity S is described as the 'apparent power', and is the term used in establishing the rating of a circuit. S has units of VA.

3.4.4 Single and Polyphase Systems

A system is single or polyphase depending upon whether the sources feeding it are single or polyphase. A source is single or polyphase according to whether there are one or several driving voltages associated with it. For example, a three-phase source is a source containing three alternating driving voltages that are assumed to reach a maximum in phase order, A, B, C. Each phase driving voltage is associated with a phase branch of the system network as shown in Figure 3.7(a).

If a polyphase system has balanced voltages, that is, equal in magnitude and reaching a maximum at equally displaced time intervals, and the phase branch impedances are identical, it is called a 'balanced' system. It will become 'unbalanced' if any of the above conditions are not satisfied. Calculations using a balanced polyphase system are simplified, as it is only necessary to solve for a single phase, the solution for the remaining phases being obtained by symmetry.

The power system is normally operated as a three-phase, balanced, system. For this reason the phase voltages are equal in magnitude and can be represented by three vectors spaced 120° or 2π/3 radians apart, as shown in Figure 3.7(b).

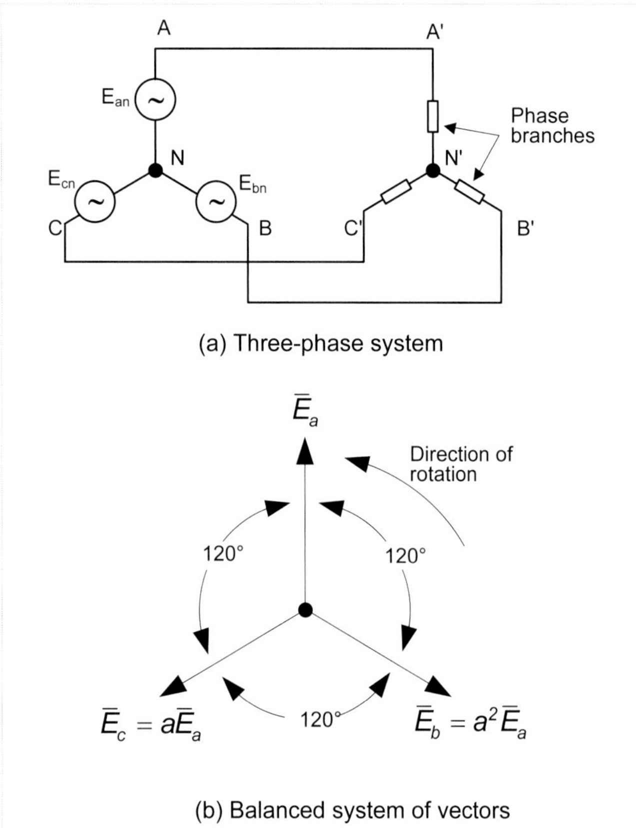

Figure 3.7: Three phase systems

Since the voltages are symmetrical, they may be expressed in terms of one, that is:

$$\bar{E}_a = \bar{E}_a$$

$$\bar{E}_b = a^2\bar{E}_a$$

$$\bar{E}_c = a\bar{E}_a$$

Equation 3.17

where a is the vector operator $e^{j\frac{2\pi}{3}}$. Further, if the phase branch impedances are identical in a balanced system, it follows that the resulting currents are also balanced.

3.5 THEOREMS AND NETWORK REDUCTION

Most practical power system problems are solved by using steady state analytical methods. These methods make the assumption that circuit parameters are linear, bilateral, and constant for constant frequency circuit variables. When analysing initial values, it is necessary to study the behaviour of a circuit in the transient state. This can be achieved using operational methods. In some problems, which fortunately are rare, the assumption of linear, bilateral circuit parameters is no longer valid. Such problems are solved using advanced mathematical techniques that are beyond the scope of this book.

3.5.1 Circuit Laws

In linear, bilateral circuits, there are three basic network laws. These laws apply, regardless of the state of the circuit, and at any particular instant of time. These laws are the branch, junction and mesh laws, derived from Ohm and Kirchhoff, and are stated below, using steady state a.c. nomenclature.

Branch law

The current $\bar{I}$ in a given branch of impedance $\bar{Z}$ is proportional to the potential difference $\bar{V}$ appearing across the branch, that is:

$$\bar{V} = \overline{IZ}$$

Junction law

The algebraic sum of all currents entering any junction (or node) in a network is zero, that is:

$$\Sigma\bar{I} = 0$$

Mesh law

The algebraic sum of all the driving voltages in any closed path (or mesh) in a network is equal to the algebraic sum of all the passive voltages (products of the impedances and the currents) in the component branches, that is:

$$\Sigma\bar{E} = \Sigma\overline{IZ}$$

Alternatively, the total change in potential around a closed loop is zero.

3.5.2 Circuit Theorems

From the above network laws, many theorems have been derived for the rationalisation of networks, either to reach a quick, simple, solution to a problem or to represent a complicated circuit by an equivalent. These theorems are divided into two classes: those concerned with the general properties of networks and those concerned with network reduction.

Of the many theorems that exist, the three most important are given. These are: the Superposition Theorem, Thévenin's Theorem and Kennelly's Star/Delta Theorem.

3.5.2.1 Superposition Theorem (general network theorem)

The resultant current that flows in any branch of a network due to the simultaneous action of several driving voltages is equal to the algebraic sum of the component currents due to each driving voltage acting alone with the remainder short-circuited.

3.5.2.2 Thévenin's Theorem (active network reduction theorem)

Any active network that may be viewed from two terminals can be replaced by single driving voltage acting in series with single impedance. The driving voltage is the open-circuit voltage between the two terminals and the impedance is the impedance of the network viewed from the terminals with all sources short-circuited.

3.5.2.3 Kennelly's Star/Delta Theorem (passive network reduction theorem)

Any three-terminal network can be replaced by a delta or star impedance equivalent without disturbing the external network. The formulae relating the replacement of a delta network by the equivalent star network is as follows:

$$\overline{Z}_{10} = \frac{\overline{Z}_{12}\overline{Z}_{31}}{\overline{Z}_{12} + \overline{Z}_{23} + \overline{Z}_{31}}$$

and so on.

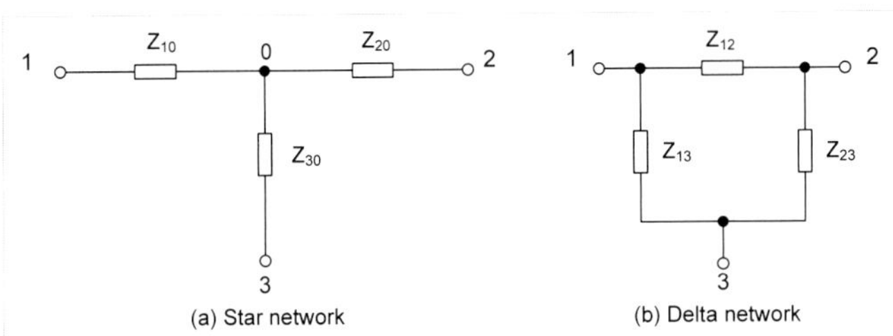

Figure 3.8: Star/Delta network reduction

The impedance of a delta network corresponding to and replacing any star network is:

$$\overline{Z}_{12} = \overline{Z}_{10} + \overline{Z}_{20} + \frac{\overline{Z}_{10}\overline{Z}_{20}}{\overline{Z}_{30}}$$

and so on.

3.5.3 Network Reduction

The aim of network reduction is to reduce a system to a simple equivalent while retaining the identity of that part of the system to be studied.

For example, consider the system shown in Figure 3.9. The network has two sources E' and E'', a line AOB shunted by an impedance, which may be regarded as the reduction of a further network connected between A and B, and a load connected between O and N. The object of the reduction is to study the effect of opening a breaker at A or B during normal system operations or of a fault at A or B. Thus the identity of nodes A and B must be retained together with the sources, but the branch ON can be eliminated, simplifying the study. Proceeding, A, B, N, forms a star branch and can therefore be converted to an equivalent delta.

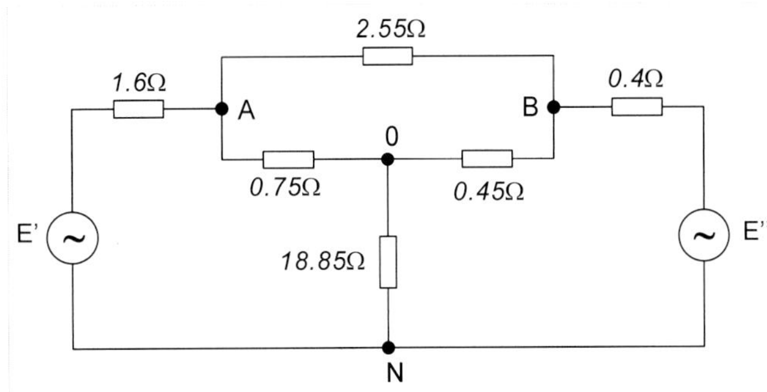

Figure 3.9: Typical power system

$$Z_{AN} = Z_{AO} + Z_{NO} + \frac{Z_{AO}Z_{BO}}{Z_{BO}}$$

$$= 0.75 + 18.85 + \frac{0.75 \times 18.85}{0.45}$$

$$= 51\Omega$$

$$Z_{BN} = Z_{BO} + Z_{NO} + \frac{Z_{BO}Z_{BO}}{Z_{AO}}$$

$$= 0.45 + 18.85 + \frac{0.45 \times 18.85}{0.75}$$

$$= 30.6\Omega$$

$$Z_{AB} = Z_{AO} + Z_{BO} + \frac{Z_{AO}Z_{BO}}{Z_{NO}}$$

$$= 1.2\Omega$$

(since $Z_{NO} >> Z_{AO}Z_{BO}$)

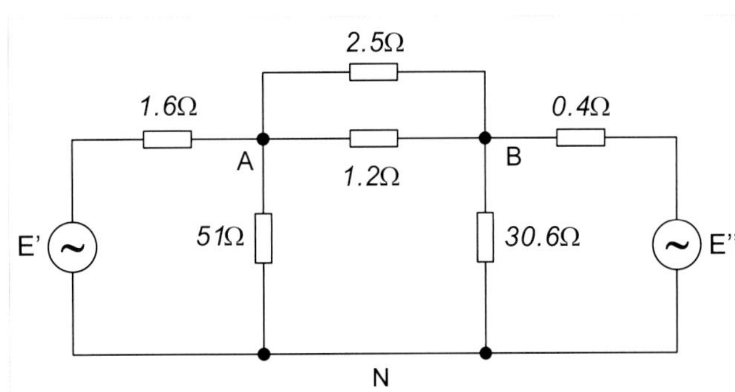

Figure 3.10: Reduction using star/delta transform

The network is now reduced as shown in Figure 3.10.

By applying Thévenin's theorem to the active loops, these can be replaced by a single driving voltage in series with impedance, as shown in Figure 3.11.

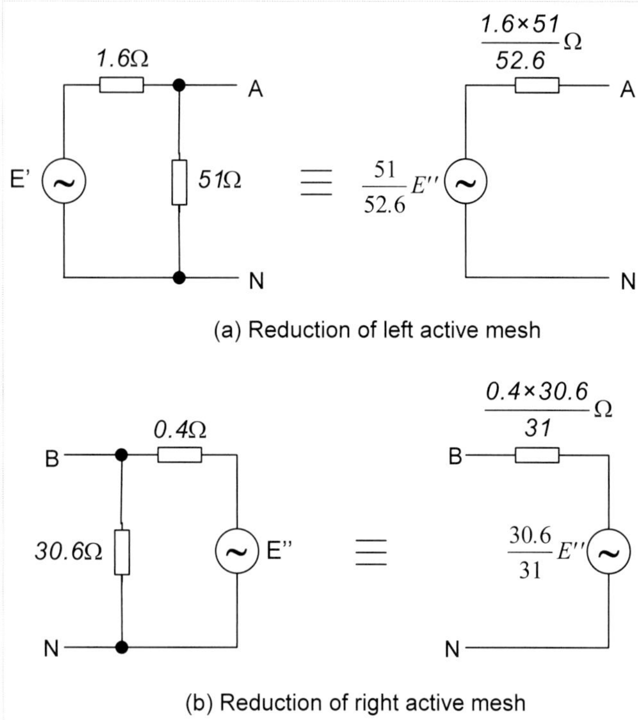

Figure 3.11: Reduction of active meshes: Thévenin's theorem

The network shown in Figure 3.9 is now reduced to that shown in Figure 3.12 with the nodes *A* and *B* retaining their identity. Further, the load impedance has been completely eliminated.

The network shown in Figure 3.12 may now be used to study system disturbances, for example power swings with and without faults.

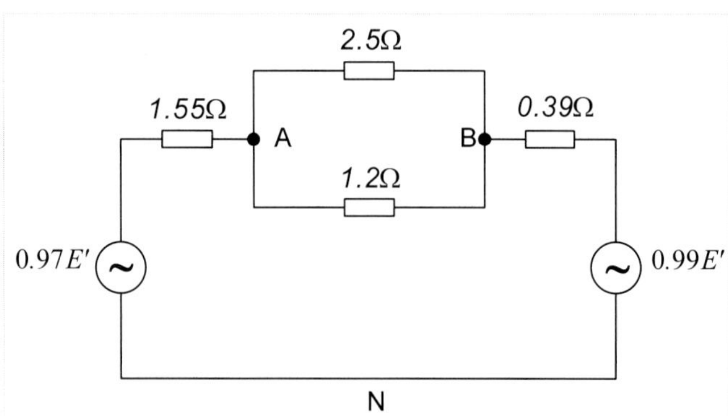

Figure 3.12: Reduction of typical power system

Most reduction problems follow the same pattern as the example above. The rules to apply in practical network reduction are:

- decide on the nature of the disturbance or disturbances to be studied
- decide on the information required, for example the branch currents in the network for a fault at a particular location
- reduce all passive sections of the network not directly involved with the section under examination
- reduce all active meshes to a simple equivalent, that is, to a simple source in series with a single impedance

With the widespread availability of computer-based power system simulation software, it is now usual to use such software on a routine basis for network calculations without significant network reduction taking place. However, the network reduction techniques given above are still valid, as there will be occasions where such software is not immediately available and a hand calculation must be carried out.

In certain circuits, for example parallel lines on the same towers, there is mutual coupling between branches. Correct circuit reduction must take account of this coupling.

Three cases are of interest. These are:

- Case a: two branches connected together at their nodes
- Case b: two branches connected together at one node only
- Case c: two branches that remain unconnected

Considering each case in turn:

Case a

Consider the circuit shown in Figure 3.13(a).

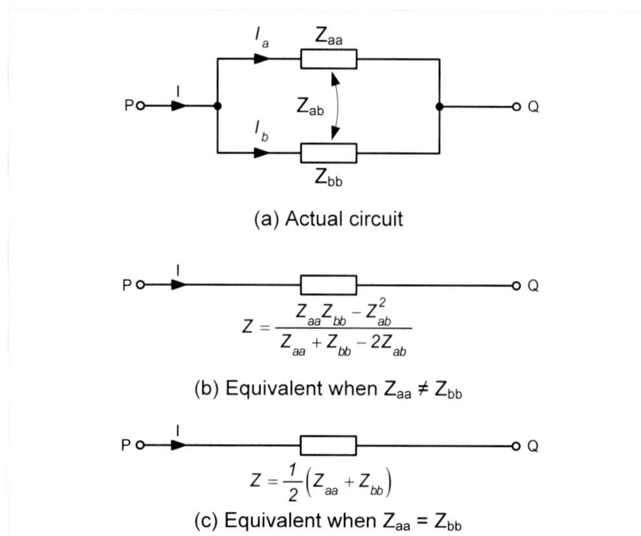

Figure 3.13: Reduction of two branches with mutual coupling

The application of a voltage *V* between the terminals *P* and *Q* gives:

$$V = I_a Z_{aa} + I_b Z_{ab}$$

$$V = I_a Z_{ab} + I_b Z_{bb}$$

where I_a and I_b are the currents in branches a and b, respectively and $I = I_a + I_b$, the total current entering at terminal P and leaving at terminal Q.

Solving for I_a and I_b :

$$I_a = \frac{(Z_{bb} - Z_{ab})V}{Z_{aa}Z_{bb} - Z_{ab}^2}$$

from which

$$I_b = \frac{(Z_{aa} - Z_{ab})V}{Z_{aa}Z_{bb} - Z_{ab}^2}$$

and

$$I = I_a + I_b = \frac{V(Z_{aa} + Z_{bb} - 2Z_{ab})}{Z_{aa}Z_{bb} - Z_{ab}^2}$$

so that the equivalent impedance of the original circuit is:

$$Z = \frac{Z_{aa}Z_{bb} - Z_{ab}^2}{Z_{aa} + Z_{bb} - 2Z_{ab}}$$

Equation 3.18

(Figure 3.13(b)), and, if the branch impedances are equal, the usual case, then:

$$Z = \frac{1}{2}(Z_{aa} + Z_{ab})$$

Equation 3.19 (see Figure 3.13c)

Case b

Consider the circuit in Figure 3.14(a).

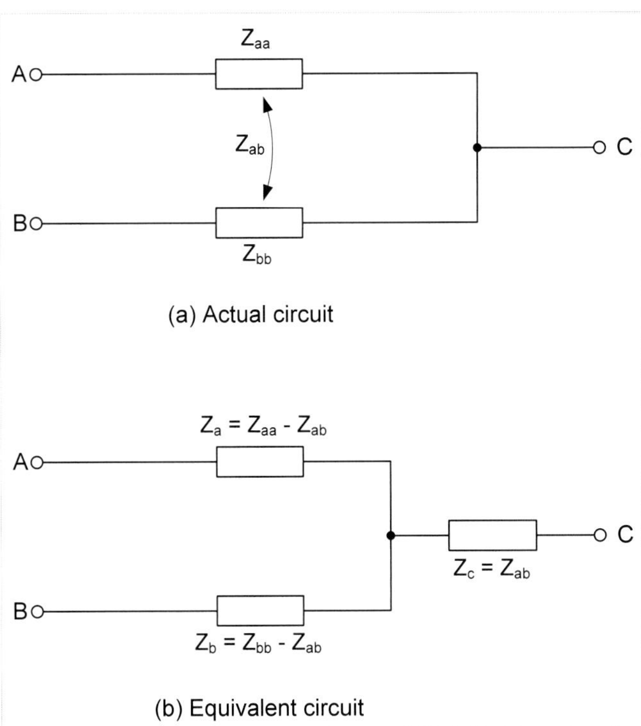

Figure 3.14: Reduction of mutually-coupled branches with a common terminal

The assumption is made that an equivalent star network can replace the network shown. From inspection with one terminal isolated in turn and a voltage V impressed across the remaining terminals it can be seen that:

$$Z_a + Z_c = Z_{aa}$$
$$Z_b + Z_c = Z_{bb}$$
$$Z_a + Z_b = Z_{aa} + Z_{bb} - 2Z_{ab}$$

Solving these equations gives:

$$Z_a = Z_{aa} - Z_{ab}$$

$$Z_b = Z_{bb} - Z_{ab}$$

$$Z_c = Z_{ab} - Z_{ab}$$

Equation 3.20 - see Figure 3.14(b).

Case c

Consider the four-terminal network given in Figure 3.15(a), in which the branches 11' and 22' are electrically separate except for a mutual link. The equations defining the network are:

$$V_1 = Z_{11}I_1 + Z_{12}I_2$$

$$V_2 = Z_{21}I_1 + Z_{22}I_2$$

$$I_1 = Y_{11}V_1 + Y_{12}V_2$$

$$I_2 = Y_{21}V_1 + Y_{22}V_2$$

where $Z_{12} = Z_{21}$ and $Y_{12} = Y_{21}$, if the network is assumed to be reciprocal. Further, by solving the above equations it can be shown that:

$$Y_{11} = Z_{22} / \Delta$$

$$Y_{22} = Z_{11} / \Delta$$

$$Y_{12} = Z_{12} / \Delta$$

$$\Delta = Z_{11}Z_{22} - Z_{12}^2$$

Equation 3.21

There are three independent coefficients, namely Z_{12}, Z_{11}, Z_{22} so the original circuit may be replaced by an equivalent mesh containing four external terminals, each terminal being connected to the other three by branch impedances as shown in Figure 3.15(b).

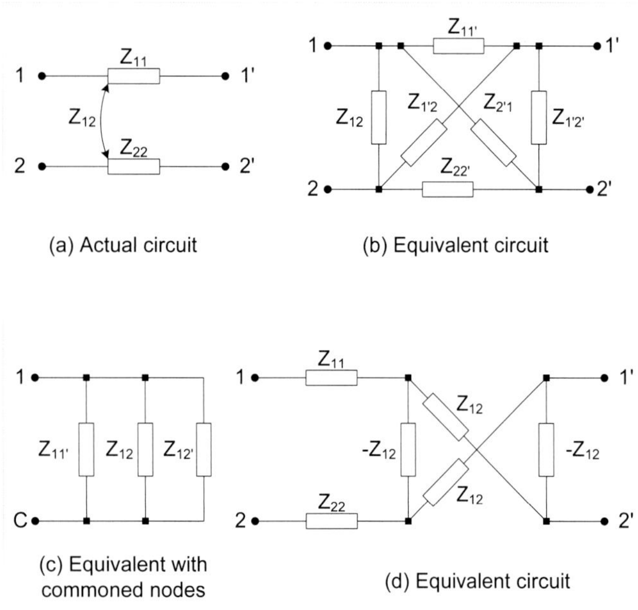

Figure 3.15: equivalent circuits for four terminal network with mutual coupling

In order to evaluate the branches of the equivalent mesh let all points of entry of the actual circuit be commoned except node 1 of circuit 1, as shown in Figure 3.15(c). Then all impressed voltages except V_1 will be zero and:

$$I_1 = Y_{11}V_1$$

$$I_2 = Y_{12}V_1$$

If the same conditions are applied to the equivalent mesh, then:

$$I_1 = \frac{-V_1}{Z_{11'}}$$

$$I_2 = \frac{-V_1}{Z_{12}} = \frac{V_1}{Z_{12'}}$$

These relations follow from the fact that the branch connecting nodes *1* and *1'* carries current I_1 and the branches connecting nodes *1* and *2'* and *1'* and *2* carry current I_2. This must be true since branches between pairs of commoned nodes can carry no current.

By considering each node in turn with the remainder commoned, the following relationships are found:

$$Z_{11'} = \frac{1}{Y_{11}}$$

$$Z_{22'} = \frac{1}{Y_{22}}$$

$$Z_{12} = \frac{-1}{Y_{12}}$$

$$Z_{12} = Z_{1'2'} = -Z_{21'} = -Z_{12'}$$

Hence:

$$Z_{11'} = \frac{Z_{11}Z_{22} - Z_{12}^2}{Z_{22}}$$

$$Z_{22'} = \frac{Z_{11}Z_{22} - Z_{12}^2}{Z_{11}}$$

$$Z_{12} = \frac{Z_{11}Z_{22} - Z_{12}^2}{Z_{12}}$$

Equation 3.22

A similar but equally rigorous equivalent circuit is shown in Figure 3.15(d). This circuit [3.2] follows from the reasoning that since the self-impedance of any circuit is independent of all other circuits it need not appear in any of the mutual branches if it is lumped as a radial branch at the terminals. So putting Z_{11} and Z_{22}, equal to zero in Equation 3.22, defining the equivalent mesh in Figure 3.15(b), and inserting radial branches having impedances equal to Z_{11} and Z_{22} in terminals 1 and 2, results in Figure 3.15(d).

3.6 IMPEDANCE NOTATION

It can be seen by inspection of any power system diagram that:

- several voltage levels exist in a system
- it is common practice to refer to plant MVA in terms of per unit or percentage values
- transmission line and cable constants are given in ohms/km

Before any system calculations can take place, the system parameters must be referred to 'base quantities' and represented as a unified system of impedances in either ohmic, percentage, or per unit values.

The base quantities are power and voltage. Normally, they are given in terms of the three-phase power in MVA and the line voltage in kV. The base impedance resulting from the above base quantities is:

$$Z_b = \frac{(kV)^2}{MVA}\Omega$$

Equation 3.23

and, provided the system is balanced, the base impedance may be calculated using either single-phase or three-phase quantities.

The per unit or percentage value of any impedance in the system is the ratio of actual to base impedance values.

Hence:

$$Z(p.u.) = Z(\Omega) \times \frac{MVA_b}{(kV_b)^2}$$

$$Z(\%) = Z(p.u.) \times 100$$

Equation 3.24

where:

MVA_b=$baseMVA$

kVA_b=$basekV$

Transferring per unit quantities from one set of base values to another can be done using the equation:

$$Z_{p.u.2} = Z_{p.u.1} \times \frac{MVA_{b2}}{MVA_{b1}} \left(\frac{kV_{b1}}{kV_{b2}} \right)^2$$

where:

- suffix $b1$ denotes the value to the original base
- suffix $b2$ denotes the value to new base

The choice of impedance notation depends upon the complexity of the system, plant impedance notation and the nature of the system calculations envisaged.

If the system is relatively simple and contains mainly transmission line data, given in ohms, then the ohmic method can be adopted with advantage. However, the per unit method of impedance notation is the most common for general system studies since:

- impedances are the same referred to either side of a transformer if the ratio of base voltages on the two sides of a transformer is equal to the transformer turns ratio
- confusion caused by the introduction of powers of 100 in percentage calculation is avoided
- by a suitable choice of bases, the magnitudes of the data and results are kept within a predictable range, and hence errors in data and computations are easier to spot

Most power system studies are carried out using software in per unit quantities. Irrespective of the method of calculation, the choice of base voltage, and unifying system impedances to this base, should be approached with caution, as shown in the following example.

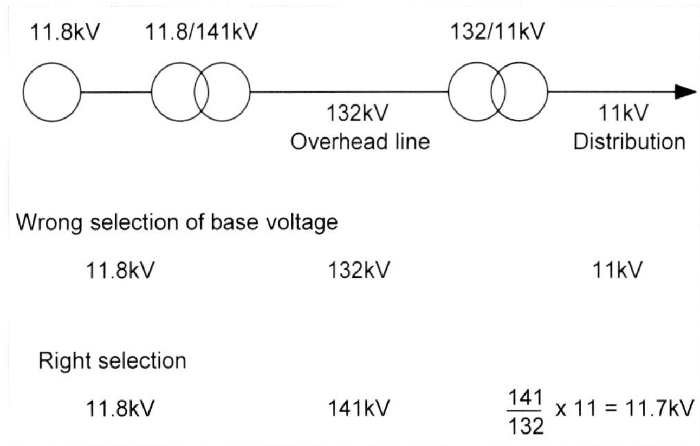

Figure 3.16: Selection of base voltages

From Figure 3.16 it can be seen that the base voltages in the three circuits are related by the turns ratios of the intervening transformers. Care is required as the nominal transformation ratios of the transformers quoted may be different from the turns ratios- e.g. a 110/33kV (nominal) transformer may have a turns ratio of 110/34.5kV. Therefore, the rule for hand calculations is: 'to refer impedance in ohms from one circuit to another multiply the given impedance by the square of the turn's ratio (open circuit voltage ratio) of the intervening transformer'.

Where power system simulation software is used, the software normally has calculation routines built in to adjust transformer parameters to take account of differences between the nominal primary and secondary voltages and turns ratios. In this case, the choice of base voltages may be more conveniently made as the nominal voltages of each section of the power system. This approach avoids confusion when per unit or percent values are used in calculations in translating the final results into volts, amps, etc.

For example, in Figure 3.17, generators G_1 and G_2 have a sub-transient reactance of 26% on 66.6MVA rating at 11kV, and transformers T_1 and T_2 a voltage ratio of 11/145kV and an impedance of 12.5% on 75MVA. Choosing 100MVA as base MVA and 132kV as base voltage, find the percentage impedances to new base quantities.

- generator reactances to new bases are:

$$26 \times \frac{100}{66.6} \times \frac{11^2}{132^2} = 0.27\%$$

- transformer reactances to new bases are:

$$12.5 \times \frac{100}{75} \times \frac{145^2}{132^2} = 20.1\%$$

NOTE: The base voltages of the generator and circuits are 11kV and 145kV respectively, that is, the turns ratio of the transformer. The corresponding per unit values can be found by dividing by 100, and the ohmic value can be found by using Equation 3.19.

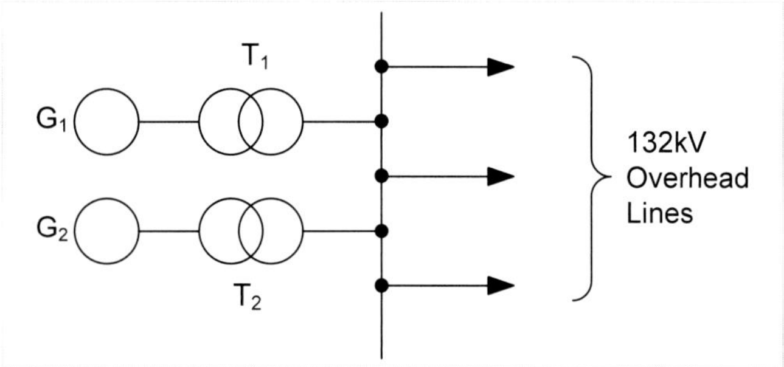

Figure 3.17: Section of a power system

3.7 REFERENCES

[3.1] Power System Analysis. J. R. Mortlock and M. W. Humphrey Davies. Chapman & Hall.

[3.2] Equivalent Circuits I. Frank M. Starr, Proc. A.I.E.E. Vol. 51. 1932, pp. 287-298.

Chapter 4
Fault Calculations

4.1 INTRODUCTION

A power system is normally treated as a balanced symmetrical three-phase network. When a fault occurs, the symmetry is normally upset, resulting in unbalanced currents and voltages appearing in the network. The only exception is the three-phase fault, where all three phase equally at the same location. This is described as a symmetrical fault. By using symmetrical component analysis and replacing the normal system sources by a source at the fault location, it is possible to analyse these fault conditions.

For the correct application of protection equipment, it is essential to know the fault current distribution throughout the system and the voltages in different parts of the system due to the fault. Further, boundary values of current at any relaying point must be known if the fault is to be cleared with discrimination. The information normally required for each kind of fault at each relaying point is:

- maximum fault current
- minimum fault current
- maximum through fault current

To obtain this information, the limits of stable generation and possible operating conditions, including the system earthing method, must be known. Faults currents are always assumed to be through zero fault impedance.

4.2 THREE-PHASE FAULT CALCULATIONS

Three-phase faults are unique in that they are balanced, that is, symmetrical in the three phases, and can be calculated from the single-phase impedance diagram and the operating conditions existing prior to the fault.

A fault condition is a sudden abnormal alteration to the normal circuit arrangement. The circuit quantities, current and voltage, will alter, and the circuit will pass through a transient state to a steady state. In the transient state, the initial magnitude of the fault current will depend upon the point on the voltage wave at which the fault occurs. The decay of the transient condition, until it merges into steady state, is a function of the parameters of the circuit elements. The transient current may be regarded as a d.c. exponential current superimposed on the symmetrical steady state fault current. In a.c. machines, owing to armature reaction, the machine reactances pass through 'sub transient' and 'transient' stages before reaching their steady state synchronous values. For this reason, the resultant fault current during the transient period, from fault inception to steady state also depends on the location of the fault in the network relative to that of the rotating plant.

In a system containing many voltage sources, or having a complex network arrangement, it is tedious to use the normal system voltage sources to evaluate the fault current in the faulty branch or to calculate the fault current distribution in the system. A more practical method [Reference 4.1] is to replace the system voltages by a single driving voltage at the fault point. This driving voltage is the voltage existing at the fault point before the fault occurs.

Consider the circuit given in Figure 4.1 where the driving voltages are $\overline{E}'$ and $\overline{E}''$, the impedances on either side of fault point F are $\overline{Z_1}'$ and $\overline{Z_1}''$, and the current through point F before the fault occurs is $\overline{I}$.

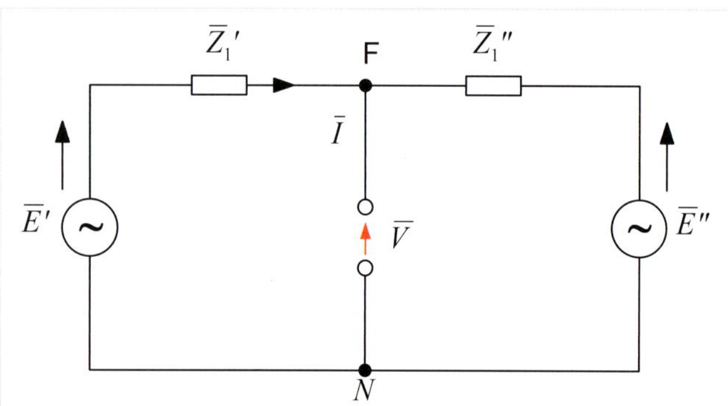

Figure 4.1: Network with fault at F

The voltage $\overline{V}$ at F before fault inception is:

$$\overline{V} = \overline{E}' - \overline{I}\overline{Z}' = \overline{E}'' + \overline{I}\overline{Z}''$$

Assuming zero fault impedance, the fault voltage $\overline{V}$ will be zero after the fault inception, and a large fault current will flow to earth. The change in voltage at the fault point is therefore $-\overline{V}$. The change in the current flowing into the network from F is thus:

$$\Delta\overline{I} = -\frac{\overline{V}}{\overline{Z}_1} = -\overline{V}\frac{(\overline{Z}_1' + \overline{Z}_1'')}{\overline{Z}_1'\overline{Z}_1''}$$

and, since no current was flowing into the network from F prior to the fault, the fault current flowing from the network into the fault is:

$$\overline{I}_f = -\Delta\overline{I} = \overline{V}\frac{(\overline{Z}_1' + \overline{Z}_1'')}{\overline{Z}_1'\overline{Z}_1''}$$

By applying the principle of superposition, the load currents circulating in the system prior to the fault may be added to the currents circulating in the system due to the fault, to give the total current in any branch of the system at the time of fault inception. However, in most problems, the load current is small in comparison to the fault current and is usually ignored.

In a practical power system, the system regulation is such that the load voltage at any point in the system is within 10% of the declared open-circuit voltage at that point. For this reason, it is usual to regard the pre-fault voltage at the fault as being the open-circuit voltage, and this assumption is also made in a number of the standards dealing with fault level calculations.

The section on Network Reduction in chapter 3, provided an example of how to reduce a three-phase network. We will use this circuit for an example of some practical three-phase fault calculations. With the network reduced as shown in Figure 4.2, the load voltage at A before the fault occurs is:

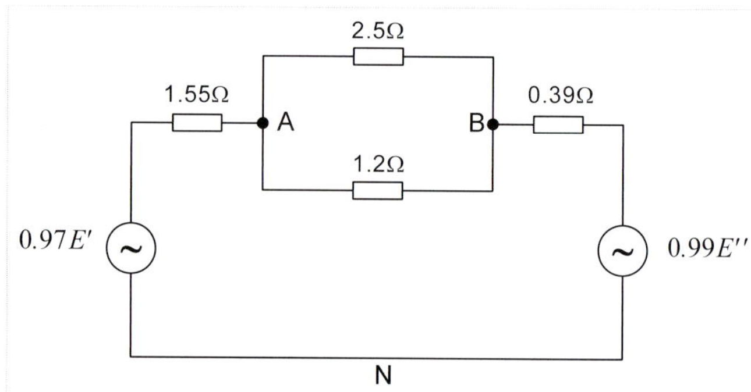

Figure 4.2: Reduction of typical power system network

$$\overline{V} = 0.97\overline{E}' - 1.55\overline{I}$$

$$\overline{V} = 0.99\overline{E}'' + \left(\frac{1.2 \times 2.5}{2.5 + 1.2} + 0.39\right)\overline{I} = 0.99\overline{E}'' + 1.2\overline{I}$$

For practical working conditions, $\overline{E}' >>> 1.55\overline{I}$ and $\overline{E}'' >>> 1.2\overline{I}$. Hence $\overline{E}' \cong \overline{E}'' \cong \overline{V}$

Replacing the driving voltages $\overline{E}'$ and $\overline{E}''$ by the load voltage $\overline{V}$ between A and N modifies the circuit as shown in Figure 4.3(a).

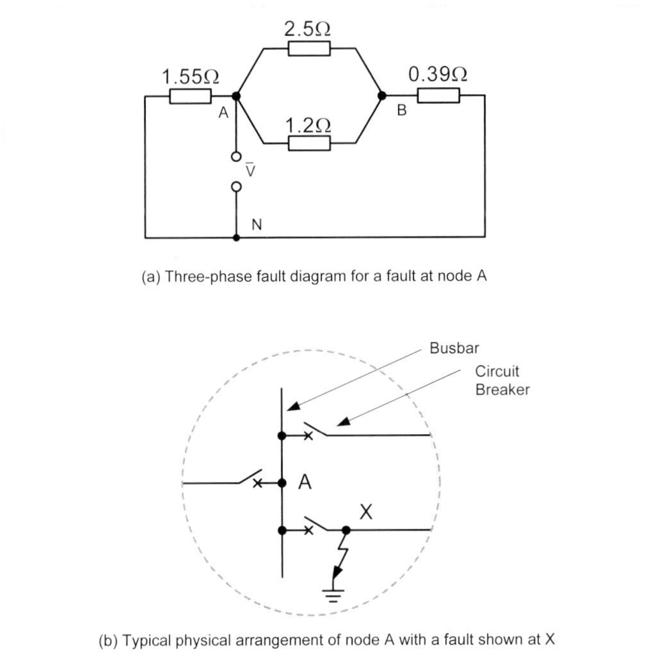

Figure 4.3: Network with fault at node A

The node A is the junction of three branches. In practice, the node would be a busbar, and the branches are feeders radiating from the bus via the closed circuit breakers, as shown in Figure 4.3(b). There are two possible locations for a fault at A; the busbar side of the breakers or the line side of one of the breakers. In this example, let us assumed that the fault is at X, and we wish to calculate the current flowing from the bus to X.

The network viewed from AN has a driving point impedance:

$$|Z_1| = \frac{1.5 \times 1.201}{1.5 + 1.201} = 0.68\Omega$$

The current in the fault is: $\left|\frac{V}{Z_1}\right| = \frac{V}{0.68}$

Let this current be 1.0 per unit. It is now necessary to find the fault current distribution in the various branches of the network and in particular the current flowing from A to X on the assumption that a relay at X is to detect the fault condition. The equivalent impedances viewed from either side of the fault are shown in Figure 4.4(a).

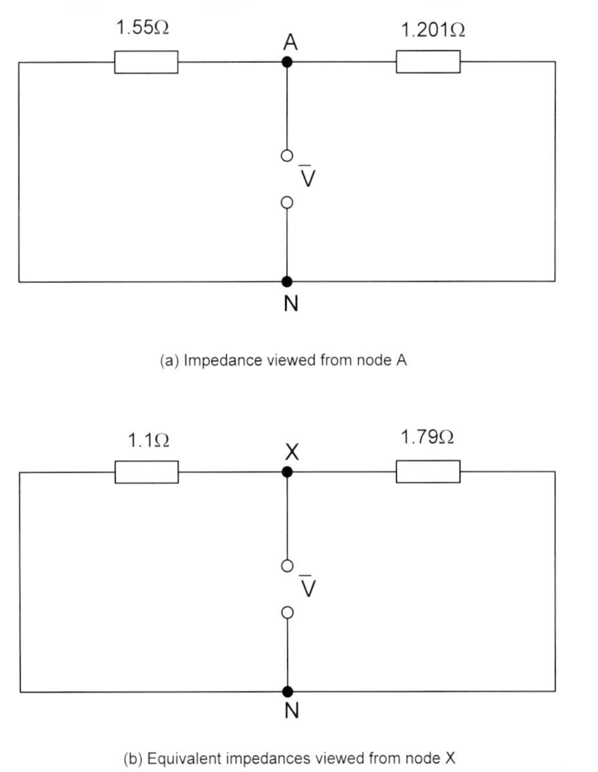

Figure 4.4: Impedances viewed from fault

The currents from Figure 4.4(a) are as follows:

From the right: $\frac{1.55}{1.55+1.201} = \frac{1.55}{2.751} = 0.563\,p.u.$

From the left: $\frac{1.201}{1.55+1.201} = \frac{1.201}{2.751} = 0.437\,p.u.$

There is a parallel branch to the right of A.

The current in the 2.5 ohm branch is:

$$\frac{1.2 \times 0.562}{2.5 + 1.2} = 0.182\,p.u.$$

and the current in 1.2 ohm branch

$$\frac{2.5 \times 0.562}{2.5 + 1.2} = 0.38\,p.u.$$

The total current entering from A to X, is $0.437+0.182 = 0.62\ p.u.$ and from B to X is $0.38 p.u.$ The equivalent network as viewed from the relay is as shown in Figure 4.4(b). The impedances on either side are:

$$\frac{0.68}{0.62} = 1.1\Omega \text{ and } \frac{0.68}{0.38} = 1.79\Omega$$

The circuit of Figure 4.4(b) has been included because the Protection Engineer is interested in these equivalent parameters when applying certain types of protection relay.

4.3 SYMMETRICAL COMPONENT ANALYSIS OF A THREE-PHASE NETWORK

It is necessary to consider the fault currents due to many different types of fault. The most common type of fault is a single-phase to earth fault, which in LV systems, can produce a higher fault current than a three-phase fault. A method of analysis that applies to unbalanced faults is required. By applying the 'Principle of Superposition', any general three-phase system of vectors may be replaced by three sets of balanced (symmetrical) vectors; two sets being three-phase but having opposite phase rotation and one set being co-phasal. These vector sets are described as the positive, negative and zero sequence sets respectively.

The equations between phase and sequence voltages are given below:

$$\overline{E}_a = \overline{E}_1 + \overline{E}_2 + \overline{E}_0$$

$$\overline{E}_b = a^2\overline{E}_1 + a\overline{E}_2 + \overline{E}_0$$

$$\overline{E}_c = a\overline{E}_1 + a^2\overline{E}_2 + \overline{E}_0$$

Equation 4.1

$$\overline{E}_1 = \frac{1}{3}\left(\overline{E}_a + a\overline{E}_b + a^2\overline{E}_c\right)$$

$$\overline{E}_2 = \frac{1}{3}\left(\overline{E}_a + a^2\overline{E}_b + a\overline{E}_c\right)$$

$$\overline{E}_0 = \frac{1}{3}\left(\overline{E}_a + \overline{E}_b + \overline{E}_c\right)$$

Equation 4.2

where all quantities are referred to the reference phase *A*. A similar set of equations can be written for phase and sequence currents. Figure 4.5 illustrates the resolution of a system of unbalanced vectors.

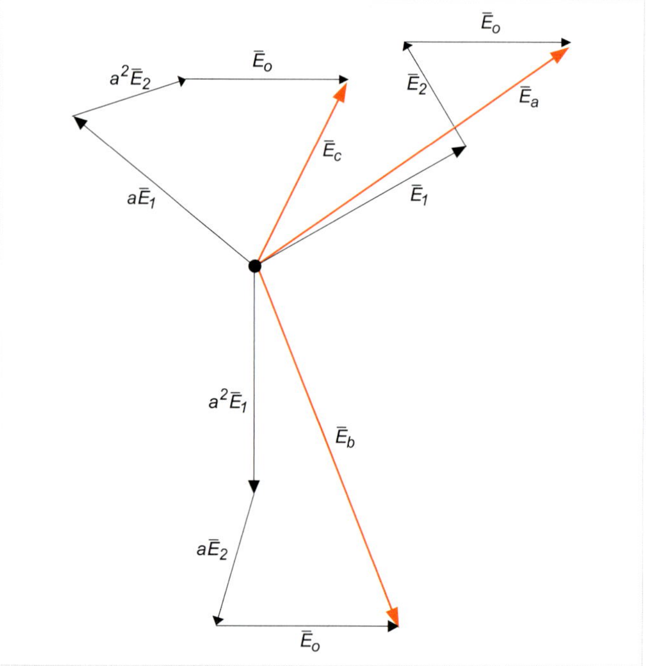

Figure 4.5: Resolution of a system of unbalanced vectors

When a fault occurs in a power system, the phase impedances are no longer identical (except in the case of three-phase faults) and the resulting currents and voltages are unbalanced, the point of greatest unbalance being at the fault point. We have shown in Chapter 3 that the fault may be studied by short-circuiting all normal driving voltages in the system and replacing the fault connection by a source whose driving voltage is equal to the pre-fault voltage at the fault point. Hence, the system impedances remain symmetrical, viewed from the fault, and the fault point may now be regarded as the point of injection of unbalanced voltages and currents into the system.

This is a most important approach in defining the fault conditions since it allows the system to be represented by sequence networks [4.3] using the method of symmetrical components

4.3.1 Positive Sequence Network

During normal balanced system conditions, only positive sequence currents and voltages can exist in the system, and therefore the normal system impedance network is a positive sequence network

When a fault occurs the current in the fault branch changes from 0 to $\overline{I}_1$ and the positive sequence voltage across the branch changes from $\overline{V}$ to $\overline{V}_1$; replacing the fault branch by a source equal to the change in voltage and short-circuiting all normal driving voltages in the system results in a current $\Delta\overline{I}$ flowing into the system, and:

$$\Delta\bar{I} = -\frac{\bar{V}-\bar{V}_1}{\bar{Z}_1}$$

Equation 4.3

where $\bar{Z}_1$ is the positive sequence impedance of the system viewed from the fault.

As before the fault no current was flowing from the fault into the system, it follows that $\bar{I}_1$, the fault current flowing from the system into the fault must equal $-\Delta\bar{I}$. Therefore:

$$\bar{V}_1 = \bar{V} - \bar{I}_1\bar{Z}_1$$

Equation 4.4

is the relationship between positive sequence currents and voltages in the fault branch during a fault.

In Figure 4.6, which represents a simple system, the voltage drops $\bar{I}'_1\bar{Z}'_1$ and $\bar{I}''_1\bar{Z}''_1$ are equal to $(\bar{V}-\bar{V}_1)$ where the currents $\bar{I}_1'$ and $\bar{I}_1''$ enter the fault from the left and right respectively and impedances $\bar{Z}_1'$ and $\bar{Z}_1''$ are the total system impedances viewed from either side of the fault branch. The voltage $\bar{V}$ is equal to the open-circuit voltage in the system, and it has been shown that $\bar{V} \cong \bar{E}' \cong \bar{E}''$ (see chapter 3). So the positive sequence voltages in the system due to the fault are greatest at the source, as shown in the gradient diagram, Figure 4.6(b).

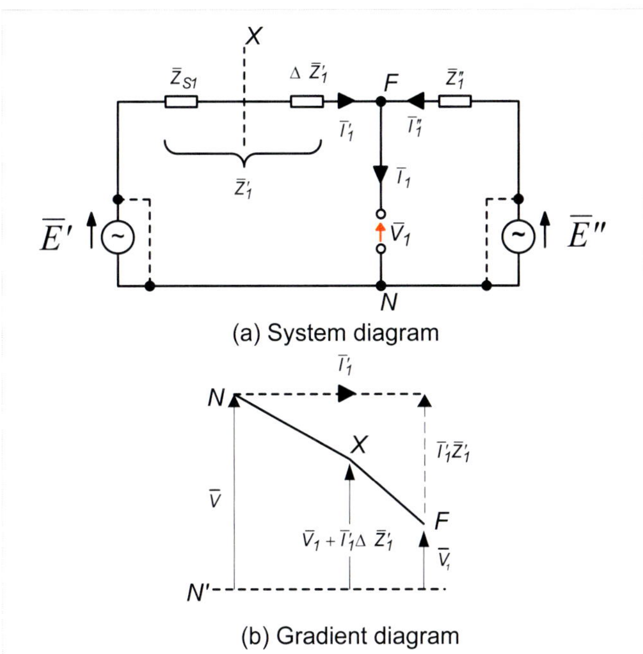

Figure 4.6: Fault at F: Positive sequence diagrams

4.3.2 Negative Sequence Network

If only positive sequence quantities appear in a power system under normal conditions, then negative sequence quantities can only exist during an unbalanced fault.

If no negative sequence quantities are present in the fault branch prior to the fault, then, when a fault occurs, the change in voltage is $\bar{V}_2$, and the resulting current $\bar{I}_2$ flowing from the network into the fault is:

$$\bar{I}_2 = \frac{-\bar{V}_2}{\bar{Z}_2}$$

Equation 4.5

The impedances in the negative sequence network are generally the same as those in the positive sequence network. In machines $\bar{Z}_1 \neq \bar{Z}_2$, but the difference is generally ignored, particularly in large networks.

The negative sequence diagrams, shown in Figure 4.7, are similar to the positive sequence diagrams, with two important differences; no driving voltages exist before the fault and the negative sequence voltage $\bar{V}_2$ is greatest at the fault point.

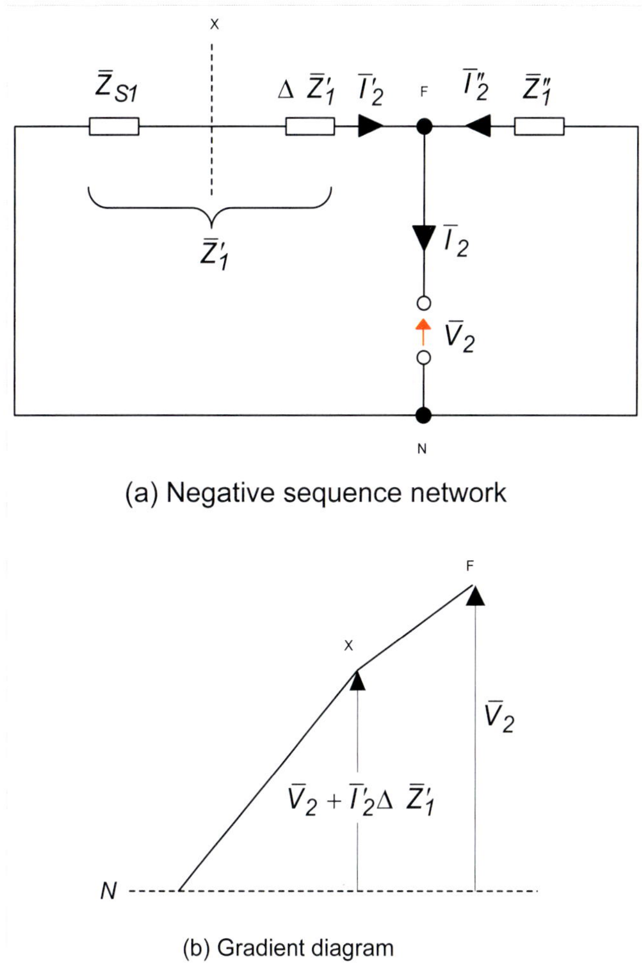

Figure 4.7: Fault at F: Negative sequence diagrams

4.3.3 Zero Sequence Network

The zero sequence current and voltage relationships during a fault condition are the same as those in the negative sequence network. Hence:

$$\overline{V}_0 = -\overline{I}_0\overline{Z}_0$$

Equation 4.6

Also, the zero sequence diagram is that of Figure 4.7, substituting $\overline{I}_0$ for $\overline{I}_2$, and so on.

The currents and voltages in the zero sequence networks are co-phasal, that is, all the same phase. For zero sequence currents to flow in a system there must be a return connection through either a neutral conductor or the general mass of earth. Note must be taken of this fact when determining zero sequence equivalent circuits. Further, in general $\overline{Z}_1 \neq \overline{Z}_0$ and the value of $\overline{Z}_0$ varies according to the type of plant, the winding arrangement and the method of earthing.

4.4 EQUATIONS AND NETWORK CONNECTIONS FOR VARIOUS TYPES OF FAULTS

The most important types of faults are as follows:

- single-phase to earth
- phase to phase
- phase-phase-earth
- three-phase (with or without earth)

The above faults are described as single shunt faults because they occur at one location and involve a connection between one phase and another or to earth.

In addition, the Protection Engineer often studies two other types of fault:

- single-phase open circuit
- cross-country fault

By determining the currents and voltages at the fault point, it is possible to define the fault and connect the sequence networks to represent the fault condition. From the initial equations and the network diagram, the nature of the fault currents and voltages in different branches of the system can be determined.

For shunt faults of zero impedance, and neglecting load current, the equations defining the first four of the above faults (using phase-neutral values) can be written down as follows:

Single-phase-earth (A-E)

$$\overline{I}_b = 0$$

$$\overline{I}_c = 0$$

$$\overline{V}_a = 0$$

Equation 4.7

Phase-phase (B-C)

$$\overline{I}_a = 0$$

$$\overline{I}_b = -\overline{I}_c$$

$$\overline{V}_b = \overline{V}_c$$

Equation 4.8

Phase-phase-earth (B-C-E)

$$\overline{I}_a = 0$$

$$\overline{V}_b = 0$$

$$\overline{V}_c = 0$$

Equation 4.9

Three-phase (A-B-C or A-B-C-E)

$$\overline{I}_a + \overline{I}_b + \overline{I}_c = 0$$

$$\overline{V}_a = \overline{V}_b$$

$$\overline{V}_b = \overline{V}_c$$

Equation 4.10

It should be noted from the above that for any type of fault there are three equations that define the fault conditions.

When there is fault impedance, this must be taken into account when writing down the equations. For example, with a single-phase earth fault through fault impedance $\overline{Z}_f$, the equations are re-written:

$$\overline{I}_b = 0$$

$$\overline{I}_c = 0$$

$$\overline{V}_a = \overline{I}_a\overline{Z}_f$$

Equation 4.11

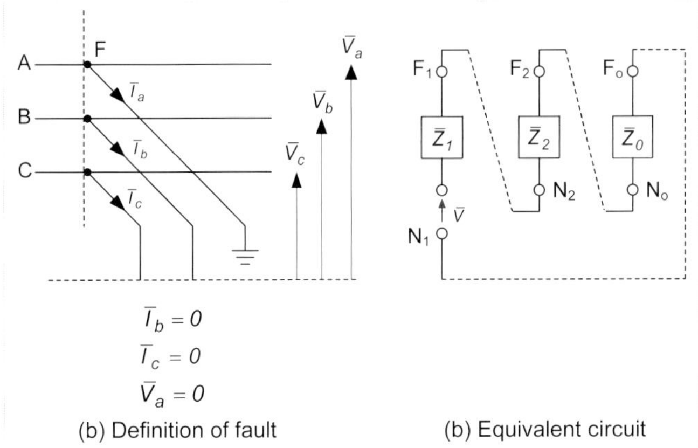

Figure 4.8: Single-phase-earth fault at F

4.4.1 Single-phase-earth Fault (A-E)

Consider a fault defined by Equation 4.7 and by Figure 4.8(a). Converting Equation 4.7 into sequence quantities by using Equation 4.1 and Equation 4.2, then:

$$\overline{I_1} = \overline{I_2} = \overline{I_0} = \frac{1}{3}\overline{I_a}$$

Equation 4.12

$$\overline{V}_1 = -\left(\overline{V}_2 + \overline{V}_0\right)$$

Equation 4.13

Substituting for $\overline{V_1}$, $\overline{V_2}$ and $\overline{V_0}$ in Equation 4.13 from Equation 4.4, Equation 4.5 and Equation 4.6:

$$\overline{V} - \overline{I}_1\overline{Z}_1 = \overline{I}_2\overline{Z}_2 + \overline{I}_0\overline{Z}_0$$

but, $\overline{I_1} = \overline{I_2} = \overline{I_0}$, therefore:

$$\overline{V} = \overline{I}_1\left(\overline{Z}_1 + \overline{Z}_2 + \overline{Z}_3\right)$$

Equation 4.14

The constraints imposed by Equation 4.12 and Equation 4.14 and indicate that the equivalent circuit for the fault is obtained by connecting the sequence networks in series, as shown in Equation 4.8(b)

4.4.2 Phase-phase Fault (B-C)

From Equation 4.8 and using Equation 4.1 and Equation 4.2:

$$\overline{I_1} = -\overline{I_2}$$

Equation 4.15

$$\overline{I_0} = 0$$

$$\overline{V_1} = \overline{V_2}$$

Equation 4.16

From network Equation 4.4, Equation 4.5 and Equation 4.16 can be re-written:

$$\overline{V} - \overline{I}_1\overline{Z}_1 = -\overline{I}_2\overline{Z}_2$$

and substituting for $\overline{I_2}$ from Equation 4.15:

$$\overline{V} = \overline{I}_1\left(\overline{Z}_1 + \overline{Z}_2\right)$$

Equation 4.17

The constraints imposed by Equations 4.15 and 4.17 indicate that there is no zero sequence network connection in the equivalent circuit and that the positive and negative sequence networks are connected in parallel. Figure 4.9 shows the defining and equivalent circuits satisfying the above equations

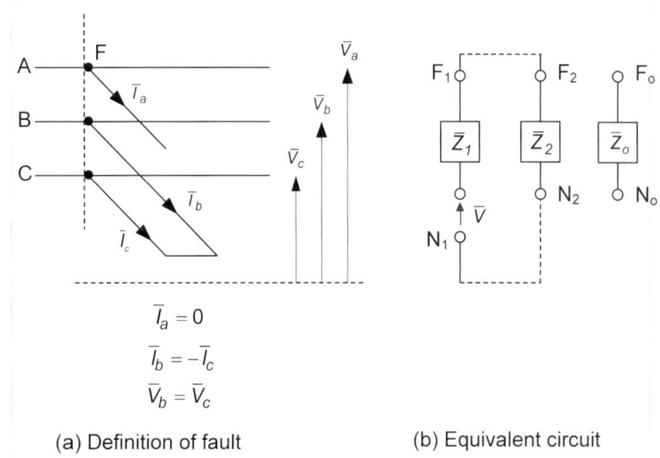

Figure 4.9: Phase-phase fault at F

4.4.3 Phase-phase-earth Fault (B-C-E)

Again, from Equation 4.9 and Equations 4.1 and 4.2:

$$\overline{I}_1 = -\left(\overline{I}_2 + \overline{I}_0\right)$$

Equation 4.18

and

$$\overline{V_1} = \overline{V_2} = \overline{V_0}$$

Equation 4.19

Substituting for $\overline{V_2}$ and $\overline{V_0}$ using Equation 4.5 and Equation 4.6:

$$\overline{I}_2\overline{Z}_2 = \overline{I}_0\overline{Z}_0$$

Thus, using Equation 4.18:

$$\overline{I}_0 = -\frac{\overline{Z}_2\overline{I}_1}{\overline{Z}_0 + \overline{Z}_2}$$

Equation 4.20

and

$$\bar{I}_2 = -\frac{\bar{Z}_0 \bar{I}_1}{\bar{Z}_0 + \bar{Z}_2}$$

Equation 4.21

Now equating $\bar{V}_1$ and $\bar{V}_2$ and using Equation 4.4 gives:

$$\bar{V} - \bar{I}_1 \bar{Z}_1 = -\bar{I}_2 \bar{Z}_2$$

or

$$\bar{V} = \bar{I}_1 \bar{Z}_1 - \bar{I}_2 \bar{Z}_2$$

Substituting for $\bar{I}_2$ from Equation 4.21:

$$\bar{V} = \left[\bar{Z}_1 + \frac{\bar{Z}_0 \bar{Z}_2}{\bar{Z}_0 + \bar{Z}_2} \right] \bar{I}_1$$

or

$$\bar{I}_1 = \bar{V} \frac{\left(\bar{Z}_0 + \bar{Z}_2\right)}{\bar{Z}_1 \bar{Z}_0 + \bar{Z}_1 \bar{Z}_2 + \bar{Z}_0 \bar{Z}_2}$$

Equation 4.22

From the above equations it follows that connecting the three sequence networks in parallel as shown in Equation 4.10(b) may represent a phase-phase-earth fault

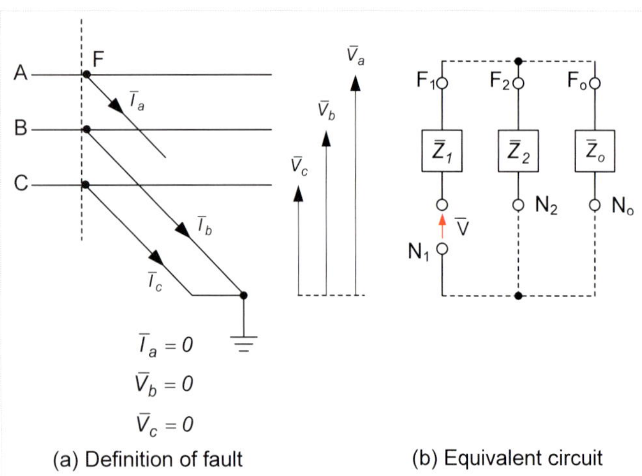

Figure 4.10: Phase-phase-earth fault

4.4.4 Three-phase Fault (A-B-C or A-B-C-E)

Assuming that the fault includes earth, then, from Equation 4.1 Equation 4.2 and Equation 4.10, it follows that:

$$\bar{V}_0 = \bar{V}_A$$

$$\bar{V}_1 = \bar{V}_2 = 0$$

Equation 4.23

and

$$\bar{I}_0 = 0$$

Equation 4.24

Substituting $\bar{V}_2 = 0$ in Equation 4.5 gives:

$$\bar{I}_2 = 0$$

Equation 4.25

and substituting $\bar{V}_1 = 0$ in Equation 4.4 gives:

$$0 = \bar{V} - \bar{I}_1 \bar{Z}_1$$

or

$$\bar{V} = \bar{I}_1 \bar{Z}_1$$

Equation 4.26

Further, since from Equation 4.24 $\bar{I}_0 = 0$, it follows from Equation 4.6 that $\bar{V}_0$ is zero when $\bar{Z}_0$ is finite. The equivalent sequence connections for a three-phase fault are shown in Figure 4.11.

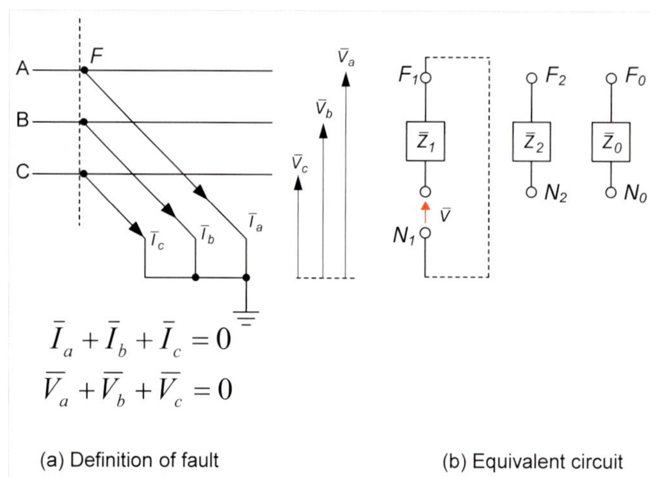

Figure 4.11: Three-phase-earth fault at F

4.4.5 Single-phase Open Circuit Fault

The single-phase open circuit fault is shown diagrammatically in Figure 4.12(a). At the fault point, the boundary conditions are:

$$\bar{I}_a = 0$$

$$\bar{V}_b = \bar{V}_c = 0$$

Equation 4.27

Hence, from Equation 4.2,

$$\overline{V_0} = \overline{V_1} = \overline{V_2} = \frac{1}{3}\overline{V_a}$$

$$\overline{I_a} = \overline{I_1} + \overline{I_2} + \overline{I_0} = 0$$

Equation 4.28

From Equation 4.8, it can be concluded that the sequence networks are connected in parallel, as shown in Figure 4.12(b).

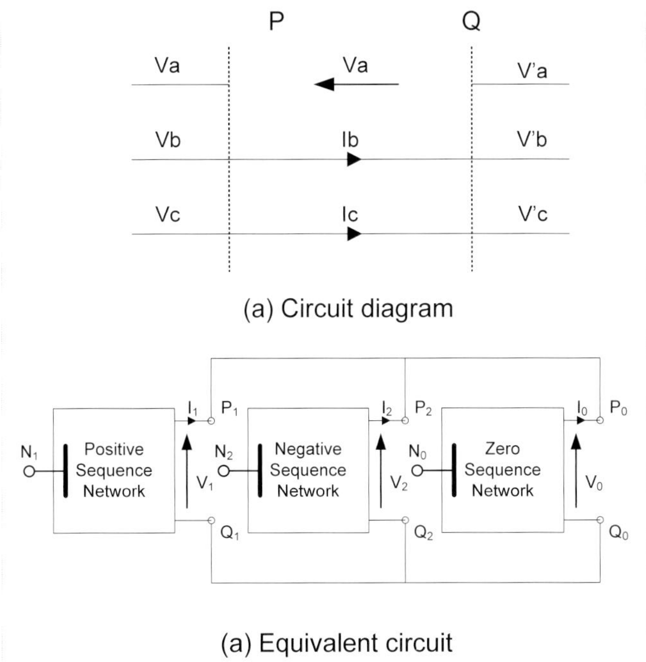

Figure 4.12: Open circuit on phase A

4.4.6 Cross-Country Faults

A cross-country fault is one where there are two faults affecting the same circuit, but in different locations and possibly involving different phases. Figure 4.13(a) illustrates this.

The constraints expressed in terms of sequence quantities are as follows:

a) at point F

$$\overline{I_b} + \overline{I_c} = 0$$

$$\overline{V_A} = 0$$

Equation 4.29

Therefore:

$$\overline{I_{a1}} = \overline{I_{a2}} = \overline{I_{a0}}$$

$$\overline{V_{a1}} + \overline{V_{a2}} + \overline{V_{a0}} = 0$$

Equation 4.30

b) at point F'

$$\bar{I'}_a = \bar{I'}_c = 0$$

$$\overline{V'}_b = 0$$

Equation 4.31

and therefore:

$$\bar{I'}_{b1} = \bar{I'}_{b2} = \bar{I'}_{b0}$$

Equation 4.32

To solve, it is necessary to convert the currents and voltages at point F' to the sequence currents in the same phase as those at point F. From Equation 4.32,

$$a^2\bar{I'}_{a1} = a\bar{I'}_{a2} = \bar{I'}_{a0}$$

or

$$\bar{I'}_{a1} = a^2\bar{I'}_{a2} = a\bar{I'}_{a0}$$

Equation 4.33

and, for the voltages

$$\overline{V'}_{b1} + \overline{V'}_{b2} + \overline{V'}_{b0} = 0$$

Converting:

$$a^2\overline{V'}_{a1} + a\overline{V'}_{a2} + \overline{V'}_{a0} = 0$$

or

$$\overline{V'}_{a1} + a^2\overline{V'}_{a2} + a\overline{V'}_{a0} = 0$$

Equation 4.34

The fault constraints involve phase shifted sequence quantities. To construct the appropriate sequence networks, it is necessary to introduce phase-shifting transformers to couple the sequence networks. This is shown in Figure 4.13(b).

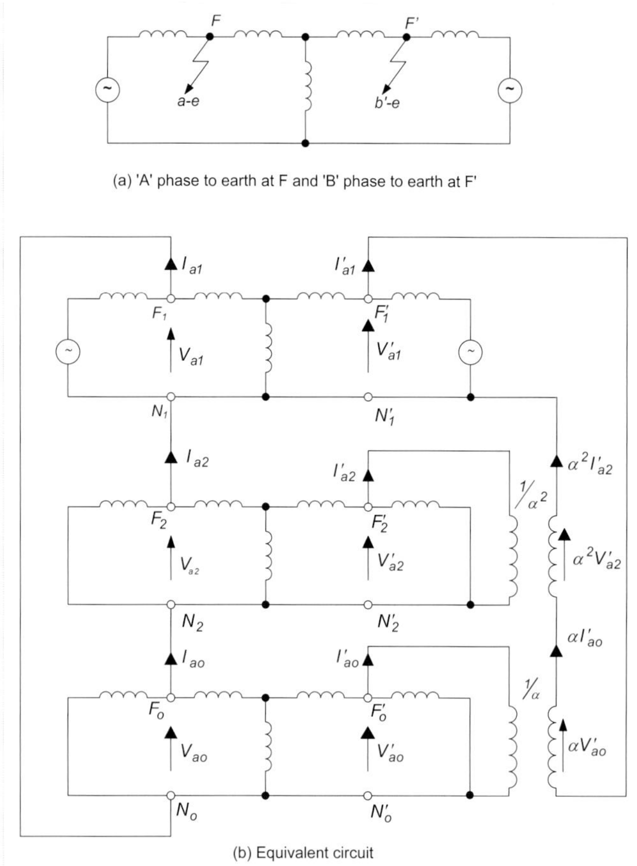

Figure 4.13: Cross-country fault: phase A to phase B

4.5 CURRENT AND VOLTAGE DISTRIBUTION IN A SYSTEM DUE TO A FAULT

Practical fault calculations involve the examination of the effect of a fault in branches of network other than the faulted branch, so that protection can be applied correctly to isolate the section of the system directly involved in the fault. It is therefore not enough to calculate the fault current in the fault itself; the fault current distribution must also be established. Further, abnormal voltage stresses may appear in a system because of a fault, and these may affect the operation of the protection. Knowledge of current and voltage distribution in a network due to a fault is essential for the application of protection.

The approach to network fault studies for assessing the application of protection equipment may be summarised as follows:

- from the network diagram and accompanying data, assess the limits of stable generation and possible operating conditions for the system

NOTE: When full information is not available assumptions may have to be made

- with faults assumed to occur at each relaying point in turn, maximum and minimum fault currents are calculated for each type of fault

NOTE: The fault is assumed to be through zero impedance

- by calculating the current distribution in the network for faults applied at different points in the network the maximum through fault currents at each relaying point are established for each type of fault
- at this stage more or less definite ideas on the type of protection to be applied are formed. Further calculations for establishing voltage variation at the relaying point, or the stability limit of the system with a fault on it, are now carried out in order to determine the class of protection necessary, such as high or low speed, unit or non-unit, etc.

4.5.1 Current Distribution

The phase current in any branch of a network is determined from the sequence current distribution in the equivalent circuit of the fault. The sequence currents are expressed in per unit terms of the sequence current in the fault branch.

In power system calculations, the positive sequence and negative sequence impedances are normally equal. Thus, the division of sequence currents in the two networks will also be identical.

The impedance values and configuration of the zero sequence network are usually different from those of the positive and negative sequence networks, so the zero sequence current distribution is calculated separately.

If C_0 and C_1 are described as the zero and positive sequence distribution factors then the actual current in a sequence branch is given by multiplying the actual current in the sequence fault branch by the appropriate distribution factor. For this reason, if $\bar{I}_1$, $\bar{I}_2$ and $\bar{I}_0$ are sequence currents in an arbitrary branch of a network due to a fault at some point in the network, then the phase currents in that branch may be expressed in terms of the distribution constants and the sequence currents in the fault. These are given below for the various common shunt faults, using Equation 4.1 and the appropriate fault equations:

a. Single-phase-earth (A-E)

$$\bar{I}'_a = (2C_1 + C_0)\bar{I}_0$$

$$\bar{I}'_b = -(C_1 - C_0)\bar{I}_0$$

$$\bar{I}'_c = -(C_1 + C_0)\bar{I}_0$$

Equation 4.35

b. Phase-phase (B-C)

$$\bar{I}'_a = 0$$

$$\bar{I}'_b = \left(a^2 - a\right)C_1\bar{I}_1$$

$$\bar{I}'_c = \left(a - a^2\right)C_1\bar{I}_1$$

Equation 4.36

c. Phase-phase-earth (B-C-E)

$$\bar{I}'_a = -\left(C_1 - C_0\right)\bar{I}_0$$

$$\bar{I}'_b = \left[\left(a - a^2\right)C_1\frac{\bar{Z}_0}{\bar{Z}_1} - a^2C_1 + C_0\right]\bar{I}_0$$

$$\bar{I}'_c = \left[\left(a^2 - a\right)C_1\frac{\bar{Z}_0}{\bar{Z}_1} - aC_1 + C_0\right]\bar{I}_0$$

Equation 4.37

d. Three-phase (A-B-C or A-B-C-E)

$$\bar{I}'_a = C_1\bar{I}_1$$

$$\bar{I}'_b = a^2C_1\bar{I}_1$$

$$\bar{I}'_c = aC_1\bar{I}_1$$

Equation 4.38

As an example of current distribution technique, consider the system in Figure 4.14(a). The equivalent sequence networks are given in Figure 4.14(b) and Figure 4.14(c), together with typical values of impedances. A fault is assumed at A and it is desired to find the currents in branch OB due to the fault. In each network, the distribution factors are given for each branch, with the current in the fault branch taken as $1.0p.u.$ From the diagram, the zero sequence distribution factor C_0 in branch OB is 0.112 and the positive sequence factor C_1 is 0.373. For an earth fault at A the phase currents in branch OB from Equation 4.35 are:

$$\bar{I}'_a = \left(0.746 + 0.112\right)\bar{I}_0 = 0.858\bar{I}_0$$

And

$$\bar{I}'_b = \bar{I}'_c = -\left(0.373 - 0.112\right)\bar{I}_0 = -0.261\bar{I}_0$$

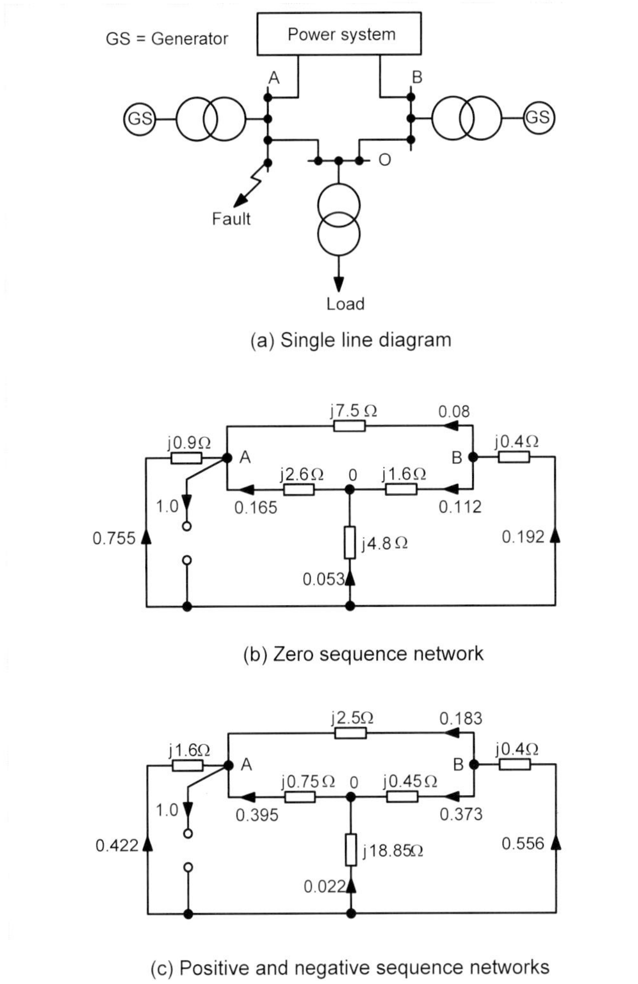

Figure 4.14: Typical power system

By using network reduction methods and assuming that all impedances are reactive, it can be shown that $\bar{Z}_1 = \bar{Z}_0 = j0.68\Omega$.

Therefore, from Equation 4.14, the current in the fault branch is $\left|I_a\right| = \frac{|V|}{0.68}$.

Assuming that $|V| = 63.5V$, then:

$$\left|I_0\right| = \frac{1}{3}\left|I_a\right| = \frac{63.5}{3 \times 0.68} = 31.2A$$

If $\bar{V}$ is taken as the reference vector, then:

$$I'_a = 26.8\angle -90°A$$

$$I'_b = I'_c = 81.5\angle 90°A$$

The vector diagram for the above fault condition is shown in Figure 4.15.

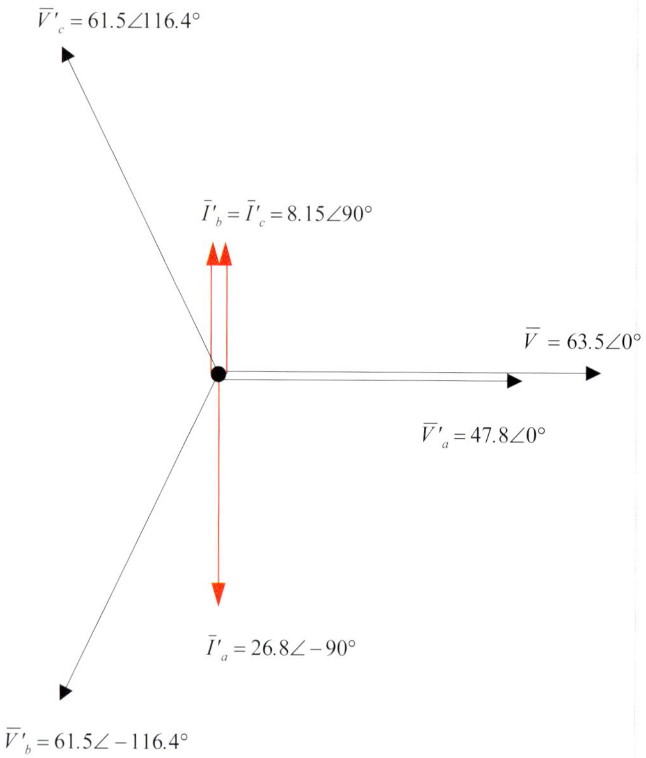

Figure 4.15: Vector diagram: Fault currents and voltages in branch OB due to Phase-to-Earth (P-E) fault at bus A

4.5.2 Voltage Distribution

The voltage distribution in any branch of a network is determined from the sequence voltage distribution. As shown by Equation 4.5, Equation 4.6 and Equation 4.7 and the gradient diagrams Figure 4.6(b) and Figure 4.7(b), the positive sequence voltage is a minimum at the fault, whereas the zero and negative sequence voltages are a maximum. Thus, the sequence voltages in any part of the system may be given generally as:

$$\overline{V'_1} = \bar{V} - \bar{I}_1\left(\bar{Z}_1 - \sum_1^n C_{1n}\Delta\bar{Z}_{1n}\right)$$

$$\overline{V'_2} = -\bar{I}_2\left(\bar{Z}_1 - \sum_1^n C_{1n}\Delta\bar{Z}_{1n}\right)$$

$$\overline{V'_0} = -\bar{I}_0\left(\bar{Z}_0 - \sum_1^n C_{0n}\Delta\bar{Z}_{0n}\right)$$

Equation 4.39

Using the above equation, the fault voltages at bus B in the previous example can be found.

From the positive sequence distribution diagram Figure 4.8(c):

$$\overline{V'_1} = \bar{V} - \bar{I}_1\left[\bar{Z}_1 - j(0.395\times 0.75) + (0.373\times 0.45)\right]$$

$$= \bar{V} - \bar{I}_1\left[\bar{Z}_1 - j0.464\right]$$

and

$$\overline{V'_2} = -\bar{I}_2\left[\bar{Z}_1 - j0.464\right]$$

From the zero sequence distribution diagram Figure 4.8(b):

$$\overline{V'_0} = I_0\left[\bar{Z}_0 - j(0.165\times 2.6) + (0.112\times 1.6)\right]$$

therefore

$$\overline{V'_0} = \bar{I}_0\left[\bar{Z}_0 - j0.608\right]$$

For earth faults, at the fault $\overline{I_1} = \overline{I_2} = \overline{I_0} = j31.2A$, when $|\bar{V}| = 63.5$ volts and is taken as the reference vector. Further,

$$\overline{Z_1} = \overline{Z_0} = j0.68\Omega$$

Hence:

$$\overline{V'_1} = 63.5 - (0.216\times 31.2) = 56.76\angle 0°V$$

$$\overline{V'_2} = 6.74\angle 180°V$$

$$\overline{V'_0} = 2.25\angle 180°V$$

and, using Equation 4.1:

$$\overline{V'_a} = \overline{V'_1} + \overline{V'_2} + \overline{V'_0} = 56.76 - (6.74 + 2.25)$$

Therefore $\overline{V'_a} = 47.8\angle 0°V$

$$\overline{V'_b} = a^2\overline{V'_1} + a\overline{V'_2} + \overline{V'_0} = 56.76a^2 - (6.74a + 2.25)$$

Therefore $\overline{V'_b} = 61.5\angle -116.4°V$

$$\overline{V'_c} = a\overline{V'_1} + a^2\overline{V'_2} + \overline{V'_0} = 56.75a - (6.74a^2 + 2.25)$$

Therefore $\overline{V'_c} = 61.5\angle 116.4°V$

These voltages are shown on the vector diagram, Figure 4.15.

4.6 EFFECT OF SYSTEM EARTHING ON ZERO SEQUENCE QUANTITIES

It has been shown previously that zero sequence currents flow in the earth path during earth faults, and it follows that the nature of these currents will be influenced by the method of earthing. Because these quantities are unique in their association with earth faults they can be utilised in protection, provided their measurement and character are understood for all practical system conditions.

4.6.1 Residual Current and Voltage

Residual currents and voltages depend for their existence on two factors:

a. a system connection to earth at two or more points

b. a potential difference between the earth points resulting in a current flow in the earth paths

Under normal system operation there is a capacitance between the phases and between phase and earth; these capacitances may be regarded as being symmetrical and distributed uniformly through the system. So even when (a) above is satisfied, if the driving voltages are symmetrical the vector sum of the currents will equate to zero and no current will flow between any two earth points in the system. When a fault to earth occurs in a system, an unbalance results in condition (b) being satisfied. From the definitions given above it follows that residual currents and voltages are the vector sum of phase currents and phase voltages respectively.

Hence:

$$\overline{I}_R = \overline{I}_a + \overline{I}_b + \overline{I}_c$$

And

$$\overline{V}_R = \overline{V}_{ae} + \overline{V}_{be} + \overline{V}_{ce}$$

Equation 4.40

Also, from Equation 4.2:

$$\overline{I}_R = 3\overline{I}_0$$

$$\overline{V}_R = 3\overline{V}_0$$

Equation 4.41

It should be further noted that:

$$\overline{V}_{ae} = \overline{V}_{an} + \overline{V}_{ne}$$

$$\overline{V}_{be} = \overline{V}_{bn} + \overline{V}_{ne}$$

$$\overline{V}_{ce} = \overline{V}_{cn} + \overline{V}_{ne}$$

Equation 4.42

and since $\overline{V}_{bn} = a^2\overline{V}_{an}$ and $\overline{V}_{cn} = a\overline{V}_{an}$

then:

$$\overline{V}_R = 3\overline{V}_{ne}$$

Equation 4.43

where $\overline{V}_{ne}$ is the neutral displacement voltage.

Measurements of residual quantities are made using current and voltage transformer connections as shown in Figure 4.16. If relays are connected into the circuits in place of the ammeter and voltmeter, earth faults in the system can be detected.

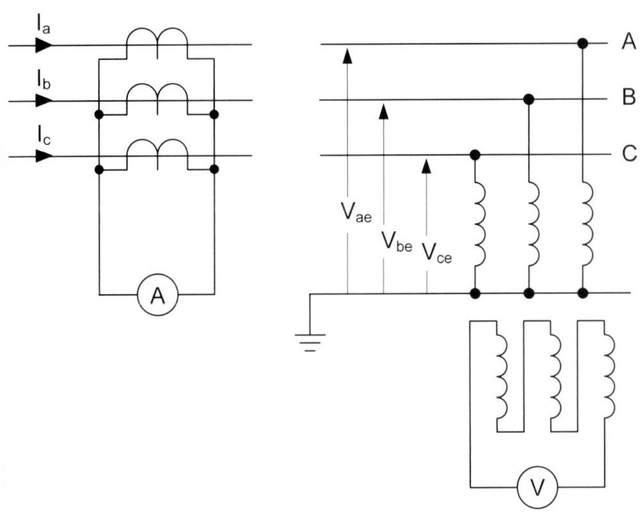

(a) Residual current (b) Residual voltage

Figure 4.16: Measurement of residual quantities

4.6.2 System Ratio

The system $\overline{Z}_0 / \overline{Z}_1$ ratio is defined as the ratio of zero sequence and positive sequence impedances viewed from the fault; it is a variable ratio, dependent upon the method of earthing, fault position and system operating arrangement.

When assessing the distribution of residual quantities through a system, it is convenient to use the fault point as the reference as it is the point of injection of unbalanced quantities into the system. The residual voltage is measured in relation to the normal phase-neutral system voltage and the residual current is compared with the three-phase fault current at the fault point. It can be shown [4.4/4.5] that the character of these quantities can be expressed in terms of the system $\overline{Z}_0 / \overline{Z}_1$ ratio.

The positive sequence impedance of a system is mainly reactive, whereas the zero sequence impedance being affected by the method of earthing may contain both resistive and reactive components of comparable magnitude. Thus the expression for the $\overline{Z}_0 / \overline{Z}_1$ ratio approximates to:

$$\frac{\overline{Z}_0}{\overline{Z}_1} = \frac{\overline{X}_0}{\overline{X}_1} - j\frac{\overline{R}_0}{\overline{X}_1}$$

Equation 4.44

Expressing the residual current in terms of the three-phase current and $\overline{Z}_0 / \overline{Z}_1$ ratio:

a. Single-phase-earth (A-E)

$$\bar{I}_R = \frac{3\bar{V}}{2\bar{Z}_1 + \bar{Z}_0} = \frac{3}{2+\bar{K}}\frac{\bar{V}}{\bar{Z}_1}$$

where $\bar{K} = \frac{\bar{Z}_0}{\bar{Z}_1}$ and $\bar{I}_{3\phi} = \frac{\bar{V}}{\bar{Z}_1}$

Thus:

$$\frac{\bar{I}_R}{\bar{I}_{3\phi}} = \frac{3}{2+\bar{K}}$$

Equation 4.45

b. Phase-phase-earth (B-C-E)

$$\bar{I}_R = 3\bar{I}_0 = -\frac{3\bar{Z}_1}{\bar{Z}_1 + \bar{Z}_0}\bar{I}_1$$

$$\bar{I}_1 = \frac{V\left(\bar{Z}_1 + \bar{Z}_0\right)}{2\bar{Z}_1\bar{Z}_0 + \bar{Z}_1^{\,2}}$$

Hence:

$$\bar{I}_R = -\frac{3\bar{V}\bar{Z}_1}{2\bar{Z}_1\bar{Z}_0 + \bar{Z}_1^{\,2}} = -\frac{3}{\left(2\bar{K}+1\right)}\frac{\bar{V}}{\bar{Z}_1}$$

Therefore:

$$\frac{\bar{I}_R}{\bar{I}_{3\phi}} = -\frac{3}{2\bar{K}+1}$$

Equation 4.46

Similarly, the residual voltages are found by multiplying Equation 4.45 and Equation 4.46 by $-\bar{K}\bar{V}$.

c. Single-phase-earth (A-E)

$$\bar{V}_R = -\frac{3\bar{K}}{2+\bar{K}}\bar{V}$$

Equation 4.47

d. Phase-phase-earth (B-C-E)

$$\bar{V}_R = -\frac{3\bar{K}}{2\bar{K}+1}\bar{V}$$

Equation 4.48

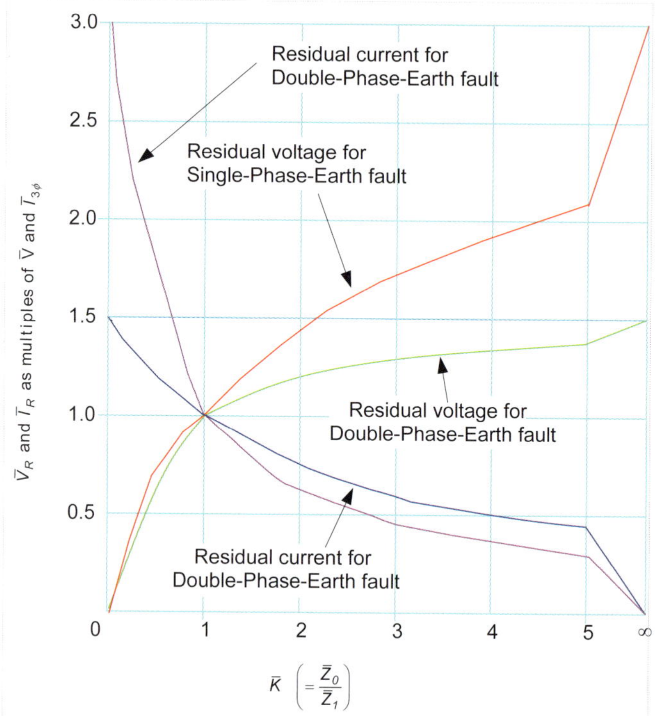

Figure 4.17: Variation of residual quantities at fault point

The curves in Figure 4.17 illustrate the variation of the above residual quantities with the $\bar{Z}_0/\bar{Z}_1$ ratio. The residual current in any part of the system can be obtained by multiplying the current from the curve by the appropriate zero sequence distribution factors. Similarly, the residual voltage is calculated by subtracting from the voltage curve three times the zero sequence voltage drops between the measuring point in the system and the fault.

4.6.3 Variation of Residual Quantities

The variation of residual quantities in a system due to different earth arrangements can be most readily understood by using vector diagrams. Three examples have been chosen, namely solid fault-isolated neutral, solid fault-resistance neutral, and resistance fault-solid neutral. These are illustrated in Figure 4.18, Figure 4.19 and Figure 4.20 respectively.

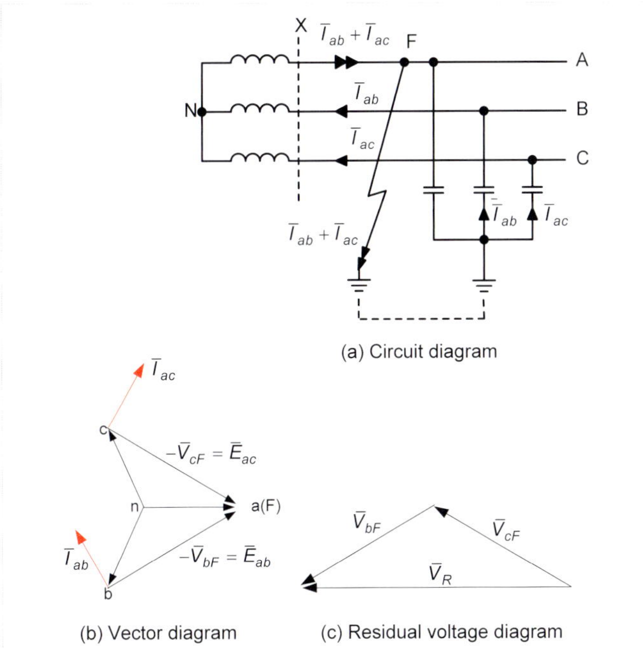

Figure 4.18: Solid fault - isolated neutral

4.6.3.1 Solid fault-isolated neutral

From Figure 4.18 it can be seen that the capacitance to earth of the faulted phase is short circuited by the fault and the resulting unbalance causes capacitance currents to flow into the fault, returning via sound phases through sound phase capacitances to earth.

At the fault point:

$$\overline{V}_{aF} = 0$$

and

$$\overline{V}_R = \overline{V}_{bF} + \overline{V}_{cF} = -3\overline{E}_{an}$$

At source:

$$\overline{V}_R = 3\overline{V}_{ne} = -3\overline{E}_{an}$$

Since

$$\overline{E}_{an} + \overline{E}_{bn} + \overline{E}_{cn} = 0$$

Thus, with an isolated neutral system, the residual voltage is three times the normal phase-neutral voltage of the faulted phase and there is no variation between $\overline{V}_R$ at source and $\overline{V}_R$ at fault.

In practice, there is some leakage impedance between neutral and earth and a small residual current would be detected at X if a very sensitive relay were employed.

4.6.3.2 Solid fault-resistance neutral

Figure 4.19 shows that the capacitance of the faulted phase is short-circuited by the fault and the neutral current combines with the sound phase capacitive currents to give $\overline{I}_a$ in the faulted phase.

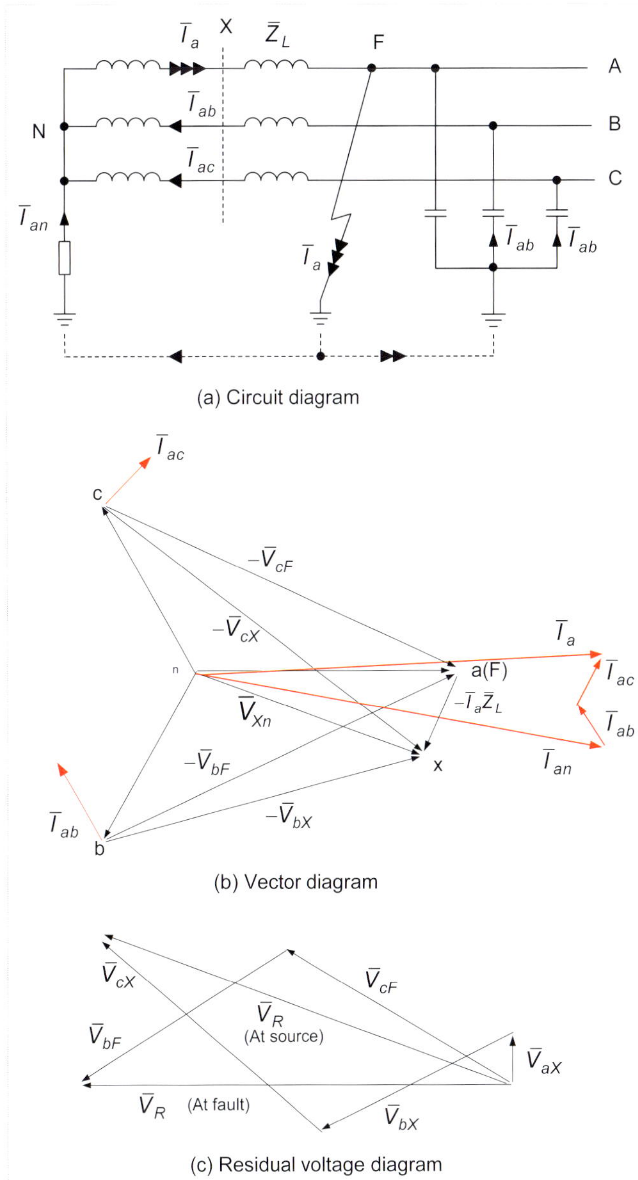

Figure 4.19: Solid fault - resistance neutral

With a relay at X, residually connected as shown in Figure 4.16, the residual current will be $\overline{I}_{an}$, that is, the neutral earth loop current.

At the fault point:

$$\overline{V}_R = \overline{V}_{bF} + \overline{V}_{cF}$$

since $\overline{V}_{Fe} = 0$

At source:

$$\overline{V}_R = \overline{V}_{aX} + \overline{V}_{bX} + \overline{V}_{cX}$$

From the residual voltage diagram it is clear that there is little variation in the residual voltages at source and fault, as most residual voltage is dropped across the neutral resistor. The degree of variation in residual quantities is therefore dependent on the neutral resistor value.

4.6.3.3 Resistance-fault-solid neutral

Capacitance can be neglected because, since the capacitance of the faulted phase is not short-circuited, the circulating capacitance currents will be negligible.

At the fault point:

$$\overline{V}_R = \overline{V}_{Fn} + \overline{V}_{bn} + \overline{V}_{cn}$$

At relaying point X:

$$\overline{V}_R = \overline{V}_{Xn} + \overline{V}_{bn} + \overline{V}_{cn}$$

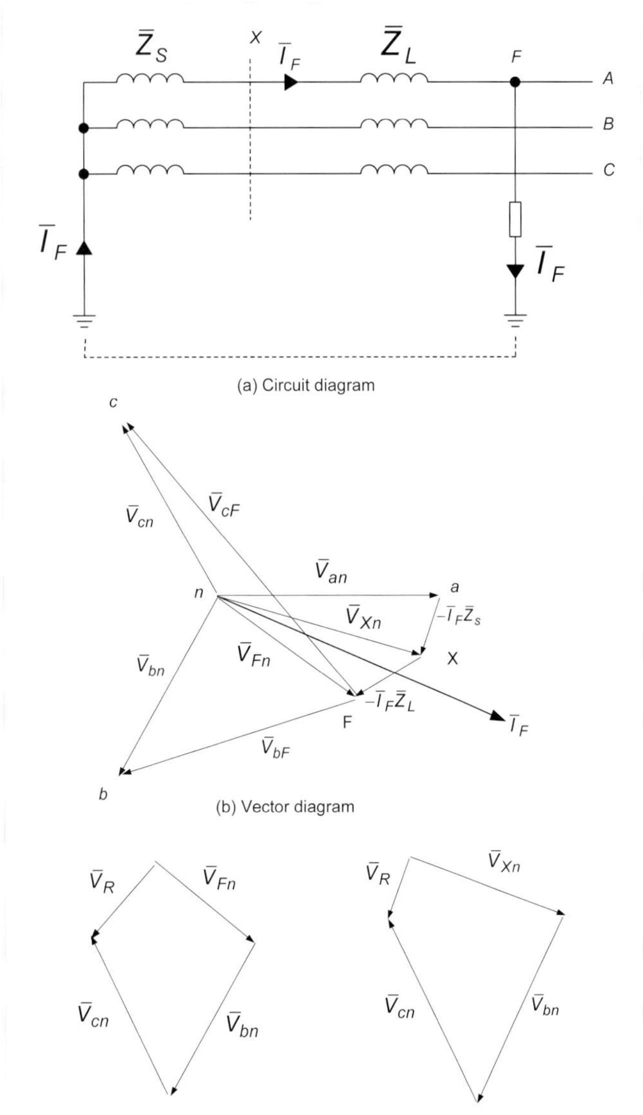

Figure 4.20: Resistance fault -solid neutral

From the residual voltage diagrams shown in Figure 4.20, it is apparent that the residual voltage is greatest at the fault and reduces towards the source. If the fault resistance approaches zero, that is, the fault becomes solid, then $\overline{V_{Fn}}$ approaches zero and the voltage drops in $\overline{Z_S}$ and $\overline{Z_L}$ become greater. The ultimate value of $\overline{V_{Fn}}$ will depend on the effectiveness of the earthing, and this is a function of the system $\overline{Z_0}/\overline{Z_1}$ ratio.

4.7 REFERENCES

[4.1] Circuit Analysis of A.C. Power Systems, Volume I. Edith Clarke. John Wiley & Sons.

[4.2] Method of Symmetrical Co-ordinates Applied to the Solution of Polyphase Networks. C.L. Fortescue. Trans. A.I.E.E.,Vol. 37, Part II, 1918, pp 1027-40.

[4.3] Power System Analysis. J.R. Mortlock and M.W. Humphrey Davies. Chapman and Hall.

[4.4] Neutral Groundings. R Willheim and M. Waters. Elsevier.

[4.5] Fault Calculations. F.H.W. Lackey. Oliver & Boyd

Chapter 5
Equivalent circuits and parameters of power system plant

5.1 INTRODUCTION

Knowledge of the behaviour of the principal electrical system plant items under normal and fault conditions is a prerequisite for the proper application of protection. This chapter summarises basic synchronous machine, transformer and transmission line theory and gives equivalent circuits and parameters so that a fault study can be successfully completed before the selection and application of the protection systems described in later chapters. Only what might be referred to as 'traditional' synchronous machine theory is covered because calculations for fault level studies generally only require this. Readers interested in more advanced models of synchronous machines are referred to the numerous papers on the subject, of which reference [5.1] is a good starting point.

Power system plant can be divided into two broad groups: static and rotating.

The modelling of static plant for fault level calculations provides few difficulties, as plant parameters generally do not change during the period of interest after a fault occurs. The problem in modelling rotating plant is that the parameters change depending on the response to a change in power system conditions.

5.2 SYNCHRONOUS MACHINES

There are two main types of synchronous machine: cylindrical rotor and salient pole. In general, the former is confined to 2 and 4 pole turbine generators, while salient pole types are built with 4 poles upwards and include most classes of duty. Both classes of machine are similar in that each has a stator carrying a three-phase winding distributed over its inner periphery. The rotor is within the stator bore and is magnetised by a d.c. current winding.

The main difference between the two classes of machine is in the rotor construction. The cylindrical rotor type has a cylindrical rotor with the excitation winding distributed over several slots around its periphery. This construction is not suited to multi-polar machines but it is very mechanically sound. It is therefore particularly well suited for the highest speed electrical machines and is universally used for 2 pole units, plus some 4 pole units.

The salient pole type has poles that are physically separate, each carrying a concentrated excitation winding. This type of construction is complementary to that of the cylindrical rotor and is used in machines of 4 poles or more. Except in special cases its use is exclusive in machines of more than 6 poles. Figure 5.1 shows a typical large cylindrical rotor generator installed in a power plant.

Two and four pole generators are most often used in applications where steam or gas turbines are used as the driver. This is because the steam turbine tends to be suited to high rotational speeds. Four pole steam turbine generators are most often found in nuclear power stations as the relative wetness of the steam makes the high rotational speed of a two-pole design unsuitable. Most generators with gas turbine drivers are four pole machines to obtain enhanced mechanical strength in the rotor - since a gearbox is often used to couple the power turbine to the generator, the choice of synchronous speed of the generator is not subject to the same constraints as with steam turbines.

Generators with diesel engine drivers are invariably of four or more pole design, to match the running speed of the driver without using a gearbox. Four-stroke diesel engines usually have a higher running speed than two-stroke engines, so generators having four or six poles are most common. Two-stroke diesel engines are often derivatives of marine designs with relatively large outputs (circa 30MW is possible) and may have running speeds of the order of 125rpm. This requires a generator with a large number of poles (48 for a 125rpm, 50Hz generator) and consequently is of large diameter and short axial length. This is a contrast to turbine-driven machines that are of small diameter and long axial length.

Figure 5.1: Large synchronous generator

5.3 ARMATURE REACTION

Armature reaction has the greatest effect on the operation of a synchronous machine with respect both to the load angle at which it operates and to the amount of excitation that it needs. The phenomenon is most easily explained by considering a simplified ideal generator with full pitch winding operating at unity p.f., zero lag p.f. and zero lead p.f. When operating at unity p.f., the voltage and current in the stator are in phase, the stator current producing a cross magnetising magneto-motive force (m.m.f.) which interacts with that of the rotor, resulting in a distortion of flux across the pole face. As can be seen from Figure 5.2(a) the tendency is to weaken the flux at the leading edge or distort the field in a manner equivalent to a shift against the direction of rotation.

If the power factor is reduced to zero lagging, the current in the stator reaches its maximum 90° after the voltage. The rotor is then in the position shown in Figure 5.2(b) and the stator m.m.f. is acting in direct opposition to the field.

Similarly, for operation at zero leading power factor, the stator m.m.f. directly assists the rotor m.m.f. This m.m.f. arising from current flowing in the stator is known as 'armature reaction'.

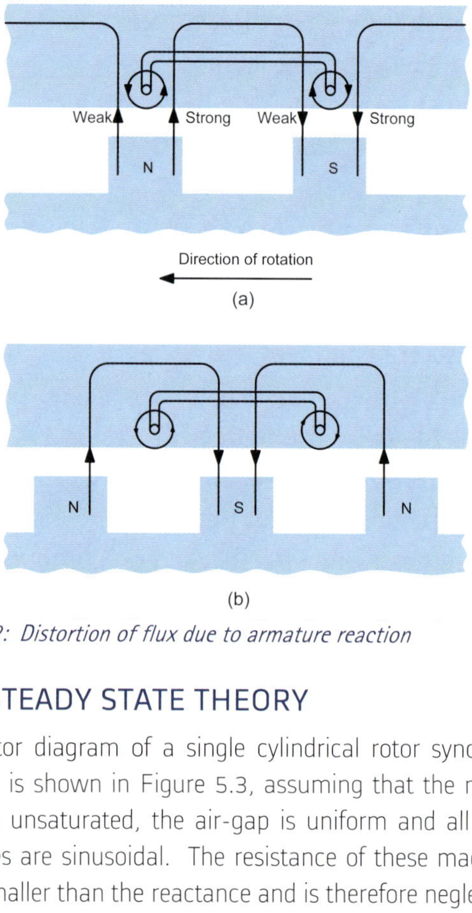

Figure 5.2: Distortion of flux due to armature reaction

5.4 STEADY STATE THEORY

The vector diagram of a single cylindrical rotor synchronous machine is shown in Figure 5.3, assuming that the magnetic circuit is unsaturated, the air-gap is uniform and all variable quantities are sinusoidal. The resistance of these machines is much smaller than the reactance and is therefore neglected.

The excitation ampere-turns AT_e produces a flux Φ across the air-gap which induces a voltage E_t in the stator. This voltage drives a current I at a power factor $cos\ \Phi$ and produces an armature reaction m.m.f. AT_{ar} in phase with it. The m.m.f. AT_f resulting from the combination of these two m.m.f. vectors (see Figure 5.3(a)) is the excitation which must be provided on the rotor to maintain flux Φ across the air gap. Rotating the rotor m.m.f. diagram, Figure 5.3(a), clockwise until AT_e coincides with E_t and changing the scale of the diagram so that AT_e becomes the basic unit, where $AT_e=E_t=1$ results in Figure 5.3(b). The m.m.f. vectors therefore become voltage vectors. For example AT_{ar}/AT_e is a unit of voltage that is directly proportional to the stator load current. This vector can be fully represented by a reactance and in practice this is called 'armature reaction reactance' and is denoted by X_{ad}. Similarly, the remaining side of the triangle becomes AT_f/AT_e which is the per unit voltage produced on open circuit by ampere-turns AT_f. It can be considered as the internal generated voltage of the machine and is designated E_0.

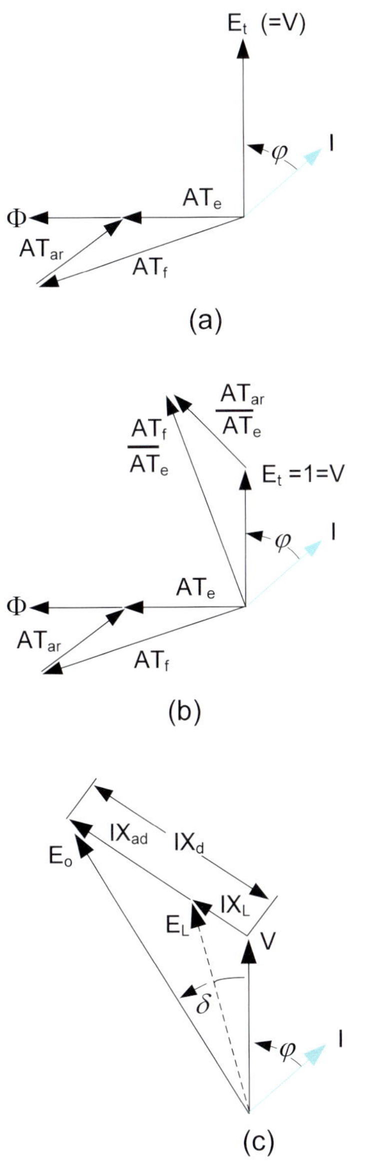

Figure 5.3: Vector diagram of synchronous machine

The true leakage reactance of the stator winding which gives rise to a voltage drop or regulation has been neglected. This reactance is designated X_L (or X_a in some texts) and the voltage drop occurring in it IX_L is the difference between the terminal voltage V and the voltage behind the stator leakage reactance E_L.

IX_L is exactly in phase with the voltage drop due to X_{ad} as shown on the vector diagram Figure 5.3(c). X_{ad} and X_L can be combined to give a simple equivalent reactance; known as the 'synchronous reactance'and denoted by X_d.

The power generated by the machine is given by:

$$P = VI\cos\phi = \frac{VE_o}{X_d}\sin\delta$$

Equation 5.1

where δ is the angle between the internal voltage and the terminal voltage and is known as the load angle of the machine.

It follows from the above analysis that, for steady state performance, the machine may be represented by the equivalent circuit shown in Figure 5.4, where X_L is a true reactance associated with flux leakage around the stator winding and X_{ad} is a fictitious reactance, being the ratio of armature reaction and open-circuit excitation magneto-motive forces.

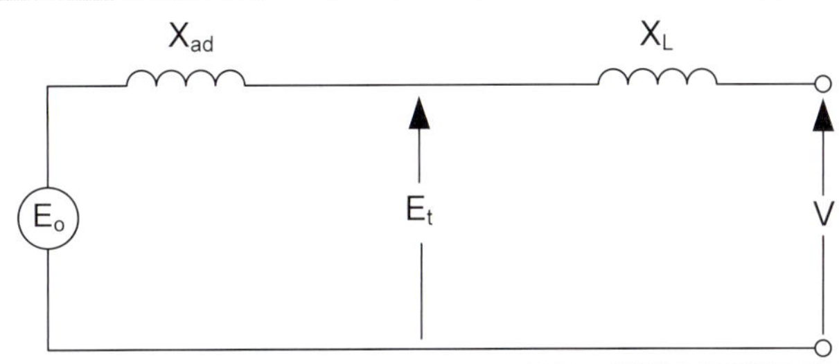

Figure 5.4: Equivalent circuit of elementary synchronous machine

In practice, due to necessary constructional features of a cylindrical rotor to accommodate the windings, the reactance X_a is not constant irrespective of rotor position, and modelling proceeds as for a generator with a salient pole rotor. However, the numerical difference between the values of X_{ad} and X_{aq} is small, much less than for the salient pole machine.

5.5 SALIENT POLE ROTOR

The preceding theory is limited to the cylindrical rotor generator. For a salient pole rotor, the air gap cannot be considered as uniform.. The effect of this is that the flux produced by armature reaction m.m.f. depends on the position of the rotor at any instant, as shown in Figure 5.5.

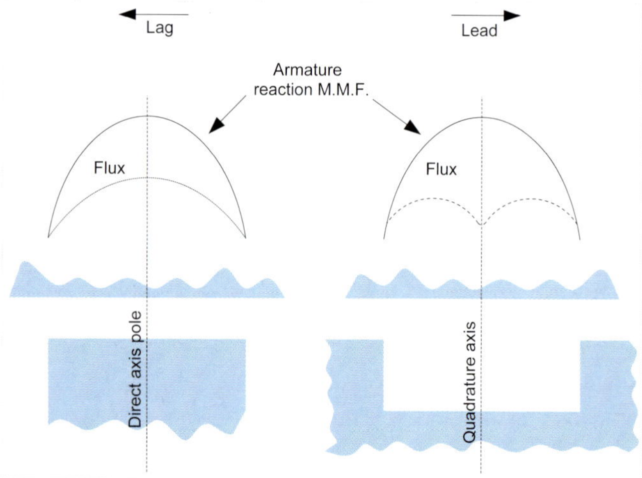

Figure 5.5: Variation of armature reaction m.m.f. with pole position

When a pole is aligned with the assumed sine wave m.m.f. set up by the stator, a corresponding sine wave flux is set up but when an inter-polar gap is aligned very severe distortion is caused. The difference is treated by considering these two axes, that is those corresponding to the pole and the inter-polar gap, separately. They are designated the 'direct' and 'quadrature' axes respectively, and the general theory is known as the 'two axis' theory.

The vector diagram for the salient pole machine is similar to that for the cylindrical rotor except that the reactance and currents associated with them are split into two components. The synchronous reactance for the direct axis is $X_d=X_{ad}+X_L$, while that in the quadrature axis is $X_q=X_{aq}+X_L$. The vector diagram is constructed as before but the appropriate quantities in this case are resolved along two axes. The resultant internal voltage is E_0, as shown in Figure 5.6.

Note that E'_0 is the internal voltage which would be given, in cylindrical rotor theory, by vectorially adding the simple vectors IX_d and V. There is very little difference in magnitude between E'_0 and E_0 but there is a substantial difference in internal angle. The simple theory is perfectly adequate for calculating excitation currents but not for stability considerations where load angle is significant.

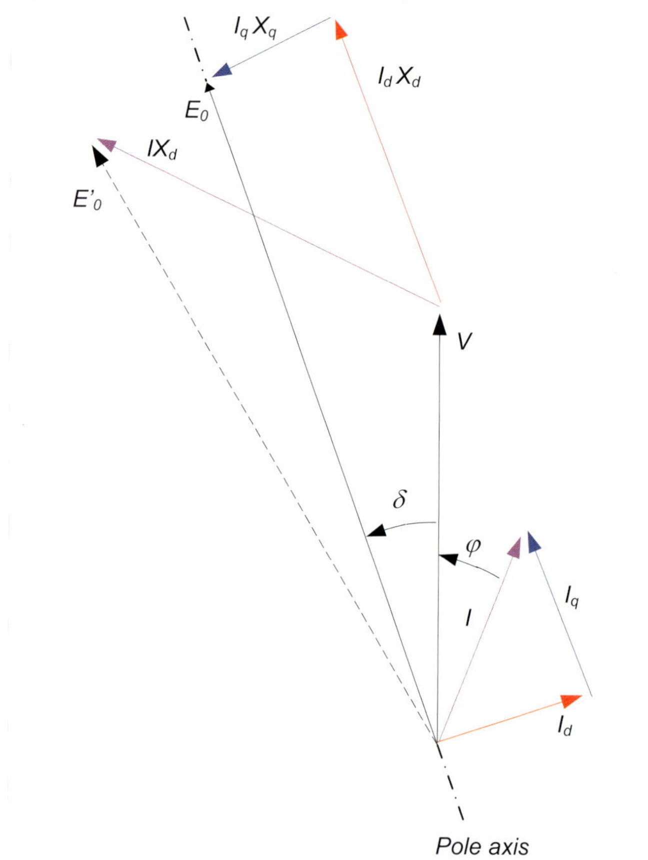

Figure 5.6: Vector diagram for salient pole machine

5.6 TRANSIENT ANALYSIS

For normal changes in load conditions, steady state theory is perfectly adequate. However, there are occasions when almost instantaneous changes are involved, such as faults or switching operations. When this happens new factors are introduced within the machine and to represent these adequately a corresponding new set of machine characteristics is required.

The generally accepted and most simple way to appreciate the meaning and derivation of these characteristics is to consider a sudden three-phase short circuit applied to a machine initially running on open circuit and excited to normal voltage E_0.

This voltage is generated by a flux crossing the air-gap. It is not possible to confine the flux to one path exclusively in any machine so there is a leakage flux Φ_L that leaks from pole to pole and across the inter-polar gaps without crossing the main air-gap as shown in Figure 5.7. The flux in the pole is $\Phi+\Phi_L$.

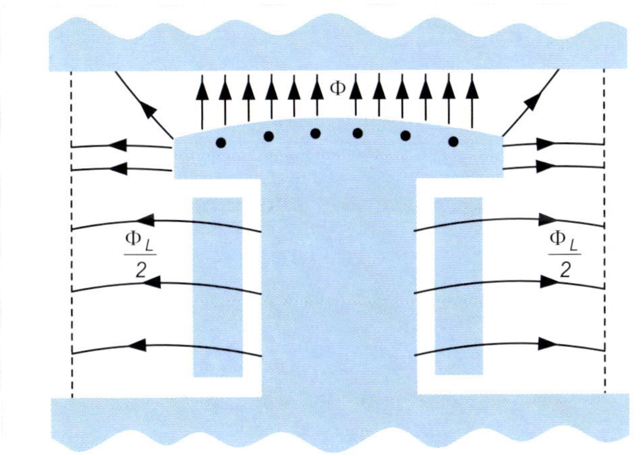

Figure 5.7: Flux paths of salient pole machine

If the stator winding is then short-circuited, the power factor in it is zero. A heavy current tends to flow as the resulting armature reaction m.m.f. is demagnetising. This reduces the flux and conditions settles until the armature reaction nearly balances the excitation m.m.f., the remainder maintaining a very much reduced flux across the air-gap which is just sufficient to generate the voltage necessary to overcome the stator leakage reactance (resistance neglected). This is the simple steady state case of a machine operating on short circuit and is fully represented by the equivalent of Figure 5.8(a); see also Figure 5.4.

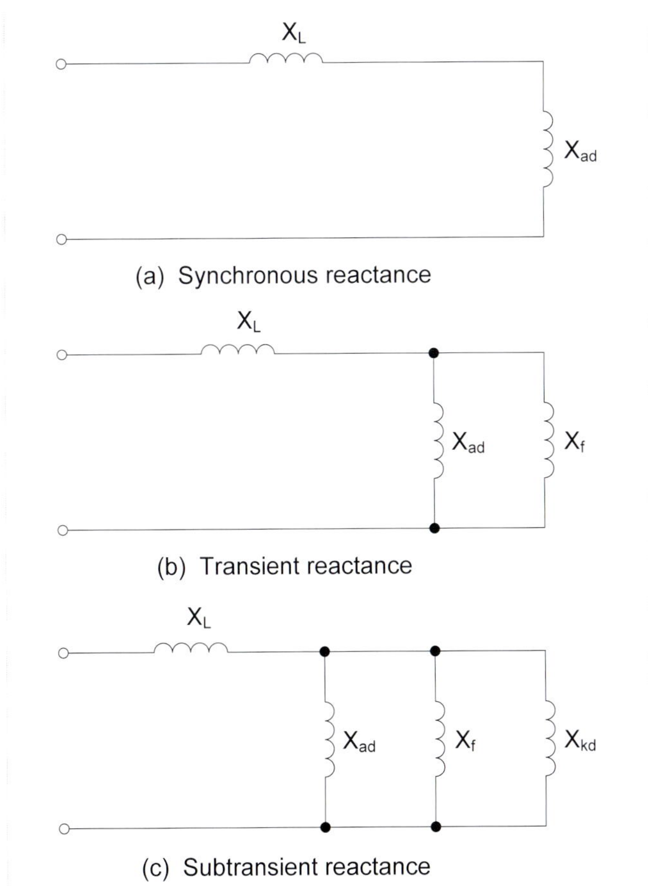

Figure 5.8: Synchronous machine reactances

It might be expected that the fault current would be given by $E_0/(X_L+X_{ad})$ equal to E_0/X_d, but this is very much reduced, and the machine is operating with no saturation. For this reason, the value of voltage used is the value read from the air-gap line corresponding to normal excitation and is higher than the normal voltage. The steady state current is given by:

$$I_d = \frac{E_g}{X_d}$$

Equation 5.2

where E_g = voltage on air gap line

Between the initial and final conditions there has been a severe reduction of flux. The rotor carries a highly inductive winding which links the flux so the rotor flux linkages before the short circuit are produced by $\Phi+\Phi_L$. In practice the leakage flux is distributed over the whole pole and all of it does not link all the winding. Φ_L is an equivalent concentrated flux imagined to link all the winding and of such a magnitude that the total linkages are equal to those actually occurring. It is a fundamental principle that any attempt to change the flux linked with such a circuit causes current to flow in a direction that opposes the change. In the present case the flux is being reduced and so the induced currents tend to sustain it.

For the position immediately following the application of the short circuit, it is valid to assume that the flux linked with the rotor remains constant, this being brought about by an induced current in the rotor which balances the heavy demagnetising effect set up by the short-circuited armature. So $\Phi+\Phi_L$ remains constant, but owing to the increased m.m.f. involved, the flux leakage increases considerably. With a constant total rotor flux, this can only increase at the expense of that flux crossing the air-gap. Consequently, this generates a reduced voltage, which, acting on the leakage reactance X_L, gives the short circuit current.

It is more convenient for machine analysis to use the rated voltage E_0 and to invent a fictitious reactance that gives rise to the same current. This reactance is called the 'transient reactance' X'_d and is defined by the equation:

$$\text{Transient current } I'_d = \frac{E_0}{X'_d}$$

Equation 5.3

It is greater than X_L and the equivalent circuit is represented by Figure 5.8(b) where:

$$X'_d = \frac{X_{ad}X_f}{X_{ad}+X_f} + X_L$$

Equation 5.4

and X_f is the leakage reactance of the field winding

Equation 5.4 may also be written as:

$$X'_d = X_L + X'_f$$

where:

X'_f = effective leakage reactance of field winding

The flux is only be sustained at its relatively high value while the induced current flows in the field winding. As this current decays, conditions approach the steady state. Consequently the duration of this phase is determined by the time constant of the excitation winding. This is usually one second or less - hence the term 'transient' applied to characteristics associated with it.

A further point now arises. All synchronous machines have what is usually called a 'damper winding' or windings. In some cases, this may be a physical winding (like a field winding, but of fewer turns and located separately), or an 'effective' one (for instance, the solid iron rotor of a cylindrical rotor machine). Sometimes, both physical and effective damper windings may exist (as in some designs of cylindrical rotor generators, having both a solid iron rotor and a physical damper winding located in slots in the pole faces).

Under short circuit conditions there is a transfer of flux from the main air-gap to leakage paths. To a small extent this diversion is opposed by the excitation winding and the main transfer is experienced towards the pole tips.

The damper winding(s) is subjected to the full effect of flux transfer to leakage paths and carries an induced current tending to oppose it. As long as this current can flow, the air-gap flux is held at a value slightly higher than would be the case if only the excitation winding were present, but still less than the original open circuit flux Φ.

As before, it is convenient to use rated voltage and to create another fictitious reactance that is considered to be effective over this period. This is known as the 'sub-transient reactance' X''_d and is defined by the equation:

$$\text{Sub-transient current } I''_d = \frac{E_0}{X''_d}$$

Equation 5.5

where:

$$X''_d = X_L + \frac{X_{ad}X_fX_{kd}}{X_{ad}X_f + X_{kd}X_f + X_{ad}X_{kd}}$$

or

$$X''_d = X_L + X'_{kd}$$

and

X_{kd} = leakage reactance of damper winding(s)

X'_{kd} = effective leakage reactance of damper winding(s)

It is greater than X_L but less than X'_d and the corresponding equivalent circuit is shown in Figure 5.8(c).

Again, the duration of this phase depends upon the time constant of the damper winding. In practice this is approximately 0.05 seconds - very much less than the transient - hence the term 'sub-transient'.

Figure 5.9 shows the envelope of the symmetrical component of an armature short circuit current indicating the values described in the preceding analysis. The analysis of the stator current waveform resulting from a sudden short circuit test is traditionally the method by which these reactances are measured. However, the major limitation is that only direct axis parameters are measured. Detailed test methods for synchronous machines are given in references [5.2] and [5.3], and include other tests that are capable of providing more detailed parameter information.

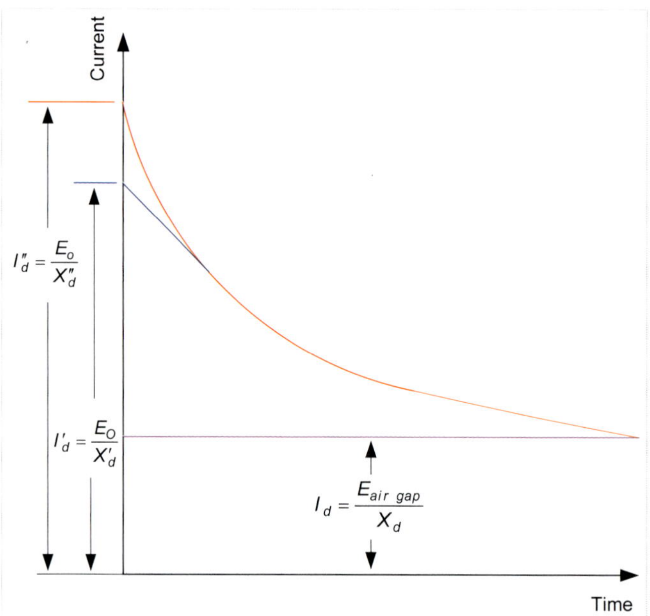

Figure 5.9: Transient decay envelope of short-circuit current

5.7 ASYMMETRY

The exact instant at which the short circuit is applied to the stator winding is of significance. If resistance is negligible compared with reactance, the current in a coil lags the voltage by 90°, that is, at the instant when the voltage wave attains a maximum, any current flowing through would be passing through zero. If a short circuit were applied at this instant, the resulting current would rise smoothly and would be a simple a.c. component. However, at the moment when the induced voltage is zero, any current flowing must pass through a maximum (owing to the 90° lag). If a fault occurs at this moment, the resulting current assumes the corresponding relationship; it is at its peak and in the ensuing 180° goes through zero to maximum in the reverse direction and so on. In fact the current must actually start from zero so it follows a sine wave that is completely asymmetrical. Intermediate positions give varying degrees of asymmetry. This asymmetry can be considered to be due to a d.c. component of current which dies away because resistance is present.

The d.c. component of stator current sets up a d.c. field in the stator which causes a supply frequency ripple on the field current, and this alternating rotor flux has a further effect on the stator. This is best shown by considering the supply frequency flux as being represented by two half magnitude waves each rotating in opposite directions at supply frequency relative to the rotor. So, as viewed from the stator, one is stationary and the other rotating at twice supply frequency. The latter sets up second harmonic currents in the stator. Further development along these lines is possible but the resulting harmonics are usually negligible and normally neglected.

5.8 MACHINE REACTANCES

Table 5.1 gives values of machine reactances for salient pole and cylindrical rotor machines typical of latest design practice. Also included are parameters for synchronous compensators – such machines are now rarely built, but significant numbers can still be found in operation.

5.8.1 Synchronous Reactance $X_d=X_L+X_{ad}$

The order of magnitude of X_L is normally 0.1-0.25p.u., while that of X_{ad} is 1.0-2.5p.u. The leakage reactance X_L can be reduced by increasing the machine size (derating), or increased by artificially increasing the slot leakage, but X_L is only about 10% of the total value of X_d and does not have much influence.

The armature reaction reactance can be reduced by decreasing the armature reaction of the machine, which in design terms means reducing the ampere conductor or electrical (as distinct from magnetic) loading - this often means a physically larger machine. Alternatively the excitation needed to generate open-circuit voltage may be increased; this is simply achieved by increasing the machine air-gap, but is only possible if the excitation system is modified to meet the increased requirements.

In general, control of X_d is obtained almost entirely by varying X_{ad} and in most cases a reduction in X_d means a larger and more costly machine. It is also worth noting that X_L normally changes in sympathy with X_{ad} but that it is completely overshadowed by it.

The value $1/X_d$ has a special significance as it approximates to the short circuit ratio (S.C.R.), the only difference being that the S.C.R. takes saturation into account whereas X_d is derived from the air-gap line.

Type of machine	Salient pole synchronous condensers		Cylindrical rotor turbine generators			Salient pole generators	
			Air Cooled	Hydrogen Cooled	Hydrogen or Water Cooled	4 Pole	Multi-pole
Short circuit ratio	0.5-0.7	1.0-1.2	0.4-0.6	0.4-0.6	0.4-0.6	0.4-0.6	0.6-0.8
Direct axis synchronous reactance Xd (p.u.)	1.6-2.0	0.8-1.0	2.0-2.8	2.1-2.4	2.1-2.6	1.75-3.0	1.4-1.9
Quadrature axis synchronous reactance Xq (p.u.)	1.0-1.23	0.5-0.65	1.8-2.7	1.9-2.4	2.0-2.5	0.9-1.5	0.8-1.0
Direct axis transient reactance X'd (p.u.)	0.3-0.5	0.2-0.35	0.2-0.3	0.27-0.33	0.3-0.36	0.26-0.35	0.24-0.4
Direct axis sub-transient reactance X''d (p.u.)	0.2-0.4	0.12-0.25	0.15-0.23	0.19-0.23	0.21-0.27	0.19-0.25	0.16-0.25
Quadrature axis sub-transient reactance X''q (p.u.)	0.25-0.6	0.15-0.25	0.16-0.25	0.19-0.23	0.21-0.28	0.19-0.35	0.18-0.24
Negative sequence reactance X2 (p.u.)	0.25-0.5	0.14-0.35	0.16-0.23	0.19-0.24	0.21-0.27	0.16-0.27	0.16-0.23
Zero sequence reactance X0 (p.u.)	0.12-0.16	0.06-0.10	0.06-0.1	0.1-0.15	0.1-0.15	0.01-0.1	0.045-0.23
Direct axis short circuit transient time constant T'd (s)	1.5-2.5	1.0-2.0	0.6-1.3	0.7-1.0	0.75-1.0	0.4-1.1	0.25-1
Direct axis open circuit transient time constant T'd (s)	5-10	3-7	6-12	6-10	6-9.5	3.0-9.0	1.7-4.0
Direct axis short circuit sub-transient- time constant T''d (s)	0.04-0.9	0.05-0.10	0.013-0.022	0.017-0.025	0.022-0.03	0.02-0.04	0.02-0.06
Direct axis open circuit sub-transient time constant T''d' (s)	0.07-0.11	0.08-0.25	0.018-0.03	0.023-0.032	0.025-0.035	0.035-0.06	0.03-0.1
Quadrature axis short circuit sub-transient time constant T''q (s)	0.04-0.6	0.05-0.6	0.013-0.022	0.018-0.027	0.02-0.03	0.025-0.04	0.025-0.08
Quadrature axis open circuit sub-transient time constant T''q (s)	0.1-0.2	0.2-0.9	0.026-0.045	0.03-0.05	0.04-0.065	0.13-0.2	0.1-0.35

Table 5.1: Typical values of machine characteristics (all reactance values are unsaturated)

5.8.2 Transient Reactance $X'_d=X_L+X'_f$

The transient reactance covers the behaviour of a machine in the period 0.1-3.0 seconds after a disturbance. This generally corresponds to the speed of changes in a system and therefore X'_d has a major influence in transient stability studies.

Generally, the leakage reactance X_L is equal to the effective field leakage reactance X'_f, about 0.1-0.25p.u. The principal factor determining the value of X'_f is the field leakage. This is largely beyond the control of the designer, in that other considerations are at present more significant than field leakage and hence take precedence in determining the field design. X_L can be varied as already outlined and, in practice, control of transient reactance is usually achieved by varying X_L.

5.8.3 Sub-Transient Reactance $X''_d=X_L+X'_{kd}$

The sub-transient reactance determines the initial current peaks following a disturbance and in the case of a sudden fault is of importance for selecting the breaking capacity of associated circuit breakers. The mechanical stresses on the machine reach maximum values that depend on this constant. The effective damper winding leakage reactance X'_{kd} is largely determined by the leakage of the damper windings and control of this is only possible to a limited extent. X'_{kd} normally has a value between 0.05 and 0.15p.u. The major factor is X_L which, as indicated previously, is of the order of 0.1-0.25p.u., and control of the sub-transient reactance is normally achieved by varying X_L.

Good transient stability is obtained by keeping the value of X'_d low, which therefore also implies a low value of X''_d. The fault rating of switchgear, etc. is therefore relatively high. It is not normally possible to improve transient stability performance in a generator without adverse effects on fault levels, and vice versa.

5.9 NEGATIVE SEQUENCE REACTANCE

Negative sequence currents can arise whenever there is any unbalance present in the system. Their effect is to set up a field rotating in the opposite direction to the main field generated by the rotor winding, so subjecting the rotor to double frequency flux pulsations. This gives rise to parasitic currents and heating; most machines are quite limited in the amount of such current which they are able to carry, both in the steady–state and transiently.

An accurate calculation of the negative sequence current capability of a generator involves consideration of the current paths in the rotor body. In a turbine generator rotor, for instance, they include the solid rotor body, slot wedges, excitation winding and end-winding retaining rings. There is a tendency for local over-heating to occur and, although possible for the stator, continuous local temperature measurement is not practical in the rotor. Calculation requires complex mathematical techniques to be applied, and involves specialist software.

In practice an empirical method is used, based on the fact that a given type of machine is capable of carrying, for short periods, an amount of heat determined by its thermal capacity, and for a long period, a rate of heat input which it can dissipate continuously. Synchronous machines are designed to operate continuously on an unbalanced system so that with none of the phase currents exceeding the rated current, the ratio of the negative sequence current I_2 to the rated current I_N does not exceed the values given in Table 5.2. Under fault conditions, the machine can also operate with the product of $\left(\frac{I_2}{I_N}\right)^2$ and time in seconds (t) not exceeding the values given.

Rotor construction	Rotor Cooling	Machine Type/Rating (SN) (MVA)	Maximum I_2/I_N for continuous operation	Maximum $(I_2/I_N)^2t$ for operation during faults
Salient	indirect	motors	0.1	20
		generators	0.08	20
		synchronous condensers	0.1	20
	direct	motors	0.08	15
		generators	0.05	15
		synchronous condensers	0.08	15
Cylindrical	indirectly cooled (air)	all	0.1	15
	indirectly cooled (hydrogen)	all	0.1	10
	directly cooled	<=350	0.08	8
		351-900	Note 1	Note 2
		901-1250	Note 1	5
		1251-1600	0.05	5

Note 1: Calculate as $\frac{I_2}{I_N}=0.08-\frac{S_N-350}{3\times10^4}$

Note 2: Calculate as $\left[\frac{I_2}{I_N}\right]^2 t = 8-0.00545\left(S_N-350\right)$

Table 5.2: Unbalanced operating conditions for synchronous machines (with acknowledgement to IEC 60034-1)

5.10 ZERO SEQUENCE REACTANCE

If a machine operates with an earthed neutral, a system earth fault gives rise to zero sequence currents in the machine. This reactance represents the machines' contribution to the total impedance offered to these currents. In practice it is generally low and often outweighed by other impedances present in the circuit.

5.11 DIRECT AND QUADRATURE AXIS VALUES

The transient reactance is associated with the field winding and since on salient pole machines this is concentrated on the direct axis, there is no corresponding quadrature axis value. The value of reactance applicable in the quadrature axis is the synchronous reactance, that is $X'_q = X_q$.

The damper winding (or its equivalent) is more widely spread and hence the sub-transient reactance associated with this has a definite quadrature axis value X''_q, which differs significantly in many generators from X''_d.

5.12 EFFECT OF SATURATION ON MACHINE REACTANCES

In general, any electrical machine is designed to avoid severe saturation of its magnetic circuit. However, it is not economically possible to operate at such low flux densities as to reduce saturation to negligible proportions, and in practice a moderate degree of saturation is accepted.

Since the armature reaction reactance X_{ad} is a ratio AT_{ar}/AT_e it is evident that AT_e does not vary in a linear manner for different voltages, while AT_{ar} remains unchanged. The value of X_{ad} varies with the degree of saturation present in the machine, and for extreme accuracy should be determined for the particular conditions involved in any calculation.

All the other reactances, namely X_L, X'_d and X''_d, are true reactances and actually arise from flux leakage. Much of this leakage occurs in the iron parts of the machines and hence must be affected by saturation. For a given set of conditions, the leakage flux exists as a result of the net m.m.f. which causes it. If the iron circuit is unsaturated its reactance is low and leakage flux is easily established. If the circuits are highly saturated the reverse is true and the leakage flux is relatively lower, so the reactance under saturated conditions is lower than when unsaturated.

Most calculation methods assume infinite iron permeability and for this reason lead to somewhat idealised unsaturated reactance values. The recognition of a finite and varying permeability makes a solution extremely laborious and in practice a simple factor of approximately 0.9 is taken as representing the reduction in reactance arising from saturation.

It is necessary to distinguish which value of reactance is being measured when on test. The normal instantaneous short-circuit test carried out from rated open-circuit voltage gives a current that is usually several times full load value, so that saturation is present and the reactance measured is the saturated value. This value is also known as the 'rated voltage' value since it is measured by a short circuit applied with the machine excited to rated voltage.

In some cases, the test may be made from a suitably reduced voltage so that the initial current is approximately full load value. This may be the case where the severe mechanical strain that occurs when the test is performed at rated voltage has to be avoided. Saturation is very much reduced and the reactance values measured are virtually unsaturated values. They are also known as 'rated current' values, for obvious reasons.

5.13 TRANSFORMERS

A transformer may be replaced in a power system by an equivalent circuit representing the self-impedance of, and the mutual coupling between, the windings. A two-winding transformer can be simply represented as a 'T' network in which the cross member is the short-circuit impedance, and the column the excitation impedance. It is rarely necessary in fault studies to consider excitation impedance as this is usually many times the magnitude of the short-circuit impedance. With these simplifying assumptions a three-winding transformer becomes a star of three impedances and a four-winding transformer a mesh of six impedances.

The impedances of a transformer, in common with other plant, can be given in ohms and qualified by a base voltage, or in per unit or percentage terms and qualified by a base MVA. Care should be taken with multi-winding transformers to refer all impedances to a common base MVA or to state the base on which each is given. The impedances of static apparatus are independent of the phase sequence of the applied voltage; in consequence, transformer negative sequence and positive sequence impedances are identical. In determining the impedance to zero phase sequence currents, account must be taken of the winding connections, earthing, and, in some cases, the construction type. The existence of a path for zero sequence currents implies a fault to earth and a flow of balancing currents in the windings of the transformer.

Practical three-phase transformers may have a phase shift between primary and secondary windings depending on the connections of the windings – delta or star. The phase shift that occurs is generally of no significance in fault level calculations as all phases are shifted equally. It is therefore ignored. It is normal to find delta-star transformers at the transmitting end of a transmission system and in distribution systems for the following reasons:

- At the transmitting end, a higher step-up voltage ratio is possible than with other winding arrangements, while the insulation to ground of the star secondary winding does not increase by the same ratio.

- In distribution systems, the star winding allows a neutral connection to be made, which may be important in considering system earthing arrangements.
- The delta winding allows circulation of zero sequence currents within the delta, thus preventing transmission of these from the secondary (star) winding into the primary circuit. This simplifies protection considerations.

5.14 TRANSFORMER POSITIVE SEQUENCE EQUIVALENT CIRCUITS

The transformer is a relatively simple device. However, the equivalent circuits for fault calculations need not necessarily be quite so simple, especially where earth faults are concerned. The following two sections discuss the positive sequence equivalent circuits of various types of transformers.

5.14.1 Two-Winding Transformers

The two-winding transformer has four terminals, but in most system problems, two-terminal or three-terminal equivalent circuits as shown in Figure 5.10 can represent it. In Figure 5.10(a), terminals A' and B' are assumed to be at the same potential. Hence if the per unit self-impedances of the windings are Z_{11} and Z_{22} respectively and the mutual impedance between them Z_{12}, the transformer may be represented by Figure 5.10(b). The circuit in Figure 5.10(b) is similar to that shown in Figure 3.14(a), and can therefore be replaced by an equivalent 'T' as shown in Figure 5.10(c) where:

$$Z_1 = Z_{11} - Z_{12}$$
$$Z_2 = Z_{22} - Z_{12}$$
$$Z_3 = Z_{12}$$

Equation 5.6

Z_1 is described as the leakage impedance of winding AA' and Z_2 the leakage impedance of winding BB'. Impedance Z_3 is the mutual impedance between the windings, usually represented by X_M, the magnetising reactance paralleled with the hysteresis and eddy current loops as shown in Figure 5.10(d).

If the secondary of the transformers is short-circuited, and Z_3 is assumed to be large with respect to Z_1 and Z_2, the short-circuit impedance viewed from the terminals AA' is $Z_T=Z_1+Z_2$ and the transformer can be replaced by a two-terminal equivalent circuit as shown in Figure 5.10(e).

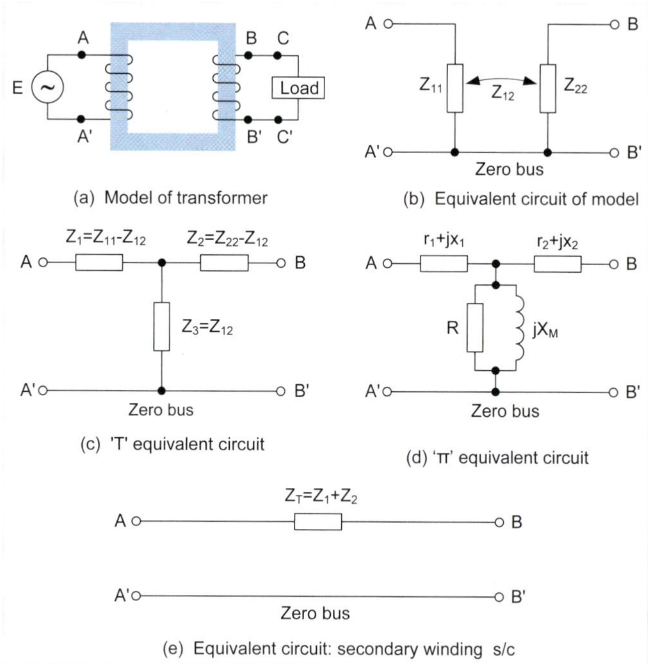

Figure 5.10: Equivalent circuits for a two-winding transformer

The relative magnitudes of Z_T and X_M are 10% and 2000% respectively. Z_T and X_M rarely have to be considered together, so that the transformer may be represented either as a series impedance or as an excitation impedance, according to the problem being studied.

Figure 5.11 shows a typical high voltage power transformer.

Figure 5.11: Testing a high voltage transformer

5.14.1 Three-Winding Transformers

If excitation impedance is neglected the equivalent circuit of a three-winding transformer may be represented by a star of impedances, as shown in Figure 5.12, where P, T and S are the primary, tertiary and secondary windings respectively. The impedance of any of these branches can be determined by considering the short-circuit impedance between pairs of windings with the third open.

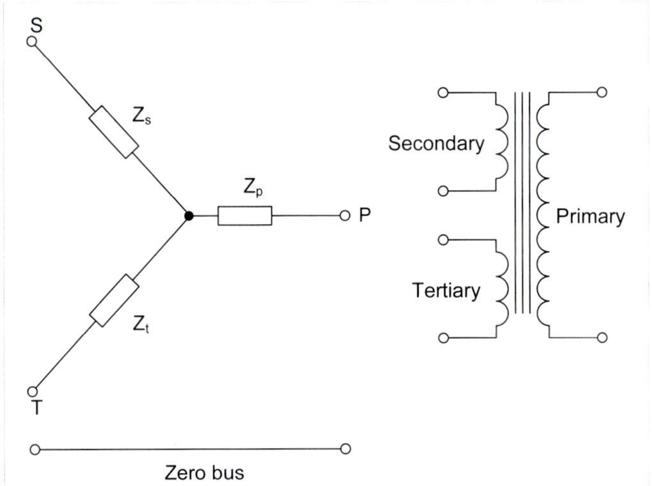

Figure 5.12: Equivalent circuit for a three-winding transformer

5.15 TRANSFORMER ZERO SEQUENCE EQUIVALENT CIRCUITS

The flow of zero sequence currents in a transformer is only possible when the transformer forms part of a closed loop for uni-directional currents and ampere-turn balance is maintained between windings.

The positive sequence equivalent circuit is still maintained to represent the transformer but now there are certain conditions attached to its connection into the external circuit. The order of excitation impedance is much lower than for the positive sequence circuit and is roughly between 1 and 4 per unit but still high enough to be neglected in most fault studies.

The mode of connection of a transformer to the external circuit is determined by taking account of each winding arrangement and its connection or otherwise to ground. If zero sequence currents can flow into and out of a winding, the winding terminal is connected to the external circuit (that is, link a is closed in Figure 5.13). If zero sequence currents can circulate in the winding without flowing in the external circuit, the winding terminal is connected directly to the zero bus (that is, link b is closed in Figure 5.13). Table 5.3 gives the zero sequence connections of some common two- and three-winding transformer arrangements applying the above rules.

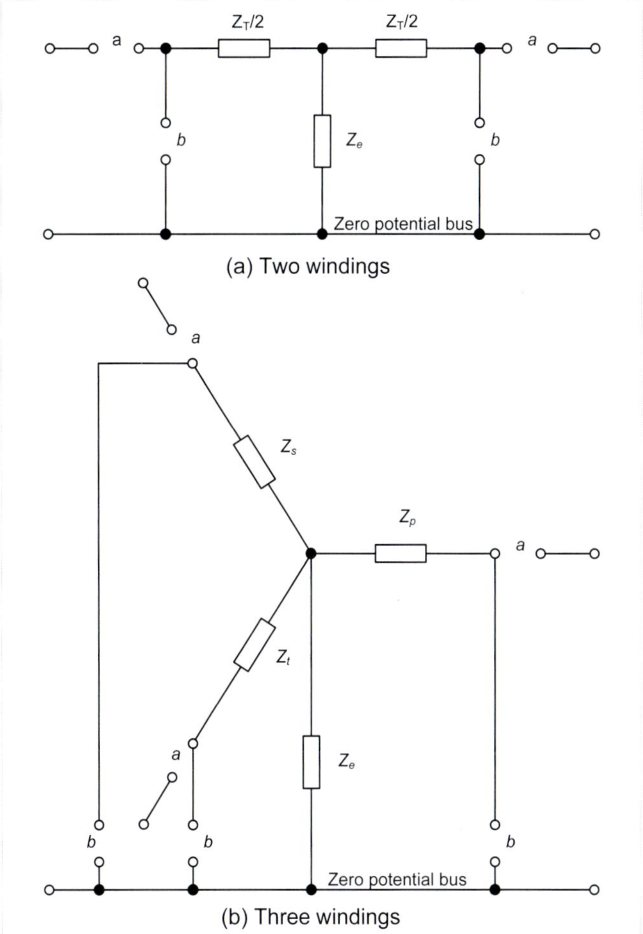

Figure 5.13: Zero sequence equivalent circuits

The exceptions to the general rule of neglecting magnetising impedance occur when the transformer is star/star and either or both neutrals are earthed. In these circumstances the transformer is connected to the zero bus through the magnetising impedance. Where a three-phase transformer bank is arranged without interlinking magnetic flux (that is a three-phase shell type, or three single-phase units) and provided there is a path for zero sequence currents, the zero sequence impedance is equal to the positive sequence impedance. In the case of three-phase core type units, the zero sequence fluxes produced by zero sequence currents can find a high reluctance path, the effect being to reduce the zero sequence impedance to about 90% of the positive sequence impedance.

However, in hand calculations, it is usual to ignore this variation and consider the positive and zero sequence impedances to be equal. It is common when using software to perform fault calculations to enter a value of zero-sequence impedance in accordance with the above guidelines, if the manufacturer is unable to provide a value.

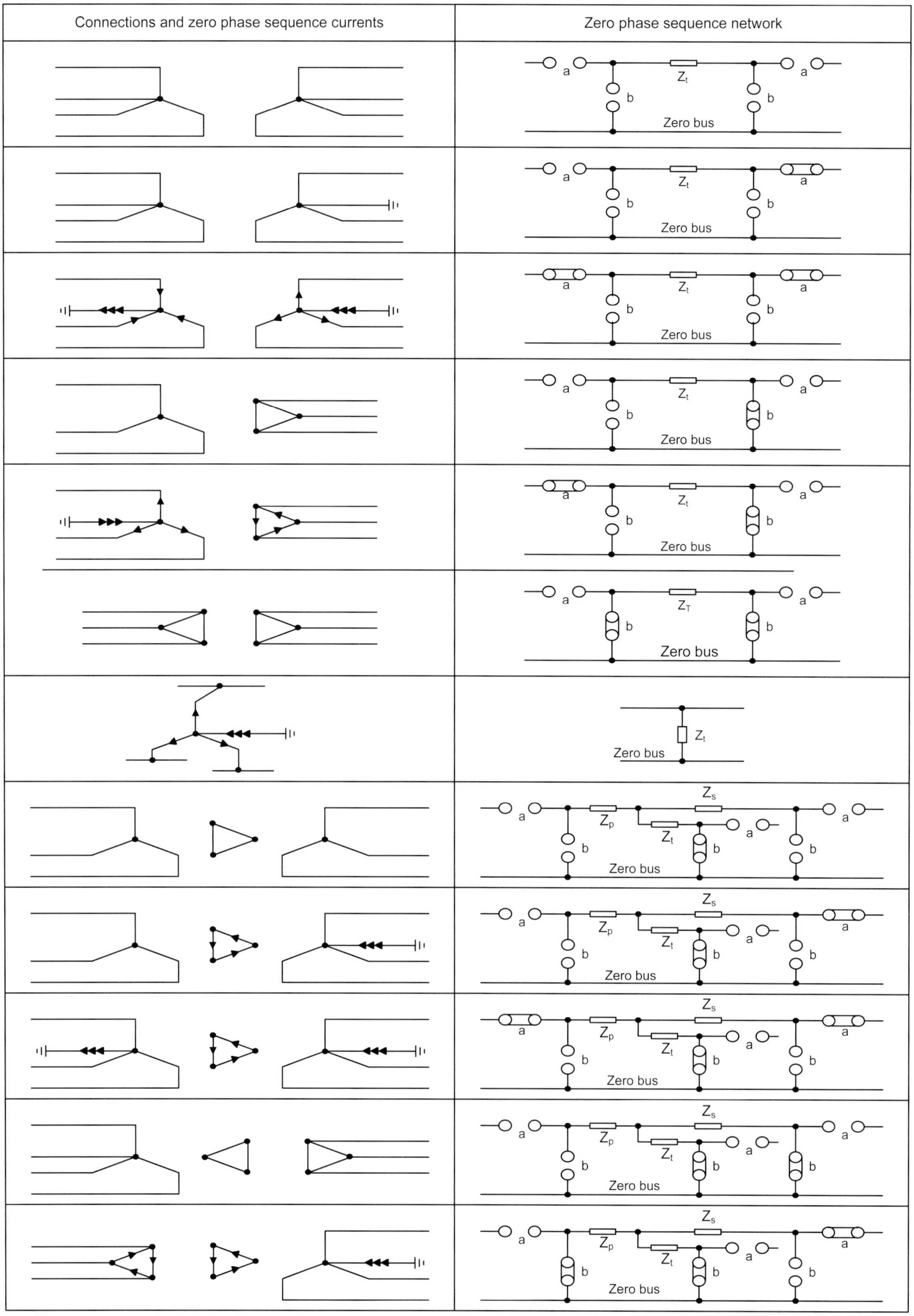

Table 5.3: Zero sequence equivalent circuit connections

5.16 AUTO-TRANSFORMERS

The auto-transformer is characterised by a single continuous winding, part of which is shared by both the high and low voltage circuits, as shown in Figure 5.14(a). The 'common' winding is the winding between the low voltage terminals whereas the remainder of the winding, belonging exclusively to the high voltage circuit, is designated the 'series' winding, and, combined with the 'common' winding, forms the 'series-common' winding between the high voltage terminals. The advantage of using an auto-transformer as opposed to a two-winding transformer is that the auto-transformer is smaller and lighter for a given rating. The disadvantage is that galvanic isolation between the two windings does not exist, giving rise to the possibility of large overvoltages on the lower voltage system in the event of major insulation breakdown.

Three-phase auto-transformer banks generally have star connected main windings, the neutral of which is normally connected solidly to earth. In addition, it is common practice to include a third winding connected in delta called the tertiary winding, as shown in Figure 5.14(b).

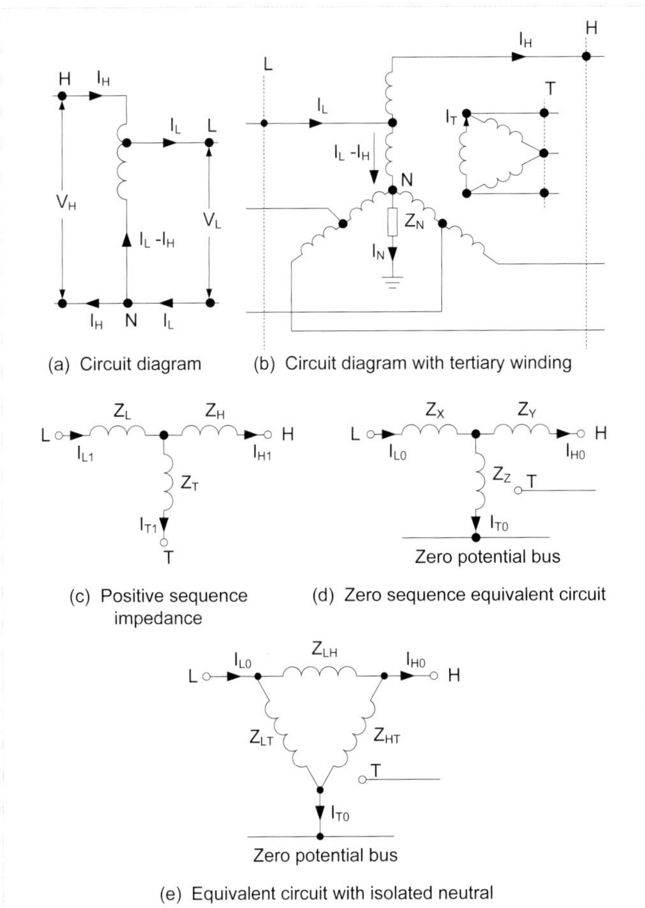

Figure 5.14: Equivalent circuits of auto-transformers

5.16.1 Positive Sequence Equivalent Circuit

The positive sequence equivalent circuit of a three-phase auto-transformer bank is the same as that of a two- or three-winding transformer. The star equivalent for a three-winding transformer, for example, is obtained in the same manner, with the difference that the impedances between windings are designated as follows:

$$Z_L = \frac{1}{2}\left(Z_{sc-c} + Z_{c-t} - Z_{sc-t}\right)$$

$$Z_H = \frac{1}{2}\left(Z_{sc-c} + Z_{sc-t} - Z_{c-t}\right)$$

$$Z_T = \frac{1}{2}\left(Z_{sc-t} + Z_{c-t} - Z_{sc-c}\right)$$

Equation 5.7

where:

$Z_{sc\text{-}t}$ = impedance between 'series-common' and tertiary windings.

$Z_{sc\text{-}c}$ = impedance between 'series-common' and 'common' windings.

$Z_{c\text{-}t}$ = impedance between 'common' and tertiary windings

When no load is connected to the delta tertiary, the point T is open-circuited and the short-circuit impedance of the transformer becomes $Z_L + Z_H = Z_{SC\text{-}C}$, similar to the equivalent circuit of a two-winding transformer, with magnetising impedance neglected; see Figure 5.14(c).

5.16.2 Zero Sequence Equivalent Circuit

The zero sequence equivalent circuit is derived in a similar manner to the positive sequence circuit, except that, as there is no identity for the neutral point, the current in the neutral and the neutral voltage cannot be given directly. Furthermore, in deriving the branch impedances, account must be taken of an impedance in the neutral Z_n, as shown in Equation 5.8, where Z_x, Z_y and Z_z are the impedances of the low, high and tertiary windings respectively and N is the ratio between the series and common windings.

$$Z_x = Z_L + 3Z_n \frac{N}{(N+1)}$$

$$Z_y = Z_H - 3Z_n \frac{N}{(N+1)^2}$$

$$Z_z = Z_T + 3Z_n \frac{1}{(N+1)}$$

Equation 5.8

Figure 5.14(d) shows the equivalent circuit of the transformer bank. Currents I_{L0} and I_{H0} are those circulating in the low and high voltage circuits respectively. The difference between these currents, expressed in amperes, is the current in the common winding. The current in the neutral impedance is three times the current in the common winding.

5.16.3 Special Conditions of Neutral Earthing

With a solidly grounded neutral, $Z_n=0$, the branch impedances Z_x, Z_y, Z_z, become Z_L, Z_H, Z_T, that is, identical to the corresponding positive sequence equivalent circuit, except that the equivalent impedance Z_T of the delta tertiary is connected to the zero potential bus in the zero sequence network.

When the neutral is ungrounded $Z_T=\infty$ and the impedances of the equivalent star also become infinite because there are apparently no paths for zero sequence currents between the windings, although a physical circuit exists and ampere-turn balance can be obtained. A solution is to use an equivalent delta circuit (see Figure 5.14(e)), and evaluate the elements of the delta directly from the actual circuit. The method requires three equations corresponding to three assumed operating conditions. Solving these equations relates the delta impedances to the impedance between the series and tertiary windings as follows:

$$Z_{LH} = Z_{s-t}\frac{N^2}{(N+1)}$$
$$Z_{LT} = -Z_{s-t}N$$
$$Z_{HT} = Z_{s-t}\frac{N}{(N+1)}$$

Equation 5.9

With the equivalent delta replacing the star impedances in the autotransformer zero sequence equivalent circuit the transformer can be combined with the system impedances in the usual manner to obtain the system zero sequence diagram.

5.17 TRANSFORMER IMPEDANCES

In most fault calculations the protection engineer is only concerned with the transformer leakage impedance; the magnetising impedance is neglected as it is very much higher. Impedances for transformers rated at 200MVA or less are given in IEC 60076 and repeated in Table 5.4, together with an indication of X/R values (not part of IEC 60076). These impedances are commonly used for transformers installed in industrial plants. Some variation is possible to assist in controlling fault levels or motor starting, and typically up to ±10% variation of the impedance values given in the table is possible without incurring a significant cost penalty. For these transformers, the tapping range is small, and the variation of impedance with tap position is normally neglected in fault level calculations.

For transformers used in electricity distribution networks, the situation is more complex, due to an increasing trend to assign importance to the standing (or no-load) losses represented by the magnetising impedance. This can be adjusted at the design stage but there is often an impact on the leakage reactance in consequence. In addition, it may be more important to control fault levels on the LV side than to improve motor starting voltage drops. Therefore, departures from the IEC 60076 values are commonplace.

IEC 60076 does not make recommendations of nominal impedance in respect of transformers rated over 200MVA, while generator transformers and a.c. traction supply transformers have impedances that are usually specified as a result of Power Systems Studies to ensure satisfactory performance. Typical values of transformer impedances covering a variety of transformer designs are given in Table 5.4 to Table 5.10. Where appropriate, they include an indication of the impedance variation at the extremes of the taps given. Transformers designed to work at 60Hz have much the same impedance as their 50Hz counterparts.

MVA	Z% HV/LV	X/R	Tolerance on Z%
<0.630	4.00	1.5	±10
0.631-1.25	5.00	3.5	±10
1.251 - 3.15	6.25	6.0	±10
3.151 - 6.3	7.15	8.5	±10
6.301-12.5	8.35	13.0	±10
12.501- 25.0	10.00	20.0	±7.5
25.001 - 200	12.50	45.0	±7.5
>200	by agreement		

Table 5.4: Transformer impedances IEC 60076

MVA	Primary kV	Primary Taps	Secondary kV	Z% HV/LV	X/R ratio
7.5	33	+5.72% -17.16%	11	7.5	15
7.5	33	+5.72% -17.16%	11	7.5	17
8	33	+5.72% -17.16%	11	8	9
11.5	33	+5.72% -17.16%	6.6	11.5	24
11.5	33	+5.72% -17.16%	6.6	11.5	24
11.5	33	+5.72% -17.16%	11	11.5	24
11.5	33	+5.72% -17.16%	11	11.5	26
11.5	33	+4.5% -18%	6.6	11.5	24
12	33	+5% -15%	11.5	12	27
12	33	±10%	11.5	12	27
12	33	±10%	11.5	12	25
15	66	+9% -15%	11.5	15	14
15	66	+9% -15%	11.5	15	16
16	33	±10%	11.5	16	16

MVA	Primary kV	Primary Taps	Secondary kV	Z% HV/LV	X/R ratio
16	33	+5.72% -17.16%	11	16	30
16	33	+5.72% -17.16%	6.6	16	31
19	33	+5.72% -17.16%	11	19	37
30	33	+5.72% -17.16%	11	30	40
24	33	±10%	6.9	24	25
30	33	+5.72% -17.16%	11	30	40
30	132	+10% -20%	11	21.3	43
30	132	+10% -20%	11	25	30
30	132	+10% -20%	11	23.5	46
40	132	+10% -20%	11	27.9	37
45	132	+10% -20%	33	11.8	18
60	132	+10% -20%	33	16.7	28
60	132	+10% -20%	33	17.7	26
60	132	+10% -20%	33	14.5	25
60	132	+10% -20%	66	11	25
60	132	+10% -20%	11/11	35.5	52
60	132	+9.3% -24%	11/11	36	75
60	132	+9.3% -24%	11/11	35.9	78
65	140	+7.5% -15%	11	12.3	28
90	132	+10% -20%	33	24.4	60
90	132	+10% -20%	66	15.1	41

Table 5.5: Impedances of two winding distribution transformers – Primary voltage <200kV

MVA	Primary kV	Primary Taps	Secondary kV	Tertiary kV	Z% HV/LV	X/R ratio
20	220	+12.5% -7.5%	6.9	-	9.9	18
20	230	+12.5% -7.5%	6.9	-	10-14	13
57	275	±10%	11.8	7.2	18.2	34
74	345	+14.4% -10%	96	12	8.9	25
79.2	220	+10% -15%	11.6	11	18.9	35
120	275	+10% -15%	34.5	-	22.5	63
125	230	±16.8%	66	-	13.1	52
125	230	not known	150	-	10-14	22
180	275	±15%	66	13	22.2	38
255	230	+10%	16.5	-	14.8	43

Table 5.6: Impedances of two winding distribution transformers – Primary voltage >200kV

MVA	Primary kV	Primary Taps	Secondary kV	Z% HV/LV	X/R ratio
95	132	±10%	11	13.5	46
140	157.5	±10%	11.5	12.7	41
141	400	±5%	15	14.7	57
151	236	±5%	15	13.6	47
167	145	+7.5% -16.5%	15	25.7	71
180	289	±5%	16	13.4	34
180	132	±10%	15	13.8	40
247	432	+3.75% -16.25%	15.5	15.2	61
250	300	+11.2% -17.6%	15	28.6	70
290	420	±10%	15	15.7	43
307	432	+3.75% -16.25%	15.5	15.3	67
346	435	+5% -15%	17.5	16.4	81
420	432	+5.55% -14.45%	22	16	87
437.8	144.1	+10.8% -21.6%	21	14.6	50
450	132	±10%	19	14	49
600	420	±11.25%	21	16.2	74
716	525	±10%	19	15.7	61
721	362	+6.25% -13.75%	22	15.2	83
736	245	+7% -13%	22	15.5	73
900	525	+7% -13%	23	15.7	67

Table 5.7: Impedances of generator transformers (three-phase units)

MVA/ phase	Primary kV	Primary Taps	Secondary kV	Z% HV/LV	X/R ratio
266.7	432/√3	+6.67% -13.33%	23.5	15.8	92
266.7	432/√3	+6.6% -13.4%	23.5	15.7	79
277	515/√3	±5%	22	16.9	105
375	525/√3	+6.66% -13.32%	26	15	118
375	420/√3	+6.66% -13.32%	26	15.1	112

Table 5.8: Impedances of generator transformers (single-phase units)

MVA	Primary kV	Primary Taps	Secondary kV	Secondary Taps	Tertiary kV	Z% HV/LV	X/R ratio
100	66	-	33	-	-	10.7	28
180	275	-	132	±15%	13	15.5	55
240	400	-	132	+15% -5%	13	20.2	83
240	400	-	132	+15% -5%	13	20.0	51
240	400	-	132	+15% -5%	13	20.0	61
250	400	-	132	+15% -5%	13	10-13	50
500	400	-	132	+0% -15%	22	14.3	51
750	400	-	275	-	13	12.1	90
1000	400	-	275	-	13	15.8	89
1000	400	-	275	-	33	17.0	91
333.3	500/√3	±10%	230/√3	-	22	18.2	101

Table 5.9: Autotransformer data

MVA	Primary kV	Primary Taps	Secondary kV	Secondary Taps	Z% HV/LV	X/R ratio
10	132	-	25	+10% -2.5%	7.7	14
12	132	±5.5%	27.5	-	12.3	21
26.5	132	±7.5%	25	-	19	63

Table 5.10: Traction supply transformer data

5.18 OVERHEAD LINES AND CABLES

In this section a description of common overhead lines and cable systems is given, together with tables of their important characteristics. The formulae for calculating the characteristics are developed to give a basic idea of the factors involved, and to enable calculations to be made for systems other than those tabulated.

A transmission circuit may be represented by an equivalent π or T network using lumped constants as shown in Figure 5.15. Z is the total series impedance $(R+jX)L$ and Y is the total shunt admittance $(G+jB)L$, where L is the circuit length. The terms inside the brackets in Figure 5.15 are correction factors that allow for the fact that in the actual circuit the parameters are distributed over the whole length of the circuit and not lumped, as in the equivalent circuits.

With short lines it is usually possible to ignore the shunt admittance, which greatly simplifies calculations, but on longer lines it must be included. Another simplification that can be made is that of assuming the conductor configuration to be symmetrical. The self-impedance of each conductor becomes Z_p, and the mutual impedance between conductors becomes Z_m. However, for rigorous calculations a detailed treatment is necessary, with account being taken of the spacing of a conductor in relation to its neighbour and earth.

5.19 CALCULATION OF SERIES IMPEDANCE

The self impedance of a conductor with an earth return and the mutual impedance between two parallel conductors with a common earth return are given by the Carson equations:

$$Z_p = R + 0.000988f + j0.0029f\log_{10}\frac{D_e}{dc}$$

$$Z_n = 0.000988f + j0.0029f\log_{10}\frac{D_e}{D}$$

Equation 5.10

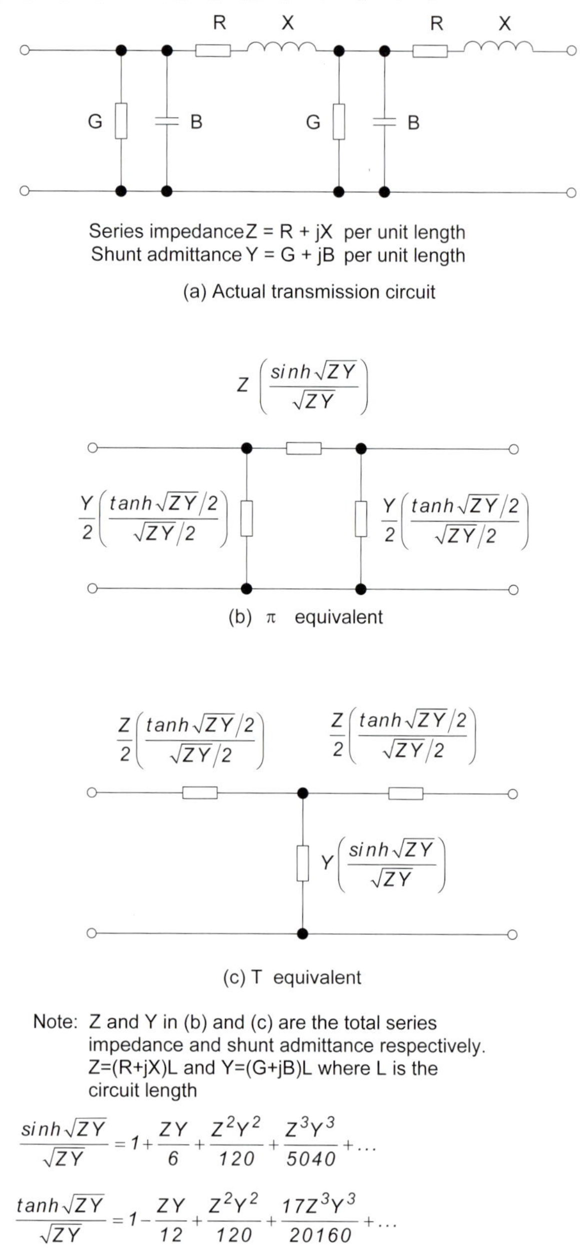

Figure 5.15: Transmission circuit equivalents

where:

R = conductor ac resistance (ohms/km)

dc = geometric mean radius of a single conductor

D = spacing between the parallel conductors

f = system frequency

D_e = equivalent spacing of the earth return path

= $216\sqrt{\rho/f}$ where ρ is earth resistivity (ohms/cm^3)

Equation 5.10 gives the impedances in ohms/km. The last terms in Equation 5.10 are very similar to the classical inductance formulae for long straight conductors.

The geometric means radius (GMR) of a conductor is an equivalent radius that allows the inductance formula to be reduced to a single term. It arises because the inductance of a solid conductor is a function of the internal flux linkages in addition to those external to it. If the original conductor can be replaced by an equivalent that is a hollow cylinder with infinitesimally thin walls, the current is confined to the surface of the conductor, and there can be no internal flux. The geometric mean radius is the radius of the equivalent conductor. If the original conductor is a solid cylinder having a radius r its equivalent has a radius of 0.779r.

It can be shown that the sequence impedances for a symmetrical three-phase circuit are:

$$Z_1 = Z_2 = Z_p - Z_m$$
$$Z_0 = Z_p + 2Z_m$$

Equation 5.11

where:

Z_p and Z_m are given by Equation 5.10. Substituting Equation 5.10 in Equation 5.11 gives:

$$Z_1 = Z_2 = R + j0.0029 f \log_{10} \frac{D}{dc}$$
$$Z_0 = R + 0.00296 f + j0.00869 f \log_{10} \frac{D_e}{\sqrt[3]{dcD^2}}$$

Equation 5.12

In the formula for Z_0 the expression $\sqrt[3]{dcD^2}$ is the geometric mean radius of the conductor group.

Typically circuits are not symmetrical. In this case symmetry can be maintained by transposing the conductors so that each conductor is in each phase position for one third of the circuit length. If A, B and C are the spacings between conductors bc, ca and ab then D in the above equations becomes the geometric mean distance between conductors, equal to $\sqrt[3]{ABC}$.

Writing $D_c = \sqrt[3]{dcD^2}$, the sequence impedances in ohms/km at 50Hz become:

$$Z_1 = Z_2 = R + j0.145 \log_{10} \frac{\sqrt[3]{ABC}}{dc}$$
$$Z_0 = (R + 0.148) + j0.434 \log_{10} \frac{D_e}{D_c}$$

Equation 5.13

5.20 CALCULATION OF SHUNT IMPEDANCE

It can be shown that the potential of a conductor a above ground due to its own charge qa and a charge $-qa$ on its image is:

$$V_a = 2qa \log_e \frac{2h}{r}$$

Equation 5.14

where:

h is the height above ground of the conductor

r is the radius of the conductor, as shown in Figure 5.16

Similarly, it can be shown that the potential of a conductor a due to a charge qb on a neighbouring conductor b and the charge $-qb$ on its image is:

$$V_a' = 2qb \log_e \frac{D'}{D}$$

Equation 5.15

where D is the spacing between conductors a and b and D' is the spacing between conductor b and the image of conductor a as shown in Figure 5.16.

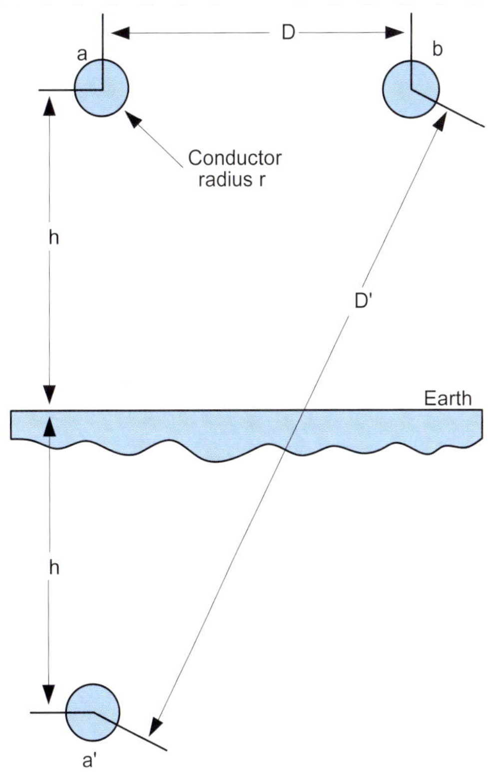

Figure 5.16: Geometry of two parallel conductors a and b and the image of a (a')

Since the capacitance $C=q/V$ and the capacitive reactance $Xc=1/\omega C$, it follows that the self and mutual capacitive reactance of the conductor system in Figure 5.16 can be obtained directly from Equation 5.14 and Equation 5.15. Further, as leakage can usually be neglected, the self and mutual shunt impedances Z'_p and Z'_m in megohm-km at a system frequency of 50Hz are:

$$Z'_p = -j0.132\log_{10}\frac{2h}{r}$$

$$Z'_m = -j0.132\log_{10}\frac{D'}{D}$$

Equation 5.16

Where the distances above ground are great in relation to the conductor spacing, which is the case with overhead lines. From Equation 5.11, the sequence impedances of a symmetrical three-phase circuit are:

$$Z_1 = Z_2 = -j0.132\log_{10}\frac{D}{r}$$

$$Z_0 = -j0.396\log_{10}\frac{D'}{\sqrt[3]{rD^2}}$$

Equation 5.17

The logarithmic terms in Equation 5.17 are similar to those in Equation 5.12 except that r is the actual radius of the conductors and D' is the spacing between the conductors and their images.

Where the conductors are transposed and not symmetrically spaced, Equation 5.17 can be rewritten using the geometric mean distance between conductors $\sqrt[3]{ABC}$, giving the distance of each conductor above ground, $h_a\ h_b\ h_c$ as follows:

$$Z_1 = Z_2 = -j0.132\log_{10}\frac{\sqrt[3]{ABC}}{r}$$

$$Z_0 = -j0.132\log_{10}\frac{8h_a h_b h_c}{r\sqrt[3]{A^2B^2C^2}}$$

Equation 5.18

5.21 OVERHEAD LINE CIRCUITS WITH OR WITHOUT EARTH WIRES

Typical configurations of overhead line circuits are given in Figure 5.18. Tower heights are not given as they vary considerably according to the design span and nature of the ground. As indicated in some of the tower outlines, some tower designs are designed with a number of base extensions for this purpose. Figure 5.17 shows a typical tower.

Figure 5.17: Double circuit 132kV overhead line tower

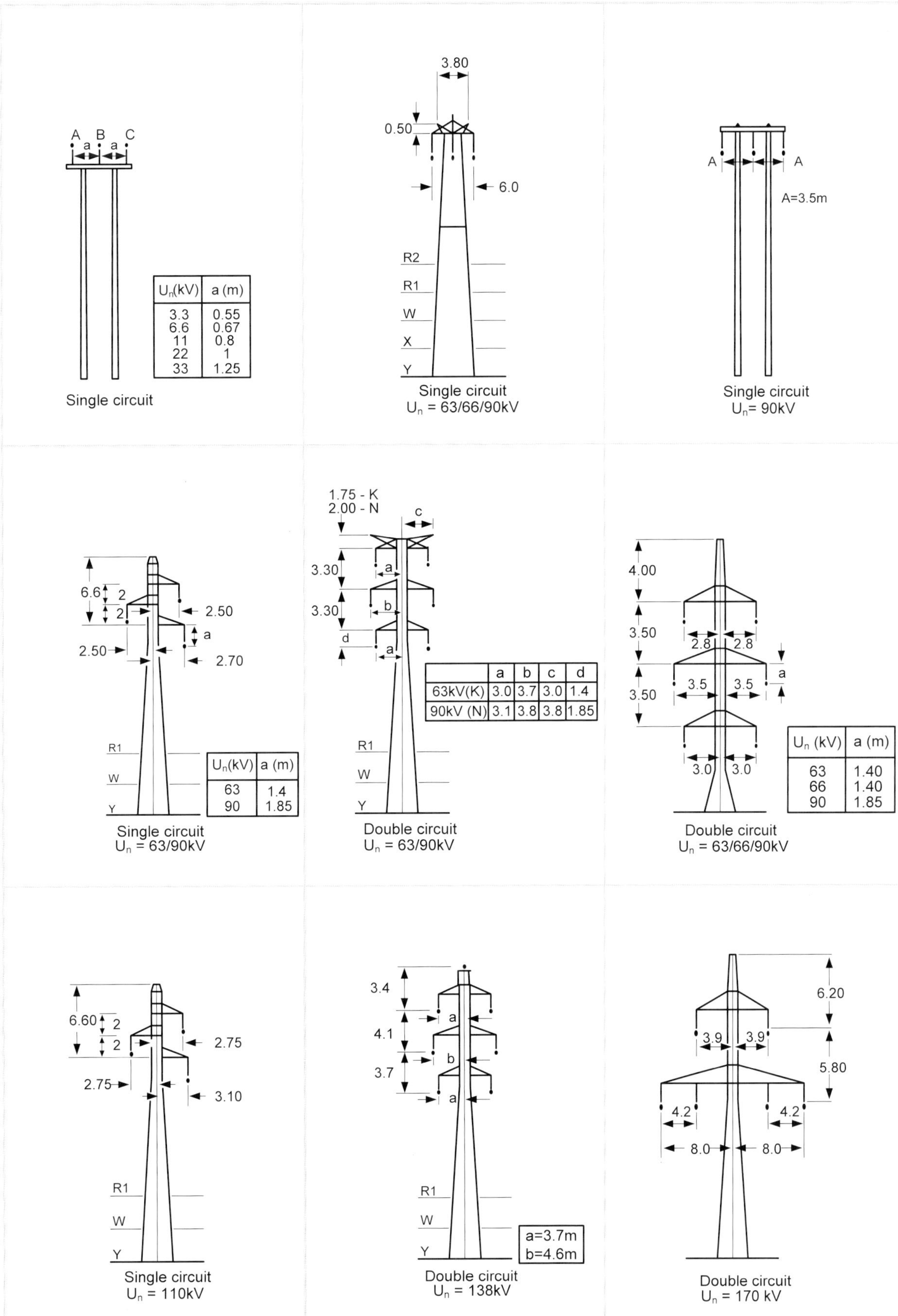

Figure 5.18: Typical overhead line tower outlines

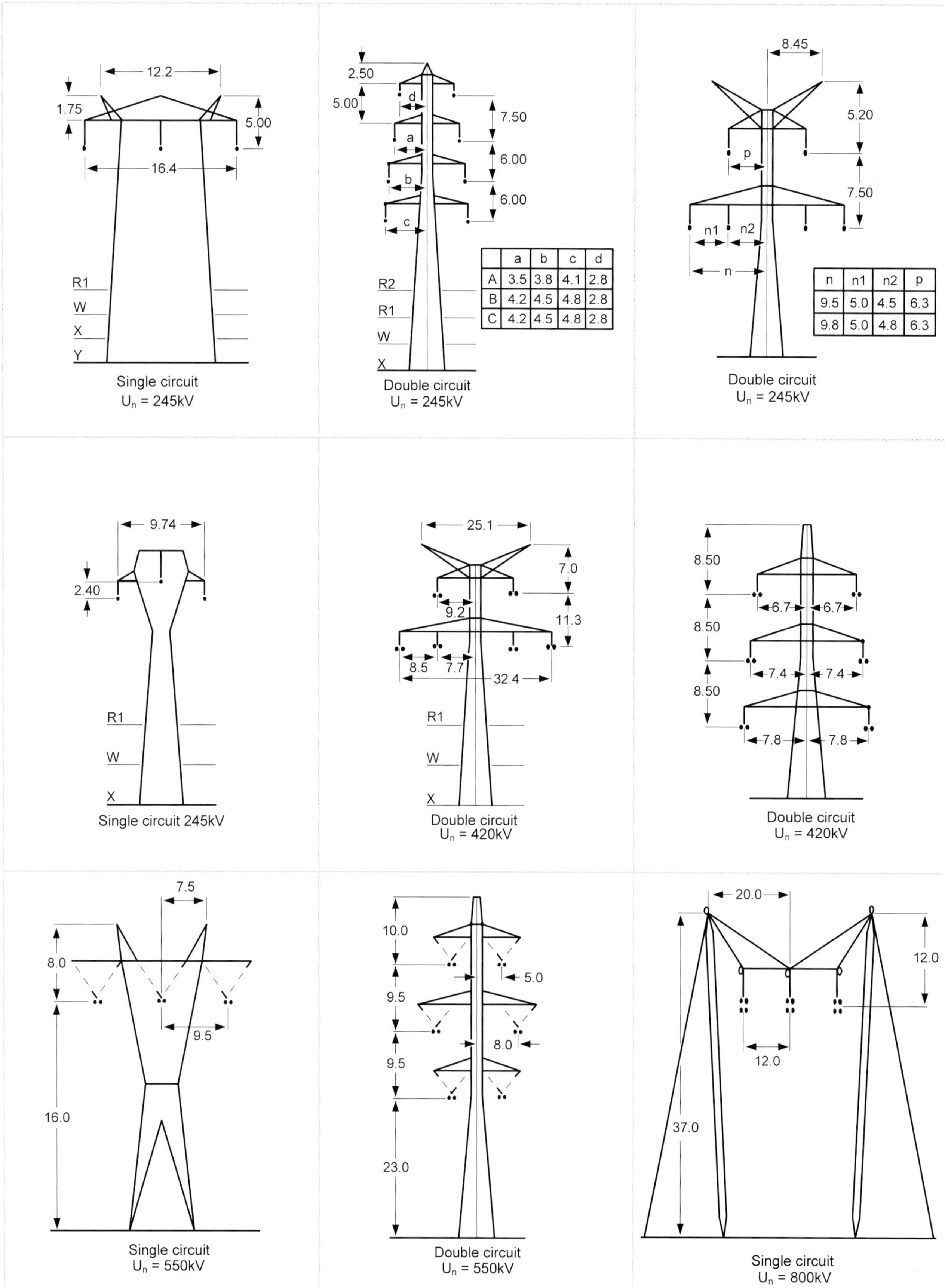

	a	b	c	d
A	3.5	3.8	4.1	2.8
B	4.2	4.5	4.8	2.8
C	4.2	4.5	4.8	2.8

n	n1	n2	p
9.5	5.0	4.5	6.3
9.8	5.0	4.8	6.3

Figure 5.19: Typical overhead line tower outlines

In some cases the phase conductors are not symmetrically disposed to each other and therefore, as previously indicated, electrostatic and electromagnetic unbalance result, which can be largely eliminated by transposition. Modern practice is to build overhead lines without transposition towers to reduce costs; this must be taken into account in rigorous calculations of the unbalances. In other cases, lines are formed of bundled conductors, that is conductors formed of two, three or four separate conductors. This arrangement minimises losses when voltages of 220kV and above are involved.

The line configuration and conductor spacings are influenced, not only by voltage, but also by many other factors including type of insulators, type of support, span length, conductor sag and the nature of terrain and external climatic loadings. Therefore there can be large variations in spacings between different line designs for the same voltage level, so those depicted in Figure 5.17 are only typical examples.

When calculating the phase self and mutual impedances, Equation 5.10 and Equation 5.16 may be used. However, in this case Z_p is calculated for each conductor and Z_m for each pair of conductors. This section is not intended to give a detailed analysis but rather to show the general method of formulating the equations, taking the calculation of series impedance as an example and assuming a single circuit line with a single earth wire.

The phase voltage drops V_a V_b V_c of a single circuit line with a single earth wire due to currents I_a I_b I_c flowing in the phases and I_e in the earth wire are:

$$\begin{aligned} V_a &= Z_{aa}I_a + Z_{ab}I_b + Z_{ac}I_c + Z_{ae}I_e \\ V_b &= Z_{ba}I_a + Z_{bb}I_b + Z_{bc}I_c + Z_{be}I_e \\ V_c &= Z_{ca}I_a + Z_{cb}I_b + Z_{cc}I_c + Z_{ce}I_e \\ 0 &= Z_{ea}I_a + Z_{eb}I_b + Z_{ec}I_c + Z_{ee}I_e \end{aligned}$$

Equation 5.19

where:

$$Z_{aa} = R + 0.000988f + j0.0029f\log_{10}\frac{D_e}{dc}$$

$$Z_{ab} = 0.000988f + j0.0029f\log_{10}\frac{D_e}{D}$$

and so on.

The equation required for the calculation of shunt voltage drops is identical to Equation 5.19 in form, except that primes must be included, the impedances being derived from Equation 5.16.

From Equation 5.19 it can be seen that:

$$-I_e = \frac{Z_{ea}}{Z_{ee}}I_a + \frac{Z_{eb}}{Z_{ee}}I_b + \frac{Z_{ec}}{Z_{ee}}I_c$$

Making use of this relation, the self and mutual impedances of the phase conductors can be modified using the following formula:

$$J_{nm} = Z_{nm} - \frac{Z_{ne}Z_{me}}{Z_{ee}}$$

Equation 5.20

For example:

$$J_{aa} = Z_{aa} - \frac{Z^2{}_{ae}}{Z_{ee}}$$

$$J_{ab} = Z_{ab} - \frac{Z_{ae}Z_{be}}{Z_{ee}}$$

and so on.

Equation 5.19 can be simplified while still accounting for the effect of the earth wire. This is done by deleting the fourth row and fourth column and substituting J_{aa} for Z_{aa}, J_{ab} for Z_{ab}, and so on, calculated using Equation 5.20. The single circuit line with a single earth wire can therefore be replaced by an equivalent single circuit line having phase self and mutual impedances J_{aa}, J_{ab} and so on.

It can be shown from the symmetrical component theory given in Chapter 4 that the sequence voltage drops of a general three-phase circuit are:

$$\begin{aligned} V_0 &= Z_{00}I_0 + Z_{01}I_1 + Z_{02}I_2 \\ V_1 &= Z_{10}I_0 + Z_{11}I_1 + Z_{12}I_2 \\ V_2 &= Z_{20}I_0 + Z_{21}I_1 + Z_{22}I_2 \end{aligned}$$

Equation 5.21

And, from Equation 5.19 modified as indicated above and Equation 5.21, the sequence impedances are:

$$Z_{00} = \frac{1}{3}(J_{aa} + J_{bb} + J_{cc}) + \frac{2}{3}(J_{ab} + J_{bc} + J_{ac})$$

$$Z_{11} = Z_{22} = \frac{1}{3}(J_{aa} + J_{bb} + J_{cc}) - \frac{1}{3}(J_{ab} + J_{bc} + J_{ac})$$

$$Z_{12} = \frac{1}{3}(J_{aa} + a^2 J_{bb} + aJ_{cc}) + \frac{2}{3}(aJ_{ab} + a^2 J_{ac} + J_{bc})$$

$$Z_{21} = \frac{1}{3}(J_{aa} + aJ_{bb} + a^2 J_{cc}) + \frac{2}{3}(a^2 J_{ab} + aJ_{ac} + J_{bc})$$

$$Z_{20} = Z_{01} = \frac{1}{3}(J_{aa} + a^2 J_{bb} + aJ_{cc}) - \frac{1}{3}(aJ_{ab} + a^2 J_{ac} + J_{bc})$$

$$Z_{10} = Z_{02} = \frac{1}{3}(J_{aa} + aJ_{bb} + a^2 J_{cc}) - \frac{1}{3}(a_2 J_{ab} + aJ_{ac} + J_{bc})$$

Equation 5.22

The development of these equations for double circuit lines with two earth wires is similar except that more terms are involved.

The sequence mutual impedances are very small and can usually be neglected; this also applies for double circuit lines except for the mutual impedance between the zero sequence circuits, namely $Z_{00'} = Z_{0'0}$. Table 5.11 and Table 5.12 give typical values of all sequence self and mutual impedances some single and double circuit lines with earth wires. All conductors are 400mm^2 ACSR, except for the 132kV double circuit example where they are 200mm^2.

Sequence impedance	132kV Single circuit line (400 mm^2)	380kV Single circuit line (400 mm^2)
$Z_{00} = (Z_{0'0'})$	1.0782 ∠73°54′	0.8227 ∠70°36′
$Z_{11} = Z_{22} = (Z_{1'1'})$	0.3947 ∠78°54′	0.3712 ∠75°57′
$(Z_{0'0} = Z_{00'})$	-	-
$Z_{01} = Z_{20} = (Z_{0'1'} = Z_{2'0'})$	0.0116 ∠-166°52′	0.0094 ∠-39°28′
$Z_{02} = Z_{10} = (Z_{0'2'} = Z_{1'0'})$	0.0185 ∠ 5°8′	0.0153 ∠28°53′
$Z_{12} = (Z_{1'2'})$	0.0255 ∠-40°9′	0.0275 ∠147°26′
$Z_{21} = (Z_{2'1'})$	0.0256 ∠-139°1′	0.0275 ∠27°29′
$(Z_{11'} = Z_{1'1} = Z_{22'} = Z_{2'2})$	-	-
$(Z_{02'} = Z_{0'2} = Z_{1'0} = Z_{10'})$	-	-
$(Z_{02'} = Z_{0'2} = Z_{1'0} = Z_{10'}$	-	-
$(Z_{1'2} = Z_{12'})$	-	-
(Z21′ = Z2′1)	-	-

Table 5.11: Sequence self and mutual impedances for various lines

Sequence impedance	132kV Double circuit line (200 mm^2)	275kV Double circuit line (400 mm^2)
$Z_{00} = (Z_{0'0'})$	1.1838 ∠71°6′	0.9520 ∠76°46′
$Z_{11} = Z_{22} = (Z_{1'1'})$	0.4433 ∠66°19′	0.3354 ∠74°35′
$(Z_{0'0} = Z_{00'})$	0.6334 ∠71°2′	0.5219 ∠75°43′
$Z_{01} = Z_{20} = (Z_{0'1'} = Z_{2'0'})$	0.0257 ∠-63°25′	0.0241 ∠-72°14′
$Z_{02} = Z_{10} = (Z_{0'2'} = Z_{1'0'})$	0.0197 ∠-94°58′	0.0217 ∠-100°20′
$Z_{12} = (Z_{1'2'})$	0.0276 ∠161°17′	0.0281 ∠149°46′
$Z_{21} = (Z_{2'1'})$	0.0277 ∠37°13′	0.0282 ∠29°6′
$(Z_{11'} = Z_{1'1} = Z_{22'} = Z_{2'2})$	0.0114 ∠88°6′	0.0129 ∠88°44′
$(Z_{02'} = Z_{0'2} = Z_{1'0} = Z_{10'})$	0.0140 ∠-93°44′	0.0185 ∠-91°16′
$(Z_{02'} = Z_{0'2} = Z_{1'0} = Z_{10'}$	0.0150 ∠-44°11′	0.0173 ∠-77°2′
$(Z_{1'2} = Z_{12'})$	0.0103 ∠145°10′	0.0101 ∠149°20′
(Z21′ = Z2′1)	0.0106 ∠30°56′	0.0102 ∠27°31′

Table 5.12: Sequence self and mutual impedances for various lines

5.22 OHL EQUIVALENT CIRCUITS

Consider an earthed, infinite busbar source behind a length of transmission line as shown in Figure 5.20(a). An earth fault involving phase A is assumed to occur at F. If the driving voltage is E and the fault current is I_a then the earth fault impedance is Z_e. From symmetrical component theory (see Chapter 4):

$$I_a = \frac{3E}{Z_1 + Z_2 + Z_0}$$

thus

$$Z_e = \frac{2Z_1 + Z_0}{3}$$

Equation 5.23

since, as shown, $Z_1 = Z_2$ for a transmission circuit. From Equation 5.11, $Z_1 = Z_p - Z_m$ and $Zo = Z_p + 2Z_m$. Thus, substituting these values in Equation 5.23 gives $Z_e = Z_p$. This relation is physically valid because Z_p is the self-impedance of a single conductor with an earth return. Similarly, for a phase fault between phases B and C at F:

$$I_b = -I_c = \frac{\sqrt{3}E}{2Z_1}$$

where $\sqrt{3}E$ is the voltage between phases and $2Z_1$ is the impedance of the fault loop.

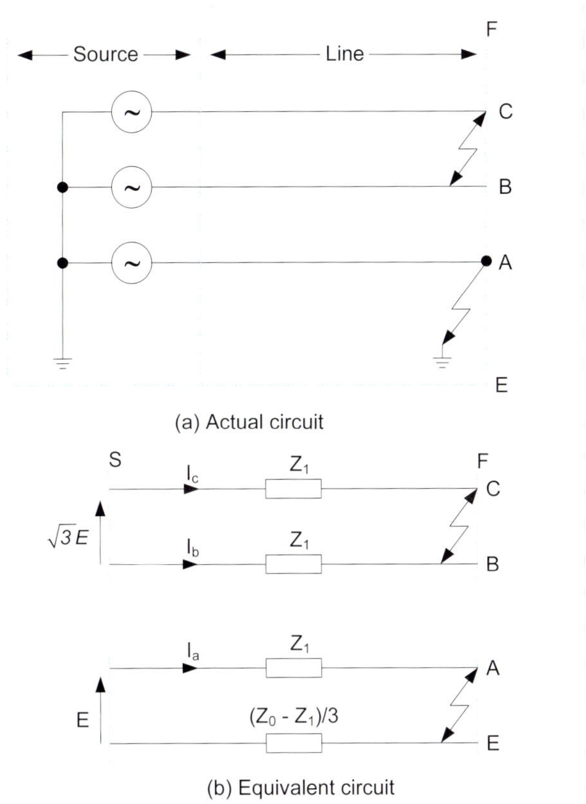

Figure 5.20: Three-phase equivalent of a transmission circuit

Making use of the above relations, a transmission circuit may be represented, generally without any loss, by the equivalent of Figure 5.20(b), where Z_1 is the phase impedance to the fault and $(Z_0-Z_1)/3$ is the impedance of the earth path, there being no mutual impedance between the phases or between phase and earth. The equivalent is valid for single and double circuit lines except that for double circuit lines there is zero sequence mutual impedance, hence $Z_0=Z_{00}-Z_{0'0}$.

The equivalent circuit of Figure 5.20(b) is valuable in distance relay applications because the phase and earth fault relays are set to measure Z_1 and are compensated for the earth return impedance $(Z_0-Z_1)/3$.

It is customary to quote the impedances of a transmission circuit in terms of Z_1 and the ratio Z_0/Z_1, since in this form they are most directly useful. By definition, the positive sequence impedance Z_1 is a function of the conductor spacing and radius, whereas the Z_0/Z_1 ratio is dependent mainly on the level of earth resistivity ρ. Further details may be found in Chapter 12.

5.23 CABLE CIRCUITS

The basic formulae for calculating the series and shunt impedances of a transmission circuit, Equation 5.10 and Equation 5.16 may be applied for evaluating cable parameters; since the conductor configuration is normally symmetrical GMD and GMR values can be used without risk of appreciable errors. However, the formulae must be modified by the inclusion of empirical factors to take account of sheath and screen effects. A useful general reference on cable formulae is given in reference [5.4]; more detailed information on particular types of cables should be obtained direct from the manufacturers. The equivalent circuit for determining the positive and negative sequence series impedances of a cable is shown in Figure 5.21. From this circuit it can be shown that:

$$Z_1 = Z_2 = \left(R_c + R_s \frac{X_{cs}^{\ 2}}{R_s^{\ 2} + X_s^{\ 2}} \right) + j\left(X_c - X_s \frac{X_{cs}^{\ 2}}{R_s^{\ 2} + X_s^{\ 2}} \right)$$

Equation 5.24

where R_c and R_s are the core and sheath (screen) resistances per unit length, X_c and X_s core and sheath (screen) reactances per unit length and X_{cs} the mutual reactance between core and sheath (screen) per unit length. X_{cs} is generally equal to X_s.

The zero sequence series impedances are obtained directly using Equation 5.10 and account can be taken of the sheath in the same way as an earth wire in the case of an overhead line.

The shunt capacitances of a sheathed cable can be calculated from the simple formula:

$$C = 0.0241\varepsilon \left(\frac{1}{\log \dfrac{d+2T}{d}} \right) \mu F / km$$

Equation 5.25

where d is the overall diameter for a round conductor, T core insulation thickness and ε permittivity of dielectric. When the conductors are oval or shaped an equivalent diameter d' may be used where $d'=(1/\pi)$ *x periphery of conductor*. No simple formula exists for belted or unscreened cables, but an empirical formula that gives reasonable results is:

$$C = \frac{0.0555\varepsilon}{G} \mu F / km$$

Equation 5.26

where G is a geometric factor which is a function of core and belt insulation thickness and overall conductor diameter.

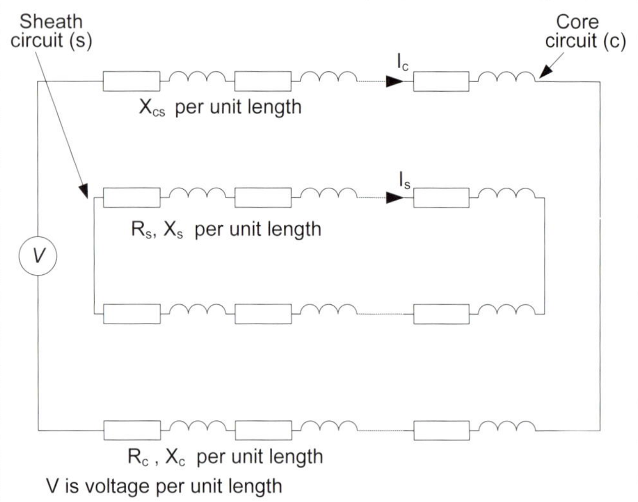

Figure 5.21: Equivalent circuit for determining positive or negative sequence impedance of cables

5.24 OVERHEAD LINE AND CABLE DATA

The following tables show typical data on overhead lines and cables that can be used with the equations in this text. The data shown is only a guide and where the results of calculations are important, data should be sourced directly from a manufacturer or supplier.

Number of Strands	GMR
7	0.726r
19	0.758r
37	0.768r
61	0.772r
91	0.774r
127	0.776r
169	0.776r
Solid	0.779r

Table 5.13: GMR for stranded copper, aluminium and aluminium alloy conductors (r = conductor radius)

Number of Layers	Number of Al Strands	GMR
1	6	0.5r*
1	12	0.75r*
2	18	0.776r
2	24	0.803r
2	26	0.812r
2	30	0.826r
2	32	0.833r
3	36	0.778r
3	45	0.794r
3	48	0.799r
3	54	0.81r
3	66	0.827r
4	72	0.789r
4	76	0.793r
4	84	0.801r

* - Indicative values only, since GMR for single layer conductors is affected by cyclic magnetic flux, which depends on various factors.

Table 5.14: GMR for aluminium conductor steel reinforced (ACSR) (r = conductor radius)

Sectional area (mm^2)	Stranding	Wire Diameter (mm)	Overall Diameter (mm)	R_{DC} (20°C) (Ohm/km)
10.6	7	1.38	4.17	1.734
21.2	7	1.96	5.89	0.865
26.7	7	2.20	6.60	0.686
33.6	7	7.00	7.42	0.544
42.4	7	2.77	8.33	0.431
53.5	7	3.12	9.35	0.342
67.4	7	3.50	10.52	0.271
85.0	7	3.93	11.79	0.215
107.2	7	4.42	13.26	0.171
126.6	19	2.91	14.58	0.144
152.0	19	3.19	15.98	0.120
177.3	19	3.45	17.25	0.103
202.7	19	3.69	18.44	0.090
228.0	37	2.80	19.61	0.080
253.3	37	2.95	20.65	0.072
278.7	37	3.10	21.67	0.066
304.3	37	3.23	22.63	0.060
329.3	61	2.62	23.60	0.056
354.7	61	2.72	24.49	0.052
380.0	61	2.82	25.35	0.048
405.3	61	2.91	26.19	0.045
456.0	61	3.09	27.79	0.040
506.7	61	3.25	29.26	0.036

Table 5.15: Overhead line conductor - hard drawn copper ASTM Standards

Sectional area (mm^2)	Stranding	Wire Diameter (mm)	Overall Diameter (mm)	R_{DC} (20°C) (Ohm/km)
11.0	1	3.73	3.25	1.617
13.0	1	4.06	4.06	1.365
14.0	1	4.22	4.22	1.269
14.5	7	1.63	4.88	1.231
16.1	1	4.52	4.52	1.103
18.9	1	4.90	4.90	0.938
23.4	1	5.46	5.46	0.756
32.2	1	6.40	6.40	0.549
38.4	7	2.64	7.92	0.466
47.7	7	2.95	8.84	0.375

Sectional area (mm²)	Stranding	Wire Diameter (mm)	Overall Diameter (mm)	R_{DC} (20°C) (Ohm/km)
65.6	7	3.45	10.36	0.273
70.1	1	9.45	9.45	0.252
97.7	7	4.22	12.65	0.183
129.5	19	2.95	14.73	0.139
132.1	7	4.90	14.71	0.135
164.0	7	5.46	16.38	0.109
165.2	19	3.33	16.64	0.109

Table 5.16: Overhead line conductor - hard drawn copper BS Standards

Designation	Stranding and wire diameter (mm)				Sectional area (mm²)		Total area (mm²)	Approximate overall diameter (mm)	R_{DC} at 20 °C (ohm/km)
	Aluminium		Steel		Aluminium	Steel			
Sparrow	6	2.67	1	2.7	33.6	5.6	39.2	8.01	0.854
Robin	6	3	1	3	42.4	7.1	49.5	9	0.677
Raven	6	3.37	1	3.4	53.5	8.9	62.4	10.11	0.536
Quail	6	3.78	1	3.8	67.4	11.2	78.6	11.34	0.426
Pigeon	6	4.25	1	4.3	85.0	14.2	99.2	12.75	0.337
Penguin	6	4.77	1	4.8	107.2	17.9	125.1	14.31	0.268
Partridge	26	2.57	7	2	135.2	22.0	157.2	16.28	0.214
Ostrich	26	2.73	7	2.2	152.0	26.9	178.9	17.28	0.191
Merlin	18	3.47	1	3.5	170.5	9.5	179.9	17.35	0.169
Lark	30	2.92	7	2.9	201.4	46.9	248.3	20.44	0.144
Hawk	26	3.44	7	2.7	241.7	39.2	280.9	21.79	0.120
Dove	26	3.72	7	2.9	282.0	45.9	327.9	23.55	0.103
Teal	30	3.61	19	2.2	306.6	69.6	376.2	25.24	0.095
Swift	36	3.38	1	3.4	322.3	9.0	331.2	23.62	0.089
Tern	45	3.38	7	2.3	402.8	27.8	430.7	27.03	0.072
Canary	54	3.28	7	3.3	456.1	59.1	515.2	29.52	0.064
Curlew	54	3.52	7	3.5	523.7	68.1	591.8	31.68	0.055
Finch	54	3.65	19	2.3	565.0	78.3	643.3	33.35	0.051
Bittern	45	4.27	7	2.9	644.5	44.7	689.2	34.17	0.045
Falcon	54	4.36	19	2.6	805.7	102.4	908.1	39.26	0.036
Kiwi	72	4.41	7	2.9	1100.0	47.5	1147.5	44.07	0.027

Table 5.17: Overhead line conductor data - aluminium conductors steel reinforced (ACSR), to ASTM B232

Designation	Stranding and wire diameter (mm)				Sectional area (mm²)		Total area (mm²)	Approximate overall diameter (mm)	R_{DC} at 20 °C (ohm/km)
	Aluminium		Steel		Aluminium	Steel			
Gopher	6	2.36	1	2.4	26.2	4.4	30.6	7.08	1.093
Weasel	6	2.59	1	2.6	31.6	5.3	36.9	7.77	0.908
Ferret	6	3	1	3	42.4	7.1	49.5	9	0.676
Rabbit	6	3.35	1	3.4	52.9	8.8	61.7	10.05	0.542
Horse	12	2.79	7	2.8	73.4	42.8	116.2	13.95	0.393
Dog	6	4.72	7	1.6	105.0	13.6	118.5	14.15	0.273
Tiger	30	2.36	7	2.4	131.2	30.6	161.9	16.52	0.220
Wolf	30	2.59	7	2.6	158.1	36.9	194.9	18.13	0.182
Dingo	18	3.35	1	3.4	158.7	8.8	167.5	16.75	0.181
Lynx	30	2.79	7	2.8	183.4	42.8	226.2	19.53	0.157
Caracal	18	3.61	1	3.6	184.2	10.2	194.5	18.05	0.156
Jaguar	18	3.86	1	3.9	210.6	11.7	222.3	19.3	0.137
Panther	30	3	7	3	212.1	49.5	261.5	21	0.136
Zebra	54	3.18	7	3.2	428.9	55.6	484.5	28.62	0.067

Table 5.18: Overhead line conductor data - aluminium conductors steel reinforced (ACSR), to BS215.2

Designation	Stranding and wire diameter (mm)				Sectional area (mm²)		Total area (mm²)	Approximate overall diameter (mm)	R_{DC} at 20 °C (ohm/km)
	Aluminium		Steel		Aluminium	Steel			
35/6	6	2.7	1	2.7	34.4	5.7	40.1	8.1	0.834
44/32	14	2	7	2.4	44.0	31.7	75.6	11.2	0.652
50/8	6	3.2	1	3.2	48.3	8.0	56.3	9.6	0.594
70/12	26	1.85	7	1.4	69.9	11.4	81.3	11.7	0.413
95/15	26	2.15	7	1.7	94.4	15.3	109.7	13.6	0.305
95/55	12	3.2	7	3.2	96.5	56.3	152.8	16	0.299
120/70	12	3.6	7	3.6	122.1	71.3	193.4	18	0.236
150/25	26	2.7	7	2.1	148.9	24.2	173.1	17.1	0.194
170/40	30	2.7	7	2.7	171.8	40.1	211.8	18.9	0.168
185/30	26	3	7	2.3	183.8	29.8	213.6	19	0.157
210/50	30	3	7	3	212.1	49.5	261.5	21	0.136
265/35	24	3.74	7	2.5	263.7	34.1	297.7	22.4	0.109
305/40	54	2.68	7	2.7	304.6	39.5	344.1	24.1	0.095
380/50	54	3	7	3	381.7	49.5	431.2	27	0.076
550/70	54	3.6	7	3.6	549.7	71.3	620.9	32.4	0.052
560/50	48	3.86	7	3	561.7	49.5	611.2	32.2	0.051
650/45	45	4.3	7	2.9	653.5	45.3	698.8	34.4	0.044
1045/45	72	4.3	7	2.9	1045.6	45.3	1090.9	43	0.028

Table 5.19: Overhead line conductor data - aluminium conductors steel reinforced (ACSR), to DIN48204

Designation	Stranding and wire diameter (mm)				Sectional area (mm^2)		Total area (mm^2)	Approximate overall diameter (mm)	R_{DC} at 20 °C (ohm/km)
	Aluminium		Steel		Aluminium	Steel			
CANNA 59.7	12	2	7	2	37.7	22.0	59.7	10	0.765
CANNA 75.5	12	2.25	7	2.3	47.7	27.8	75.5	11.25	0.604
CANNA 93.3	12	2.5	7	2.5	58.9	34.4	93.3	12.5	0.489
CANNA 116.2	30	2	7	2	94.2	22.0	116.2	14	0.306
CROCUS 116.2	30	2	7	2	94.2	22.0	116.2	14	0.306
CANNA 147.1	30	2.25	7	2.3	119.3	27.8	147.1	15.75	0.243
CROCUS 181.6	30	2.5	7	2.5	147.3	34.4	181.6	17.5	0.197
CROCUS 228	30	2.8	7	2.8	184.7	43.1	227.8	19.6	0.157
CROCUS 297	36	2.8	19	2.3	221.7	75.5	297.2	22.45	0.131
CANNA 288	30	3.15	7	3.2	233.8	54.6	288.3	22.05	0.124
CROCUS 288	30	3.15	7	3.2	233.8	54.6	288.3	22.05	0.124
CROCUS 412	32	3.6	19	2.4	325.7	86.0	411.7	26.4	0.089
CROCUS 612	66	3.13	19	2.7	507.8	104.8	612.6	32.03	0.057
CROCUS 865	66	3.72	19	3.2	717.3	148.1	865.4	38.01	0.040

Table 5.20: Overhead line conductor data - aluminium conductors steel reinforced (ACSR), to NF C34-120

Standard	Designation	No. of Al Strands	Wire diameter (mm)	Sectional area (mm^2)	Overall diameter (mm)	R_{DC} at 20°C (Ohm/km)
ASTM B-397	Kench	7	2.67	39.2	8.0	0.838
ASTM B-397	Kibe	7	3.37	62.4	10.1	0.526
ASTM B-397	Kayak	7	3.78	78.6	11.4	0.418
ASTM B-397	Kopeck	7	4.25	99.3	12.8	0.331
ASTM B-397	Kittle	7	4.77	125.1	14.3	0.262
ASTM B-397	Radian	19	3.66	199.9	18.3	0.164
ASTM B-397	Rede	19	3.78	212.6	18.9	0.155
ASTM B-397	Ragout	19	3.98	236.4	19.9	0.140
ASTM B-397	Rex	19	4.14	255.8	19.9	0.129
ASTM B-397	Remex	19	4.36	283.7	21.8	0.116
ASTM B-397	Ruble	19	4.46	296.8	22.4	0.111
ASTM B-397	Rune	19	4.7	330.6	23.6	0.100
ASTM B-397	Spar	37	3.6	376.6	25.2	0.087
ASTM B-397	Solar	37	4.02	469.6	28.2	0.070
ASTM B-399	-	19	3.686	202.7	18.4	0.165
ASTM B-399	-	19	3.909	228.0	19.6	0.147
ASTM B-399	-	19	4.12	253.3	20.6	0.132
ASTM B-399	-	37	3.096	278.5	21.7	0.120
ASTM B-399	-	37	3.233	303.7	22.6	0.110
ASTM B-399	-	37	3.366	329.2	23.6	0.102
ASTM B-399	-	37	3.493	354.6	24.5	0.094
ASTM B-399	-	37	3.617	380.2	25.3	0.088
ASTM B-399	-	37	3.734	405.2	26.1	0.083
ASTM B-399	-	37	3.962	456.2	27.7	0.073
ASTM B-399	-	37	4.176	506.8	29.2	0.066

Table 5.21: Overhead line conductor data - aluminium alloy (ASTM)

Standard	Designation	No. of Al Strands	Wire diameter (mm)	Sectional area (mm^2)	Overall diameter (mm)	R_{DC} at 20°C (Ohm/km)
BS 3242	Box	7	1.85	18.8	5.6	1.750
BS 3242	Acacia	7	2.08	23.8	6.2	1.384
BS 3242	Almond	7	2.34	30.1	7.0	1.094
BS 3242	Cedar	7	2.54	35.5	7.6	0.928
BS 3242	Fir	7	2.95	47.8	8.9	0.688
BS 3242	Hazel	7	3.3	59.9	9.9	0.550
BS 3242	Pine	7	3.61	71.6	10.8	0.460
BS 3242	Willow	7	4.04	89.7	12.1	0.367
BS 3242	-	7	4.19	96.5	12.6	0.341
BS 3242	-	7	4.45	108.9	13.4	0.302
BS 3242	Oak	7	4.65	118.9	14.0	0.277
BS 3242	Mullberry	19	3.18	150.9	15.9	0.219
BS 3242	Ash	19	3.48	180.7	17.4	0.183
BS 3242	Elm	19	3.76	211.0	18.8	0.157
BS 3242	Poplar	37	2.87	239.4	20.1	0.139
BS 3242	Sycamore	37	3.23	303.2	22.6	0.109
BS 3242	Upas	37	3.53	362.1	24.7	0.092
BS 3242	Yew	37	4.06	479.0	28.4	0.069
BS 3242	Totara	37	4.14	498.1	29.0	0.067
BS 3242	Rubus	61	3.5	586.9	31.5	0.057
BS 3242	Araucaria	61	4.14	821.1	28.4	0.040

Table 5.22: Overhead line conductor data - aluminium alloy (BS)

Standard	Designation	No. of Al Strands	Wire diameter (mm)	Sectional area (mm²)	Overall diameter (mm)	R_{DC} at 20°C (Ohm/km)
CSA C49.1-M87	10	7	1.45	11.5	4.3	2.863
CSA C49.1-M87	16	7	1.83	18.4	5.5	1.788
CSA C49.1-M87	25	7	2.29	28.8	6.9	1.142
CSA C49.1-M87	40	7	2.89	46.0	8.7	0.716
CSA C49.1-M87	63	7	3.63	72.5	10.9	0.454
CSA C49.1-M87	100	19	2.78	115.1	13.9	0.287
CSA C49.1-M87	125	19	3.1	143.9	15.5	0.230
CSA C49.1-M87	160	19	3.51	184.2	17.6	0.180
CSA C49.1-M87	200	19	3.93	230.2	19.6	0.144
CSA C49.1-M87	250	19	4.39	287.7	22.0	0.115
CSA C49.1-M87	315	37	3.53	362.1	24.7	0.092
CSA C49.1-M87	400	37	3.98	460.4	27.9	0.072
CSA C49.1-M87	450	37	4.22	517.9	29.6	0.064
CSA C49.1-M87	500	37	4.45	575.5	31.2	0.058
CSA C49.1-M87	560	37	4.71	644.5	33.0	0.051
CSA C49.1-M87	630	61	3.89	725.0	35.0	0.046
CSA C49.1-M87	710	61	4.13	817.2	37.2	0.041
CSA C49.1-M87	800	61	4.38	920.8	39.5	0.036
CSA C49.1-M87	900	61	4.65	1035.8	41.9	0.032
CSA C49.1-M87	1000	91	4.01	1150.9	44.1	0.029
CSA C49.1-M87	1120	91	4.25	1289.1	46.7	0.026
CSA C49.1-M87	1250	91	4.49	1438.7	49.4	0.023
CSA C49.1-M87	1400	91	4.75	1611.3	52.2	0.021
CSA C49.1-M87	1500	91	4.91	1726.4	54.1	0.019

Table 5.23: Overhead line conductor data - aluminium alloy (CSA)

Standard	Designation	No. of Al Strands	Wire diameter (mm)	Sectional area (mm²)	Overall diameter (mm)	R_{DC} at 20°C (Ohm/km)
DIN 48201	16	7	1.7	15.9	5.1	2.091
DIN 48201	25	7	2.1	24.3	6.3	1.370
DIN 48201	35	7	2.5	34.4	7.5	0.967
DIN 48201	50	19	1.8	48.4	9.0	0.690
DIN 48201	50	7	3	49.5	9.0	0.672
DIN 48201	70	19	2.1	65.8	10.5	0.507
DIN 48201	95	19	2.5	93.3	12.5	0.358
DIN 48201	120	19	2.8	117.0	14.0	0.285
DIN 48201	150	37	2.25	147.1	15.7	0.228
DIN 48201	185	37	2.5	181.6	17.5	0.184
DIN 48201	240	61	2.25	242.5	20.2	0.138
DIN 48201	300	61	2.5	299.4	22.5	0.112
DIN 48201	400	61	2.89	400.1	26.0	0.084
DIN 48201	500	61	3.23	499.8	29.1	0.067

Table 5.24: Overhead line conductor data - aluminium alloy (DIN)

Standard	Designation	No. of Al Strands	Wire diameter (mm)	Sectional area (mm²)	Overall diameter (mm)	R_{DC} at 20°C (Ohm/km)
NF C34-125	ASTER 22	7	2	22.0	6.0	1.497
NF C34-125	ASTER 34-4	7	2.5	34.4	7.5	0.958
NF C34-125	ASTER 54-6	7	3.15	54.6	9.5	0.604
NF C34-125	ASTER 75-5	19	2.25	75.5	11.3	0.438
NF C34-125	ASTER 93,3	19	2.5	93.3	12.5	0.355
NF C34-125	ASTER 117	19	2.8	117.0	14.0	0.283
NF C34-125	ASTER 148	19	3.15	148.1	15.8	0.223
NF C34-125	ASTER 181-6	37	2.5	181.6	17.5	0.183
NF C34-125	ASTER 228	37	2.8	227.8	19.6	0.146
NF C34-125	ASTER 288	37	3.15	288.3	22.1	0.115
NF C34-125	ASTER 366	37	3.55	366.2	24.9	0.091
NF C34-125	ASTER 570	61	3.45	570.2	31.1	0.058
NF C34-125	ASTER 851	91	3.45	850.7	38.0	0.039
NF C34-125	ASTER 1144	91	4	1143.5	44.0	0.029
NF C34-125	ASTER 1600	127	4	1595.9	52.0	0.021

Table 5.25: Overhead line conductor data - aluminium alloy (NF)

Standard	Designation	Stranding and wire diameter (mm)				Sectional area (mm^2)		Total area (mm^2)	Approximate overall dia (mm)	R_{DC} at 20 °C (ohm /km)
		Alloy		Steel		Alloy	Steel			
ASTM B711		26	2.62	7	2	140.2	22.9	163.1	7.08	0.240
ASTM B711		26	2.97	7	2.3	180.1	29.3	209.5	11.08	0.187
ASTM B711		30	2.76	7	2.8	179.5	41.9	221.4	12.08	0.188
ASTM B711		26	3.13	7	2.4	200.1	32.5	232.5	13.08	0.168
ASTM B711		30	3.08	7	3.1	223.5	52.2	275.7	16.08	0.151
ASTM B711		26	3.5	7	2.7	250.1	40.7	290.8	17.08	0.135
ASTM B711		26	3.7	7	2.9	279.6	45.6	325.2	19.08	0.120
ASTM B711		30	3.66	19	2.2	315.6	72.2	387.9	22.08	0.107
ASTM B711		30	3.88	19	2.3	354.7	81.0	435.7	24.08	0.095
ASTM B711		30	4.12	19	2.5	399.9	91.0	491.0	26.08	0.084
ASTM B711		54	3.26	19	2	450.7	58.5	509.2	27.08	0.075
ASTM B711		54	3.63	19	2.2	558.9	70.9	629.8	29.08	0.060
ASTM B711		54	3.85	19	2.3	628.6	79.6	708.3	30.08	0.054
ASTM B711		54	4.34	19	2.6	798.8	100.9	899.7	32.08	0.042
ASTM B711		84	4.12	19	2.5	1119.9	91.0	1210.9	35.08	0.030
ASTM B711		84	4.35	19	2.6	1248.4	101.7	1350.0	36.08	0.027

Table 5.26: Overhead line conductor data – aluminium alloy conductors, steel re-inforced (AACSR) ASTM

Standard	Designation	Stranding and wire diameter (mm)				Sectional area (mm^2)		Total area (mm^2)	Approximate overall dia (mm)	R_{DC} at 20 °C (ohm /km)
		Alloy		Steel		Alloy	Steel			
DIN 48206	70/12	26	1.85	7	1.4	69.9	11.4	81.3	11.7	0.479
DIN 48206	95/15	26	2.15	7	1.7	94.4	15.3	109.7	13.6	0.355
DIN 48206	125/30	30	2.33	7	2.3	127.9	29.8	157.8	16.3	0.262
DIN 48206	150/25	26	2.7	7	2.1	148.9	24.2	173.1	17.1	0.225
DIN 48206	170/40	30	2.7	7	2.7	171.8	40.1	211.8	18.9	0.195
DIN 48206	185/30	26	3	7	2.3	183.8	29.8	213.6	19	0.182
DIN 48206	210/50	30	3	7	3	212.1	49.5	261.5	21	0.158
DIN 48206	230/30	24	3.5	7	2.3	230.9	29.8	260.8	21	0.145
DIN 48206	265/35	24	3.74	7	2.5	263.7	34.1	297.7	22.4	0.127
DIN 48206	305/40	54	2.68	7	2.7	304.6	39.5	344.1	24.1	0.110
DIN 48206	380/50	54	3	7	3	381.7	49.5	431.2	27	0.088
DIN 48206	450/40	48	3.45	7	2.7	448.7	39.5	488.2	28.7	0.075
DIN 48206	560/50	48	3.86	7	3	561.7	49.5	611.2	32.2	0.060
DIN 48206	680/85	54	4	19	2.4	678.6	86.0	764.5	36	0.049

Table 5.27: Overhead line conductor data – aluminium alloy conductors, steel re-inforced (AACSR) DIN

Standard	Designation	Stranding and wire diameter (mm)				Sectional area (mm^2)		Total area (mm^2)	Approximate overall dia (mm)	R_{DC} at 20 °C (ohm /km)
		Alloy		Steel		Alloy	Steel			
NF C34-125	PHLOX 116.2	18	2	19	2	56.5	59.7	116.2	14	0.591
NF C34-125	PHLOX 147.1	18	2.25	19	2.3	71.6	75.5	147.1	15.75	0.467
NF C34-125	PASTEL 147.1	30	2.25	7	2.3	119.3	27.8	147.1	15.75	0.279
NF C34-125	PHLOX 181.6	18	2.5	19	2.5	88.4	93.3	181.6	17.5	0.378
NF C34-125	PASTEL 181.6	30	2.5	7	2.5	147.3	34.4	181.6	17.5	0.226
NF C34-125	PHLOX 228	18	2.8	19	2.8	110.8	117.0	227.8	19.6	0.300
NF C34-125	PASTEL 228	30	2.8	7	2.8	184.7	43.1	227.8	19.6	0.180
NF C34-125	PHLOX 288	18	3.15	19	3.2	140.3	148.1	288.3	22.05	0.238
NF C34-125	PASTEL 288	30	3.15	7	3.2	233.8	54.6	288.3	22.05	0.142
NF C34-125	PASTEL 299	42	2.5	19	2.5	206.2	93.3	299.4	22.45	0.162
NF C34-125	PHLOX 376	24	2.8	37	2.8	147.8	227.8	375.6	26.4	0.226

Table 5.28: Overhead line conductor data – aluminium alloy conductors, steel re-inforced (AACSR) NF

			Conductor Size (mm²)								
			10	16	25	35	50	70	95	120	150
3.3kV	Series Resistance	R (Ω/km)	206	1303	825.5	595	439.9	304.9	220.4	174.5	142.3
	Series Reactance	X (Ω/km)	87.7	83.6	76.7	74.8	72.5	70.2	67.5	66.6	65.7
	Susceptance	ωC (mS/km)									
6.6kV	Series Resistance	R (Ω/km)	514.2	326	206.4	148.8	110	76.2	55.1	43.6	35.6
	Series Reactance	X (Ω/km)	26.2	24.3	22	21.2	20.4	19.6	18.7	18.3	17.9
	Susceptance	ωC (mS/km)									
11kV	Series Resistance	R (Ω/km)	-	111	0.87	0.63	0.46	0.32	0.23	0.184	0.15
	Series Reactance	X (Ω/km)	-	9.26	0.107	0.1	0.096	0.091	0.087	0.085	0.083
	Susceptance	ωC (mS/km)									
22kV	Series Resistance	R (Ω/km)	-	-	17.69	12.75	9.42	6.53	4.71	3.74	3.04
	Series Reactance	X (Ω/km)	-	-	2.89	2.71	2.6	2.46	2.36	2.25	2.19
	Susceptance	ωC (mS/km)									
33kV	Series Resistance	R (Ω/km)	-	-	-	-	4.19	2.9	2.09	0.181	0.147
	Series Reactance	X (Ω/km)	-	-	-	-	1.16	1.09	1.03	0.107	0.103
	Susceptance	ωC (mS/km)								0.104	0.116
Cables are solid type 3 core except for those marked *. Impedances are at 50Hz											

Table 5.29: Characteristics of paper insulated cables, conductor size 10 to 150 mm²

			Conductor Size (mm²)							
			185	240	300	400	*500	*630	*800	*1000
3.3kV	Series Resistance	R (Ω/km)	113.9	87.6	70.8	56.7	45.5	37.1	31.2	27.2
	Series Reactance	X (Ω/km)	64.7	63.8	62.9	62.4	73.5	72.1	71.2	69.8
	Susceptance	ωC (mS/km)								
6.6kV	Series Resistance	R (Ω/km)	28.5	21.9	17.6	14.1	11.3	9.3	7.8	6.7
	Series Reactance	X (Ω/km)	17.6	17.1	16.9	16.5	18.8	18.4	18	17.8
	Susceptance	ωC (mS/km)								
11kV	Series Resistance	R (Ω/km)	0.12	0.092	0.074	0.059	0.048	0.039	0.033	0.028
	Series Reactance	X (Ω/km)	0.081	0.079	0.077	0.076	0.085	0.083	0.081	0.08
	Susceptance	ωC (mS/km)								
22kV	Series Resistance	R (Ω/km)	2.44	1.87	1.51	1.21	0.96	0.79	0.66	0.57
	Series Reactance	X (Ω/km)	2.11	2.04	1.97	1.92	1.9	1.84	1.8	1.76
	Susceptance	ωC (mS/km)								
33kV	Series Resistance	R (Ω/km)	0.118	0.09	0.073	0.058	0.046	0.038	0.031	0.027
	Series Reactance	X (Ω/km)	0.101	0.097	0.094	0.09	0.098	0.097	0.092	0.089
	Susceptance	ωC (mS/km)	0.124	0.194	0.151	0.281	0.179	0.198	0.22	0.245
Cables are solid type 3 core except for those marked *. Impedances are at 50Hz										

Table 5.30: Characteristics of paper insulated cables, conductor size 185 to 1000 mm²

Conductor size (mm²)	3.3kV	
	R Ω/km	X Ω/km
16	1.380	0.106
25	0.870	0.100
35	0.627	0.094
50	0.463	0.091
70	0.321	0.086
95	0.232	0.084
120	0.184	0.081
150	0.150	0.079
185	0.121	0.077
240	0.093	0.076
300	0.075	0.075
400	0.060	0.075
*500	0.049	0.089
*630	0.041	0.086
*800	0.035	0.086
*1000	0.030	0.084
3 core copper conductors, 50Hz values.		
* - single core cables in trefoil		

Table 5.31: 3.3 kV PVC insulated cables

At the conceptual design stage, initial selection of overhead line conductor size is determined by four factors:

- maximum load to be carried in MVA
- length of line
- conductor material and hence maximum temperature
- cost of losses

gives indicative details of the capability of various sizes of overhead lines using the above factors, for AAAC and ACSR conductor materials. It is based on commonly used standards for voltage drop and ambient temperature. Since these factors may not be appropriate for any particular project, the Table should only be used as a guide for initial sizing, with appropriately detailed calculations carried out to arrive at a final proposal.

Voltage Level		Cross Sectional Area mm²	Conductors per phase	Surge Imped-ance Loading	Voltage Drop Loading	Indicative Thermal Loading	
Un kV	Um kV			MVA	MWkm	MVA	A
11	12	30	1	0.3	11	2.9	151
		50	1	0.3	17	3.9	204
		90	1	0.4	23	5.1	268
		120	1	0.5	27	6.2	328
		150	1	0.5	30	7.3	383
22	24	30	1	1.2	44	5.8	151
		50	1	1.2	66	7.8	204
		90	1	1.2	92	10.2	268
		120	1	1.4	106	12.5	328
		150	1	1.5	119	14.6	383
33	36	50	1	2.7	149	11.7	204
		90	1	2.7	207	15.3	268
		120	1	3.1	239	18.7	328
		150	1	3.5	267	21.9	383
66	72.5	90	1	11	827	41	268
		150	1	11	1068	59	383
		250	1	11	1240	77	502
		250	2	15	1790	153	1004
132	145	150	1	44	4070	85	370
		250	1	44	4960	115	502
		250	2	58	7160	230	1004
		400	1	56	6274	160	698
		400	2	73	9057	320	1395
220	245	400	1	130	15600	247	648
		400	2	184	22062	494	1296
		400	4	260	31200	988	2592
380	420	400	2	410	58100	850	1296
		400	4	582	82200	1700	2590
		550	2	482	68200	1085	1650
		550	3	540	81200	1630	2475

Table 5.32: OHL capabilities

Sectional area of Al	R_{DC} (20°C)	X_{AC} at 50 Hz				X_{AC} at 50 Hz and shunt capacitance																	
						33kV		66kV						132kV					
		3.3kV	6.6kV	11kV	22kV			Flat circuit		Double vertical		Triangle		Double vertical		Double triangle		Flat circuit	
						X	C	X	C	X	C	X	C	X	C	X	C	X	C
mm²	Ω/km	Ω/km	Ω/km	Ω/km	Ω/km	Ω/km	nF/km	Ω/km	nF/km	Ω/km	nF/km	Ω/km	nF/km	Ω/km	nF/km	Ω/km	nF/km	Ω/km	nF/km
13.3	2.1586	0.395	0.409	0.420	0.434	0.445	8.7	0.503	7.6	0.513	7.4	0.520	7.3	0.541	7.0	0.528	7.2	0.556	6.8
15.3	1.8771	0.391	0.405	0.415	0.429	0.441	8.8	0.499	7.7	0.508	7.5	0.515	7.4	0.537	7.1	0.523	7.3	0.552	6.9
21.2	1.3557	0.381	0.395	0.405	0.419	0.430	9.0	0.488	7.8	0.498	7.7	0.505	7.6	0.527	7.2	0.513	7.4	0.542	7.0
23.9	1.2013	0.376	0.390	0.401	0.415	0.426	9.1	0.484	7.9	0.494	7.8	0.501	7.6	0.522	7.3	0.509	7.5	0.537	7.1
26.2	1.0930	0.374	0.388	0.398	0.412	0.424	9.2	0.482	8.0	0.491	7.8	0.498	7.7	0.520	7.3	0.506	7.5	0.535	7.1
28.3	1.0246	0.352	0.366	0.377	0.391	0.402	9.4	0.460	8.2	0.470	8.0	0.477	7.8	0.498	7.5	0.485	7.7	0.513	7.3
33.6	0.8535	0.366	0.380	0.390	0.404	0.416	9.4	0.474	8.1	0.484	7.9	0.491	7.8	0.512	7.5	0.499	7.7	0.527	7.2
37.7	0.7647	0.327	0.341	0.351	0.365	0.376	9.7	0.435	8.4	0.444	8.2	0.451	8.1	0.473	7.7	0.459	7.9	0.488	7.4
42.4	0.6768	0.359	0.373	0.383	0.397	0.409	9.6	0.467	8.3	0.476	8.1	0.483	7.9	0.505	7.6	0.491	7.8	0.520	7.3
44.0	0.6516	0.320	0.334	0.344	0.358	0.369	9.9	0.427	8.5	0.437	8.3	0.444	8.2	0.465	7.8	0.452	8.0	0.481	7.5
47.7	0.6042	0.319	0.333	0.344	0.358	0.369	9.9	0.427	8.5	0.437	8.3	0.444	8.2	0.465	7.8	0.452	8.1	0.480	7.6
51.2	0.5634	0.317	0.331	0.341	0.355	0.367	10.0	0.425	8.6	0.434	8.4	0.441	8.2	0.463	7.9	0.449	8.1	0.478	7.6
58.9	0.4894	0.313	0.327	0.337	0.351	0.362	10.1	0.421	8.7	0.430	8.5	0.437	8.3	0.459	7.9	0.445	8.2	0.474	7.7
63.1	0.4545	0.346	0.360	0.371	0.385	0.396	9.9	0.454	8.5	0.464	8.3	0.471	8.2	0.492	7.8	0.479	8.0	0.507	7.5
67.4	0.4255	0.344	0.358	0.369	0.383	0.394	10.0	0.452	8.5	0.462	8.3	0.469	8.2	0.490	7.8	0.477	8.1	0.505	7.6
73.4	0.3930	0.306	0.320	0.330	0.344	0.356	10.3	0.414	8.8	0.423	8.6	0.430	8.5	0.452	8.1	0.438	8.3	0.467	7.8
79.2	0.3622	0.339	0.353	0.363	0.377	0.389	10.1	0.447	8.7	0.457	8.4	0.464	8.3	0.485	7.9	0.472	8.2	0.500	7.6
85.0	0.3374	0.337	0.351	0.361	0.375	0.387	10.2	0.445	8.7	0.454	8.5	0.461	8.4	0.483	7.9	0.469	8.2	0.498	7.7
94.4	0.3054	0.302	0.316	0.327	0.341	0.352	10.3	0.410	8.8	0.420	8.6	0.427	8.4	0.448	8.0	0.435	8.3	0.463	7.8
105.0	0.2733	0.330	0.344	0.355	0.369	0.380	10.4	0.438	8.8	0.448	8.6	0.455	8.5	0.476	8.1	0.463	8.3	0.491	7.8
121.6	0.2371	0.294	0.308	0.318	0.332	0.344	10.6	0.402	9.0	0.412	8.8	0.419	8.6	0.440	8.2	0.427	8.4	0.455	7.9
127.9	0.2254	0.290	0.304	0.314	0.328	0.340	10.7	0.398	9.0	0.407	8.8	0.414	8.7	0.436	8.2	0.422	8.5	0.451	8.0
131.2	0.2197	0.289	0.303	0.313	0.327	0.339	10.7	0.397	9.1	0.407	8.8	0.414	8.7	0.435	8.3	0.421	8.5	0.450	8.0
135.2	0.2133	0.297	0.311	0.322	0.336	0.347	10.5	0.405	9.0	0.415	8.8	0.422	8.6	0.443	8.2	0.430	8.4	0.458	7.9
148.9	0.1937	0.288	0.302	0.312	0.326	0.338	10.8	0.396	9.1	0.406	8.9	0.413	8.7	0.434	8.3	0.420	8.6	0.449	8.0
158.7	0.1814	0.292	0.306	0.316	0.330	0.342	10.7	0.400	9.1	0.410	8.9	0.417	8.7	0.438	8.3	0.425	8.5	0.453	8.0
170.5	0.1691	0.290	0.304	0.314	0.328	0.340	10.8	0.398	9.1	0.407	8.9	0.414	8.8	0.436	8.3	0.422	8.6	0.451	8.0
184.2	0.1565	0.287	0.302	0.312	0.326	0.337	10.9	0.395	9.2	0.405	9.0	0.412	8.8	0.433	8.4	0.420	8.6	0.449	8.1
201.4	0.1438	0.280	0.294	0.304	0.318	0.330	11.0	0.388	9.3	0.398	9.1	0.405	8.9	0.426	8.5	0.412	8.8	0.441	8.2
210.6	0.1366	0.283	0.297	0.308	0.322	0.333	11.0	0.391	9.3	0.401	9.1	0.408	8.9	0.429	8.4	0.416	8.7	0.444	8.1
221.7	0.1307	0.274	0.288	0.298	0.312	0.323	11.3	0.381	9.5	0.391	9.3	0.398	9.1	0.419	8.6	0.406	8.9	0.435	8.3
230.9	0.1249	0.276	0.290	0.300	0.314	0.326	11.2	0.384	9.4	0.393	9.2	0.400	9.0	0.422	8.6	0.408	8.9	0.437	8.3
241.7	0.1193	0.279	0.293	0.303	0.317	0.329	11.2	0.387	9.4	0.396	9.2	0.403	9.0	0.425	8.5	0.411	8.8	0.440	8.2
263.7	0.1093	0.272	0.286	0.296	0.310	0.321	11.3	0.380	9.5	0.389	9.3	0.396	9.1	0.418	8.6	0.404	8.9	0.433	8.3
282.0	0.1022	0.274	0.288	0.298	0.312	0.324	11.3	0.382	9.5	0.392	9.3	0.399	9.1	0.420	8.6	0.406	8.9	0.435	8.3
306.6	0.0945	0.267	0.281	0.291	0.305	0.317	11.5	0.375	9.7	0.384	9.4	0.391	9.2	0.413	8.7	0.399	9.1	0.428	8.4
322.3	0.0895	0.270	0.284	0.294	0.308	0.320	11.5	0.378	9.6	0.387	9.4	0.394	9.2	0.416	8.7	0.402	9.0	0.431	8.4
339.3	0.0850	0.265	0.279	0.289	0.303	0.315	11.6	0.373	9.7	0.383	9.5	0.390	9.3	0.411	8.8	0.398	9.1	0.426	8.5
362.6	0.0799	0.262	0.276	0.286	0.300	0.311	11.7	0.369	9.8	0.379	9.6	0.386	9.4	0.408	8.9	0.394	9.2	0.423	8.5
386.0	0.0747	0.261	0.275	0.285	0.299	0.311	11.8	0.369	9.8	0.379	9.6	0.386	9.4	0.407	8.9	0.393	9.2	0.422	8.6
402.8	0.0719	0.261	0.275	0.285	0.299	0.310	11.8	0.368	9.9	0.378	9.6	0.385	9.4	0.407	8.9	0.393	9.2	0.422	8.6
428.9	0.0671	0.267	0.281	0.291	0.305	0.316	11.5	0.374	9.7	0.384	9.4	0.391	9.2	0.413	8.7	0.399	9.0	0.428	8.4
448.7	0.0642	0.257	0.271	0.281	0.295	0.306	11.9	0.364	10.0	0.374	9.7	0.381	9.5	0.402	9.0	0.389	9.3	0.418	8.7
456.1	0.0635	0.257	0.271	0.281	0.295	0.307	12.0	0.365	10.0	0.374	9.7	0.381	9.5	0.403	9.0	0.389	9.3	0.418	8.7
483.4	0.0599	0.255	0.269	0.279	0.293	0.305	12.0	0.363	10.0	0.372	9.8	0.379	9.6	0.401	9.0	0.387	9.4	0.416	8.7
494.4	0.0583	0.254	0.268	0.279	0.293	0.304	12.1	0.362	10.0	0.372	9.8	0.379	9.6	0.400	9.0	0.387	9.4	0.415	8.7
510.5	0.0565	0.252	0.266	0.277	0.291	0.302	12.1	0.360	10.1	0.370	9.8	0.377	9.6	0.398	9.1	0.385	9.4	0.413	8.7
523.7	0.0553	0.252	0.266	0.277	0.291	0.302	12.1	0.360	10.1	0.370	9.8	0.377	9.6	0.398	9.1	0.385	9.4	0.413	8.7

Table 5.33: Overhead line feeder circuit data, ACSR Conductors, 50Hz

Sectional area of Al	R_{DC} (20°C)	R_{AC} at 60Hz & 20° C	X_{AC} at 60 Hz .				X_{AC} at 60 Hz and shunt capacitance															
			3.3kV	6.6kV	11kV	22kV	33kV		66kV						132kV					
									Flat circuit		Double vertical		Triangle		Double vertical		Double triangle		Flat circuit	
			X	X	X	X	X	C	X	C	X	C	X	C	X	C	X	C	X	C
mm²	Ω/km	Ω/km	Ω/km	Ω/km	Ω/km	Ω/km	Ω/km	nF/km	Ω/km	nF/km	Ω/km	nF/km	Ω/km	nF/km	Ω/km	nF/km	Ω/km	nF/km	Ω/km	nF/km
13.3	2.1586	2.159	0.474	0.491	0.503	0.520	0.534	8.7	0.604	7.6	0.615	7.4	0.624	7.3	0.649	7.0	0.633	7.2	0.668	6.8
15.3	1.8771	1.877	0.469	0.486	0.498	0.515	0.529	8.8	0.598	7.7	0.610	7.5	0.619	7.4	0.644	7.1	0.628	7.3	0.662	6.9
21.2	1.3557	1.356	0.457	0.474	0.486	0.503	0.516	9.0	0.586	7.8	0.598	7.7	0.606	7.6	0.632	7.2	0.616	7.4	0.650	7.0
23.9	1.2013	1.201	0.452	0.469	0.481	0.498	0.511	9.1	0.581	7.9	0.593	7.8	0.601	7.6	0.627	7.3	0.611	7.5	0.645	7.1
26.2	1.0930	1.093	0.449	0.466	0.478	0.495	0.508	9.2	0.578	8.0	0.590	7.8	0.598	7.7	0.624	7.3	0.608	7.5	0.642	7.1
28.3	1.0246	1.025	0.423	0.440	0.452	0.469	0.483	9.4	0.552	8.2	0.564	8.0	0.572	7.8	0.598	7.5	0.582	7.7	0.616	7.3
33.6	0.8535	0.854	0.439	0.456	0.468	0.485	0.499	9.4	0.569	8.1	0.580	7.9	0.589	7.8	0.614	7.5	0.598	7.7	0.633	7.2
37.7	0.7647	0.765	0.392	0.409	0.421	0.438	0.452	9.7	0.521	8.4	0.533	8.2	0.541	8.1	0.567	7.7	0.551	7.9	0.585	7.4
42.4	0.6768	0.677	0.431	0.447	0.460	0.477	0.490	9.6	0.560	8.3	0.572	8.1	0.580	7.9	0.606	7.6	0.589	7.8	0.624	7.3
44.0	0.6516	0.652	0.384	0.400	0.413	0.429	0.443	9.9	0.513	8.5	0.525	8.3	0.533	8.2	0.559	7.8	0.542	8.0	0.577	7.5
47.7	0.6042	0.604	0.383	0.400	0.412	0.429	0.443	9.9	0.513	8.5	0.524	8.3	0.533	8.2	0.558	7.8	0.542	8.1	0.576	7.6
51.2	0.5634	0.564	0.380	0.397	0.409	0.426	0.440	10.0	0.510	8.6	0.521	8.4	0.530	8.2	0.555	7.9	0.539	8.1	0.573	7.6
58.9	0.4894	0.490	0.375	0.392	0.404	0.421	0.435	10.1	0.505	8.7	0.516	8.5	0.525	8.3	0.550	7.9	0.534	8.2	0.568	7.7
63.1	0.4545	0.455	0.416	0.432	0.445	0.462	0.475	9.9	0.545	8.5	0.557	8.3	0.565	8.2	0.591	7.8	0.574	8.0	0.609	7.5
67.4	0.4255	0.426	0.413	0.430	0.442	0.459	0.473	10.0	0.543	8.5	0.554	8.3	0.563	8.2	0.588	7.8	0.572	8.1	0.606	7.6
73.4	0.3930	0.393	0.367	0.384	0.396	0.413	0.427	10.3	0.496	8.8	0.508	8.6	0.516	8.5	0.542	8.1	0.526	8.3	0.560	7.8
79.2	0.3622	0.362	0.407	0.424	0.436	0.453	0.467	10.1	0.536	8.7	0.548	8.4	0.556	8.3	0.582	7.9	0.566	8.2	0.600	7.6
85.0	0.3374	0.338	0.404	0.421	0.433	0.450	0.464	10.2	0.534	8.7	0.545	8.5	0.554	8.4	0.579	7.9	0.563	8.2	0.598	7.7
94.4	0.3054	0.306	0.363	0.380	0.392	0.409	0.423	10.3	0.492	8.8	0.504	8.6	0.512	8.4	0.538	8.0	0.522	8.3	0.556	7.8
105.0	0.2733	0.274	0.396	0.413	0.426	0.442	0.456	10.4	0.526	8.8	0.537	8.6	0.546	8.5	0.572	8.1	0.555	8.3	0.590	7.8
121.6	0.2371	0.238	0.353	0.370	0.382	0.399	0.413	10.6	0.482	9.0	0.494	8.8	0.502	8.6	0.528	8.2	0.512	8.4	0.546	7.9
127.9	0.2254	0.226	0.348	0.365	0.377	0.394	0.408	10.7	0.477	9.0	0.489	8.8	0.497	8.7	0.523	8.2	0.507	8.5	0.541	8.0
131.2	0.2197	0.220	0.347	0.364	0.376	0.393	0.407	10.7	0.476	9.1	0.488	8.8	0.496	8.7	0.522	8.3	0.506	8.5	0.540	8.0
135.2	0.2133	0.214	0.357	0.374	0.386	0.403	0.416	10.5	0.486	9.0	0.498	8.8	0.506	8.6	0.532	8.2	0.516	8.4	0.550	7.9
148.9	0.1937	0.194	0.346	0.362	0.375	0.392	0.405	10.8	0.475	9.1	0.487	8.9	0.495	8.7	0.521	8.3	0.504	8.6	0.539	8.0
158.7	0.1814	0.182	0.351	0.367	0.380	0.397	0.410	10.7	0.480	9.1	0.492	8.9	0.500	8.7	0.526	8.3	0.509	8.5	0.544	8.0
170.5	0.1691	0.170	0.348	0.365	0.377	0.394	0.408	10.8	0.477	9.1	0.489	8.9	0.497	8.8	0.523	8.3	0.507	8.6	0.541	8.0
184.2	0.1565	0.157	0.345	0.362	0.374	0.391	0.405	10.9	0.474	9.2	0.486	9.0	0.494	8.8	0.520	8.4	0.504	8.6	0.538	8.1
201.4	0.1438	0.145	0.336	0.353	0.365	0.382	0.396	11.0	0.466	9.3	0.477	9.1	0.486	8.9	0.511	8.5	0.495	8.8	0.529	8.2
210.6	0.1366	0.137	0.340	0.357	0.369	0.386	0.400	11.0	0.469	9.3	0.481	9.1	0.489	8.9	0.515	8.4	0.499	8.7	0.533	8.1
221.7	0.1307	0.132	0.328	0.345	0.357	0.374	0.388	11.3	0.458	9.5	0.469	9.3	0.478	9.1	0.503	8.6	0.487	8.9	0.522	8.3
230.9	0.1249	0.126	0.331	0.348	0.360	0.377	0.391	11.2	0.460	9.4	0.472	9.2	0.480	9.0	0.506	8.6	0.490	8.9	0.524	8.3
241.7	0.1193	0.120	0.335	0.351	0.364	0.381	0.394	11.2	0.464	9.4	0.476	9.2	0.484	9.0	0.510	8.5	0.493	8.8	0.528	8.2
263.7	0.1093	0.110	0.326	0.343	0.355	0.372	0.386	11.3	0.455	9.5	0.467	9.3	0.476	9.1	0.501	8.6	0.485	8.9	0.519	8.3
282.0	0.1022	0.103	0.329	0.346	0.358	0.375	0.389	11.3	0.458	9.5	0.470	9.3	0.478	9.1	0.504	8.6	0.488	8.9	0.522	8.3
306.6	0.0945	0.096	0.320	0.337	0.349	0.366	0.380	11.5	0.450	9.7	0.461	9.4	0.470	9.2	0.495	8.7	0.479	9.1	0.514	8.4
322.3	0.0895	0.091	0.324	0.341	0.353	0.370	0.384	11.5	0.453	9.6	0.465	9.4	0.473	9.2	0.499	8.7	0.483	9.0	0.517	8.4
339.3	0.0850	0.086	0.318	0.335	0.347	0.364	0.378	11.6	0.448	9.7	0.459	9.5	0.468	9.3	0.493	8.8	0.477	9.1	0.511	8.5
362.6	0.0799	0.081	0.314	0.331	0.343	0.360	0.374	11.7	0.443	9.8	0.455	9.6	0.463	9.4	0.489	8.9	0.473	9.2	0.507	8.5
386.0	0.0747	0.076	0.313	0.330	0.342	0.359	0.373	11.8	0.443	9.8	0.454	9.6	0.463	9.4	0.488	8.9	0.472	9.2	0.506	8.6
402.8	0.0719	0.074	0.313	0.330	0.342	0.359	0.372	11.8	0.442	9.9	0.454	9.6	0.462	9.4	0.488	8.9	0.472	9.2	0.506	8.6
428.9	0.0671	0.069	0.320	0.337	0.349	0.366	0.380	11.5	0.449	9.7	0.461	9.4	0.469	9.2	0.495	8.7	0.479	9.0	0.513	8.4
448.7	0.0642	0.066	0.308	0.325	0.337	0.354	0.367	11.9	0.437	10.0	0.449	9.7	0.457	9.5	0.483	9.0	0.467	9.3	0.501	8.7
456.1	0.0635	0.065	0.305	0.322	0.334	0.351	0.364	12.0	0.434	10.0	0.446	9.7	0.454	9.6	0.480	9.0	0.463	9.4	0.498	8.7
483.4	0.0599	0.062	0.306	0.323	0.335	0.352	0.366	12.0	0.435	10.0	0.447	9.8	0.455	9.6	0.481	9.0	0.465	9.4	0.499	8.7
494.4	0.0583	0.060	0.305	0.322	0.334	0.351	0.365	12.1	0.435	10.0	0.446	9.8	0.455	9.6	0.480	9.0	0.464	9.4	0.498	8.7
510.5	0.0565	0.059	0.303	0.320	0.332	0.349	0.362	12.1	0.432	10.1	0.444	9.8	0.452	9.6	0.478	9.1	0.462	9.4	0.496	8.7
523.7	0.0553	0.057	0.303	0.320	0.332	0.349	0.363	12.1	0.432	10.1	0.444	9.8	0.452	9.6	0.478	9.1	0.462	9.4	0.496	8.7

Table 5.34: Overhead line feeder circuit data, ACSR Conductors, 60Hz

			Conductor size mm²																
			25	35	50	70	95	120	150	185	240	300	400	*500	*630	*800	*1000	*1200	*1600
3.3kV	Series Resistance	R (Ω/km)	0.927	0.669	0.494	0.342	0.247	0.196	0.158	0.127	0.098	0.08	0.064	0.051	0.042				
	Series Reactance	X (Ω/km)	0.097	0.092	0.089	0.083	0.08	0.078	0.076	0.075	0.073	0.072	0.071	0.088	0.086				
	Susceptance	ωC (mS/km)	0.059	0.067	0.079	0.09	0.104	0.111	0.122	0.133	0.146	0.16	0.179	0.19	0.202				
6.6kV	Series Resistance	R (Ω/km)	0.927	0.669	0.494	0.342	0.247	0.196	0.158	0.127	0.098	0.08	0.064	0.057	0.042				
	Series Reactance	X (Ω/km)	0.121	0.113	0.108	0.102	0.096	0.093	0.091	0.088	0.086	0.085	0.083	0.088	0.086				
	Susceptance	ωC (mS/km)	0.085	0.095	0.104	0.12	0.136	0.149	0.16	0.177	0.189	0.195	0.204	0.205	0.228				
11kV	Series Resistance	R (Ω/km)	0.927	0.669	0.494	0.342	0.247	0.196	0.158	0.127	0.098	0.08	0.064	0.051	0.042				
	Series Reactance	X (Ω/km)	0.128	0.119	0.114	0.107	0.101	0.098	0.095	0.092	0.089	0.087	0.084	0.089	0.086				
	Susceptance	ωC (mS/km)	0.068	0.074	0.082	0.094	0.105	0.115	0.123	0.135	0.15	0.165	0.182	0.194	0.216				
22kV	Series Resistance	R (Ω/km)	-	0.669	0.494	0.348	0.247	0.196	0.158	0.127	0.098	0.08	0.064	0.051	0.042				
	Series Reactance	X (Ω/km)	-	0.136	0.129	0.121	0.114	0.11	0.107	0.103	0.1	0.094	0.091	0.096	0.093				
	Susceptance	ωC (mS/km)		0.053	0.057	0.065	0.072	0.078	0.084	0.091	0.1	0.109	0.12	0.128	0.141				
33kV	Series Resistance	R (Ω/km)	-	0.669	0.494	0.348	0.247	0.196	0.158	0.127	0.098	0.08	0.064	0.051	0.042				
	Series Reactance	X (Ω/km)	-	0.15	0.143	0.134	0.127	0.122	0.118	0.114	0.109	0.105	0.102	0.103	0.1				
	Susceptance	ωC (mS/km)		0.042	0.045	0.05	0.055	0.059	0.063	0.068	0.075	0.081	0.089	0.094	0.103				
66kV*	Series Resistance	R (Ω/km)	-	-	-	-	-	-	-	-	-	-	-	0.0387	0.031	0.0254	0.0215		
	Series Reactance	X (Ω/km)	-	-	-	-	-	-	-	-	-	-	-	0.117	0.113	0.109	0.102		
	Susceptance	ωC (mS/km)												0.079	0.082	0.088	0.11		
145kV*	Series Resistance	R (Ω/km)	-	-	-	-	-	-	-	-	-	-	-	0.0387	0.031	0.0254	0.0215		
	Series Reactance	X (Ω/km)	-	-	-	-	-	-	-	-	-	-	-	0.13	0.125	0.12	0.115		
	Susceptance	ωC (mS/km)												0.053	0.06	0.063	0.072		
245kV*	Series Resistance	R (Ω/km)											0.0487	0.0387	0.0310	0.0254	0.0215	0.0161	0.0126
	Series Reactance	X (Ω/km)											0.145	0.137	0.134	0.128	0.123	0.119	0.113
	Susceptance	ωC (mS/km)											0.044	0.047	0.05	0.057	0.057	0.063	0.072
420kV*	Series Resistance	R (Ω/km)													0.0310	0.0254	0.0215	0.0161	0.0126
	Series Reactance	X (Ω/km)													0.172	0.162	0.156	0.151	0.144
	Susceptance	ωC (mS/km)													0.04	0.047	0.05	0.057	0.063

Table 5.35: Characteristics of polyethylene insulated cables (XLPE), copper conductors (50Hz)

5.25 REFERENCES

[5.1] Physical significance of sub-subtransient quantities in dynamic behaviour of synchronous machines. I.M. Canay. Proc. IEE, Vol. 135, Pt. B, November 1988.

[5.2] IEC 60034-4. Methods for determining synchronous machine quantities from tests.

[5.3] IEEE Standards 115/115A. IEEE Test Procedures for Synchronous Machines.

[5.4] Power System Analysis. J.R. Mortlock and M.W. Humphrey Davies. Chapman & Hall, London.

Chapter 6
Current and Voltage Transformers

6.1 INTRODUCTION

If the voltage or current in a power circuit are too high to connect measuring instruments or relays directly, coupling is made through transformers. Such 'measuring' transformers are required to produce a scaled down replica of the input quantity to the accuracy expected for the particular measurement; this is made possible by the high efficiency of the transformer. During and following large instantaneous changes in the input quantity, the waveform may no longer be sinusoidal, therefore the performance of measuring transformers is important. The deviation may be a step change in magnitude, or a transient component that persists for an significant period, or both. The resulting effect on instrument performance is usually negligible, although for precision metering a persistent change in the accuracy of the transformer may be significant.

However, many protection systems are required to operate during the transient disturbance in the output of the measuring transformers following a system fault. The errors in transformer output may delay the operation of the protection or cause unnecessary operations. Therefore the functioning of such transformers must be examined analytically.

The transformer can be represented by the equivalent circuit of Figure 6.1, where all quantities are referred to the secondary side.

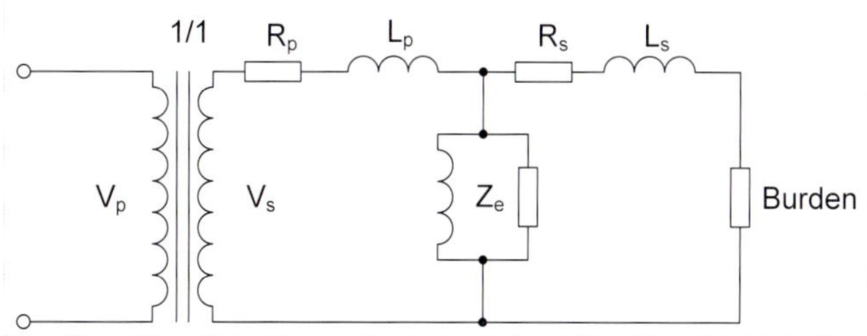

Figure 6.1: Equivalent circuit of transformer

When the transformer is not 1/1 ratio, this condition can be represented by energising the equivalent circuit with an ideal transformer of the given ratio but having no losses.

6.1.1 Measuring Transformers

Voltage and current transformers for low primary voltage or current ratings are not readily distinguishable; for higher ratings, dissimilarities of construction are usual. Nevertheless the main differences between these devices are the way they are connected into the power circuit. Voltage transformers are much like small power transformers, differing only in details of design that control ratio accuracy over the specified range of output. Current transformers have their primary windings connected in series with the power circuit, and so also in series with the system impedance. The response of the transformer is radically different in these two modes of operation.

6.2 ELECTROMAGNETIC VOLTAGE TRANSFOMERS

In the shunt mode, the system voltage is applied across the input terminals of the equivalent circuit of Figure 6.1. The vector diagram for this circuit is shown in Figure 6.2.

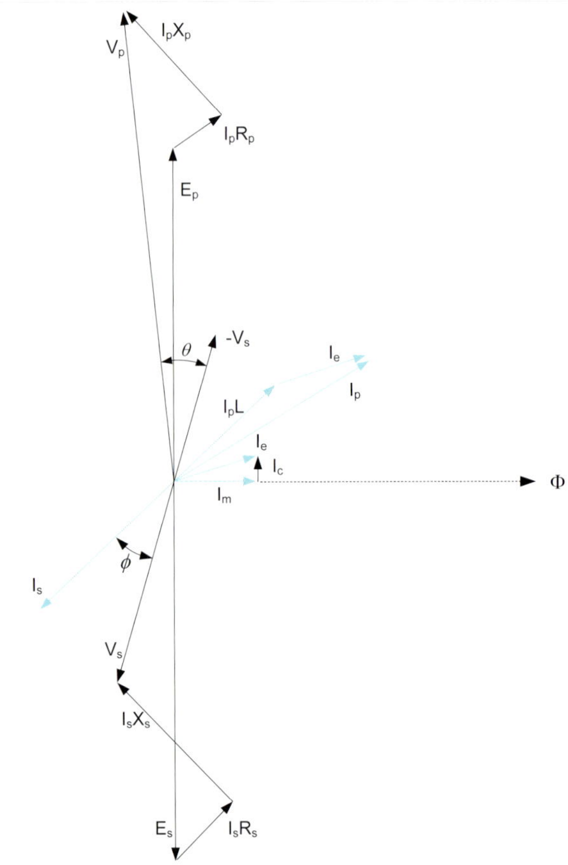

V_p = Primary applied voltage
E_p = Primary induced e.m.f.
V_s = Secondary output voltage
E_s = Secondary induced e.m.f.
Φ = Flux
I_e = Exciting current
I_m = Magnetising component
I_c = Iron loss component
θ = Phase angle error
ϕ = Secondary burden angle
I_pR_p = Primary resistance voltage drop
I_pX_p = Primary reactance voltage drop
I_sR_s = Secondary resistance voltage drop
I_sX_s = Secondary reactance voltage drop
I_s = Secondary current
I_pL = Load component of primary current
I_p = Primary current

Figure 6.2: Vector diagram of voltage transformer

The secondary output voltage V_s is required to be an accurate scaled replica of the input voltage V_p over a specified range of output. Therefore the winding voltage drops are made small and the normal flux density in the core is designed to be well below the saturation density, so the exciting current can be low and the exciting impedance substantially constant with a variation of applied voltage over the desired operating range including some degree of overvoltage. These limitations in design result in a VT for a given burden being much larger than a typical power transformer of similar rating. Consequently the exciting current is not as small, relative to the rated burden, as it would be for a typical power transformer.

6.2.1 Errors

The ratio and phase errors of the transformer can be calculated using the vector diagram of Figure 6.2.

The ratio error is defined as:

$$\frac{\left(K_n V_s - V_p\right)}{V_p} \times 100\%$$

where:

K_n is the nominal ratio

V_p is the primary voltage

V_s is the secondary voltage

If the error is positive, the secondary voltage is greater than the nominal value. If the error is negative, the secondary voltage is less than the nominal value. The turns ratio of the transformer need not be equal to the nominal ratio and a small turns compensation is usually used so the error is positive for low burdens and negative for high burdens.

The phase error is the phase difference between the reversed secondary and the primary voltage vectors. It is positive when the reversed secondary voltage leads the primary vector. Requirements in this respect are set out in IEC 60044-2. All voltage transformers are required to comply with one of the classes in Table 6.1.

For protection purposes, accuracy of voltage measurement may be important during fault conditions, as the system voltage might be reduced by the fault to a low value. Voltage transformers for such types of service must comply with the extended range of requirements set out in Table 6.2.

Accuracy Class	0.8 - 1.2 x rated voltage 0.25 - 1.0 x rated burden at 0.8pf	
	voltage ratio error (%)	phase displacement (minutes)
0.1	+/- 0.1	+/- 5
0.2	+/- 0.2	+/- 10
0.5	+/- 0.5	+/- 20
1.0	+/- 1.0	+/- 40
3.0	+/- 3.0	not specified

Table 6.1: Measuring Voltage Transformer error limits

Accuracy Class	0.25 - 1.0 x rated burden at 0.8pf 0.05 - Vf x rated primary voltage	
	Voltage ratio error (%)	Phase displacement (minutes)
3P	+/- 3.0	+/- 120
6P	+/- 6.0	+/- 240

Table 6.2: Additional limits for protection Voltage Transformers

6.2.2 Voltage Factors

The quantity V_f in Table 6.2 is an upper limit of operating voltage, expressed in per unit of rated voltage. This is important for correct relay operation and operation under unbalanced fault conditions on unearthed or impedance earthed systems, resulting in a rise in the voltage on the healthy phases.

Voltage factor Vf	Time rating	Primary winding connection/system earthing conditions
1.2	continuous	Between lines in any network.
		Between transformer star point and earth in any network
1.2	continuous	Between line and earth in an effectively earthed network
1.5	30 sec	
1.2	continuous	Between line and earth in a non-effectively earthed neutral system with automatic earth fault tripping
1.9	30 sec	
1.2	continuous	Between line and earth in an isolated neutral system without automatic earth fault tripping, or in a resonant earthed system without automatic earth fault tripping
1.9	8 hours	

Table 6.3: Voltage transformers permissible duration of maximum voltage

6.2.3 Secondary Leads

Voltage transformers are designed to maintain the specified accuracy in voltage output at their secondary terminals. To maintain this if long secondary leads are required, a distribution box can be fitted close to the VT to supply relay and metering burdens over separate leads. If necessary, allowance can be made for the resistance of the leads to individual burdens when the particular equipment is calibrated

6.2.4 Protection of Voltage Transformers

Voltage Transformers can be protected by High Rupturing Capacity (H.R.C.) fuses on the primary side for voltages up to 66kV. Fuses do not usually have a sufficient interrupting capacity for use with higher voltages. Practice varies, and in some cases protection on the primary is omitted.

The secondary of a Voltage Transformer should always be protected by fuses or a miniature circuit breaker (MCB). The device should be located as near to the transformer as possible. A short circuit on the secondary circuit wiring produces a current of many times the rated output and causes excessive heating. Even where primary fuses can be fitted, these usually do not clear a secondary side short circuit because of the low value of primary current and the minimum practicable fuse rating.

6.2.5 Construction of Voltage Transformers

The construction of a voltage transformer differs from that of a power transformer in that different emphasis is placed on cooling, insulation and mechanical design. The rated output seldom exceeds a few hundred VA and therefore the heat generated normally presents no problem. The size of a VT is largely determined by the system voltage and the insulation of the primary winding often exceeds the winding in volume.

A VT should be insulated to withstand overvoltages, including impulse voltages, of a level equal to the withstand value of the switchgear and the high voltage system. To achieve this in a compact design the voltage must be distributed uniformly across the winding, which requires uniform distribution of the winding capacitance or the application of electrostatic shields.

Voltage transformers are commonly used with switchgear so the physical design must be compact and adapted for mounting in or near to the switchgear. Three-phase units are common up to 36kV but for higher voltages single-phase units are usual. Voltage transformers for medium voltage circuits have dry type insulation, but high and extra high voltage systems still use oil immersed units. Figure 6.3 shows an Alstom OTEF 36.5kV to 765kV high voltage electromagnetic transformer.

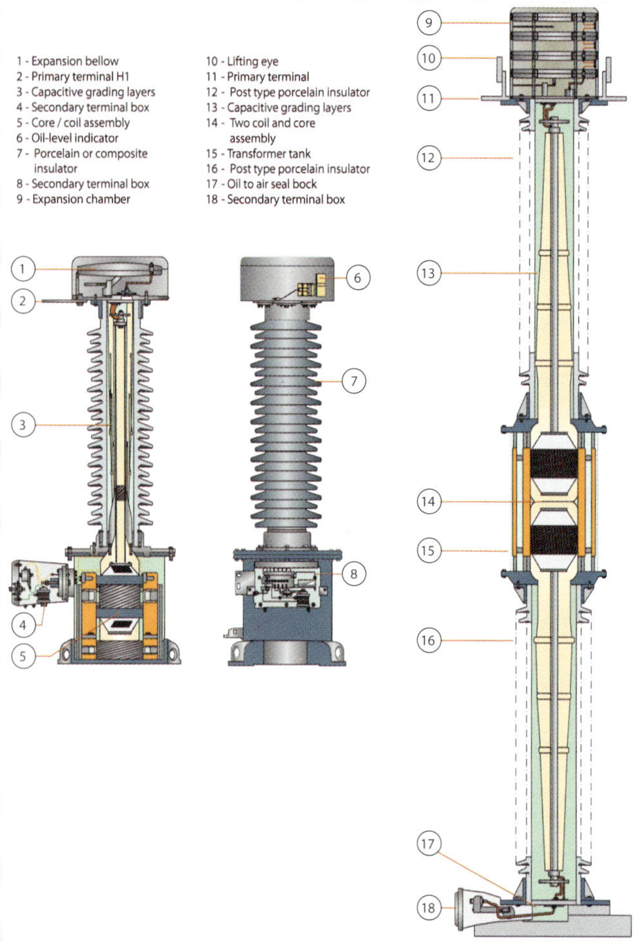

Figure 6.3: Alstom OTEF electromagnetic 36.6kV to 765kV high voltage transformer

6.2.6 Residually connected Voltage Transformers

The three voltages of a balanced system summate to zero, but this is not so when the system is subject to a single-phase earth fault. The residual voltage of a system is measured by connecting the secondary windings of a VT in 'broken delta' as shown in Figure 6.4.

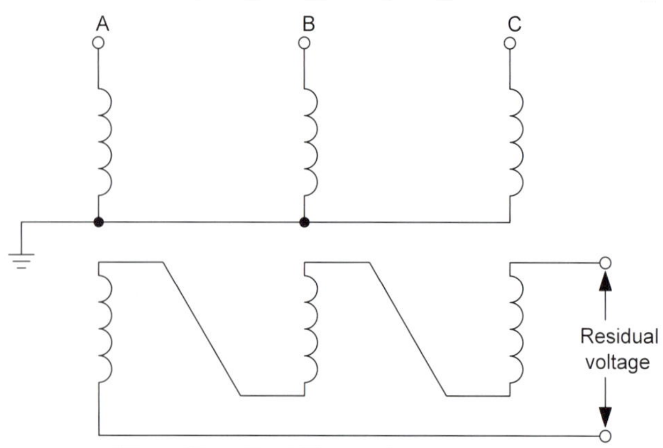

Figure 6.4: Residual voltage connection

The output of the secondary windings connected in broken delta is zero when balanced sinusoidal voltages are applied, but under conditions of imbalance a residual voltage equal to three times the zero sequence voltage of the system is developed. To measure this component it is necessary for a zero sequence flux to be set up in the VT, and for this to be possible there must be a return path for the resultant summated flux. The VT core must have one or more unwound limbs linking the yokes in addition to the limbs carrying windings. Usually the core is made symmetrically, with five limbs, the two outermost ones being unwound. Alternatively, three single-phase units can be used. It is equally necessary for the primary winding neutral to be earthed, for without an earth, zero sequence exciting current cannot flow.

A VT should be rated to have an appropriate voltage factor as described in Section 6.2.2 and Table 6.3, to cater for the voltage rise on healthy phases during earth faults.

Voltage transformers are often provided with a normal star-connected secondary winding and a broken-delta connected 'tertiary' winding. Alternatively the residual voltage can be extracted by using a star/broken-delta connected group of auxiliary voltage transformers energised from the secondary winding of the main unit, providing the main voltage transformer fulfils all the requirements for handling a zero sequence voltage as previously described. The auxiliary VT must also be suitable for the appropriate voltage factor. It should be noted that third harmonics in the primary voltage wave, which are of zero sequence, summate in the broken-delta winding.

6.2.7 Transient Performance

Transient errors cause few difficulties in the use of conventional voltage transformers although some do occur. Errors are generally limited to short time periods following the sudden application or removal of voltage from the VT primary.

If a voltage is suddenly applied, an inrush transient occurs, as with power transformers. However, the effect is less severe than for power transformers because of the lower flux density for which the VT is designed. If the VT is rated to have a fairly high voltage factor, there is little inrush effect. An error appears in the first few cycles of the output current in proportion to the inrush transient that occurs.

When the supply to a voltage transformer is interrupted, the core flux does not immediately collapse. The secondary winding maintains the magnetising force to sustain this flux and circulates a current through the burden, which decays more or less exponentially. There may also be a superimposed audio-frequency oscillation due to the capacitance of the winding. If the exciting quantity in ampere-turns exceeds the burden, the transient current may be significant.

6.2.8 Cascade Voltage Transformer

The capacitor VT (section 6.3) was developed because of the high cost of conventional electromagnetic voltage transformers but, as shown in Section 6.3.2, the frequency and transient responses are less satisfactory than those of the orthodox voltage transformers. Another solution to the problem is the cascade VT shown in Figure 6.5.

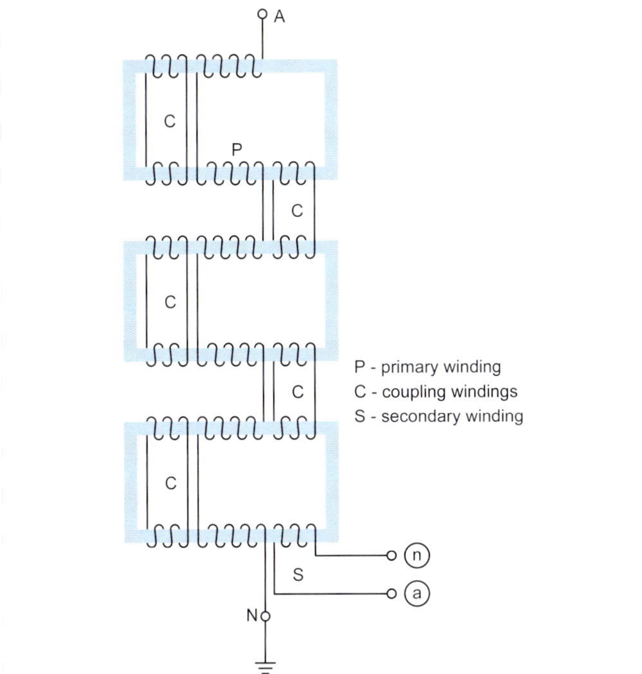

Figure 6.5: Schematic diagram of typical cascade voltage transformer

The conventional type of VT has a single primary winding, the insulation of which presents a problem for voltages above about 132kV. The cascade VT avoids these difficulties by breaking down the primary voltage in several distinct and separate stages.

The complete VT is made up of several individual transformers, the primary windings of which are connected in series as shown in Figure 6.5. Each magnetic core has primary windings (P) on two opposite sides. The secondary winding (S) consists of a single winding on the last stage only. Coupling windings (C) connected in pairs between stages, provide low impedance circuits for the transfer of load ampere-turns between stages and ensure that the power frequency voltage is equally distributed over the several primary windings.

The potentials of the cores and coupling windings are fixed at definite values by connecting them to selected points on the primary windings. The insulation of each winding is sufficient for the voltage developed in that winding, which is a fraction of the total according to the number of stages. The individual transformers are mounted on a structure built of insulating material, which provides the interstage insulation, accumulating to a value able to withstand the full system voltage across the complete height of the stack. The entire assembly is contained in a hollow cylindrical porcelain housing with external weather-sheds; the housing is filled with oil and sealed, an expansion bellows being included to maintain hermetic sealing and to permit expansion with temperature change.

6.3 CAPACITOR VOLTAGE TRANSFOMERS

The size of electromagnetic voltage transformers for the higher voltages is largely proportional to the rated voltage; the cost tends to increase at a disproportionate rate. The capacitor voltage transformer (CVT) is often more economic.

This device is basically a capacitance potential divider. As with resistance-type potential dividers, the output voltage is seriously affected by load at the tapping point. The capacitance divider differs in that its equivalent source impedance is capacitive and can therefore be compensated by a reactor connected in series with the tapping point. With an ideal reactor, such an arrangement would have no regulation and could supply any value of output.

A reactor possesses some resistance, which limits the output that can be obtained. For a secondary output voltage of 110V, the capacitors would have to be very large to provide a useful output while keeping errors within the usual limits. The solution is to use a high secondary voltage and further transform the output to the normal value using a relatively inexpensive electromagnetic transformer. The successive stages of this reasoning are shown in Figure 6.6: Development of capacitor voltage transformer.

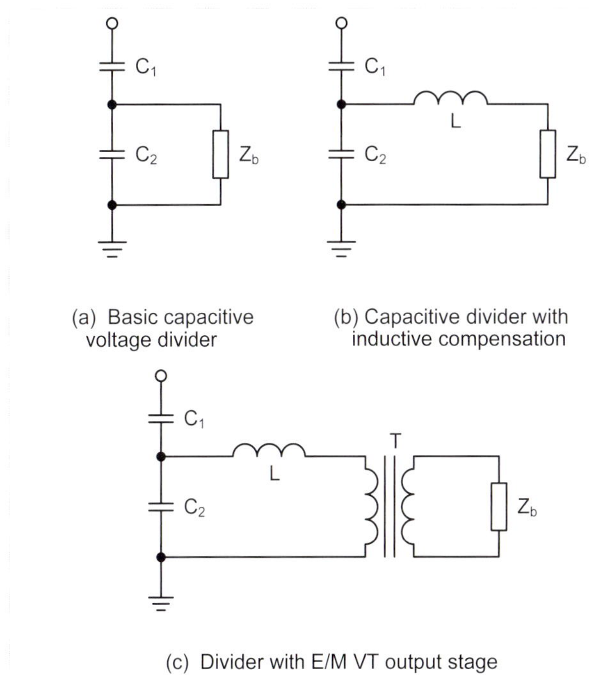

Figure 6.6: Development of capacitor voltage transformer

There are numerous variations of this basic circuit. The inductance L may be a separate unit or it may be incorporated in the form of leakage reactance in the transformer T. Capacitors C_1 and C_2 cannot conveniently be made to close tolerances, so tappings are provided for ratio adjustment, either on the transformer T, or on a separate auto-transformer in the secondary circuit. Adjustment of the tuning inductance L is also needed; this can be done with tappings, a separate tapped inductor in the secondary circuit, by adjustment of gaps in the iron cores, or by shunting with variable capacitance. A simplified equivalent circuit is shown in Figure 6.7.

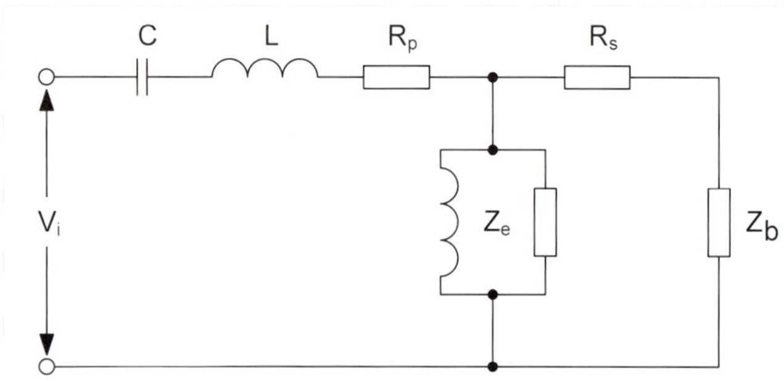

L = tuning inductance
R_p = primary winding resistance (plus losses)
Z_e = exciting impedance of transformer T
R_s = secondary circuit resistance
Z_b = burden impedance
C = C_1+C_2

Figure 6.7: Simplified equivalent circuit of capacitor voltage transformer

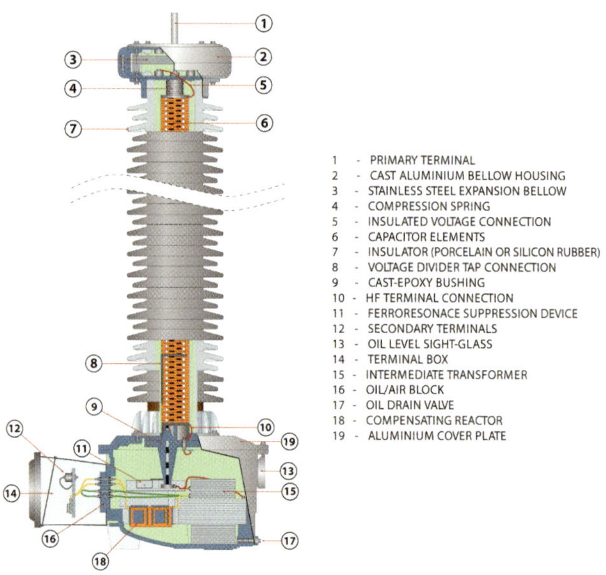

Figure 6.8: Section view of an Alstom OTCF 72.5kV to 765kV coupling capacitor voltage transformer

The main difference between Figure 6.7 and Figure 6.1 is the presence of C and L. At normal frequency when C and L resonate and therefore cancel, the circuit behaves in a similar way to a conventional VT. However, at other frequencies a reactive component exists which modifies the errors.

Standards generally require a CVT used for protection to conform to accuracy requirements of Table 6.2 within a frequency range of 97-103% of nominal. The corresponding frequency range of measurement CVTs is much less, 99%-101%, as reductions in accuracy for frequency deviations outside this range are less important than for protection applications.

6.3.1 Voltage Protection of Auxiliary Capacitor

If the burden impedance of a CVT is short-circuited, the rise in the reactor voltage is limited only by the reactor losses and possible saturation to $Q \times E_2$ where E_2 is the no-load tapping point voltage and Q is the amplification factor of the resonant circuit. This value would be excessive and is therefore limited by a spark gap connected across the auxiliary capacitor. The voltage on the auxiliary capacitor is higher at full rated output than at no load, and the capacitor is rated for continuous service at this raised value. The spark gap is set to flash over at about twice the full load voltage.

The spark gap limits the short-circuit current which the VT delivers and fuse protection of the secondary circuit is carefully designed with this in mind. Usually the tapping point can be earthed either manually or automatically before making any adjustments to tappings or connections.

6.3.2 Transient Behaviour of Capacitor Voltage Transformers

A CVT is a series resonant circuit. The introduction of the electromagnetic transformer between the intermediate voltage and the output makes further resonance possible involving the exciting impedance of this unit and the capacitance of the divider stack. When a sudden voltage step is applied, oscillations in line with these different modes take place and persist for a period governed by the total resistive damping that is present. Any increase in resistive burden reduces the time constant of a transient oscillation, although the chance of a large initial amplitude is increased.

For very high-speed protection, transient oscillations should be minimised. Modern capacitor voltage transformers are much better in this respect than their earlier counterparts. However, high performance protection schemes may still be adversely affected unless their algorithms and filters have been specifically designed with care.

6.3.3 Ferro-Resonance

The exciting impedance Z_e of the auxiliary transformer T and the capacitance of the potential divider together form a resonant circuit that usually oscillates at a sub-normal frequency. If this circuit is subjected to a voltage impulse, the resulting oscillation may pass through a range of frequencies. If the basic frequency of this circuit is slightly less than one-third of the system frequency, it is possible for energy to be absorbed from the system and cause the oscillation to build up. The increasing flux density in the transformer core reduces the inductance, bringing the resonant frequency nearer to the one-third value of the system frequency. The result is a progressive build-up until the oscillation stabilises as a third sub-harmonic of the system, which can be maintained indefinitely. Depending on the values of components, oscillations at fundamental frequency or at other sub-harmonics or multiples of the supply frequency are possible but the third sub-harmonic is the one most likely to be encountered. The principal manifestation of such an oscillation is a rise in output voltage, the r.m.s. value being perhaps 25% to 50% above the normal value. The output waveform would generally be of the form shown in Figure 6.9.

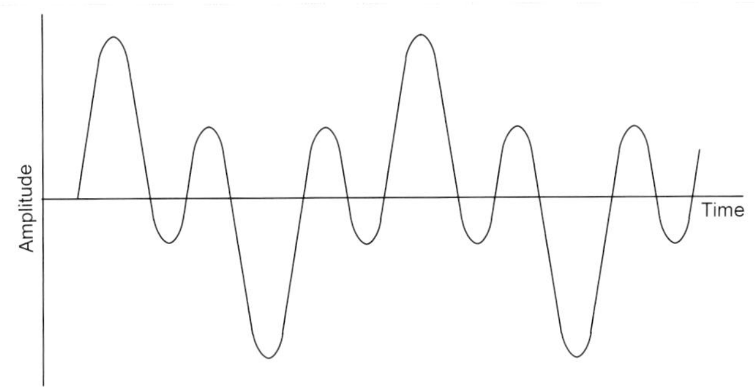

Figure 6.9: Typical secondary voltage waveform with third sub-harmonic oscillation

Such oscillations are less likely to occur when the circuit losses are high, as is the case with a resistive burden, and can be prevented by increasing the resistive burden. Special anti-ferro-resonance devices that use a parallel-tuned circuit are sometimes built into the VT. Although such arrangements help to suppress ferro-resonance, they tend to impair the transient response, so that the design is a matter of compromise.

Correct design prevents a CVT that supplies a resistive burden from exhibiting this effect, but it is possible for non-linear inductive burdens, such as auxiliary voltage transformers, to induce ferro-resonance. Auxiliary voltage transformers for use with capacitor voltage transformers should be designed with a low value of flux density that prevents transient voltages from causing core saturation, which in turn would bring high exciting currents.

6.4 CURRENT TRANSFOMERS

The primary winding of a current transformer is connected in series with the power circuit and the impedance is negligible compared with that of the power circuit. The power system impedance governs the current passing through the primary winding of the current transformer. This condition can be represented by inserting the load impedance, referred through the turns ratio, in the input connection of Figure 6.1.

This approach is developed in Figure 6.10, taking the numerical example of a 300/5A CT applied to an 11kV power system. The system is considered to be carrying rated current (300A) and the CT is feeding a burden of 10VA.

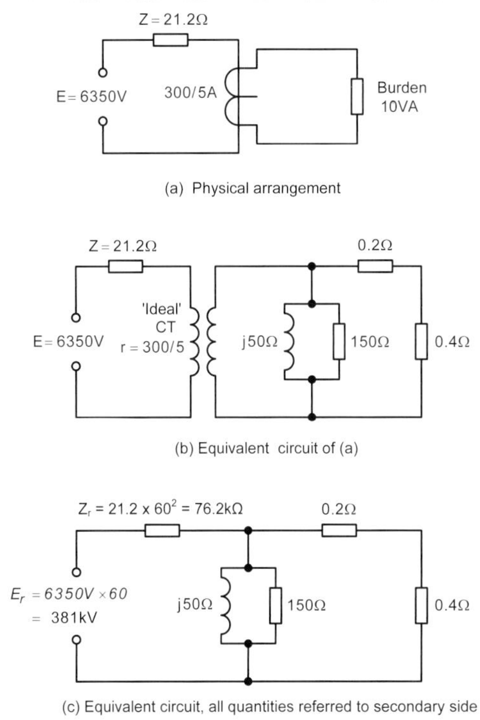

Figure 6.10: Derivation of equivalent circuit of a current transformer

A study of the final equivalent circuit of Figure 6.10(c), taking note of the typical component values, reveals all the properties of a current transformer. It can be seen that:

- The secondary current is not affected by change of the burden impedance over a considerable range.
- The secondary circuit must not be interrupted while the primary winding is energised. The induced secondary e.m.f. under these circumstances is high enough to present a danger to life and insulation.
- The ratio and phase angle errors can be calculated easily if the magnetising characteristics and the burden impedance are known.

6.4.1 Errors

The general vector diagram shown in Figure 6.2 can be simplified by omitting details that are not of interest in current measurement; see Figure 6.11. Errors arise because of the shunting of the burden by the exciting impedance. This uses a small portion of the input current for exciting the core, reducing the amount passed to the burden. So $I_s = I_p - I_e$, where I_e is dependent on Z_e, the exciting impedance and the secondary e.m.f. E_s, given by the equation $E_s=I_s(Z_s+Z_b)$, where:

Z_s = the self-impedance of the secondary winding, which can generally be taken as the resistive component R_s only

Z_b = the impedance of the burden

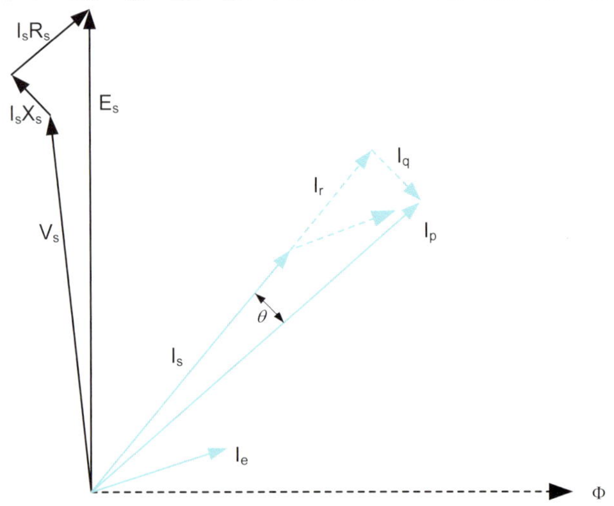

E_s = Secondary induced e.m.f.
V_s = Secondary output voltage
I_p = Primary current
I_s = Secondary current
θ = Phase angle error
Φ = Flux
I_sR_s = Secondary resistance voltage drop
I_sX_s = Secondary reactance voltage drop
I_e = Exciting current
I_r = Component of I_e in phase with I_s
I_q = Component of I_e in quadrature with I_s

Figure 6.11: Vector diagram for current transformer (referred to secondary)

6.4.1.1 Current or Ratio Error

This is the difference in magnitude between I_p and I_s and is equal to I_r, the component of I_e which is in phase with I_s.

6.4.1.2 Phase Error

This is represented by I_q, the component of I_e in quadrature with I_s and results in the phase error ϕ.

The values of the current error and phase error depend on the phase displacement between I_s and I_e, but neither current nor phase error can exceed the vectorial error I_e. With a moderately inductive burden, resulting in I_s and I_e approximately in phase, there is little phase error and the exciting component results almost entirely in ratio error.

A reduction of the secondary winding by one or two turns is often used to compensate for this. For example, in the CT corresponding to Figure 6.10, the worst error due to the use of an inductive burden of rated value would be about 1.2%. If the nominal turns ratio is 2:120, removal of one secondary turn would raise the output by 0.83% leaving the overall current error as -0.37%.

For lower value burden or a different burden power factor, the error would change in the positive direction to a maximum of +0.7% at zero burden; the leakage reactance of the secondary winding is assumed to be negligible. No corresponding correction can be made for phase error, but it should be noted that the phase error is small for moderately reactive burdens.

6.4.2 Composite Error

This is defined in IEC 60044-1 as the r.m.s. value of the difference between the ideal secondary current and the actual secondary current. It includes current and phase errors and the effects of harmonics in the exciting current. The accuracy class of measuring current transformers is shown in Table 6.4 and Table 6.5.

Accuracy Class		+/- Percentage current (ratio) error				+/- Phase displacement (minutes)			
	% current	5	20	100	120	5	20	100	120
0.1		0.4	0.2	0.1	0.1	15	8	5	5
0.2		0.75	0.35	0.2	0.2	30	15	10	10
0.5		1.5	0.75	0.5	0.5	90	45	30	30
1		3	1.5	1.0	1.0	180	90	60	60

Table 6.4: Limits of CT error for accuracy classes 0.1 to 1.0

Accuracy Class		+/- current (ratio) error, %	
	% current	50	120
3		3	3
5		5	5

Table 6.5: Limits of CT error for accuracy classes 3 and 5

6.4.3 Accuracy Limit Current of Protection Current Transformers

Protection equipment is intended to respond to fault conditions, and is for this reason required to function at current values above the normal rating. Protection class current transformers must retain a reasonable accuracy up to the largest relevant current. This value is known as the 'accuracy limit current' and may be expressed in primary or equivalent secondary terms. The ratio of the accuracy limit current to the rated current is known as the 'accuracy limit factor'. The accuracy class of protection current transformers is shown in Table 6.6.

Class	Current error at rated primary current (%)	Phase displacement at rated current (minutes)	Composite error at rated accuracy limit primary current (%)
5P	+/-1	+/-60	5
10P	+/-3	-	10
Standard accuracy limit factors are 5, 10, 15, 20, and 30			

Table 6.6: Protection CT error limits for classes 5P and 10P

Even though the burden of a protection CT is only a few VA at rated current, the output required from the CT may be considerable if the accuracy limit factor is high. For example, with an accuracy limit factor of 30 and a burden of 10VA, the CT may have to supply 9000VA to the secondary circuit.

Alternatively, the same CT may be subjected to a high burden. For overcurrent and earth fault protection, with elements of similar VA consumption at setting, the earth fault element of an electromechanical relay set at 10% would have 100 times the impedance of the overcurrent elements set at 100%. Although saturation of the relay elements somewhat modifies this aspect of the matter, the earth fault element is a severe burden, and the CT is likely to have a considerable ratio error in this case. Therefore it is not much use applying turns compensation to such current transformers; it is generally simpler to wind the CT with turns corresponding to the nominal ratio.

Current transformers are often used for the dual duty of measurement and protection. They then need to be rated according to a class selected from Table 6.4, Table 6.5 and Table 6.6. The applied burden is the total of instrument and relay burdens. Turns compensation may well be needed to achieve the measurement performance. Measurement ratings are expressed in terms of rated burden and class, for example 15VA Class 0.5. Protection ratings are expressed in terms of rated burden, class, and accuracy limit factor, for example 10VA Class 10P10.

6.4.4 Class PX Current Transformers

The classification of Table 6.6 is only used for overcurrent protection. Class PX is the definition in IEC 60044-1 for the quasi-transient current transformers formerly covered by Class X of BS 3938, commonly used with unit protection schemes.

Guidance was given in the specifications to the application of current transformers to earth fault protection, but for this and for the majority of other protection applications it is better to refer directly to the maximum useful e.m.f. that can be obtained from the CT. In this context, the 'knee-point' of the excitation curve is defined as 'that point at which a further increase of 10% of secondary e.m.f. would require an increment of exciting current of 50%'; see Figure 6.12.

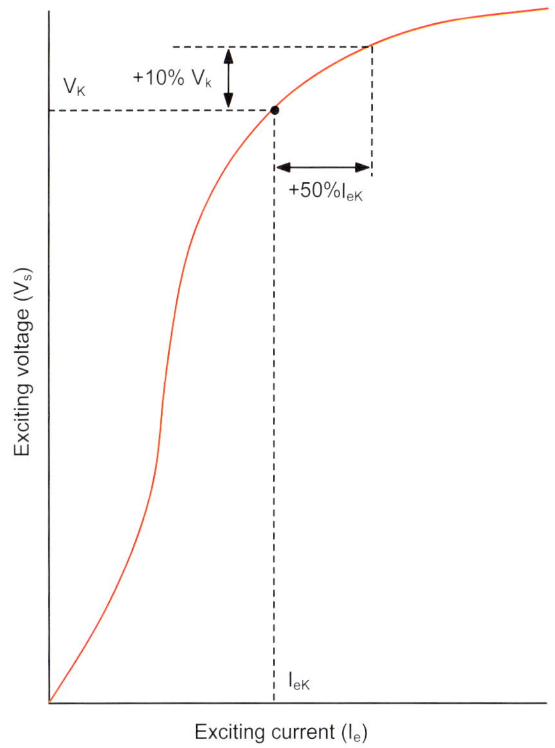

Figure 6.12: Definition of knee-point of excitation curve

Design requirements for current transformers for general protection purposes are frequently laid out in terms of knee-point e.m.f., exciting current at the knee-point (or some other specified point) and secondary winding resistance. Such current transformers are designated Class PX

6.4.5 CT Winding Arrangements

Several CT winding arrangements are used. These are described in the following sections.

6.4.5.1 Wound primary type

This type of CT has conventional windings formed of copper wire wound round a core. It is used for auxiliary current transformers and for many low or moderate ratio current transformers used in switchgear of up to 11kV rating.

6.4.5.2 Bushing or bar primary type

Many current transformers have a ring-shaped core, sometimes built up from annular stampings, but often consisting of a single length of strip tightly wound to form a close-turned spiral. The distributed secondary winding forms a toroid which should occupy the whole perimeter of the core, a small gap being left between start and finish leads for insulation.

Such current transformers normally have a single concentrically placed primary conductor, sometimes permanently built into the CT and provided with the necessary primary insulation. In other cases, the bushing of a circuit breaker or power transformer is used for this purpose. At low primary current ratings it may be difficult to obtain sufficient output at the desired accuracy. This is because a large core section is needed to provide enough flux to induce the secondary e.m.f. in the small number of turns, and because the exciting ampere-turns form a large proportion of the primary ampere-turns available. The effect is particularly pronounced when the core diameter has been made large to fit over large EHV bushings.

6.4.5.3 Core-Balance Current Transformers

The core-balance CT (or CBCT) is normally of the ring type, through the centre of which is passed cable that forms the primary winding. An earth fault relay, connected to the secondary winding, is energised only when there is residual current in the primary system.

The advantage in using this method of earth fault protection lies in the fact that only one CT core is used in place of three phase CTs whose secondary windings are residually connected. In this way the CT magnetising current at relay operation is reduced by approximately three-to-one, an important consideration in sensitive earth fault relays where a low effective setting is required. The number of secondary turns does not need to be related to the cable rated current because no secondary current would flow under normal balanced conditions. This allows the number of secondary turns to be chosen such as to optimise the effective primary pick-up current.

Core-balance transformers are normally mounted over a cable at a point close up to the cable gland of switchgear or other apparatus. Physically split cores ('slip-over' types) are normally available for applications in which the cables are already made up, as on existing switchgear.

6.4.5.4 Summation Current Transformers

The summation arrangement is a winding arrangement used in a measuring relay or on an auxiliary current transformer to give a single-phase output signal having a specific relationship to the three-phase current input.

6.4.5.5 Air-gapped current transformers

These are auxiliary current transformers in which a small air gap is included in the core to produce a secondary voltage output proportional in magnitude to current in the primary winding. Sometimes termed 'transactors' and 'quadrature current transformers', this form of current transformer has been used as an auxiliary component of traditional pilot-wire unit protection schemes in which the outputs into multiple secondary circuits must remain linear for and proportional to the widest practical range of input currents.

6.4.6 CT Winding Arrangements

CTs for measuring line currents fall into one of three types.

6.4.6.1 Over-Dimensioned CTs

Over-dimensioned CTs are capable of transforming fully offset fault currents without distortion. In consequence, they are very large, as can be deduced from Section 6.4.10. They are prone to errors due to remanent flux arising, for instance, from the interruption of heavy fault currents.

6.4.6.2 Anti-Remanence CTs

This is a variation of the overdimensioned current transformer and has small gap(s) in the core magnetic circuit, thus reducing the possible remanent flux from approximately 90% of saturation value to approximately 10%. These gap(s) are quite small, for example 0.12mm total, and so the excitation characteristic is not significantly changed by their presence. However, the resulting decrease in possible remanent core flux confines any subsequent d.c. flux excursion, resulting from primary current asymmetry, to within the core saturation limits. Errors in current transformation are therefore significantly reduced when compared with those with the gapless type of core.

Transient protection Current Transformers are included in IEC 60044-6 as types TPX, TPY and TPZ and this specification gives good guidance to their application and use.

6.4.6.3 Linear Current Transformers

The 'linear' current transformer constitutes an even more radical departure from the normal solid core CT in that it incorporates an appreciable air gap, for example 7.5-10mm. As its name implies the magnetic behaviour tends to linearisation by the inclusion of this gap in the magnetic circuit. However, the purpose of introducing more reluctance into the magnetic circuit is to reduce the value of magnetising reactance. This in turn reduces the secondary time-constant of the CT, thereby reducing the overdimensioning factor necessary for faithful transformation.

Figure 6.13 shows a CT for use on HV systems.

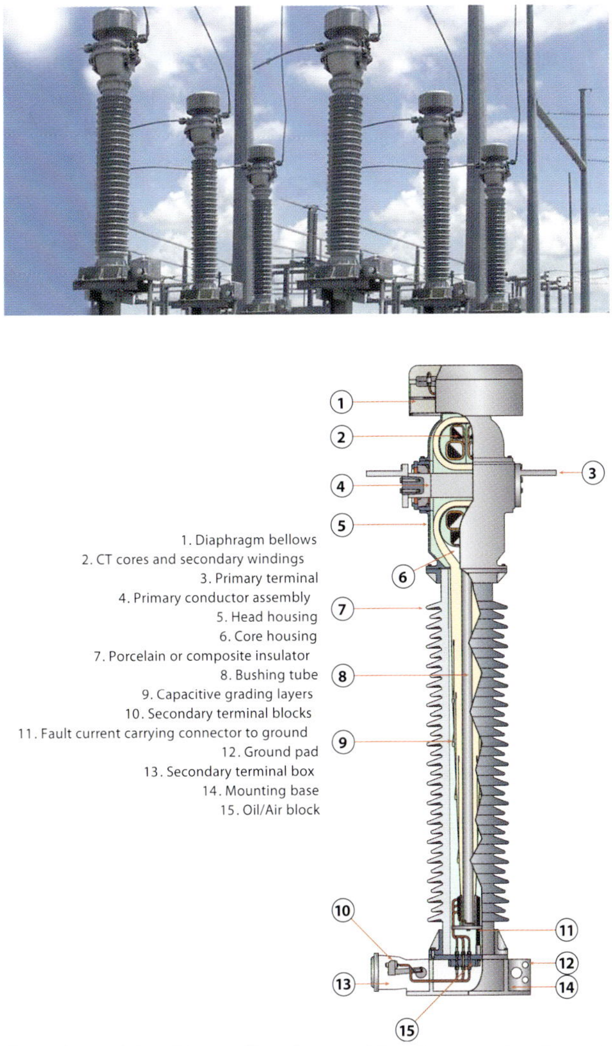

Figure 6.13: Alstom OSKF 72.5kV to 765kV high voltage current transformer

6.4.7 Secondary Winding Impedance

As a protection CT may be required to deliver high values of secondary current, the secondary winding resistance must be made as low as practicable. Secondary leakage reactance also occurs, particularly in wound primary current transformers, although its precise measurement is difficult. The non-linear nature of the CT magnetic circuit makes it difficult to assess the definite ohmic value representing secondary leakage reactance.

It is however, normally accepted that a current transformer is of the low reactance type provided that the following conditions prevail:

- The core is of the jointless ring type (including spirally wound cores).
- The secondary turns are substantially evenly distributed along the whole length of the magnetic circuit.
- The primary conductor(s) passes through the approximate centre of the core aperture or, if wound, is approximately evenly distributed along the whole length of the magnetic circuit.
- Flux equalising windings, where fitted to the requirements of the design, consist of at least four parallel-connected coils, evenly distributed along the whole length of the magnetic circuit, each coil occupying one quadrant.

Alternatively, when a current transformer does not comply with all of the above requirements, it may be proved to be of low-reactance. In this case the composite error, as measured in the accepted way, does not exceed by a factor of 1.3 that error obtained directly from the V-I excitation characteristic of the secondary winding.

6.4.8 Secondary Current Rating

The choice of secondary current rating is determined largely by the secondary winding burden and the standard practice of the user. Standard CT secondary current ratings are 5A and 1A. The burden at rated current imposed by digital or numerical relays or instruments is largely independent of the rated value of current. This is because the winding of the device has to develop a given number of ampere-turns at rated current, so that the actual number of turns is inversely proportional to the current, and the impedance of the winding varies inversely with the square of the current rating. However, electromechanical or static earth-fault relays may have a burden that varies with the current tapping used.

Interconnection leads do not share this property, however, being commonly of standard cross-section regardless of rating. Where the leads are long, their resistance may be appreciable, and the resultant burden varies with the square of the current rating. For example a CT lead run of the order of 200 metres, a typical distance for outdoor EHV switchgear, could have a loop resistance of approximately 3 ohms.

The CT lead VA burden if a 5A CT is used would be 75VA, to which must be added the relay burden (up to of perhaps 10VA for an electromechanical relay, but less than 1VA for a numerical relay), making a total of 85VA. Such a burden would require the CT to be very large and expensive, particularly if a high accuracy limit factor were also applicable.

With a 1A CT secondary rating, the lead burden is reduced to 3VA, so that with the same relay burden the total becomes a maximum of 13VA. This can be provided by a CT of normal dimensions, resulting in a saving in size, weight and cost. Hence modern CTs tend to have secondary windings of 1A rating. However, where the primary rating is high, say above 2000A, a CT of higher secondary rating may be used, to limit the number of secondary turns. In such a situation secondary ratings of 2A, 5A or, in extreme cases, 20A, might be used.

6.4.9 Rated Short-Time Current

A current transformer is overloaded while system short-circuit currents are flowing and is short-time rated. Standard times for which the CT must be able to carry rated short-time current (STC) are 0.25, 0.5, 1.0, 2.0 or 3.0 seconds.

A CT with a particular short-time current/ time rating carries a lower current for a longer time in inverse proportion to the square of the ratio of current values. The converse, however, cannot be assumed, and larger current values than the STC rating are not permissible for any duration unless justified by a new rating test to prove the dynamic capability.

6.4.10 Transient Response of a Current Transformer

When accuracy of response during very short intervals is being studied, it is necessary to examine what happens when the primary current is suddenly changed. The effects are most important, and were first observed in connection with balanced forms of protection, which were liable to operate unnecessarily when short-circuit currents were suddenly established.

6.4.10.1 Primary Current Transient

The power system, neglecting load circuits, is mostly inductive, so that when a short circuit occurs, the fault current that flows is given by:

$$i_p = \frac{E_p}{\sqrt{R^2 + \omega^2 L^2}}\left[\sin(\omega t + \beta - \alpha) + \sin(\alpha - \beta)e^{-(R/L)t)}\right]$$

Equation 6.1

where:

E_p = peak system e.m.f.

R = system resistance

L = system inductance

β = initial phase angle governed by instant of fault occurrence

α = system power factor angle

= $\tan^{-1} \omega L/R$

The first term of *Equation 6.1* represents the steady state alternating current, while the second is a transient quantity responsible for displacing the waveform asymmetrically.

$\frac{E_p}{\sqrt{R^2 + \omega^2 L^2}}$ is the steady state peak current I_p

The maximum transient occurs when $\sin(\alpha - \beta) = 1$ and no other condition need be examined.

So:

$$i_p = I_p\left[\sin\left(\omega t - \frac{\pi}{2}\right) + e^{-(R/L)t}\right]$$

Equation 6.2

When the current is passed through the primary winding of a current transformer, the response can be examined by replacing the CT with an equivalent circuit as shown in Figure 6.10(b).

As the 'ideal' CT has no losses, it transfers the entire function, and all further analysis can be carried out in terms of equivalent secondary quantities (i_s and I_s). A simplified solution is obtainable by neglecting the exciting current of the CT.

The flux developed in an inductance is obtained by integrating the applied e.m.f. through a time interval:

$$\phi = K\int_{t_1}^{t_2} v\,dt$$

Equation 6.3

For the CT equivalent circuit, the voltage is the drop on the burden resistance R_b.

Integrating for each component in turn, the steady state peak flux is given by:

$$\phi_A = KR_b I_s \int_{\pi/\omega}^{3\pi/2\omega} \sin\left(\omega t - \frac{\pi}{2}\right)dt$$

$$= \frac{KR_b I_s}{\omega}$$

Equation 6.4

The transient flux is given by:

$$\phi_B = KR_b I_s \int_0^{\infty} e^{-(R/L)t}\,dt$$

$$= \frac{KR_b I_s L}{R}$$

Equation 6.5

Hence, the ratio of the transient flux to the steady state value is:

$$\frac{\phi_B}{\phi_A} = \frac{\omega L}{R} = \frac{X}{R}$$

where X and R are the primary system reactance and resistance values.

The CT core has to carry both fluxes, so that:

$$\phi_C = \phi_A + \phi_B = \phi_A\left(1+\frac{X}{R}\right)$$

Equation 6.6

The term $(1+X/R)$ has been called the 'transient factor' (TF), the core flux being increased by this factor during the transient asymmetric current period. From this it can be seen that the ratio of reactance to resistance of the power system is an important feature in the study of the behaviour of protection relays.

Alternatively, L/R is the primary system time constant T, so that the transient factor TF can be written:

$$TF = 1+\frac{\omega L}{R} = 1+\omega T$$

Again, fT is the time constant expressed in cycles of the a.c. quantity T' so that:

$$TF = 1+2\pi fT = 1+2\pi T'$$

This latter expression is particularly useful when assessing a recording of a fault current, because the time constant in cycles can be easily estimated and leads directly to the transient factor. For example, a system time constant of three cycles results in a transient factor of $(1+6\pi)$, or 19.85; that is, the CT would be required to handle almost twenty times the maximum flux produced under steady state conditions. The above theory is sufficient to give a general view of the problem. In this simplified treatment, no reverse voltage is applied to demagnetise the CT, so that the flux would build up as shown in Figure 6.14.

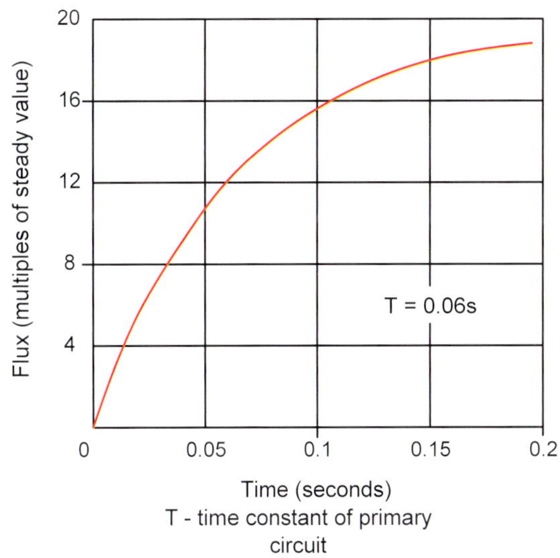

Figure 6.14: Response of a CT of infinite shunt impedance to transient asymmetric primary current

Since a CT requires a finite exciting current to maintain a flux, it does not remain magnetised (neglecting hysteresis), and for this reason a complete representation of the effects can only be obtained by including the finite inductance of the CT in the calculation. The response of a current transformer to a transient asymmetric current is shown in Figure 6.15.

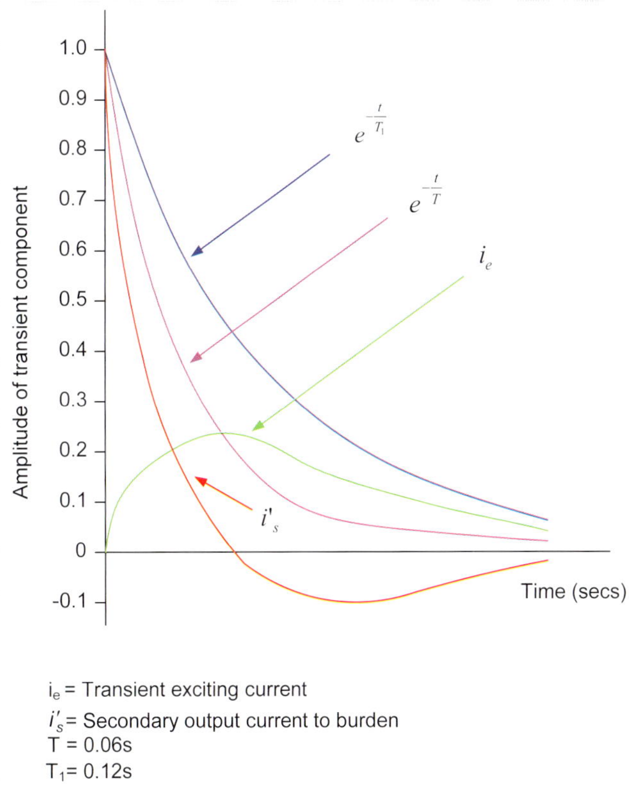

Figure 6.15: Response of a current transformer to a transient asymmetric current

Let

i_s = the nominal secondary current

i'_s = the actual secondary output current

i_e = the exciting current

then:

$$i_s = i_e + i'_s$$

Equation 6.7

also,

$$L_e \frac{di_e}{dt} = R_b i'_s$$

Equation 6.8

where:

$$\frac{di_e}{dt} = \frac{R_b i_e}{L_e} = \frac{R_b i_s}{L_e}$$

Equation 6.9

which gives for the transient term

$$i_e = I_1 \frac{T}{T_1 - T}\left(e^{-t/T_1} - e^{-t/T}\right)$$

where:
T = primary system time constant L/R

T_1 = CT secondary circuit time constant L_e/R_b

I_1 = prospective peak secondary current

6.4.10.2 Practical Conditions

Practical conditions differ from theory for the following reasons:

a. No account has been taken of secondary leakage or burden inductance. This is usually small compared with L_e so has little effect on the maximum transient flux.

b. Iron loss has not been considered. This has the effect of reducing the secondary time constant, but the value of the equivalent resistance is variable, depending upon both the sine and exponential terms. Consequently, it cannot be included in any linear theory and is too complicated for a satisfactory treatment to be evolved.

c. The theory is based upon a linear excitation characteristic. This is only approximately true up to the knee-point of the excitation curve. A precise solution allowing for non-linearity is not practicable. Solutions have been sought by replacing the excitation curve with several chords; a linear analysis can then be made for the extent of each chord.
The above theory is sufficient to give a good insight into the problem and to allow most practical issues to be decided.

d. The effect of hysteresis, apart from loss as discussed under (b) above, is not included. Hysteresis makes the inductance different for flux build up and decay, so that the secondary time constant is variable. Moreover, the ability of the core to retain a 'remanent' flux means that the value of ϕ_B developed in Equation 6.5 has to be regarded as an increment of flux from any possible remanent value positive or negative. The formula would then be reasonable provided the applied current transient did not produce saturation.

A precise calculation of the flux and excitation current is not feasible; the value of the study is to explain the observed phenomena. The asymmetric (or d.c.) component can be regarded as building up the mean flux over a period corresponding to several cycles of the sinusoidal component, during which period the latter component produces a flux swing about the varying 'mean level' established by the former. The asymmetric flux ceases to increase when the exciting current is equal to the total asymmetric input current, since beyond this point the output current, and hence the voltage drop across the burden resistance, is negative. Saturation makes the point of equality between the excitation current and the input occur at a flux level lower than would be expected from linear theory.

When the exponential component drives the CT into saturation, the magnetising inductance decreases, causing a large increase in the alternating component i_e.

The total exciting current during the transient period is of the form shown in Figure 6.16 and the corresponding resultant distortion in the secondary current output, due to saturation, is shown in Figure 6.17.

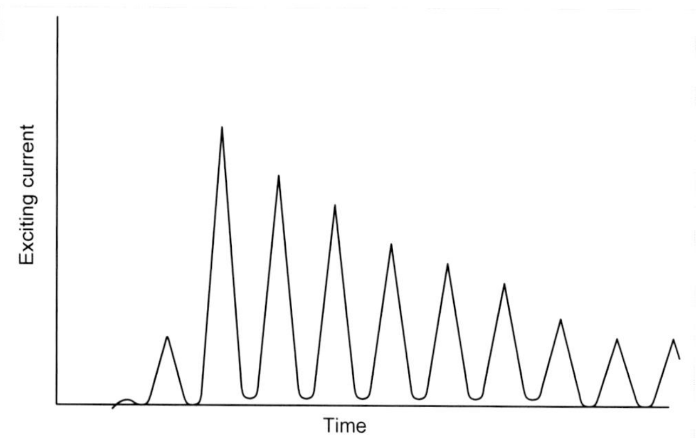

Figure 6.16: Typical exciting current of CT during transient asymmetric input current

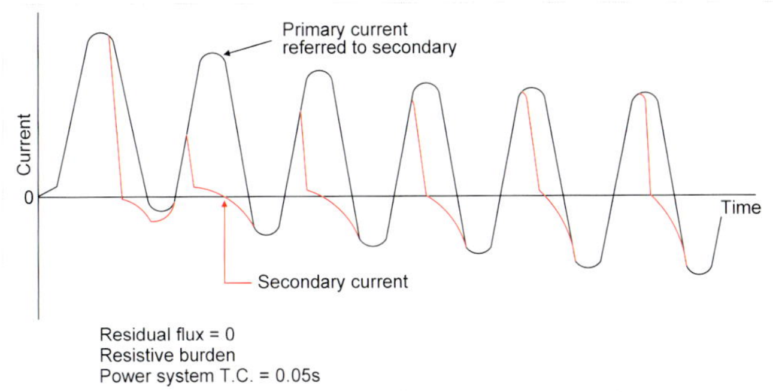

Figure 6.17: Distortion in secondary current due to saturation

The presence of residual flux varies the starting point of the transient flux excursion on the excitation characteristic. Remanence of like polarity to the transient reduces the value of symmetric current of given time constant which the CT can transform without severe saturation. Conversely, reverse remanence greatly increases the ability of a CT to transform transient current.

If the CT were the linear non-saturable device considered in the analysis, the sine current would be transformed without loss of accuracy. In practice the variation in excitation inductance caused by transferring the centre of the flux swing to other points on the excitation curve causes an error that may be very large. The effect on measurement is of little consequence, but for protection equipment that is required to function during fault conditions, the effect is more serious. The output current is reduced during transient saturation, which may prevent the relays from operating if the conditions are near to the relay setting. This must not be confused with the increased r.m.s. value of the primary current due to the asymmetric transient, a feature which sometimes offsets the increase ratio error. In the case of balanced protection, during through faults the errors of the several current transformers may differ and produce an out-of-balance quantity, causing unwanted operation.

6.4.11 Harmonics During the Transient Period

When a CT is required to develop a high secondary e.m.f. under steady state conditions, the non-linearity of the excitation impedance causes some distortion of the output waveform. In addition to the fundamental current, such a waveform contains odd harmonics only.

However, when the CT is saturated unidirectionally while being simultaneously subjected to a small a.c. quantity, as in the transient condition discussed above, the output contains both odd and even harmonics. Usually the lower numbered harmonics are of greatest amplitude and the second and third harmonic components may be of considerable value. This may affect relays that are sensitive to harmonics.

6.4.12 Test Windings

On-site conjunctive testing of current transformers and the apparatus that they energise is often required. It may be difficult, however, to pass a suitable value of current through the primary windings, because of the scale of such current and in many cases because access to the primary conductors is difficult. Additional windings can be provided to make such tests easier and these windings are usually rated at 10A. The test winding inevitably occupies appreciable space and the CT costs more. This should be weighed against the convenience achieved and often the tests can be replaced by alternative procedures.

6.4.13 Use of IEEE Standard Current Transformers

Most of this chapter has been based around IEC standards for current transformers. Parts of the world preferring IEEE specifications for CTs may require an easy method of converting requirements between the two. In reality, the fundamental technology and construction of the CTs remains the same, however the knee-point voltage is specified in a different way. IEC CT excitation curves are typically drawn on linear scales, whereas IEEE CT standards prefer to use log-log scales, defining the knee point as the excitation voltage at which the gradient of the curve is 45°. The voltage found by this definition is typically 5 to 10% different to the point on the excitation curve found by the IEC definition, as in Figure 6.12.

An additional complication is that the IEC voltage is an e.m.f., which means that it is not an actual measurable voltage at the CT terminals. The e.m.f. is the internal voltage, compounded by any voltage drop across the CT winding resistance. The IEEE specifications relate to a terminal voltage, which is often referred to as a "C class" voltage rating. This means that a C200 rating CT has a knee voltage of 200V according to the IEEE definition of the knee point.

Assume that the IEC knee point voltage required in a protection application is Vk_{IEC}. The IEEE C class standard voltage rating required is lower and the method of conversion is as follows:

$$Vc = \frac{Vk_{IEC} - (In \times R_{CT} \times ALF)}{1.05}$$

where:

Vc = IEEE C Class standard voltage rating

Vk_{IEC} = IEC Knee point voltage

In = CT rated current, usually always 5A for IEEE

R_{CT} = CT secondary winding resistance

ALF = CT accuracy limit factor, always 20 for an IEEE CT.

The factor of 1.05 accounts for the differing points on the excitation curve at which the two philosophy standards are defined.

6.5 NON-CONVENTIONAL INSTRUMENT TRANSFORMERS

The preceding types of instrument transformers have all been based on electromagnetic principles using a magnetic core. There are now available several new methods of transforming the measured quantity using optical and mass state methods.

6.5.1 Optical Instrument Transducers

Figure 6.18 shows the key features of a freestanding optical instrument transducer.

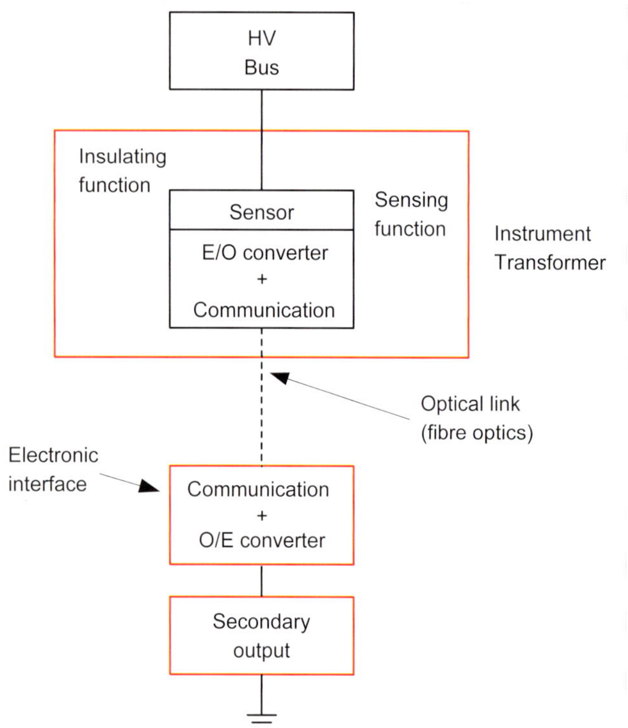

Figure 6.18: Typical architecture using optical communication between sensing unit and electronic interface

Non-conventional optical transducers lend themselves to smaller, lighter devices where the overall size and power rating of the unit does not have any significant bearing on the size and the complexity of the sensor. Small, lightweight insulator structures may be tailor-made to fit optical sensing devices as an integral part of the insulator. Additionally, the non-linear effects and electromagnetic interference problems in the secondary wiring of conventional VTs and CTs are minimised.

Optical transducers can be separated in two families: firstly the *hybrid* transducers, making use of conventional electrical circuit techniques to which are coupled various optical converter systems, and secondly the *'all-optical'* transducers that are based on fundamental, optical sensing principles.

6.5.1.1 Optical Sensor Concepts

Certain optical sensing media (glass, crystals, plastics) show a sensitivity to electric and magnetic fields and that some properties of a probing light beam can be altered when passing through them. A simple optical transducer description is shown in Figure 6.19.

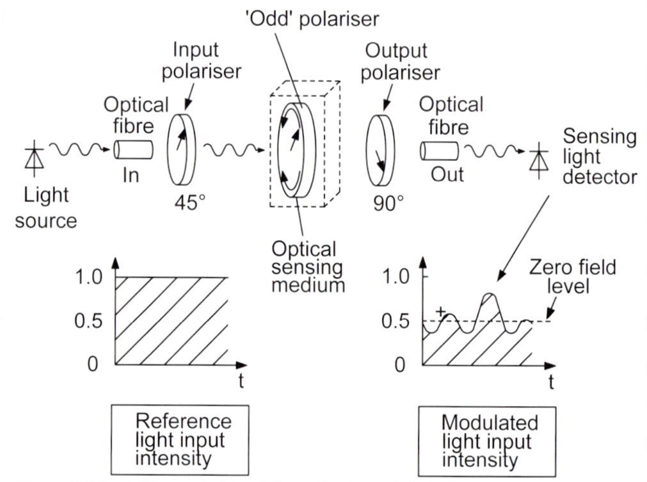

Figure 6.19: Schematic representation of the concepts behind the optical sensing of varying electric and magnetic fields

If a beam of light passes through a pair of polarising filters, and if the input and output polarising filters have their axes rotated 45° from each other, only half the light comes through. The reference light input intensity is maintained constant over time. If these two polarising filters remain fixed and a third polarising filter is placed in between them, a random rotation of this middle polariser either clockwise or anticlockwise is monitored as a varying or modulated light output intensity at the light detector.

When a block of optical sensing material (glass or crystal) is immersed in a varying magnetic or electric field, it plays the role of the 'odd' polariser. Changes in the magnetic or electric field in which the optical sensor is immersed are monitored as a varying intensity of the probing light beam at the light detector. The light output intensity fluctuates around the zero-field level equal to 50% of the reference light input. This modulation of the light intensity due to the presence of varying fields is converted back to time-varying currents or voltages.

A transducer uses a *magneto-optic effect sensor* for optical *current* measuring applications. This reflects the fact that the sensor is not basically sensitive to a current but to the magnetic field generated by this current. Solutions exist using both wrapped fibre optics and bulk glass sensors as the optical sensing medium. However, most optical voltage transducers rely on an electro-optic effect sensor. This reflects the fact that the sensor used is sensitive to the imposed electric field.

6.5.1.2 Hybrid Transducers

The hybrid family of non-conventional instrument transducers can be divided in two types: those with active sensors and those with passive sensors. The idea behind a transducer with an active sensor is to change the existing output of the conventional instrument transformer into an optically isolated output by adding an optical conversion system (Figure 6.19). This conversion system may require a power supply of its own: this is the active sensor type. The use of an optical isolating system serves to de-couple the instrument transformer output secondary voltages and currents from earthed or galvanic links. Therefore the only link that remains between the control-room and the switchyard is a fibre optic cable.

6.5.1.3 'All-optical' Transducers

These instrument transformers are based entirely on optical materials and are fully passive. The sensing function is achieved directly by the sensing material and a simple fibre optic cable running between the base of the unit and the sensor location provides the communication link.

The sensing element consists of an optical material that is positioned in the electric or magnetic field to be sensed. The sensitive element of a current measuring device is either located freely in the magnetic field (Figure 6.20(a)) or it can be immersed in a field-shaping magnetic 'gap' (Figure 6.20(b)). In the case of a voltage-sensing device (Figure 6.21) the same alternatives exist, this time for elements that are sensitive to electric fields. Both sensors can be combined in a single compact housing, providing both a CT and VT to save space in a substation.

In all cases there is an optical fibre that channels the probing reference light from a source into the medium and another fibre that channels the light back to the analysing circuitry. In sharp contrast with a conventional free-standing instrument transformer, the optical instrument transformer needs an electronic interface module to function. Therefore its sensing principle (the optical material) is passive but its operational integrity relies on a powered interface.

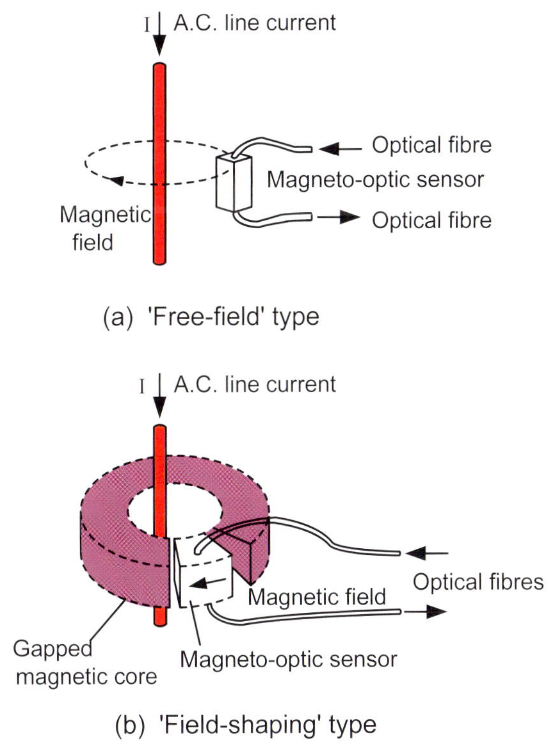

Figure 6.20: Optical current sensor based on the magnetic properties of optical materials

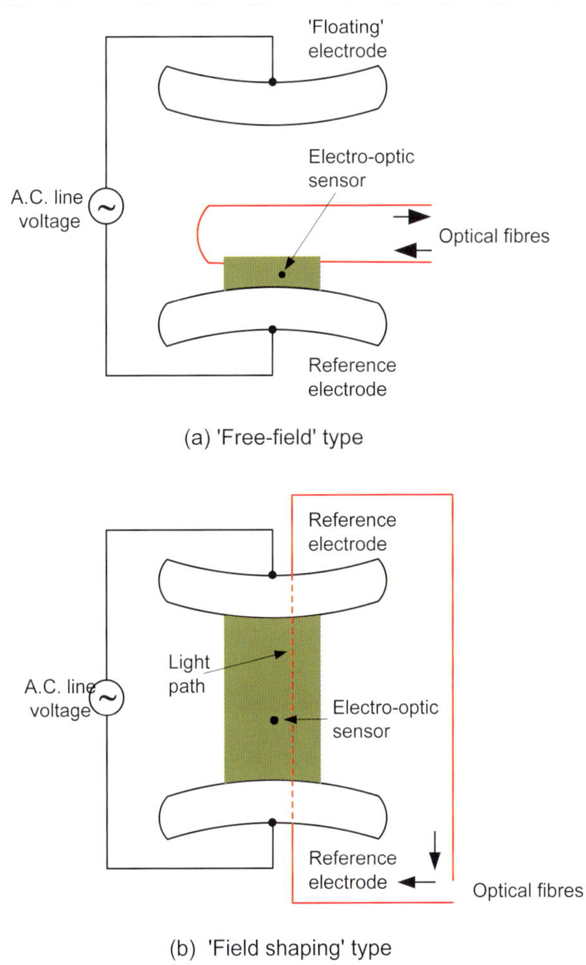

Figure 6.21: Optical voltage sensor based on the electrical properties of optical materials

Typically, current transducers take the shape of a closed loop of light-transparent material, fitted around a straight conductor carrying the line current (Figure 6.22). In this case a bulk-glass sensor unit is depicted (Figure 6.22(a)), along with a wrapped fibre sensor example, as shown in Figure 6.22(b) and Figure 6.23. Light detectors are very sensitive devices and the sensing material can be selected to scale-up readily for larger currents. However, 'all-optical' voltage transducers are not ideally suited to extremely high line voltages. Two concepts using a 'full voltage' sensor are shown in Figure 6.24.

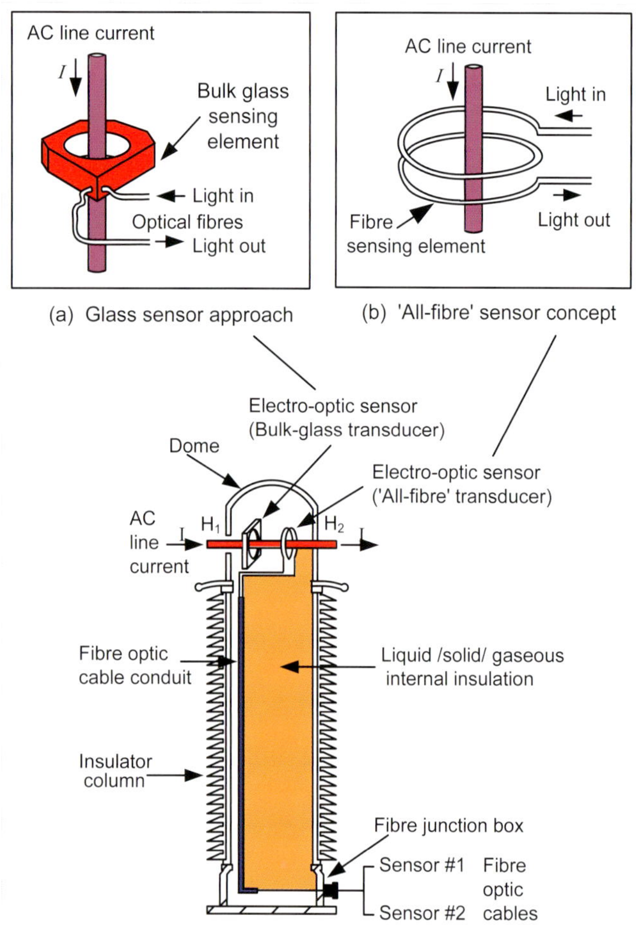

Figure 6.22: Conceptual design of a double-sensor optical CT

Figure 6.23: Alstom COSI-NXCT F3 flexible optical current transformer in a portable substation application

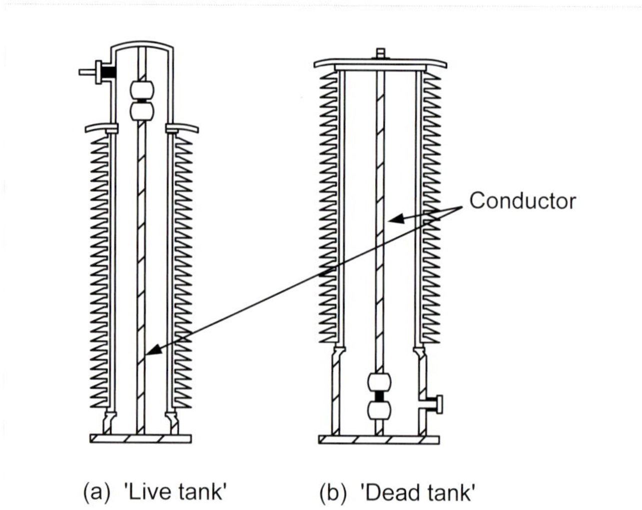

Figure 6.24: Optical voltage transducer concepts, using a 'full-voltage' sensor

Figure 6.25: Installation of a CT with an optical sensor

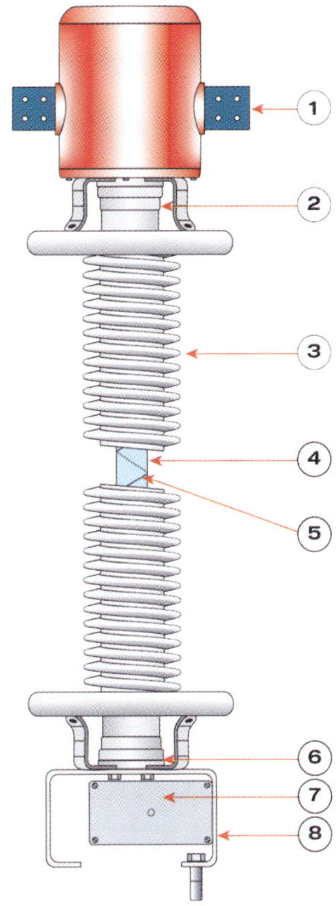

1 High voltage terminal
2 Aluminum top fitting
3 Silicone sheds
4 Fiberglass epoxy tube
5 Optical fibres
6 Aluminum bottom fitting
7 Junction box
8 Base

Figure 6.26 Cross section of an Alstom CTO 72.5kV to 765kV current transformer with an optical sensor

6.5.2 Other Sensing Systems

There are several other sensing systems that can be used, as described in the following sections.

6.5.2.1 Zero-flux (Hall Effect) Current Transformer

In this case the sensing element is a semi-conducting wafer that is placed in the gap of a magnetic concentrating ring. This type of transformer is also sensitive to d.c. currents. The transformer requires a power supply that is fed from the line or from a separate power supply. The sensing current is typically 0.1% of the current to be measured. In its simplest shape, the Hall effect voltage is directly proportional to the magnetising current to be measured. For more accurate and more sensitive applications, the sensing current is fed through a secondary, multiple-turn winding, placed around the magnetic ring to balance out the gap magnetic field. This zero-flux or null-flux version allows very accurate current measurements in both d.c. and high-frequency applications. A schematic representation of the sensing part is shown in Figure 6.27.

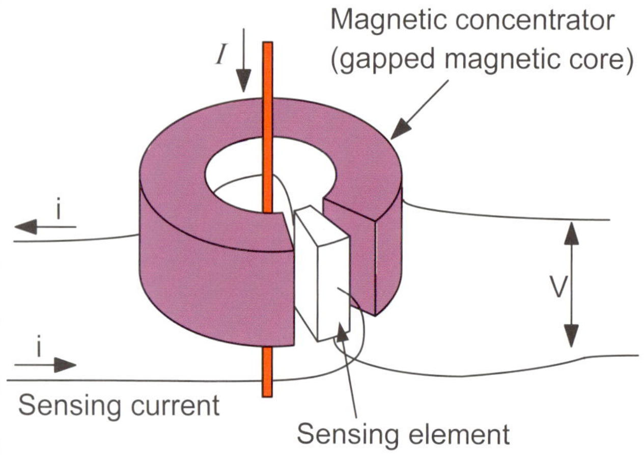

Figure 6.27: Conceptual design of a Hall-effect current sensing element fitted in a field-shaping gap

6.5.2.2 Hybrid Magnetic-Optical Sensor

This type of transformer is mostly used in applications such as series capacitive compensation of long transmission lines, where a non-grounded measurement of current is required. In this case, several current sensors are required on each phase to achieve capacitor surge protection and balance. The preferred solution is to use small toroidally wound magnetic core transformers connected to fibre optic isolating systems. These sensors are usually active sensors because the isolated systems require a power supply. This is shown in Figure 6.28.

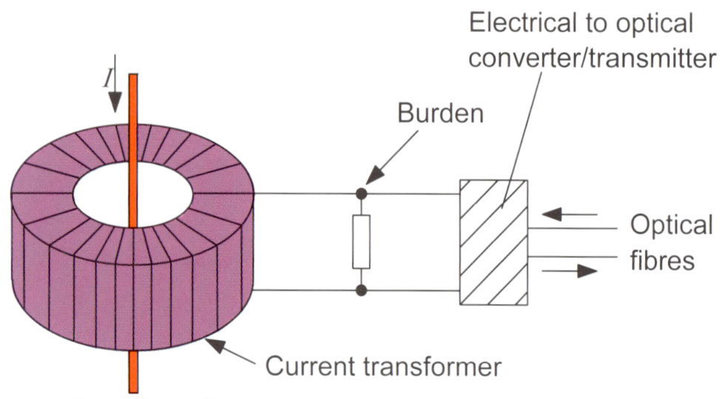

Figure 6.28: Design principle of a hybrid magnetic current transformer fitted with an optical transmitter

6.5.2.3 Rogowski Coils

The Rogowski coil is based on the principle of an air-cored current transformer with a very high load impedance. The secondary winding is wound on a toroid of insulation material. In most cases the Rogowski coil is connected to an amplifier, to deliver sufficient power to the connected measuring or protection equipment and to match the input impedance of this equipment. The Rogowski coil requires integration of the magnetic field and therefore has a time and phase delay while the integration is completed. This can be corrected for in a digital protection relay. The schematic representation of the Rogowski coil sensor is shown in Figure 6.29.

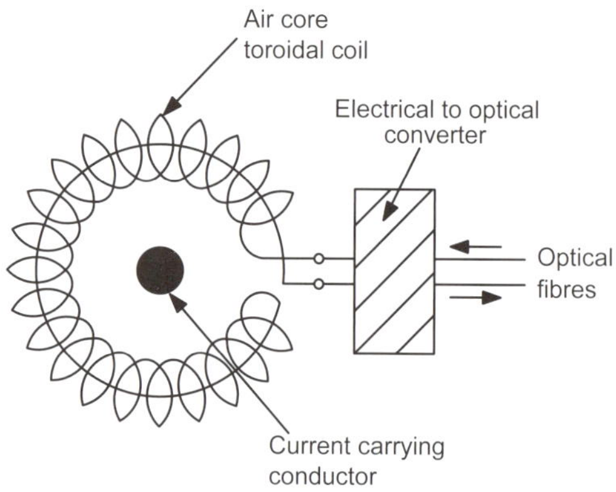

Figure 6.29: Schematic representation of a Rogowski coil, used for current sensing

Chapter 7
Relay Technology

7.1 INTRODUCTION

This chapter describes how relay technology has changed. The electromechanical relay in all of its different forms has been replaced successively by static, digital and numerical relays, each change bringing with it reductions in size and improvements in functionality. Reliability levels have also been maintained or even improved and availability significantly increased due to techniques not available with older relay types. This represents a tremendous achievement for all those involved in relay design and manufacture.

This book concentrates on modern digital and numerical protection relay technology, although the vast number of electromechanical and static relays are still giving dependable service.

7.2 ELECTROMECHANICAL RELAYS

These relays were the earliest forms of relay used for the protection of power systems, and they date back around 100 years. They work on the principle of a mechanical force operating a relay contact in response to a stimulus. The mechanical force is generated through current flow in one or more windings on a magnetic core or cores, hence the term electromechanical relay. The main advantage of such relays is that they provide galvanic isolation between the inputs and outputs in a simple, cheap and reliable form. Therefore these relays are still used for simple on/off switching functions where the output contacts carry substantial currents.

Electromechanical relays can be classified into several different types as follows:

- attracted armature
- moving coil
- induction
- thermal
- motor operated
- mechanical

However, only attracted armature types presently have significant applications while all other types have been superseded by more modern equivalents.

7.2.1 Attracted Armature Relays

These generally consist of an iron-cored electromagnet that attracts a hinged armature when energised. A restoring force is provided by a spring or gravity so that the armature returns to its original position when the electromagnet is de-energised. Typical forms of an attracted armature relay are shown in Figure 7.1. Movement of the armature opens or closes a contact. The armature either carries a moving contact that engages with a fixed one or causes a rod to move that brings two contacts together. It is easy to mount multiple contacts in rows or stacks, causing a single input to actuate several outputs. The contacts can be robust and therefore able to make, carry and break large currents under difficult conditions such as highly inductive circuits. This is still a significant advantage of this type of relay that ensures its continued use.

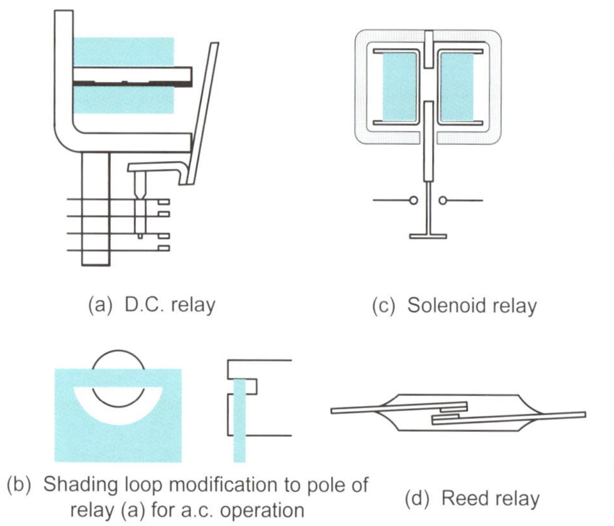

Figure 7.1: Typical attracted armature relays

The energising quantity can be either an a.c. or a d.c. current. If an a.c. current is used, there is chatter due to the flux passing through zero every half cycle. A common solution is to split the magnetic pole and provide a copper loop around one half. The flux is then phase-shifted in that pole so the total flux is never equal to zero. Conversely, for relays energised using a d.c. current, remanent flux may prevent the relay from releasing when the actuating current is removed. This can be avoided by preventing the armature from contacting the electromagnet by a non-magnetic stop, or constructing the electromagnet using a material with very low remanent flux properties.

Operating speed, power consumption and the number and type of contacts required are a function of the design. The typical attracted armature relay has an operating speed of between 100ms and 400ms, but reed relays (whose use spanned a relatively short period in the history of protection relays) with light current contacts can be designed to have an operating time of as little as 1ms. Operating power is typically 0.05-0.2 watts, but could be as large as 80 watts for a relay with several heavy-duty contacts and a high degree of resistance to mechanical shock.

Some applications need a polarised relay. This is a permanent magnet added to the basic electromagnet. Both self-reset and bistable forms can be made using the basic construction shown in Figure 7.2. An example of its use is to provide very fast operating times for a single contact, with speeds of less than 1msec. Figure 7.3 shows a typical attracted armature relay.

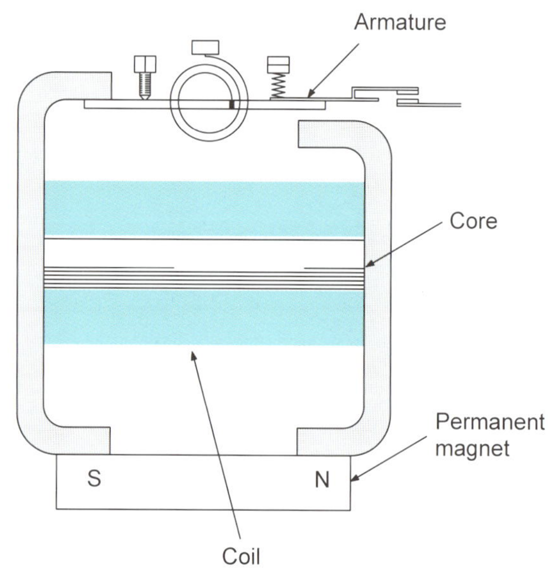

Figure 7.2: Typical polarised relay

Figure 7.3: Attracted armature relay in manufacture

7.3 STATIC RELAYS

The term 'static' implies that the relay has no moving parts. This is not strictly the case for a static relay, as the output contacts are still generally attracted armature relays. In a protection relay, the term 'static' refers to the absence of moving parts to create the relay characteristic.

Introduction of static relays began in the early 1960s. Their design is based on the use of analogue electronic devices instead of coils and magnets to create the relay characteristic. Early versions used discrete devices such as transistors and diodes with resistors, capacitors and inductors. However, advances in electronics enabled the use of linear and digital integrated circuits in later versions for signal processing and implementation of logic functions. Although basic circuits were common to several relays, each protection function had its own case, so complex functions required several cases of interconnected hardware. User programming was restricted to the basic functions of adjustment of relay characteristic curves. Therefore they can be considered as an analogue electronic replacement for electromechanical relays, with some additional flexibility in settings and some saving in space requirements. In some cases, relay burden is reduced, reducing CT/VT output requirements.

Several design problems had to be solved with static relays, such as a reliable d.c. power source and measures to prevent damage to vulnerable electronic circuits. Substation environments are particularly hostile to electronic circuits due to electrical interference of various forms that are commonly found, such as switching operations and the effect of faults. Although the d.c. supply can be generated from the measured quantities of the relay, this has the disadvantage of increasing the burden on the CTs or VTs, and there is a minimum primary current or voltage below which the relay will not operate. This directly affects the possible sensitivity of the relay. So provision of an independent, highly reliable and secure source of relay power supply was an important consideration.

To prevent maloperation or destruction of electronic devices during faults or switching operations, sensitive circuitry is housed in a shielded case to exclude common mode and radiated interference. The devices are also sensitive to electrostatic discharge (ESD), requiring special precautions during handling. ESD damage may not be immediately apparent but may cause premature failure of the relay. Therefore, radically different relay manufacturing facilities are required compared to electromechanical relays. Calibration and repair is no longer a task performed in the field without specialised equipment. Figure 7.4 shows a typical static relay.

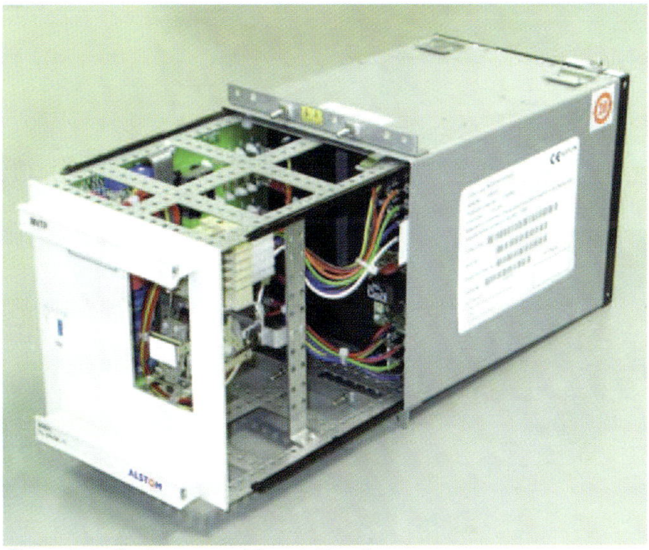

Figure 7.4: Busbar supervision static relay

7.4 DIGITAL RELAYS

Digital protection relays introduced a step change in technology. Microprocessors and microcontrollers replaced analogue circuits used in static relays to implement relay functions. Early examples were introduced around 1980 and with improvements in processing capacity are still current technology for many relay applications. However, such technology could be completely superseded by numerical relays.

Compared to static relays, digital relays use analogue to digital conversion of all measured quantities and use a microprocessor to implement the protection algorithm. The microprocessor may use a counting technique or use Discrete Fourier Transforms (DFT) to implement the algorithm. However, these microprocessors have limited processing capacity and associated memory compared to numerical relays. Therefore the functionality is limited mainly to the protection function itself. Compared to an electromechanical or static relay, digital relays have a wider range of settings, greater accuracy and a communications link to a remote computer. Figure 7.5 shows a typical digital relay.

Digital relays typically use 8 or 16-bit microprocessors that were later used in modems, hard disk controllers or early car engine management systems. The limited power of the microprocessors used in digital relays restricts the number of samples of the waveform that can be measured per cycle. This limits the speed of operation of the relay in certain applications. Therefore a digital relay for a particular protection function may have a longer operation time than the static relay equivalent. However, the extra time is insignificant compared to overall tripping time and possible effects on power system stability.

Figure 7.5: Second generation distribution digital relay (1982)

7.5 NUMERICAL RELAYS

The distinction between digital and numerical relays is particular to Protection. Numerical relays are natural developments of digital relays due to advances in technology. They use one or more digital signal processors (DSP) optimised for real time signal processing, running the mathematical algorithms for the protection functions.

Figure 7.6: First generation transmission numerical relay (1986)

Figure 7.7: First generation distribution numerical relay

The continuing reduction in the cost and size of microprocessors, memory and I/O circuitry leads to a single item of hardware for a range of functions. For faster real time processing and more detailed analysis of waveforms, several DSPs can be run in parallel. Many functions previously implemented in separate items of hardware can then be included in a single item. Table 7.1 provides a list of typical functions available, while Table 7.2 summarises the advantages of a modern numerical relay over static relay equivalents.

Figure 7.6, Figure 7.7 and Figure 7.8 show typical numerical relays, while Figure 7.9 and Figure 7.10 show typical numerical relay boards.

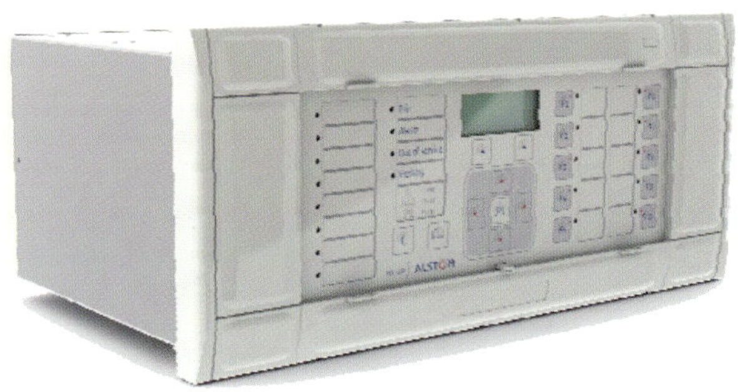

Figure 7.8: Typical modern numerical relay

Figure 7.9: Numerical relay processor board

Figure 7.10: Numerical relay redundant Ethernet board

A numerical relay has the functionality that previously required several discrete relays, therefore the relay functions such as overcurrent or earth fault are referred to as 'relay elements'. Each relay element is in software so with modular hardware the main signal processor can run a vast variety of relay elements.

The argument against putting many features into one piece of hardware centres on the issues of reliability and availability. A failure of a numerical relay may cause many more functions to be lost, compared to applications where different functions are implemented by separate hardware items. Comparison of reliability and availability between the two methods is complex as inter-dependency of elements of an application provided by separate relay elements needs to be taken into account.

With the experience gained with static and digital relays, most hardware failure mechanisms are now well understood and suitable precautions taken at the design stage. Software problems are minimised by rigorous use of software design techniques, extensive prototype testing (see Chapter 21) and the ability to download updated software. Practical experience indicates that numerical relays are as reliable as relays of earlier technologies. Modern numerical relays will have comprehensive self monitoring to alert the user to any problems.

Distance Protection- several schemes including user definable
Overcurrent Protection (directional/non-directional)
Several Setting Groups for protection values
Switch-on-to-Fault Protection
Power Swing Blocking
Voltage Transformer Supervision
Negative Sequence Current Protection
Undervoltage Protection
Overvoltage Protection
CB Fail Protection
Fault Location
CT Supervision
VT Supervision
Check Synchronisation
Autoreclose
CB Condition Monitoring
CB State Monitoring
User-Definable Logic
Broken Conductor Detection
Measurement of Power System Quantities (Current, Voltage, etc.)
Fault/Event/Disturbance recorder

Table 7.1: Numerical distance relay features

Several setting groups
Wider range of parameter adjustment
Communications built in (serial, Ethernet, teleprotection, etc.)
Internal Fault diagnosis
Power system measurements available
Distance to fault locator
Disturbance recorder
Auxiliary protection functions (broken conductor, negative sequence, etc.)
CB monitoring (state, condition)
User-definable logic
Backup protection functions in-built
Consistency of operation times - reduced grading margin

Table 7.2: Advantages of numerical relays over static relays

7.5.1 Hardware Architecture

The typical architecture of a numerical relay is shown in Figure 7.11. It consists of one or more DSPs, some memory, digital and analogue input/output (I/O), and a power supply. Where multiple processors are used, one of them is a general controller of the I/O, Human Machine Interface (HMI) and any associated logic while the others are dedicated to the protection relay algorithms. By organising the I/O on a set of plug-in printed circuit boards (PCBs), additional I/O up to the limits of the hardware/software can be easily added. The internal communications bus links the hardware and therefore is a critical component in the design. It must work at high speed, use low voltages, yet be immune to conducted and radiated interference from the electrically noisy substation environment. Excellent shielding of the relevant areas is therefore required. Digital inputs are optically isolated to prevent transients being transmitted to the internal circuitry. Analogue inputs are isolated using precision transformers to maintain measurement accuracy while removing harmful transients. Additionally, the input signals must be amplitude limited to avoid them exceeding the measurement range, otherwise the waveform is clipped, introducing harmonics. See Figure 7.12.

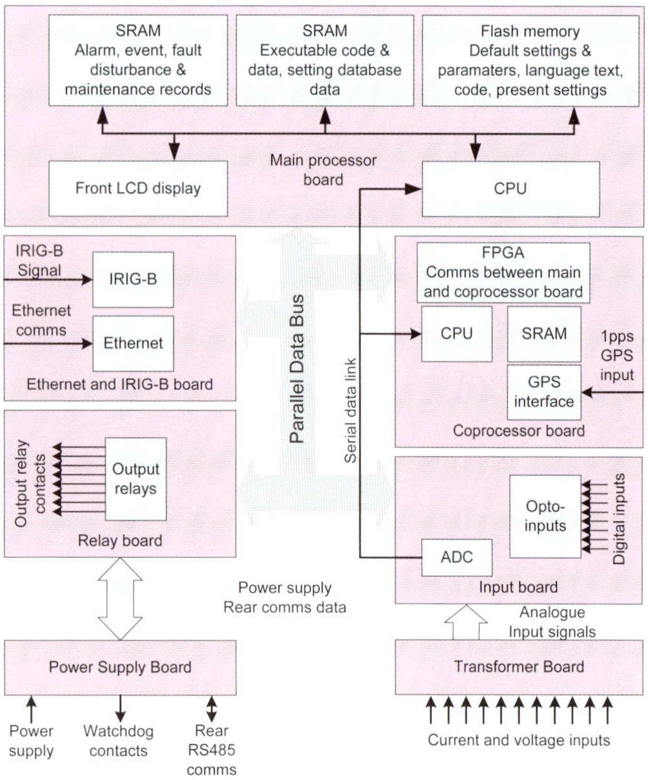

Figure 7.11: Typical numerical relay hardware architecture

Analogue signals are converted to digital form using an A/D converter. The cheapest method is to use a single A/D converter, preceded by a multiplexer to connect each of the input signals in turn to the converter. The signals may be initially input to several simultaneous sample-and-hold circuits before multiplexing, or the time relationship between successive samples must be known if the phase relationship between signals is important. The alternative is to provide each input with a dedicated A/D converter and logic to ensure that all converters perform the measurement simultaneously.

The frequency of sampling must be carefully considered, as the Nyquist criterion applies:

$$f_s \geq 2 f_h$$

where:

f_s = sampling rate

f_h = highest frequency of interest

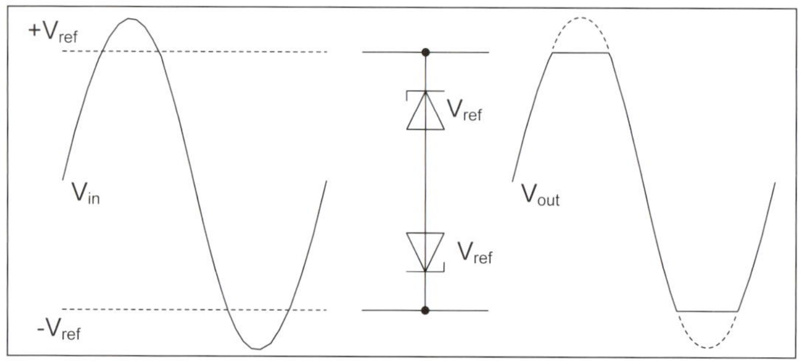

Figure 7.12: Clipping due to excessive amplitude

If the sampling frequency is too low, aliasing of the input signal can occur (see Figure 7.13) so that high frequencies can appear as part of the signal in the frequency range of interest. Incorrect results are then obtained. The solution is to use an anti-aliasing filter and the correct sampling frequency on the analogue signal, filtering out the frequency components that could cause aliasing. Digital sine and cosine filters (Figure 7.12) extract the real and imaginary components of the signal; the frequency response of the filters is shown in Figure 7.13. Frequency tracking of the input signals is applied to adjust the sampling frequency so that the desired number of samples/cycle is always obtained. A modern numerical relay can sample each analogue input quantity at typically between 24 and 80 samples per cycle.

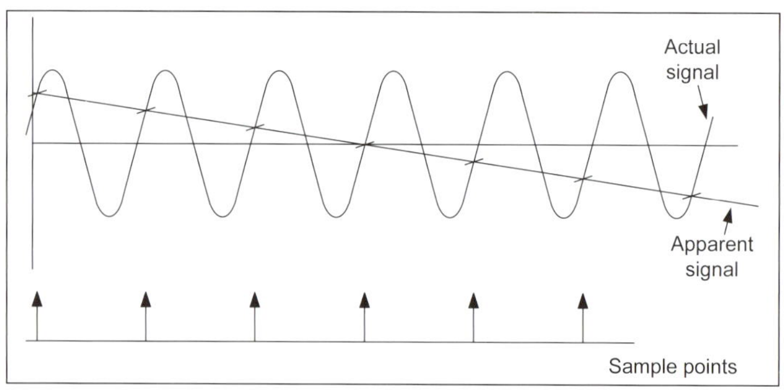

Figure 7.13: Signal aliasing problem

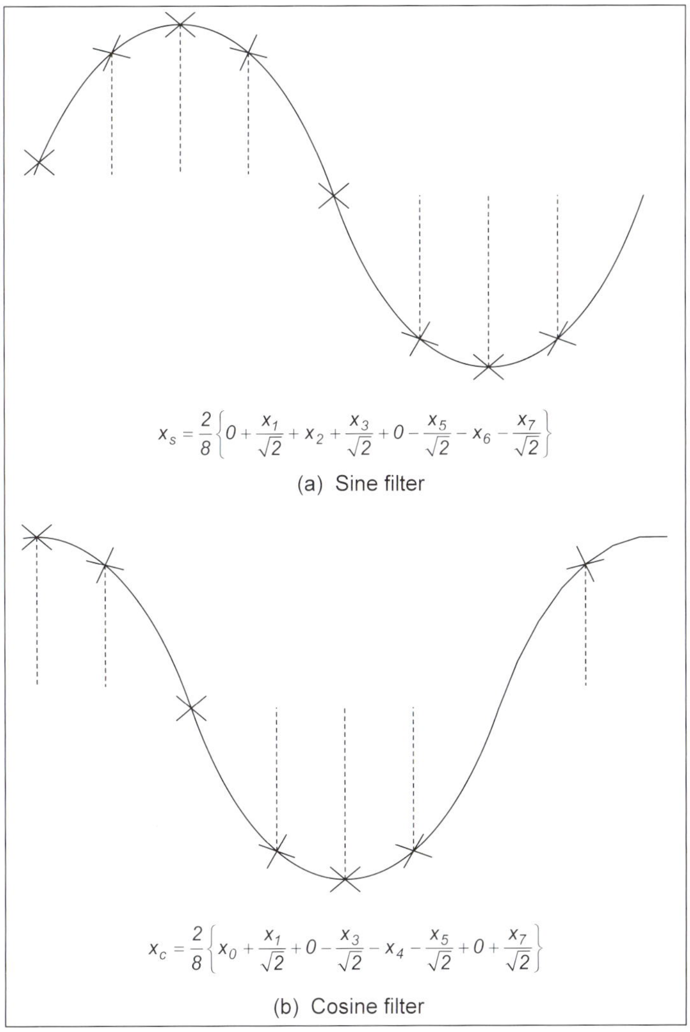

Figure 7.14: Sine and cosine filters (simple 8 sample per cycle example)

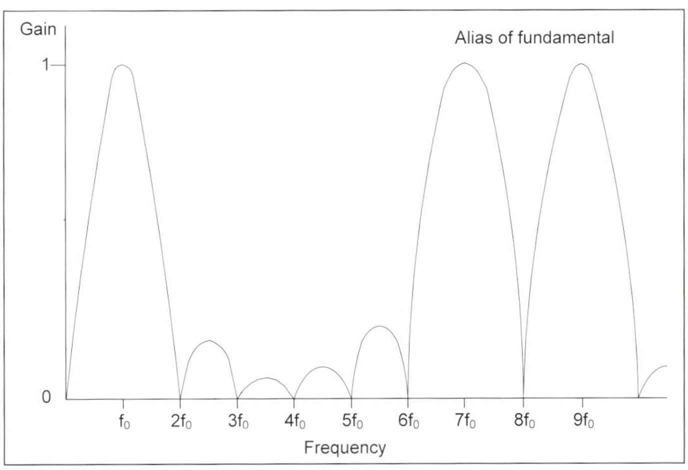

Figure 7.15: Filter frequency response

All subsequent signal processing is carried out in software. The final digital outputs use relays to provide isolation or are sent through an external communications bus to other devices.

7.5.2 Relay Operating System Software

The software provided is commonly organised into a series of tasks operating in real time. An essential component is the Real Time Operating System (RTOS) which ensures that the other tasks are executed when required, in the correct priority.

Other software depends on the function of the relay, but can be generalised as follows:

- system services software – this is comparable with the BIOS of an ordinary PC and controls the low-level I/O for the relay such as drivers for the relay hardware and boot-up sequence.
- HMI interface software – this is the high level software for communicating with a user on the front panel controls or through a data link to another computer to store data such as settings or event records.
- application software – this is the software that defines the protection function of the relay
- auxiliary functions – software to implement other features in the relay, often structured as a series of modules to reflect the options offered by the manufacturer.

7.5.3 Relay Application Software

The relevant software algorithm is then applied. Firstly the quantities of interest are determined from the information in the data samples. This is often done using a Discrete Fourier Transform (DFT) and the result is magnitude and phase information for the selected quantity. This calculation is repeated for all of the quantities of interest. The quantities can then be compared with the relay characteristic, and a decision made in terms of the following:

- value above setting – start timers, etc.
- timer expired – action alarm/trip
- value returned below setting – reset timers, etc.
- value below setting – do nothing
- value still above setting – increment timer, etc.

Since the overall cycle time for the software is known, timers are generally implemented as counters.

7.6 ADDITIONAL FEATURES OF NUMERICAL RELAYS

The DSP in a numerical relay normally can handle both the relay protection function calculations and general management of the relay such as HMI and I/O. However, if the DSP is overloaded it cannot complete the protection algorithm calculations in the required time and the protection function is slowed.

Typical functions that may be found in a numerical relay other than protection functions are described in this section. Note that not all functions are found in a particular relay. As with earlier generations of relays and according to market segmentation, manufacturers offer different versions, each with a different set of functions. Function parameters are usually displayed on the front panel of the relay and through an external communications port.

7.6.1 Measured Values Display

This is perhaps the most obvious and simple function to implement, as it involves the least additional processor time. The values that the relay must measure to perform its protection function have already been acquired and processed. It is therefore a simple task to display them on the front panel, or transmit them to a remote computer or HMI station. Several extra quantities may be derived from the measured quantities, depending on the input signals available. These might include:

- sequence quantities (positive, negative, zero)
- power, reactive power and power factor
- energy (kWh, kVArh)
- max. demand in a period (kW, kVAr; average and peak values)
- harmonic quantities
- frequency
- temperatures/RTD status
- motor start information (start time, total no. of starts/reaccelerations, total running time
- distance to fault

The accuracy of the measured values can only be as good as the accuracy of the transducers used such as VTs CTs, and the A/D converter. As CTs and VTs for protection functions may have a different accuracy specification to those for metering functions, such data may not be sufficiently accurate for tariff purposes. However, it is sufficiently accurate for an operator to assess system conditions and make appropriate decisions.

7.6.2 VT/CT Supervision

If suitable VTs are used, supervision of the VT/CT supplies can be made available. VT supervision is made more complicated by the different conditions under which there may be no VT signal, some of which indicate VT failure and some occur because a power system fault has occurred. If only one or two phases are lost, the VT failure algorithm can be accomplished by detecting residual voltage without the presence of zero or negative phase sequence current. For loss of all three phases, a fault in the system would be accompanied by a change in the phase currents, so absence of such a change can be taken as loss of the VT signal. However, this technique fails when closing onto a dead but healthy line, so a level detector is also required to distinguish between current inrush due to line charging and that due to a fault.

CT supervision is carried out more easily. The general principle is the calculation of a level of negative sequence current that is inconsistent with the calculated value of negative sequence voltage.

7.6.3 CB Control/State Indication /Condition Monitoring

System operators normally require knowledge of the state of all circuit breakers under their control. The CB position-switch outputs can be connected to the relay digital inputs and therefore provide the indication of state through the communications bus to a remote control centre.

Circuit breakers also require periodic maintenance of their operating mechanisms and contacts to ensure they operate when required and that the fault capacity is not affected adversely. The requirement for maintenance is a function of the number of trip operations, the cumulative current broken and the type of breaker. A numerical relay can record all of these parameters and hence be configured to send an alarm when maintenance is due. If maintenance is not carried out within a predefined time or number of trips after maintenance is required, the CB can be arranged to trip and lockout or inhibit certain functions such as auto-reclose.

Finally, as well as tripping the CB as required under fault conditions, it can also be arranged for a digital output to be used for CB closure, so that separate CB close control circuits can be eliminated.

7.6.4 Disturbance Recorder (Oscillograph)

The relay memory requires a certain minimum number of cycles of measured data to be stored for correct signal processing and detection of events. The memory can easily be expanded to allow storage of a greater time period of input data, both analogue and digital, plus the state of the relay outputs. It then has the capability to act as a disturbance recorder for the circuit being monitored, so that by freezing the memory at the instant of fault detection or trip, a record of the disturbance is available for later download and analysis. It may be inconvenient to download the record immediately, so facilities may be provided to capture and store a number of disturbances. In industrial and small distribution networks, this may be all that is required. In transmission networks, it may be necessary to provide a single recorder to monitor several circuits simultaneously, and in this case, a separate disturbance recorder is still required. For more information on the different types of disturbance recording, see Chapter 22.

7.6.5 Time Synchronisation

Disturbance records and data relating to energy consumption requires time tagging to serve any useful purpose. Although there is an internal clock, this is of limited accuracy and use of this clock to provide time information may cause problems if the disturbance record has to be correlated with similar records from other sources to obtain a complete picture of an event. Many numerical relays have the facility for time synchronisation from an external clock. The standard normally used is an IRIG-B or IEEE 1588 signal, which may be derived from several sources including a GPS satellite receiver.

7.6.6 Programmable Logic

Logic functions are well suited to implementation using microprocessors. The implementation of logic in a relay is not new, as functions such as intertripping and auto-reclose require a certain amount of logic. However, by providing a substantial number of digital I/O and making the logic capable of being programmed using suitable off-line software, the functionality of such schemes can be enhanced or additional features provided. For instance, an overcurrent relay at the receiving end of a transformer feeder could use the temperature inputs provided to monitor transformer winding temperature and provide alarm or trip facilities to the operator or upstream relay, eliminating the need for a separate winding temperature relay. There may be other advantages such as different logic schemes required by different utilities that no longer need separate relay versions, or some hard-wired logic to implement, and all of these reduce the cost of manufacture. It is also easier to customise a relay for a specific application, and eliminate other devices that would otherwise be required.

7.6.7 Provision of Setting Groups

Historically, electromechanical and static relays have been provided with fixed plug settings applied to the relay. Unfortunately, power systems change their topology due to operational reasons on a regular basis, such as supply from normal or emergency generation. Different configurations may require different relay settings to maintain the desired level of network protection. Fault levels are significantly different on parts of the network that are energised under normal and emergency generation.

This problem can be overcome by the provision within the relay of several setting groups, only one of which is in use at any one time. Changeover between groups can be achieved from a remote command from the operator, or possibly through the programmable logic system. This may obviate the need for duplicate relays to be fitted with some form of switching arrangement of the inputs and outputs depending on network configuration. Also the operator can program the relay remotely with a group of settings if required.

7.6.8 Conclusions

The extra facilities in numerical relays may avoid the need for other measurement and control devices to be fitted in a substation. Also numerical relays have functionality that previously required separate equipment. The protection relay no longer performs a basic protection function but is an integral and major part of a substation automation scheme. The choice of a protection relay rather than some other device is logical as the protection relay is probably the only device that is virtually mandatory on circuits of any significant rating. Therefore the functions previously carried out by separate devices such as bay controllers, discrete metering transducers and similar devices are now found in a protection relay. It is now possible to implement a substation automation scheme using numerical relays as the main hardware provided at bay level. As the power of microprocessors continues to grow and pressure on operators to reduce costs continues, this trend will continue; one obvious development is the provision of RTU facilities in designated relays that act as local concentrators of information within the overall network automation scheme.

7.7 NUMERICAL RELAY CONSIDERATIONS

The introduction of numerical relays replaces some of the issues of previous generations of relays with new ones. Some of the new issues that must be addressed are as follows:

- software version control
- relay data management
- testing and commissioning

7.7.1 Software Version Control

Numerical relays perform their functions in software. The process used for software generation is no different in principle to that for any other device using real-time software, and includes the difficulties of developing code that is error-free. Manufacturers must therefore pay particular attention to the methodology used for software generation and testing to ensure that as far as possible, the code contains no errors.

If a manufacturer advances the functionality available in a new software version of a relay, or where problems are discovered in software after the release of a numerical relay, a field upgrade may be desirable. This process then requires a rigorous system of software version control to keep track of:

- the different software versions in existence
- the differences between each version
- the reasons for the change
- relays fitted with each of the versions

With an effective version control system, manufacturers are able to advise users in the event of reported problems if the problem is a known software related problem and what remedial action is required. With the aid of suitable software held by a user, it may be possible to download the new software version instead of requiring a visit from a service engineer.

7.7.2 Relay Data Management

A numerical relay usually provides many more features than a relay using static or electromechanical technology. To use these features, the appropriate data must be entered into the relay's memory and it is good practice to keep a backup copy. The amount of data per numerical relay may be 10-100 times that of an equivalent electromechanical relay, to which must be added the possibility of user-defined logic functions. The task of entering the data correctly into a numerical relay becomes a much more complex task than previously, which adds to the possibility of a mistake being made. Similarly, the amount of data that must be recorded is much larger, requiring larger storage.

The problems have been addressed with software to automate the preparation and download of relay setting data from a laptop PC connected to a communications port of the relay. As part of the process, the setting data can be read back from the relay and compared with the desired settings to ensure that the download has been error-free. A copy of the setting data (including user-defined logic schemes where used) can also be stored on the computer. These can then be printed or uploaded to the user's database facilities.

7.7.3 Relay Testing and Commissioning

Numerical relays perform many more functions than earlier generations of relays so testing them is more complex. Site commissioning is usually restricted to running the in-built software self-check, verifying that currents and voltages measured by the relay are correct, and exercising a subset of the protection functions. Any problems revealed by such tests require specialist equipment to resolve so it is simpler to replace a faulty relay and send it for repair.

Figure 7.16: Quality inspection of a numerical relay printed circuit board

Chapter 8
Protection Signalling and Intertripping

8.1 INTRODUCTION

Unit protection schemes can be formed by several relays located remotely from each other and some distance protectionn schemes. Such unit protection schemes need communication between each location to achieve a unit protection function. This communication is known as protection signalling. Communications facilities are also needed when remote circuit breakers need to be operated due to a local event. This communication is known as intertripping.

The communication messages involved may be quite simple, involving instructions for the receiving device to take some defined action (trip, block, etc.), or it may be the passing of measured data in some form from one device to another (as in a unit protection scheme).

Various types of communication links are available for protection signalling, for example:

- private pilot wires installed by the utility
- pilot wires or channels rented from a communications company
- carrier channels at high frequencies over the power lines
- radio channels at very high or ultra high frequencies
- optical fibres

Whether or not a particular link is used depends on factors such as the availability of an appropriate communication network, the distance between protection relaying points, the terrain over which the power network is constructed, as well as cost.

Protection signalling is used to implement unit protection schemes, provide teleprotection commands, or implement intertripping between circuit breakers.

8.2 UNIT PROTECTION SCHEMES

Phase comparison and current differential schemes use signalling to convey information concerning the relaying quantity - phase angle of current and phase and magnitude of current respectively - between local and remote relaying points. Comparison of local and remote signals provides the basis for both fault detection and discrimination of the schemes.

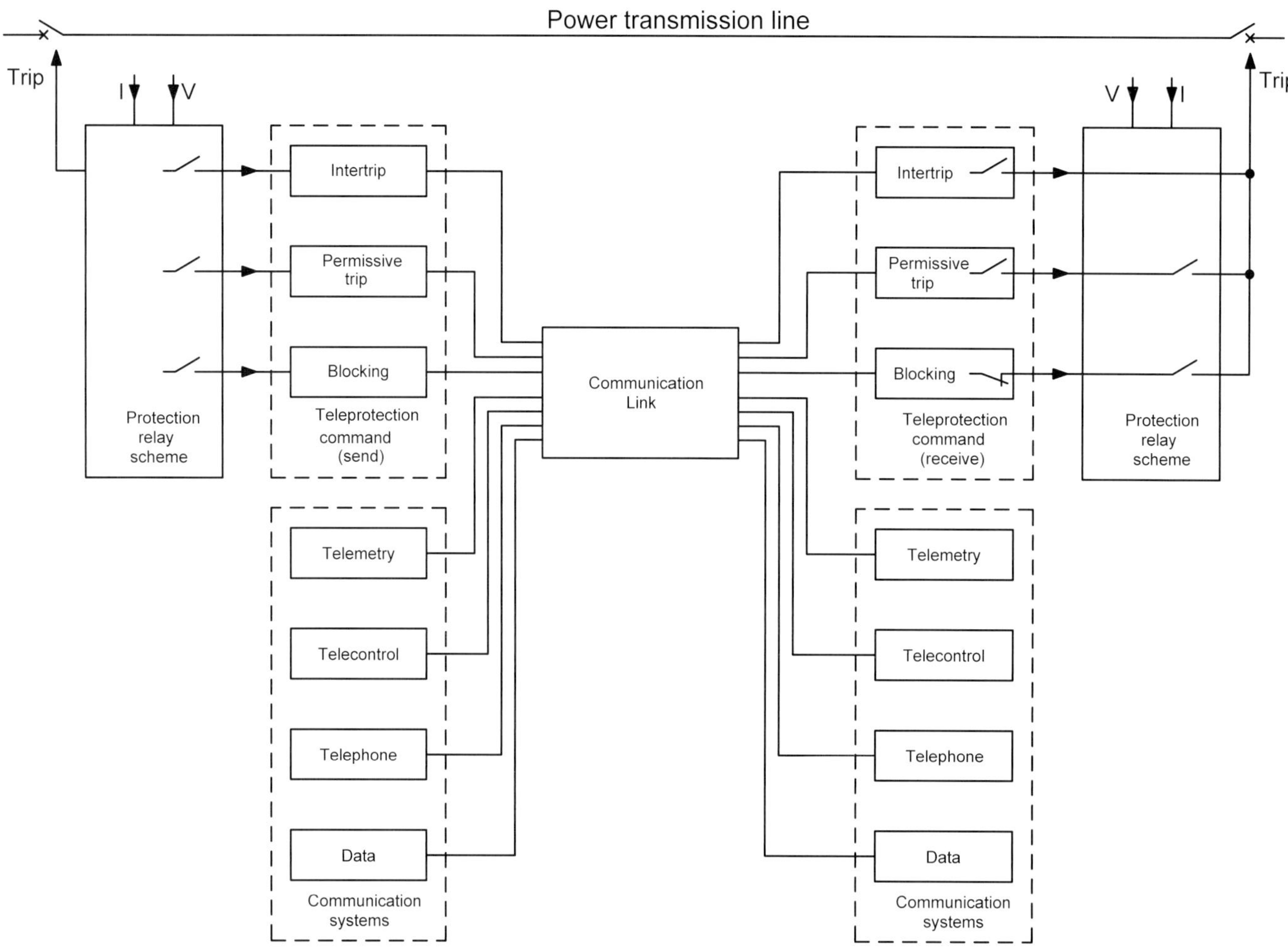

Figure 8.1: Application of protection signalling and its relationship to other systems using communication. (Shown as a unidirectional system for simplicity)

Details of Unit Protection schemes are given in Chapter 10. Communications methods are covered later in this Chapter.

8.3 TELEPROTECTION COMMANDS

Some Distance Protection schemes described in Chapter 12 use signalling to convey a command between local and remote relaying points. Receipt of the information is used to aid or speed up clearance of faults within a protected zone or to prevent tripping from faults outside a protected zone.

Teleprotection systems are often referred to by their mode of operation, or the role of the teleprotection command in the system.

8.3.1 Intertripping

Intertripping is the controlled tripping of a circuit breaker to complete the isolation of a circuit or piece of apparatus in sympathy with the tripping of other circuit breakers. The main use of such schemes is to ensure that protection at both ends of a faulted circuit isolates the equipment concerned. Possible circumstances when it may be used are:

- a feeder with a weak infeed at one end, insufficient to operate the protection for all faults
- feeder protection applied to transformer–feeder circuits. Faults on the transformer windings may operate the transformer protection but not the feeder protection. Similarly, some earth faults may not be detected due to transformer connections
- faults between the CB and feeder protection CTs, when these are located on the feeder side of the CB. Bus-zone protection does not result in fault clearance – the fault is still fed from the remote end of the feeder, while feeder unit protection may not operate as the fault is outside the protected zone
- Some distance protection schemes use intertripping to improve fault clearance times for particular kinds of fault – see Chapters 12/13

Intertripping schemes use signalling to convey a trip command to remote circuit breakers to isolate circuits. For high reliability EHV protection schemes, intertripping can be used to provide back-up to main protection, or back-tripping in the case of breaker failure. Three types of intertripping are commonly encountered, as described in the following sections.

8.3.2 Direct Tripping

In direct tripping applications, intertrip signals are sent directly to the master trip relay. Receipt of the command causes circuit breaker operation. The method of communication must be reliable and secure because any signal detected at the receiving end causes a trip of the circuit at that end. The communications system must be designed so that interference on the communication circuit does not cause spurious trips. If a spurious trip occurs, the primary system might be unnecessarily isolated.

8.3.3 Permissive Tripping

Permissive trip commands are always monitored by a protection relay. The circuit breaker is tripped when receipt of the command coincides with a 'start' condition being detected by the protection relay at the receiving end responding to a system fault. Requirements for the communications channel are less onerous than for direct tripping schemes, since receipt of an incorrect signal must coincide with a 'start' of the receiving end protection for a trip operation to take place. The intention of these schemes is to speed up tripping for faults occurring within the protected zone.

8.3.4 Blocking Scheme

Blocking commands are initiated by a protection element that detects faults external to the protected zone. Detection of an external fault at the local end of a protected circuit results in a blocking signal being transmitted to the remote end. At the remote end, receipt of the blocking signal prevents the remote end protection operating if it had detected the external fault. Loss of the communications channel is less serious for this scheme than in others as loss of the channel does not result in a failure to trip when required. However, the risk of a spurious trip is higher.

Figure 8.2 shows the typical trade-off triangle that applies when determining the required performance for teleprotection.

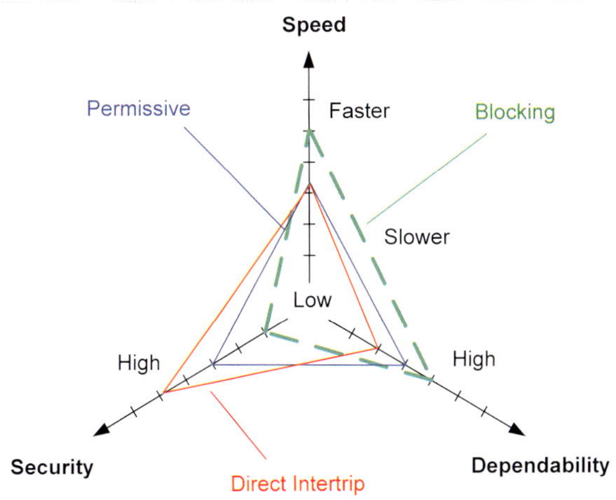

Figure 8.2: Teleprotection performance trade-off triangle

High security means that an intertrip command does not spuriously pick up due to a noisy channel. High dependability is the means by which a blocking or permissive command may easily pass through noise and still be received at the remote line end.

Figure 8.1 shows the typical applications of protection signalling and their relationship to other signalling systems commonly required for control and management of a power system. Of course, not all of the protection signals shown are required in any particular scheme.

8.4 PERFORMANCE REQUIREMENTS

Overall fault clearance time is the sum of:

- signalling time
- protection relay operating time
- trip relay operating time
- circuit breaker operating time

The overall time must be less than the maximum time for which a fault can remain on the system for minimum plant damage, loss of stability, etc. Fast operation is therefore a pre-requisite of most signalling systems.

Typically the time allowed for the transfer of a command is of the same order as the operating time of the associated protection relays. Nominal operating times range from 4 to 40ms dependent on the mode of operation of the teleprotection system.

Protection signals are subjected to the noise and interference associated with each communication medium. If noise reproduces the signal used to convey the command, unwanted commands may be produced, whilst if noise occurs when a command signal is being transmitted, the command may be retarded or missed completely. Performance is expressed in terms of security, dependability and speed. Security is assessed by the probability of an unwanted command occurring, and dependability is assessed by the probability of missing a command.

Poor security is indicated by a high probability of an unwanted command being received (Puc), therefore a lower Puc figure is generally preferable. Poor dependability is indicated by a high probability of a missing command (Pmc). Generally a lower Pmc figure is also preferable.

The required degree of security and dependability is related to the mode of operation, the characteristics of the communication medium and the operating standards of the particular utility.

Typical design objectives for teleprotection systems are not more than one incorrect trip per 500 equipment years and less than one failure to trip in every 1000 attempts, or a delay of more than 50msec should not occur more than once per 10 equipment years. To achieve these objectives, special emphasis may be attached to the security and dependability of the teleprotection command for each mode of operation in the system, as follows.

8.4.1 Performance Requirements – Intertripping

Since any unwanted command causes incorrect tripping, very high security is required at all noise levels up to the maximum that might ever be encountered.

8.4.2 Performance Requirements – Permissive Tripping

Security somewhat lower than that required for intertripping is usually satisfactory, since incorrect tripping can occur only if an unwanted command happens to coincide with operation of the protection relay for an out-of-zone fault.

For permissive over-reach schemes, resetting after a command should be highly dependable to avoid any chance of maloperations during current reversals.

8.4.3 Performance Requirements – Blocking Schemes

Low security is usually adequate since an unwanted command can never cause an incorrect trip. High dependability is required since absence of the command could cause incorrect tripping if the protection relay operates for an out-of-zone fault.

8.5 TRANSMISSION MEDIA, INTERFERENCE AND NOISE

The transmission media that provide the communication links involved in protection signalling are:

- private pilots
- rented pilots or channels
- power line carrier
- radio
- optical fibres

Historically, pilot wires and channels (discontinuous pilot wires with isolation transformers or repeaters along the route between signalling points) have been the most widely used due to their availability, followed by Power Line Carrier Communications (PLCC) techniques and radio. In recent years, fibre-optic systems have become the usual choice for new installations, primarily due to their complete immunity from electrical interference. The use of fibre-optic cables also greatly increases the number of communication channels available for each physical fibre connection and thus enables more comprehensive monitoring of the power system to be achieved by the provision of a large number of communication channels.

8.5.1 Private Pilot Wires and Channels

Pilot wires are continuous copper connections between signalling stations, while pilot channels are discontinuous pilot wires with isolation transformers or repeaters along the route between signalling stations. They may be laid in a trench with high voltage cables, laid by a separate route or strung as an open wire on a separate wood pole route.

Distances over which signalling is required vary considerably. At one end of the scale, the distance may be only a few tens of metres, where the devices concerned are located in the same substation. For applications on EHV lines, the distance between devices may be between 10-100km or more. For short distances, no special measures are required against interference, but over longer distances, special send and receive relays may be required to boost signal levels and provide immunity against induced voltages from power circuits, lightning strikes to ground adjacent to the route, etc. Isolation transformers may also have to be provided to guard against rises in substation ground potential due to earth faults.

The capacity of a link can be increased if multiplexing techniques are used to run parallel signalling systems but some utilities prefer the link to be used only for protection signalling.

Private pilot wires or channels can be attractive to a utility running a very dense power system with short distances between stations.

8.5.2 Rented Pilot Wires and Channels

These are rented from national communication authorities and, apart from the connection from the relaying point to the nearest telephone exchange, the routing is through cables forming part of the national communication network.

An economic decision has to be made between the use of private or rented pilots. If private pilots are used, the owner has complete control, but bears the cost of installation and maintenance. If rented pilots are used, most of these costs are eliminated, but fees must be paid to the owner of the pilots and the signal path may be changed without warning. This may be a problem in protection applications where signal transmission times are critical.

The chance of voltages being induced in rented pilots is smaller than for private pilots, as the pilot route is normally not related to the route of the power line with which it is associated. However, some degree of security and protection against induced voltages must be built into signalling systems.

Station earth potential rise is a significant factor to be taken into account and isolation must be provided to protect both the personnel and equipment of the communication authority.

The most significant hazard to be withstood by a protection signalling system using this medium arises when a linesman inadvertently connects a low impedance test oscillator across the pilot link that can generate signalling tones. Transmissions by such an oscillator may simulate the operating code or tone sequence that, in the case of direct intertripping schemes, would result in incorrect operation of the circuit breaker.

Communication between relaying points may be over a two-wire or four-wire link. Consequently the effect of a particular human action, for example an incorrect disconnection, may disrupt communication in one direction or both.

The signals transmitted must be limited in both level and bandwidth to avoid interference with other signalling systems. The owner of the pilots impose standards in this respect that may limit transmission capacity or transmission distance, or both.

With a power system operating at, say, 132kV, where relatively long protection signalling times are acceptable, signalling has been achieved above speech together with metering and control signalling on an established control network. Consequently the protection signalling was achieved at very low cost. High voltage systems; (220kV and above), have demanded shorter operating times and improved security, which has led to the renting of pilot links exclusively for protection signalling purposes.

8.5.3 Power Line Carrier Communications Techniques

Where long line sections are involved, or if the route involves installation difficulties, the expense of providing physical pilot connections or operational restrictions associated with the route length require that other means of providing signalling facilities are required.

Power Line Carrier Communications (PLCC) is a technique that involves high frequency signal transmission along the overhead power line, typically in the 300Hz to 3400Hz band. It is robust and therefore reliable, constituting a low loss transmission path that is fully controlled by the Utility.

High voltage capacitors are used, along with drainage coils, for the purpose of injecting the signal to and extracting it from the line. Injection can be carried out by impressing the carrier signal voltage between one conductor and earth or between any two phase conductors. The basic units can be built up into a high pass or band pass filter as shown in Figure 8.3.

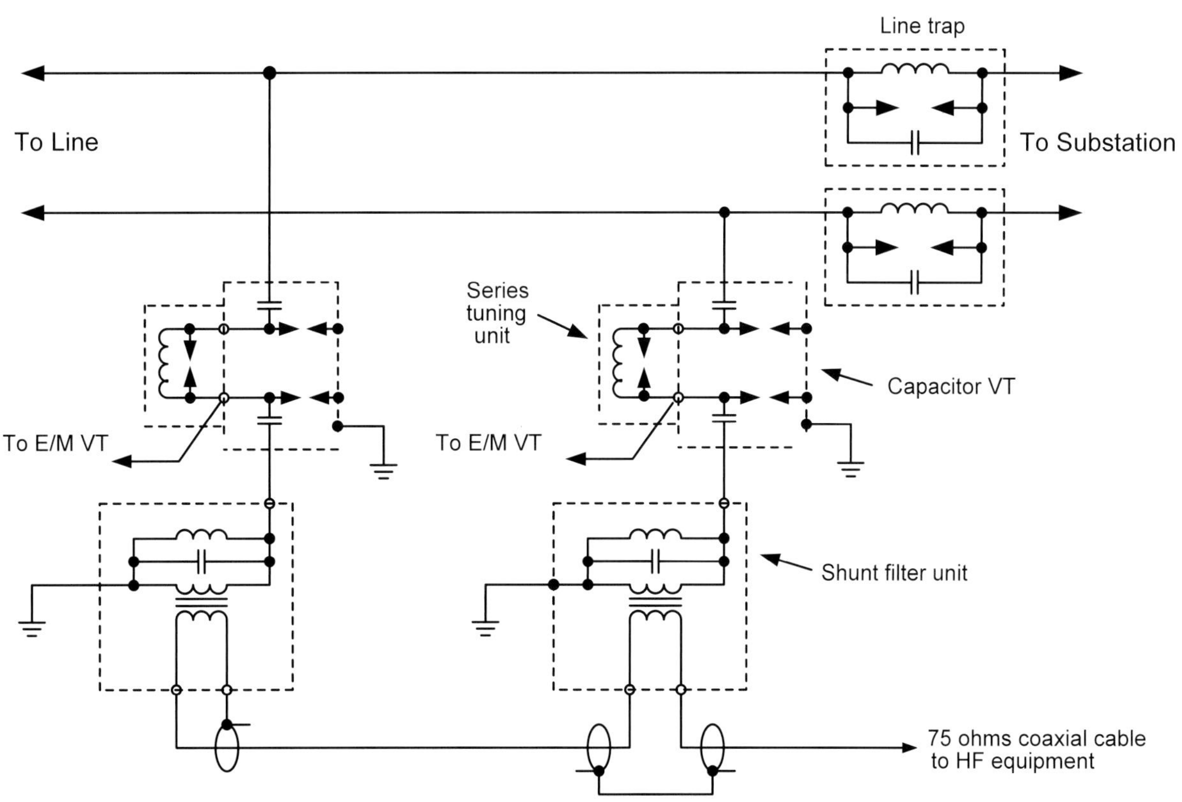

Figure 8.3: Typical phase-to-phase coupling equipment

The high voltage capacitor is tuned by a tuning coil to present a low impedance at the signal frequency; the parallel circuit presents a high impedance at the signal frequency while providing a path for the power frequency currents passed by the capacitor.

Figure 8.4: Carrier coupling equipment

It is necessary to minimise the loss of signal into other parts of the power system, to allow the same frequency to be used on another line. This is done with a 'line trap' or 'wave trap', which in its simplest form is a parallel circuit tuned to present a very high impedance to the signal frequency. It is connected in the phase conductor on the station side of the injection equipment.

The single frequency line trap can be treated as an integral part of the complete injection equipment to accommodate two or more carrier systems. However, difficulties may arise in an overall design because at certain frequencies the actual station reactance, which is normally capacitive, tunes with the trap which is inductive below its resonant frequency. The result is a low impedance across the transmission path, preventing operation at these frequencies. This situation can be avoided by using an independent 'double frequency' or 'broad-band' trap.

The attenuation of a channel is of prime importance in the application of carrier signalling because it determines the amount of transmitted energy available at the receiving end to overcome noise and interfering voltages. The loss of each line terminal is 1 to 2dB through the coupling filter, a maximum of 3dB through its broad-band trap and not more than 0.5dB per 100 metres through the high frequency cable.

The high frequency transmission characteristics of power circuits are good the loss amounting to 0.02 to 0.2dB per kilometre depending upon line voltage and frequency. Line attenuation is not affected appreciably by rain, but serious increase in loss may occur when the phase conductors are thickly coated with hoar-frost or ice. Attenuations of up to three times the fair weather value have been experienced. Receiving equipment commonly incorporates automatic gain control (AGC) to compensate for variations in attenuation of signals.

High noise levels arise from lightning strikes and system fault inception or clearance. Although these are of short duration, lasting only a few milliseconds at the most, they may cause overloading of power line carrier receiving equipment.

8.5.3.1 Application Considerations

Signalling systems used for intertripping in particular must incorporate appropriate security features to avoid maloperation. The most severe noise levels are encountered with operation of the line isolators, and these may last for some seconds. Although maloperation of the associated teleprotection scheme may have little operational significance, since the circuit breaker at one end at least is normally already open, high security is generally required to cater for noise coupled between parallel lines or passed through line traps from adjacent lines.

Signalling for permissive intertrip applications needs special consideration, as this involves signalling through a power system fault. The increase in channel attenuation due to the fault varies according to the type of fault, but most utilities select a nominal value, usually between 20 and 30dB, as an application guide. A protection signal boost facility can be employed to cater for an increase in attenuation of this order of magnitude, to maintain an acceptable signal-to-noise ratio at the receiving end, so that the performance of the service is not impaired.

Most direct intertrip applications require signalling over a healthy power system, so boosting is not normally needed. In fact, if a voice frequency intertrip system is operating over a carrier bearer channel, the dynamic operating range of the receiver must be increased to accommodate a boosted signal. This makes it less inherently secure in the presence of noise during a quiescent signalling condition.

8.5.3.2 Digital Power Line Carrier

The latest power line carrier equipment allows analogue, digital and mixed-mode communication. Digital communication up to 128kbits/s can be achieved using a 16kHz bandwidth. Low-latency Ethernet bridging facilities are a cost-effective communications solution for substations that have no access to any fibre network.

8.5.4 Radio Channels

At first consideration, the wide bandwidth associated with radio frequency transmissions could allow the use of modems operating at very high data rates. Protection signalling commands could be sent by serial coded messages of sufficient length and complexity to give high security, but still achieve fast operating times. In practice, it is seldom economic to provide radio equipment exclusively for protection signalling, so standard general-purpose telecommunications channel equipment is normally adopted.

Typical radio bearer equipment operates at the microwave frequencies of 0.2 to 10GHz. Because of the relatively short range and directional nature of the transmitter and receiver aerial systems at these frequencies, large bandwidths can be allocated without much chance of mutual interference with other systems.

Multiplexing techniques allow several channels to share the common bearer medium and exploit the large bandwidth. In addition to voice frequency channels, wider bandwidth channels or data channels may be available, dependent on the particular system. For instance, in analogue systems using frequency division multiplexing, normally up to 12 voice frequency channels are grouped together in basebands at 12-60kHz or 60-108kHz, but alternatively the baseband may be used as a 48kHz signal channel. Modern digital systems employing pulse code modulation and time division multiplexing usually provide the voice frequency channels by sampling at 8kHz and quantising to 8 bits; alternatively, access may be available for data at 64kbits/s (equivalent to one voice frequency channel) or higher data rates.

Radio systems are well suited to the bulk transmission of information between control centres and are widely used for this. When the route of the trunk data network coincides with that of transmission lines, channels can often be allocated for protection signalling. More generally, radio communication is between major stations rather than the ends of individual lines, because of the need for line-of-sight operation between aerials and other requirements of the network. Roundabout routes involving repeater stations and the addition of pilot channels to interconnect the radio installation and the relay station may be possible, but overall dependability is normally much lower than for PLCC systems in which the communication is direct from one end of the line to the other.

Radio channels are not affected by increased attenuation due to power system faults, but fading has to be taken into account when the signal-to-noise ratio of a particular installation is being considered.

Most of the noise in such a protection signalling system is generated in the radio equipment.

A polluted atmosphere can cause radio beam refraction that interferes with efficient signalling. The height of aerial tower should be limited, so that winds and temperature changes have the minimum effect on their position.

8.5.5 Optical Fibre Channels

Optical fibres are fine strands of glass, which behave as wave guides for light. This ability to transmit light over considerable distances can be used to provide optical communication links with enormous information carrying capacity and an inherent immunity to electromagnetic interference.

A practical optical cable consists of a central optical fibre which comprises core, cladding and protective buffer coating surrounded by a protective plastic oversheath containing strength members which, in some cases, are enclosed by a layer of armouring.

To communicate information a beam of light is modulated in accordance with the signal to be transmitted. This modulated beam travels along the optical fibre and is subsequently decoded at the remote terminal into the received signal. On/off modulation of the light source is normally preferred to linear modulation since the distortion caused by non-linearities in the light source and detectors, as well as variations in received light power, are largely avoided.

The light transmitter and receiver are usually laser or LED devices capable of emitting and detecting narrow beams of light at selected frequencies in the low attenuation 850, 1300 and 1550 nanometre spectral windows. The distance over which effective communications can be established depends on the attenuation and dispersion of the communication link and this depends on the type and quality of the fibre and the wavelength of the optical source. Within the fibre there are many modes of propagation with different optical paths that cause dispersion of the light signal and result in pulse broadening. The degrading of the signal in this way can be reduced by the use of 'graded index' fibres that cause the various modes to follow nearly equal paths. The distance over which signals can be transmitted is significantly increased by the use of 'monomode' fibres that support only one mode of propagation.

Optical fibre channels allow communication at data rates of hundreds of megahertz over a few tens of kilometres, however, repeaters are needed for greater distances. An optical fibre can be used as a dedicated link between two items of terminal equipment or as a multiplexed link that carries all communication traffic such as voice, telecontrol and protection signalling. For protection signalling, the available bandwidth of a link is divided by time division multiplexing (T.D.M.) techniques into several channels, each of 64kbits/s. Each 64kbits/s channel is equivalent to one voice frequency channel, which typically uses 8-bit analogue-to-digital conversion at a sampling rate of 8kHz. Several utilities sell surplus capacity on their links to telecommunications operators, or they may take the opportunity to increase the data rate in end-to-end applications up to 2Mbits/s. The trend of using rented pilot circuits is being reversed as the utilities revert to ownership of the communication circuits that carry protection signalling.

The equipment that carries out this multiplexing at each end of a line is known as 'Pulse Code Modulation' (P.C.M.) terminal equipment. This approach is the one adopted by telecommunications authorities and some utilities favour its adoption on their private systems, for economic considerations.

Optical fibre communications are well established in the electrical supply industry. They are the preferred means for the communications link between a substation and a telephone exchange when rented circuits are used, as trials have shown that this link is particularly susceptible to interference from power system faults if copper conductors are used. Whilst such fibres can be laid in cable trenches, there is a strong trend to associate them with the conductors themselves by producing composite cables comprising optical fibres embedded in the conductors, either earth or phase. For overhead lines use of OPGW (Optical Ground Wire) earth conductors is very common, while an alternative is to wrap the optical cable helically around a phase or earth conductor. This latter technique can be used without restringing of the line.

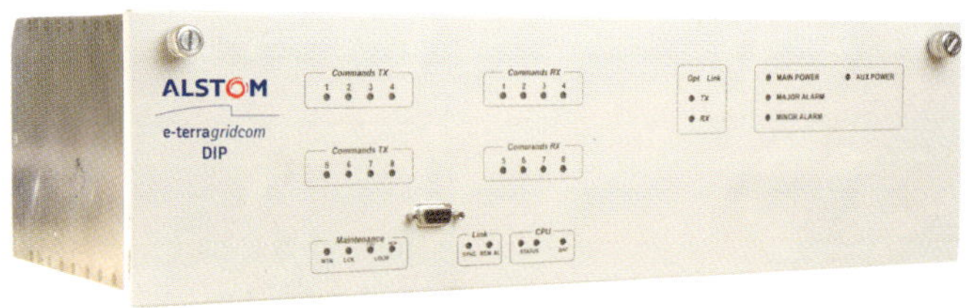

Figure 8.6: Example digital teleprotection, e-terra*gridcom* DIP

8.6 SIGNALLING METHODS

Various methods are used in protection signalling; not all need be suited to every transmission medium. The methods to be considered briefly are:

- d.c. voltage step or d.c. voltage reversals
- plain tone keyed signals at high and voice frequencies
- frequency shift keyed signals involving two or more tones at high and voice frequencies

General purpose telecommunications equipment operating

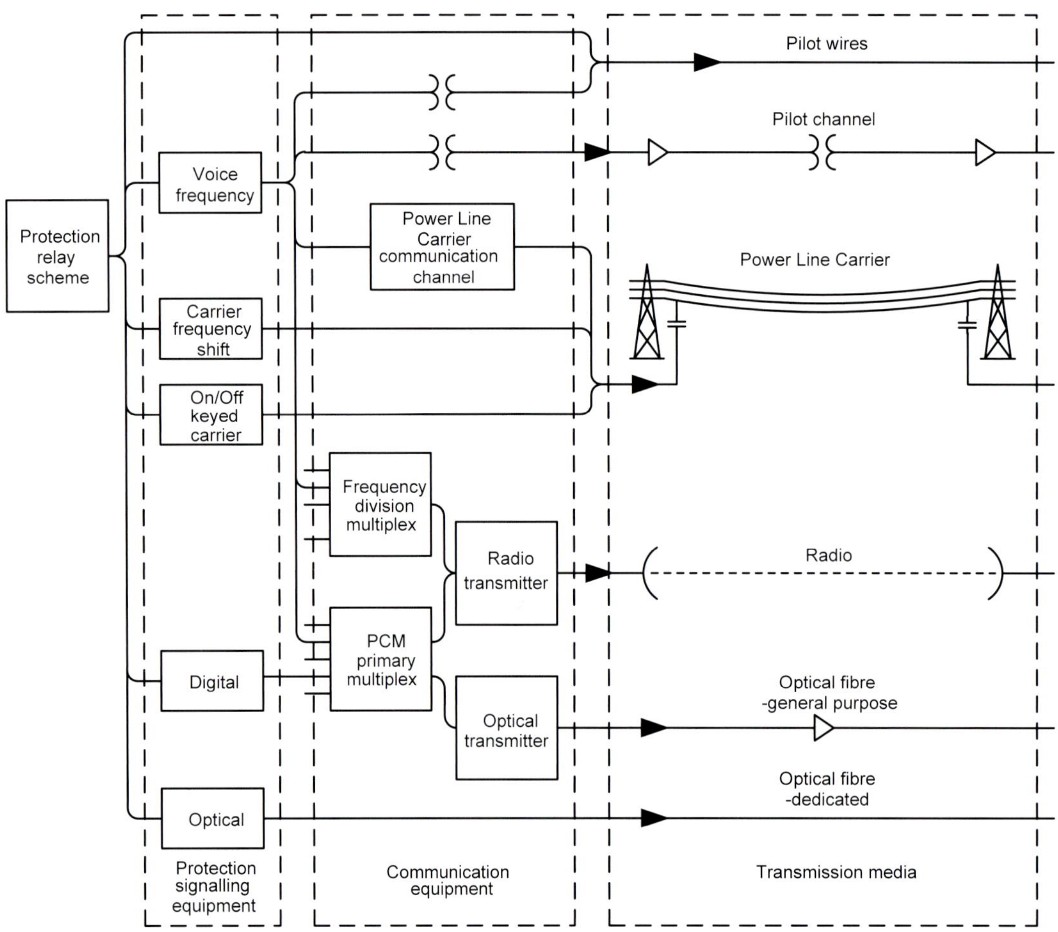

Figure 8.5: Communication arrangements commonly encountered in protection signalling

over power line carrier, radio or optical fibre media incorporate frequency translating or multiplexing techniques to provide the user with standardised communication channels. They have a nominal bandwidth/channel of 2kHz to 4kHz and are often referred to as voice frequency (vf) channels. Protection signalling equipment operating at voice frequencies exploits the standardisation of the communication interface. Where voice frequency channels are not available or suitable, protection signalling may make use of media or specialised equipment dedicated entirely to the signalling requirements.

Figure 8.5 shows the communication arrangements commonly encountered in protection signalling.

8.6.1 D.C. Voltage Signalling

A d.c. voltage step or d.c. voltage reversals may be used to convey a signalling instruction between protection relaying points in a power system, but these are suited only to private pilot wires, where low speed signalling is acceptable, with its inherent security. In all applications there is the need to ensure that no maloperation can occur due to power frequency interference.

8.6.2 Plain Tone Signals

Plain high frequency signals can be used successfully for the signalling of blocking information over a power line. A normally quiescent power line carrier equipment can be dedicated entirely to the transfer to teleprotection blocking commands. Phase comparison power line carrier unit protection schemes often use such equipment and take advantage of the very high speed and dependability of the signalling system. The special characteristics of dedicated 'on/off' keyed carrier systems are discussed later. A relatively insensitive receiver is used to discriminate against noise on an amplitude basis, and for some applications the security may be satisfactory for permissive tripping, particularly if the normal high-speed operation of about 7ms or less is sacrificed by the addition of delays. The need for regular reflex testing of a normally quiescent channel usually precludes any use for intertripping.

Plain tone power line carrier signalling systems are particularly suited to providing the blocking commands often associated with the protection of multi-ended feeders, as described in Chapter 13. A blocking command sent from one end can be received simultaneously at all the other ends using a single power line carrier channel. Other signalling systems usually require discrete communication channels between each of the ends or involve repeaters, leading to decreased dependability of the blocking command.

8.6.3 Frequency Shift Keyed Signals

Frequency shift keyed high frequency signals can be used over a power line carrier link to give short operating times (15 milliseconds for blocking and permissive intertripping, 20 milliseconds for direct intertripping) for all applications of protection signalling. The required amount of security can be achieved by using a broadband noise detector to monitor the actual operational signalling equipment.

Frequency shift keyed voice frequency signals can be used for all protection signalling applications over all transmission media. Frequency modulation techniques make possible an improvement in performance, because amplitude limiting rejects the amplitude modulation component of noise, leaving only the phase modulation components to be detected.

The operational protection signal may consist of tone sequence codes with, say, three tones, or a multi-bit code using two discrete tones for successive bits, or of a single frequency shift.

Modern high-speed systems use multi-bit code or single frequency shift techniques. Complex codes are used to give the required degree of security in direct intertrip schemes: the short operating times needed may result in uneconomical use of the available voice frequency spectrum, particularly if the channel is not exclusively employed for protection signalling. As noise power is directly proportional to bandwidth, a large bandwidth causes an increase in the noise level admitted to the detector, making operation in the presence of noise more difficult. So, again, it is difficult to obtain both high dependability and high security.

The signal frequency shift technique has advantages where fast signalling is needed for blocked distance and permissive intertrip applications. It has little inherent security, but additional algorithms responsive to every type of interference can give acceptable security. This system does not require a high transmission rate channel as the frequency changes once only. The bandwidth can therefore be narrower than in coded systems, giving better noise rejection as well as being advantageous if the channel is shared with telemetry and control signalling, which is inevitably the case if a power line carrier bearer is used.

Chapter 9
Overcurrent Protection for Phase and Earth Faults

9.1 INTRODUCTION

Protection against excess current was naturally the earliest protection system to evolve. From this basic principle, the graded overcurrent system, a discriminative fault protection, has been developed. This should not be confused with 'overload' protection, which normally makes use of relays that operate in a time related in some degree to the thermal capability of the plant to be protected. Overcurrent protection, on the other hand, is directed entirely to the clearance of faults, although with the settings usually adopted some measure of overload protection may be obtained.

9.2 CO-ORDINATION PROCEDURE

Correct overcurrent relay application requires knowledge of the fault current that can flow in each part of the network. Since large-scale tests are normally impracticable, system analysis must be used – see Chapter 4 for details. The data required for a relay setting study are:

- a one-line diagram of the power system involved, showing the type and rating of the protection devices and their associated current transformers
- the impedances in ohms, per cent or per unit, of all power transformers, rotating machine and feeder circuits
- the maximum and minimum values of short circuit currents that are expected to flow through each protection device
- the maximum load current through protection devices
- the starting current requirements of motors and the starting and locked rotor/stalling times of induction motors
- the transformer inrush, thermal withstand and damage characteristics
- decrement curves showing the rate of decay of the fault current supplied by the generators
- performance curves of the current transformers

The relay settings are first determined to give the shortest operating times at maximum fault levels and then checked to see if operation will also be satisfactory at the minimum fault current expected. It is always advisable to plot the curves of relays and other protection devices, such as fuses, that are to operate in series, on a common scale. It is usually more convenient to use a scale corresponding to the current expected at the lowest voltage base, or to use the predominant voltage base. The alternatives are a common MVA base or a separate current scale for each system voltage.

The basic rules for correct relay co-ordination can generally be stated as follows:

- whenever possible, use relays with the same operating characteristic in series with each other
- make sure that the relay farthest from the source has current settings equal to or less than the relays behind it, that is, that the primary current required to operate the relay in front is always equal to or less than the primary current required to operate the relay behind it

9.3 PRINCIPLES OF TIME/CURRENT GRADING

Among the various possible methods used to achieve correct relay co-ordination are those using either time or overcurrent, or a combination of both. The common aim of all three methods is to give correct discrimination. That is to say, each one must isolate only the faulty section of the power system network, leaving the rest of the system undisturbed.

9.3.1 Discrimination by Time

In this method, an appropriate time setting is given to each of the relays controlling the circuit breakers in a power system to ensure that the breaker nearest to the fault opens first. A simple radial distribution system is shown in Figure 9.1, to illustrate the principle.

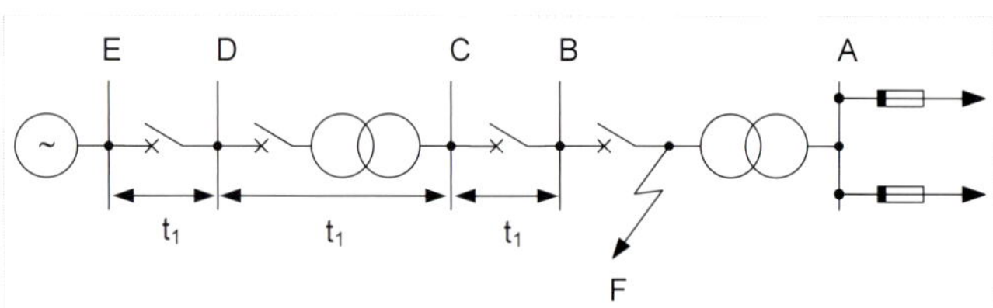

Figure 9.1: Radial system with time discrimination

Overcurrent protection is provided at *B*, *C*, *D* and *E*, that is, at the infeed end of each section of the power system. Each protection unit comprises a definite-time delay overcurrent relay in which the operation of the current sensitive element simply initiates the time delay element. Provided the setting of the current element is below the fault current value, this element plays no part in the achievement of discrimination. For this reason, the relay is sometimes described as an 'independent definite-time delay relay', since its operating time is for practical purposes independent of the level of overcurrent.

It is the time delay element, therefore, which provides the means of discrimination. The relay at *B* is set at the shortest time delay possible to allow the fuse to blow for a fault at *A* on the secondary side of the transformer. After the time delay has expired, the relay output contact closes to trip the circuit breaker. The relay at *C* has a time delay setting equal to t_1 seconds, and similarly for the relays at *D* and *E*.

If a fault occurs at *F*, the relay at *B* will operate in *t* seconds and the subsequent operation of the circuit breaker at *B* will clear the fault before the relays at *C*, *D* and *E* have time to operate. The time interval t_1 between each relay time setting must be long enough to ensure that the upstream relays do not operate before the circuit breaker at the fault location has tripped and cleared the fault.

The main disadvantage of this method of discrimination is that the longest fault clearance time occurs for faults in the section closest to the power source, where the fault level (MVA) is highest.

9.3.2 Discrimination by Current

Discrimination by current relies on the fact that the fault current varies with the position of the fault because of the difference in impedance values between the source and the fault. Hence, typically, the relays controlling the various circuit breakers are set to operate at suitably tapered values of current such that only the relay nearest to the fault trips its breaker. Figure 9.2 illustrates the method.

For a fault at F_1, the system short-circuit current is given by:

$$I = \frac{6350}{Z_S + Z_{L1}} A$$

where:

Z_S = source impedance = $\frac{11^2}{250}$ = 0.485Ω

Z_{L1} = cable impedance between C and B = 0.24Ω

Hence $I = \frac{6350}{0.725}$ = 8800A

So, a relay controlling the circuit breaker at *C* and set to operate at a fault current of 8800A would in theory protect the whole of the cable section between *C* and *B*. However, there are two important practical points that affect this method of co-ordination:

- it is not practical to distinguish between a fault at F_1 and a fault at F_2, since the distance between these points may be only a few metres, corresponding to a change in fault current of approximately 0.1%

- in practice, there would be variations in the source fault level, typically from 250MVA to 130MVA. At this lower fault level the fault current would not exceed 6800A, even for a cable fault close to *C*. A relay set at 8800A would not protect any part of the cable section concerned

Discrimination by current is therefore not a practical proposition for correct grading between the circuit breakers at *C* and *B*. However, the problem changes appreciably when there is significant impedance between the two circuit breakers concerned. Consider the grading required between the circuit breakers at *C* and *A* in Figure 9.2. Assuming a fault at F_4, the short-circuit current is given by:

$$I = \frac{6350}{Z_S + Z_{L1} + Z_{L2} + Z_T}$$

where

Z_S = source impedance = $\frac{11^2}{250}$ = 0.485Ω

Z_{L1} = cable impedance between C and B = 0.24Ω

Z_{L2} = cable impedance between B and 4MVA transformer = 0.04Ω

Z_T = transformer impedance = $0.07\left(\frac{11^2}{4}\right)$ =2.12Ω

Hence $I = \frac{6350}{2.885}$ = 2200A

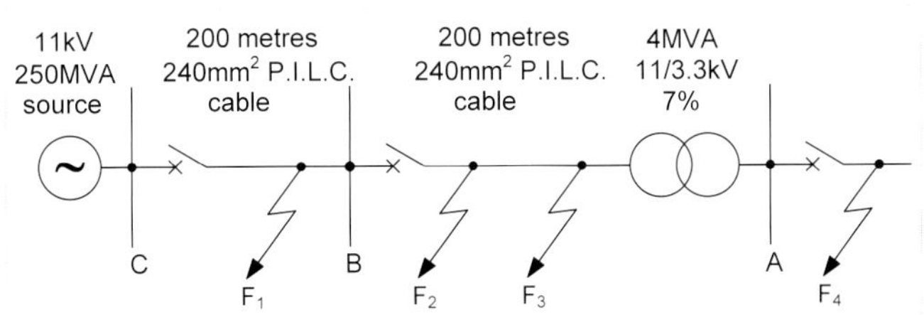

Figure 9.2: Radial system with current discrimination

For this reason, a relay controlling the circuit breaker at *B* and set to operate at a current of 2200A plus a safety margin would not operate for a fault at F_4 and would thus discriminate with the relay at *A*. Assuming a safety margin of 20% to allow for relay errors and a further 10% for variations in the system impedance values, it is reasonable to choose a relay setting of 1.3 x 2200A, that is, 2860A, for the relay at *B*. Now, assuming a fault at F_3, at the end of the 11kV cable feeding the 4MVA transformer, the short-circuit current is given by:

$$I = \frac{6350}{Z_S + Z_{L1} + Z_{L2}}$$

Thus, assuming a 250MVA source fault level:

$$I = \frac{6350}{0.485 + 0.24 + 0.04} = 8300\text{A}$$

Alternatively, assuming a source fault level of 130MVA:

$$I = \frac{6350}{0.93 + 0.214 + 0.04} = 5250\text{A}$$

For either value of source level, the relay at *B* would operate correctly for faults anywhere on the 11kV cable feeding the transformer.

9.3.3 Discrimination by both Time and Current

Each of the two methods described so far has a fundamental disadvantage. In the case of discrimination by time alone, the disadvantage is due to the fact that the more severe faults are cleared in the longest operating time. On the other hand, discrimination by current can be applied only where there is appreciable impedance between the two circuit breakers concerned.

It is because of the limitations imposed by the independent use of either time or current co-ordination that the inverse time overcurrent relay characteristic has evolved. With this characteristic, the time of operation is inversely proportional to the fault current level and the actual characteristic is a function of both 'time' and 'current' settings. Figure 9.3 shows the characteristics of two relays given different current/time settings. For a large variation in fault current between the two ends of the feeder, faster operating times can be achieved by the relays nearest to the source, where the fault level is the highest. The disadvantages of grading by time or current alone are overcome.

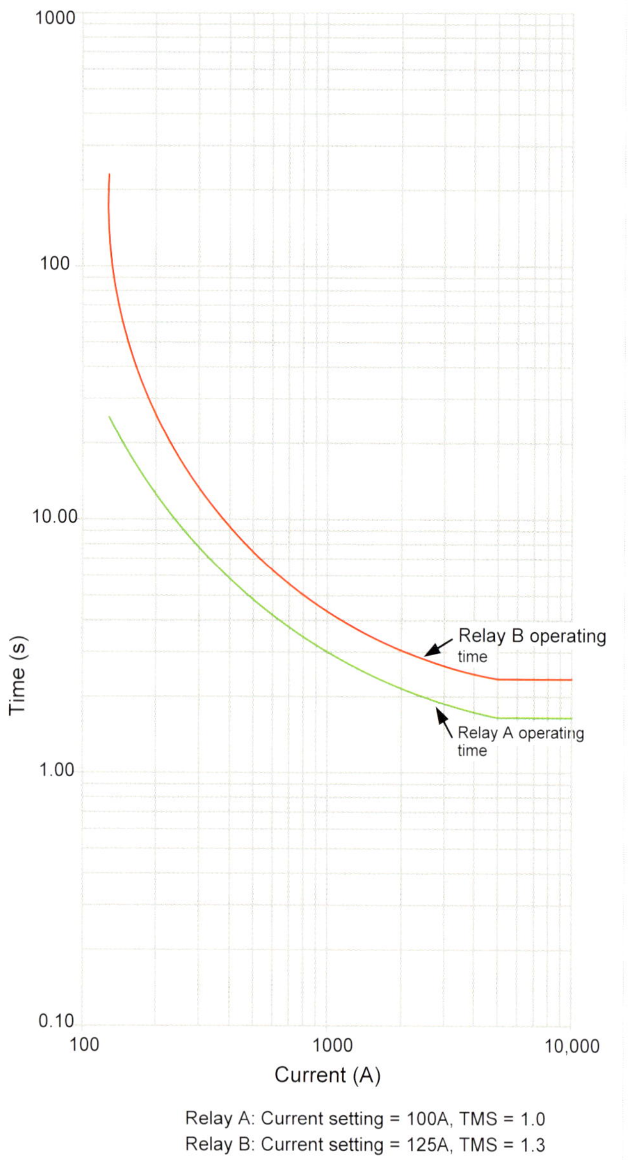

Figure 9.3: Relay characteristics for different settings

9.4 STANDARD IDMT OVERCURRENT RELAYS

The current/time tripping characteristics of IDMT relays may need to be varied according to the tripping time required and the characteristics of other protection devices used in the network. For these purposes, IEC 60255 defines a number of standard characteristics as follows:

- Standard Inverse (SI)
- Very Inverse (VI)
- Extremely Inverse (EI)
- Definite Time (DT)

The mathematical descriptions of the curves are given in Table 9.1, and the curves based on a common setting current and time multiplier setting of 1 second are shown in Figure 9.4. The tripping characteristics for different TMS settings using the SI curve are shown in Figure 9.6.

Relay Characteristic	Equation (IEC 60255)
Standard Inverse (SI)	$t = TMS \times \frac{0.14}{I_r^{0.02} - 1}$
Very Inverse (VI)	$t = TMS \times \frac{13.5}{I_r - 1}$
Extremely Inverse (EI)	$t = TMS \times \frac{80}{I_r^2 - 1}$
Long time standby earth fault	$t = TMS \times \frac{120}{I_r - 1}$

Table 9.1: Definitions of standard relay characteristics

Characteristic	Equation
IEEE Moderately Inverse	$t = \frac{TD}{7}\left[\left(\frac{0.0515}{I_r^{0.02} - 1}\right) + 0.114\right]$
IEEE Very Inverse	$t = \frac{TD}{7}\left[\left(\frac{19.61}{I_r^2 - 1}\right) + 0.491\right]$
IEEE Extremely Inverse	$t = \frac{TD}{7}\left[\left(\frac{28.2}{I_r^2 - 1}\right) + 0.1217\right]$
US CO8 Inverse	$t = \frac{TD}{7}\left[\left(\frac{5.95}{I_r^2 - 1}\right) + 0.18\right]$
US CO2 Short Time Inverse	$t = \frac{TD}{7}\left[\left(\frac{0.02394}{I_r^{0.02} - 1}\right) + 0.01694\right]$

Table 9.2: North American IDMT definitions of standard relay characteristics

For Table 9.1 and Table 9.2:

$Ir = I / Is$

Where:

I = Measured current

Is = Relay setting current

TMS = Time Multiplier Setting

TD = Time Dial setting

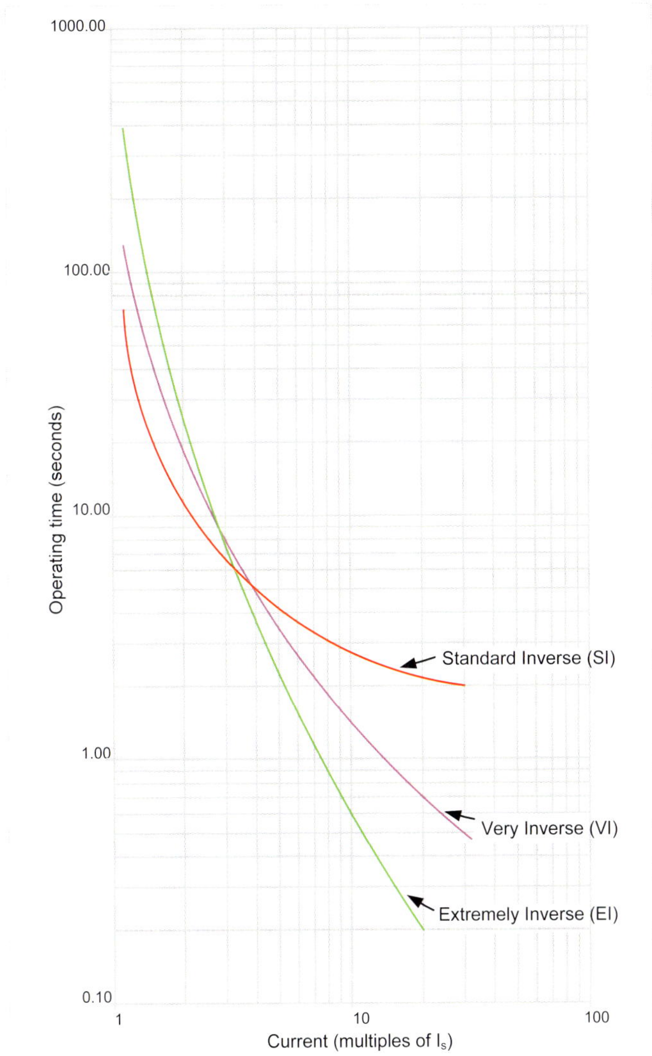

Figure 9.4: IEC 60255 IDMT relay characteristics; TMS=1.0

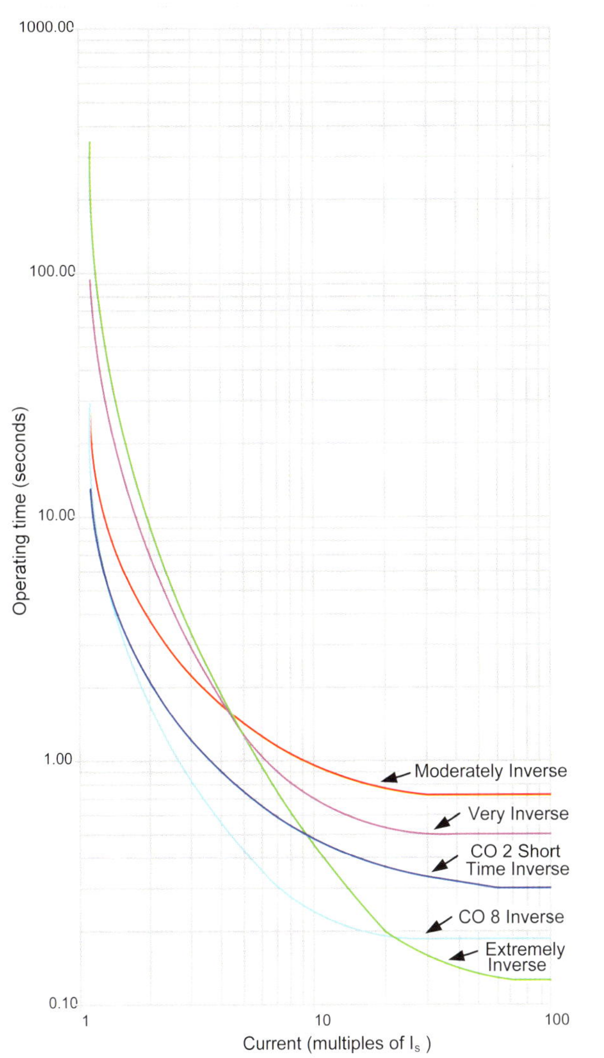

Figure 9.5: North American IDMT relay characteristics; TD=7

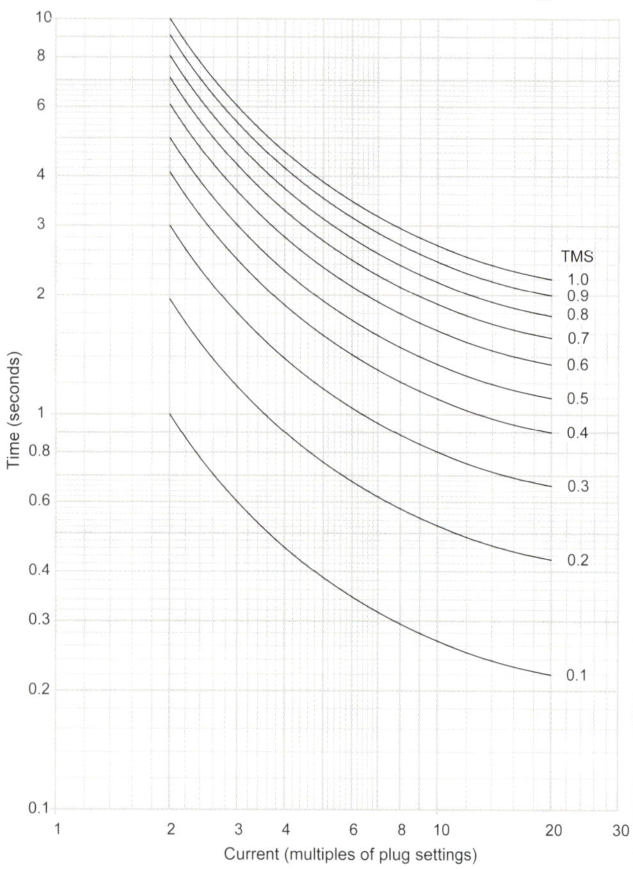

Figure 9.6: Typical time/current characteristics of standard IDMT relay

Although the curves are only shown for discrete values of TMS, continuous adjustment may be possible in an electromechanical relay. For other relay types, the setting steps may be so small as to effectively provide continuous adjustment. In addition, almost all overcurrent relays are also fitted with a high-set instantaneous element. In most cases, use of the standard SI curve proves satisfactory, but if satisfactory grading cannot be achieved, use of the VI or EI curves may help to resolve the problem. When digital or numeric relays are used, other characteristics may be provided, including the possibility of user-definable curves. More details are provided in the following sections.

Relays for power systems designed to North American practice utilise ANSI/IEEE curves. Table 9.2 gives the mathematical description of these characteristics and Figure 9.5 shows the curves standardised to a time dial setting of 7.

Take great care that different vendors may standardise their curves at different settings other than TD=7. The protection engineer must ensure whether the factor of 7, or some other nominal, is applied.

9.5 COMBINED IDMT AND HIGH SET INSTANTANEOUS OVERCURRENT RELAYS

A high-set instantaneous element can be used where the source impedance is small in comparison with the protected circuit impedance. This makes a reduction in the tripping time at high fault levels possible. It also improves the overall system grading by allowing the 'discriminating curves' behind the high set instantaneous elements to be lowered.

As shown in Figure 9.7, one of the advantages of the high set instantaneous elements is to reduce the operating time of the circuit protection by the shaded area below the 'discriminating curves'. If the source impedance remains constant, it is then possible to achieve high-speed protection over a large section of the protected circuit. The rapid fault clearance time achieved helps to minimise damage at the fault location. Figure 9.7 also shows a further important advantage gained by the use of high set instantaneous elements. Grading with the relay immediately behind the relay that has the instantaneous elements enabled is carried out at the current setting of the instantaneous elements and not at the maximum fault level. For example, in Figure 9.7 relay R_2 is graded with relay R_3 at 500A and not 1100A, allowing relay R_2 to be set with a TMS of 0.15 instead of 0.2 while maintaining a grading margin between relays of 0.4s. Similarly, relay R_1 is graded with R_2 at 1400A and not at 2300A.

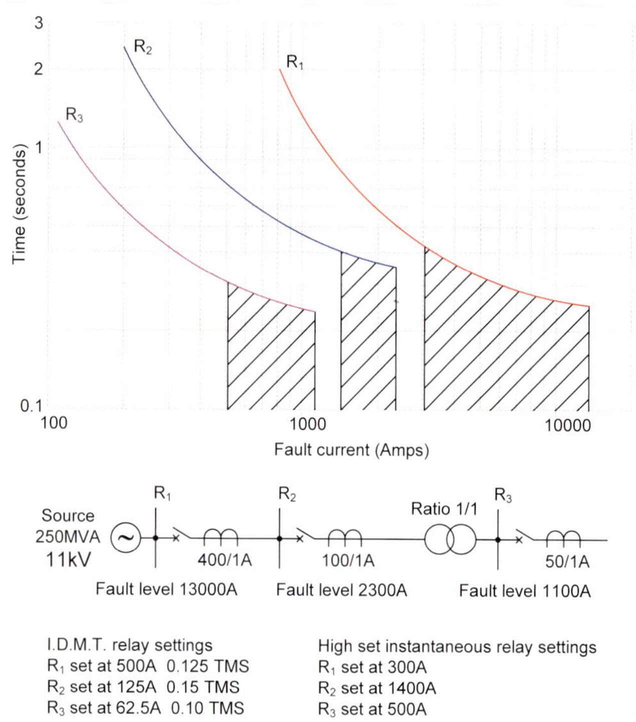

Figure 9.7: Characteristics of combined IDMT and high-set instantaneous overcurrent relays

9.5.1 Transient Overreach

The *reach* of a relay is that part of the system protected by the relay if a fault occurs. A relay that operates for a fault that lies beyond the intended zone of protection is said to overreach.

When using instantaneous overcurrent elements, care must be exercised in choosing the settings to prevent them operating for faults beyond the protected section. The initial current due to a d.c. offset in the current wave may be greater than the relay pick-up value and cause it to operate. This may occur even though the steady state r.m.s. value of the fault current for a fault at a point beyond the required reach point may be less than the relay setting. This phenomenon is called transient over-reach, and is defined as:

$$\% \text{ Transient over-reach} = \frac{I_1 - I_2}{I_2} \times 100\%$$

Equation 9.1

where:

I_1 = r.m.s steady state pickup current

I_2 = steady state r.m.s current which when fully offset just causes relay pickup

When applied to power transformers, the high set instantaneous overcurrent elements must be set above the maximum through fault current than the power transformer can supply for a fault across its LV terminals, to maintain discrimination with the relays on the LV side of the transformer.

9.6 VERY INVERSE (VI) OVERCURRENT RELAYS

Very inverse overcurrent relays are particularly suitable if there is a substantial reduction of fault current as the distance from the power source increases, i.e. there is a substantial increase in fault impedance. The VI operating characteristic is such that the operating time is approximately doubled for reduction in current from 7 to 4 times the relay current setting. This permits the use of the same time multiplier setting for several relays in series.

Figure 9.8 compares the SI and VI curves for a relay. The VI curve is much steeper and therefore the operation increases much faster for the same reduction in current compared to the SI curve. This enables the requisite grading margin to be obtained with a lower TMS for the same setting current, and hence the tripping time at source can be minimised.

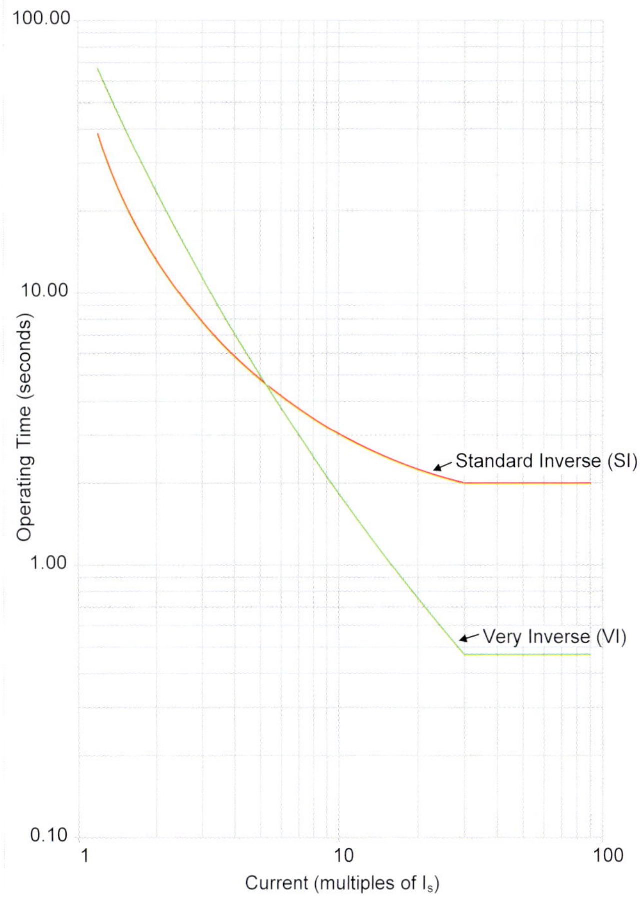

Figure 9.8: Comparison of SI and VI relay characteristics

9.7 EXTREMELY INVERSE (EI) OVERCURRENT RELAYS

With this characteristic, the operation time is approximately inversely proportional to the square of the applied current. This makes it suitable for the protection of distribution feeder circuits in which the feeder is subjected to peak currents on switching in, as would be the case on a power circuit supplying refrigerators, pumps, water heaters and so on, which remain connected even after a prolonged interruption of supply. The long time operating characteristic of the extremely inverse relay at normal peak load values of current also makes this relay particularly suitable for grading with fuses. Figure 9.9 shows typical curves. The EI characteristic gives a satisfactory grading margin, but the VI or SI characteristics at the same settings does not. Another application of this relay is in conjunction with auto-reclosers in low voltage distribution circuits. The majority of faults are transient in nature and unnecessary blowing and replacing of the fuses present in final circuits of such a system can be avoided if the auto-reclosers are set to operate before the fuse blows. If the fault persists, the auto-recloser locks itself in the closed position after one opening and the fuse blows to isolate the fault.

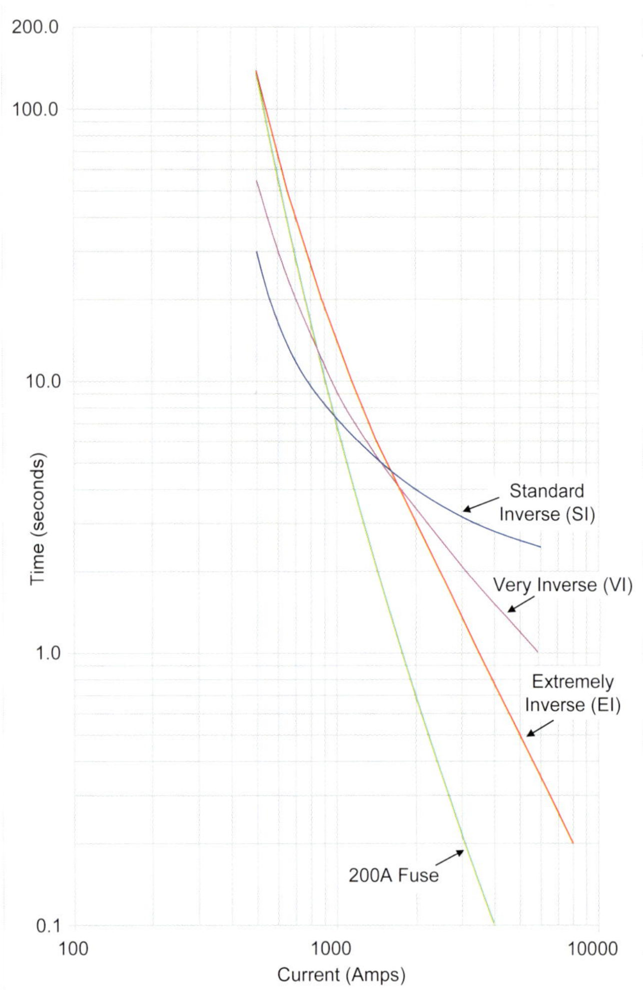

Figure 9.9: Comparison of relay and fuse characteristics

9.8 OTHER RELAY CHARACTERISTICS

User definable curves may be provided on some types of digital or numerical relays. The general principle is that the user enters a series of current/time co-ordinates that are stored in the memory of the relay. Interpolation between points is used to provide a smooth trip characteristic. Such a feature, if available, may be used in special cases if none of the standard tripping characteristics is suitable. However, grading of upstream protection may become more difficult, and it is necessary to ensure that the curve is properly documented, along with the reasons for use. Since the standard curves provided cover most cases with adequate tripping times, and most equipment is designed with standard protection curves in mind, the need to utilise this form of protection is relatively rare.

Digital and numerical relays may also include pre-defined logic schemes utilising digital (relay) I/O provided in the relay to implement standard schemes such as CB failure and trip circuit supervision. This saves the provision of separate relay hardware to perform these functions.

9.9 INDEPENDENT (DEFINITE) TIME OVERCURRENT RELAYS

Overcurrent relays are normally also provided with elements having independent or definite time characteristics. These characteristics provide a ready means of co-ordinating several relays in series in situations in which the system fault current varies very widely due to changes in source impedance, as there is no change in time with the variation of fault current. The time/current characteristics of this curve are shown in Figure 9.10, together with those of the standard IDMT characteristic, to indicate that lower operating times are achieved by the inverse relay at the higher values of fault current, whereas the definite time relay has lower operating times at the lower current values.

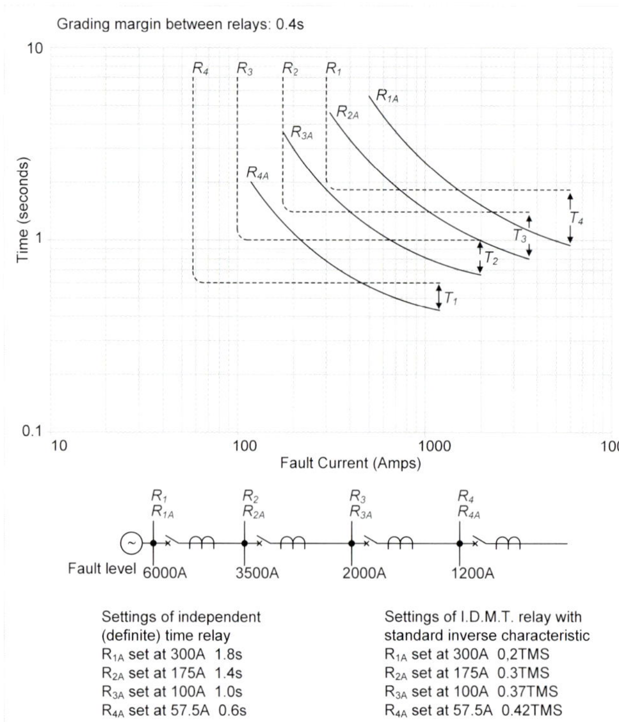

Figure 9.10: Comparison of definite time and standard IDMT relay

Vertical lines T_1, T_2, T_3, and T_4 indicate the reduction in operating times achieved by the inverse relay at high fault levels.

9.10 RELAY CURRENT SETTING

An overcurrent relay has a minimum operating current, known as the current setting of the relay. The current setting must be chosen so that the relay does not operate for the maximum load current in the circuit being protected, but does operate for a current equal or greater to the minimum expected fault current. Although by using a current setting that is only just above the maximum load current in the circuit a certain degree of protection against overloads as well as faults may be provided, the main function of overcurrent protection is to isolate primary system faults and not to provide overload protection. In general, the current setting will be selected to be above the maximum short time rated current of the circuit involved. Since all relays have hysteresis in their current settings, the setting must be sufficiently high to allow the relay to reset when the rated current of the circuit is being carried. The amount of hysteresis in the current setting is denoted by the pick-up/drop-off ratio of a relay – the value for a modern relay is typically 0.95. Thus, a relay minimum current setting of at least 1.05 times the short-time rated current of the circuit is likely to be required.

9.11 RELAY TIME GRADING MARGIN

The time interval that must be allowed between the operation of two adjacent relays to achieve correct discrimination between them is called the *grading margin*. If a grading margin is not provided, or is insufficient, more than one relay will operate for a fault, leading to difficulties in determining the location of the fault and unnecessary loss of supply to some consumers.

The grading margin depends on a number of factors:

1. the fault current interrupting time of the circuit breaker
2. relay timing errors
3. the overshoot time of the relay
4. CT errors
5. final margin on completion of operation

Factors (2) and (3) depend on the relay technology used. For example, an electromechanical relay has a larger overshoot time than a numerical relay.

Grading is initially carried out for the maximum fault level at the relaying point under consideration, but a check is also made that the required grading margin exists for all current levels between relay pick-up current and maximum fault level.

9.11.1 Circuit Breaker Interrupting Time

The circuit breaker interrupting the fault must have completely interrupted the current before the discriminating relay ceases to be energised. The time taken is dependent on the type of circuit breaker used and the fault current to be interrupted. Manufacturers normally provide the fault interrupting time at rated interrupting capacity and this value is invariably used in the calculation of grading margin.

9.11.2 Relay Timing Error

All relays have errors in their timing compared to the ideal characteristic as defined in IEC 60255. For a relay specified to IEC 60255, a relay error index is quoted that determines the maximum timing error of the relay. The timing error must be taken into account when determining the grading margin.

9.11.3 Overshoot

When the relay is de-energised, operation may continue for a little longer until any stored energy has been dissipated. For example, an induction disc relay will have stored kinetic energy in the motion of the disc; static relay circuits may have energy stored in capacitors. Relay design is directed to minimising and absorbing these energies, but some allowance is usually necessary.

The overshoot time is defined as the difference between the operating time of a relay at a specified value of input current and the maximum duration of input current, which when suddenly reduced below the relay operating level, is insufficient to cause relay operation.

9.11.4 CT Errors

Current transformers have phase and ratio errors due to the exciting current required to magnetise their cores. The result is that the CT secondary current is not an identical scaled replica of the primary current. This leads to errors in the operation of relays, especially in the operation time. CT errors are not relevant for independent definite-time delay overcurrent relays.

9.11.5 Final Margin

After allowances have been made for circuit breaker interrupting time, relay timing error, overshoot and CT errors, the discriminating relay must just fail to complete its operation. Some extra safety margin is required to ensure that relay operation does not occur.

9.11.6 Overall Accuracy

The overall limits of accuracy according to IEC 60255-4 for an IDMT relay with standard inverse characteristic are shown in Figure 9.11.

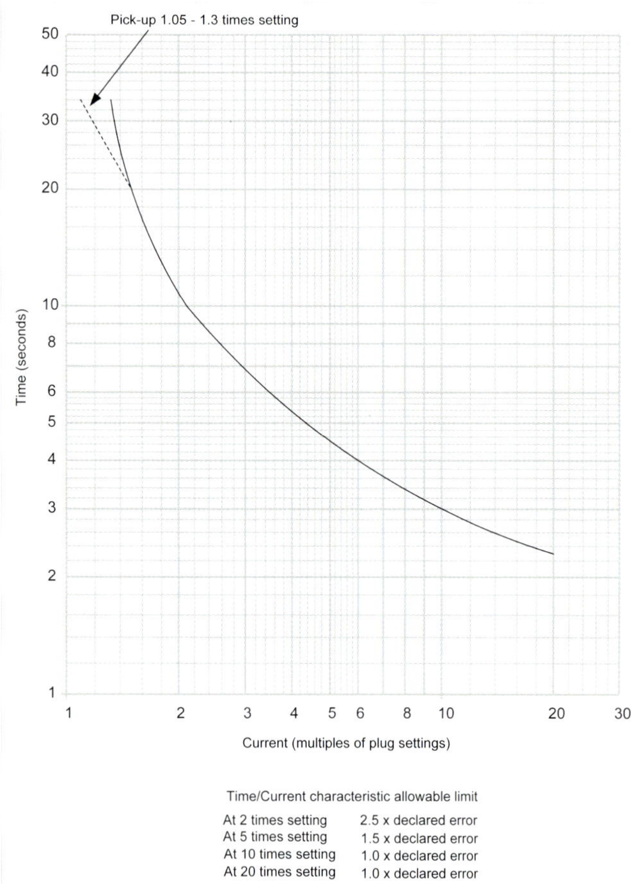

Figure 9.11: Typical limits of accuracy from IEC 60255-4 for an inverse definite minimum time overcurrent relay

9.12 RECOMMENDED GRADING MARGINS

The following sections give the recommended overall grading margins for between different protection devices.

9.12.1 Grading: Relay to Relay

The total interval required to cover circuit breaker interrupting time, relay timing error, overshoot and CT errors, depends on the operating speed of the circuit breakers and the relay performance. At one time 0.5s was a normal grading margin. With faster modern circuit breakers and a lower relay overshoot time, 0.4s is reasonable, while under the best conditions even lower intervals may be practical.

The use of a fixed grading margin is popular, but it may be better to calculate the required value for each relay location. This more precise margin comprises a fixed time, covering circuit breaker fault interrupting time, relay overshoot time and a safety margin, plus a variable time that allows for relay and CT errors. Table 9.3 gives typical relay errors according to the technology used.

It should be noted that use of a fixed grading margin is only appropriate at high fault levels that lead to short relay operating times. At lower fault current levels, with longer operating times, the permitted error specified in IEC 60255 (7.5% of operating time) may exceed the fixed grading margin, resulting in the possibility that the relay fails to grade correctly while remaining within specification. This requires consideration when considering the grading margin at low fault current levels.

A practical solution for determining the optimum grading margin is to assume that the relay nearer to the fault has a maximum possible timing error of +2E, where E is the basic timing error. To this total effective error for the relay, a further 10% should be added for the overall current transformer error.

	Relay Technology			
	Electro-mechanical	Static	Digital	Numerical
Typical basic timing error (%)	7.5	5	5	5
Overshoot time (s)	0.05	0.03	0.02	0.02
Safety margin (s)	0.1	0.05	0.03	0.03
Typical overall grading margin - relay to relay(s)	0.4	0.35	0.3	0.3

Table 9.3: Typical relay timing errors – standard IDMT relays

A suitable minimum grading time interval, t', may be calculated as follows:

$$t' = \left[\frac{2E_R + E_{CT}}{100}\right] t + t_{CB} + t_o + t_S \text{ seconds}$$

Equation 9.2

where:

E_R = relay timing error (IEC60255-4)
E_{CT} = allowance for CT ratio error (%)
t = nominal operating time of relay nearer to fault (sec)
t_{CB} = CB interrupting time (sec)
t_O = relay overshoot time (sec)
t_S = safety margin (sec)

if, for example *t=0.5s*, the time interval for an electromechanical relay tripping a conventional circuit breaker would be 0.375s, whereas, at the lower extreme, for a static relay tripping a vacuum circuit breaker, the interval could be as low as 0.25s.

When the overcurrent relays have independent definite time delay characteristics, it is not necessary to include the allowance for CT error. Hence:

$$t' = \left[\frac{2E_R}{100}\right] t + t_{CB} + t_o + t_S \text{ seconds}$$

Equation 9.3

Calculation of specific grading times for each relay can often be tedious when performing a protection grading calculation on a power system. Table 9.3 also gives practical grading times at high fault current levels between overcurrent relays for different technologies. Where relays of different technologies are used, the time appropriate to the technology of the downstream relay should be used.

9.12.2 Grading: Fuse to Fuse

The operating time of a fuse is a function of both the pre-arcing and arcing time of the fusing element, which follows an I^2t law. So, to achieve proper co-ordination between two fuses in series, it is necessary to ensure that the total I^2t taken by the smaller fuse is not greater than the pre-arcing I^2t value of the larger fuse. It has been established by tests that satisfactory grading between the two fuses will generally be achieved if the current rating ratio between them is greater than two.

9.12.3 Grading: Fuse to Relay

For grading inverse time relays with fuses, the basic approach is to ensure whenever possible that the relay backs up the fuse and not vice versa. If the fuse is upstream of the relay, it is very difficult to maintain correct discrimination at high values of fault current because of the fast operation of the fuse.

The relay characteristic best suited for this co-ordination with fuses is normally the extremely inverse (EI) characteristic as it follows a similar I^2t characteristic. To ensure satisfactory co-ordination between relay and fuse, the primary current setting of the relay should be approximately three times the current rating of the fuse. The grading margin for proper co-ordination, when expressed as a fixed quantity, should not be less than 0.4s or, when expressed as a variable quantity, should have a minimum value of:

$$t' = 0.4t + 0.15 \text{ seconds}$$

Equation 9.4

where t is the nominal operating time of the fuse.

Section 9.20.1 gives an example of fuse to relay grading.

9.13 CALCULATION OF PHASE FAULT OVERCURRENT RELAY SETTINGS

The correct co-ordination of overcurrent relays in a power system requires the calculation of the estimated relay settings in terms of both current and time. The resultant settings are then traditionally plotted in suitable log/log format to show pictorially that a suitable grading margin exists between the relays at adjacent substations. Plotting is usually done using suitable software although it can be done by hand.

The information required at each relaying point to allow a relay setting calculation to proceed is given in Section 9.2. The main relay data can be recorded in a table such as that shown in Table 9.4, populating the first five columns.

Location	Fault Current (A)		Maximum Load Current (A)	CT Ratio	Relay Current Setting		Relay Time Multiplier Setting
	Maximum	Minimum			Per Cent	Primary Current (A)	

Table 9.4: Typical relay data table

It is usual to plot all time/current characteristics to a common voltage/MVA base on log/log scales. The plot includes all relays in a single path, starting with the relay nearest the load and finishing with the relay nearest the source of supply. A separate plot is required for each independent path The settings of any relays that lie on multiple paths must be carefully considered to ensure that the final setting is appropriate for all conditions. Earth faults are considered separately from phase faults and require separate plots.

After relay settings have been finalised they are entered into a table such as that shown in Table 9.4, populating the last three columns. This also assists in record keeping during commissioning of the relays at site.

9.13.1 Independent (Definite) Time Relays

The selection of settings for independent (definite) time relays presents little difficulty. The overcurrent elements must be given settings that are lower, by a reasonable margin, than the fault current that is likely to flow to a fault at the remote end of the system up to which back-up protection is required, with the minimum plant in service. The settings must be high enough to avoid relay operation with the maximum probable load, a suitable margin being allowed for large motor starting currents or transformer inrush transients. Time settings will be chosen to allow suitable grading margins, as discussed in Section 9.12.

9.13.2 Inverse Time Relays

When the power system consists of a series of short sections of cable, so that the total line impedance is low, the value of fault current will be controlled principally by the impedance of transformers or other fixed plant and will not vary greatly with the location of the fault. In such cases, it may be possible to grade the inverse time relays in very much the same way as definite time relays. However, when the prospective fault current varies substantially with the location of the fault, it is possible to make use of this fact by employing both current and time grading to improve the overall performance of the relay.

The procedure begins by selection of the appropriate relay characteristics. Current settings are then chosen, with finally the time multiplier settings to give appropriate grading margins between relays. Otherwise, the procedure is similar to that for definite time delay relays. An example of a relay setting study is given in Section 9.20.1.

9.14 DIRECTIONAL PHASE FAULT OVERCURRENT RELAYS

When fault current can flow in both directions through the relay location, it may be necessary to make the response of the relay directional by the introduction of a directional control facility. The facility is provided by use of additional voltage inputs to the relay.

9.14.1 Relay Connections

There are many possibilities for a suitable connection of voltage and current inputs. The various connections are dependent on the phase angle, at unity system power factor, by which the current and voltage applied to the relay are displaced. Reference [9.1] details all of the connections that have been used. However, only very few are used in current practice and these are described below.

In a digital or numerical relay, the phase displacements are obtained by software, while electromechanical and static relays generally obtain the required phase displacements by connecting the input quantities to the relay. The history of the topic results in the relay connections being defined as if they were obtained by suitable connection of the input quantities, irrespective of the actual method used.

9.14.2 90° Relay Quadrature Connection

This is the standard connection for static, digital or numerical relays. Depending on the angle by which the applied voltage is shifted to produce maximum relay sensitivity (the Relay Characteristic Angle, or RCA), two types are available.

9.14.2.1 90°-30° Characteristic (30° RCA)

The *A* phase relay element is supplied with I_a current and V_{bc} voltage displaced by 30° in an anti-clockwise direction. In this case, the relay maximum sensitivity is produced when the current lags the system phase to neutral voltage by 60°. This connection gives a correct directional tripping zone over the current range of 30° leading to 150° lagging; see Figure 9.12. The relay sensitivity at unity power factor is 50% of the relay maximum sensitivity and 86.6% at zero power factor lagging. This characteristic is recommended when the relay is used for the protection of plain feeders with the zero sequence source behind the relaying point.

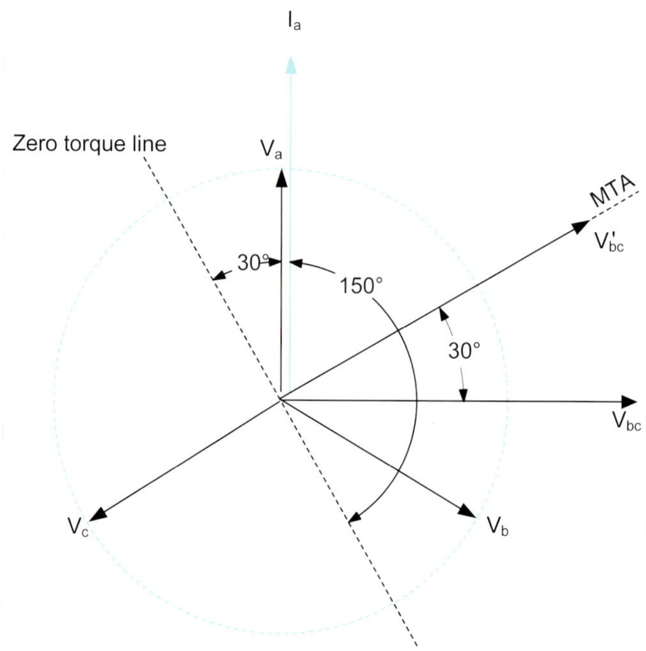

A phase element connected I_a V_{bc}
B phase element connected I_b V_{ca}
C phase element connected I_c V_{ab}

Figure 9.12: Vector diagram for the 90°-30° connection (phase A element)

9.14.2.2 90°-45° characteristic (45° RCA)

The *A* phase relay element is supplied with current I_a and voltage V_{bc} displaced by 45° in an anti-clockwise direction. The relay maximum sensitivity is produced when the current lags the system phase to neutral voltage by 45°. This connection gives a correct directional tripping zone over the current range of 45° leading to 135° lagging. The relay sensitivity at unity power factor is 70.7% of the maximum torque and the same at zero power factor lagging; see Figure 9.13.

This connection is recommended for the protection of transformer feeders or feeders that have a zero sequence source in front of the relay. It is essential in the case of parallel transformers or transformer feeders, to ensure correct relay operation for faults beyond the star/delta transformer.

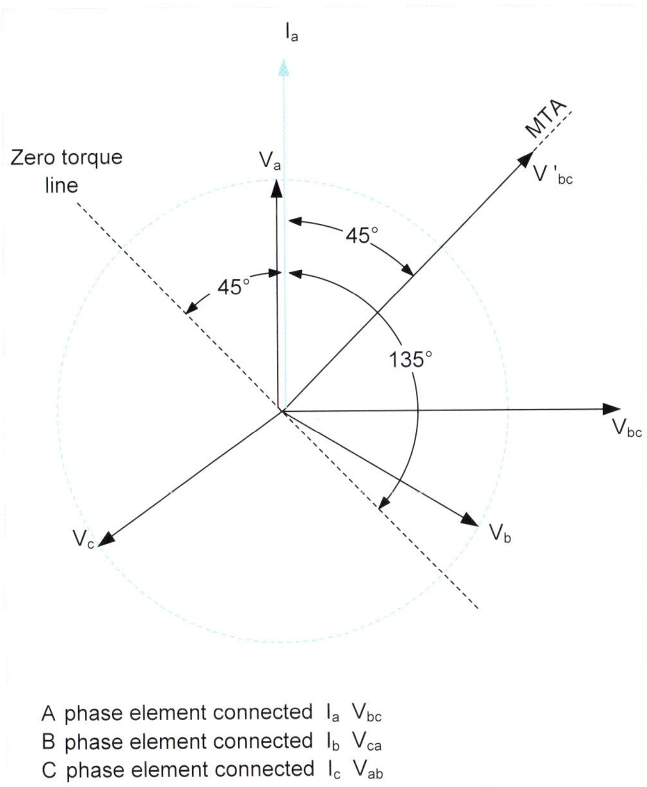

Figure 9.13: Vector diagram for the 90°-45° connection (phase A element)

For a digital or numerical relay, it is common to allow user-selection of the RCA within a wide range.

Theoretically, three fault conditions can cause maloperation of the directional element:

- a phase-phase-ground fault on a plain feeder
- a phase-ground fault on a transformer feeder with the zero sequence source in front of the relay
- a phase-phase fault on a power transformer with the relay looking into the delta winding of the transformer

These conditions are assumed to establish the maximum angular displacement between the current and voltage quantities at the relay. The magnitude of the current input to the relay is insufficient to cause the overcurrent element to operate. The possibility of maloperation with the 90^0-45^0 connection is non-existent.

9.14.3 Application of Directional Relays

If non-unit, non-directional relays are applied to parallel feeders having a single generating source, any faults that might occur on any one line will, regardless of the relay settings used, isolate both lines and completely disconnect the power supply. With this type of system configuration, it is necessary to apply directional relays at the receiving end and to grade them with the non-directional relays at the sending end, to ensure correct discriminative operation of the relays during line faults. This is done by setting the directional relays R'_1 and R'_2 in Figure 9.14 with their directional elements looking into the protected line, and giving them lower time and current settings than relays R_1 and R_2. The usual practice is to set relays R'_1 and R'_2 to 50% of the normal full load of the protected circuit and 0.1TMS, but care must be taken to ensure that the continuous thermal rating of the relays of twice rated current is not exceeded. An example calculation is given in Section 9.20.3

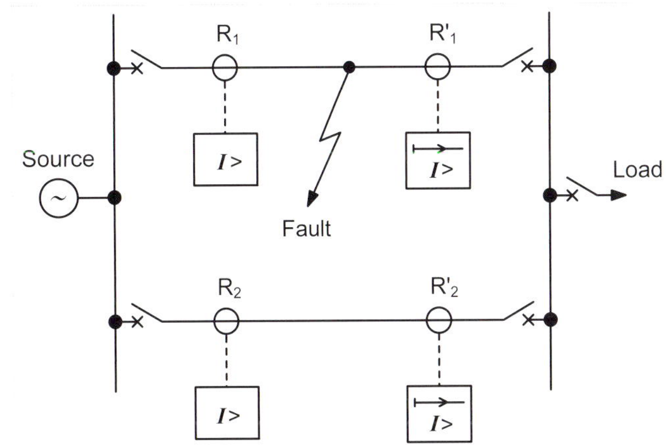

Figure 9.14: Directional relays applied to parallel feeders

9.15 RING MAINS

A particularly common arrangement within distribution networks is the Ring Main. The primary reason for its use is to maintain supplies to consumers in case of fault conditions occurring on the interconnecting feeders. A typical ring main with associated overcurrent protection is shown in Figure 9.15. Current may flow in either direction through the various relay locations, and therefore directional overcurrent relays are applied.

In the case of a ring main fed at one point only, the settings of the relays at the supply end and at the mid-point substation are identical. They can therefore be made non-directional, if, in the latter case, the relays are located on the same feeder, that is, one at each end of the feeder.

It is interesting to note that when the number of feeders round the ring is an even number, the two relays with the same operating time are at the same substation. They will therefore have to be directional. When the number of feeders is an odd number, the two relays with the same operating time are at different substations and therefore do not need to be directional. It may also be noted that, at intermediate substations, whenever the operating time of the relays at each substation are different, the difference between their operating times is never less than the grading margin, so the relay with the longer operating time can be non-directional. With modern numerical relays, a directional facility is often available for little or no extra cost, so that it may be simpler in practice to apply directional relays at all locations. Also, in the event of an additional feeder being added subsequently, the relays that can be non-directional need to be re-determined and will not necessarily be the same – giving rise to problems of changing a non-directional relay for a directional one. If a VT was not provided originally, this may be very difficult to install at a later date.

9.15.1 Grading of Ring Mains

The usual grading procedure for relays in a ring main circuit is to open the ring at the supply point and to grade the relays first clockwise and then anti-clockwise. That is, the relays looking in a clockwise direction around the ring are arranged to operate in the sequence 1-2-3-4-5-6 and the relays looking in the anti-clockwise direction are arranged to operate in the sequence 1'-2'-3'-4'-5'-6', as shown in Figure 9.15.

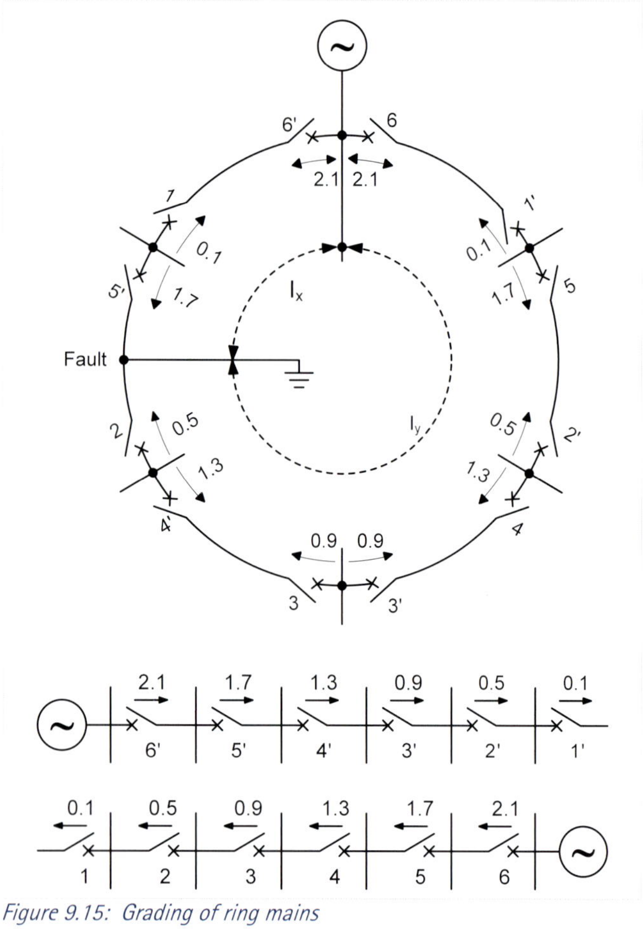

Figure 9.15: Grading of ring mains

The arrows associated with the relaying points indicate the direction of current flow that will cause the relay to operate. A double-headed arrow is used to indicate a non-directional relay, such as those at the supply point where the power can flow only in one direction. A single-headed arrow is used to indicate a directional relay, such as those at intermediate substations around the ring where the power can flow in either direction. The directional relays are set in accordance with the invariable rule, applicable to all forms of directional protection, that the current in the system must flow from the substation busbars into the protected line so the relays may operate.

Disconnection of the faulted line is carried out according to time and fault current direction. As in any parallel system, the fault current has two parallel paths and divides itself in the inverse ratio of their impedances. Thus, at each substation in the ring, one set of relays will be made inoperative because of the direction of current flow, and the other set operative. It will also be found that the operating times of the relays that are inoperative are faster than those of the operative relays, with the exception of the mid-point substation, where the operating times of relays 3 and 3' happen to be the same.

The relays that are operative are graded downwards towards the fault and the last to be affected by the fault operates first. This applies to both paths to the fault. Consequently, the faulted line is the only one to be disconnected from the ring and the power supply is maintained to all the substations.

When two or more power sources feed into a ring main, time graded overcurrent protection is difficult to apply and full discrimination may not be possible. With two sources of supply, two solutions are possible. The first is to open the ring at one of the supply points, whichever is more convenient, by means of a suitable high set instantaneous overcurrent relay. The ring is then graded as in the case of a single infeed. The second method is to treat the section of the ring between the two supply points as a continuous bus separate from the ring and to protect it with a unit protection system, and then proceed to grade the ring as in the case of a single infeed. Section 9.20.4 provides a worked example of ring main grading.

9.16 EARTH FAULT PROTECTION

In the foregoing description, attention has been principally directed towards phase fault overcurrent protection. More sensitive protection against earth faults can be obtained by using a relay that responds only to the residual current of the system, since a residual component exists only when fault current flows to earth. The earth fault relay is therefore completely unaffected by load currents, whether balanced or not, and can be given a setting which is limited only by the design of the equipment and the presence of unbalanced leakage or capacitance currents to earth. This is an important consideration if settings of only a few percent of system rating are considered, since leakage currents may produce a residual quantity of this order.

On the whole, the low settings permissible for earth fault relays are very useful, as earth faults are not only by far the most frequent of all faults, but may be limited in magnitude by the neutral earthing impedance, or by earth contact resistance.

The residual component is extracted by connecting the line current transformers in parallel as shown in Figure 9.16. The simple connection shown in Figure 9.16(a) can be extended by connecting overcurrent elements in the individual phase leads, as shown in Figure 9.16(b), and inserting the earth fault relay between the star points of the relay group and the current transformers.

Phase fault overcurrent relays are often provided on only two phases since these will detect any interphase fault; the connections to the earth fault relay are unaffected by this consideration. The arrangement is shown in Figure 9.16(c).

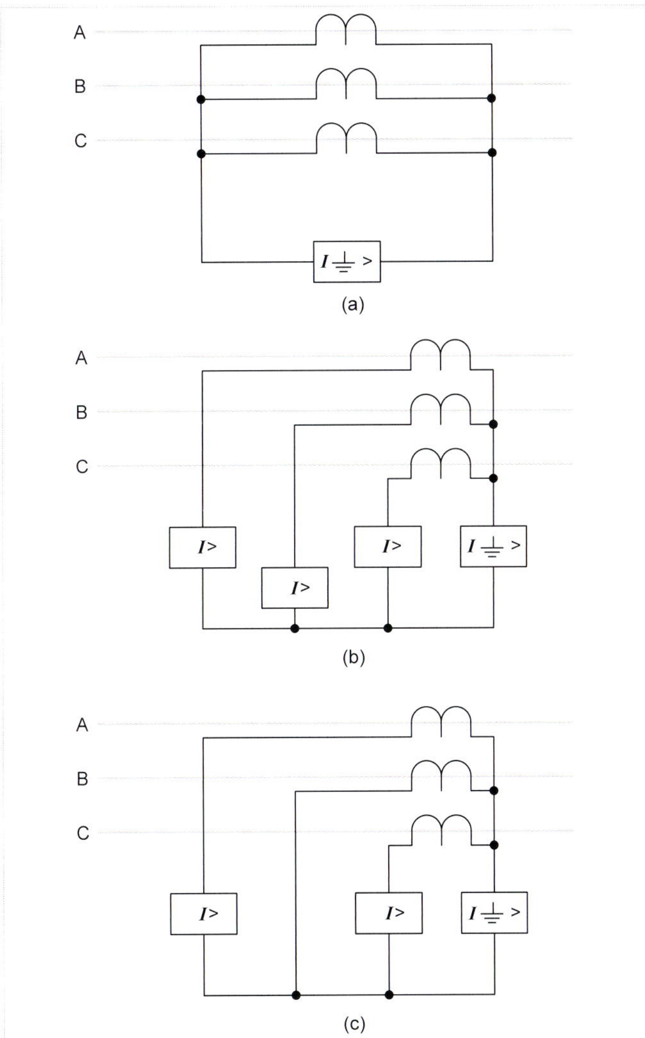

Figure 9.16: Residual connection of current transformers to earth fault relays

The typical settings for earth fault relays are 30%-40% of the full-load current or minimum earth fault current on the part of the system being protected. However, account may have to be taken of the variation of setting with relay burden as described in Section 9.16.1. If greater sensitivity than this is required, one of the methods described in Section 9.16.3 for obtaining sensitive earth fault protection must be used.

9.16.1 Effective Setting of Earth Fault Relays

The primary setting of an overcurrent relay can usually be taken as the relay setting multiplied by the CT ratio. The CT can be assumed to maintain a sufficiently accurate ratio so that, expressed as a percentage of rated current, the primary setting is directly proportional to the relay setting. However, this may not be true for an earth fault relay. The performance varies according to the relay technology used.

9.16.1.1 Static, digital and numerical relays

When static, digital or numerical relays are used the relatively low value and limited variation of the relay burden over the relay setting range results in the above statement holding true. The variation of input burden with current should be checked to ensure that the variation is sufficiently small. If not, substantial errors may occur, and the setting procedure will have to follow that for electromechanical relays.

9.16.1.2 Electromechanical Relays

When using an electromechanical relay, the earth fault element generally will be similar to the phase elements. It will have a similar VA consumption at setting, but will impose a far higher burden at nominal or rated current, because of its lower setting. For example, a relay with a setting of 20% will have an impedance of 25 times that of a similar element with a setting of 100%. Very frequently, this burden will exceed the rated burden of the current transformers. It might be thought that correspondingly larger current transformers should be used, but this is considered to be unnecessary. The current transformers that handle the phase burdens can operate the earth fault relay and the increased errors can easily be allowed for.

Not only is the exciting current of the energising current transformer proportionately high due to the large burden of the earth fault relay, but the voltage drop on this relay is impressed on the other current transformers of the paralleled group, whether they are carrying primary current or not. The total exciting current is therefore the product of the magnetising loss in one CT and the number of current transformers in parallel. The summated magnetising loss can be appreciable in comparison with the operating current of the relay, and in extreme cases where the setting current is low or the current transformers are of low performance, may even exceed the output to the relay. The 'effective setting current' in secondary terms is the sum of the relay setting current and the total excitation loss. Strictly speaking, the effective setting is the vector sum of the relay setting current and the total exciting current, but the arithmetic sum is near enough, because of the similarity of power factors. It is instructive to calculate the effective setting for a range of setting values of a relay, a process that is set out in Table 9.5, with the results shown in Figure 9.17.

The effect of the relatively high relay impedance and the summation of CT excitation losses in the residual circuit is augmented still further by the fact that, at setting, the flux density in the current transformers corresponds to the bottom bend of the excitation characteristic. The exciting impedance under this condition is relatively low, causing the ratio error to be high. The current transformer actually improves in performance with increased primary current, while the relay impedance decreases until, with an input current several times greater than the primary setting, the multiple of setting current in the relay is appreciably higher than the multiple of primary setting current which is applied to the primary circuit. This causes the relay operating time to be shorter than might be expected.

At still higher input currents, the CT performance falls off until finally the output current ceases to increase substantially. Beyond this value of input current, operation is further complicated by distortion of the output current waveform.

Relay Plug Setting		Coil voltage at Setting (V)	Exciting Current Ie	Effective Setting	
%	Current (A)			Current (A)	%
5	0.25	12	0.583	2	40
10	0.5	6	0.405	1.715	34.3
15	0.75	4	0.3	1.65	33
20	1	3	0.27	1.81	36
40	2	1.5	0.17	2.51	50
60	3	1	0.12	3.36	67
80	4	0.75	0.1	4.3	86
100	5	0.6	0.08	5.24	105

Table 9.5: Calculation of effective settings

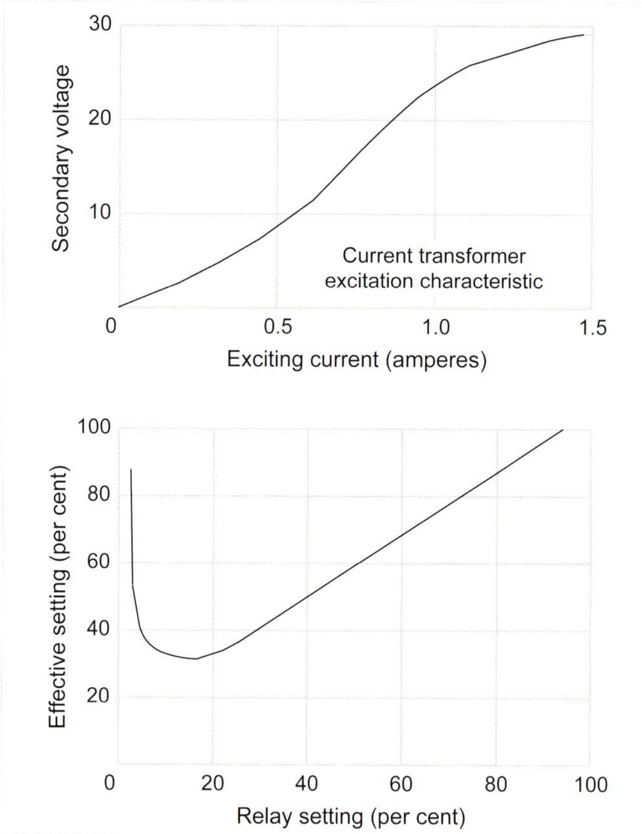

Figure 9.17: Effective setting of earth fault relay

9.16.1.3 Time Grading of Electromechanical Earth Fault Relays

The time grading of earth fault relays can be arranged in the same manner as for phase fault relays. The time/primary current characteristic for electromechanical relays cannot be kept proportionate to the relay characteristic with anything like the accuracy that is possible for phase fault relays. As shown above, the ratio error of the current transformers at relay setting current may be very high. It is clear that time grading of electromechanical earth fault relays is not such a simple matter as the procedure adopted for phase relays in Table 9.4. Either the above factors must be taken into account with the errors calculated for each current level, making the process much more tedious, or longer grading margins must be allowed. However, for other types of relay, the procedure adopted for phase fault relays can be used.

9.16.2 Sensitive Earth Fault Protection

LV systems are not normally earthed through an impedance, due to the resulting overvoltages that may occur and consequential safety implications. HV systems may be designed to accommodate such overvoltages, but not the majority of LV systems.

However, it is quite common to earth HV systems through an impedance that limits the earth fault current. Further, in some countries, the resistivity of the earth path may be very high due to the nature of the ground itself (e.g. desert or rock). A fault to earth not involving earth conductors may result in the flow of only a small current, insufficient to operate a normal protection system. A similar difficulty also arises in the case of broken line conductors, which, after falling on to hedges or dry metalled roads, remain energised because of the low leakage current, and therefore present a danger to life.

To overcome the problem, it is necessary to provide an earth fault protection system with a setting that is considerably lower than the normal line protection. This presents no difficulty to a modern digital or numerical relay. However, older electromechanical or static relays may present difficulties due to the high effective burden they may present to the CT.

The required sensitivity cannot normally be provided by means of conventional CTs. A core balance current transformer (CBCT) will normally be used. The CBCT is a current transformer mounted around all three phase (and neutral if present) conductors so that the CT secondary current is proportional to the residual (i.e. earth) current. Such a CT can be made to have any convenient ratio suitable for operating a sensitive earth fault relay element. By use of such techniques, earth fault settings down to 10% of the current rating of the circuit to be protected can be obtained.

Care must be taken to position a CBCT correctly in a cable circuit. If the cable sheath is earthed, the earth connection from the cable gland/sheath junction must be taken through the CBCT primary to ensure that phase-sheath faults are detected. Figure 9.18 shows the correct and incorrect methods. With the incorrect method, the fault current in the sheath is not seen as an unbalance current and hence relay operation does not occur.

The normal residual current that may flow during healthy conditions limits the application of non-directional sensitive earth fault protection. Such residual effects can occur due to unbalanced leakage or capacitance in the system.

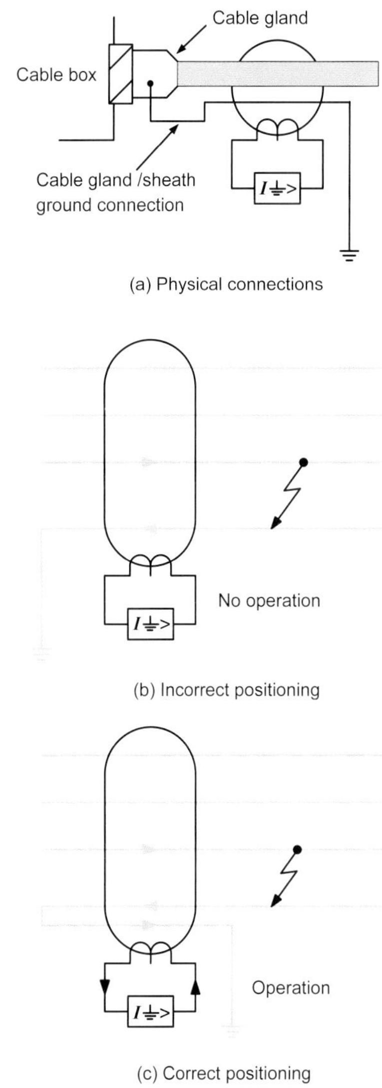

Figure 9.18: Positioning of core balance current transformers

9.17 DIRECTIONAL EARTH FAULT OVERCURRENT PROTECTION

Directional earth fault overcurrent may need to be applied in the following situations:

- for earth fault protection where the overcurrent protection is by directional relays
- in insulated-earth networks
- in Petersen coil earthed networks
- where the sensitivity of sensitive earth fault protection is insufficient – use of a directional earth fault relay may provide greater sensitivity

The relay elements previously described as phase fault elements respond to the flow of earth fault current, and it is important that their directional response be correct for this condition. If a special earth fault element is provided as described in Section 9.16 (which will normally be the case), a related directional element is needed.

9.17.1 Relay Connections

The residual current is extracted as shown in Figure 9.16. Since this current may be derived from any phase, to obtain a directional response it is necessary to obtain an appropriate quantity to polarise the relay. In digital or numerical relays there are usually two choices provided.

9.17.2 Residual Voltage

A suitable quantity is the residual voltage of the system. This is the vector sum of the individual phase voltages. If the secondary windings of a three-phase, five limb voltage transformer or three single-phase units are connected in broken delta, the voltage developed across its terminals will be the vector sum of the phase to ground voltages and hence the residual voltage of the system, as shown in Figure 9.19.

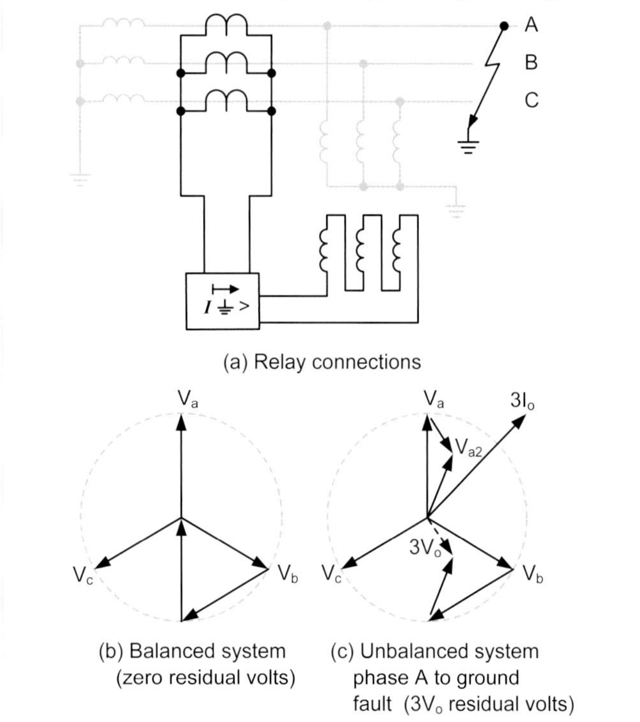

Figure 9.19: Voltage polarised directional earth fault relay

The primary star point of the VT must be earthed. However, a three-phase, three limb VT is not suitable, as there is no path for the residual magnetic flux.

When the main voltage transformer associated with the high voltage system is not provided with a broken delta secondary winding to polarise the directional earth fault relay, it is permissible to use three single-phase interposing voltage transformers. Their primary windings are connected in star and their secondary windings are connected in broken delta. For satisfactory operation, however, it is necessary to ensure that the main voltage transformers are of a suitable construction to reproduce the residual voltage and that the star point of the primary winding is solidly earthed. In addition, the star point of the primary windings of the interposing voltage transformers must be connected to the star point of the secondary windings of the main voltage transformers.

The residual voltage will be zero for balanced phase voltages. For simple earth fault conditions, it will be equal to the depression of the faulted phase voltage. In all cases the residual voltage is equal to three times the zero sequence voltage drop on the source impedance and is therefore displaced from the residual current by the characteristic angle of the source impedance. The residual quantities are applied to the directional element of the earth fault relay.

The residual current is phase offset from the residual voltage and hence angle adjustment is required. Typically, the current will lag the polarising voltage. The method of system earthing also affects the Relay Characteristic Angle (RCA), and the following settings are usual:

- Resistance-earthed system: 0° RCA
- Distribution system, solidly-earthed: -45° RCA
- Transmission system, solidly-earthed: -60° RCA

The different settings for distribution and transmission systems arise from the different X/R ratios found in these systems.

9.17.3 Negative Sequence Current

The residual voltage at any point in the system may be insufficient to polarise a directional relay, or the voltage transformers available may not satisfy the conditions for providing residual voltage. In these circumstances, negative sequence current can be used as the polarising quantity. The fault direction is determined by comparison of the negative sequence voltage with the negative sequence current. The RCA must be set based on the angle of the negative phase sequence source voltage.

9.18 EARTH FAULT PROTECTION ON INSULATED NETWORKS

Occasionally, a power system is run completely insulated from earth. The advantage of this is that a single phase-earth fault on the system does not cause any earth fault current to flow, and so the whole system remains operational. The system must be designed to withstand high transient and steady-state overvoltages however, so its use is generally restricted to low and medium voltage systems.

It is vital that detection of a single phase-earth fault is achieved, so that the fault can be traced and rectified. While system operation is unaffected for this condition, the occurrence of a second earth fault allows substantial currents to flow.

The absence of earth fault current for a single phase-earth fault clearly presents some difficulties in fault detection. Two methods are available using modern relays.

9.18.1 Residual Voltage

When a single phase-earth fault occurs, the healthy phase voltages rise by a factor of $\sqrt{3}$ and the three phase voltages no longer have a phasor sum of zero. Hence, a residual voltage element can be used to detect the fault. However, the method does not provide any discrimination, as the unbalanced voltage occurs on the whole of the affected section of the system. One advantage of this method is that no CTs are required, as voltage is being measured. However, the requirements for the VTs as given in Section 9.17.1.1 apply.

Grading is a problem with this method, since all relays in the affected section will see the fault. It may be possible to use definite-time grading, but in general, it is not possible to provide fully discriminative protection using this technique.

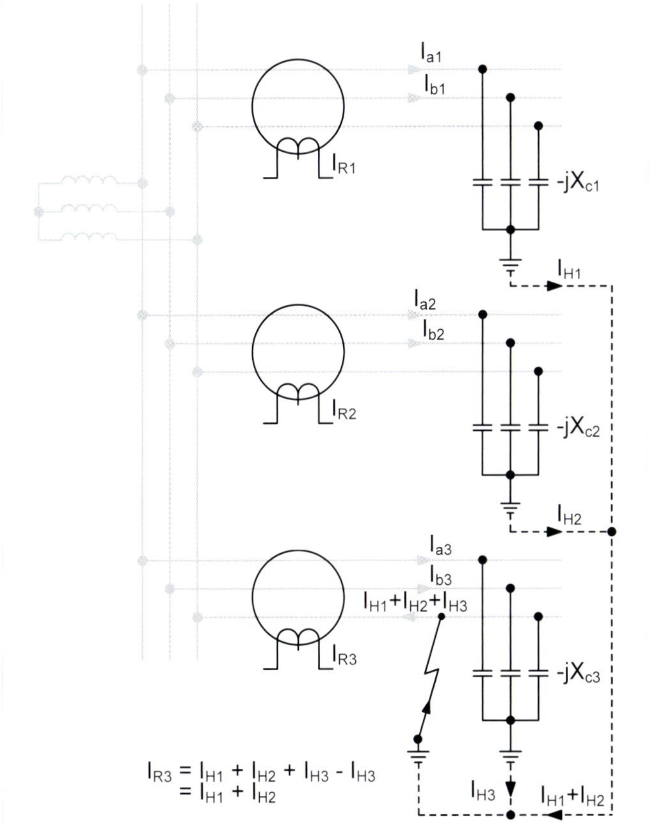

Figure 9.20: Current distribution in an insulated system with a C phase–earth fault

9.18.2 Sensitive Earth Fault

This method is principally applied to MV systems, as it relies on detection of the imbalance in the per-phase charging currents that occurs.

Figure 9.20 shows the situation that occurs when a single phase-earth fault is present. The relays on the healthy feeders see the unbalance in charging currents for their own feeders. The relay in the faulted feeder sees the charging currents in the rest of the system, with the current of its' own feeders cancelled out. Figure 9.21 shows the phasor diagram.

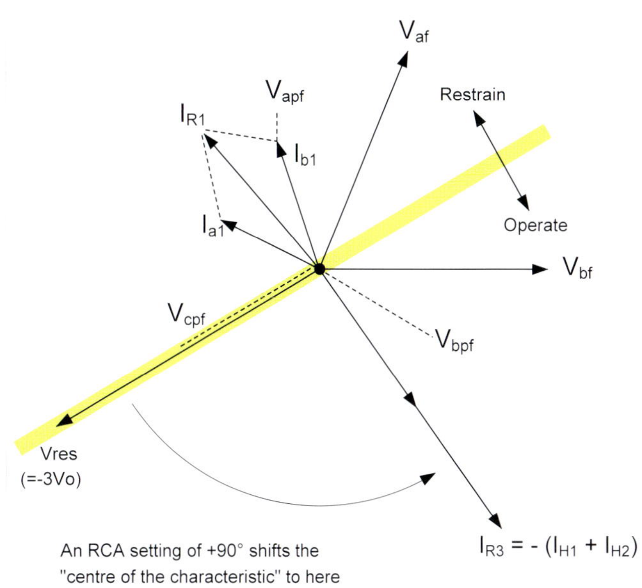

Figure 9.21: Phasor diagram for insulated system with C phase-earth fault

Use of Core Balance CTs is essential. With reference to Figure 9.21, the unbalance current on the healthy feeders lags the residual voltage by 90°

. The charging currents on these feeders will be √3 times the normal value, as the phase-earth voltages have risen by this amount. The magnitude of the residual current is therefore three times the steady-state charging current per phase. As the residual currents on the healthy and faulted feeders are in antiphase, use of a directional earth fault relay can provide the discrimination required.

The polarising quantity used is the residual voltage. By shifting this by 90°, the residual current seen by the relay on the faulted feeder lies within the 'operate' region of the directional characteristic, while the residual currents on the healthy feeders lie within the 'restrain' region. Thus, the RCA required is 90°. The relay setting has to lie between one and three times the per-phase charging current.

This may be calculated at the design stage, but confirmation by means of tests on-site is usual. A single phase-earth fault is deliberately applied and the resulting currents noted, a process made easier in a modern digital or numeric relay by the measurement facilities provided. As noted earlier, application of such a fault for a short period does not involve any disruption to the network, or fault currents, but the duration should be as short as possible to guard against a second such fault occurring.

It is also possible to dispense with the directional element if the relay can be set at a current value that lies between the charging current on the feeder to be protected and the charging current of the rest of the system.

9.19 EARTH FAULT PROTECTION ON PETERSEN COIL EARTHED NETWORKS

Petersen Coil earthing is a special case of high impedance earthing. The network is earthed via a reactor, whose reactance is made nominally equal to the total system capacitance to earth. Under this condition, a single phase-earth fault does not result in any earth fault current in steady-state conditions. The effect is therefore similar to having an insulated system. The effectiveness of the method is dependent on the accuracy of tuning of the reactance value – changes in system capacitance (due to system configuration changes for instance) require changes to the coil reactance. In practice, perfect matching of the coil reactance to the system capacitance is difficult to achieve, so that a small earth fault current will flow. Petersen Coil earthed systems are commonly found in areas where the system consists mainly of rural overhead lines, and are particularly beneficial in locations subject to a high incidence of transient faults.

To understand how to correctly apply earth fault protection to such systems, system behaviour under earth fault conditions must first be understood. Figure 9.22 shows a simple network earthed through a Petersen Coil. The equations clearly show that, if the reactor is correctly tuned, no earth fault current will flow.

Figure 9.23 shows a radial distribution system earthed using a Petersen Coil. One feeder has a phase-earth fault on phase C. Figure 9.24 shows the resulting phasor diagrams, assuming that no resistance is present. In Figure 9.24(a), it can be seen that the fault causes the healthy phase voltages to rise by a factor of √3 and the charging currents lead the voltages by 90°.

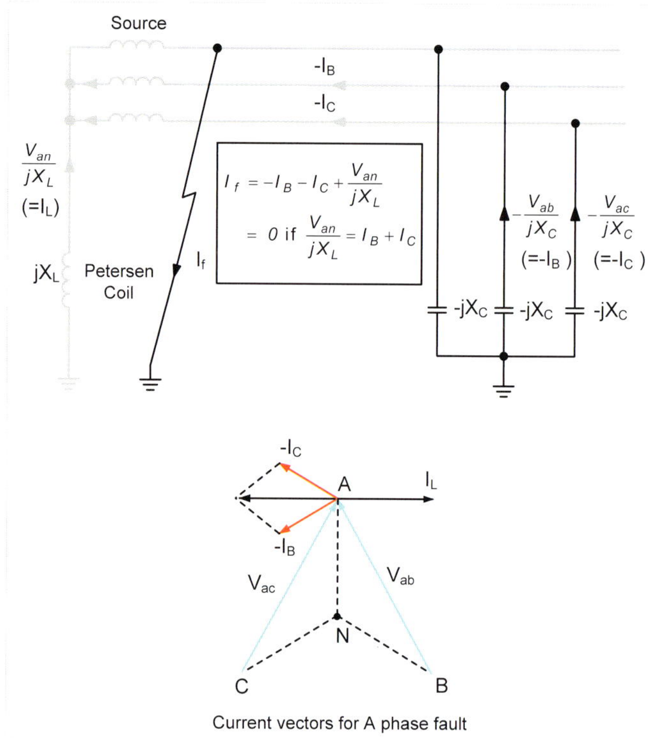

Figure 9.22: Earth fault in Petersen Coil earthed system

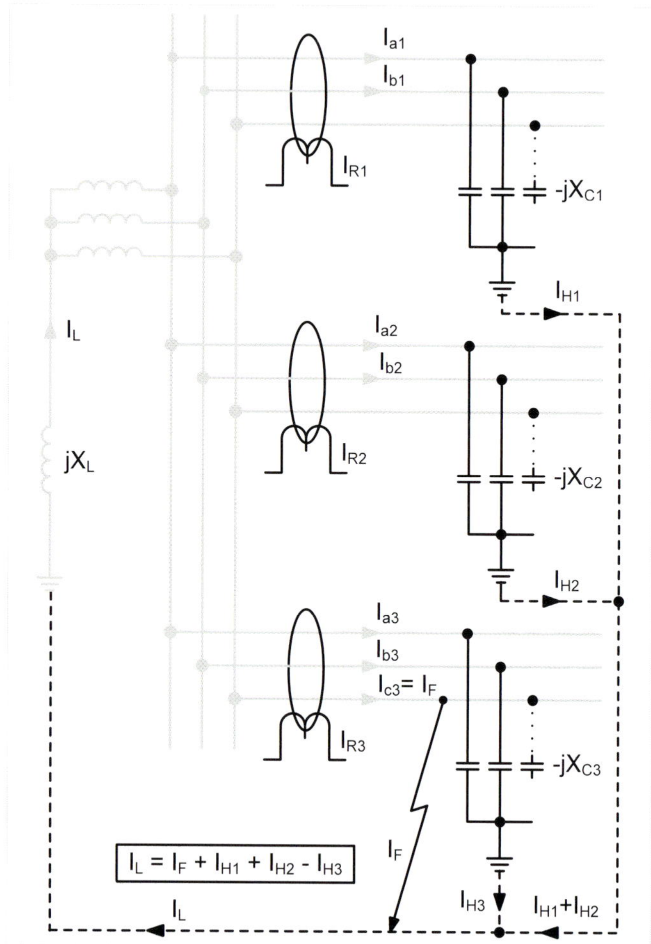

Figure 9.23: Distribution of currents during a C phase-earth fault – radial distribution system

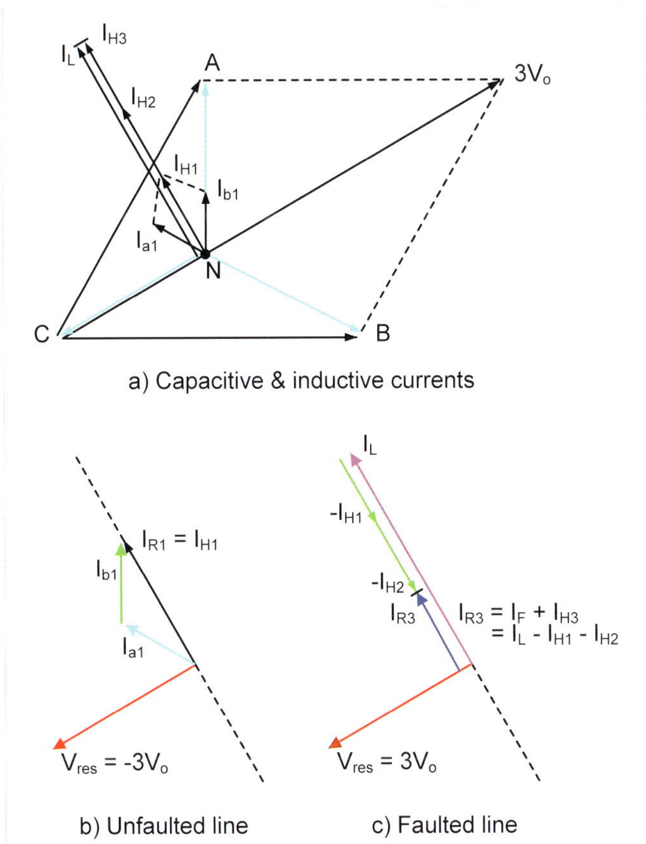

Figure 9.24: C phase-earth fault in Petersen Coil earthed network: theoretical case –no resistance present in X_L or X_C

Using a CBCT, the unbalance currents seen on the healthy feeders can be seen to be a simple vector addition of I_{a1} and I_{b1} and this lies at exactly 90° lagging to the residual voltage (Figure 9.24(b)). The magnitude of the residual current I_{R1} is equal to three times the steady-state charging current per phase. On the faulted feeder, the residual current is equal to $I_L - I_{H1} - I_{H2}$, as shown in Figure 9.24(c) and more clearly by the zero sequence network of Figure 9.25.

However, in practical cases, resistance is present and Figure 9.26 shows the resulting phasor diagrams. If the residual voltage V_{res} is used as the polarising voltage, the residual current is phase shifted by an angle less than 90° on the faulted feeder and greater than 90° on the healthy feeders. Hence a directional relay can be used, and with an RCA of 0°, the healthy feeder residual current will fall in the 'restrain' area of the relay characteristic while the faulted feeder residual current falls in the 'operate' area.

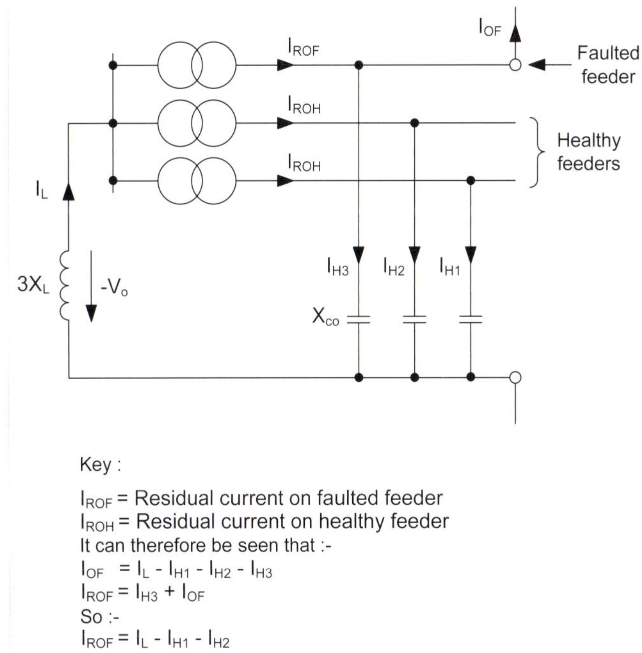

Figure 9.25: Zero sequence network showing residual currents

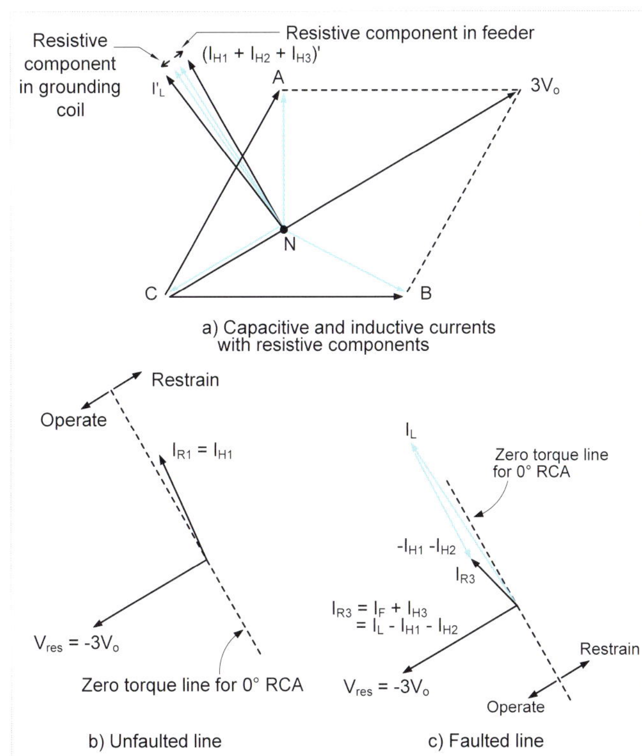

Figure 9.26: C phase-earth fault in Petersen Coil earthed network: practical case with resistance present in X_L or X_C

Often, a resistance is deliberately inserted in parallel with the Petersen Coil to ensure a measurable earth fault current and increase the angular difference between the residual signals to aid relay application.

Having established that a directional relay can be used, two possibilities exist for the type of protection element that can be applied – sensitive earth fault and zero sequence wattmetric.

9.19.1 Sensitive Earth Fault Protection

To apply this form of protection, the relay must meet two requirements:

- current measurement setting capable of being set to very low values
- an RCA of 0°, and capable of fine adjustment around this value

The sensitive current element is required because of the very low current that may flow – so settings of less than 0.5% of rated current may be required. However, as compensation by the Petersen Coil may not be perfect, low levels of steady-state earth fault current will flow and increase the residual current seen by the relay. An often used setting value is the per phase charging current of the circuit being protected. Fine tuning of the RCA is also required about the 0° setting, to compensate for coil and feeder resistances and the performance of the CT used. In practice, these adjustments are best carried out on site through deliberate application of faults and recording of the resulting currents.

9.19.2 Sensitive Wattmetric Protection

It can be seen in Figure 9.26 that a small angular difference exists between the spill current on the healthy and faulted feeders. Figure 9.27 shows how this angular difference gives rise to active components of current which are in antiphase to each other.

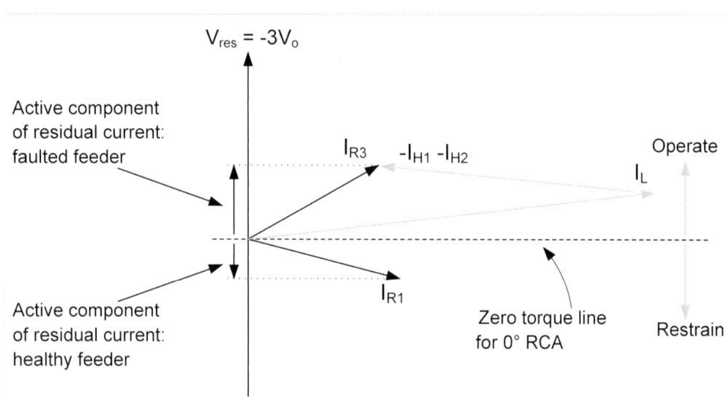

Figure 9.27: Resistive components of spill current

Consequently, the active components of zero sequence power will also lie in similar planes and a relay capable of detecting active power can make a discriminatory decision. If the wattmetric component of zero sequence power is detected in the forward direction, it indicates a fault on that feeder, while a power in the reverse direction indicates a fault elsewhere on the system. This method of protection is more popular than the sensitive earth fault method, and can provide greater security against false operation due to spurious CBCT output under non-earth fault conditions.

Wattmetric power is calculated in practice using residual quantities instead of zero sequence ones. The resulting values are therefore nine times the zero sequence quantities as the residual values of current and voltage are each three times the corresponding zero sequence values. The equation used is:

$$V_{res} \times I_{res} \times \cos(\varphi - \varphi_c)$$
$$= 9 \times V_O \times I_O \times \cos(\varphi - \varphi_c)$$

Equation 9.5

where:

V_{res} = residual voltage

I_{res} = residual current

V_O = zero sequence voltage

I_O = zero sequence current

φ = angle between V_{res} and I_{res}

φ_C = relay characteristic angle setting

The current and RCA settings are as for a sensitive earth fault relay.

9.20 EXAMPLES OF TIME AND CURRENT GRADING

This section provides details of the time/current grading of some example networks, to illustrate the process of relay setting calculations and relay grading. They are based on the use of a modern numerical overcurrent relay shown in Figure 9.28 with setting data taken from this relay.

Figure 9.28: MiCOM numerical overcurrent relay

9.20.1 Relay Phase Fault Setting Example – IDMT Relays/Fuses

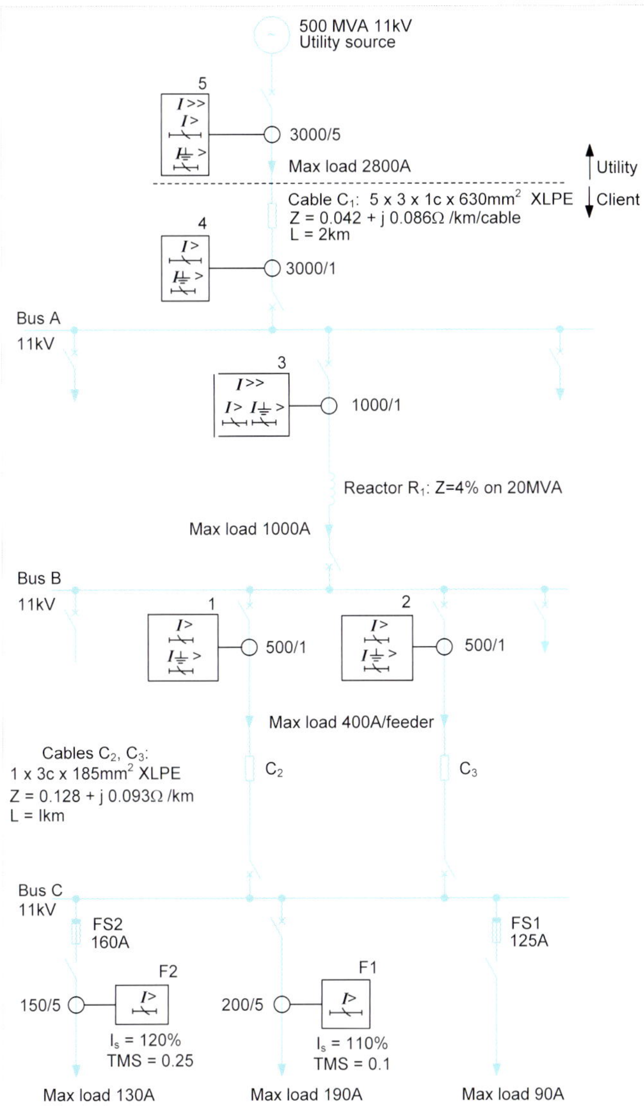

Figure 9.29: IDMT relay grading example

In the system shown in Figure 9.29., the problem is to calculate appropriate relay settings for relays *1-5* inclusive. Because the example is concerned with grading, considerations such as bus-zone protection and CT knee-point voltage requirements are not dealt with. All curves are plotted to an 11kV base. The contactors in series with fuses *FS1/FS2* have a maximum breaking capacity of 3kA, and relay *F2* has been set to ensure that the fuse operates before the contactor for currents in excess of this value. CTs for relays F1, *F2* and *5* are existing CTs with 5A secondaries, while the remaining CTs are new with 1A secondaries. Relay *5* is the property of the supply utility, and is required to be set using an SI characteristic to ensure grading with upstream relays.

9.20.1.1 Impedance Calculations

All impedances must first be referred to a common base, taken as 500MVA, as follows:

- Reactor R_1

$$Z_{R1} = \frac{4 \times 500}{20} = 100\%$$

- Cable C_1

$$Z_{C1} = \frac{0.096}{5} \times 2 = 0.038\Omega$$

On 500MVA base,

$$Z_{C1} = \frac{0.038 \times 100 \times 500}{11^2} = 15.7\%$$

- Cables C_2,C_3

$$Z_{C2}, Z_{C3} = 0.158\Omega$$

On 500MVA base,

$$Z_{C2}, Z_{C3} = \frac{0.158 \times 100 \times 500}{11^2} = 65.3\%$$

- Source Impedance (500MVA base)

$$Z_S = \frac{500}{500} \times 100\% = 100\%$$

9.20.1.2 Fault Levels

The fault levels are calculated as follows:

- (i) At bus C, For 2 feeders,

$$\text{Fault Level} = \frac{500 \times 100}{\dfrac{Z_{R1} + Z_S + Z_{C1} + Z_{C2}}{2}} MVA$$

= 201MVA = 10.6kA on 11kV base. For a single feeder, fault level = 178MVA = 9.33kA

- (ii) At bus B

$$\text{Fault Level} = \frac{500 \times 100}{Z_S + Z_{C1} + Z_{R1}} MVA$$

= 232MVA = 12.2kA

- (iii) At bus A

$$\text{Fault Level} = \frac{500 \times 100}{Z_S + Z_{C1}} MVA$$

= 432MVA = 22.7kA

- (iv) Source

Fault Level = 500MVA = 26.3kA

9.20.1.3 CT Ratio Selection

This requires consideration not only of the maximum load current, but also of the maximum secondary current under fault conditions.

CT secondaries are generally rated to carry a short-term current equal to 100 x rated secondary current. Therefore, a check is required that none of the new CT secondaries has a current of more than 100A when maximum fault current is flowing in the primary. Using the calculated fault currents, this condition is satisfied, so modifications to the CT ratios are not required.

9.20.1.4 Relay Overcurrent Settings – Relays 1/2

These relays perform overcurrent protection of the cable feeders, Busbar *C* and backup-protection to relays *F1*, *F2* and their associated fuses FS1 and *FS2*. The settings for Relays *1* and *2* will be identical, so calculations will only be performed for Relay *1*. Consider first the current setting of the relay.

Relay *1* must be able to reset at a current of 400A – the rating of the feeder. The relay has a drop-off/pick-up ratio of 0.95, so the relay current setting must not be less than 400/0.95A, or 421A. A suitable setting that is greater than this value is 450A. However, Section 9.12.3 also recommends that the current setting should be three times the largest fuse rating (i.e. 3 x 160A, the rating of the largest fuse on the outgoing circuits from Busbar *C*), leading to a current setting of 480A, or 96% of relay rated primary current. Note that in this application of relays to a distribution system, the question of maximum and minimum fault levels are probably not relevant as the difference between maximum and minimum fault levels will be very small. However in other applications where significant differences between maximum and minimum fault levels exist, it is essential to ensure that the selection of a current setting that is greater than full load current does not result in the relay failing to operate under minimum fault current conditions. Such a situation may arise for example in a self-contained power system with its own generation. Minimum generation may be represented by the presence of a single generator and the difference between minimum fault level and maximum load level may make the choice of relay current settings difficult.

The grading margin now has to be considered. For simplicity, a fixed grading margin of 0.3s between relays is used in the calculations, in accordance with Table 9.3. Between fuse and relay, Equation 9.4 is applied, and with fuse *FS2* pre-arcing time of 0.01s (from Figure 9.30), the grading margin is 0.154s.

Consider first the IDMT overcurrent protection. Select the EI characteristic, as fuses exist downstream, to ensure grading. The relay must discriminate with the longest operating time between relays *F1*, *F2* and fuse *FS2* (being the largest fuse) at the maximum fault level seen by relays *1* and *2*. The maximum fault current seen by relay *1* for a fault at Busbar *C* occurs when only one of cables C_2, C_3 is in service. This is because the whole of the fault current then flows through the feeder that is in service. With two feeders in service, although the fault level at Busbar *C* is higher, each relay only sees half of the total fault current, which is less than the fault current with a single feeder in service. With EI characteristics used for relays *F1* and *F2*, the operating time for relay *F1* is 0.02s at TMS=0.1 because the fault current is greater than 20 times relay setting, at which point the EI characteristic becomes definite time (Figure 9.4) and 0.05s for relay *F2* (TMS=0.25).

Hence select relay *1* operating time = 0.3+0.05=0.35s, to ensure grading with relay *F2* at a fault current of 9.33kA.

With a primary setting of 480A, a fault current of 9.33kA represents

9330/480 = 19.44 times setting

Thus relay *1* operating time at TMS=1.0 is 0.21s. The required TMS setting is given by the formula:

Operation Time Required / Actual Operation Time @ TMS=1

$$\therefore TMS = \frac{0.35}{0.21} = 1.66$$

If this value of TMS is outside the settable range of the relay (maximum setting was 1.2 in historical variants), changes must be made to the relay current setting. This is necessary to bring the value of TMS required into the range available, provided this does not result in the inability of the relay to operate at the minimum fault level.

By re-arrangement of the formula for the EI characteristic:

$$I_{srIf} = \sqrt{\left(\frac{80}{t} + 1\right)}$$

Where

t= the required operation time (sec)

I_{srIf}= setting of relay at fault current

Hence, with t= 0.35, I_{srIf}= 15.16

$$\text{Or, } I_{srI} = \frac{9330}{15.16} = 615.4A$$

$$\text{So, } I_{srI} = \frac{616}{500} = 1.232$$

Use 1.24 = 620A nearest available value

At a TMS of 1.0, operation time at 9330A

$$= \frac{80}{\left(\frac{9330}{620}\right)^2 - 1} = 0.355s$$

Hence, required TMS = 0.35/0.355 = 0.99, for convenience, use a TMS of 1.0, slightly greater than the required value.

From the grading curves of Figure 9.30, it can be seen that there are no grading problems with fuse *FS1* or relays *F1* and *F2*.

9.20.1.5 Relay Overcurrent Settings - Relay 3

This relay provides overcurrent protection for reactor R_1, and backup overcurrent protection for cables C_2 and C_3. The overcurrent protection also provides busbar protection for Busbar *B*.

Again, the EI characteristic is used to ensure grading with relays *1* and *2*. The maximum load current is 1000A. Relay *3* current setting is therefore:

$$I_{sr3} > \frac{feederFLC}{CTprimary \times 0.95}$$

Substituting values, $I_{sr3} > 1052A$

Use a setting of 106% or 1060A, nearest available setting above 1052A.

Relay *3* has to grade with relays *1/2* under two conditions:

- Condition 1: for a fault just beyond relays *1* and *2* where the fault current is the busbar fault current of 12.2kA
- Condition 2: for a fault at Bus *C* where the fault current seen by either relay *1* or *2* is half the total Bus *C* fault current of 10.6kA, i.e. 5.3kA

Examining first condition 1. With a current setting of 620A, a TMS of 1.0 and a fault current of 12.2kA, relay 1 operates in 0.21s. Using a grading interval of 0.3s, relay 3 must therefore operate in

$0.3+0.21=0.51s$ *at a fault current of 12.2kA*

12.2kA represents 12200/1060 = 11.51 times setting for relay *3* and thus the time multiplier setting of relay *3* should be 0.84 to give an operating time of 0.51s at 11.51 times setting.

Consider now condition 2. With settings of 620A and TMS of 1.0 and a fault current of 5.3kA, relay *1* operates in 1.11s. Using a grading interval of 0.3s, relay 3 must therefore operate in

$0.3+1.11=1.41s$ *at a fault current of 5.3kA*

5.3kA represents 5300/1060 = 5 times setting for relay *3*, and thus the time multiplier setting of relay *3* should be 0.33 to give an operating time of 1.11s at 5 times setting. Thus condition 1 represents the worst case and the time multiplier setting of relay *3* should be set at 0.84. In practice, a value of 0.85 is used as the nearest available setting on the relay.

Relay *3* also has an instantaneous element. This is set so that it does not operate for the maximum through-fault current seen by the relay and a setting of 130% of this value is satisfactory. The setting is therefore:

$1.3 \times 12.2kA = 15.86kA$

This is equal to a current setting of 14.96 times the setting of relay *3*.

9.20.1.6 Relay 4

This must grade with relay *3* and relay *5*. The supply authority requires that relay *5* use an SI characteristic to ensure grading with relays further upstream, so the SI characteristic will be used for relay 4 also. Relay 4 must grade with relay 3 at Bus A maximum fault level of 22.7kA. However with the use of an instantaneous high set element for relay 3, the actual grading point becomes the point at which the high set setting of relay 3 operates, i.e. 15.86kA. At this current, the operation time of relay 3 is

$$= \frac{80}{(14.96)^2 - 1} \times 0.85s = 0.305s$$

Thus, relay 4 required operating time is

0.305 + 0.3 = 0.605s at a fault level of 15.86kA.

Relay *4* current setting must be at least

$$= \frac{2800}{3000 \times 0.95} = 98\%$$

For convenience, use a value of 100% (=3000A). Thus relay *4* must operate in 0.605s at 15860/3000 = 5.29 times setting. Thus select a time multiplier setting of 0.15, giving a relay operating time of 0.62s for a normal inverse type characteristic.

At this stage, it is instructive to review the grading curves, which are shown in Figure 9.30(a). While it can be seen that there are no grading problems between the fuses and relays *1/2*, and between relays *F1/2* and relays *1/2*, it is clear that relay 3 and relay *4* do not grade over the whole range of fault current. This is a consequence of the change in characteristic for relay *4* to SI from the EI characteristic of relay *3* to ensure grading of relay *4* with relay *5*. The solution is to increase the TMS setting of relay *4* until correct grading is achieved. The alternative is to increase the current setting, but this is undesirable unless the limit of the TMS setting is reached, because the current setting should always be as low as possible to help ensure positive operation of the relay and provide overload protection. Trial and error is often used, but suitable software can speed the task – for instance it is not difficult to construct a spreadsheet with the fuse/relay operation times and grading margins calculated. Satisfactory grading can be found for relay *4* setting values of:

$I_{srl4} = 1.0 \text{ or } 3000A;\ TMS = 0.275$

At 22.7kA, the operation time of relay *4* is 0.93s. The revised grading curves are shown in Figure 9.30(b).

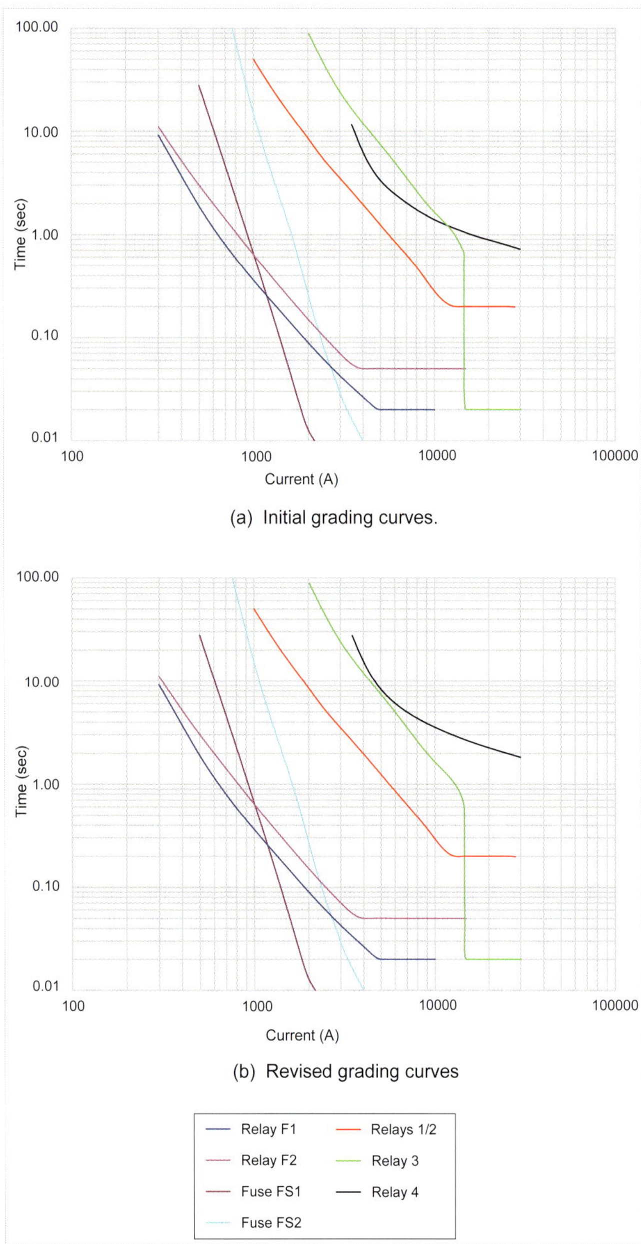

Figure 9.30: Overcurrent grading exercise - initial relay grading curves

9.20.1.7 Relay 5

Relay *5* must grade with relay *4* at a fault current of 22.7kA. At this fault current, relay *4* operates in 0.93s and thus relay *5* must operate in

$$0.3 + 0.93 = 1.23s \text{ at } 22.7kA$$

A current setting of 110% of relay *4* current setting (i.e. 110% or 3300A) is chosen to ensure relay *4* picks up prior to relay *5*. Thus 22.7kA represents 6.88 times the setting of relay *5*. Relay *5* must grade with relay *4* at a fault current of 22.7kA, where the required operation time is 1.23s. At a TMS of 1.0, relay *5* operation time is

$$= \frac{0.14}{(6.88)^{0.02} - 1} = 03.56s$$

Therefore, the required TMS is 1.23/3.56 = 0.345, use 0.35 nearest available value.

The protection grading curves that result are shown in Figure 9.31 and the setting values in Table 9.6. Grading is now satisfactory.

In situations where one of the relays to be graded is provided by a third party, it is common for the settings of the relay to be specified and this may lead to a lack of co-ordination between this relay and others (usually those downstream). Negotiation is then required to try and achieve acceptable settings, but it is often the case that no change to the settings of the relay provided by the third party is allowed. A lack of co-ordination between relays then has to be accepted over at least part of the range of fault currents.

Relay or Fuse	Relay Settings							
	Load current (A)	Max Fault Current kA	CT Ratio	Fuse Rating	Characteristic	Current Setting Primary Amps	Current Setting Per Cent	TMS
F1	190	10.6	200/5		EI	100	100	0.1
F2	130	10.6	150/5		EI	150	120	0.25
FS1	90	10.6	-	125A				
FS2	130	10.6	-	160A		-	-	-
1	400	12.2	500/1		EI	620	124	1
2	400	12.2	500/1		EI	620	124	1
3	1000	22.7	1000/1		EI	1060	106	0.85
					Instantaneous	15860	14.96	-
4	3000	22.7	3000/1		SI	3000	100	0.275
5	3000	26.25	3000/5		SI	3300	110	0.35

Table 9.6: Relay settings for overcurrent relay example

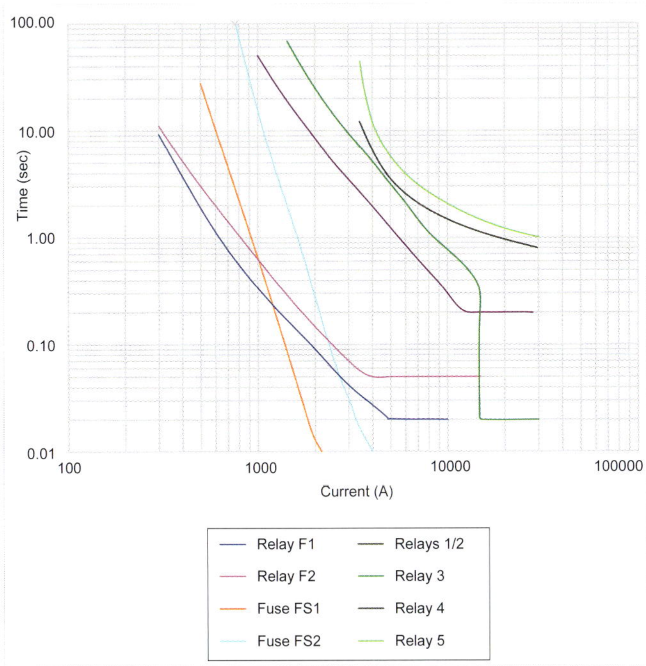

Figure 9.31: Final relay grading curves for overcurrent relay example

9.20.2 Relay Earth Fault Settings

The procedure for setting the earth fault elements is identical to that for the overcurrent elements, except that zero sequence impedances must be used if available and different from positive sequence impedances to calculate fault levels. However, such impedances are frequently not available, or known only approximately and the phase fault current levels have to be used. Note that earth fault levels can be higher than phase fault levels if the system contains multiple earth points or if earth fault levels are considered on the star side of a delta-star transformer when the star winding is solidly earthed.

On the circuit with fuse *F2*, low-level earth faults may not be of sufficient magnitude to blow the fuse. Attempting to grade the earth fault element of the upstream relay with fuse *F2* will not be possible. Similarly, relays *F1* and *F2* have phase fault settings that do not provide effective protection against earth faults. The remedy would be to modify the downstream protection, but such considerations lie outside the scope of this example. In general therefore, the earth fault elements of relays upstream of circuits with only phase fault protection (i.e. relays with only phase fault elements or fuses) will have to be set with a compromise that they will detect downstream earth faults but will not provide a discriminative trip. This illustrates the practical point that it is rare in anything other than a very simple network to achieve satisfactory grading for all faults throughout the network.

In the example of Figure 9.29, the difference in fault levels between phase to phase and phase to earth faults is probably very small so the only function of earth fault elements is to detect and isolate low level earth faults not seen by the phase fault elements. Following the guidelines of Section 9.16, relays *1/2* can use a current setting of 30% (150A) and a TMS of 0.2, using the EI characteristic. Grading of relays *3/4/5* follows the same procedure as described for the phase-fault elements of these relays.

9.20.3 Protection of Parallel Feeders

Figure 9.32(a) shows two parallel transformer feeders forming part of a supply circuit. Impedances are as shown in the diagram. The example shows that unless relays 2 and 3 are made directional, they maloperate for a fault at *F3*. Also shown is how to calculate appropriate relay settings for all six relays to ensure satisfactory protection for faults at locations *F1-F4*.

Figure 9.32(b) shows the impedance diagram, to 100MVA, 110kV base. The fault currents for faults with various system configurations are shown in Table 9.7.

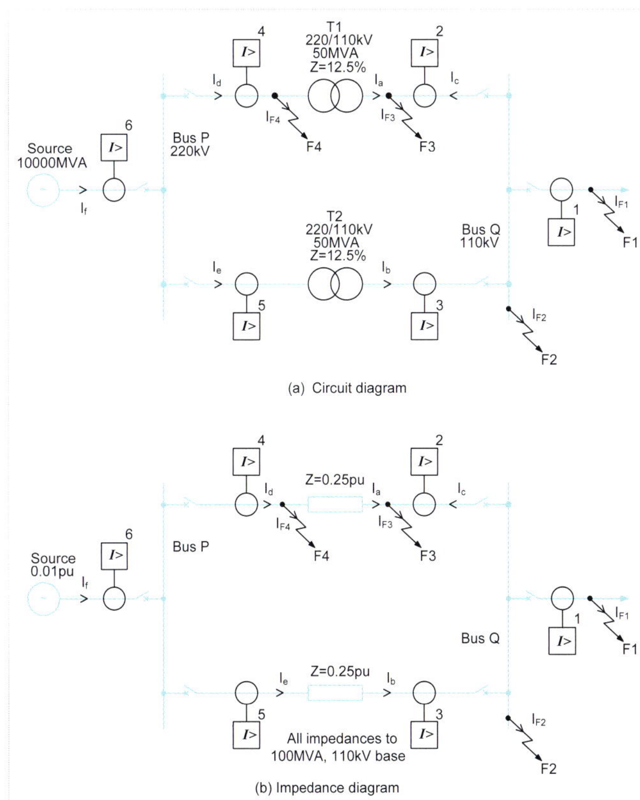

Figure 9.32: System diagram: Parallel feeder example

Fault Position	System Configuration	Currents (A)						
		Fault	Ia	Ib	Ic	Id	Ie	If
F1	2 fdrs	3888	1944	1944	0	972	972	1944
F1/F2	1 fdr	2019	2019	0	0	1009	0	1009
F2	2 fdrs	3888	1944	1944	0	972	972	1944
F3	2 fdrs	3888	1944	1944	1944	972	972	1944
F4	1 fdr	26243	0	0	0	26243	0	26243

Table 9.7: Fault currents for parallel feeder example

If relays 2 and 3 are non-directional, then, using SI relay characteristics for all relays, grading of the relays is dictated by the following:

- fault at location *F1*, with 2 feeders in service
- fault at location *F4*, with one feeder in service

The settings shown in Figure 9.33(a) can be arrived at, with the relay operation times shown in Figure 9.33 (b). It is clear that for a fault at *F3* with both transformer feeders in service, relay *3* operates at the same time as relay *2* and results in total disconnection of Bus *Q* and all consumers supplied solely from it. This is undesirable – the advantages of duplicated 100% rated transformers have been lost.

Relay	CT Primary (A)	Current Setting	TMS	Characteristic
1	300	1	0.2	SI
2	300	1.1	0.3	SI
3	300	1.1	0.3	SI
4	300	0.61	0.425	SI
5	300	0.61	0.425	SI
6	300	0.7	0.7	SI

(a) Relay settings - non-directional relays

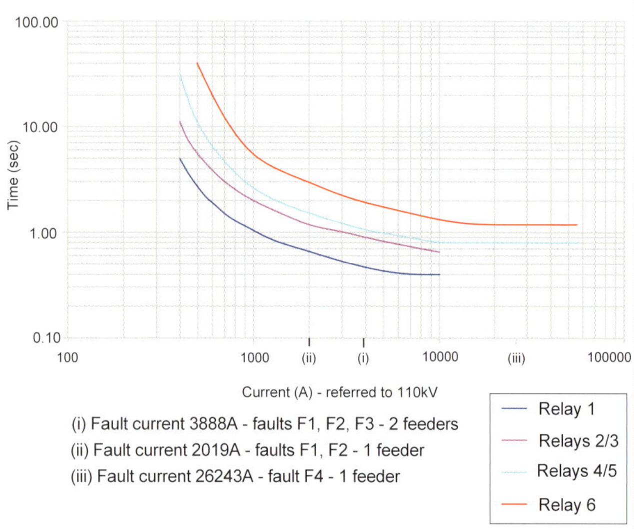

(b) Relay grading curves - non-directional relays

Figure 9.33: Relay grading for parallel feeder example – non-directional relays

By making relays *2* and *3* directional as shown in Figure 9.34 (a), lower settings for these relays can be adopted – they can be set as low as reasonably practical but normally a current setting of about 50% of feeder full load current is used, with a TMS of 0.1. Grading rules can be established as follows:

- relay *4* is graded with relay *1* for faults at location *F1* with one transformer feeder in service
- relay *4* is graded with relay *3* for faults at location *F3* with two transformer feeders in service
- relay *6* grades with relay *4* for faults at F4
- relay *6* also has to grade with relay *4* for faults at *F1* with both transformer feeders in service – relay 6 sees the total fault current but relay *4* only 50% of this current

Normal rules about calculating current setting values of relays in series apply. The settings and resulting operation times are given in Figure 9.34(b) and(c) respectively.

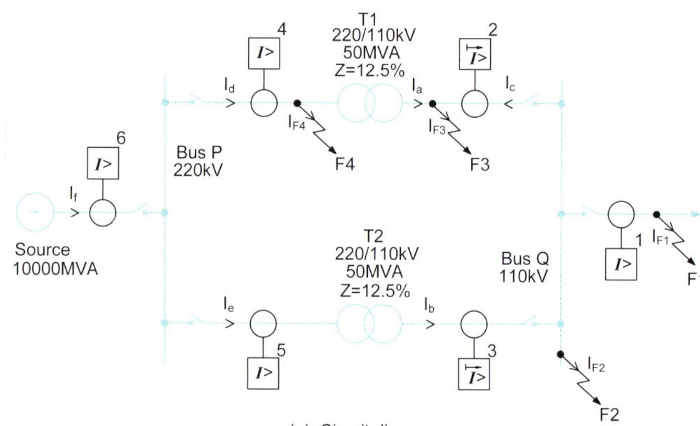

(a) Circuit diagram

Relay	CT Primary	Current Setting	TMS	Characteristic
1	300	1	0.2	SI
2	300	0.42	0.1	SI
3	300	0.42	0.1	SI
4	300	0.6	0.275	SI
5	300	0.6	0.275	SI
6	300	0.7	0.475	SI

(b) Relay settings

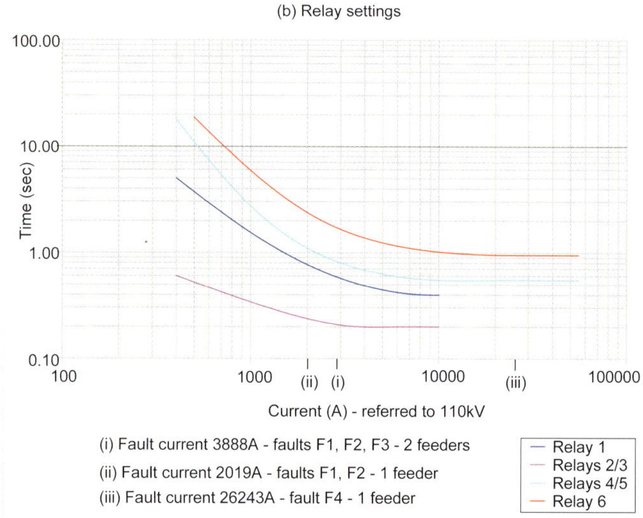

(c) Relay characteristics

Figure 9.34: Relay grading for parallel feeder example – directional relays

In practice, a complete protection study would include instantaneous elements on the primary side of the transformers and analysis of the situation with only one transformer in service. These have been omitted from this example, as the purpose is to illustrate the principles of parallel feeder protection in a simple fashion.

9.20.4 Grading of a Ring Main

Figure 9.35 shows a simple ring main, with a single infeed at Bus *A* and three load busbars. Settings for the directional relays *R2-R7* and non-directional relays *R1/R8* are required. Maximum load current in the ring is 785A (maximum continuous current with one transformer out of service), so 1000/1A CTs are chosen. The relay considered is a MiCOM P140 series.

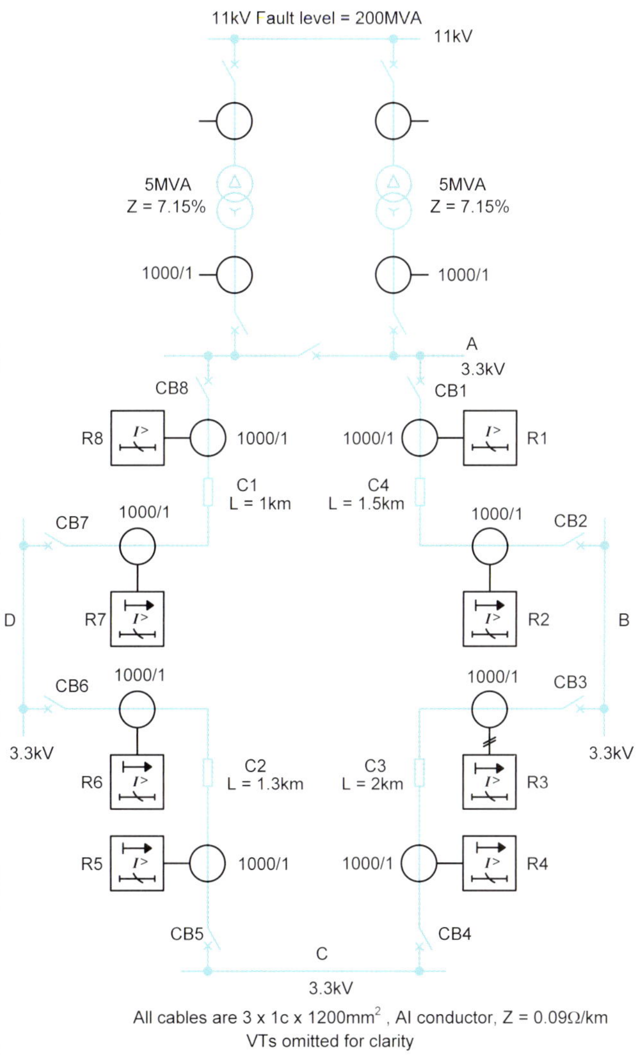

Figure 9.35: Ring main grading example – circuit diagram

The first step is to establish the maximum fault current at each relay location. Assuming a fault at Bus *B* (the actual location is not important), two possible configurations of the ring have to be considered, firstly a closed ring and secondly an open ring. For convenience, the ring will be considered to be open at *CB1* (*CB8* is the other possibility to be considered, but the conclusion will be the same).

Figure 9.36 shows the impedance diagram for these two cases. Three-phase fault currents I_1 and I_1' can be calculated as 2.13kA and 3.67kA respectively, so that the worst case is with the ring open (this can also be seen from consideration of the impedance relationships, without the necessity of performing the calculation). Table 9.8 shows the fault currents at each bus for open points at *CB1* and *CB8*.

For grading of the relays, consider relays looking in a clockwise direction round the ring, i.e. relays *R1/R3/R5/R7*.

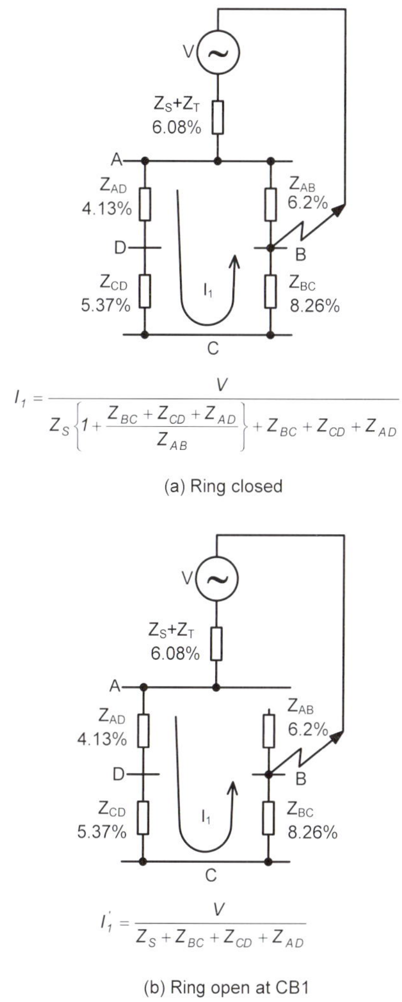

Figure 9.36: Impedance diagrams with ring open

Clockwise		Anticlockwise	
Open Point CB8		Open Point CB1	
Bus	Fault Current kA	Bus	Fault Current kA
D	7.124	B	3.665
C	4.259	C	5.615
B	3.376	D	8.568

Table 9.8: Fault current tabulation with ring open

9.20.4.1 Relay R7

Load current cannot flow from Bus *D* to Bus *A* since Bus *A* is the only source. Hence low relay current and TMS settings can be chosen to ensure a rapid fault clearance time. These can be chosen arbitrarily, so long as they are above the cable charging current and within the relay setting characteristics. Select a relay current setting of 0.8 (i.e. 800A CT primary current) and TMS of 0.05. This ensures that the other relays will not pick up under conditions of normal load current. At a fault current of 3376A, relay operating time on the SI characteristic is:

$$0.05 \times \left[\frac{0.14}{(4.22)^{0.02} - 1} \right] s = 0.24s$$

9.20.4.2 Relay R5

This relay must grade with Relay *R7* at 3376A and have a minimum operation time of 0.54s. The current setting for Relay *R5* must be at least 110% that of relay *R7* to prevent unwanted pickup. Therefore select relay *R5* current setting of 0.88 (880A CT primary current).

Relay *R5* operating time at TMS = 1.0

$$\left[\frac{0.14}{(3.84)^{0.02} - 1} \right] s = 5.14s$$

Hence, relay *R5* TMS = 0.54/5.14 = 0.105

Let's assume that we have, for example, an earlier version of the relay whose available TMS settings were less granular in steps of 0.025. In this case use the nearest settable value of TMS of 0.125. Table 9.9 summarises the relay settings while Figure 9.37 and Figure 9.38 show the relay grading curves.

Bus	Relay	Relay Characteristic	CT Ratio	Max Load Current (A)	Max Fault Current (A) (3.3kV base)	Current Setting p.u.	TMS
D	R7	SI	1000/1	874	3376	0.8	0.05
C	R5	SI	1000/1	874	4259	0.88	0.125
B	R3	SI	1000/1	874	7124	0.97	0.2
A	R1	SI	1000/1	874	14387	1.07	0.275
A	R8	SI	1000/1	874	14387	1.07	0.3
D	R6	SI	1000/1	874	8568	0.97	0.2
C	R4	SI	1000/1	874	5615	0.88	0.125
B	R2	SI	1000/1	874	3665	0.8	0.05

Table 9.9: Ring main example relay settings

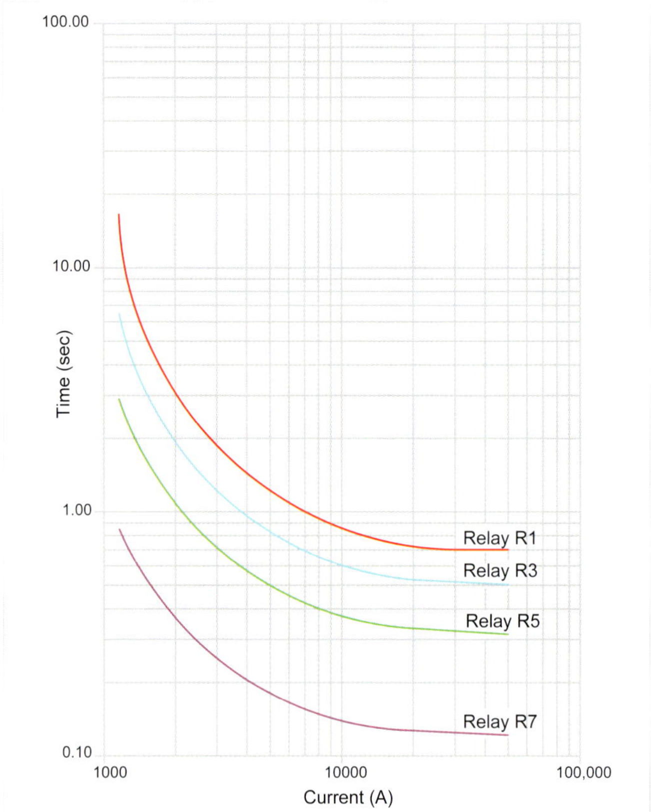

Figure 9.37: Ring main example – relay grading curves. Clockwise grading of relays (ring open at CB8)

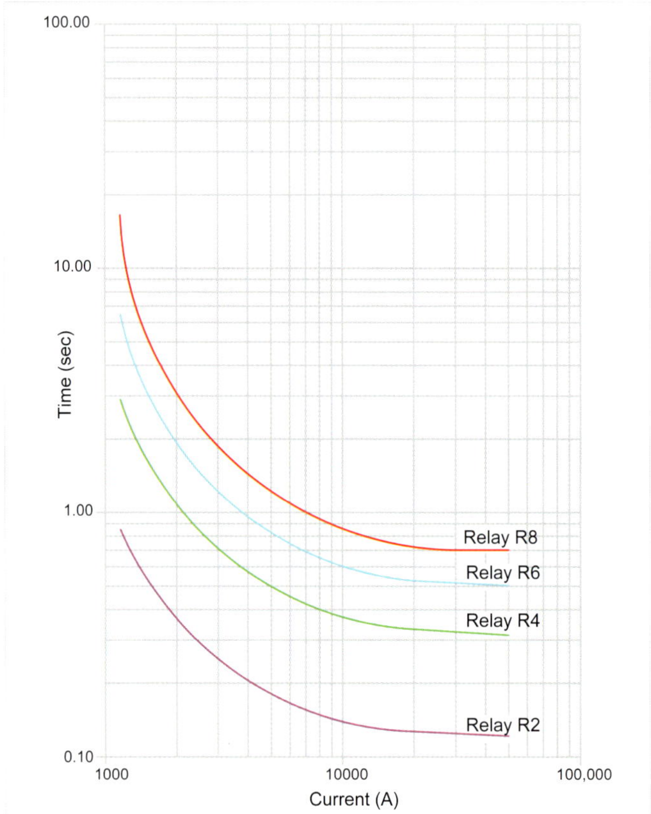

Figure 9.38: Ring main example – relay grading curves. Anticlockwise grading of relays (ring open at CB1)

9.21 HI-Z - HIGH IMPEDANCE DOWNED CONDUCTOR PROTECTION

High impedance ("Hi-Z") faults are generally defined as the unwanted contact of an electrical conductor with a non-conductive surface like asphalt road, concrete, tree limbs, sand, wooden fences or some other surface or object which restricts the fault current to a level below that which can be reliably detected by conventional overcurrent and earth fault relays. In some cases even sensitive earth fault protection cannot reliably detect such low levels of fault current flow. Undetected high impedance faults such as downed conductors are dangerous for nearby staff, the public, and livestock. The primary objective of clearing such faults is therefore towards protection of human and animal life, and property, and not towards the integrity and selectivity of the power system. Therefore, high-impedance fault detection is becoming increasingly important for utilities and protection engineers, as moral and legal challenges press them to take a greater duty of care and social responsibility for all that may be in the proximity of their power assets.

The typical fault scenario is where a distribution overhead line conductor has fallen, for example where corrosion or wind and/or ice loading over time have caused the conductor to break free from its retaining clamp at the tower or pole. Repeat failures of jumpers or previously-repaired spliced sections may also give rise to a downed conductor, which then will fall under a combination of gravity and any remaining tension to rest on whatever surface lies below it.

As previously described, non-conductive surfaces will tend to limit the fault current which flows. This is due to their high resistivity, and the need for the earth fault current to flow back to the source of supply, and a legitimate zero sequence current source. Typically, this will necessitate the current returning to the nearest adjacent earthed tower, for the return current then to flow in any aerial earth wire. If the circuit has no earth wire, the fault current will need to return typically to the earthed star-point of the upstream distribution transformer. In the case of a conductor falling onto rock or sand, the challenge is made all the harder in that the initial contact surface, and many metres of fault current flow in the same material composition which can drastically limit the prospective fault current. In the case of a conductor falling onto a fence, if the wood is dry this may have a high resistivity, but that high resistance may only apply for a few metres, until the current can flow in moist soil underground.

Sand poses a particular problem, because once a conductor falls onto it and arc current strikes, the heat in the arc can cause clumps of the surrounding sand to turn to glass, which partially insulates the conductor from the earth. Asphalt roads too can pose fault detection problems due to the natural good insulating properties, especially if the road is dry.

It has been explained that such a downed conductor will tend to strike a fault arc. This offers a real advantage in terms of detection of such faults, because arcs have particular characteristics:

- Arcs are rich in harmonics, with a persistence and randomness of the harmonic profile which is not typically seen in normal load current.
- The heat and energy in the arc, plus any wind and remaining tension in the wire, tend to cause movement of the conductor – this leads to randomness in the flow of fundamental current too.
- On certain surfaces, the heat from the arc will affect the insulating properties, and any moisture in the contact area – this will further induce randomness in harmonics, fundamental current, and the persistence of any fault current flow.

Modern feeder management relays offer numerical algorithms which react to (1) prolonged intermittency in current flow, and (2) unusual levels or prevalence of harmonics, to be used independently, or in combination, as a reliable method to detect high impedance faults such as downed conductors. These techniques can be directionalised using power-based techniques.

9.22 REFERENCES

[9.1]. *Directional Element Connections for Phase Relays*. W.K Sonnemann, Transactions A.I.E.E. 1950.

Chapter 10
Unit Protection of Feeders

10.1 INTRODUCTION

The graded overcurrent systems described in Chapter 9, though attractively simple in principle, do not meet all the protection requirements of a power system. Application difficulties are encountered for two reasons: firstly, satisfactory grading cannot always be arranged for a complex network, and secondly, the settings may lead to maximum tripping times at points in the system that are too long to prevent excessive disturbances occurring.

These problems led to the concept of 'Unit Protection', whereby sections of the power system are protected individually as a complete unit without reference to other sections. One form of 'Unit Protection' is also known as 'Differential Protection', as the principle is to sense the difference in currents between the incoming and outgoing terminals of the unit being protected. Other forms can be based on directional comparison, or distance teleprotection schemes, which are covered in Chapter 12, or phase comparison protection, which is discussed later in this chapter. The configuration of the power system may lend itself to unit protection; for instance, a simple earth fault relay applied at the source end of a transformer-feeder can be regarded as unit protection provided that the transformer winding associated with the feeder is not earthed. In this case the protection coverage is restricted to the feeder and transformer winding because the transformer cannot transmit zero sequence current to an out-of-zone fault.

In most cases, however, a unit protection system involves the measurement of fault currents (and possibly voltages) at each end of the zone, and the transmission of information between the equipment at zone boundaries. It should be noted that a stand-alone distance relay, although nominally responding only to faults within their setting zone, does not satisfy the conditions for a unit system because the zone is not clearly defined; it is defined only within the accuracy limits of the measurement. Also, to cater for some conditions, the setting of a stand-alone distance relay may also extend outside of the protected zone to cater for some conditions.

Merz and Price [10.1] first established the principle of current differential unit systems; their fundamental differential systems have formed the basis of many highly developed protection arrangements for feeders and numerous other items of plant. In one arrangement, an auxiliary 'pilot' circuit interconnects similar current transformers at each end of the protected zone, as shown in Figure 10.1. Current transmitted through the zone causes secondary current to circulate round the pilot circuit without producing any current in the relay. For a fault within the protected zone the CT secondary currents will not balance, compared with the through-fault condition, and the difference between the currents will flow in the relay.

An alternative arrangement is shown in Figure 10.2, in which the CT secondary windings are opposed for through-fault conditions so that no current flows in the series connected relays. The former system is known as a 'Circulating Current' system, while the latter is known as a 'Balanced Voltage' system.

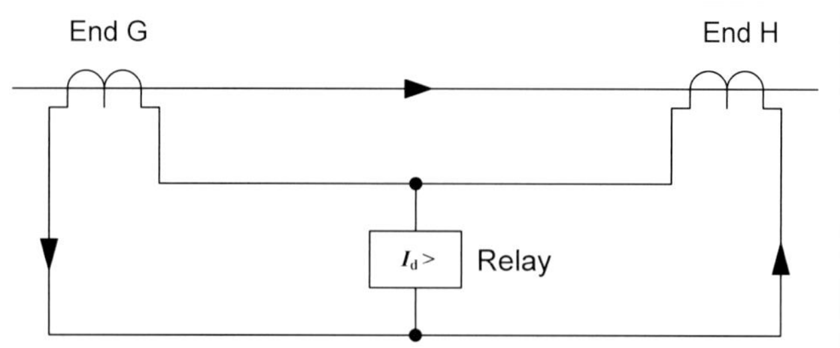

Figure 10.1: Circulating current system

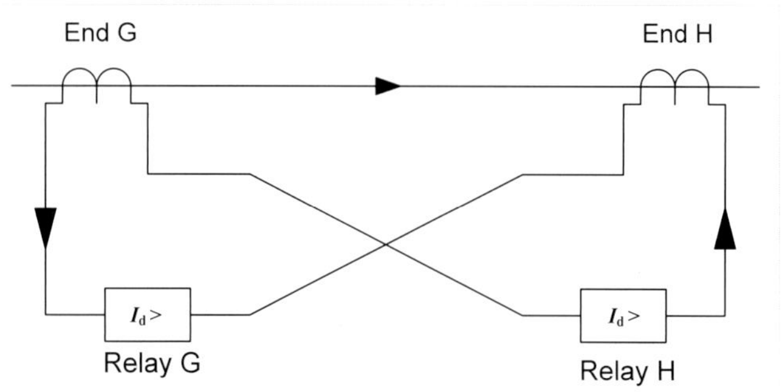

Figure 10.2: Balanced voltage system

Most systems of unit protection function through the determination of the relative direction of the fault current. This direction can only be expressed on a comparative basis, and such a comparative measurement is the common factor of many systems, including directional comparison protection and distance teleprotection schemes with directional impedance measurement.

A major factor in consideration of unit protection is the method of communication between the relays. This is covered in detail in Chapter 8 in respect of the latest fibre-optic based digital techniques.

10.2 CONVENTION OF DIRECTION

It is useful to establish a convention of direction of current flow; for this purpose, the direction measured from a busbar outwards along a feeder is taken as positive. Hence the notation of current flow shown in Figure 10.3; the section *GH* carries a through current which is counted positive at *G* but negative at *H*, while the infeeds to the faulted section *HJ* are both positive.

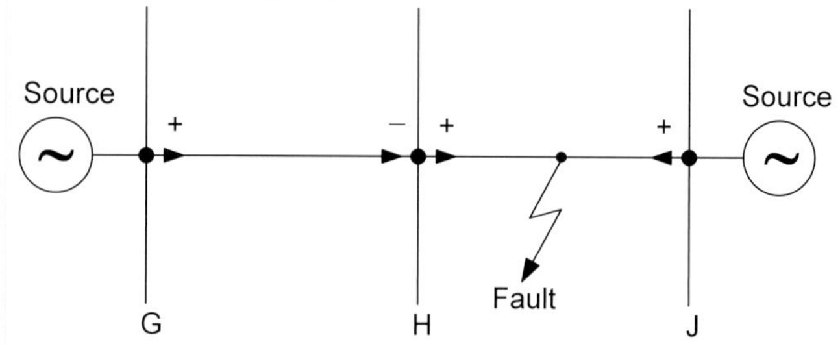

Figure 10.3: Convention of current direction

Neglect of this rule has often led to anomalous arrangements of equipment or difficulty in describing the action of a complex system. When applied, the rule will normally lead to the use of identical equipments at the zone boundaries, and is equally suitable for extension to multi-ended systems. It also conforms to the standard methods of network analysis

10.3 CONDITIONS FOR DIRECTION COMPARISON

The circulating current and balanced voltage systems of Figure 10.1 and Figure 10.2 perform full vectorial comparison of the zone boundary currents. Such systems can be treated as analogues of the protected zone of the power system, in which CT secondary quantities represent primary currents and the relay operating current corresponds to an in-zone fault current.

These systems are simple in concept; they are nevertheless applicable to zones having any number of boundary connections and for any pattern of terminal currents.

To define a current requires that both magnitude and phase be stated. Comparison in terms of both of these quantities is performed in the Merz-Price systems, but it is not always easy to transmit all this information over some pilot channels. Chapter 8 provides a detailed description of modern methods that may be used.

10.4 CIRCULATING CURRENT SYSTEM

The principle of this system is shown in outline in Figure 10.1. If the current transformers are ideal, the functioning of the system is straightforward. The transformers will, however, have errors arising from both Wattmetric and magnetising current losses that cause deviation from the ideal, and the interconnections between them may have unequal impedances. This can give rise to a 'spill' current through the relay even without a fault being present, thus limiting the sensitivity that can be obtained. Figure 10.4 shows the equivalent circuit of the circulating current scheme. If a high impedance relay is used, then unless the relay is located at point J in the circuit a current will flow through the relay even with currents I_{Pg} and I_{Ph} being identical. If a low impedance relay is used, voltage FF' will be very small, but the CT exciting currents will be unequal due to the unequal burdens and relay current I_R will still be non-zero.

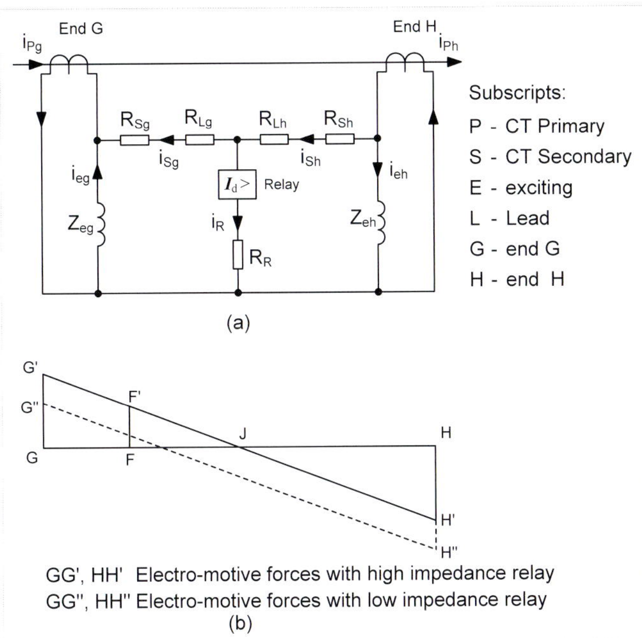

Figure 10.4: Equivalent circuit and potential diagram for circulating current scheme

10.4.1 Transient Instability

It is shown in Section 6.4.10 that an asymmetrical current applied to a current transformer will induce a flux that is greater than the peak flux corresponding to the steady state alternating component of the current. It may take the CT into saturation, with the result that the dynamic exciting impedance is reduced and the exciting current greatly increased.

When the balancing current transformers of a unit protection system differ in excitation characteristics, or have unequal burdens, the transient flux build-ups will differ and an increased 'spill' current will result. There is a consequent risk of relay operation on a healthy circuit under transient conditions, which is clearly unacceptable. One solution is to include a stabilising resistance in series with the relay. Details of how to calculate the value of the stabilising resistor are usually included in the instruction manuals of all relays that require one.

When a stabilising resistor is used, the relay current setting can be reduced to any practical value, the relay now being a voltage-measuring device. There is obviously a lower limit, below which the relay element does not have the sensitivity to pick up.

10.4.2 Bias

The 'spill' current in the relay arising from these various sources of error is dependent on the magnitude of the through current, being negligible at low values of through-fault current but sometimes reaching a disproportionately large value for more severe faults. Setting the operating threshold of the protection above the maximum level of spill current produces poor sensitivity. By making the differential setting approximately proportional to the fault current, the low-level fault sensitivity is greatly improved. Figure 10.5 shows a typical bias characteristic for a modern relay that overcomes the problem. At low currents, the bias is small, thus enabling the relay to be made sensitive. At higher currents, such as would be obtained from inrush or through fault conditions, the bias used is higher, and thus the spill current required to cause operation is higher. The relay is therefore more tolerant of spill current at higher fault currents and therefore less likely to maloperate, while still being sensitive at lower current levels.

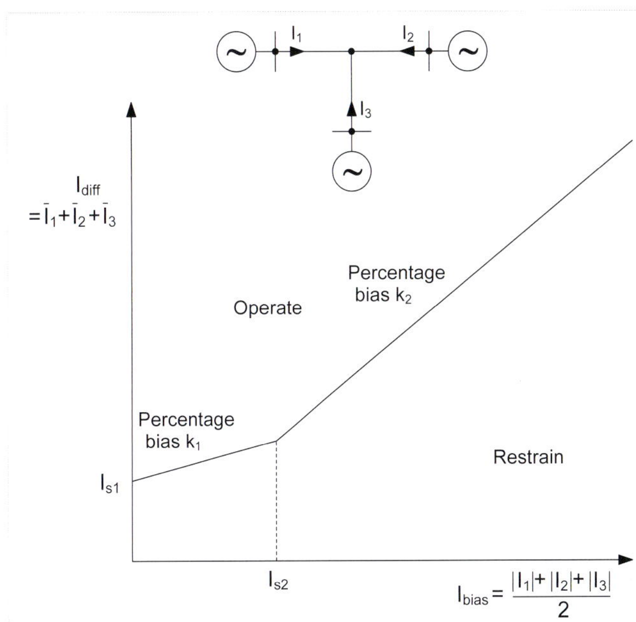

Figure 10.5: Typical bias characteristic of relay

10.5 BALANCED VOLTAGE SYSTEM

This section is included for historical reasons, mainly because of the number of such schemes still to be found in service – for new installations it has been almost completely superseded by circulating current schemes. It is the dual of the circulating current protection, and is summarised in Figure 10.2 as used in the 'MHOR' scheme.

With primary through current, the secondary e.m.f.s of the current transformers are opposed, and provide no current in the interconnecting pilot leads or the series connected relays. An in-zone fault leads to a circulating current condition in the CT secondaries and hence to relay operation.

An immediate consequence of the arrangement is that the current transformers are in effect open-circuited, as no secondary current flows for any primary through-current conditions. To avoid excessive saturation of the core and secondary waveform distortion, the core is provided with non-magnetic gaps sufficient to absorb the whole primary m.m.f. at the maximum current level, the flux density remaining within the linear range. The secondary winding therefore develops an e.m.f. and can be regarded as a voltage source. The shunt reactance of the transformer is relatively low, so the device acts as a transformer loaded with a reactive shunt; hence the name of transactor. The equivalent circuit of the system is as shown in Figure 10.6.

The series connected relays are of relatively high impedance; because of this the CT secondary winding resistances are not of great significance and the pilot resistance can be moderately large without significantly affecting the operation of the system. This is why the scheme was developed for feeder protection.

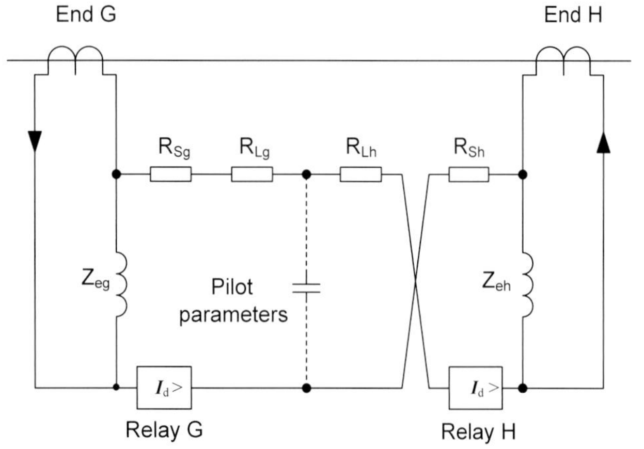

Figure 10.6: Equivalent circuit for balanced voltage system

10.5.1 Stability Limit of the Voltage Balance System

Unlike normal current transformers, transactors are not subject to errors caused by the progressive build-up of exciting current, because the whole of the primary current is expended as exciting current. In consequence, the secondary e.m.f. is an accurate measure of the primary current within the linear range of the transformer. Provided the transformers are designed to be linear up to the maximum value of fault current, balance is limited only by the inherent limit of accuracy of the transformers, and as a result of capacitance between the pilot cores. A broken line in the equivalent circuit shown in Figure 10.6 indicates such capacitance. Under through-fault conditions the pilots are energised to a proportionate voltage, the charging current flowing through the relays. The stability ratio that can be achieved with this system is only moderate and a bias technique is used to overcome the problem.

10.6 SUMMATION ARRANGEMENTS

Schemes have so far been discussed as though they were applied to single-phase systems. A polyphase system could be provided with independent protection for each phase. Modern digital or numerical relays communicating via fibre-optic links operate on this basis, since the amount of data to be communicated is not a major constraint. For older relays, use of this technique over pilot wires may be possible for relatively short distances, such as would be found with industrial and urban power distribution systems. Clearly, each phase would require a separate set of pilot wires if the protection was applied on a per phase basis. The cost of providing separate pilot-pairs and also separate relay elements per phase is generally prohibitive. Summation techniques can be used to combine the separate phase currents into a single relaying quantity for comparison over a single pair of pilot wires.

10.6.1 Summation Transformer Principle

A winding, either within a measuring relay, or an auxiliary current transformer, is arranged as in Figure 10.7.

The interphase sections of this winding, A-B, and B-C, often have a relatively similar number of turns, with the neutral end of the winding (C-N) generally having a greater number of turns.

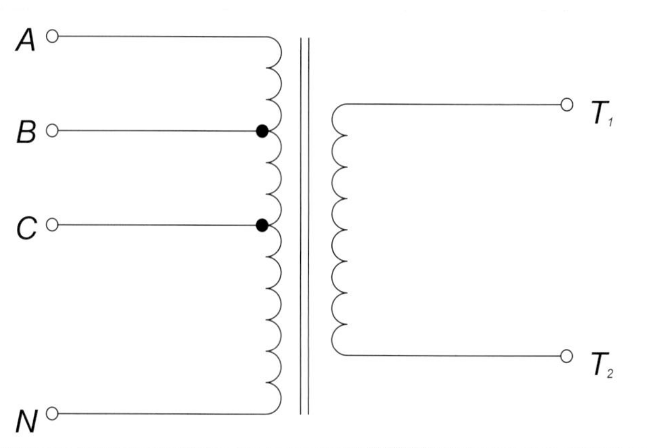

Figure 10.7: Typical summation winding

This winding has a number of special properties, which are discussed in the next section.

10.6.2 Sensitivity Using Summation Transformers

This section shows the performance of the summation transformer in a typical MBCI 'Translay' scheme.

In the MBCI relay, the number of turns on the input side of the summation CT is in the ratio:

- A-B 1.25
- B-C 1
- C-N 3 (or 6)

Boosting the turns ratio to 6 for the neutral end of the winding serves to increase the earth fault sensitivity of the scheme, as is shown in the bracketed performance below.

Unbalanced fault currents will energise different numbers of turns, according to which phase(s) is/are faulted. This leads to relay settings which are in inverse ratio to the number of turns involved. If the relay has a setting of 100% for a B-C fault, the following proportionate trip thresholds will apply:

- A-B 80%
- B-C 100%
- C-A 44%
- A-B-C 50%
- A-N 19% (or 12% for N=6)
- B-N 25% (or 14% for N=6)
- C-N 33% (or 17% for N=6)

10.7 EXAMPLES OF ELECTROMECHANICAL AND STATIC UNIT PROTECTION SYSTEMS

As mentioned above, the basic balanced voltage principle of protection evolved to biased protection systems. Several of these have been designed, some of which appear to be quite different from others. These dissimilarities are, however, superficial. A number of these systems that are still in common use are described below.

10.7.1 'Translay' Balanced Voltage Electromechanical System

A typical biased, electromechanical balanced voltage system, trade name 'Translay', still giving useful service on distribution systems is shown in Figure 10.8.

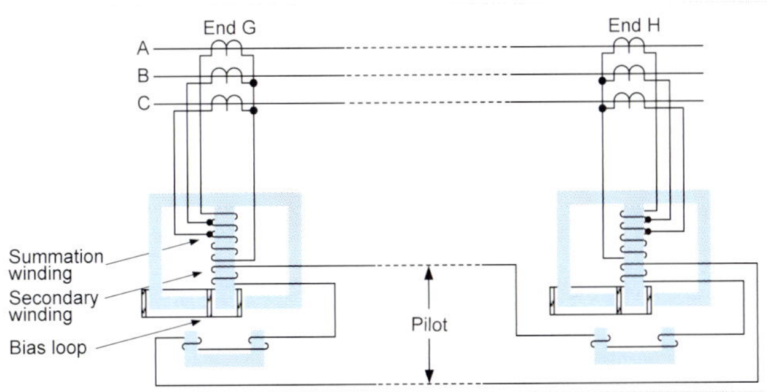

Figure 10.8: 'Translay' biased electromechanical differential protection system

The electromechanical design derives its balancing voltages from the transactor incorporated in the measuring relay at each line end. The latter are based on the induction-type meter electromagnet as shown in Figure 10.8.

The upper magnet carries a summation winding to receive the output of the current transformers, and a secondary winding which delivers the reference e.m.f. The secondary windings of the conjugate relays are interconnected as a balanced voltage system over the pilot channel, the lower electromagnets of both relays being included in this circuit.

Through current in the power circuit produces a state of balance in the pilot circuit and zero current in the lower electromagnet coils. In this condition, no operating torque is produced.

An in-zone fault causing an inflow of current from each end of the line produces circulating current in the pilot circuit and the energisation of the lower electromagnets. These co-operate with the flux of the upper electromagnets to produce an operating torque in the discs of both relays. An infeed from one end only will result in relay operation at the feeding end, but no operation at the other, because of the absence of upper magnet flux.

Bias is produced by a copper shading loop fitted to the pole of the upper magnet, thereby establishing a Ferraris motor action that gives a reverse or restraining torque proportional to the square of the upper magnet flux value.

Typical settings achievable with such a relay are:

- Least sensitive earth fault - 40% of rating
- Least sensitive phase-phase fault - 90% of rating
- Three-phase fault - 52% of rating

10.7.2 Static Circulating Current Unit Protection System - 'MBCI Translay'

A typical static modular pilot wire unit protection system, operating on the circulating current principle is shown in Figure 10.9. This uses summation transformers with a neutral section that is tapped, to provide alternative earth fault sensitivities. Phase comparators tuned to the power frequency are used for measurement and a restraint circuit gives a high level of stability for through faults and transient charging currents. High-speed operation is obtained with moderately sized current transformers and where space for current transformers is limited and where the lowest possible operating time is not essential, smaller current transformers may be used. This is made possible by a special adjustment (Kt) by which the operating time of the differential protection can be selectively increased if necessary, thereby enabling the use of current transformers having a correspondingly decreased knee-point voltage, whilst ensuring that through-fault stability is maintained to greater than 50 times the rated current.

Internal faults give simultaneous tripping of relays at both ends of the line, providing rapid fault clearance irrespective of whether the fault current is fed from both line ends or from only one line end.

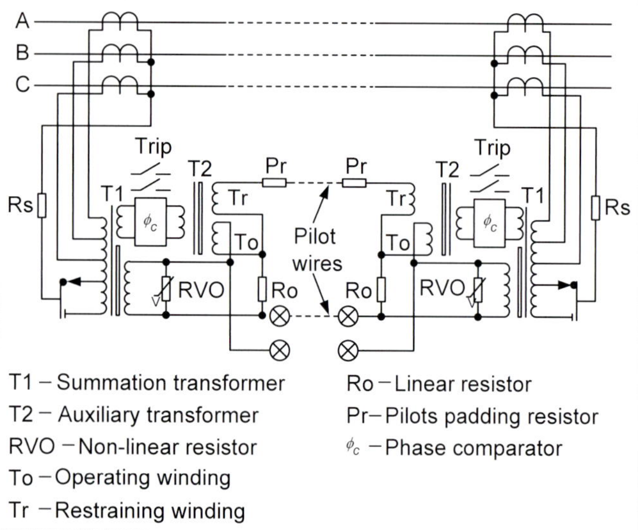

Figure 10.9: Typical static circulating current feeder unit protection circuit diagram

10.8 DIGITAL/NUMERICAL CURRENT DIFFERENTIAL PROTECTION SYSTEMS

A digital or numerical unit protection relay may typically provide phase-segregated current differential protection. This means that the comparison of the currents at each relay is done on a per phase basis. For digital data communication between relays, it is usual that a direct optical connection is used (for short distances) or a multiplexed link. Link speeds of 64kbit/s (56kbit/s in N. America) are normal, and up to 2 Mbit/s in some cases. Through current bias is typically applied to provide through fault stability in the event of CT saturation. A dual slope bias technique (Figure 10.5) is used to enhance stability for through faults. A typical trip criterion is as follows:

For $|I_{bias}| < I_{S2}$

$$|I_{diff}| > k1|I_{bias}| + I_{S1}$$

For $|I_{bias}| > I_{S2}$

$$|I_{diff}| > k2|I_{bias}| - (k2 - k1)I_{s2} + I_{S1}$$

Once the relay at one end of the protected section has determined that a trip condition exists, an intertrip signal is transmitted to the relay at the other end. Relays that are supplied with information on line currents at all ends of the line may not need to implement intertripping facilities. However, it is usual to provide intertripping in any case to ensure the protection operates in the event of any of the relays detecting a fault.

A facility for vector/ratio compensation of the measured currents, so that transformer feeders can be included in the unit protection scheme without the use of interposing CTs or defining the transformer as a separate zone increases versatility. Any interposing CTs required are implemented in software. Maloperation on transformer inrush is prevented by second harmonic detection. Care must be taken if the transformer has a wide-ratio on-load tap changer, as this results in the current ratio departing from nominal and may cause maloperation, depending on the sensitivity of the relays. The initial bias slope should be set taking this into consideration.

Tuned measurement of power frequency currents provides a high level of stability with capacitance inrush currents during line energisation. The normal steady-state capacitive charging current can be allowed for if a voltage signal can be made available and the susceptance of the protected zone is known.

Where an earthed transformer winding or earthing transformer is included within the zone of protection, some form of zero sequence current filtering is required. This is because there will be an in-zone source of zero sequence current for an external earth fault. The differential protection will see zero sequence differential current for an external fault and it could incorrectly operate as a result. In older protection schemes, the problem was eliminated by delta connection of the CT secondary windings. For a digital or numerical relay, a selectable software zero sequence filter is typically employed.

The minimum setting that can be achieved with such techniques while ensuring good stability is 20% of CT primary current.

The problem remains of compensating for the time difference between the current measurements made at the ends of the feeder, since small differences can upset the stability of the scheme, even when using fast direct fibre-optic links. The problem is overcome by either time synchronisation of the measurements taken by the relays, or calculation of the propagation delay of the link continuously.

10.8.1 Time Synchronisation of Relays

Fibre-optic media allow direct transmission of the signals between relays for distances of up to several km without the need for repeaters. For longer distances repeaters will be required. Where a dedicated fibre pair is not available, multiplexing techniques can be used. As phase comparison techniques are used on a per phase basis, time synchronisation of the measurements is vitally important. This requires knowledge of the transmission delay between the relays. Four techniques are possible for this:

a. assume a value

b. measurement during commissioning only

c. continuous online measurement

d. GPS time signal

Method (a) is not used, as the error between the assumed and actual value will be too great.

Method (b) provides reliable data if direct communication between relays is used. As signal propagation delays may change over a period of years, repeat measurements may be required at intervals and relays re-programmed accordingly. There is some risk of maloperation due to changes in signal propagation time causing incorrect time synchronisation between measurement intervals. The technique is less suitable if rented fibre-optic pilots are used, since the owner may perform circuit re-routing for operational reasons without warning, resulting in the propagation delay being outside of limits and leading to scheme maloperation. Where re-routing is limited to a few routes, it may be possible to measure the delay on all routes and pre-program the relays accordingly, with the relay digital inputs and ladder logic being used to detect changes in route and select the appropriate delay accordingly.

Method (c), continuous sensing of the signal propagation delay, also known as 'ping-pong', is a robust technique. One method of achieving this is shown in Figure 10.10.

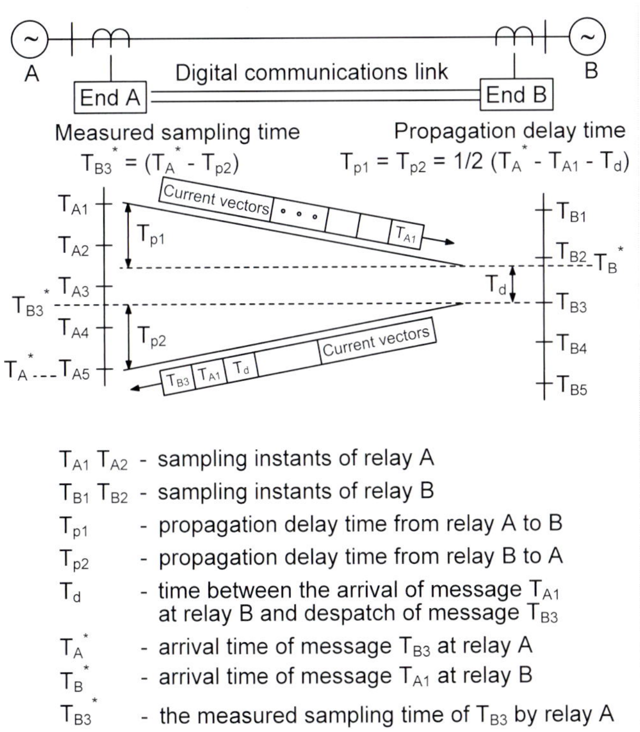

Figure 10.10: Signal propagation delay measurement

Relays at ends A and B sample signals at time T_{A1}, T_{A2} ... and T_{B1}, T_{B2} ... respectively. The times will not be coincident, even if they start coincidentally, due to slight differences in sampling frequencies. At time T_{A1} relay A transmits its data to relay B, containing a time tag and other data. Relay B receives it at time $T_{A1}+T_{p1}$ where T_{p1} is the propagation time from relay A to relay B. Relay B records this time as time T_{B^*}. Relay B also sends messages of identical format to relay A. It transmits such a message at time T_{B3}, received by relay A at time $T_{B3}+T_{p2}$ (say time T_{A^*}), where T_{p2} is the propagation time from relay B to relay A. The message from relay B to relay A includes the time T_{B3}, the last received time tag from relay A (T_{A1}) and the delay time between the arrival time of the message from A (T_{B^*}) and T_{B3} – call this the delay time T_d. The total elapsed time is therefore:

$$(T_A{*} - T_{A1}) = (T_d + T_{p1} + T_{p2})$$

If it is assumed that $T_{p1} = T_{p2}$, then the value of T_{p1} and T_{p2} can be calculated, and hence also T_{B3}. The relay B measured data as received at relay A can then be adjusted to enable data comparison to be performed. Relay B performs similar computations in respect of the data received from relay A (which also contains similar time information). Therefore, continuous measurement of the propagation delay is made, thus reducing the possibility of maloperation due to this cause to a minimum. Comparison is carried out on a per-phase basis, so signal transmission and the calculations are required for each phase. A variation of this technique is available that can cope with unequal propagation delays in the two communication channels under well-defined conditions.

The technique can also be used with all types of pilots, subject to provision of appropriate interfacing devices.

Method (d) is also a robust technique. It involves both relays being capable of receiving a time signal from a GPS clock source. The propagation delay on each communication channel is no longer required to be known or calculated as both relays are synchronised to a common time signal. For the protection scheme to meet the required performance in respect of availability and maloperation, the GPS signal must be capable of reliable receipt under all atmospheric conditions. There is extra satellite signal receiving equipment required at both ends of the line, which implies extra cost.

In applications where SONET (synchronous digital hierarchy) communication links are used, it cannot be assumed that T_{P1} is equal to T_{P2}, as split path routings which differ from A to B, and B to A are possible. In this scenario, method (d) is highly recommended, as it is able to calculate the independent propagation delays in each direction.

The minimum setting that can be achieved with such techniques while ensuring good stability is 20% of CT primary current.

10.8.2 Application to Mesh Corner and 1 1/2 Breaker Switched Substations

These substation arrangements are quite common, and the arrangement for the latter is shown in Figure 10.11. Problems exist in protecting the feeders due to the location of the line CTs, as either Bus 1 or Bus 2 or both can supply the feeder. Two alternatives are used to overcome the problem, and they are shown in the Figure. The first is to common the line CT inputs (as shown for Feeder *A*) and the alternative is to use a second set of CT inputs to the relay (as shown for Feeder *B*).

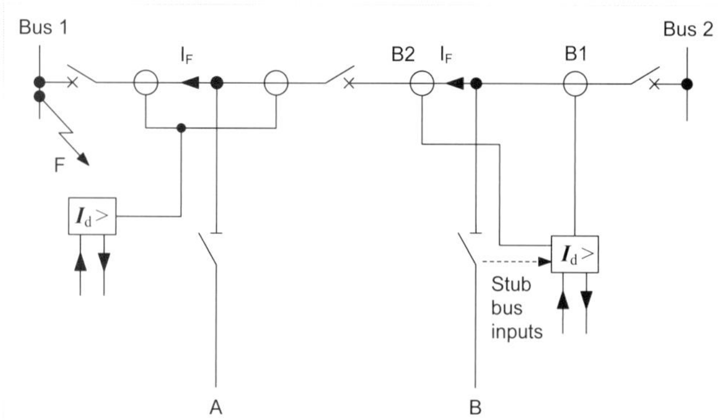

Figure 10.11: Breaker and a half switched substation

In the case of a through fault as shown, the relay connected to Feeder *A* theoretically sees no unbalance current, and hence will be stable. However, with the line disconnect switch open, no bias is produced in the relay, so CTs need to be well matched and equally loaded if maloperation is to be avoided.

For Feeder *B*, the relay also theoretically sees no differential current, but it will see a large bias current even with the line disconnect switch open. This provides a high degree of stability, in the event of transient asymmetric CT saturation. Therefore, this technique is preferred.

Sensing of the state of the line isolator through auxiliary contacts enables the current values transmitted to and received from remote relays to be set to zero when the isolator is open. Hence, stub-bus protection for the energised part of the bus is then possible, with any fault resulting in tripping of the relevant CB.

10.9 CARRIER UNIT PROTECTION SCHEMES

In earlier sections, the pilot links between relays have been treated as an auxiliary wire circuit that interconnects relays at the boundaries of the protected zone. In many circumstances, such as the protection of longer line sections or where the route involves installation difficulties, it is too expensive to provide an auxiliary cable circuit for this purpose, and other means are sought.

In all cases (apart from private pilots and some short rented pilots) power system frequencies cannot be transmitted directly on the communication medium. Instead a relaying quantity may be used to vary the higher frequency associated with each medium (or the light intensity for fibre-optic systems), and this process is normally referred to as modulation of a carrier wave. Demodulation or detection of the variation at a remote receiver permits the relaying quantity to be reconstituted for use in conjunction with the relaying quantities derived locally, and forms the basis for all carrier systems of unit protection.

Carrier systems are generally insensitive to induced power system currents since the systems are designed to operate at much higher frequencies, but each medium may be subjected to noise at the carrier frequencies that may interfere with its correct operation. Variations of signal level, restrictions of the bandwidth available for relaying and other characteristics unique to each medium influence the choice of the most appropriate type of scheme. Methods and media for communication are discussed in Chapter 8.

10.10 CURRENT DIFFERENTIAL SCHEME – ANALOGUE TECHNIQUES

The carrier channel is used in this type of scheme to convey both the phase and magnitude of the current at one relaying point to another for comparison with the phase and magnitude of the current at that point. Transmission techniques may use either voice frequency channels using FM modulation or A/D converters and digital transmission. Signal propagation delays still need to be taken into consideration by introducing a deliberate delay in the locally derived signal before a comparison with the remote signal is made.

A further problem that may occur concerns the dynamic range of the scheme. As the fault current may be up to 30 times the rated current, a scheme with linear characteristics requires a wide dynamic range, which implies a wide signal transmission bandwidth. In practice, bandwidth is limited, so either a non-linear modulation characteristic must be used or detection of fault currents close to the setpoint will be difficult.

10.10.1 Phase Comparison Scheme

The carrier channel is used to convey the phase angle of the current at one relaying point to another for comparison with the phase angle of the current at that point.

The principles of phase comparison are shown in Figure 10.12. The carrier channel transfers a logic or 'on/off' signal that switches at the zero crossing points of the power frequency waveform. Comparison of a local logic signal with the corresponding signal from the remote end provides the basis for the measurement of phase shift between power system currents at the two ends and hence discrimination between internal and through faults.

Load or through fault currents at the two ends of a protected feeder are in antiphase (using the normal relay convention for direction), whilst during an internal fault the (conventional) currents tend towards the in-phase condition. Hence, if the phase relationship of through fault currents is taken as a reference condition, internal faults cause a phase shift of approximately 180° with respect to the reference condition.

Phase comparison schemes respond to any phase shift from the reference conditions, but tripping is usually permitted only when the phase shift exceeds an angle of typically 30 to 90 degrees, determined by the time delay setting of the measurement circuit, and this angle is usually referred to as the Stability Angle. Figure 10.13 is a polar diagram that shows the discrimination characteristics that result from the measurement techniques used in phase comparison schemes.

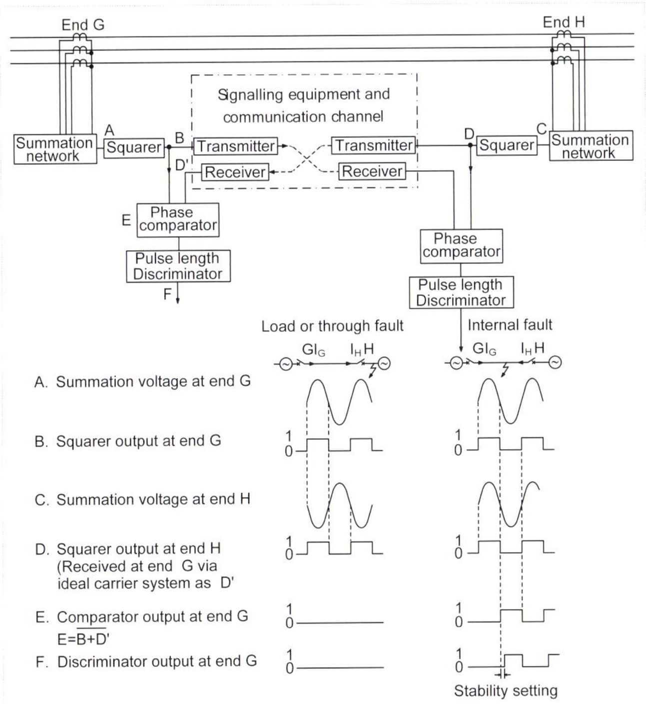

Figure 10.12: Principles of phase comparison protection

Since the carrier channel is required to transfer only binary information, the techniques associated with sending teleprotection commands. Blocking or permissive trip modes of operation are possible, however Figure 10.12 shows the more usual blocking mode, since the comparator provides an output when neither squarer is at logic '1'. A permissive trip scheme can be realised if the comparator is arranged to give an output when both squarers are at logic '1'. Performance of the scheme during failure or disturbance of the carrier channel and its ability to clear single-end-fed faults depends on the mode of operation, the type and function of fault detectors or starting units, and the use of any additional signals or codes for channel monitoring and transfer tripping.

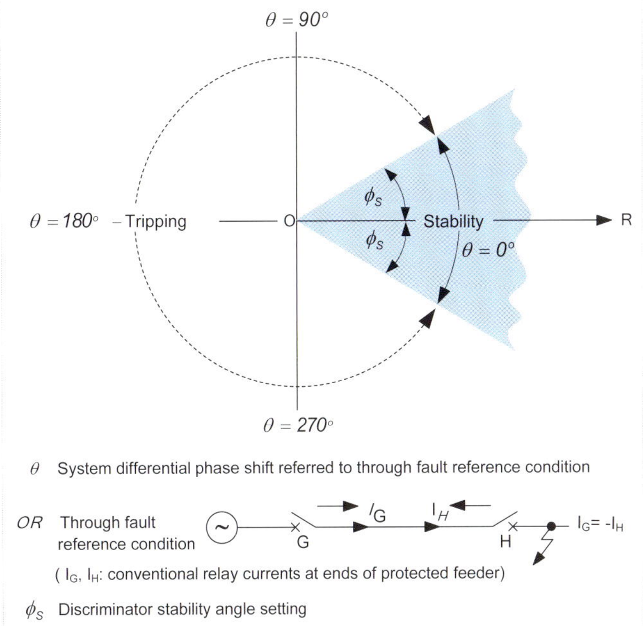

Figure 10.13: Polar diagram for phase comparison scheme

Signal transmission is usually performed by voice frequency channels using frequency shift keying (FSK) or PLC techniques.

Voice frequency channels involving FSK use two discrete frequencies either side of the middle of the voice band. This arrangement is less sensitive to variations in delay or frequency response than if the full bandwidth was used. Blocking or permissive trip modes of operation may be implemented. In addition to the two frequencies used for conveying the squarer information, a third tone is often used, either for channel monitoring or transfer tripping dependent on the scheme.

For a sensitive phase comparison scheme, accurate compensation for channel delay is required. However, since both the local and remote signals are logic pulses, simple time delay circuits can be used, in contrast to the analogue delay circuitry usually required for current differential schemes.

The principles of the Power Line Carrier channel technique are shown in Figure 10.14. The scheme operates in the blocking mode. The 'squarer' logic is used directly to turn a transmitter 'on' or 'off' at one end, and the resultant burst (or block) of carrier is coupled to and propagates along the power line which is being protected to a receiver at the other end. Carrier signals above a threshold are detected by the receiver, and hence produce a logic signal corresponding to the block of carrier. In contrast to Figure 10.12, the signalling system is a 2-wire rather than 4-wire arrangement, in which the local transmission is fed directly to the local receiver along with any received signal. The transmitter frequencies at both ends are nominally equal, so the receiver responds equally to blocks of carrier from either end. Through-fault current results in transmission of blocks of carrier from both ends, each lasting for half a cycle, but with a phase displacement of half a cycle, so that the composite signal is continuously above the threshold level and the detector output logic is continuously '1'. Any phase shift relative to the through fault condition produces a gap in the composite carrier signal and hence a corresponding '0' logic level from the detector. The duration of the logic '0' provides the basis for discrimination between internal and external faults, tripping being permitted only when a time delay setting is exceeded. This delay is usually expressed in terms of the corresponding phase shift in degrees at system frequency φ_s in Figure 10.13.

The advantages generally associated with the use of the power line as the communication medium apply namely, that a power line provides a robust, reliable, and low-loss interconnection between the relaying points. In addition dedicated 'on/off' signalling is particularly suited for use in phase comparison blocking mode schemes, as signal attenuation is not a problem. This is in contrast to permissive or direct tripping schemes, where high power output or boosting is required to overcome the extra attenuation due to the fault.

The noise immunity is also very good, making the scheme very reliable. Signal propagation delay is easily allowed for in the stability angle setting, making the scheme very sensitive as well.

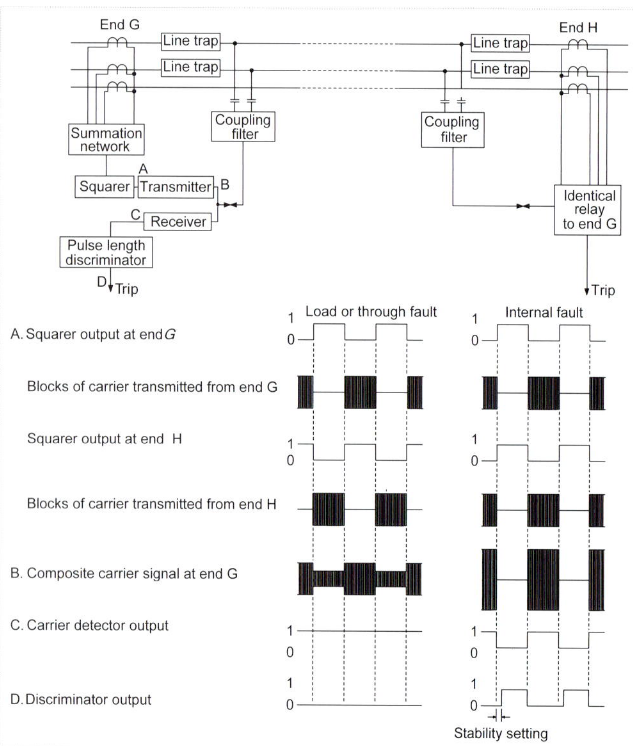

Figure 10.14: Principles of power line carrier phase comparison

10.11 PHASE COMPARISON PROTECTION SCHEME CONSIDERATIONS

One type of unit protection that uses carrier techniques for communication between relays is phase comparison protection. Communication between relays commonly uses PLCC or frequency modulated carrier modem techniques. There are a number of considerations that apply only to phase comparison protection systems, which are discussed in this section.

10.11.1 Lines with Shunt Capacitance

A problem can occur with the shunt capacitance current that flows from an energising source. Since this current is in addition to the load current that flows out of the line, and typically leads it by more than 90°, significant differential phase shifts between the currents at the ends of the line can occur, particularly when load current is low.

The system differential phase shift may encroach into the tripping region of the simple discriminator characteristic, regardless of how large the stability angle setting may be. Figure 10.15 shows the effect and indicates techniques that are commonly used to ensure stability.

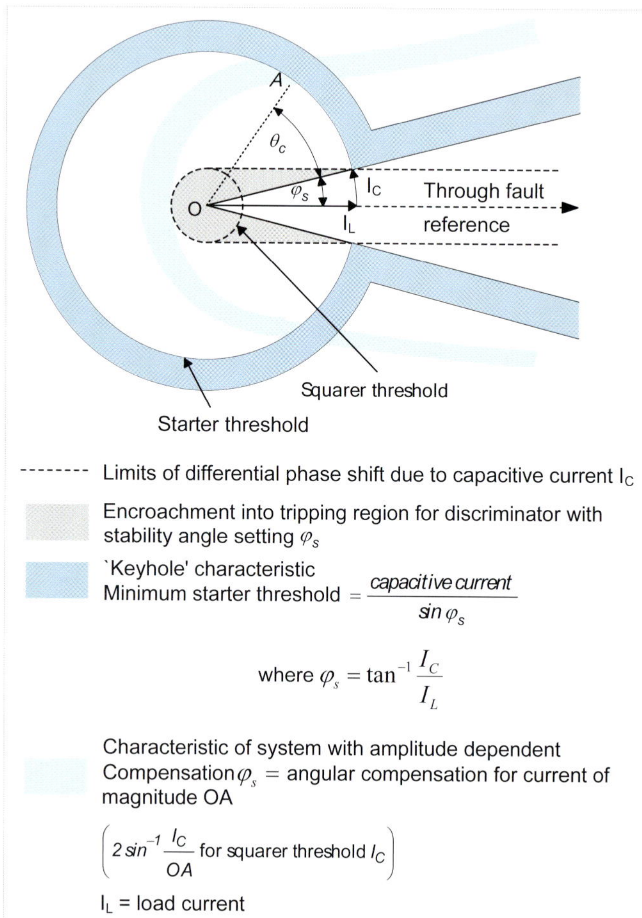

Figure 10.15: Capacitive current in phase comparison schemes and techniques used to avoid instability

Operation of the discriminator can be permitted only when current is above some threshold, so that measurement of the large differential phase shifts which occur near the origin of the polar diagram is avoided. By choice of a suitable threshold and stability angle, a 'keyhole' characteristic can be provided so the capacitive current characteristic falls within the resultant stability region. Fast resetting of the fault detector is required to ensure stability following the clearance of a through fault when the currents tend to fall towards the origin of the polar diagram.

The mark-space ratio of the squarer (or modulating) waveform can be made dependent on the current amplitude. Any decrease in the mark-space ratio will permit a corresponding differential phase shift to occur between the currents before any output is given from the comparator for measurement in the discriminator. A squarer circuit with an offset or bias can provide a decreasing mark-space ratio at low currents, and with a suitable threshold level the extra phase shift θ_c which is permitted can be arranged to equal or exceed the phase shift due to capacitive current. At high current levels the capacitive current compensation falls towards zero and the resultant stability region on the polar diagram is usually smaller than on the keyhole characteristic, giving improvements in sensitivity and/or dependability of the scheme. Since the stability region encompasses all through-fault currents, the resetting speed of any fault detectors or starter (which may still be required for other purposes, such as the control of a normally quiescent scheme) is much less critical than with the keyhole characteristic.

10.11.2 System Tripping Angles

For the protection scheme to trip correctly on internal faults the change in differential phase shift, θ_o, from the through-fault condition taken as reference, must exceed the effective stability angle of the scheme. Hence:

$$\theta_0 \geq \phi_s + \theta_c$$

Equation 10.1

Where:

ϕ_s = stability angle setting

θ_c = capacitive current compensation (when applicable)

The currents at the ends of a transmission line I_G and I_H may be expressed in terms of magnitude and phase shift θ with respect a common system voltage.

$$I_G = |I_G| \angle \theta_G$$

$$I_H = |I_H| \angle \theta_H$$

Using the relay convention described in Section 10.2, the reference through-fault condition is

$$I_G = -I_H$$

$$\therefore I_G \angle \theta_G = -I_H \angle \theta_H = I_H \angle \theta_H \pm 180^\circ$$

$$\therefore |\theta_G - \theta_H| = 180^\circ$$

During internal faults, the system tripping angle θ_o is the differential phase shift relative to the reference condition.

$$\therefore \theta_o = 180 - |\theta_G - \theta_H|$$

Substituting θ_o in Equation 10.1, the conditions for tripping are:

$$\theta_o = 180 - |\theta_G - \theta_H| \geq \phi_s + \theta_c$$

$$\therefore |\theta_G - \theta_H| \leq 180 - (\phi_s + \theta_c)$$

Equation 10.2

The term $(Ø_s + \theta_c)$ is the effective stability angle setting of the scheme. Substituting a typical value of 60° in Equation 10.2, gives the tripping condition as

$$|\theta_G - \theta_H| \leq 120^\circ$$

Equation 10.3

In the absence of pre-fault load current, the voltages at the two ends of a line are in phase. Internal faults are fed from both ends with fault contributions whose magnitudes and angles are determined by the position of the fault and the system source impedances. Although the magnitudes may be markedly different, the angles (line plus source) are similar and seldom differ by more than about 20°

Hence $|\theta_G - \theta_H| \leq 20^\circ$ and the requirements of Equation 10.3 are very easily satisfied. The addition of arc or fault resistance makes no difference to the reasoning above, so the scheme is inherently capable of clearing such faults.

10.11.3 Effect of Load Current

When a line is heavily loaded prior to a fault the e.m.f.s of the sources which cause the fault current to flow may be displaced by up to about 50°, that is, the power system stability limit. To this the differential line and source angles of up to 20° mentioned above need to be added. So $|\theta_G - \theta_H| \leq 70^\circ$ and the requirements of Equation 10.3 are still easily satisfied.

For three phase faults, or solid earth faults on phase-by-phase comparison schemes, through load current falls to zero during the fault and so need not be considered. For all other faults, load current continues to flow in the healthy phases and may therefore tend to increase $|\theta_G - \theta_H|$ towards the through fault reference value. For low resistance faults the fault current usually far exceeds the load current and so has little effect. High resistance faults or the presence of a weak source at one end can prove more difficult, but high performance is still possible if the modulating quantity is chosen with care and/or fault detectors are added.

10.11.4 Modulating Quantity

Phase-by-phase comparison schemes usually use phase current for modulation of the carrier. Load and fault currents are almost in antiphase at an end with a weak source. Correct performance is possible only when fault current exceeds load current, or

for $I_F < I_L, |\theta_G - \theta_H| \approx 180^\circ$

for $I_F > I_L, |\theta_G - \theta_H| \approx 0^\circ$

Equation 10.4

where

I_F = fault current contribution from weak source

I_L = load current flowing towards weak source

To avoid any risk of failure to operate, fault detectors with a setting greater than the maximum load current may be applied, but they may limit the sensitivity of scheme. When the fault detector is not operated at one end, fault clearance invariably involves sequential tripping of the circuit breakers.

Most phase comparison schemes use summation techniques to produce a single modulating quantity, responsive to faults on any of the three phases. Phase sequence components are often used and a typical modulating quantity is

$$I_m = MI_2 + NI_1$$

Equation 10.5

Where:

I_1 = Positive Phase Sequence Component

I_2 = Negative Phase Sequence Component

M, N = constants (N typically negative)

With the exception of three phase faults all internal faults give rise to negative phase sequence (NPS) currents, I_2, which are approximately in phase at the ends of the line and therefore could form an ideal modulating quantity. To provide a modulating signal during three phase faults, which give rise to positive phase sequence (PPS) currents, I_1, only, a practical modulating quantity must include some response to I_1 in addition to I_2.

Typical values of the ratio $M : N$ exceed 5:1, so that the modulating quantity is weighted heavily in favour of NPS, and any PPS associated with load current tends to be swamped out on all but the highest resistance faults.

For a high resistance phase-earth fault, the system remains well balanced so that load current I_L is entirely positive sequence. The fault contribution I_F provides equal parts of positive, negative and zero sequence components $\frac{I_F}{3}$.

Assuming the fault is on 'A' phase and the load is resistive, all sequence components are in phase at the infeed end G.

$$\therefore I_{mG} = NI_L + \frac{MI_{FG}}{3} + \frac{NI_{FG}}{3} \quad \therefore \theta_G \approx 0$$

At the outfeed end load current is negative,

$$\therefore I_{mH} = -NI_L + \frac{MI_{FH}}{3} + \frac{NI_{FH}}{3}$$

for

$$I_{mH} > 0, \theta_H \approx 0 \therefore |\theta_G - \theta_H| = 0^\circ$$

for

$$I_{mH} < 0, \theta_H \approx 180^\circ \therefore |\theta_G - \theta_H| = 180^\circ$$

Hence for correct operation $I_{mH} \geq 0$

let $I_{mH} = 0$

then

$$I_{FH} = \frac{3I_L}{\left(\frac{M}{N} + 1\right)} = I_E$$

Equation 10.6

The fault current in Equation 10.6 is the effective earth fault sensitivity I_E of the scheme. For the typical values of

$$M = 6 \text{ and } N = -1, \frac{M}{N} = -6$$

$$\therefore I_E = -\frac{3}{5} I_L$$

Comparing this with Equation 10.4, a scheme using summation is potentially 1.667 times more sensitive than one using phase current for modulation.

Even though the use of a negative value of M gives a lower value of I_E than if it were positive, it is usually preferred since the limiting condition of $I_m = 0$ then applies at the load infeed end. Load and fault components are additive at the outfeed end so that a correct modulating quantity occurs there, even with the lowest fault levels. For operation of the scheme it is sufficient therefore that the fault current contribution from the load infeed end exceeds the effective setting.

For faults on B or C phases, the NPS components are displaced by 120° or 240° with respect to the PPS components. No simple cancellation can occur, but instead a phase displacement is introduced. For tripping to occur, Equation 10.2 must be satisfied, and to achieve high dependability under these marginal conditions, a small effective stability angle is essential. Figure 10.16 shows operation near to the limits of earth fault sensitivity.

Very sensitive schemes may be implemented by using high values of $\frac{M}{N}$ but the scheme then becomes more sensitive to differential errors in NPS currents such as the unbalanced components of capacitive current or spill from partially saturated CTs.

Techniques such as capacitive current compensation and reduction of $\frac{M}{N}$ at high fault levels may be required to ensure stability of the scheme.

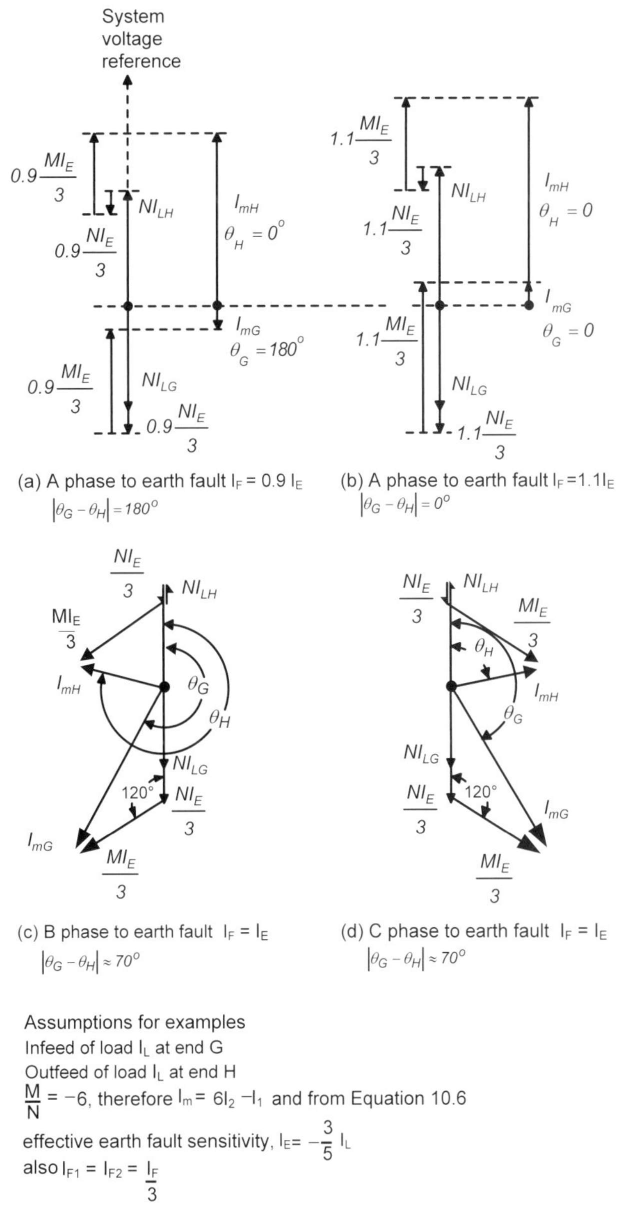

Figure 10.16: Effect of load current on differential phase shift $|\theta_g - \theta_H|$ for resistive earth faults at the effective earth fault sensitivity I_E

10.11.5 Fault Detection and Starting

For a scheme using a carrier system that continuously transmits the modulating quantity, protecting an ideal line (capacitive current = 0) in an interconnected transmission system, measurement of current magnitude might be unnecessary. In practice, fault detector or starting elements are invariably provided and the scheme then becomes a permissive tripping scheme in which both the fault detector and the discriminator must operate to provide a trip output, and the fault detector may limit the sensitivity of the scheme. Requirements for the fault detectors vary according to the type of carrier channel used, mode of operation used in the phase angle measurement, that is, blocking or permissive, and the features used to provide tolerance to capacitive current.

10.11.6 Normally Quiescent Power Line Carrier (Blocking Mode)

To ensure stability of through faults, it is essential that carrier transmission starts before any measurement of the width of the gap is permitted. To allow for equipment tolerances and the difference in magnitude of the two currents due to capacitive current, two starting elements are used, usually referred to as 'Low Set' and 'High Set' respectively. Low Set controls the start-up of transmission whilst High Set, having a setting typically 1.5 to 2 times that of the Low Set element, permits the phase angle measurement to proceed.

The use of impulse starters that respond to the change in current level enables sensitivities of less than rated current to be achieved. Resetting of the starters occurs naturally after a swell time or at the clearance of the fault. Dwell times and resetting characteristics must ensure that during through faults, a High Set is never operated when a Low Set has reset and potential race conditions are often avoided by the transmitting of an demodulated (and therefore blocking) carrier for a short time following the reset of low set; this feature is often referred to as 'Marginal Guard.'

10.11.7 Scheme without Capacitive Current Compensation

The 'keyhole' discrimination characteristic of depends on the inclusion of a fault detector to ensure that no measurements of phase angle can occur at low current levels, when the capacitive current might cause large phase shifts. Resetting must be very fast to ensure stability following the shedding of through load.

10.11.8 Scheme with Capacitive Current Compensation (Blocking Mode)

When the magnitude of the modulating quantity is less than the threshold of the squarer, transmission if it occurred, would be a continuous blocking signal. This might occur at an end with a weak source, remote from a fault close to a strong source. A fault detector is required to permit transmission only when the current exceeds the modulator threshold by some multiple (typically about 2 times) so that the effective stability angle is not excessive. For PLCC schemes, the low set element referred to in Section 10.11.6 is usually used for this purpose. If the fault current is insufficient to operate the fault detector, circuit breaker tripping will normally occur sequentially.

10.11.9 Fault Detector Operating Quantities

Most faults cause an increase in the corresponding phase current(s) so measurement of current increase could form the basis for fault detection. However, when a line is heavily loaded and has a low fault level at the outfeed end, some faults can be accompanied by a fall in current, which would lead to failure of such fault detection, resulting in sequential tripping (for blocking mode schemes) or no tripping (for permissive schemes). Although fault detectors can be designed to respond to any disturbance (increase or decrease of current), it is more usual to use phase sequence components. All unbalanced faults produce a rise in the NPS components from the zero level associated with balanced load current, whilst balanced faults produce an increase in the PPS components from the load level (except at ends with very low fault level) so that the use of NPS and PPS fault detectors make the scheme sensitive to all faults. For schemes using summation of NPS and PPS components for the modulating quantity, the use of NPS and PPS fault detectors is particularly appropriate since, in addition to any reductions in hardware, the scheme may be characterised entirely in terms of sequence components. Fault sensitivities I_F for PPS and NPS impulse starter settings I_{1s} and I_{2s} respectively are as follows:

Three phase fault $I_F = I_{1S}$

Phase-phase fault $I_F = \sqrt{3}\, I_{2S}$

Phase-earth fault $I_F = 3\, I_{2S}$

10.12 EXAMPLES

This section gives examples of setting calculations for simple unit protection schemes. It cannot and is not intended to replace a proper setting calculation for a particular application. It is intended to show the principles of the calculations required. The examples use the Alstom MiCOM P54x Current Differential relay, which has the setting ranges given in Table 10.1 for differential protection. The relay also has backup distance, high-set instantaneous, and earth-fault protection included in the basic model to provide a complete 'one-box' solution of main and backup protection.

Parameter	Setting Range
Differential Current Setting I_{s1}	0.2 - 2.0 I_n
Bias Current Threshold Setting I_{s2}	1.0 - 30.0 I_n
Lower Percentage Bias Setting k_1	0.3 - 1.5
Higher Percentage Bias Setting k_2	0.3 - 1.5
I_n = CT Rated Secondary Current	

Table 10.1: Relay Setting Ranges

10.12.1 Unit Protection of a Plain Feeder

The circuit to be protected is shown in Figure 10.17. It consists of a plain feeder circuit formed of an overhead line 25km long. The relevant properties of the line are:

Line voltage: 33kV

$$Z = 0.156 + j0.337\Omega / km$$

Shunt charging current $= 0.065A / km$

To arrive at the correct settings, the characteristics of the relays to be applied must be considered.

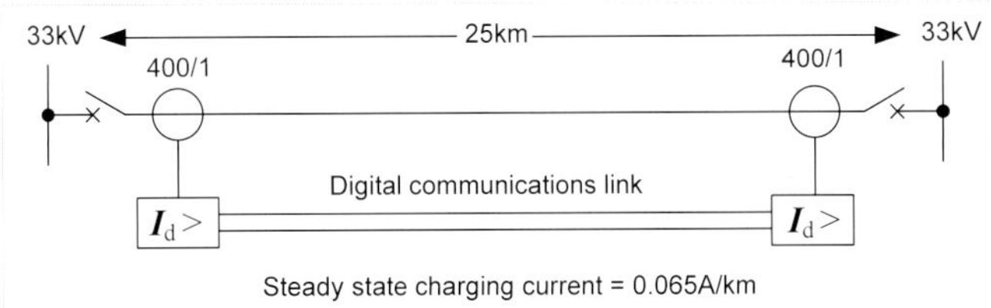

Figure 10.17: Protection of a plain feeder

The recommended settings for three of the adjustable values (taken from the relay manual) are:

$$I_{s2} = 2.0\,pu$$
$$k_1 = 30\%$$
$$k_2 = 150\%$$

To provide immunity from the effects of line charging current, the setting of I_{s1} must be at least 2.5 times the steady-state charging current, i.e. 4.1A or 0.01p.u., after taking into consideration the CT ratio of 400/1. The nearest available setting above this is 0.20p.u. This gives the points on the relay characteristic as shown in Figure 10.18.

The minimum operating current $I_{d\,\min}$ is related to the value of I_{s1} by the formula

$$I_{d\min} = \frac{(k_1 I_L + I_{S1})}{(1 - 0.5k_1)}$$

for $I_{bias} =< I_{s2}$

and

$$I_{d\min} = \frac{(k_2 I_L - (k_2 - k_1) I_{S2} - I_{S1})}{1 - 0.5k_2}$$

for $I_{bias} => I_{s2}$

where I_L = load current and hence the minimum operating current at no load is 0.235p.u. or 94A.

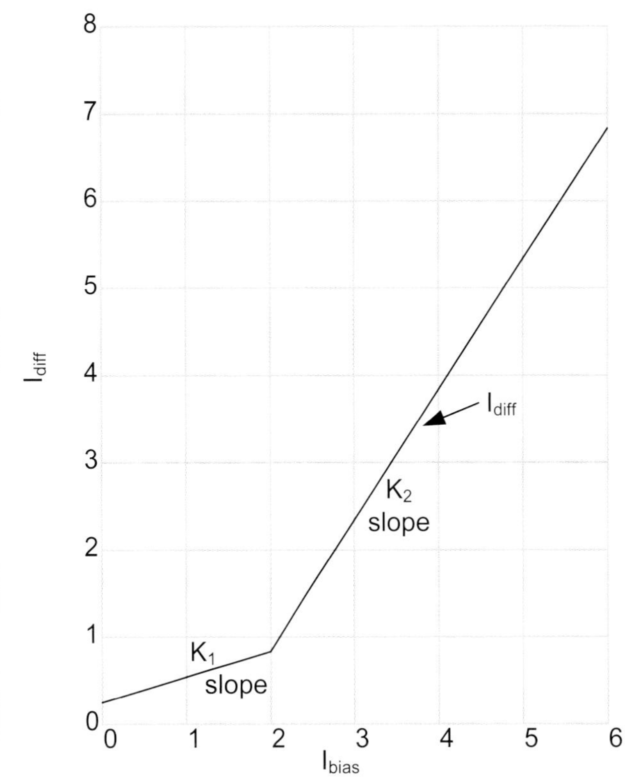

Figure 10.18: Relay characteristic; plain feeder example

In cases where the capacitive charging current is very large and hence the minimum tripping current needs to be set to an unacceptably high value, some relays offer the facility of subtracting the charging current from the measured value. Use of this facility depends on having a suitable VT input and knowledge of the shunt capacitance of the circuit.

10.12.2 Unit Protection of a Transformer Feeder

Figure 10.19 shows unit protection applied to a transformer feeder. The feeder is assumed to be a 100m length of cable, such as might be found in some industrial plants or where a short distance separates the 33kV and 11kV substations. While 11kV cable capacitance will exist, it can be regarded as negligible for the purposes of this example.

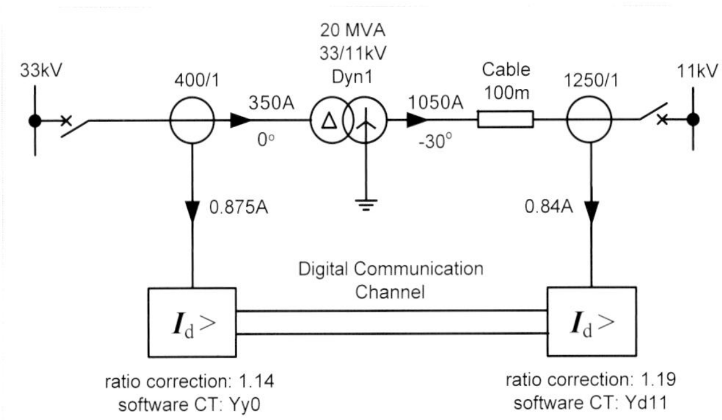

Figure 10.19: Unit Protection of a transformer feeder

The delta/star transformer connection requires phase shift correction of CT secondary currents across the transformer, and in this case software equivalents of interposing CTs are used.

Since the LV side quantities lag the HV side quantities by 30°, it is necessary to correct this phase shift by using software CT settings that produce a 30° phase shift. There are two obvious possibilities:

- HV side: Yd1, LV side: Yy0
- HV side: Yy0, LV side: Yd11

Only the second combination is satisfactory, since only this one provides the necessary zero-sequence current trap to avoid maloperation of the protection scheme for earth faults on the LV side of the transformer outside of the protected zone.

Ratio correction must also be applied, to ensure that the relays see currents from the primary and secondary sides of the transformer feeder that are well balanced under full load conditions. This is not always inherently the case, due to selection of the main CT ratios. For the example of Figure 10.19, transformer turns ratio at nominal tap

$$=\frac{11}{33}=0.3333$$

Required turns ratio according to the CT ratios used

$$=\frac{400/1}{1250/1}=0.32$$

Spill current that will arise due to the incompatibility of the CT ratios used with the power transformer turns ratio may cause relay maloperation. This has to be eliminated by using the facility in the relay for CT ratio correction factors. For this particular relay, the correction factors are chosen so the full load current seen by the relay software is equal to 1A.

The appropriate correction factors are:

HV: $\frac{400}{350}=1.14$

LV: $\frac{1250}{1050}=1.19$

where:

transformer rated primary current = 350A

transformer rated secondary current = 1050A

With the line charging current being negligible, the following relay settings are then suitable, and allow for transformer efficiency and mismatch due to tap-changing:

$I_{S1}=20\%$ (Minimum possible)

$$I_{S1}=20\%$$
$$k_1=30\%$$
$$k_2=150\%$$

10.12.3 Unit Protection of Transmission Circuits

Application of current differential to transmission circuits is similar to that described in Section 10.12.1, except that:

- Lines will often be longer, and hence have higher charging current.
- Signal attenuation in fibre optic channels will become larger.
- The relay may be expected to single-pole trip and reclose.
- High-speed backup distance protection elements may be brought into service automatically, in instances where signalling channel failure has been detected.

10.13 REFERENCE

[10.1] Merz-Price Protective Gear. K. Faye-Hansen and G. Harlow. IEE Proceedings, 1911.

Chapter 11
Distance Protection

11.1 INTRODUCTION

The problem of combining fast fault clearance with selective tripping of plant is a key aim for the protection of power systems. To meet these requirements, high-speed protection systems for transmission and primary distribution circuits that are suitable for use with the automatic reclosure of circuit breakers are under continuous development and are very widely applied.

Distance protection, in its basic form, is a non-unit system of protection offering considerable economic and technical advantages. Unlike phase and neutral overcurrent protection, the key advantage of distance protection is that its fault coverage of the protected circuit is virtually independent of source impedance variations. This is illustrated in Figure 11.1, where it can be seen that overcurrent protection cannot be applied satisfactorily.

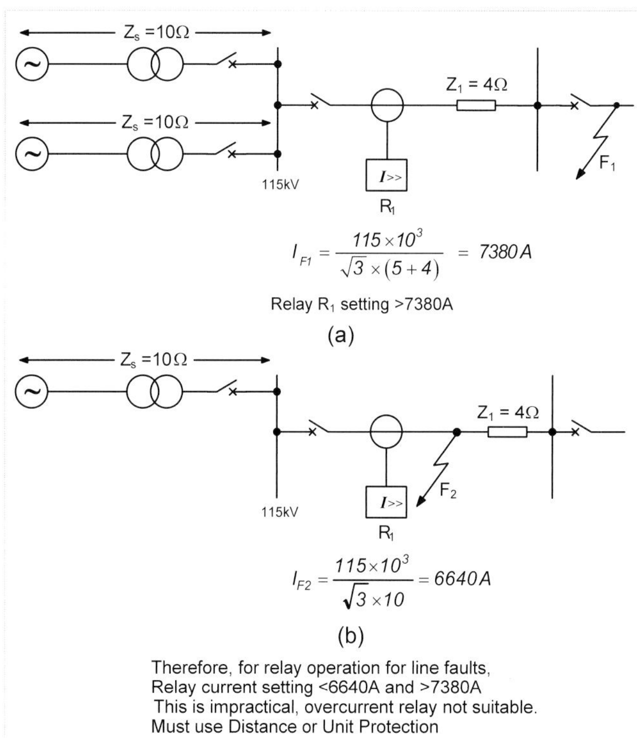

Figure 11.1: Advantages of distance over overcurrent protection

Distance protection is comparatively simple to apply and it can be fast in operation for faults located along most of a protected circuit. It can also provide both primary and remote back-up functions in a single scheme. It can easily be adapted to create a unit protection scheme when applied with a signalling channel. In this form it is eminently suitable for application with high-speed auto-reclosing, for the protection of critical transmission lines.

11.2 PRINCIPLES OF DISTANCE RELAYS

Since the impedance of a transmission line is proportional to its length, for distance measurement it is appropriate to use a relay capable of measuring the impedance of a line up to a predetermined point (the reach point). Such a relay is described as a distance relay and is designed to operate only for faults occurring between the relay location and the selected reach point, thus giving discrimination for faults that may occur in different line sections.

The basic principle of distance protection involves the division of the voltage at the relaying point by the measured current. The apparent impedance so calculated is compared with the reach point impedance. If the measured impedance is less than the reach point impedance, it is assumed that a fault exists on the line between the relay and the reach point.

The reach point of a relay is the point along the line impedance locus that is intersected by the boundary characteristic of the relay. Since this is dependent on the ratio of voltage and current and the phase angle between them, it may be plotted on an *R/X* diagram. The loci of power system impedances as seen by the relay during faults, power swings and load variations may be plotted on the same diagram and in this manner the performance of the relay in the presence of system faults and disturbances may be studied.

11.3 RELAY PERFORMANCE

Distance relay performance is defined in terms of reach accuracy and operating time. Reach accuracy is a comparison of the actual ohmic reach of the relay under practical conditions with the relay setting value in ohms. Reach accuracy particularly depends on the level of voltage presented to the relay under fault conditions. The impedance measuring techniques employed in particular relay designs also have an impact.

Operating times can vary with fault current, with fault position relative to the relay setting, and with the point on the voltage wave at which the fault occurs. Depending on the measuring techniques employed in a particular relay design, measuring signal transient errors, such as those produced by Capacitor Voltage Transformers or saturating CTs, can also adversely delay relay operation for faults close to the reach point. It is usual for electromechanical and static distance relays to claim both maximum and minimum operating times. However, for modern digital or numerical distance relays, the variation between these is small over a wide range of system operating conditions and fault positions.

11.3.1 Electromechanical/Static Distance Relays

With electromechanical and earlier static relay designs, the magnitude of input quantities particularly influenced both reach accuracy and operating time. It was customary to present information on relay performance by voltage/reach curves, as shown in Figure 11.2, and operating time/fault position curves for various values of system impedance ratios (*S.I.R.s*) as shown in Figure 11.3, where:

$$S.I.R = \frac{Z_S}{Z_L}$$

and

Z_S = system source impedance behind the relay location

Z_L = line impedance equivalent to relay reach setting

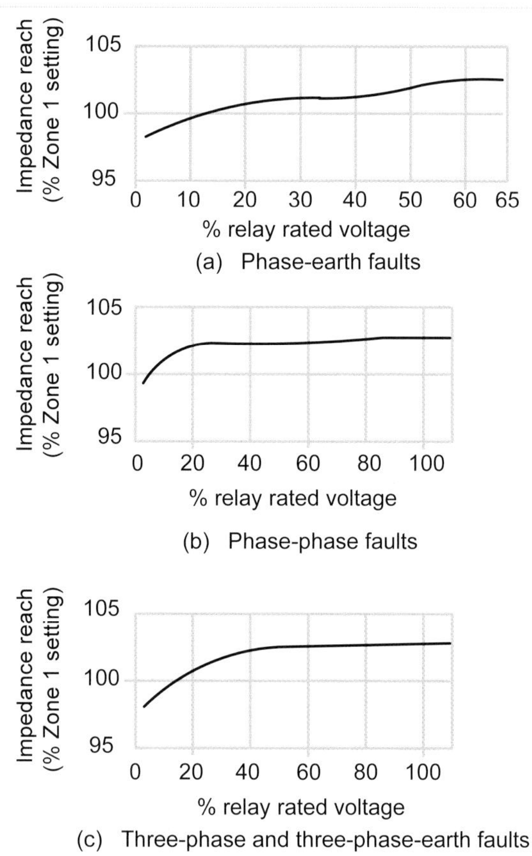

Figure 11.2: Typical impedance reach accuracy characteristics for Zone 1

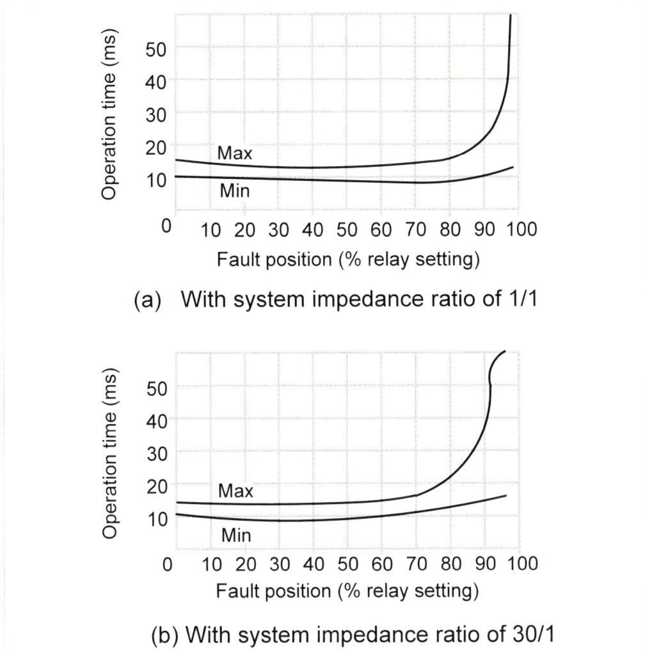

Figure 11.3: Typical operation time characteristics for Zone 1 phase-phase faults

Alternatively the above information was combined in a family of contour curves, where the fault position expressed as a percentage of the relay setting is plotted against the source to line impedance ratio, as illustrated in Figure 11.4.

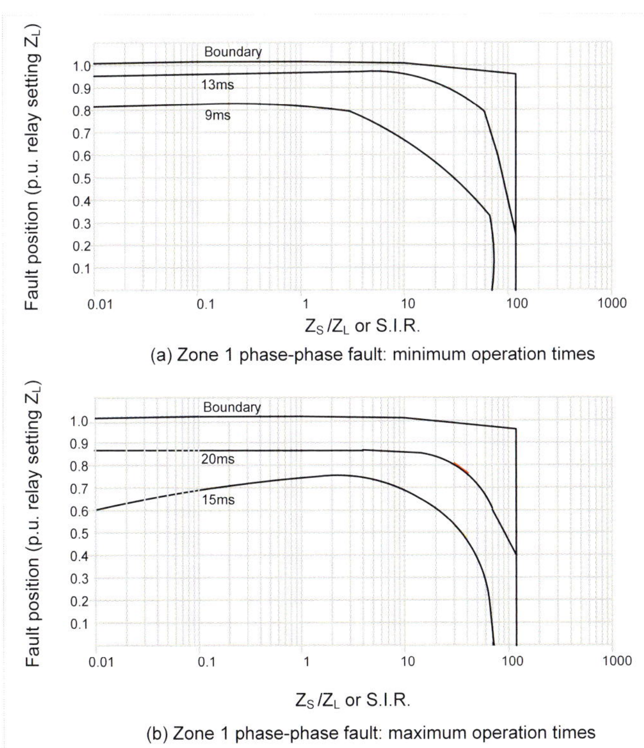

Figure 11.4: Typical operation-time contours

11.3.2 Digital/Numerical Distance Relays

Digital/Numerical distance relays tend to have more consistent operating times. The best transmission-class relays can achieve subcycle tripping, and the digital filtering techniques available ensure optimum performance under adverse waveform conditions or for boundary fault conditions.

11.4 RELATIONSHIP BETWEEN RELAY VOLTAGE AND ZS/ZL RATIO

A single, generic, equivalent circuit, as shown in Figure 11.5, may represent any fault condition in a three-phase power system. The voltage V applied to the impedance loop is the open circuit voltage of the power system. Point R represents the relay location; I_R and V_R are the current and voltage measured by the relay, respectively.

The impedances Z_s and Z_L are described as source and line impedances because of their position with respect to the relay location. Source impedance Z_s is a measure of the fault level at the relaying point. For faults involving earth it is dependent on the method of system earthing behind the relaying point. Line impedance Z_L is a measure of the impedance of the protected section. The voltage V_R applied to the relay is, therefore, $I_R Z_L$. For a fault at the reach point, this may be alternatively expressed in terms of source to line impedance ratio Z_S/Z_L using the following expressions:

$$V_R = I_R Z_L$$

where:

$$I_R = \frac{V}{Z_S + Z_L}$$

Therefore:

$$V_R = \frac{Z_L}{Z_S + Z_L} V$$

Or

$$V_R = \frac{1}{\left(\frac{Z_S}{Z_L}\right) + 1} V$$

Equation 11.1

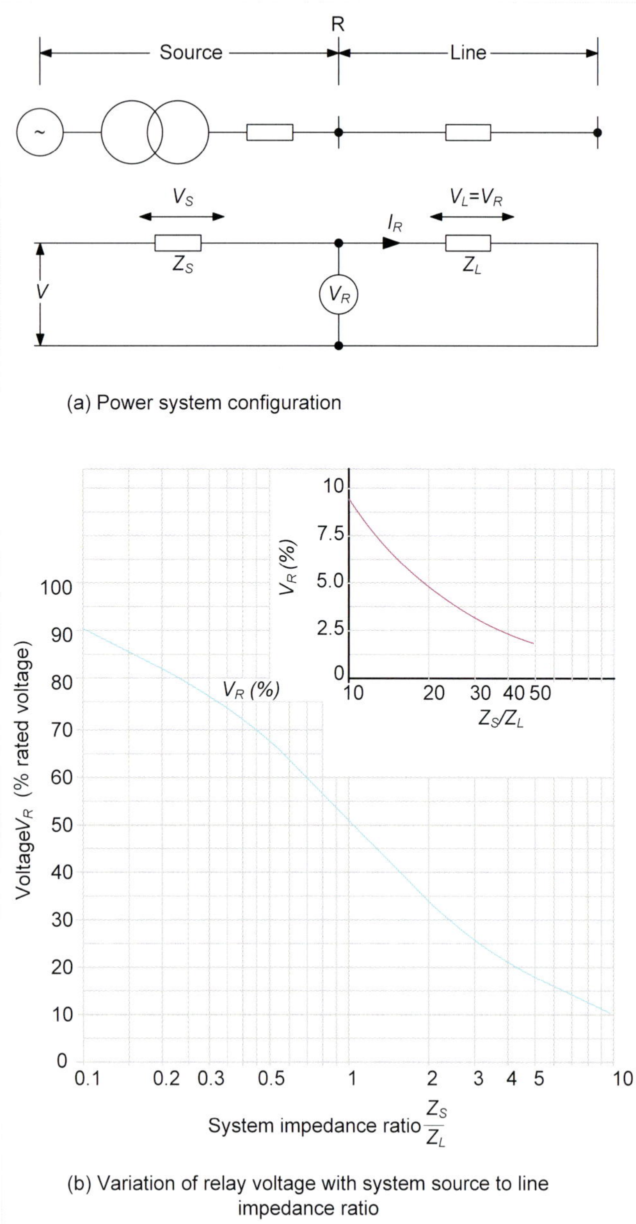

(a) Power system configuration

(b) Variation of relay voltage with system source to line impedance ratio

Figure 11.5: Relationship between source to line ratio and relay voltage

The above generic relationship between V_R and Z_S/Z_L, illustrated in Figure 11.5, is valid for all types of short circuits provided a few simple rules are observed. These are:

- for phase faults, V_Δ is the phase-phase source voltage and Z_S/Z_L is the positive sequence source to line impedance ratio. V_R is the phase-phase relay voltage and I_R is the phase-phase relay current, for the faulted phases

$$V_R = \frac{1}{\left(\frac{Z_S}{Z_L}\right)+1} V_\Delta$$

Equation 11.2

- for earth faults, $V_{l\text{-}n}$ is the phase-neutral source voltage and Z_S/Z_L, is a composite ratio involving the positive and zero sequence impedances. V_R is the phase-neutral relay voltage and I_R is the relay current for the faulted phase

$$V_R = \frac{1}{\left(\frac{Z_S}{Z_L}\right)\left(\frac{2+p}{2+q}\right)+1} V_{l-n}$$

Equation 11.3

where:

$$Z_S = 2Z_{S1} + Z_{S0} = Z_{S1}(2+p)$$

$$Z_L = 2Z_{L1} + Z_{L0} = Z_{L1}(2+q)$$

and

$$p = \frac{Z_{S0}}{Z_{S1}}$$

$$q = \frac{Z_{L0}}{Z_{L1}}$$

11.5 VOLTAGE LIMIT FOR ACCURATE REACH POINT MEASUREMENT

The ability of a distance relay to measure accurately for a reach point fault depends on the minimum voltage at the relay location under this condition being above a declared value. This voltage, which depends on the relay design, can also be quoted in terms of an equivalent maximum Z_S/Z_L or *S.I.R.*

Distance relays are designed so that, provided the reach point voltage criterion is met, any increased measuring errors for faults closer to the relay will not prevent relay operation. Most modern relays are provided with healthy phase voltage polarisation and/or memory voltage polarisation. The prime purpose of the relay polarising voltage is to ensure correct relay directional response for close-up faults, in the forward or reverse direction, where the fault-loop voltage measured by the relay may be very small.

11.6 ZONES OF PROTECTION

Careful selection of the reach settings and tripping times for the various zones of measurement enables correct co-ordination between distance relays on a power system. Basic distance protection will comprise instantaneous directional Zone 1 protection and one or more time-delayed zones. Typical reach and time settings for a 3-zone distance protection are shown in Figure 11.6. Digital and numerical distance relays may have up to five or six zones, some set to measure in the reverse direction. Typical settings for three forward-looking zones of basic distance protection are given in the following sub-sections. To determine the settings for a particular relay design or for a particular distance teleprotection scheme, involving end-to-end signalling, the relay manufacturer's instructions should be referred to.

11.6.1 Zone 1 Setting

Electromechanical/static relays usually have a reach setting of up to 80% of the protected line impedance for instantaneous Zone 1 protection. For digital/numerical distance relays, settings of up to 85% may be safe. The resulting 15-20% safety margin ensures that there is no risk of the Zone 1 protection over-reaching the protected line due to errors in the current and voltage transformers, inaccuracies in line impedance data provided for setting purposes and errors of relay setting and measurement. Otherwise, there would be a loss of discrimination with fast operating protection on the following line section. Zone 2 of the distance protection must cover the remaining 15-20% of the line.

11.6.2 Zone 2 Setting

To ensure full coverage of the line with allowance for the sources of error already listed in the previous section, the reach setting of the Zone 2 protection should be at least 120% of the protected line impedance. In many applications it is common practice to set the Zone 2 reach to be equal to the protected line section +50% of the shortest adjacent line. Where possible, this ensures that the resulting maximum effective Zone 2 reach does not extend beyond the minimum effective Zone 1 reach of the adjacent line protection. This avoids the need to grade the Zone 2 time settings between upstream and downstream relays. In electromechanical and static relays, Zone 2 protection is provided either by separate elements or by extending the reach of the Zone 1 elements after a time delay that is initiated by a fault detector. In most digital and numerical relays, the Zone 2 elements are implemented in software.

Zone 2 tripping must be time-delayed to ensure grading with the primary relaying applied to adjacent circuits that fall within the Zone 2 reach. Thus complete coverage of a line section is obtained, with fast clearance of faults in the first 80-85% of the line and somewhat slower clearance of faults in the remaining section of the line.

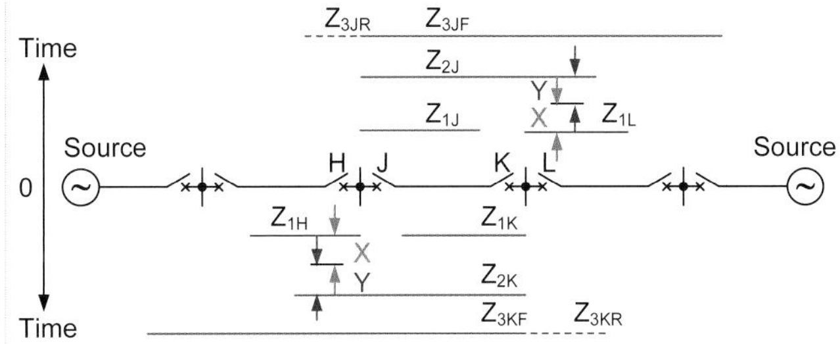

Zone 1 = 80-85% of protected line impedance
Zone 2 (minimum) = 120% of protected line
Zone 2 (maximum) < Protected line + 50% of shortest second line
Zone 3F = 1.2 (protected line + longest second line)
Zone 3R = 20% of protected line
X = Circuit breaker tripping time
Y = Discriminating time

Figure 11.6: Typical time/distance characteristics for three zone distance protection

11.6.3 Zone 3 Setting

Remote back-up protection for all faults on adjacent lines can be provided by a third zone of protection that is time delayed to discriminate with Zone 2 protection plus circuit breaker trip time for the adjacent line. Zone 3 reach should be set to at least 1.2 times the impedance presented to the relay for a fault at the remote end of the second line section.

On interconnected power systems, the effect of fault current infeed at the remote busbars will cause the impedance presented to the relay to be much greater than the actual impedance to the fault and this needs to be taken into account when setting Zone 3. In some systems, variations in the remote busbar infeed can prevent the application of remote back-up Zone 3 protection but on radial distribution systems with single end infeed, no difficulties should arise.

11.6.4 Settings for Reverse Reach and Other Zones

Modern digital or numerical relays may have additional impedance zones that can be utilised to provide additional protection functions. For example, where the first three zones are set as above, Zone 4 might be used to provide back-up protection for the local busbar, by applying a reverse reach setting of the order of 25% of the Zone 1 reach. Alternatively, one of the forward-looking zones (typically Zone 3) could be set with a small reverse offset reach from the origin of the R/X diagram, in addition to its forward reach setting. An offset impedance measurement characteristic is non-directional. One advantage of a non-directional zone of impedance measurement is that it is able to operate for a close-up, zero-impedance fault, in situations where there may be no healthy phase voltage signal or memory voltage signal available to allow operation of a directional impedance zone. With the offset-zone time delay bypassed, there can be provision of

'Switch-on-to-Fault' (SOTF) protection. This is required where there are line voltage transformers, to provide fast tripping in the event of accidental line energisation with maintenance earthing clamps left in position. Additional impedance zones may be deployed as part of a distance protection scheme used in conjunction with a teleprotection signalling channel.

11.7 DISTANCE RELAY CHARACTERISTICS

Some numerical relays measure the absolute fault impedance and then determine whether operation is required according to impedance boundaries defined on the R/X diagram. Traditional distance relays and numerical relays that emulate the impedance elements of traditional relays do not measure absolute impedance. They compare the measured fault voltage with a replica voltage derived from the fault current and the zone impedance setting to determine whether the fault is within zone or out-of-zone. Distance relay impedance comparators or algorithms which emulate traditional comparators are classified according to their polar characteristics, the number of signal inputs they have, and the method by which signal comparisons are made. The common types compare either the relative amplitude or phase of two input quantities to obtain operating characteristics that are either straight lines or circles when plotted on an R/X diagram. At each stage of distance relay design evolution, the development of impedance operating characteristic shapes and sophistication has been governed by the technology available and the acceptable cost. Since many traditional relays are still in service and since some numerical relays emulate the techniques of the traditional relays, a brief review of impedance comparators is justified.

11.7.1 Amplitude and Phase Comparison

Relay measuring elements whose functionality is based on the comparison of two independent quantities are essentially either amplitude or phase comparators. For the impedance elements of a distance relay, the quantities being compared are the voltage and current measured by the relay. There are numerous techniques available for performing the comparison, depending on the technology used. They vary from balanced-beam (amplitude comparison) and induction cup (phase comparison) electromagnetic relays, through diode and operational amplifier comparators in static-type distance relays, to digital sequence comparators in digital relays and to algorithms used in numerical relays.

Any type of impedance characteristic obtainable with one comparator is also obtainable with the other. The addition and subtraction of the signals for one type of comparator produces the required signals to obtain a similar characteristic using the other type. For example, comparing V and I in an amplitude comparator results in a circular impedance characteristic centred at the origin of the *R/X* diagram. If the sum and difference of V and I are applied to the phase comparator the result is a similar characteristic.

11.7.2 Plain Impedance Characteristic

This characteristic takes no account of the phase angle between the current and the voltage applied to it; for this reason its impedance characteristic when plotted on an *R/X* diagram is a circle with its centre at the origin of the co-ordinates and of radius equal to its setting in ohms. Operation occurs for all impedance values less than the setting, that is, for all points within the circle. The relay characteristic, shown in Figure 11.7, is therefore non-directional, and in this form would operate for all faults along the vector *AL* and also for all faults behind the busbars up to an impedance *AM*. *A* is the relaying point and *RAB* is the angle by which the fault current lags the relay voltage for a fault on the line *AB* and *RAC* is the equivalent leading angle for a fault on line *AC*. Vector *AB* represents the impedance in front of the relay between the relaying point *A* and the end of line *AB*. Vector *AC* represents the impedance of line *AC* behind the relaying point. *AL* represents the reach of instantaneous Zone 1 protection, set to cover 80% to 85% of the protected line.

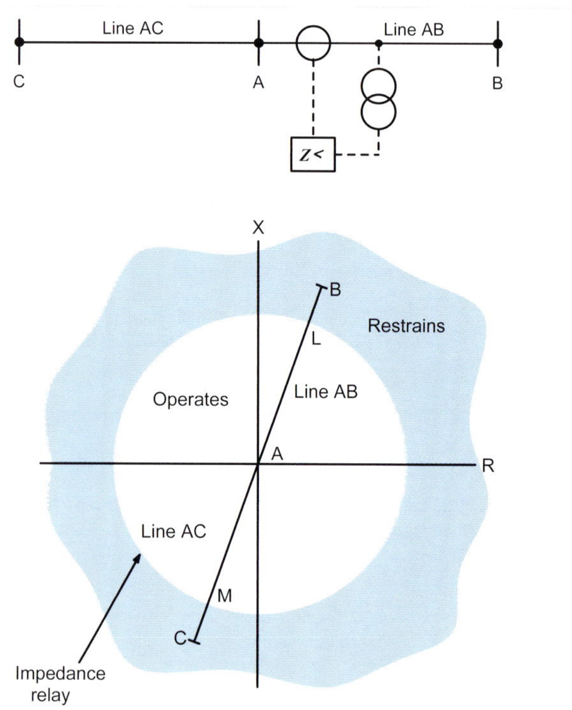

Figure 11.7: Plain impedance relay characteristic

A relay using this characteristic has three important disadvantages:

- it is non-directional; it will see faults both in front of and behind the relaying point, and therefore requires a directional element to give it correct discrimination
- it has non-uniform fault resistance coverage
- it is susceptible to power swings and heavy loading of a long line because of the large area covered by the impedance circle

Directional control is an essential discrimination quality for a distance relay, to make the relay non-responsive to faults outside the protected line. This can be obtained by the addition of a separate directional control element. The impedance characteristic of a directional control element is a straight line on the R/X diagram, so the combined characteristic of the directional and impedance relays is the semi-circle *APLQ* shown in Figure 11.8.

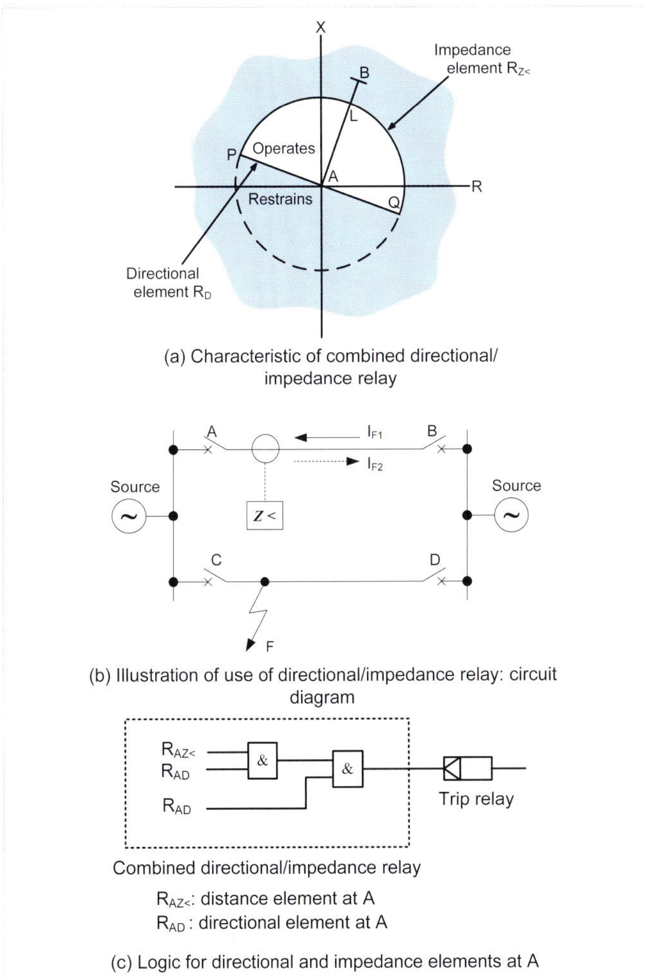

Figure 11.8: Combined directional and impedance relays

If a fault occurs at *F* close to *C* on the parallel line *CD*, the directional unit R_D at *A* will restrain due to current I_{F1}. At the same time, the impedance unit is prevented from operating by the inhibiting output of unit R_D. If this control is not provided, the under impedance element could operate prior to circuit breaker *C* opening. Reversal of current through the relay from I_{F1} to I_{F2} when *C* opens could then result in incorrect tripping of the healthy line if the directional unit R_D operates before the impedance unit resets. This is an example of the need to consider the proper co-ordination of multiple relay elements to attain reliable relay performance during evolving fault conditions. In older relay designs, the type of problem to be addressed was commonly referred to as one of 'contact race'.

11.7.3 Self-Polarised Mho Relay

The mho impedance element is generally known as such because its characteristic is a straight line on an admittance diagram. It cleverly combines the discriminating qualities of both reach control and directional control, thereby eliminating the 'contact race' problems that may be encountered with separate reach and directional control elements. This is achieved by the addition of a polarising signal. Mho impedance elements were particularly attractive for economic reasons where electromechanical relay elements were employed. As a result, they have been widely deployed worldwide for many years and their advantages and limitations are now well understood. For this reason they are still emulated in the algorithms of some modern numerical relays.

The characteristic of a mho impedance element, when plotted on an *R/X* diagram, is a circle whose circumference passes through the origin, as illustrated in Figure 11.9. This demonstrates that the impedance element is inherently directional and such that it will operate only for faults in the forward direction along line *AB*.

The impedance characteristic is adjusted by setting Zn, the impedance reach, along the diameter and φ, the angle of displacement of the diameter from the R axis. Angle φ is known as the Relay Characteristic Angle (RCA). The relay operates for values of fault impedance ZF within its characteristic.

The self-polarised mho characteristic can be obtained using a phase comparator circuit which compares input signals S_2 and S_1 and operates whenever S_2 lags S_1 by between 90° and 270°, as shown in the voltage diagram of Figure 11.9(a).

The two input signals are:

$$S_2 = V\text{-}IZ_n$$

$$S_1 = V$$

where:

V = fault voltage from VT secondary

I = fault current from CT secondary

Z_n = impedance setting of the zone

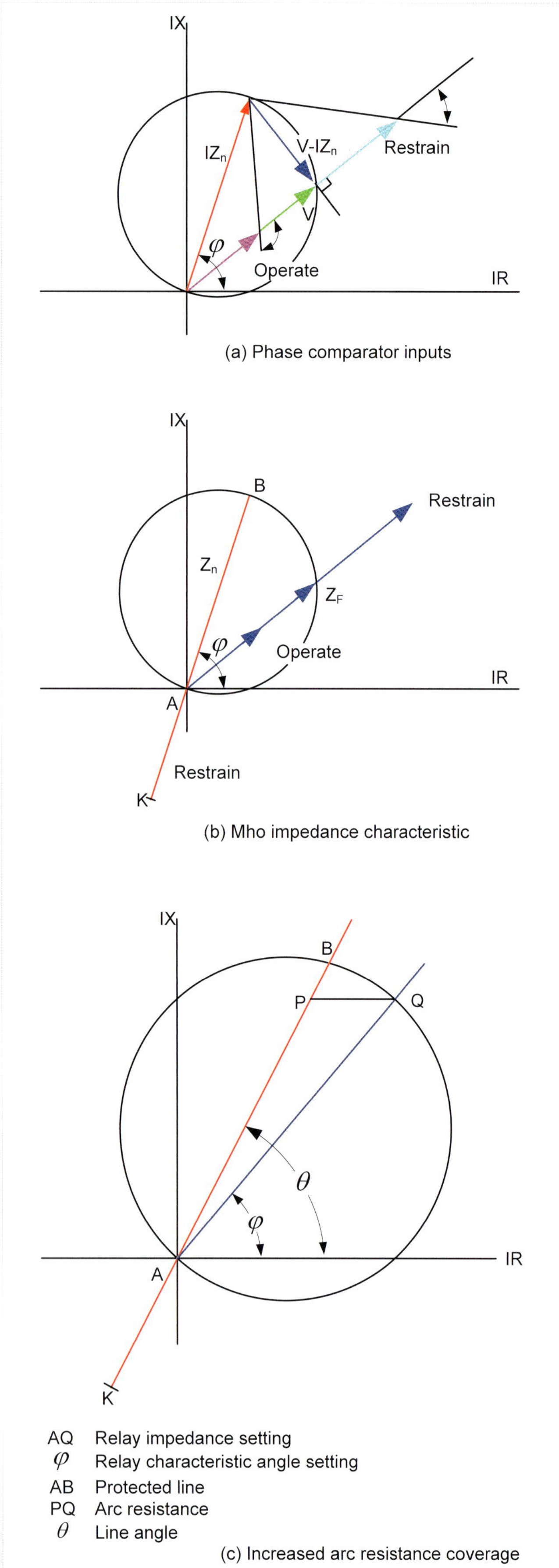

Figure 11.9: Mho relay characteristic

The characteristic of Figure 11.9(a) can be converted to the impedance plane of Figure 11.9(b) by dividing each voltage by I.

The impedance reach varies with fault angle. As the line to be protected is made up of resistance and inductance, its fault angle will be dependent upon the relative values of R and X at the system operating frequency. Under an arcing fault condition, or an earth fault involving additional resistance, such as tower footing resistance or fault through vegetation, the value of the resistive component of fault impedance will increase to change the impedance angle. Thus a relay having a characteristic angle equivalent to the line angle will under-reach under resistive fault conditions.

Some users set the RCA less than the line angle, so that it is possible to accept a small amount of fault resistance without causing under-reach. However, when setting the relay, the difference between the line angle θ and the relay characteristic angle $\varnothing$ must be known. The resulting characteristic is shown in Figure 11.9 where GL corresponds to the length of the line to be protected. With $\varnothing$ set less than θ, the actual amount of line protected, AB, would be equal to the relay setting value AQ multiplied by cosine $(\theta\text{-}\varnothing)$. Therefore the required relay setting AQ is given by:

$$AQ = \frac{AB}{\cos(\theta - \phi)}$$

Due to the physical nature of an arc, there is a non-linear relationship between arc voltage and arc current, which results in a non-linear resistance. Using the empirical formula derived by A.R. van C. Warrington, [11.1] the approximate value of arc resistance can be assessed as:

$$R_a = \frac{28{,}710}{I^{1.4}} L$$

Equation 11.4

where:

R_a = arc resistence (ohms)

L = length of arc (metres)

I = arc current (A)

On long overhead lines carried on steel towers with overhead earth wires the effect of arc resistance can usually be neglected. The effect is most significant on short overhead lines and with fault currents below 2000A (i.e. minimum plant condition), or if the protected line is of wood-pole construction without earth wires. In the latter case, the earth fault resistance reduces the effective earth-fault reach of a 'mho' Zone 1 element to such an extent that the majority of faults are detected in Zone 2 time. This problem can usually be overcome by using a relay with a cross-polarised mho or a polygonal characteristic.

Where a power system is resistance-earthed, it should be appreciated that this does not need to be considered with regard to the relay settings other than the effect that reduced fault current may have on the value of arc resistance seen. The earthing resistance is in the source behind the relay and only modifies the source angle and source to line impedance ratio for earth faults. It would therefore be taken into account only when assessing relay performance in terms of system impedance ratio.

11.7.4 Offset Mho/Lenticular Characteristics

Under close up fault conditions, when the relay voltage falls to zero or near-zero, a relay using a self-polarised mho characteristic or any other form of self-polarised directional impedance characteristic may fail to operate when it is required to do so. Methods of covering this condition include the use of non-directional impedance characteristics, such as offset mho, offset lenticular, or cross-polarised and memory polarised directional impedance characteristics.

If current bias is employed, the mho characteristic is shifted to embrace the origin, so that the measuring element can operate for close-up faults in both the forward and the reverse directions. The offset mho relay has two main applications:

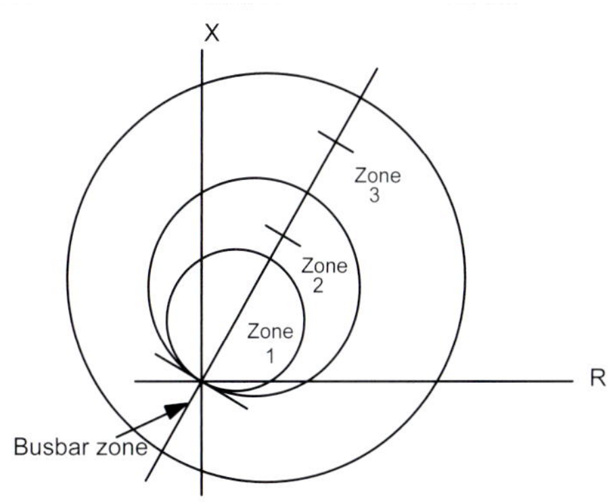

(a) Busbar zone back-up using an offset mho relay

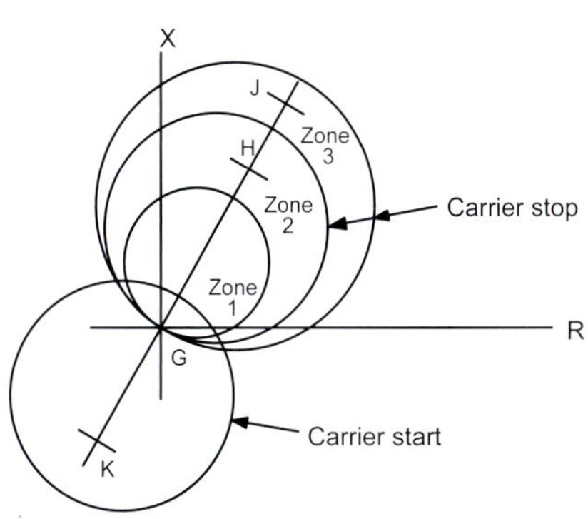

(b) Carrier starting in distance blocking schemes

Figure 11.10: Typical applications for the offset mho relay

11.7.4.1 Third Zone and Busbar Back-Up Zone

In this application it is used in conjunction with mho measuring units as a fault detector and/or Zone 3 measuring unit. So, with the reverse reach arranged to extend into the busbar zone, as shown in Figure 11.10, it will provide back-up protection for busbar faults. This facility can also be provided with quadrilateral characteristics. A further benefit of the Zone 3 application is for Switch-on-to-Fault (SOTF) protection, where the Zone 3 time delay would be bypassed for a short period immediately following line energisation to allow rapid clearance of a fault anywhere along the protected line.

11.7.4.2 Carrier Starting Unit in Distance Schemes With Carrier Blocking

If the offset mho unit is used for starting carrier signalling, it is arranged as shown in Figure 11.10. The carrier is transmitted if the fault is external to the protected line but inside the reach of the offset mho relay, to prevent accelerated tripping of the second or third zone relay at the remote station. Transmission is prevented for internal faults by operation of the local mho measuring units, which allows high-speed fault clearance by the local and remote end circuit breakers.

11.7.4.3 Application of Lenticular Characteristic

There is a danger that the offset mho relay shown in Figure 11.10 may operate under maximum load transfer conditions if Zone 3 of the relay has a large reach setting. A large Zone 3 reach may be required to provide remote back-up protection for faults on the adjacent feeder. To avoid this, a shaped type of characteristic may be used, where the resistive coverage is restricted. With a 'lenticular' characteristic, the aspect ratio of the lens $\left(\frac{a}{b}\right)$ is adjustable, enabling it to be set to provide the maximum fault resistance coverage consistent with non-operation under maximum load transfer conditions.

Figure 11.11 shows how the lenticular characteristic can tolerate much higher degrees of line loading than offset mho and plain impedance characteristics. Reduction of load impedance from Z_{D3} to Z_{D1} will correspond to an equivalent increase in load current.

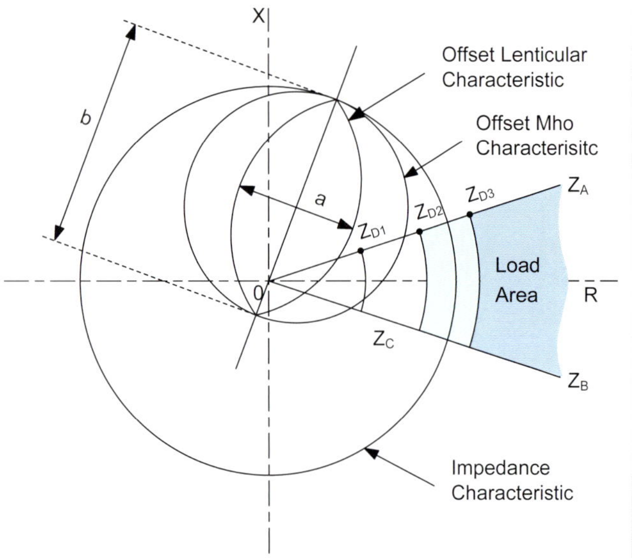

Figure 11.11: Minimum load impedance permitted with lenticular, offset mho and impedance relays

It can be observed in Figure 11.11 how the load area is defined according to a minimum impedance arc, constrained by straight lines which emanate from the origin, O. Modem numerical relays typically do not use lenticular characteristic shaping, but instead use load encroachment (load blinder) detection. This allows a full mho characteristic to be used, but with tripping prevented in the region of the impedance plane known to be frequented by load (Z_A-Z_B-Z_C-Z_D).

11.7.5 Fully Cross-Polarised Mho Characteristic

The previous section showed how the non-directional offset mho characteristic is inherently able to operate for close-up zero voltage faults, where there would be no polarising voltage to allow operation of a plain mho directional element. One way of ensuring correct mho element response for zero-voltage faults is to add a percentage of voltage from the healthy phase(s) to the main polarising voltage as a substitute phase reference. This technique is called cross-polarising, and it has the advantage of preserving and indeed enhancing the directional properties of the mho characteristic. By the use of a phase voltage memory system, that provides several cycles of pre-fault voltage reference during a fault, the cross-polarisation technique is also effective for close-up three-phase faults. For this type of fault, no healthy phase voltage reference is available.

Early memory systems were based on tuned, resonant, analogue circuits, but problems occurred when applied to networks where the power system operating frequency could vary. More modern digital or numerical systems can offer a synchronous phase reference for variations in power system frequency before or even during a fault.

As described in Section 11.7.3, a disadvantage of the self-polarised, plain mho impedance characteristic, when applied to overhead line circuits with high impedance angles, is that it has limited coverage of arc or fault resistance. The problem is aggravated in the case of short lines, since the required Zone 1 ohmic setting is low. The amount of the resistive coverage offered by the mho circle is directly related to the forward reach setting. Hence, the resulting resistive coverage may be too small in relation to the expected values of fault resistance.

One additional benefit of applying cross-polarisation to a mho impedance element is that its resistive coverage will be enhanced. This effect is illustrated in Figure 11.12, for the case where a mho element has 100% cross-polarisation. With cross-polarisation from the healthy phase(s) or from a memory system, the mho resistive expansion will occur during a balanced three-phase fault as well as for unbalanced faults. The expansion will not occur under load conditions, when there is no phase shift between the measured voltage and the polarising voltage. The degree of resistive reach enhancement depends on the ratio of source impedance to relay reach (impedance) setting as can be deduced by reference to Figure 11.13.

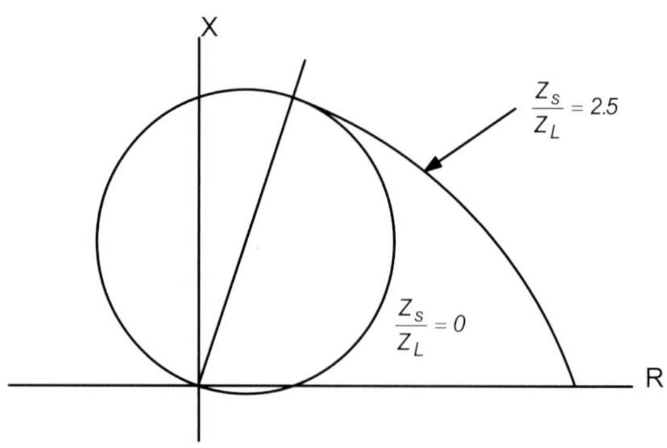

Figure 11.12: Fully cross-polarised mho relay characteristic with variations of Z_S/Z_L ratio

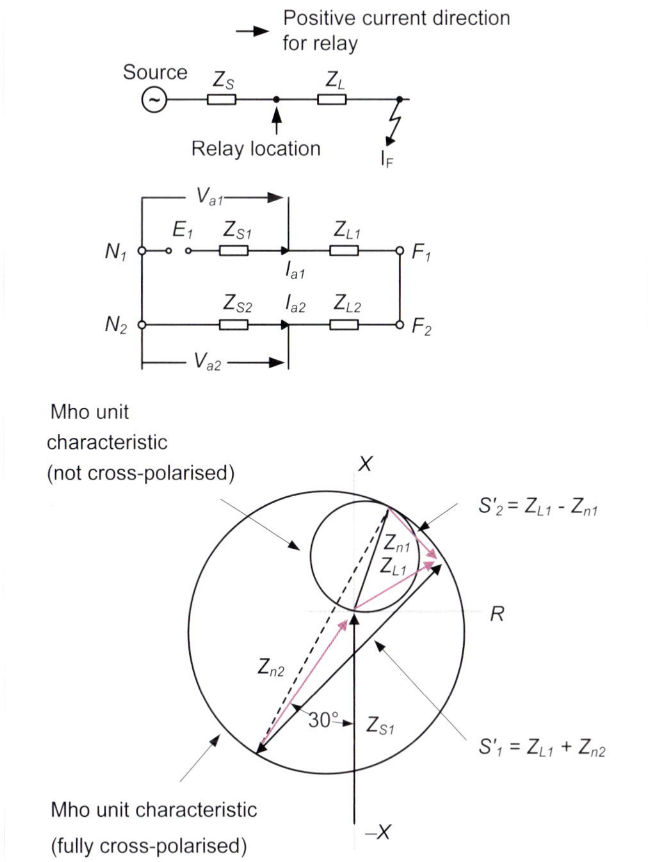

Figure 11.13: Illustration of improvement in relay resistive coverage for fully cross-polarised characteristic

It must be emphasised that the apparent extension of a fully cross-polarised impedance characteristic into the negative reactance quadrants of Figure 11.13 does not imply that there would be operation for reverse faults. With cross-polarisation, the relay characteristic expands to encompass the origin of the impedance diagram for forward faults only. For reverse faults, the effect is to exclude the origin of the impedance diagram, thereby ensuring proper directional responses for close-up forward or reverse faults.

Fully cross-polarised characteristics have now largely been superseded, due to the tendency of comparators connected to healthy phases to operate under heavy fault conditions on another phase. This is of no consequence in a switched distance relay, where a single comparator is connected to the correct fault loop impedance by starting units before measurement begins. However, modern relays offer independent impedance measurement for each of the three earth-fault and three phase-fault loops. For these types of relay, mal-operation of healthy phases is undesirable, especially when single-pole tripping is required for single-phase faults.

11.7.6 Partially Cross-Polarised Mho Characteristic

Where a reliable, independent method of faulted phase selection is not provided, a modern non-switched distance relay may only employ a relatively small percentage of cross polarisation. The level selected must be sufficient to provide reliable directional control in the presence of CVT transients for close-up faults, and also attain reliable faulted phase selection. By employing only partial cross-polarisation, the disadvantages of the fully cross-polarised characteristic are avoided, while still retaining the advantages. Figure 11.14 shows a typical characteristic that can be obtained using this technique (reference Micromho, Quadramho and Optimho family).

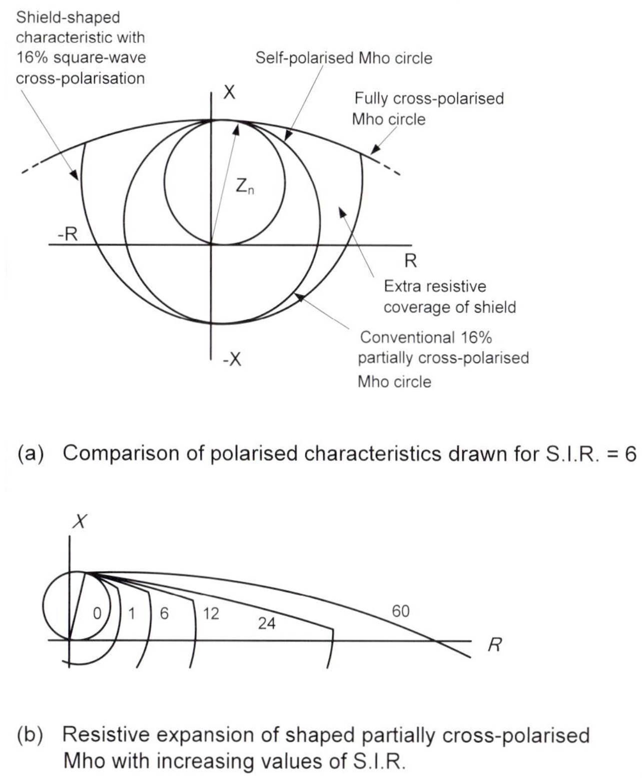

Figure 11.14: Partially cross-polarised characteristic with 'shield' shape

11.7.7 Quadrilateral Characteristic

This form of polygonal impedance characteristic is shown in Figure 11.15. The characteristic is provided with forward reach and resistive reach settings that are independently adjustable. It therefore provides better resistive coverage than any mho-type characteristic for short lines. This is especially true for earth fault impedance measurement, where the arc resistances and fault resistance to earth contribute to the highest values of fault resistance. To avoid excessive errors in the zone reach accuracy, it is common to impose a maximum resistive reach in terms of the zone impedance reach. Recommendations in this respect can usually be found in the appropriate relay manuals.

Quadrilateral elements with plain reactance reach lines can introduce reach error problems for resistive earth faults where the angle of total fault current differs from the angle of the current measured by the relay. This will be the case where the local and remote source voltage vectors are phase shifted with respect to each other due to pre-fault power flow. This can be overcome by selecting an alternative to use of a phase current for polarisation of the reactance reach line. Polygonal impedance characteristics are highly flexible in terms of fault impedance coverage for both phase and earth faults. For this reason, most digital and numerical distance relays now offer this form of characteristic. A further factor is that the additional cost implications of implementing this characteristic using discrete component electromechanical or early static relay technology do not arise.

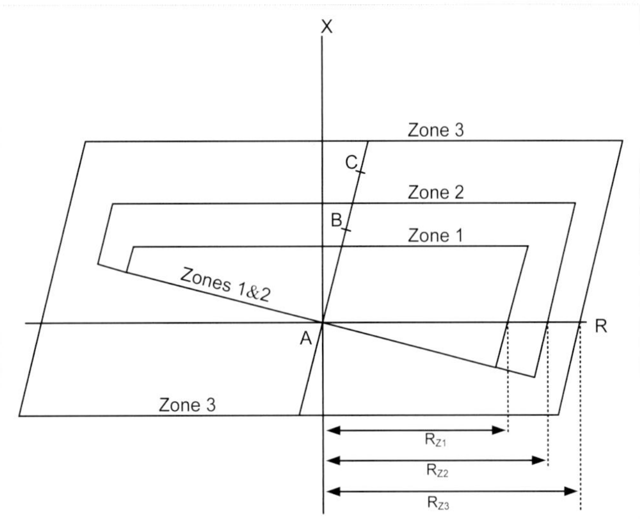

Figure 11.15: Quadrilateral characteristic

11.7.8 Protection against Power Swings – Use of the Ohm Characteristic

During severe power swing conditions from which a system is unlikely to recover, stability might only be regained if the swinging sources are separated. Where such scenarios are identified, power swing, or out-of-step, tripping protection can be deployed, to strategically split a power system at a preferred location. Ideally, the split should be made so that the plant capacity and connected loads on either side of the split are matched.

This type of disturbance cannot normally be correctly identified by an ordinary distance protection. As previously mentioned, it is often necessary to prevent distance protection schemes from operating during stable or unstable power swings, to avoid cascade tripping. To initiate system separation for a prospective unstable power swing, an out-of-step tripping scheme employing ohm impedance measuring elements can be deployed.

Ohm impedance characteristics are applied along the forward and reverse resistance axes of the R/X diagram and their operating boundaries are set to be parallel to the protected line impedance vector, as shown in Figure 11.16.

The ohm impedance elements divide the R/X impedance diagram into three zones, A, B and C. As the impedance changes during a power swing, the point representing the impedance moves along the swing locus, entering the three zones in turn and causing the ohm units to operate in sequence. When the impedance enters the third zone the trip sequence is completed and the circuit breaker trip coil can be energised at a favourable angle between system sources for arc interruption with little risk of restriking.

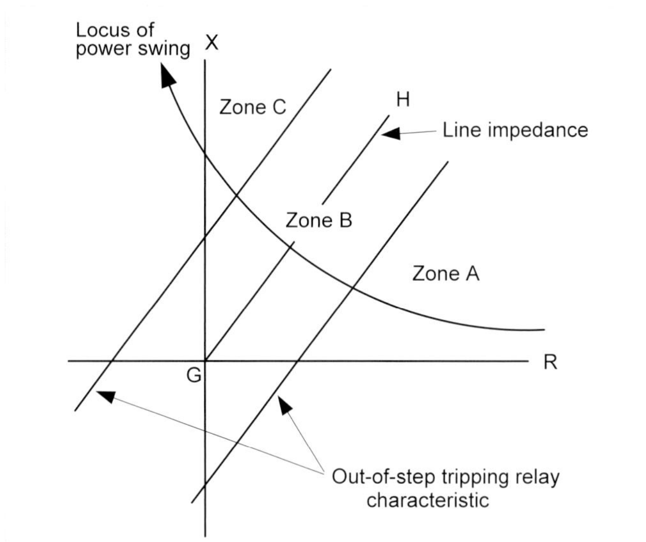

Figure 11.16: Application of out-of-step tripping relay characteristic

Only an unstable power swing condition can cause the impedance vector to move successively through the three zones. Therefore, other types of system disturbance, such as power system fault conditions, will not result in relay element operation.

11.7.9 Other Characteristics

The execution time for the algorithm for traditional distance protection using quadrilateral or similar characteristics may result in a relatively long operation time, possibly up to 40ms in some relay designs. To overcome this, some numerical distance relays also use alternative algorithms that can be executed significantly faster. These algorithms are based generally on detecting changes in current and voltage that are in excess of what is expected, often known as the 'Delta' algorithm.

This algorithm detects a fault by comparing the measured values of current and voltage with the values sampled previously. If the change between these samples exceeds a predefined amount (the 'delta'), it is assumed a fault has occurred. In parallel, the distance to fault is also computed. Provided the computed distance to fault lies within the Zone reach of the relay, a trip command is issued. This algorithm can be executed significantly faster than the conventional distance algorithm, resulting in faster overall tripping times. Faulted phase selection can be carried out by comparing the signs of the changes in voltage and current.

Relays that use the 'Delta' algorithm generally run both this and conventional distance protection algorithms in parallel, as some types of fault (e.g. high-resistance faults) may not fall within the fault detection criteria of the 'Delta' algorithm.

11.8 DISTANCE RELAY IMPLEMENTATION

Discriminating zones of protection can be achieved using distance relays, provided that fault distance is a simple function of impedance. While this is true in principle for transmission circuits, the impedances actually measured by a distance relay also depend on the following factors:

- the magnitudes of current and voltage (the relay may not see all the current that produces the fault voltage)
- the fault impedance loop being measured
- the type of fault
- the fault resistance
- the symmetry of line impedance
- the circuit configuration (single, double or multi-terminal circuit)

It is impossible to eliminate all of the above factors for all possible operating conditions. However, considerable success can be achieved with a suitable distance relay. This may comprise relay elements or algorithms for starting, distance measuring and for scheme logic. The distance measurement elements may produce impedance characteristics selected from those described in Section 11.7. Various distance relay formats exist, depending on the operating speed required and cost considerations related to the relaying hardware, software or numerical relay processing capacity required. The most common formats are:

- a single measuring element for each phase is provided, that covers all phase faults
- a more economical arrangement is for 'starter' elements to detect which phase or phases have suffered a fault. The starter elements switch a single measuring element or algorithm to measure the most appropriate fault impedance loop. This is commonly referred to as a switched distance relay
- a single set of impedance measuring elements for each impedance loop may have their reach settings progressively increased from one zone reach setting to another. The increase occurs after zone time delays that are initiated by operation of starter elements. This type of relay is commonly referred to as a reach-stepped distance relay
- each zone may be provided with independent sets of impedance measuring elements for each impedance loop. This is known as a full distance scheme, capable of offering the highest performance in terms of speed and application flexibility

Furthermore, protection against earth faults may require different characteristics and/or settings to those required for phase faults, resulting in additional units being required. A total of 18 impedance-measuring elements or algorithms would be required in a full scheme distance relay for three-zone protection for all types of fault. With electromechanical or static technology, each of the measuring elements would have been a separate relay housed in its own case, so that the distance relay comprised a panel-mounted assembly of the required relays with suitable inter-unit wiring. Figure 11.17(a) shows an example of such a relay scheme.

Figure 11.17a: Static distance relay

Figure 11.17b: MiCOM P440 series numerical distance relay

Digital/numerical distance relays (Figure 11.17(b) are likely to have all of the above functions implemented in software. Starter units may not be necessary. The complete distance relay is housed in a single unit, making for significant economies in space, wiring and increased dependability, through the increased availability that stems from the provision of continuous self-supervision. When the additional features detailed in Section 11.11 are taken into consideration, such equipment offers substantial user benefits.

11.8.1 Starters for Switched Distance Protection

Electromechanical and static distance relays do not normally use an individual impedance-measuring element per phase. The cost and the resulting physical scheme size made this arrangement impractical, except for the most demanding EHV transmission applications. To achieve economy for other applications, only one measuring element was provided, together with 'starter' units that detected which phases were faulted, to switch the appropriate signals to the single measuring function. A distance relay using this technique is known as a switched distance relay. A number of different types of starters have been used, the most common being based on overcurrent, undervoltage or under-impedance measurement.

Numerical distance relays permit direct detection of the phases involved in a fault. This is called faulted phase selection, often abbreviated to phase selection. Several techniques are available for faulted phase selection, which then permits the appropriate distance-measuring zone to trip. Without phase selection, the relay risks having over or underreach problems, or tripping three-phase when single-pole fault clearance is required. Several techniques are available for faulted phase selection, such as:

- superimposed current comparisons, comparing the step change of level between pre-fault load, and fault current (the 'delta' algorithm). This enables very fast detection of the faulted phases, within only a few samples of the analogue current inputs
- change in voltage magnitude
- change in current magnitude

Numerical phase selection is much faster than traditional starter techniques used in electromechanical or static distance relays. It does not impose a time penalty as the phase selection and measuring zone algorithms run in parallel. It is possible to build a full-scheme relay with these numerical techniques. The phase selection algorithm provides faulted phase selection, together with a segregated measuring algorithm for each phase-ground and phase to phase fault loop (AN, BN, CN, AB, BC, CA), thus ensuring full-scheme operation.

However, there may be occasions where a numerical relay that mimics earlier switched distance protection techniques is desired. The reasons may be economic (less software required – thus cheaper than a relay that contains a full-scheme implementation) and/or technical. Some applications may require the numerical relay characteristics to match those of earlier generations already installed on a network, to aid selectivity. Such relays are available, often with refinements such as multi-sided polygonal impedance characteristics that assist in avoiding tripping due to heavy load conditions.

With electromechanical or static switched distance relays, a selection of available starters often had to be made. The choice of starter was dependent on power system parameters such as maximum load transfer in relation to maximum reach required and power system earthing arrangements.

Where overcurrent starters are used, care must be taken to ensure that, with minimum generating plant in service, the setting of the overcurrent starters is sensitive enough to detect faults beyond the third zone. Furthermore, these starters require a high drop-off to pick-up ratio, to ensure that they will drop off under maximum load conditions after a second or third zone fault has been cleared by the first zone relay in the faulty section. Without this feature, indiscriminate tripping may result for subsequent faults in the second or third zone. For satisfactory operation of the overcurrent starters in a switched distance scheme, the following conditions must be fulfilled:

- the current setting of the overcurrent starters must be not less than 1.2 times the maximum full load current of the protected line
- the power system minimum fault current for a fault at the Zone 3 reach of the distance relay must not be less than 1.5 times the setting of the overcurrent starters

On multiple-earthed systems where the neutrals of all the power transformers are solidly earthed, or in power systems where the fault current is less than the full load current of the protected line, it is not possible to use overcurrent starters. In these circumstances under-impedance starters are typically used.

The type of under-impedance starter used is mainly dependent on the maximum expected load current and equivalent minimum load impedance in relation to the required relay setting to cover faults in Zone 3. This is illustrated in Figure 11.11 where Z_{D1}, Z_{D2}, and Z_{D3} are respectively the minimum load impedances permitted when lenticular, offset mho and impedance relays are used.

11.9 EFFECT OF SOURCE IMPEDANCE AND EARTHING METHODS

For correct operation, distance relays must be capable of measuring the distance to the fault accurately. To ensure this, it is necessary to provide the correct measured quantities to the measurement elements. It is not always the case that use of the voltage and current for a particular phase will give the correct result, or that additional compensation is required.

11.9.1 Phase Fault Impedance Measurement

Figure 11.18 shows the current and voltage relations for the different types of fault. If Z_{S1} and Z_{L1} are the source and line positive sequence impedances, viewed from the relaying point, the currents and voltages at this point for double phase faults are dependent on the source impedance as well as the line impedance. The relationships are given in Figure 11.19.

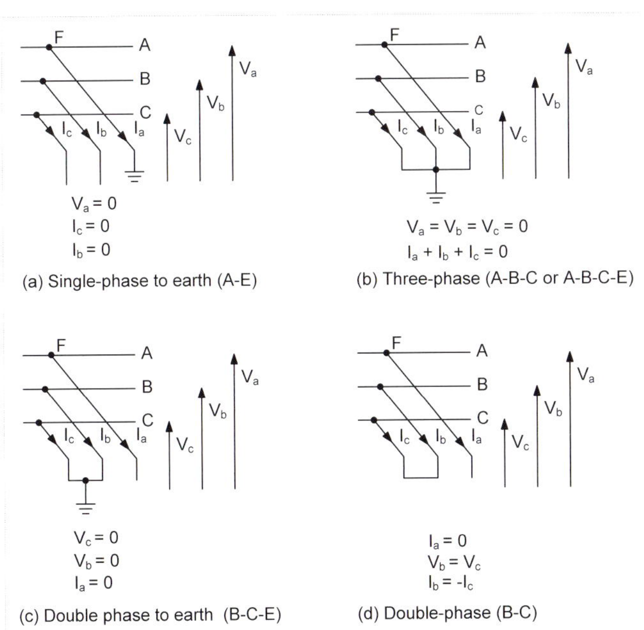

Figure 11.18: Current and voltage relationships for some shunt faults

Fault quantity	Three-phase (A-B-C)	Double-phase (B-C)
I'_a	I'_1	0
I'_b	$a^2 I'_1$	$(a^2 - a) I'_1$
I'_c	$a I'_1$	$(a - a^2) I'_1$
V'_a	$Z_{L1} I'_1$	$2(Z_{S1} + Z_{L1}) I'_1$
V'_b	$a^2 Z_{L1} I'_1$	$(2a^2 Z_{L1} - Z_{S1}) I'_1$
V'_c	$a Z_{L1} I'_1$	$(2a Z_{L1} - Z_{S1}) I'_1$

NOTE: $I'_1 = \frac{1}{3}(I'_a + aI'_b + a^2 I'_c)$

I' and V' are at relay location

Figure 11.19: Phase currents and voltages at relaying point for 3-phase and double-phase faults

Applying the difference of the phase voltages to the relay eliminates the dependence on Z_{S1}. For example:

$$V'_{bc} = (a^2 - a) Z_{L1} I'_1 \text{ for 3 phase faults}$$

$$V'_{bc} = 2(a^2 - a) Z_{L1} I'_1 \text{ for double phase faults}$$

Distance measuring elements are usually calibrated in terms of the positive sequence impedance. Correct measurement for both phase-phase and three-phase faults is achieved by supplying each phase-phase measuring element with its corresponding phase-phase voltage and difference of phase currents. Thus, for the B-C element, the current measured will be:

$$I'_b - I'_c = (a^2 - a) I'_1 \text{ for 3 phase faults}$$

$$I'_b - I'_c = 2(a^2 - a) I'_1 \text{ for double phase faults}$$

and the relay will measure Z_{L1} in each case.

11.9.2 Earth Fault Impedance Measurement

When a phase-earth fault occurs, the phase-earth voltage at the fault location is zero. It would appear that the voltage drop to the fault is simply the product of the phase current and line impedance. However, the current in the fault loop depends on the number of earthing points, the method of earthing and sequence impedances of the fault loop. Unless these factors are taken into account, the impedance measurement will be incorrect.

The voltage drop to the fault is the sum of the sequence voltage drops between the relaying point and the fault. The voltage drop to the fault and current in the fault loop are:

$$V'_a = I'_1 Z_{L1} + I'_2 Z_{L1} + I'_o Z_{Lo}$$

$$I'_a = I'_1 + I'_2 + I'_o$$

and the residual current I'_N at the relaying point is given by:

$$I'_N = I'_a + I'_b + I'_c = 3I'_o$$

Where I'_a, I'_b, I'_c are the phase currents at the relaying point. From the above expressions, the voltage at the relaying point can be expressed in terms of:

- the phase currents at the relaying point,
- the ratio of the transmission line zero sequence to positive sequence impedance, K, $\left(= Z_{LO}/Z_{L1}\right)$
- the transmission line positive sequence impedance Z_{L1}:

$$V'_a = Z_{L1}\left\{I'_a + \left(I'_a + I'_b + I'_c\right)\frac{K-1}{3}\right\}$$

Equation 11.5

The voltage appearing at the relaying point, as previously mentioned, varies with the number of infeeds, the method of system earthing and the position of the relay relative to the infeed and earthing points in the system. Figure 11.20 illustrates the three possible arrangements that can occur in practice with a single infeed. In Figure 11.20(a), the healthy phase currents are zero, so that the phase currents I_a, I_b and I_c have a 1-0-0 pattern. The impedance seen by a relay comparing I_a and V_a is:

$$Z = \left\{1 + \frac{(K-1)}{3}\right\} Z_{L1}$$

Equation 11.6

In Figure 11.20(b), the currents entering the fault from the relay branch have a 2-1-1 distribution, so:

$$Z = Z_{L1}$$

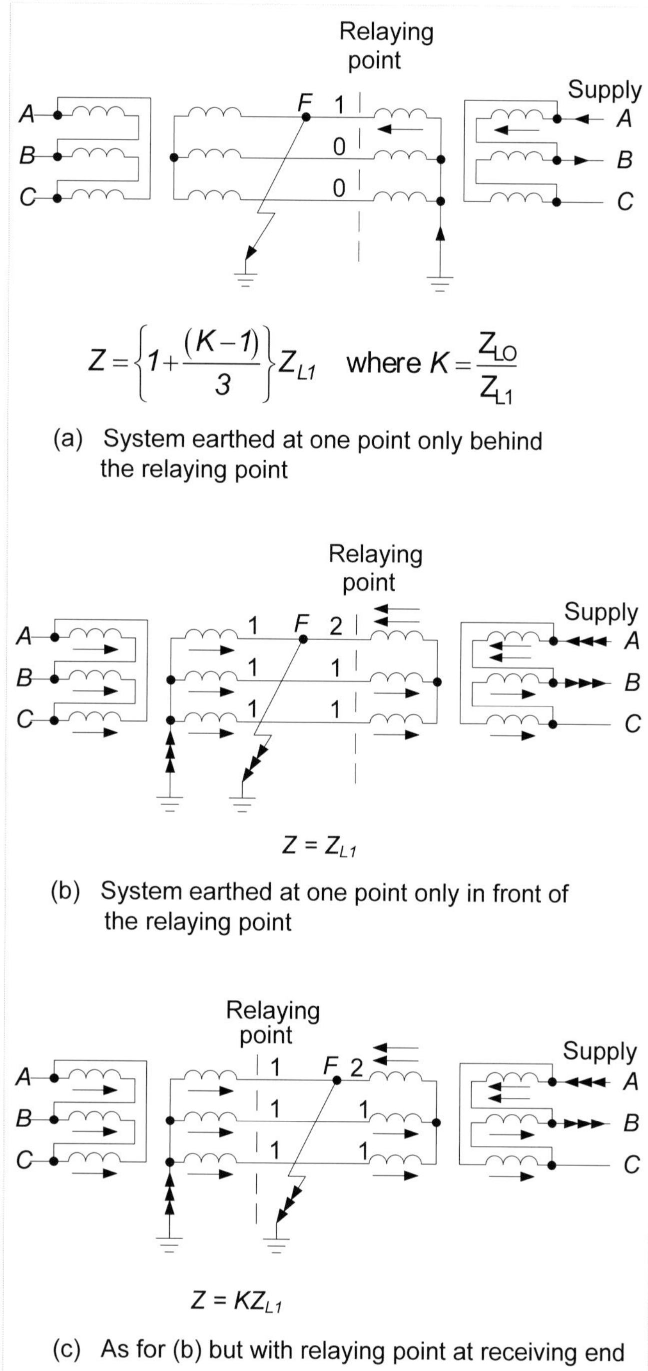

Figure 11.20: Effect of infeed and earthing arrangements on earth fault distance measurement

In Figure 11.20(c), the phase currents have a 1-1-1 distribution, and hence:

$$Z = KZ_{L1}$$

If there were infeeds at both ends of the line, the impedance measured would be a superposition of any two of the above examples, with the relative magnitudes of the infeeds taken into account.

This analysis shows that the relay can only measure an impedance which is independent of infeed and earthing arrangements if a proportion $K_N = \frac{(K-1)}{3}$ of the residual current $I_N = I_a + I_b + I_c$ is added to the phase current I_a. This technique is known as '*residual compensation*'.

Static distance relays compensate for the earth fault conditions by using an additional replica impedance Z_N within the measuring circuits. This compensation is implemented in software in numerical relays. Whereas the phase replica impedance Z_1 is fed with the phase current at the relaying point, Z_N is fed with the full residual current. The value of Z_N is adjusted so that for a fault at the reach point, the sum of the voltages developed across Z_1 and Z_N equals the measured phase to neutral voltage in the faulted phase.

The required setting for Z_N can be determined by considering an earth fault at the reach point of the relay. This is illustrated with reference to the A-N fault with single earthing point behind the relay as in Figure 11.20(a)

Voltage supplied from the VT's:

$$= I_1(Z_1 + Z_2 + Z_O) = I_1(2Z_1 + Z_O)$$

Voltage across replica impedances:

$$= I_A Z_1 + I_N Z_N$$

$$= I_A(Z_1 + Z_N)$$

$$= I_A(Z_1 + Z_N) = 3I_1(Z_1 + Z_N)$$

Hence, the required setting of Z_N for balance at the reach point is given by equating the above two expressions:

$$3I_1(Z_1 + Z_N) = I_1(2Z_1 + Z_0)$$

$$\therefore Z_N = \frac{Z_O - Z_1}{3}$$

$$= \frac{(Z_O - Z_1)}{3Z_1} Z_1 = K_N Z_1$$

Equation 11.7

With the replica impedance set to $\frac{Z_O - Z_1}{3}$, earth fault measuring elements will measure the fault impedance correctly, irrespective of the number of infeeds and earthing points on the system.

11.10 DISTANCE RELAY APPLICATION PROBLEMS

Distance relays may suffer from a number of difficulties in their application. Many of them have been overcome in the latest numerical relays. Nevertheless, an awareness of the problems is useful where a protection engineer has to deal with older relays that are already installed and not due for replacement.

11.10.1 Minimum Voltage at Relay Terminals

To attain their claimed accuracy, distance relays that do not employ voltage memory techniques require a minimum voltage at the relay terminals under fault conditions. This voltage should be declared in the data sheet for the relay. With knowledge of the sequence impedances involved in the fault, or alternatively the fault MVA, the system voltage and the earthing arrangements, it is possible to calculate the minimum voltage at the relay terminals for a fault at the reach point of the relay. It is then only necessary to check that the minimum voltage for accurate reach measurement can be attained for a given application. Care should be taken that both phase and earth faults are considered.

11.10.2 Minimum Length of Line

To determine the minimum length of line that can be protected by a distance relay, it is necessary to check first that any minimum voltage requirement of the relay for a fault at the Zone 1 reach is within the declared sensitivity for the relay. Secondly, the ohmic impedance of the line (referred if necessary to VT/CT secondary side quantities) must fall within the ohmic setting range for Zone 1 reach of the relay. For very short lines and especially for cable circuits, it may be found that the circuit impedance is less than the minimum setting range of the relay. In such cases, an alternative method of protection will be required. A suitable alternative might be current differential protection, as the line length will probably be short enough for the cost-effective provision of a high bandwidth communication link between the relays fitted at the ends of the protected circuit. However, the latest numerical distance relays have a very wide range of impedance setting ranges and good sensitivity with low levels of relaying voltage, so such problems are now rarely encountered. Application checks are still essential, though. When considering earth faults, particular care must be taken to ensure that the appropriate earth fault loop impedance is used in the calculation.

11.10.3 Under-Reach - Effect of Remote Infeed

A distance relay is said to under-reach when the impedance presented to it is apparently greater than the impedance to the fault.

Percentage under-reach is defined as:

$$\frac{Z_R - Z_F}{Z_R} \text{x} 100\%$$

where:

Z_R = intended relay reach (relay reach setting)

Z_F = effective reach

The main cause of under-reaching is the effect of fault current infeed at remote busbars. This is best illustrated by an example.

In Figure 11.21, the relay at *A* will not measure the correct impedance for a fault on line section Z_C due to current infeed I_B.

For a fault at point *F*, the relay is presented with an impedance:

$$Z_A + \frac{I_A + I_B}{I_A} Z_C$$

So, for relay balance:

$$Z_A + Z_C = Z_A + \frac{(I_A + I_B)}{I_A} Z_C$$

Therefore the apparent impedance is

$$Z_A + \left(\frac{I_A + I_B}{I_A} \right) Z_C$$

Equation 11.8

It is clear from Equation 11.8 that the relay will underreach. It is relatively easy to compensate for this by increasing the reach setting of the relay, but care has to be taken. Should there be a possibility of the remote infeed being reduced or zero, the relay will then reach further than intended. For example, setting Zone 2 to reach a specific distance into an adjacent line section under parallel circuit conditions may mean that Zone 2 reaches beyond the Zone 1 reach of the adjacent line protection under single circuit operation. If $I_B = 9I_A$ and the relay reach is set to see faults at F, then in the absence of the remote infeed, the relay effective setting becomes $Z_A + 10Z_C$.

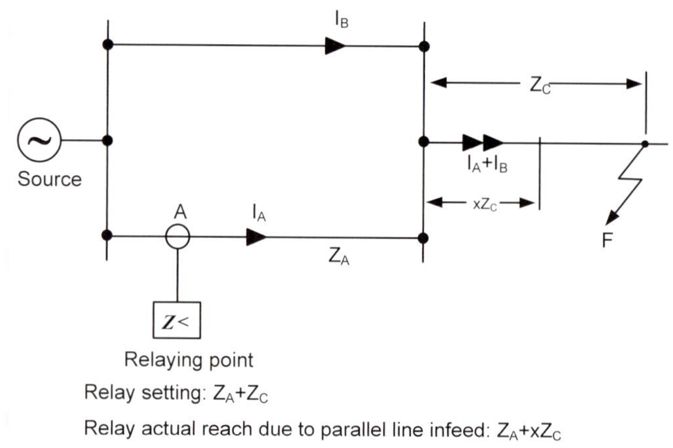

Figure 11.21: Effect on distance relays of infeed at the remote busbar

Care should also be taken that large forward reach settings will not result in operation of healthy phase relays for reverse earth faults, see Section 11.10.5.

11.10.4 Over-Reach

A distance relay is said to over-reach when the apparent impedance presented to it is less than the impedance to the fault.

Percentage over-reach is defined by the equation:

$$\frac{Z_F - Z_R}{Z_R} \text{x} 100\%$$

Equation 11.9

where:

Z_R = relay reach setting

Z_F = effective reach

An example of the over-reaching effect is when distance relays are applied on parallel lines and one line is taken out of service and earthed at each end. This is covered in Section 13.2.3.

11.10.5 Forward Reach Limitations

There are limitations on the maximum forward reach setting that can be applied to a distance relay. For example, with reference to Figure 11.6, Zone 2 of one line section should not reach beyond the Zone 1 coverage of the next line section relay. Where there is a link between the forward reach setting and the relay resistive coverage (e.g. a Mho Zone 3 element), a relay must not operate under maximum load conditions. Also, if the relay reach is excessive, the healthy phase-earth fault units of some relay designs may be prone to operation for heavy reverse faults. This problem only affected older relays applied to three-terminal lines that have significant line section length asymmetry. A number of the features offered with modern relays can eliminate this problem.

11.10.6 Power Swing Blocking

Power swings are variations in power flow that occur when the internal voltages of generators at different points of the power system slip relative to each other. The changes in load flows that occur as a result of faults and their subsequent clearance are one cause of power swings.

A power swing may cause the impedance presented to a distance relay to move away from the normal load area and into the relay characteristic. In the case of a stable power swing it is especially important that the distance relay should not trip to allow the power system to return to a stable conditions. For this reason, most distance protection schemes applied to transmission systems have a power swing blocking facility available. Different relays may use different principles for detection of a power swing, but all involve recognising that the movement of the measured impedance in relation to the relay measurement characteristics is at a rate that is significantly less than the rate of change that occurs during fault conditions. When the relay detects such a condition, operation of the relay elements can be blocked. Power swing blocking may be applied individually to each of the relay zones, or on an all zones applied/inhibited basis, depending on the particular relay used.

Various techniques are used in different relay designs to inhibit power swing blocking in the event of a fault occurring while a power swing is in progress. This is particularly important, for example, to allow the relay to respond to a fault that develops on a line during the dead time of a single pole autoreclose cycle.

Some Utilities may designate certain points on the network as split points, where the network should be split in the event of an unstable power swing or pole-slipping occurring. A dedicated power swing tripping relay may be employed for this purpose (see Section 11.7.8). Alternatively, it may be possible to achieve splitting by strategically limiting the duration for which the operation a specific distance relay is blocked during power swing conditions.

11.10.7 Voltage Transformer Supervision

Fuses or sensitive miniature circuit breakers normally protect the secondary wiring between the voltage transformer secondary windings and the relay terminals.

Distance relays having:

- self-polarised offset characteristics encompassing the zero impedance point of the R/X diagram
- sound phase polarisation
- voltage memory polarisation

may maloperate if one or more voltage inputs are removed due to operation of these devices.

For these types of distance relay, supervision of the voltage inputs is recommended. The supervision may be provided by external means, e.g. separate voltage supervision circuits, or it may be incorporated into the distance relay itself. On detection of VT failure, tripping of the distance relay can be inhibited and/or an alarm is given. Modern distance protection relays employ voltage supervision that operates from sequence voltages and currents. Zero or negative sequence voltages and corresponding zero or negative sequence currents are derived. Discrimination between primary power system faults and wiring faults or loss of supply due to individual fuses blowing or MCB's being opened is obtained by blocking the distance protection only when zero or negative sequence voltage is detected without the presence of zero or negative sequence current. This arrangement will not detect the simultaneous loss of all three voltages and additional detection is required that operates for loss of voltage with no change in current, or a current less than that corresponding to the three phase fault current under minimum fault infeed conditions. If fast-acting miniature circuit breakers are used to protect the VT secondary circuits, contacts from these may be used to inhibit operation of the distance protection elements and prevent tripping.

11.11 OTHER DISTANCE RELAY FEATURES

A modern digital or numerical distance relay will often incorporate additional features that assist the protection engineer in providing a comprehensive solution to the protection requirements of a particular part of a network. Table 11.1 provides an indication of the additional features that may be provided in such a relay. The combination of features that are actually provided is manufacturer and relay model dependent, but it can be seen from the Table that steady progression is being made towards a 'one-box' solution that incorporates all the protection and control requirements for a line or cable. However, at the highest transmission voltages, the level of dependability required for rapid clearance of any protected circuit fault will still demand the use of two independent protection systems.

Fault location (distance to fault)
Instantaneous overcurrent protection
Tee'd feeder protection
Alternative setting groups
CT supervision
Check synchroniser
Auto-reclose
CB state monitoring
CB condition monitoring
CB control
Measurement of voltages, currents, etc.
Event recorder
Disturbance recorder
CB failure detection/logic
Directional/Non-directional phase fault overcurrent protection (backup to distance protection)
Directional/Non-directional earth fault overcurrent protection (backup to distance protection)
Negative sequence protection
Under/overvoltage protection
Stub-bus protection
Broken conductor detection
User-programmable scheme logic

Table 11.1: Listing of possible additional features in a numerical distance relay

11.12 DISTANCE RELAY APPLICATION EXAMPLE

The system diagram shown in Figure 11.22 shows a simple 230kV network. The following example shows the calculations necessary to apply three-zone distance protection to the line interconnecting substations ABC and XYZ. All relevant data for this exercise are given in the diagram. The MiCOM P441 relay with quadrilateral characteristics is considered in this example. Relay parameters used in the example are listed in Table 11.2.

Relay Parameter	Parameter Description	Parameter Value	Units
ZL1 (mag)	Line positive sequence impedance (magnitude)	48.42	Ω
ZL1 (ang)	Line positive sequence impedance (phase angle)	79.41	deg
ZL0 (mag)	Line zero sequence impedance (magnitude)	163.26	Ω
ZL0 (ang)	Line zero sequence impedance (phase angle)	74.87	deg
KZ0 (mag)	Default residual compensation factor (magnitude)	0.79	-
KZ0 (ang)	Default residual compensation factor (phase angle)	-6.5	deg
Z1 (mag)	Zone 1 reach impedance setting (magnitude)	38.74	Ω
Z1 (ang)	Zone 1 reach impedance setting (phase angle)	80	deg
Z2 (mag)	Zone 2 reach impedance setting (magnitude)	62.95	Ω
Z2 (ang)	Zone 2 reach impedance setting (phase angle)	80	deg
Z3 (mag)	Zone 3 reach impedance setting (magnitude)	83.27	Ω
Z3 (ang)	Zone 3 reach impedance setting (phase angle)	80	deg
R1ph	Phase fault resistive reach value - Zone 1	78	Ω
R2ph	Phase fault resistive reach value - Zone 2	78	Ω
R3ph	Phase fault resistive reach value - Zone 3	78	Ω
TZ1	Time delay - Zone 1	0	sec
TZ2	Time delay - Zone 2	0.35	sec
TZ3	Time delay - Zone 3	0.8	sec
R1G	Ground fault resistive reach value - Zone 1	104	Ω
R2G	Ground fault resistive reach value - Zone 2	104	Ω
R3G	Ground fault resistive reach value - Zone 3	104	Ω

Table 11.2: Distance relay parameters for example

Calculations are carried out in terms of primary system impedances in ohms, rather than the traditional practice of using secondary impedances. With numerical relays, where the CT and VT ratios may be entered as parameters, the scaling between primary and secondary ohms can be performed by the relay. This simplifies the example by allowing calculations to be carried out in primary quantities and eliminates considerations of VT/CT ratios.

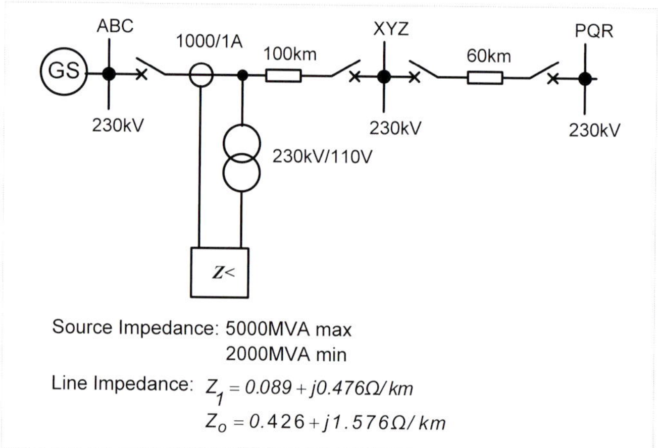

Figure 11.22: Example network for distance relay setting calculation

For simplicity it is assumed that only a conventional 3-zone distance protection is to be set and that there is no teleprotection scheme to be considered. In practice, a teleprotection scheme would normally be applied to a line at this voltage level.

11.12.1 Line Impedance

The line impedance is:

$$Z_L = (0.089 + j0.476) \text{ x } 100\Omega$$

$$= 8.9 + j47.6\Omega$$

$$= 48.42\angle 79.41^{\circ}\Omega$$

Use values of 48.42Ω (magnitude) and 80° (angle) as nearest settable values.

11.12.2 Residual Compensation

The relays used are calibrated in terms of the positive sequence impedance of the protected line. Since the zero sequence impedance of the line between substations ABC and XYZ is different from the positive sequence impedance, the impedance seen by the relay in the case of an earth fault, involving the passage of zero sequence current, will be different to that seen for a phase fault. Hence, the earth fault reach of the relay requires zero sequence compensation (see Section 11.9.2).

For the relay used, this adjustment is provided by the residual (or neutral) compensation factor K_{ZO}, set equal to:

$$|K_{ZO}| = \left|\frac{(Z_0 - Z_1)}{3Z_1}\right|$$

$$\angle K_{Z0} = \angle\frac{(Z_0 - Z_1)}{3Z_1}$$

For each of the transmission lines:

$$Z_{L1} = 0.089 + j0.476\Omega\ (0.484\angle 79.41^\circ\Omega)$$

$$Z_{L0} = 0.426 + j1.576\Omega\ (1.632\angle 74.87^\circ\Omega)$$

Hence,

$$|K_{Z0}| = 0.792$$

$$\angle K_{Z0} = -6.5^\circ$$

11.12.3 Zone 1 Phase Reach

The required Zone 1 reach is 80% of the line impedance. Therefore,

$$0.8\text{x}(48.42\angle 79.41^\circ) = 38.74\angle 79.41^\circ$$

Use $38.74\angle 80^\circ\Omega$ nearest settable value.

11.12.4 Zone 2 Phase Reach

Ideally, the requirements for setting Zone 2 reach are:

- at least 120% of the protected line
- less than the protected line + 50% of the next line

Sometimes, the two requirements are in conflict. In this case, both requirements can be met. A setting of the whole of the line between substations ABC and XYZ, plus 50% of the adjacent line section to substation PQR is used. Hence, Zone 2 reach:

$$= 48.42\angle 79.41^\circ + (0.5\,\text{x}\,60\,\text{x}\,(0.484\angle 79.41^\circ))\,\Omega$$

$$= 62.95\angle 79.41^\circ\Omega$$

Use $62.95\angle 80^\circ\Omega$ nearest available setting.

11.12.5 Zone 3 Phase Reach

Zone 3 is set to cover 120% of the sum of the lines between substations ABC and PQR, provided this does not result in any transformers at substation XYZ being included. It is assumed that this constraint is met. Hence, Zone 3 reach:

$$= 48.42\angle 79.41^\circ + (1.2\,\text{x}\,60\,\text{x}\,(0.484\angle 79.41^\circ))\,\Omega$$

$$= 83.275\angle 79.41^\circ\Omega$$

Use a setting of $83.27\angle 80^\circ\Omega$, nearest available setting.

11.12.6 Zone Time Delay Settings

Proper co-ordination of the distance relay settings with those of other relays is required. Independent timers are available for the three zones to ensure this.

For Zone 1, instantaneous tripping is normal. A time delay is used only in cases where large d.c. offsets occur and old circuit breakers, incapable of breaking the instantaneous d.c. component, are involved.

The Zone 2 element has to grade with the relays protecting the line between substations XYZ and PQR since the Zone 2 element covers part of these lines. Assuming that this line has distance, unit or instantaneous high-set overcurrent protection applied, the time delay required is that to cover the total clearance time of the downstream relays. To this must be added the reset time for the Zone 2 element following clearance of a fault on the adjacent line, and a suitable safety margin. A typical time delay is 350ms, and the normal range is 200-500ms.

The considerations for the Zone 3 element are the same as for the Zone 2 element, except that the downstream fault clearance time is that for the Zone 2 element of a distance relay or IDMT overcurrent protection. Assuming distance relays are used, a typical time is 800ms. In summary:

$$T_{Z1} = 0ms$$

$$T_{Z2} = 250ms$$

$$T_{Z3} = 800ms$$

11.12.7 Phase Fault Resistive Reach Settings

With the use of a quadrilateral characteristic, the resistive reach settings for each zone can be set independently of the impedance reach settings. The resistive reach setting represents the maximum amount of additional fault resistance (in excess of the line impedance) for which a zone will trip, regardless of the fault within the zone.

Two constraints are imposed upon the settings, as follows:

- it must be greater than the maximum expected phase-phase fault resistance (principally that of the fault arc)
- it must be less than the apparent resistance measured due to the heaviest load on the line, unless load blinding (load encroachment) is applied

The minimum fault current at Substation ABC is of the order of 1.8kA, leading to a typical arc resistance R_{arc} using the van Warrington formula (Equation 11.4) of 8Ω. Using the current transformer ratio as a guide to the maximum expected load current, the minimum load impedance Z_{lmin} will be 130Ω. Typically, the resistive reaches will be set to avoid the minimum load impedance by a 40% margin for the phase elements, leading to a maximum resistive reach setting of 78Ω.

Therefore, the resistive reach setting lies between 8Ω and 78Ω. Allowance should be made for the effects of any remote fault infeed, by using the maximum resistive reach possible. While each zone can have its own resistive reach setting, for this simple example they can all be set equal. This need not always be the case, it depends on the particular distance protection scheme used and the need to include Power Swing Blocking.

Suitable settings are chosen to be 80% of the load resistance:

$$R_{1ph} = 78\Omega$$

$$R_{2ph} = 78\Omega$$

$$R_{3ph} = 78\Omega$$

11.12.8 Earth Fault Impedance Reach Settings

By default, the residual compensation factor as calculated in Section 11.12.2 is used to adjust the phase fault reach setting in the case of earth faults, and is applied to all zones.

11.12.9 Earth Fault Resistive Reach Settings

The margin for avoiding the minimum load impedance need only be 20%. Hence the settings are:

$$R_{1G} = 104\Omega$$

$$R_{2G} = 104\Omega$$

$$R_{3G} = 104\Omega$$

This completes the setting of the relay. Table 11.2 also shows the settings calculated.

11.13 REFERENCE

[11.1] Protective Relays – their Theory and Practice. A.R. van C. Warrington. Chapman and Hall, 1962

Chapter 12
Distance Protection Schemes

12.1 INTRODUCTION

Conventional time-stepped distance protection is illustrated in Figure 12.1. One of the main disadvantages of this scheme is that the instantaneous Zone 1 protection at each end of the protected line cannot be set to cover the whole of the feeder length and is usually set to about 80%. This leaves two 'end zones', each being about 20% of the protected feeder length. Faults in these zones are cleared in Zone 1 time by the protection at one end of the feeder and in Zone 2 time (typically 0.25 to 0.4 seconds) by the protection at the other end of the feeder.

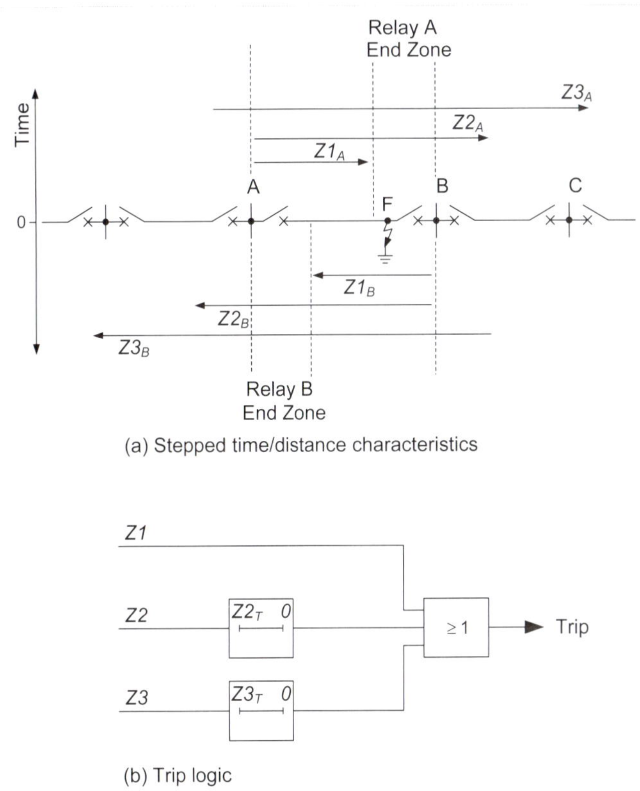

Figure 12.1: Conventional distance scheme

This situation cannot be tolerated in some applications, for two main reasons:

- faults remaining on the feeder for Zone 2 time may cause the system to become unstable
- where high-speed auto-reclosing is used, the non-simultaneous opening of the circuit breakers at both ends of the faulted section results in no 'dead time' during the auto-reclose cycle for the fault to be extinguished and for ionised gases to clear. This results

in the possibility that a transient fault will cause permanent lockout of the circuit breakers at each end of the line section

Even where instability does not occur, the increased duration of the disturbance may give rise to power quality problems, and may result in increased plant damage.

Unit schemes of protection that compare the conditions at the two ends of the feeder simultaneously positively identify whether the fault is internal or external to the protected section and provide high-speed protection for the whole feeder length. This advantage is balanced by the fact that the unit scheme does not provide the back up protection for adjacent feeders given by a distance scheme.

The most desirable scheme is obviously a combination of the best features of both arrangements, that is, instantaneous tripping over the whole feeder length plus back-up protection to adjacent feeders. This can be achieved by interconnecting the distance protection relays at each end of the protected feeder by a communications channel. Communication techniques are described in detail in Chapter 8.

The purpose of the communications channel is to transmit information about the system conditions from one end of the protected line to the other, including requests to initiate or prevent tripping of the remote circuit breaker. The former arrangement is generally known as a 'transfer tripping scheme' while the latter is generally known as a 'blocking scheme'. However, the terminology of the various schemes varies widely, according to local custom and practice.

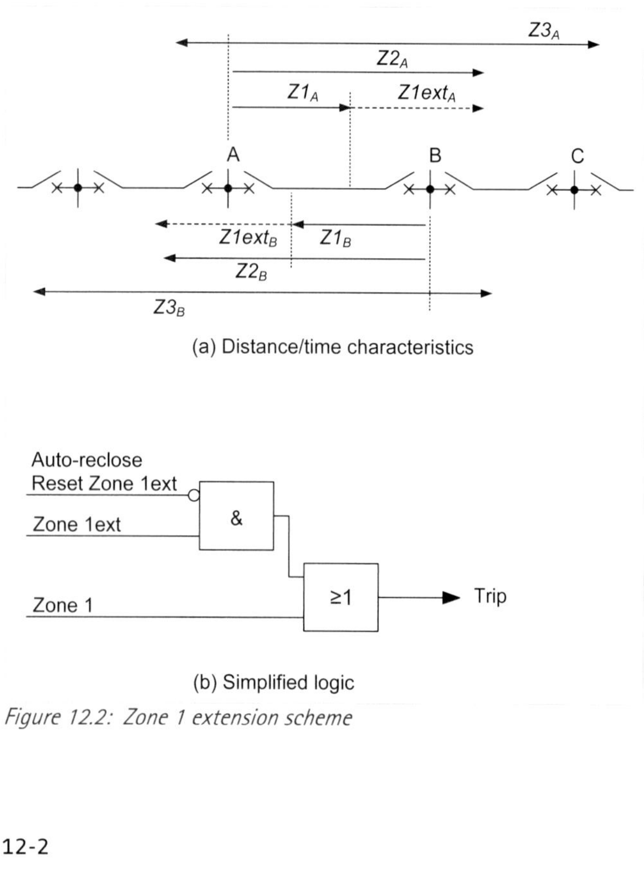

Figure 12.2: Zone 1 extension scheme

12.2 ZONE 1 EXTENSION SCHEME (Z1X SCHEME)

This scheme is intended for use with an auto-reclose facility, or where no communications channel is available, or the channel has failed. Thus it may be used on radial distribution feeders, or on interconnected lines as a fallback when no communications channel is available, e.g. due to maintenance or temporary fault. The scheme is shown in Figure 12.2.

The Zone 1 elements of the distance relay have two settings. One is set to cover 80% of the protected line length as in the basic distance scheme. The other, known as 'Extended Zone 1' or 'Z1X', is set to overreach the protected line, a setting of 120% of the protected line being common. The Zone 1 reach is normally controlled by the Z1X setting and is reset to the basic Zone 1 setting when a command from the auto-reclose relay is received.

On occurrence of a fault at any point within the Z1X reach, the relay operates in Zone 1 time, trips the circuit breaker and initiates auto-reclosure. The Zone 1 reach of the distance relay is also reset to the basic value of 80%, prior to the auto-reclose closing pulse being applied to the breaker. This should also occur when the auto-reclose facility is out of service. Reversion to the Z1X reach setting occurs only at the end of the reclaim time. For interconnected lines, the Z1X scheme is established (automatically or manually) upon loss of the communications channel by selection of the appropriate relay setting (setting group in a numerical relay). If the fault is transient, the tripped circuit breakers will reclose successfully, but otherwise further tripping during the reclaim time is subject to the discrimination obtained with normal Zone 1 and Zone 2 settings.

The disadvantage of the Zone 1 extension scheme is that external faults within the Z1X reach of the relay result in tripping of circuit breakers external to the faulted section, increasing the amount of breaker maintenance needed and needless transient loss of supply to some consumers. This is illustrated in Figure 12.3(a) for a single circuit line where three circuit breakers operate and in Figure 12.3(b) for a double circuit line, where five circuit breakers operate.

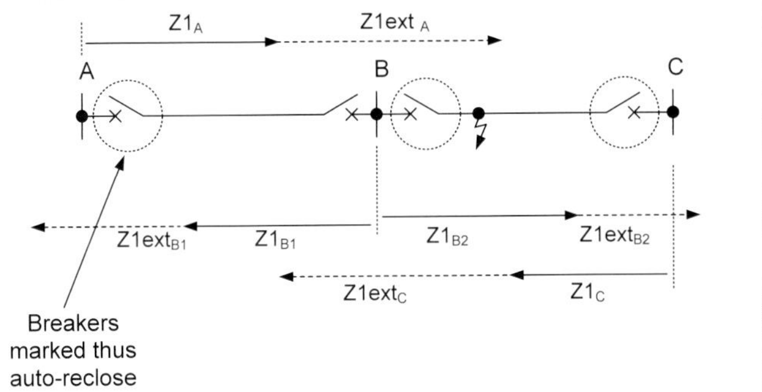

(a) Fault within Zone 1 extension reach of distance relays (single circuit lines)

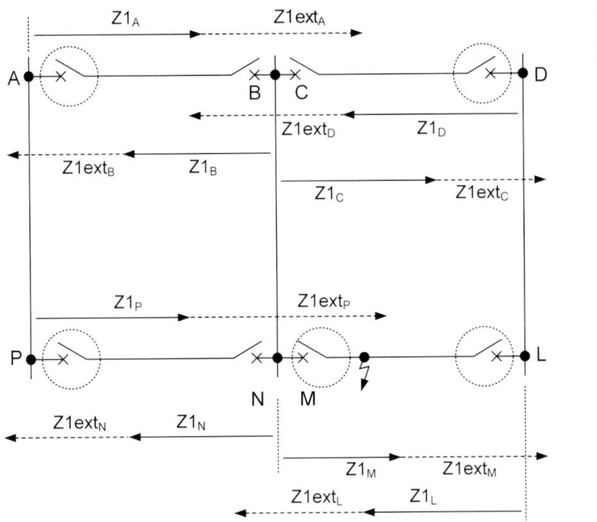

(b) Fault within Zone 1 extension reach of distance relays (double circuit lines)

Figure 12.3: Performance of Zone 1 extension scheme in conjunction with auto-reclose relays

12.3 TRANSFER TRIPPING SCHEMES

A number of these schemes are available, as described below. Selection of an appropriate scheme depends on the requirements of the system being protected.

12.3.1 Direct Under-reach Transfer Tripping Scheme

The simplest way of reducing the fault clearance time at the terminal that clears an end zone fault in Zone 2 time is to adopt a direct transfer trip or intertrip technique, the logic of which is shown in Figure 12.4. A contact operated by the Zone 1 relay element is arranged to send a signal to the remote relay requesting a trip. The scheme may be called a 'direct under-reach transfer tripping scheme', 'transfer trip under-reaching scheme', or 'intertripping underreach distance protection scheme', as the Zone 1 relay elements do not cover the whole of the line.

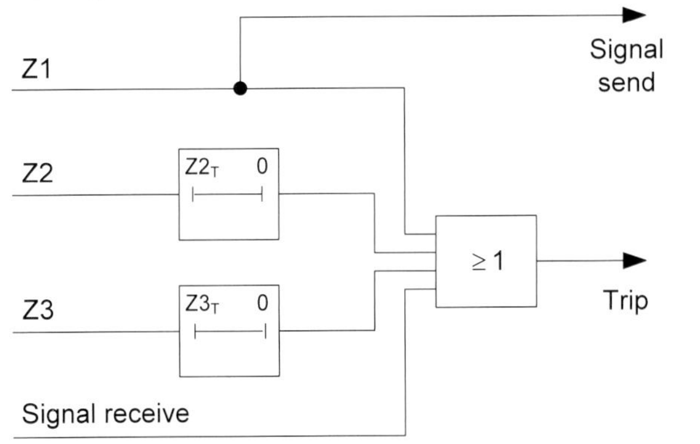

Figure 12.4: Logic for direct underreach transfer tripping scheme

A fault F in the end zone at end *B* in Figure 12.1(a) results in operation of the Zone 1 relay and tripping of the circuit breaker at end *B*. A request to trip is also sent to the relay at end *A*. The receipt of a signal at A initiates tripping immediately because the receive relay contact is connected directly to the trip relay. The disadvantage of this scheme is the possibility of undesired tripping by accidental operation or maloperation of signalling equipment, or interference on the communications channel. As a result, it is not commonly used.

12.3.2 Permissive Under-reach Transfer Tripping (PUP) Scheme

The direct under-reach transfer tripping scheme described above is made more secure by supervising the received signal with the operation of the Zone 2 relay element before allowing an instantaneous trip, as shown in Figure 12.5. The scheme is then known as a 'permissive under-reach transfer tripping scheme' (sometimes abbreviated as a PUTT, PUR or PUP Z2 scheme) or 'permissive underreach distance protection', as both relays must detect a fault before the remote end relay is permitted to trip in Zone 1 time.

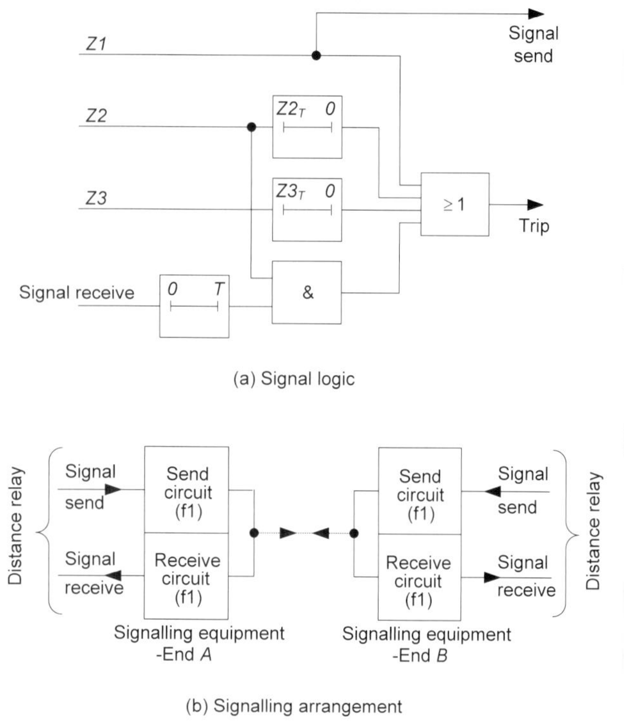

Figure 12.5: Permissive under-reach transfer tripping scheme

A variant of this scheme, found on some relays, allows tripping by Zone 3 element operation as well as Zone 2, provided the fault is in the forward direction. This is sometimes called the PUP-Fwd scheme.

Time delayed resetting of the 'signal received' element is required to ensure that the relays at both ends of a single-end fed faulted line of a parallel feeder circuit have time to trip when the fault is close to one end. Consider a fault F in a double circuit line, as shown in Figure 12.6. The fault is close to end A, so there is negligible infeed from end B when the fault at F occurs. The protection at B detects a Zone 2 fault only after the breaker at end A has tripped. It is possible for the Zone 1 element at A to reset, thus removing the permissive signal to B and causing the 'signal received' element at B to reset before the Zone 2 unit at end B operates. It is therefore necessary to delay the resetting of the 'signal received' element to ensure high speed tripping at end B.

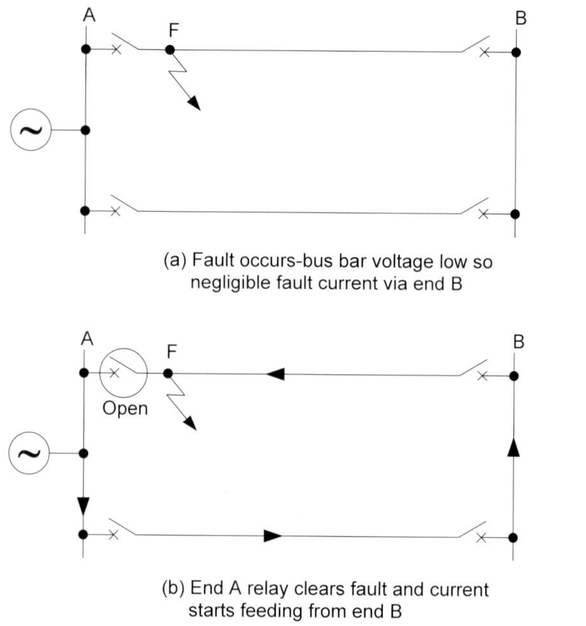

Figure 12.6: PUP scheme: Single-end fed close-up fault on double circuit line

The PUP schemes require only a single communications channel for two-way signalling between the line ends, as the channel is keyed by the under-reaching Zone 1 elements.

When the circuit breaker at one end is open, or there is a weak infeed such that the relevant relay element does not operate, instantaneous clearance cannot be achieved for end-zone faults near the 'breaker open' terminal unless special features are included, as detailed in Section 12.3.5.

12.3.3 Permissive Under-reaching Acceleration Scheme

This scheme is applicable only to zone switched distance relays that share the same measuring elements for both Zone 1 and Zone 2. In these relays, the reach of the measuring elements is extended from Zone 1 to Zone 2 by means of a range change signal immediately, instead of after Zone 2 time. It is also called an 'accelerated underreach distance protection scheme'.

The under-reaching Zone 1 unit is arranged to send a signal to the remote end of the feeder in addition to tripping the local circuit breaker. The receive relay contact is arranged to extend the reach of the measuring element from Zone 1 to Zone 2. This accelerates the fault clearance at the remote end for faults that lie in the region between the Zone 1 and Zone 2 reaches. The scheme is shown in Figure 12.7. Most quality modern distance relays do not employ switched measuring elements, so the scheme is likely to fall into disuse.

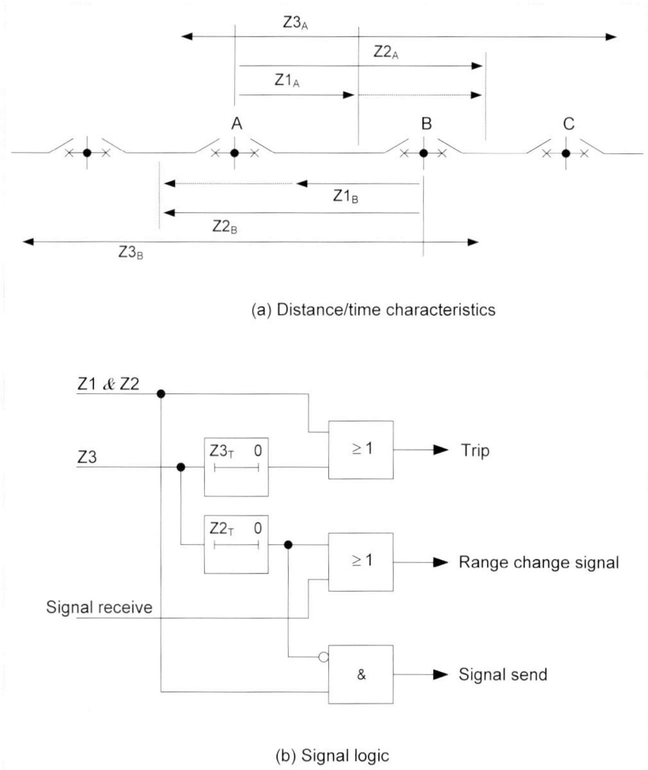

Figure 12.7: Permissive under-reaching acceleration scheme

12.3.4 Permissive Over-reach Transfer Tripping (POP) Scheme

In this scheme, a distance relay element set to reach beyond the remote end of the protected line is used to send an intertripping signal to the remote end. However, it is essential that the receive relay contact is monitored by a directional relay contact to ensure that tripping does not take place unless the fault is within the protected section; see Figure 12.8. The instantaneous contacts of the Zone 2 unit are arranged to send the signal, and the received signal, supervised by Zone 2 operation, is used to energise the trip circuit. The scheme is then known as a 'permissive over-reach transfer tripping scheme' (sometimes abbreviated to POTT, POR or POP), 'directional comparison scheme', or 'permissive overreach distance protection scheme'.

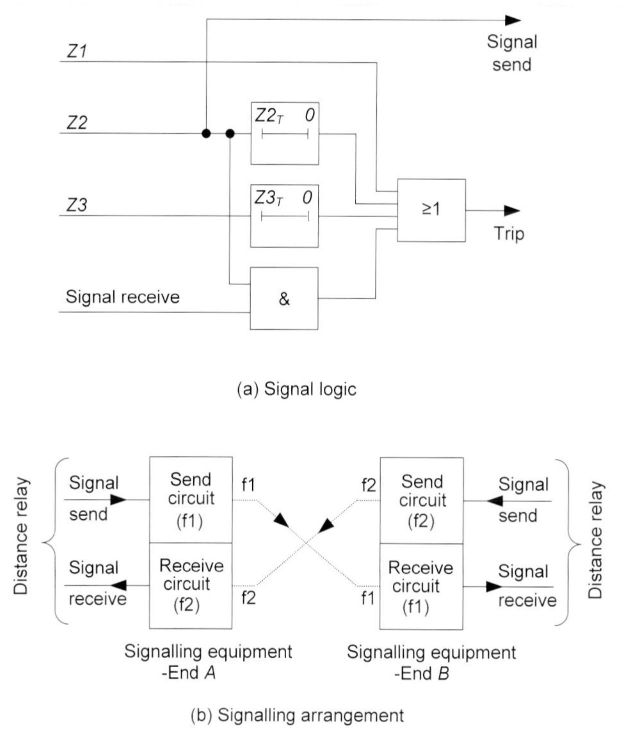

Figure 12.8: Permissive over-reach transfer tripping scheme

Since the signalling channel is keyed by over-reaching Zone 2 elements, the scheme requires duplex communication channels - one frequency for each direction of signalling.

If distance relays with mho characteristics are used, the scheme may be more advantageous than the permissive under-reaching scheme for protecting short lines, because the resistive coverage of the Zone 2 unit may be greater than that of Zone 1.

To prevent operation under current reversal conditions in a parallel feeder circuit, it is necessary to use a current reversal guard timer to inhibit the tripping of the forward Zone 2 elements. Otherwise maloperation of the scheme may occur under current reversal conditions, see Section 12.3.2 for more details. It is necessary only when the Zone 2 reach is set greater than 150% of the protected line impedance.

The timer is used to block the permissive trip and signal send circuits as shown in Figure 12.9. The timer is energised if a signal is received and there is no operation of Zone 2 elements. An adjustable time delay on pick-up (t_p) is usually set to allow instantaneous tripping to take place for any internal faults, taking into account a possible slower operation of Zone 2. The timer will have operated and blocked the 'permissive trip' and 'signal send' circuits by the time the current reversal takes place.

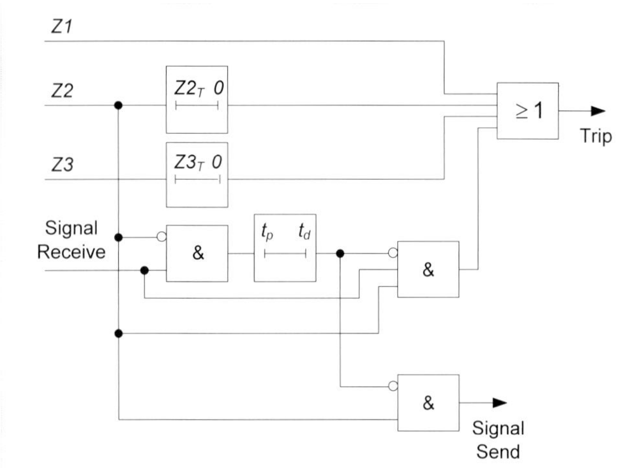

Figure 12.9: Current reversal guard logic – permissive over-reach scheme

The timer is de-energised if the Zone 2 elements operate or the 'signal received' element resets. The reset time delay (t_d) of the timer is set to cover any overlap in time caused by Zone 2 elements operating and the signal resetting at the remote end, when the current in the healthy feeder reverses. Using a timer in this manner means that no extra time delay is added in the permissive trip circuit for an internal fault.

The above scheme using Zone 2 relay elements is often referred to as a POP Z2 scheme. An alternative exists that uses Zone 1 elements instead of Zone 2, and this is referred to as the POP Z1 scheme. However POP Z1 is unusual as it requires Zone 1 to be set overreaching, which is not usual practice.

12.3.5 Weak Infeed Conditions

In the standard permissive over-reach scheme, as with the permissive under-reach scheme, instantaneous clearance cannot be achieved for end-zone faults under weak infeed or breaker open conditions. To overcome this disadvantage, two possibilities exist.

The Weak Infeed Echo feature available in some protection relays allows the remote relay to echo the trip signal back to the sending relay even if the appropriate remote relay element has not operated. This caters for conditions of the remote end having a weak infeed or circuit breaker open condition, so that the relevant remote relay element does not operate. Fast clearance for these faults is now obtained at both ends of the line. The logic is shown in Figure 12.10. A time delay (T_1) is required in the echo circuit to prevent tripping of the remote end breaker when the local breaker is tripped by the busbar protection or breaker fail protection associated with other feeders connected to the busbar. The time delay ensures that the remote end Zone 2 element will reset by the time the echoed signal is received at that end.

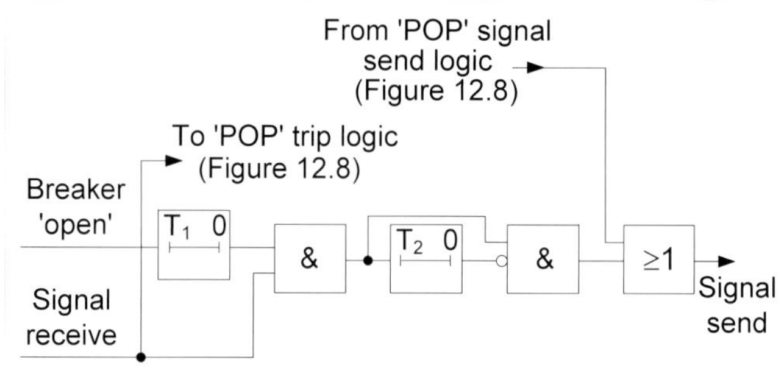

Figure 12.10: Weak Infeed Echo logic circuit

Signal transmission can take place even after the remote end breaker has tripped. This gives rise to the possibility of continuous signal transmission due to lock-up of both signals. Timer T_2 is used to prevent this. After this time delay, 'signal send' is blocked.

A variation on the Weak Infeed Echo feature is to allow tripping of the remote relay under the circumstances described above, providing that an undervoltage condition exists, due to the fault. This is known as the Weak Infeed Trip feature and ensures that both ends are tripped if the conditions are satisfied.

12.4 BLOCKING OVERREACHING SCHEMES

The arrangements described so far have used the signalling channel(s) to transmit a tripping instruction. If the signalling channel fails or there is no Weak Infeed feature provided, end-zone faults may take longer to be cleared.

Blocking over-reaching schemes use an over-reaching distance scheme and inverse logic. Signalling is initiated only for external faults and signalling transmission takes place over healthy line sections. Fast fault clearance occurs when no signal is received and the over-reaching Zone 2 distance measuring elements looking into the line operate. The signalling channel is keyed by reverse-looking distance elements (Z3 in the diagram, though which zone is used depends on the particular relay, with a dedicated reverse zone such as Zone 4 now commonly-used in numerical relays). An ideal blocking scheme is shown in Figure 12.11.

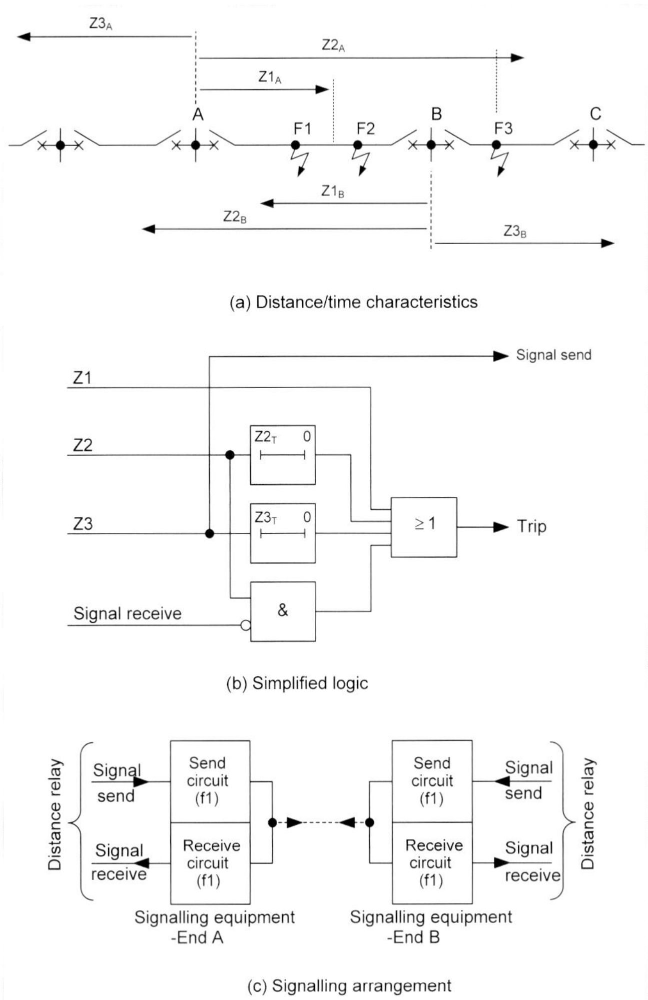

Figure 12.11: Ideal distance protection blocking scheme

The single frequency signalling channel operates both local and remote receive relays when a block signal is initiated at any end of the protected section.

12.4.1 Practical Blocking Schemes

A blocking instruction has to be sent by the reverse-looking relay elements to prevent instantaneous tripping of the remote relay for Zone 2 faults external to the protected section. To achieve this, the reverse-looking elements and the signalling channel must operate faster than the forward-looking elements. In practice, this is seldom the case and to ensure discrimination, a short time delay is generally introduced into the blocking mode trip circuit. Either the Zone 2 or Zone 1 element can be used as the forward-looking element, giving rise to two variants of the scheme.

12.4.1.1 Blocking over-reaching protection scheme using Zone 2 element

This scheme (sometimes abbreviated to 'BLOCKING' or BOP Z2) is based on the ideal blocking scheme of Figure 12.11, but has the signal logic illustrated in Figure 12.12. It is also known as a 'directional comparison blocking scheme' or a 'blocking over-reach distance protection scheme'.

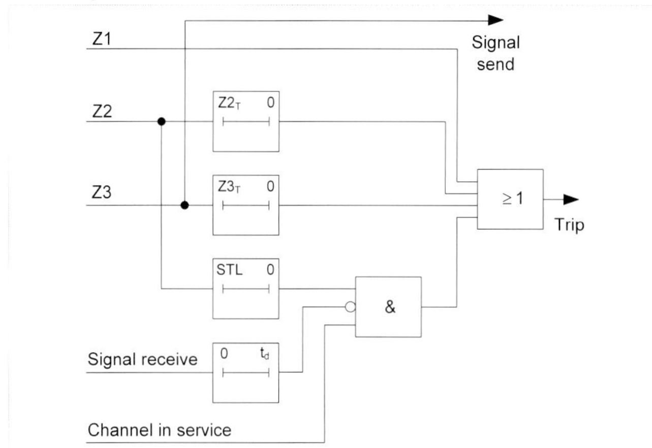

Figure 12.12: Signal logic for BOP Z2 scheme

Operation of the scheme can be understood by considering the faults shown at F1, F2 and F3 in Figure 12.11 along with the signal logic of Figure 12.12.

A fault at F1 is seen by the Zone 1 relay elements at both ends A and B; as a result, the fault is cleared instantaneously at both ends of the protected line. Signalling is controlled by the Z3 elements looking away from the protected section, so no transmission takes place, thus giving fast tripping via the forward-looking Zone 1 elements.

A fault at F2 is seen by the forward-looking Zone 2 elements at ends A and B and by the Zone 1 elements at end B. No signal transmission takes place, since the fault is internal and the fault is cleared in Zone 1 time at end B and after the short time lag (STL) at end A.

A fault at F3 is seen by the reverse-looking Z3 elements at end B and the forward looking Zone 2 elements at end A. The Zone 1 relay elements at end B associated with line section B-C would normally clear the fault at F3. To prevent the Z2 elements at end A from tripping, the reverse-looking Zone 3 elements at end B send a blocking signal to end A. If the fault is not cleared instantaneously by the protection on line section B-C, the trip signal will be given at end B for section A-B after the Z3 time delay.

The setting of the reverse-looking Zone 3 elements must be greater than that of the Zone 2 elements at the remote end of the feeder, otherwise there is the possibility of Zone 2 elements initiating tripping and the reverse looking Zone 3 elements failing to see an external fault. This would result in instantaneous tripping for an external fault. When the signalling channel is used for a stabilising signal, as in the above case, transmission takes place over a healthy line section if power line carrier is used. The signalling channel should then be more reliable when used in the blocking mode than in tripping mode.

It is essential that the operating times of the various relays be skilfully co-ordinated for all system conditions, so that sufficient time is always allowed for the receipt of a blocking signal from the remote end of the feeder. If this is not done accurately, the scheme may trip for an external fault or alternatively, the end zone tripping times may be delayed longer than is necessary.

If the signalling channel fails, the scheme must be arranged to revert to conventional basic distance protection. Normally, the blocking mode trip circuit is supervised by a 'channel-in-service' contact so that the blocking mode trip circuit is isolated when the channel is out of service, as shown in Figure 12.12.

In a practical application, the reverse-looking relay elements may be set with a forward offset characteristic to provide back-up protection for busbar faults after the zone time delay. It is then necessary to stop the blocking signal being sent for internal faults. This is achieved by making the 'signal send' circuit conditional upon non-operation of the forward-looking Zone 2 elements, as shown in Figure 12.13.

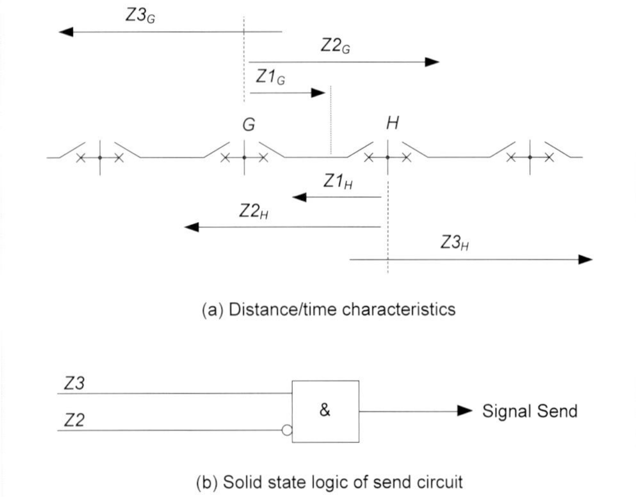

Figure 12.13: Blocking scheme using reverse-looking relays

Blocking schemes, like the permissive over-reach scheme, are also affected by the current reversal in the healthy feeder due to a fault in a double circuit line. If current reversal conditions occur, as described in Section 12.3.2, it may be possible for the maloperation of a breaker on the healthy line to occur. To avoid this, the resetting of the 'signal received' element provided in the blocking scheme is time delayed.

The timer with delayed resetting (t_d) is set to cover the time difference between the maximum resetting time of reverse-looking Zone 3 elements and the signalling channel. So, if there is a momentary loss of the blocking signal during the current reversal, the timer does not have time to reset in the blocking mode trip circuit and no false tripping takes place.

12.4.1.2 Blocking over-reaching protection scheme using Zone 1 element

This is similar to the BOP Z2 scheme described above, except that an over-reaching Zone 1 element is used in the logic, instead of the Zone 2 element. It may also be known as the BOP Z1 scheme, although it is rarely used.

12.4.2 Weak Infeed Conditions

The protection at the strong infeed terminal will operate for all internal faults, since a blocking signal is not received from the weak infeed terminal end. In the case of external faults behind the weak infeed terminal, the reverse-looking elements at that end will see the fault current fed from the strong infeed terminal and operate, initiating a block signal to the remote end. The relay at the strong infeed end operates correctly without the need for any additional circuits. The relay at the weak infeed end cannot operate for internal faults, and so tripping of that breaker is possible only by means of direct intertripping from the strong source end.

12.5 DIRECTIONAL COMPARISON UNBLOCKING SCHEME

The permissive over-reach scheme described in Section 12.3.4 can be arranged to operate on a directional comparison unblocking principle by providing additional circuitry in the signalling equipment. In this scheme (also called a 'deblocking overreach distance protection scheme'), a continuous block (or guard) signal is transmitted. When the over-reaching distance elements operate, the frequency of the signal transmitted is shifted to an 'unblock' (trip) frequency. The receipt of the unblock frequency signal and the operation of over-reaching distance elements allow fast tripping to occur for faults within the protected zone. In principle, the scheme is similar to the permissive over-reach scheme.

The scheme is made more dependable than the standard permissive over-reach scheme by providing additional circuits in the receiver equipment. These allow tripping to take place for internal faults even if the transmitted unblock signal is short-circuited by the fault. This is achieved by allowing aided tripping for a short time interval, typically 100 to 150 milliseconds, after the loss of both the block and the unblock frequency signals. After this time interval, aided tripping is permitted only if the unblock frequency signal is received.

This arrangement gives the scheme improved security over a blocking scheme, since tripping for external faults is possible only if the fault occurs within the above time interval of channel failure. Weak Infeed terminal conditions can be catered for by the techniques detailed in 12.3.5 Weak Infeed Conditions.

In this way, the scheme has the dependability of a blocking scheme and the security of a permissive over-reach scheme. This scheme is generally preferred when power line carrier is used, except when continuous transmission of signal is not acceptable.

12.6 COMPARISON OF TRANSFER TRIP AND BLOCKING RELAYING SCHEMES

On normal two-terminal lines the main deciding factors in the choice of the type of scheme, apart from the reliability of the signalling channel previously discussed, are operating speed and the method of operation of the system. Table 12.1 compares the important characteristics of the various types of scheme.

Criterion	Transfer Tripping Scheme	Blocking Scheme
Speed of operation	Fast	Not as fast
Speed with in-service testing	Slower	As fast
Suitable for auto-reclose	Yes	Yes
Security against maloperation due to:		
Current reversal	Special features required	Special features required
Loss of communications	Poor	Good
Weak Infeed/Open CB	Special features required	Special features required

Table 12.1: Comparison of different distance protection schemes

Modern digital or numerical distance relays are provided with a choice of several schemes in the same relay. Thus scheme selection is now largely independent of relay selection, and the user is assured that a relay is available with all the required features to cope with changing system conditions.

Chapter 13
Protection of Complex Transmission Circuits

13.1 INTRODUCTION

Chapters 10-12 have covered the basic principles of protection for two terminal, single circuit lines whose circuit impedance is due solely to the conductors used. However parallel transmission circuits are often installed, either as duplicate circuits on a common structure, or as separate lines connecting the same two terminal points via different routes. Also, circuits may be multi-ended, a three-ended circuit being the most common.

For economic reasons, transmission and distribution lines can be much more complicated, maybe having three or more terminals (multi-ended feeder), or with more than one circuit carried on a common structure (parallel feeders), as shown in Figure 13.1. Other possibilities are the use of series capacitors or direct-connected shunt reactors. The protection of such lines is more complicated and requires the basic schemes described in the above chapters to be modified.

The purpose of this chapter is to explain the special requirements of some of these situations in respect of protection and identify which protection schemes are particularly appropriate for use in these situations.

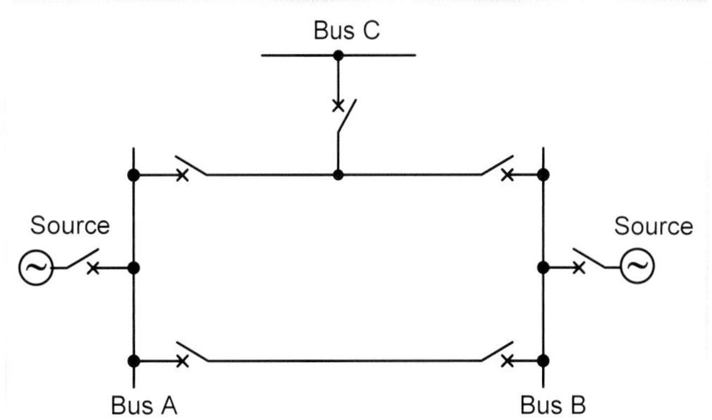

Figure 13.1: Parallel and Multi-ended feeders

13.2 PARALLEL FEEDERS

If two overhead lines are supported on the same structures or are otherwise in close proximity over part or whole of their length, there is a mutual coupling between the two circuits. The positive and negative sequence coupling between the two circuits is small and is usually neglected. The zero sequence coupling can be strong and its effect cannot be ignored.

The other situation that requires mutual effects to be taken into account is when there is an earth fault on a feeder when the parallel feeder is out of service and earthed at both ends. An earth fault in the feeder that is in service can induce current in the earth loop of the earthed feeder, causing a misleading mutual compensation signal.

13.2.1 Unit Protection Systems

Types of protection that use current only, for example unit protection systems, are not affected by the coupling between the feeders. Therefore, compensation for the effects of mutual coupling is not required for the relay tripping elements.

If the relay has a distance-to-fault feature, mutual compensation is required for an accurate measurement. Refer to Section 13.2.2.3 for how this is achieved.

13.2.2 Distance Protection

There are a number of problems applicable to distance relays, as described in the following sections.

13.2.2.1 Current Reversal on Double Circuit Lines

When a fault is cleared sequentially on one circuit of a double circuit line with generation sources at both ends of the circuit, the current in the healthy line can reverse for a short time. Unwanted tripping of CBs on the healthy line can then occur if a Permissive Overreach or Blocking distance scheme (see Chapter 12) is used. Figure 13.2 shows how the situation can arise. The CB at D clears the fault at F faster than the CB at C. Before CB D opens, the Zone 2 elements at A may see the fault and operate, sending a permissive trip signal to the relay for CB B. The reverse looking element of the relay at CB B also sees the fault and inhibits tripping of CBs A and B. However, once CB D opens, the relay element at A starts to reset, while the forward looking elements at B pick up (due to current reversal) and could initiate tripping. If the reset time of the forward-looking elements of the relay at A is longer than the operating time of the forward-looking elements at B, the relays trip the healthy line. The solution is to incorporate a blocking time delay that prevents the tripping of the forward-looking elements of the relays and is initiated by the reverse-looking element. This time delay must be longer than the reset time of the relay elements at A.

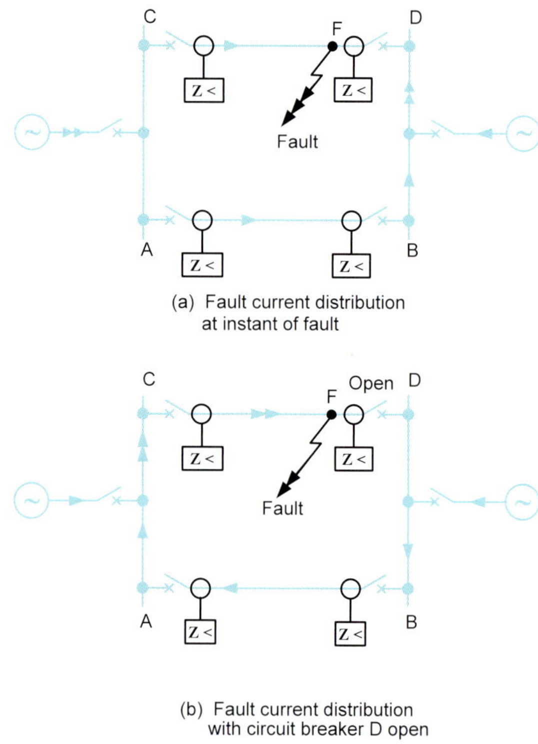

Figure 13.2: Fault current distribution in double-circuit line

13.2.2.2 Under-Reach on Parallel Lines

If a fault occurs on a line that lies beyond the remote terminal end of a parallel line circuit, the distance relay under-reaches for those zones set to reach into the affected line. Analysis shows that under these conditions, because the relay sees only 50% (for two parallel circuits) of the total fault current for a fault in the adjacent line section, the relay sees the impedance of the affected section as twice the correct value. This may have to be allowed for in the settings of Zones 2 and 3 of conventionally set distance relays.

Since the requirement for the minimum reach of Zone 2 is to the end of the protected line section and the under-reach effect only occurs for faults in the following line section(s), it is not usually necessary to adjust Zone 2 impedance settings to compensate. However, Zone 3 elements are intended to provide backup protection to adjacent line sections and hence the under-reaching effect must be allowed for in the impedance calculations.

13.2.2.3 Behaviour of Distance Relays with Earth Faults on the Protected Feeder

When an earth fault occurs in the system, the voltage applied to the earth fault element of the relay in one circuit includes an induced voltage proportional to the zero sequence current in the other circuit.

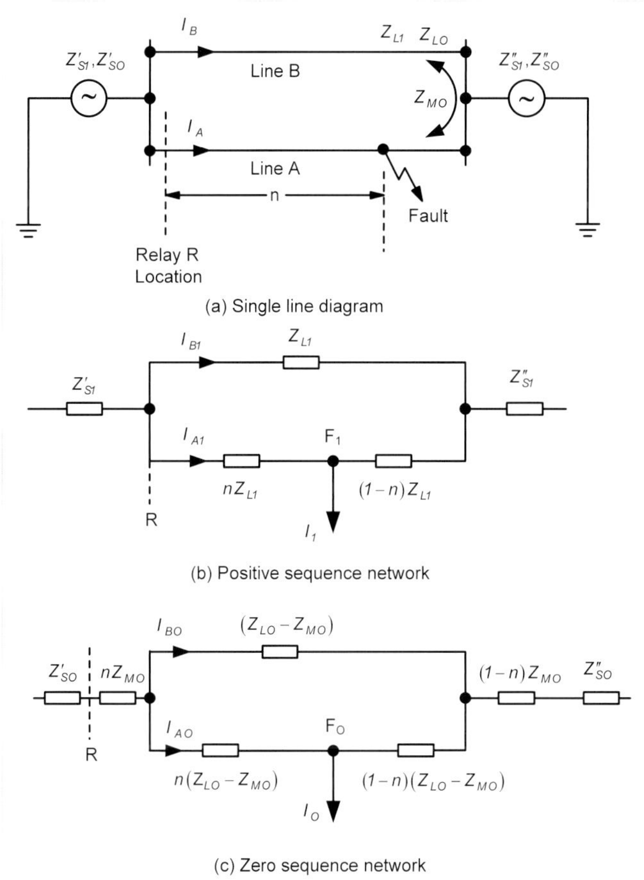

Figure 13.3: General parallel circuit fed from both ends

As the current distribution in the two circuits is unaffected by the presence of mutual coupling, no similar variation in the current applied to the relay element takes place and, consequently, the relay measures the impedance to the fault incorrectly. Whether the apparent impedance to the fault is greater or less than the actual impedance depends on the direction of the current flow in the healthy circuit. For the common case of two circuits, *A* and *B*, connected at the local and remote busbars, as shown in Figure 13.3, the impedance of Line *A* measured by a distance relay, with the normal zero sequence current compensation from its own feeder, is given by:

$$Z_A = nZ_{L1}\left\{1 + \frac{\left(\frac{I_{B0}}{I_{A0}}\right)M}{2\left(\frac{I_{A1}}{I_{A0}}\right)+K}\right\}$$

Equation 13.1

where:

$$M = \frac{Z_{M0}}{Z_{L1}}$$

The true impedance to the fault is nZ_{L1} where n is the per unit fault position measured from R and Z_{L1} is the positive sequence impedance of a single circuit. The 'error' in measurement is determined from the fraction inside the bracket; this varies with the positive and zero sequence currents in circuit *A* and the zero sequence current in circuit *B*. These currents are expressed below in terms of the line and source parameters:

$$\frac{I_{B0}}{I_{A0}} = \frac{nZ''_{S0} - (1-n)Z'_{S0}}{(2-n)Z''_{S0} + (1-n)(Z'_{S0} + Z_{L0} + Z_{M0})}$$

$$I_{A1} = \frac{(2-n)Z''_{S1} + (1-n)(Z'_{S1} + Z_{L1})}{2(Z'_{S1} + Z''_{S1}) + Z_{L1}} I_1$$

$$I_{A0} = \frac{(2-n)Z''_{S0} + (1-n)(Z'_{S0} + Z_{L0} + Z_{M0})}{2(Z'_{S0} + Z''_{S0}) + Z_{L0} + Z_{M0}} I_0$$

and

Z_{M0} = zero sequence mutual impedance between two circuits

NOTE: For earth faults $I_1 = I_0$

All symbols in the above expressions are either self-explanatory from Figure 13.3 or have been introduced in Chapter 11. Using the above formulae, families of reach curves may be constructed, of which Figure 13.4 is typical. In this figure, n' is the effective per unit reach of a relay set to protect 80% of the line. It has been assumed that an infinite busbar is located at each line end, that is, Z'_{S1} and Z''_{S1} are both zero. A family of curves of constant n' has been plotted for variations in the source zero sequence impedances Z'_{S0} and Z''_{S0}.

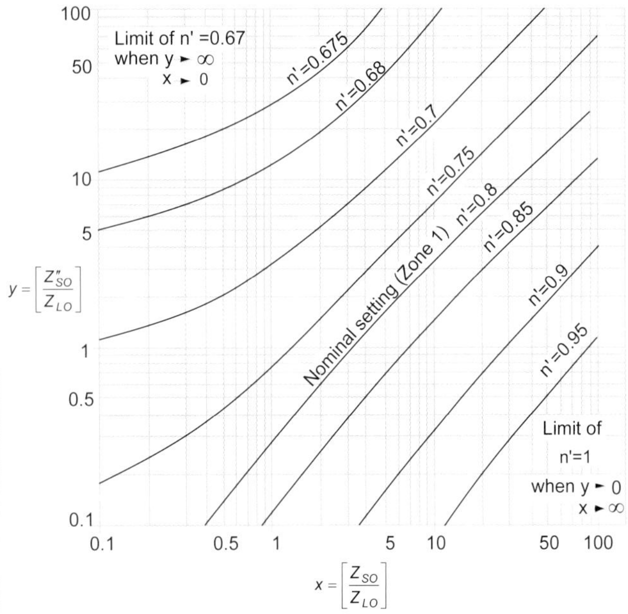

Figure 13.4: Typical reach curves showing the effect of mutual coupling

Figure 13.4 shows that relay R can under-reach or over-reach, according to the relative values of the zero sequence source to line impedance ratios; the extreme effective per unit reaches for the relay are 0.67 and 1. Relay over-reach is not a problem, as the condition being examined is a fault in the protected feeder, for which relay operation is desirable. In Figure 13.4, relay R tends to under-reach whereas the relay at the opposite line end tends to over-reach. As a result, the Zone 1 characteristic of the relays at both ends of the feeder overlap for an earth fault anywhere in the feeder – see Section 13.2.2.5 for more details.

Satisfactory protection can be obtained with a transfer trip, under-reach type distance scheme. Further, compensation for the effect of zero sequence mutual impedance is not necessary unless a distance-to-fault facility is provided. Some manufacturers compensate for the effect of the mutual impedance in the distance relay elements, while others may restrict the application of compensation to the distance-to-fault function only. The latter is easy to implement in software for a digital/numerical relay but is impractical in relays using older technologies. Compensation is achieved by injecting a proportion of the zero sequence current flowing in the parallel feeder into the relay. However, some Utilities do not permit this due to the potential hazards associated with feeding a relay protecting one circuit from a CT located in a different circuit.

For the relay to measure the line impedance accurately, the following condition must be met:

$$\frac{V_R}{I_R} = Z_{L1}$$

For a solid phase to earth fault at the theoretical reach of the relay, the voltage and current in the faulty phase at the relaying point are given by:

$$V_A = I_{A1}Z_{L1} + I_{A2}Z_{L2} + I_{A0}Z_{L0} + I_{B0}Z_{M0}$$

$$I_A = I_{A1} + I_{A2} + I_{A0}$$

Equation 13.2

The voltage and current fed into the relay are given by:

$$V_R = V_A$$

$$I_R = I_A + K_R I_{A0} + K_M I_{B0}$$

Equation 13.3

where:

K_R = the residual compensation factor

K_M = the mutual compensation factor

Thus:

$$K_R = \frac{Z_{L0} - Z_{L1}}{Z_{L1}}$$

$$K_M = \frac{Z_{M0}}{Z_{L1}}$$

13.2.2.4 Distance Relay Behaviour with Earth Faults on the Parallel Feeder

Although distance relays with mutual compensation measure the correct distance to the fault, they may not operate correctly if the fault occurs in the adjacent feeder. Davison and Wright [13.1] have shown that while distance relays without mutual compensation do not over-reach for faults outside the protected feeder, the relays may see faults in the adjacent feeder if mutual compensation is provided. With reference to Figure 13.3, the amount of over-reach is highest when $Z''_{S1} = Z''_{S2} = Z''_{S0} = \infty$. Under these conditions, faults occurring in the first 43% of feeder A appear to the distance relay in feeder B to be in its Zone 1 reach. The solution is to limit the mutual compensation applied to 150% of the zero sequence compensation.

13.2.2.5 Distance Relay Behaviour with Single-Circuit Operation

If only one of the parallel feeders is in service, the protection in the remaining feeder measures the fault impedance correctly, except when the feeder that is not in service is earthed at both ends. In this case, the zero sequence impedance network is as shown in Figure 13.5.

Humpage and Kandil [13.2] have shown that the apparent impedance presented to the relay under these conditions is given by:

$$Z_R = Z_{L1} - \frac{I_{A0} Z_{M0}^2}{I_R Z_{L0}}$$

Equation 13.4

where:

I_R is the current fed into the relay

$$= I_A + K_R I_{A0}$$

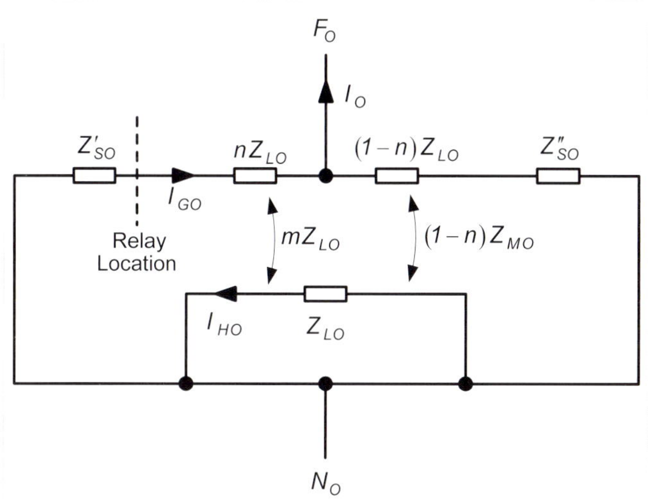

Figure 13.5: Zero sequence impedance network during single circuit operation

The ratio $\frac{I_{A0}}{I_R}$ varies with the system conditions, reaching a maximum when the system is earthed behind the relay with no generation at that end. In this case, the ratio $\frac{I_{A0}}{I_R}$ is equal to $\frac{Z_{L1}}{Z_{L0}}$, and the apparent impedance presented to the relay is:

$$Z_R = Z_{L1}\left(1 - \frac{Z_{M0}^2}{Z_{L0}^2}\right)$$

It is apparent from the above formulae that the relay has a tendency to over-reach. Care should be taken when Zone 1 settings are selected for the distance protection of lines in which this condition may be encountered. To overcome this possible over-reaching effect, some Utilities reduce the reach of earth fault relays to around $0.65Z_{L1}$ when lines are taken out of service for maintenance. However, the probability of having a fault on the first section of the following line while one line is out of service is very small, and many Utilities do not reduce the setting under this condition. It should be noted that the use of mutual compensation would not overcome the over-reaching effect since earthing clamps are normally placed on the line side of the current transformers.

Typical values of zero sequence line impedances for HV lines in the United Kingdom are given in Table 13.1, where the maximum per unit over-reach error $\left(\frac{Z_{M0}}{Z_{L0}}\right)^2$ is also given. The over-reach values quoted in this table are maxima, and are found only in rare cases. In most cases, there is generation at both ends of the feeder and the amount of over-reach is therefore be reduced. In the calculations carried out by Humpage and Kandil, with more realistic conditions, the maximum error found in a 400kV double circuit line was 18.6%.

Line volts	Conductor size		Zero sequence mutual impedance ZMO		Zero sequence line impedance ZLO		Per unit over-reach error $(Z_{MO}/Z_{LO})^2$
	(in^2)	Metric equiv. (mm^2)	Ohms /mile	Ohms /km	Ohms /mile	Ohms /km	
132kV	0.4	258	0.3 + j0.81	0.19+ j0.5	0.41+j 1.61	0.25+ j1.0	0.264
275kV	2 x 0.4	516	0.18+ j0.69	0.11+ j0.43	0.24+ j1.3	0.15+ j0.81	0.292
400kV	4 x 0.4	1032	0.135+ j0.6	0.80+ j0.37	0.16+ j1.18	0.1+ j0.73	0.2666

Table 13.1: Maximum over-reach errors found during single circuit working

13.3 MULTI-ENDED FEEDERS – UNIT PROTECTION SCHEMES

A multi-ended feeder is defined as one having three or more terminals, with either load or generation, or both, at any terminal. Those terminals with load only are usually known as 'taps'.

The simplest multi-terminal feeders are three-ended, and are generally known as tee'd feeders. This is the type most commonly found in practice.

The protection schemes described previously for the protection of two-ended feeders can also be used for multi-ended feeders. However, the problems involved in the application of these schemes to multi-ended feeders are much more complex and require special attention.

The protection schemes that can be used with multi-ended feeders are unit protection and distance schemes. Each uses some form of signalling channel, such as fibre-optic cable, power line carrier or pilot wires. The specific problems that may be met when applying these protections to multi-ended feeders are discussed in the following sections.

13.3.1 Balanced Voltage Schemes for Tee'd Circuits

Although pilot wire schemes are uncommon in the protection of transmission circuits (as per the strict title of this chapter), they are discussed here for completeness.

The balanced voltage scheme is a modification of the MHOA04 / HOA4 scheme described in Section 10.7.1. Since it is necessary to maintain linearity in the balancing circuit, though not in the sending element, the voltage reference is derived from separate quadrature transformers, as shown in Figure 13.6. These are auxiliary units with summation windings energised by the main current transformers in series with the upper electromagnets of the sensing elements. The secondary windings of the quadrature current transformers at all ends are interconnected by the pilots in a series circuit that also includes the lower electromagnets of the relays. Secondary windings on the relay elements are not used, but these elements are fitted with bias loops in the usual way.

The plain feeder settings are increased in the tee'd scheme by 50% for one tee and 75% for two.

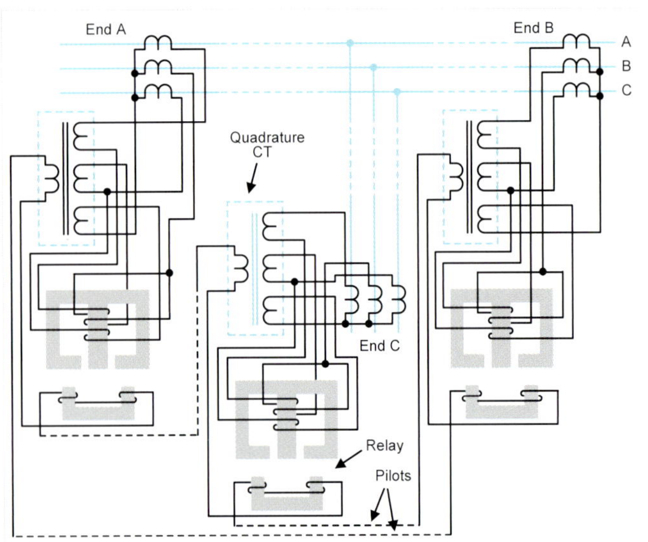

Figure 13.6: Balanced voltage Tee'd feeder scheme

13.3.2 Power Line Carrier Phase Comparison Schemes

The operating principle of these protection schemes has already been covered in detail in Section 10.9. It involves comparing the phase angles of signals derived from a combination of the sequence currents at each end of the feeder. When the phase angle difference exceeds a pre-set value, the 'trip angle', a trip signal is sent to the corresponding circuit breakers. To prevent incorrect operation for external faults, two different detectors, set at different levels, are used. The low-set detector starts the transmission of carrier signal, while the high-set detector is used to control the trip output. Without this safeguard, the scheme could operate incorrectly for external faults because of operating tolerances of the equipment and the capacitive current of the protected feeder. This condition is worse with multi-terminal feeders, since the currents at the feeder terminals can be very dissimilar for an external fault. In the case of the three-terminal feeder in Figure 13.7, if incorrect operation is to be avoided, it is necessary to make certain that the low-set detector at end A or end B is energised when the current at end C is high enough to operate the high-set detector at that end. As only one low-set starter, at end A or end B, needs to be energised for correct operation, the most unfavourable condition is when currents I_A and I_B are equal. To maintain stability under this condition, the high-set to low-set setting ratio of the fault detectors needs to be twice as large as that required when the scheme is applied to a plain feeder. This results in a loss of sensitivity, which may make the equipment unsuitable if the minimum fault level of the power system is low.

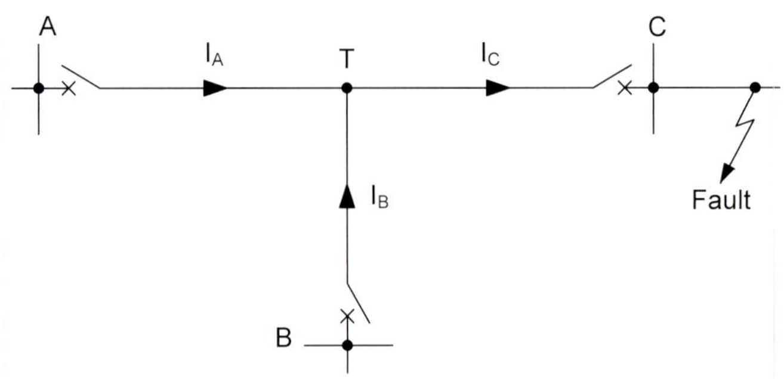

Figure 13.7: External fault conditions

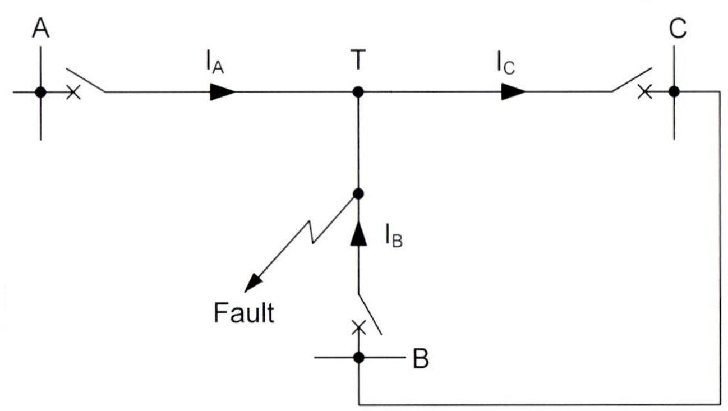

Figure 13.8: Internal fault with current flowing out at one line end

A further unfavourable condition is that shown in Figure 13.8. If an internal fault occurs near one of the ends of the feeder (end B in Figure 13.8) and there is little or no generation at end C, the current at this end may be flowing outwards. The protection is then prevented from operating, since the fault current distribution is similar to that for an external fault; see Figure 13.7. The fault can be cleared only by the back-up protection and, if high speed of operation is required, an alternative type of primary protection must be used.

A point that should also be considered when applying this scheme is the attenuation of carrier signal at the 'tee' junctions. This attenuation is a function of the relative impedances of the branches of the feeder at the carrier frequency, including the impedance of the receiving equipment. When the impedances of the second and third terminals are equal, a power loss of 50% takes place. In other words, the carrier signal sent from terminal A to terminal B is attenuated by 3dB by the existence of the third terminal C. If the impedances of the two branches corresponding to terminal B to C are not equal, the attenuation may be either greater or less than 3dB.

13.3.3 Differential Relay using Optical Fibre Signalling

Current differential relays can provide unit protection for multi-ended circuits without the restrictions associated with other forms of protection. In Section 8.6.5, the characteristics of optical fibre cables and their use in protection signalling are outlined.

Their use in a three-ended system is shown in Figure 13.9, where the relays at each line end are digital/numerical relays interconnected by optical fibre links so that each can send information to the others. In practice the optical fibre links can be dedicated to the protection system or multiplexed, in which case multiplexing equipment, not shown in Figure 13.9,is used to terminate the fibres.

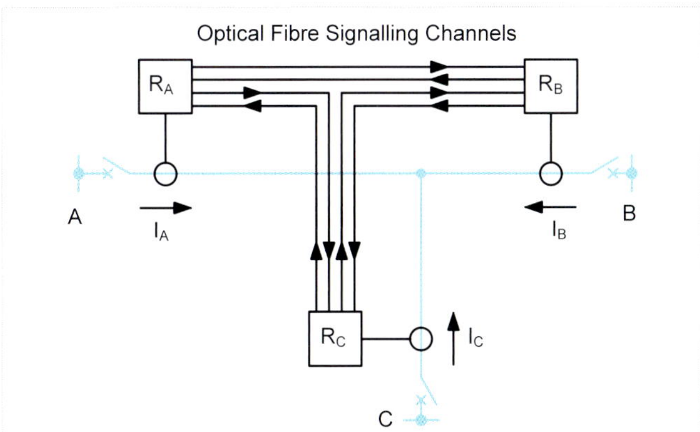

Figure 13.9: Current differential protection for tee'd feeders using optical fibre signalling

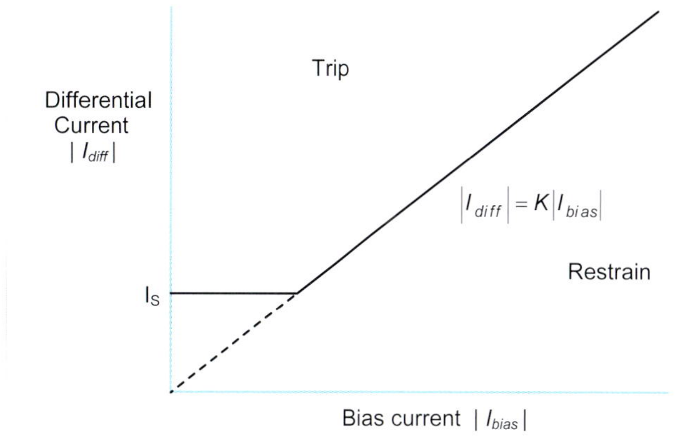

Figure 13.10: Percentage biased differential protection characteristic

If I_A, I_B, I_C are the current vector signals at line ends A, B, C, then for a healthy circuit:

$$I_A + I_B + I_C = 0$$

The basic principles of operation of the system are that each relay measures its local three phase currents and sends its values to the other relays. Each relay then calculates, for each phase, a resultant differential current and also a bias current, which is used to restrain the relay in the manner conventional for biased differential unit protection.

The bias feature is necessary in this scheme because it is designed to operate from conventional current transformers that are subject to steady-state and transient transformation errors.

The two quantities are:

$$\left|I_{diff}\right| = \left|I_A + I_B + I_C\right|$$

$$\left|I_{bias}\right| = \frac{1}{2}\left(\left|I_A\right| + \left|I_B\right| + \left|I_C\right|\right)$$

Figure 13.10 shows the percentage biased differential characteristic used, the tripping criteria being:

$$\left|I_{diff}\right| > K\left|I_{bias}\right|$$

and

$$\left|I_{diff}\right| > I_S$$

where:

K = percentage bias setting

I_S = minimum differential current setting

If the magnitudes of the differential currents indicate that a fault has occurred, the relays trip their local circuit breaker.

The relays also continuously monitor the communication channel performance and carry out self-testing and diagnostic operations. The system measures individual phase currents and so single phase tripping can be used when required. Relays are provided with software to re-configure the protection between two and three terminal lines, so that modification of the system from two terminals to three terminals does not require relay replacement. Further, loss of a single communications link only degrades scheme performance slightly. The relays can recognise this and use alternate communications paths. Only if all communication paths from a relay fail does the scheme have to revert to backup protection.

13.4 MULTI-ENDED FEEDERS - DISTANCE RELAYS

Distance protection is widely used at present for tee'd feeder protection. However, its application is not straightforward, requiring careful consideration and systematic checking of all the conditions described later in this section.

Most of the problems found when applying distance protection to tee'd feeders are common to all schemes. A preliminary discussion of these problems will assist in the assessment of the performance of the different types of distance schemes.

13.4.1 Apparent Impedance Seen by Distance Relays

The impedance seen by the distance relays is affected by the current infeeds in the branches of the feeders. Referring to Figure 13.11, for a fault at the busbars of the substation B, the voltage V_A at busbar A is given by:

$$V_A = I_A Z_{LA} + I_B Z_{LB}$$

so the impedance Z_A seen by the distance relay at terminal A is given by:

$$Z_A = \frac{V_A}{I_A} = Z_{LA} + \frac{I_B}{I_A} Z_{LB}$$

Or

$$Z_A = Z_{LA} + \frac{I_B}{I_A} Z_{LB}$$

Equation 13.5

or

$$Z_A = Z_{LA} + Z_{LB} + \frac{I_C}{I_A} Z_{LB}$$

The apparent impedance presented to the relay has been modified by the term $\left(\frac{I_C}{I_A}\right) Z_{LB}$. If the pre-fault load is zero, the currents I_A and I_C are in phase and their ratio is a real number. The apparent impedance presented to the relay in this case can be expressed in terms of the source impedances as follows:

$$Z_A = Z_{LA} + Z_{LB} + \frac{(Z_{SA} + Z_{LA})}{(Z_{SC} + Z_{LC})} Z_{LB}$$

The magnitude of the third term in this expression is a function of the total impedances of the branches A and B and can reach a relatively high value when the fault current contribution of branch C is much larger than that of branch A. Figure 13.12 shows how a distance relay with a mho characteristic located at A with a Zone 2 element set to 120% of the protected feeder AB, fails to see a fault at the remote busbar B. The 'tee' point T in this example is halfway between substations A and B $(Z_{LA} = Z_{LB})$ and the fault currents I_A and I_C have been assumed to be identical in magnitude and phase angle. With these conditions, the fault appears to the relay to be located at B' instead of at B so the relay under-reaches.

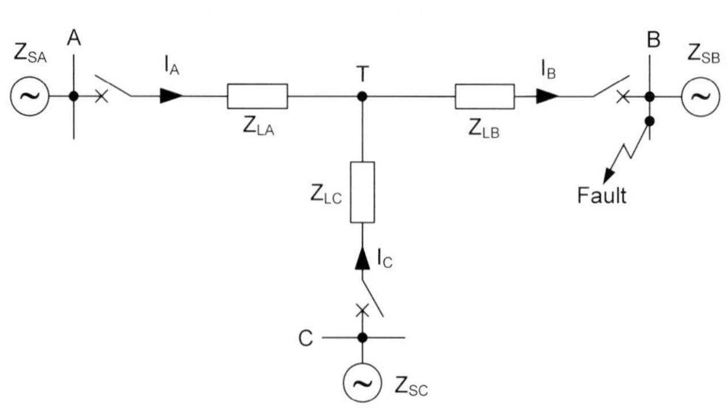

Figure 13.11: Fault at substation B busbars

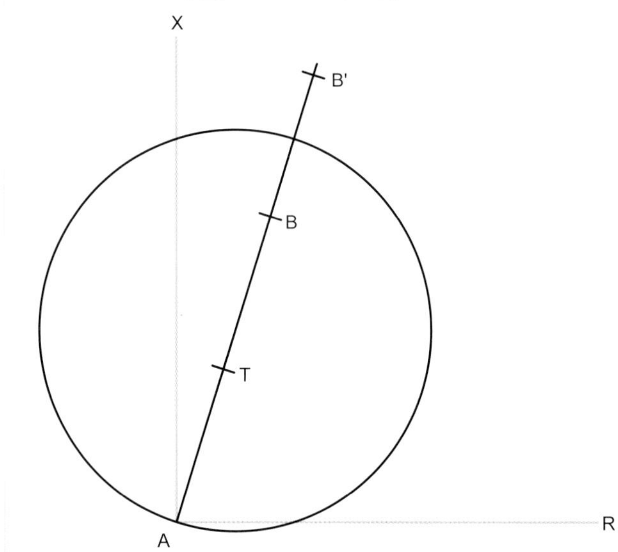

Figure 13.12: Apparent impedance presented to the relay at substation A for a fault at substation B busbars

The under-reaching effect in tee'd feeders can be found for any kind of fault. For the sake of simplicity, the equations and examples mentioned so far have been for balanced faults only. For unbalanced faults, especially those involving earth, the equations become somewhat more complicated, as the ratios of the sequence fault current contributions at terminals A and C may not be the same. An extreme example of this condition is found when the third terminal is a tap with no generation but with the star point of the primary winding of the transformer connected directly to earth, as shown in Figure 13.13. The corresponding sequence networks are shown in Figure 13.14.

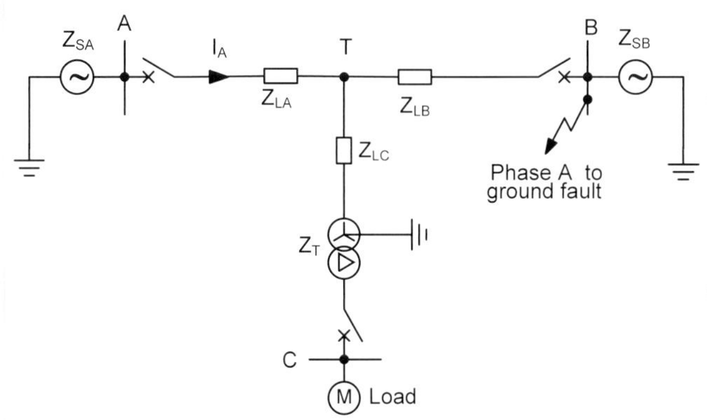

Figure 13.13: Transformer tap with primary winding solidly earthed

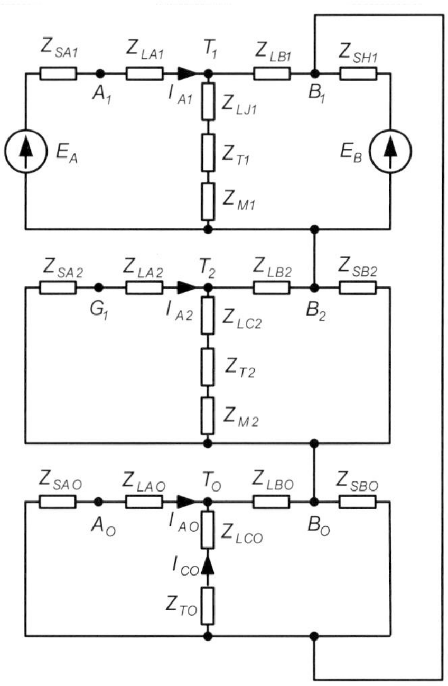

Figure 13.14: Sequence networks for a phase A to earth fault at busbar B in the system shown in Figure 13.13

Figure 13.14 shows that the presence of the tap has little effect in the positive and negative sequence networks. However, the zero sequence impedance of the branch actually shunts the zero sequence current in branch A. As a result, the distance relay located at terminal A tends to under-reach. One solution to the problem is to increase the residual current compensating factor in the distance relay, to compensate for the reduction in zero sequence current. However, the solution has two possible limitations:

- over-reach occurs when the transformer is not connected so operation for faults outside the protected zone may occur
- the inherent possibility of maloperation of the earth fault elements for earth faults behind the relay location is increased

13.4.2 Effect of Pre-fault Load

In all the previous discussions it has been assumed that the power transfer between terminals of the feeder immediately before the fault occurred was zero. If this is not the case, the fault currents I_A and I_C in Figure 13.11 may not be in phase, and the factor I_C / I_A in the equation for the impedance seen by the relay at *A*, is a complex quantity with a positive or a negative phase angle according to whether the current I_C leads or lags the current I_A. For the fault condition previously considered in Figure 13.11 and Figure 13.12, the pre-fault load current may displace the impedance seen by the distance relay to points such as $B_1^{'}$ or $B_2^{'}$, shown in Figure 13.15, according to the phase angle and the magnitude of the pre-fault load current. Humpage and Lewis [13.3] have analysed the effect of pre-fault load on the impedances seen by distance relays for typical cases. Their results and conclusions point out some of the limitations of certain relay characteristics and schemes.

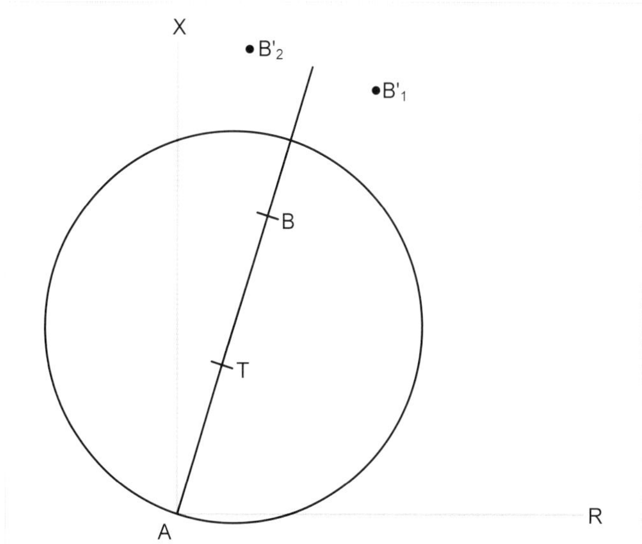

Figure 13.15: effects of the pre-fault load on the apparent impedance presented to the relay

13.4.3 Effect of the Fault Current Flowing Outwards at One Terminal

Up to this point it has been assumed that the fault currents at terminals A and C flow into the feeder for a fault at the busbar

B. Under some conditions, however, the current at one of these terminals may flow outwards instead of inwards. A typical case is shown in Figure 13.16; that of a parallel tapped feeder with one of the ends of the parallel circuit open at terminal A.

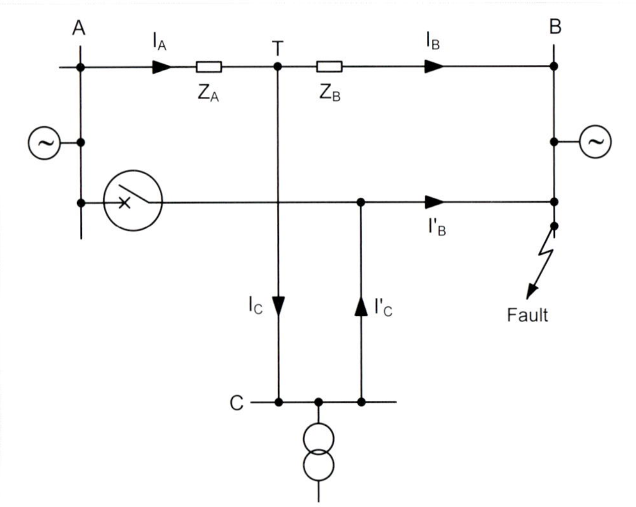

Figure 13.16: Internal Fault at busbar B with current flowing out at terminal C

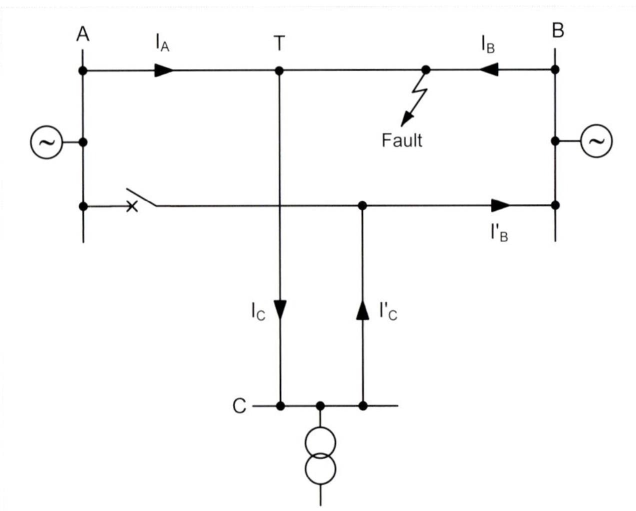

Figure 13.17: Internal fault near busbar B with current flowing out at terminal C

As the currents I_A and I_C now have different signs, the factor I_C / I_A becomes negative. Consequently, the distance relay at terminal A sees an impedance smaller than that of the protected feeder Z_A+Z_B and therefore has a tendency to over-reach. In some cases the apparent impedance presented to the relay may be as low as 50% of the impedance of the protected feeder, and even lower if other lines exist between terminals B and C.

If the fault is internal to the feeder and close to the busbar B, as shown in Figure 13.17, the current at terminal C may still flow outwards. As a result, the fault appears as an external fault to the distance relay at terminal C, which fails to operate.

13.4.4 Maloperation with Reverse Faults

Earth fault distance relays with a directional characteristic tend to lose their directional properties under reverse unbalanced fault conditions if the current flowing through the relay is high and the relay setting relatively large. These conditions arise principally from earth faults. The relay setting and the reverse fault current are now related, the first being a function of the maximum line length and the second depending mainly on the impedance of the shortest feeder and the fault level at that terminal. For instance, referring to Figure 13.18, the setting of the relay at terminal A depends on the impedance Z_A+Z_B and the fault current infeed I_C, for a fault at B, while the fault current I_A for a reverse fault may be quite large if the T point is near the terminals A and C.

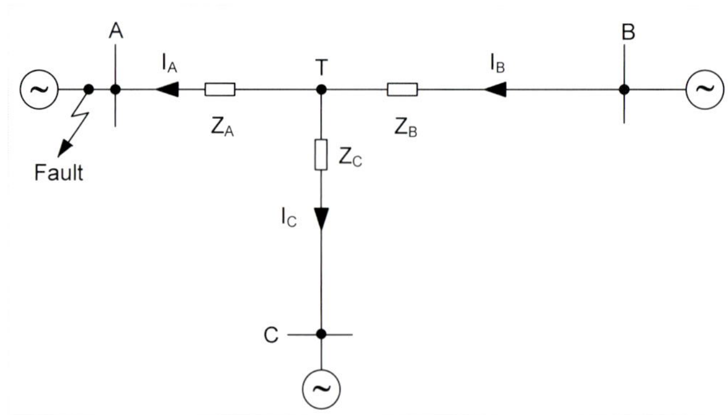

Figure 13.18: External fault behind the relay at terminal A

A summary of the main problems met in the application of distance protection to tee'd feeders is given in Table 13.2 .

Case	Description	Relevant figure number
1	Under-reaching effect for internal faults due to current infeed at the T point	13.12 to 13.15
2	Effect of pre-fault load on the impedance 'seen' by the relay	13.16
3	Over-reaching effect for external faults, due to current flowing outwards at one terminal	13.17
4	Failure to operate for an internal fault, due to current flowing out at one terminal	13.18
5	Incorrect operation for an external fault, due to high current fed from nearest terminal	13.19

Table 13.2: Main problems met in the application of distance protection to tee'd feeders.

13.5 MULTI-ENDED FEEDERS – APPLICATION OF DISTANCE PROTECTION SCHEMES

The schemes that have been described in Chapter 12 for the protection of plain feeders may also be used for tee'd feeder protection. However, the applications of some of these schemes are much more limited in this case.

Distance schemes can be subdivided into two main groups; transfer trip schemes and blocking schemes. The usual considerations when comparing these schemes are security, that is, no operation for external faults, and dependability, that is, assured operation for internal faults.

In addition, it should be borne in mind that transfer trip schemes require fault current infeed at all the terminals to achieve high-speed protection for any fault in the feeder. This is not the case with blocking schemes. While it is rare to find a plain feeder in high voltage systems where there is current infeed at one end only, it is not difficult to envisage a tee'd feeder with no current infeed at one end, for example when the tee'd feeder is operating as a plain feeder with the circuit breaker at one of the terminals open. Nevertheless, transfer trip schemes are also used for tee'd feeder protection, as they offer some advantages under certain conditions.

13.5.1 Transfer Trip Under-Reach Schemes

The main requirement for transfer trip under-reach schemes is that the Zone 1 of the protection, at one end at least, shall see a fault in the feeder. To meet this requirement, the Zone 1 characteristics of the relays at different ends must overlap, either the three of them or in pairs. Cases 1, 2 and 3 in Table 13.2. should be checked when the settings for the Zone 1 characteristics are selected. If the conditions mentioned in case 4 are found, direct transfer tripping may be used to clear the fault; the alternative is to trip sequentially at end C when the fault current I_C reverses after the circuit breaker at terminal B has opened; see Figure 13.17.

Transfer trip schemes may be applied to feeders that have branches of similar length. If one or two of the branches are very short, and this is often the case in tee'd feeders, it may be difficult or impossible to make the Zone 1 characteristics overlap. Alternative schemes are then required.

Another case for which under-reach schemes may be advantageous is the protection of tapped feeders, mainly when the tap is short and is not near one of the main terminals. Overlap of the Zone 1 characteristics is then easily achieved, and the tap does not require protection applied to the terminal.

13.5.2 Transfer Trip Over-Reach Schemes

For correct operation when internal faults occur, the relays at the three ends should see a fault at any point in the feeder. This condition is often difficult to meet, since the impedance seen by the relays for faults at one of the remote ends of the feeder may be too large, as in case 1 in Table 13.2, increasing the possibility of maloperation for reverse faults, case 5 in Table 13.2. In addition, the relay characteristic might encroach on the load impedance.

These considerations, in addition to the signalling channel requirements mentioned later on, make transfer trip over-reach schemes unattractive for multi-ended feeder protection.

13.5.3 Blocking Schemes

Blocking schemes are particularly suited to the protection of multi-ended feeders, since high-speed operation can be obtained with no fault current infeed at one or more terminals. The only disadvantage is when there is fault current outfeed from a terminal, as shown in Figure 13.17. This is case 4 in Table 13.2. The protection units at that terminal may see the fault as an external fault and send a blocking signal to the remote terminals. Depending on the scheme logic either relay operation is blocked or clearance is in Zone 2 time.

The directional unit should be set so that no maloperation can occur for faults in the reverse direction; case 5 in Table 13.2.

13.5.4 Signalling Channel Considerations

The minimum number of signalling channels required depends on the type of scheme used. With under-reach and blocking schemes, only one channel is required, whereas a permissive over-reach scheme requires as many channels as there are feeder ends. The signalling channel equipment at each terminal should include one transmitter and *(N-1)* receivers, where *N* is the total number of feeder ends. This may not be a problem if fibre-optic cables are used, but could lead to problems otherwise.

If frequency shift channels are used to improve the reliability of the protection schemes, mainly with transfer trip schemes, N additional frequencies are required for the purpose. Problems of signal attenuation and impedance matching should also be carefully considered when power line carrier frequency channels are used.

13.5.5 Directional Comparison Blocking Schemes

The principle of operation of these schemes is the same as that of the distance blocking schemes described in the previous section. The main advantage of directional comparison schemes over distance schemes is their greater capability to detect high-resistance earth faults. The reliability of these schemes, in terms of stability for through faults, is lower than that of distance blocking schemes. However, with the increasing reliability of modern signalling channels, directional comparison blocking schemes offer good solutions to the many difficult problems encountered in the protection of multi-ended feeders. For further information see Chapter 12 and specific relay manuals.

13.6 PROTECTION OF SERIES COMPENSATED LINES

The basic power transfer equation in Figure 13.19 shows that transmitted power is proportional to the system voltage level and load angle while being inversely proportional to system impedance. Series compensated lines are used in transmission networks where the required level of transmitted power can not be met, either from a load requirement or system stability requirement. Series compensated transmission lines introduce a series connected capacitor, which has the net result of reducing the overall inductive impedance of the line, hence increasing the prospective, power flow. Typical levels of compensation are 35%, 50% and 70%, where the percentage level dictates the capacitor impedance compared to the transmission line it is associated with.

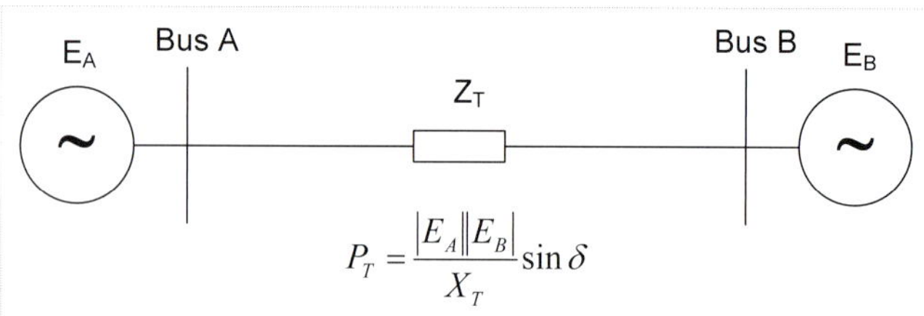

Figure 13.19: Power transfer in a transmission line

The introduction of a capacitive impedance to a network can give rise to several relaying problems. The most common of these is the situation of voltage inversion, which is shown in Figure 13.20. In this case a fault occurs on the protected line. The overall fault impedance is inductive and hence the fault current is inductive (shown lagging the system e.m.f. by 90 degrees in this case). However, the voltage measured by the relay is that across the capacitor and therefore lags the fault current by 90 degrees. The net result is that the voltage measured by the relay is in anti-phase to the system e.m.f. Whilst this view is highly simplistic, it adequately demonstrates potential relay problems, in that any protection reliant upon making a directional decision bases its decision on an inductive system i.e. one where a forward fault is indicated by fault current lagging the measured voltage. A good example of this is a distance relay, which assumes the transmission line is an evenly distributed inductive impedance. Presenting the relay with a capacitive voltage (impedance) can lead the relay to make an incorrect directional decision.

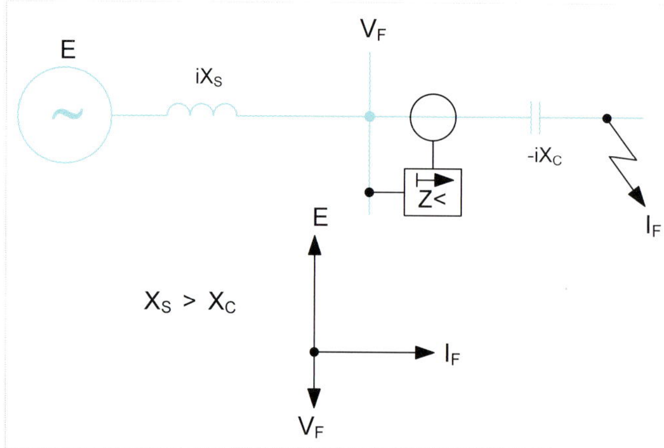

Figure 13.20: Voltage inversion on a transmission line

A second problem is that of current inversion which is demonstrated in Figure 13.21. In this case, the overall fault impedance is taken to be capacitive. The fault current therefore leads the system e.m.f. by 90° whilst the measured fault voltage remains in phase with system e.m.f. Again this condition can give rise to directional stability problems for a variety of protection devices. Practically, the case of current inversion is difficult to obtain. To protect capacitors from high over voltages during fault conditions some form of voltage limiting device (usually in the form of MOVs) is installed to bypass the capacitor at a set current level. In the case of current inversion, the overall fault impedance has to be capacitive and is generally small. This leads to high levels of fault current, which triggers the MOVs and bypasses the capacitors, leaving an inductive fault impedance and preventing the current inversion.

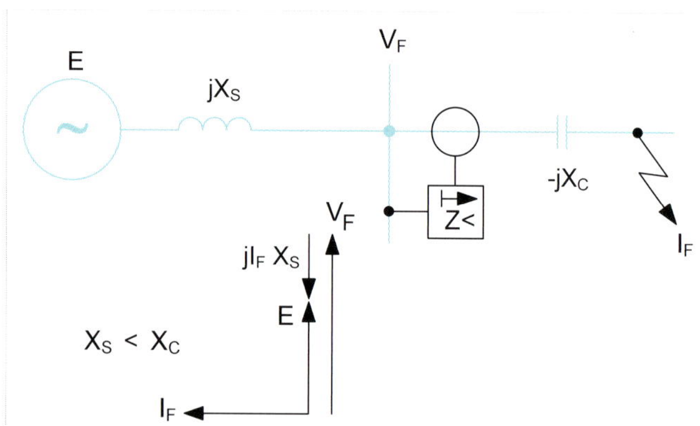

Figure 13.21: Current inversion in a transmission line

The application of protective relays in a series compensated power system needs careful evaluation. The problems associated with the introduction of a series capacitor can be overcome by a variety of relaying techniques so it is important to ensure the suitability of the chosen protection. Each particular application requires careful investigation to determine the most appropriate solution in respect of protection – there are no general guidelines that can be given.

13.7 EXAMPLE

In this section, an example calculation showing the solution to a problem mentioned in this chapter is given.

13.7.1 Distance Relay applied to Parallel Circuits

The system diagram shown in Figure 13.22 indicates a simple 110kV network supplied from a 220kV grid through two auto-transformers. The following example shows the calculations necessary to check the suitability of three zone distance protection to the two parallel feeders interconnecting substations A and B, Line 1 being selected for this purpose. All relevant data for this exercise are given in the diagram. The MiCOM P441 relay with quadrilateral characteristics is used to provide the relay data for the example. Relay quantities used in the example are listed in Table 13.3, and calculations are carried out in terms of actual system impedances in ohms, rather than CT secondary quantities. This simplifies the calculations, and enables the example to be simplified by excluding considerations of CT ratios. Most modern distance relays permit settings to be specified in system quantities rather than CT secondary quantities, but older relays may require the system quantities to be converted to impedances as seen by the relay.

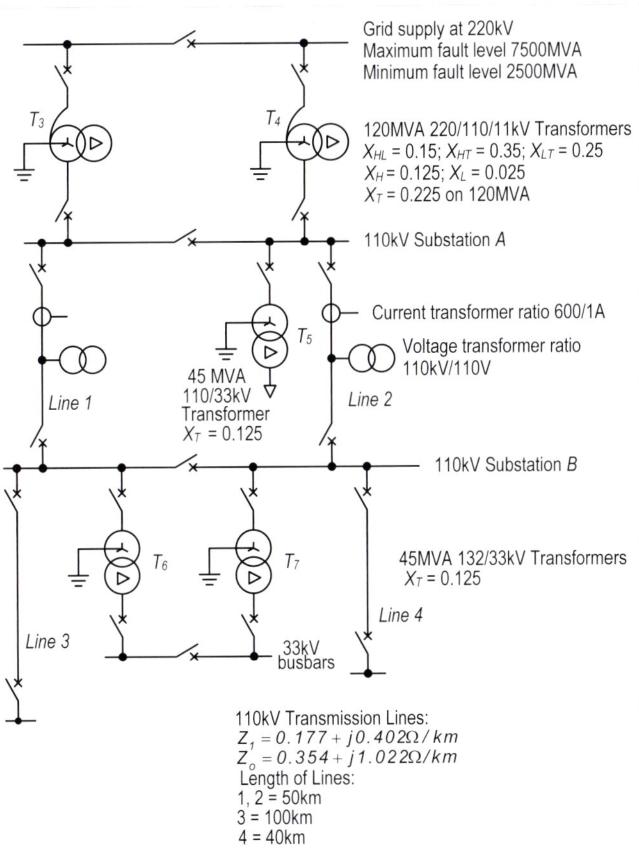

Figure 13.22: Example network for distance relay setting calculation

Relay Parameter	Parameter Description	Parameter Value	Units
ZL1 (mag)	Line positive sequence impedance (magnitude)	21.95	Ω
ZL1 (ang)	Line positive sequence impedance (phase angle)	66.236	deg
ZL0 (mag)	Line zero sequence impedance (magnitude)	54.1	Ω
ZL0 (ang)	Line zero sequence impedance (phase angle)	70.895	deg
KZ0 (mag)	Default residual compensation factor (magnitude)	0.49	-
KZ0 (ang)	Default residual compensation factor (phase angle)	7.8	deg
Z1 (mag)	Zone 1 reach impedance setting (magnitude)	17.56	Ω
Z1 (ang)	Zone 1 reach impedance setting (phase angle)	66.3	deg
Z2 (mag)	Zone 2 reach impedance setting (magnitude)	30.73	Ω
Z2 (ang)	Zone 2 reach impedance setting (phase angle)	66.3	deg
Z3 (mag)	Zone 3 reach impedance setting (magnitude)	131.8	Ω
Z3 (ang)	Zone 3 reach impedance setting (phase angle)	66.3	deg
R1ph	Phase fault resistive reach value - Zone 1	84.8	Ω
R2ph	Phase fault resistive reach value - Zone 2	84.8	Ω
R3ph	Phase fault resistive reach value - Zone 3	84.8	Ω
KZ1 (mag)	Zone 1 residual compensation factor (magnitude)	0.426	-
KZ1 (ang)	Zone 1 residual compensation factor (phase angle)	9.2	deg
KZ2 (mag)	Zone 2 residual compensation factor (magnitude)	not used	-
KZ2 (ang)	Zone 2 residual compensation factor (phase angle)	not used	deg
TZ1	Time delay - Zone 1	0	s
TZ2	Time delay - Zone 2	0.25	s
TZ3	Time delay - Zone 3	0.45	s
R1G	Earth fault resistive reach value - Zone 1	84.8	Ω
R2G	Earth fault resistive reach value - Zone 2	84.8	Ω
R3G	Earth fault resistive reach value - Zone 3	84.8	Ω

Table 13.3: Distance relay settings

13.7.1.1 Residual Compensation

The relays used are calibrated in terms of the positive sequence impedance of the protected line. Since the earth fault impedance of Line 1 is different from the positive sequence impedance, the impedance seen by the relay in the case of a fault involving earth is different to that seen for a phase fault. Therefore the reach of the earth fault elements of the relay needs to be different.

For the relay used, this adjustment is provided by the residual (or neutral) compensation factor K_{ZO}, set equal to:

$$|K_{Z0}| = \left|\frac{(Z_0 - Z_1)}{3Z_1}\right|$$

$$\angle K_{Z0} = \angle\frac{(Z_0 - Z_1)}{3Z_1}$$

For Lines 1 and 2,

$$Z_{L1} = 0.177 + j0.402\Omega$$
$$= (0.439\angle 66.236^\circ\,\Omega)$$

$$Z_{L0} = 0.354 + j1.022\Omega$$
$$= (1.082\angle 70.895^\circ\,\Omega)$$

Hence,

$$|K_{Z0}| = 0.490$$
$$\angle K_{Z0} = 7.8^\circ$$

13.7.1.2 Zone Impedance Reach Settings – Phase Faults

Firstly, the impedance reaches for the three relay zones are calculated.

13.7.1.3 Zone 1 Reach

Zone 1 impedance is set to 80% of the impedance of the protected line. Hence,

$$Z_1 = 0.8 \times 50 \times (0.439\angle 66.236^\circ)\Omega$$

$$= 0.8 \times 21.95\angle 66.236^\circ\,\Omega$$

$$= 17.56\angle 66.236^\circ\,\Omega$$

Use a value of $17.56\angle 66.3^\circ\,\Omega$

13.7.1.4 Zone 2 Reach

Zone 2 impedance reach is set to cover the maximum of:

i. 120% of Line 1 length

ii. Line 1 + 50% of shortest line from Substation B i.e. 50% of Line 4

From the line impedances given,

i. $1.2 \times 21.95\angle 66.236^\circ = 26.34\angle 66.236^\circ\,\Omega$

ii. $21.95\angle 66.236^\circ + 0.5 \times 40 \times 0.439\angle 66.236^\circ\,\Omega$

It is clear that condition (ii) governs the setting, and therefore the initial Zone 2 reach setting is:

$$Z_2 = 30.73\angle 66.3^\circ\,\Omega$$

The effect of parallel Line 2 is to make relay 1 underreach for faults on adjacent line sections, as discussed in Section 11.9.3. This is not a problem for the phase fault elements because Line 1 is always protected.

13.7.1.5 Zone 3 Reach

The function of Zone 3 is to provide backup protection for uncleared faults in adjacent line sections. The criterion used is that the relay should be set to cover 120% of the impedance between the relay location and the end of the longest adjacent line, taking account of any possible fault infeed from other circuits or parallel paths. In this case, faults in Line 3 results in the relay under-reaching due to the parallel Lines 1 and 2, so the impedance of Line 3 should be doubled to take this effect into account. Therefore,

$$Z_3 = 1.2 \times (21.95\angle 66.3^\circ + 100 \times 2 \times 0.439\angle 66.3^\circ)\Omega$$
$$= 131.8\angle 66.3^\circ\,\Omega$$

13.7.1.6 Zone Time Delay Settings

Proper co-ordination of the distance relay settings with those of other relays is required. Independent timers are available for the three zones to ensure this.

For Zone 1, instantaneous tripping is normal. A time delay is used only in cases where large d.c. offsets occur and old circuit breakers, incapable of breaking the instantaneous d.c. component, are involved.

The Zone 2 element has to grade with the relays protecting Lines 3 and 4 since the Zone 2 element covers part of these lines. Assuming that Lines 3/4 have distance, unit or instantaneous high-set overcurrent protection applied, the time delay required is that to cover the total clearance time of the downstream relays. To this must be added the reset time for the Zone 2 elements following clearance of a fault on an adjacent line, and a suitable safety margin. A typical time delay is 250ms, and the normal range is 200-300ms.

The considerations for the Zone 3 element are the same as for the Zone 2 element, except that the downstream fault clearance time is that for the Zone 2 element of a distance relay or IDMT overcurrent protection. Assuming distance relays are used, a typical time is 450ms. In summary:

T_{Z1} = 0ms (instantaneous)

T_{Z2} = 250ms

T_{Z3} = 450ms

13.7.1.7 Phase Fault Resistive Reach Settings

With the use of a quadrilateral characteristic, the resistive reach settings for each zone can be set independently of the impedance reach settings. The resistive reach setting represents the maximum amount of additional fault resistance (in excess of the line impedance) for which a zone trips, regardless of the fault in the zone.

Two constraints are imposed upon the settings, as follows:

- it must be greater than the maximum expected phase-phase fault resistance (principally that of the fault arc)
- it must be less than the apparent resistance measured due to the heaviest load on the line

The minimum fault current at Substation B is of the order of 1.5kA, leading to a typical arc resistance R_{arc} using the van Warrington formula (equation 11.6) of 9Ω.

Using the current transformer ratio on Line 1 as a guide to the maximum expected load current, the minimum load impedance Z_{lmin} is 106Ω

. Typically, the resistive reaches are set to avoid the minimum load impedance by a 20% margin for the phase elements, leading to a maximum resistive reach setting of 84.8.Ω.

Therefore, the resistive reach setting lies between 9Ω and 84.8Ω. While each zone can have its own resistive reach setting, for this simple example, all of the resistive reach settings can be set equal (depending on the particular distance protection scheme used and the need to include Power Swing Blocking, this need not always be the case).

Suitable settings are chosen to be 80% of the load resistance:

R_{3ph} = 84.8Ω

R_{2ph} = 84.8Ω

R_{1ph} = 84.8Ω

13.7.1.8 Earth Fault Reach Settings

By default, the residual compensation factor as calculated in section 13.7.1.1 is used to adjust the phase fault reach setting in the case of earth faults, and is applied to all zones. However, it is also possible to apply this compensation to zones individually. Two cases in particular require consideration, and are covered in this example

13.7.1.9 Zone 1 Earth Fault Reach

Where distance protection is applied to parallel lines (as in this example), the Zone 1 earth fault elements may sometimes overreach and therefore operate when one line is out of service and earthed at both ends

The solution is to reduce the earth fault reach of the Zone 1 element to typically 80% of the default setting. Hence:

$$K_{Z1} = 0.8 \times K_{Z0}$$
$$= 0.8 \times 0.532$$
$$= 0.426$$

In practice, the setting is selected by using an alternative setting group, selected when the parallel line is out of service and earthed.

13.7.1.10 Zone 2 Earth Fault Reach

With parallel circuits, the Zone 2 element tends to under-reach due to the zero sequence mutual coupling between the lines. Maloperation may occur, particularly for earth faults occurring on the remote busbar. The effect can be countered by increasing the Zone 2 earth fault reach setting, but first it is necessary to calculate the amount of under-reach that occurs.

Underreach = $Z_{adj} \times \dfrac{I_{fltp}}{I_{flt}}$

where:

Z_{adj} = impedance of adjacent line covered by Zone 2

I_{fltp} = fault current in parallel line

I_{flt} = total fault current

since the two parallel lines are identical, $I_{fltp} = 0.5 I_{flt}$ and hence, for Lines 1 and 2,

Underreach = $8.78\angle 66.3^\circ \times 0.5 = 4.39\angle 66.3^\circ$

% Underreach = Underreach / Reach of Protected Zone, and hence:

% Underreach =14.3%

This amount of under-reach is not significant and no adjustment need be made. If adjustment is required, this can be achieved by using the K_{Z2} relay setting, increasing it over the K_{Z0} setting by the percentage under-reach. When this is done, care must also be taken that the percentage over-reach during single circuit operation is not excessive – if it is then use can be made of the alternative setting groups provided in most modern distance relays to change the relay settings according to the number of circuits in operation.

13.7.1.11 Earth Fault Resistive Reach Settings

The same settings can be used as for the phase fault resistive reaches. Hence,

$R_{3G} = 84.8\Omega$

$R_{2G} = 84.8\Omega$

$R_{1G} = 84.8\Omega$

This completes the setting of the relay. Table 13.3 also shows the settings calculated.

13.8 REFERENCES

[13.1] Some factors affecting the accuracy of distance type protective equipment under earth fault conditions. Davison, E.B. and Wright, A. Proc. IEE Vol. 110, No. 9, Sept. 1963, pp. 1678-1688.

[13.2] Distance protection performance under conditions of single-circuit working in double-circuit transmission lines. Humpage, W.D. and Kandil, M.S. Proc. IEE. Vol. 117. No. 4, April 1970, pp. 766-770.

[13.3] Distance protection of tee'd circuits. Humpage, W.A. and Lewis, D.W. Proc. IEE, Vol. 114, No. 10, Oct. 1967, pp. 1483-1498.

Chapter 14
Auto-Reclosing

14.1 INTRODUCTION

Faults on overhead lines fall into one of three categories:

- transient
- semi-permanent
- permanent

80-90% of faults on any overhead line network are transient in nature. The remaining 10%-20% of faults are either semi-permanent or permanent.

Transient faults are commonly caused by lightning or temporary contact with foreign objects, and immediate tripping of one or more circuit breakers clears the fault. Subsequent re-energisation of the line is usually successful.

A small tree branch falling on the line could cause a semi-permanent fault. The cause of the fault would not be removed by the immediate tripping of the circuit, but could be burnt away during a time-delayed trip. HV overhead lines in forest areas are prone to this type of fault. Permanent faults, such as broken conductors, and faults on underground cable sections, must be located and repaired before the supply can be restored.

Use of an auto-reclose scheme to re-energise the line after a fault trip permits successful re-energisation of the line. Sufficient time must be allowed after tripping for the fault arc to de-energise before reclosing otherwise the arc will re-strike. Such schemes have been the cause of a substantial improvement in continuity of supply. A further benefit, particularly to EHV systems, is the maintenance of system stability and synchronism.

A typical single-shot auto-reclose scheme is shown in Figure 14.1 and Figure 14.2. Figure 14.1 shows a successful reclosure in the event of a transient fault, and Figure 14.2 an unsuccessful reclosure followed by lockout of the circuit breaker if the fault is permanent.

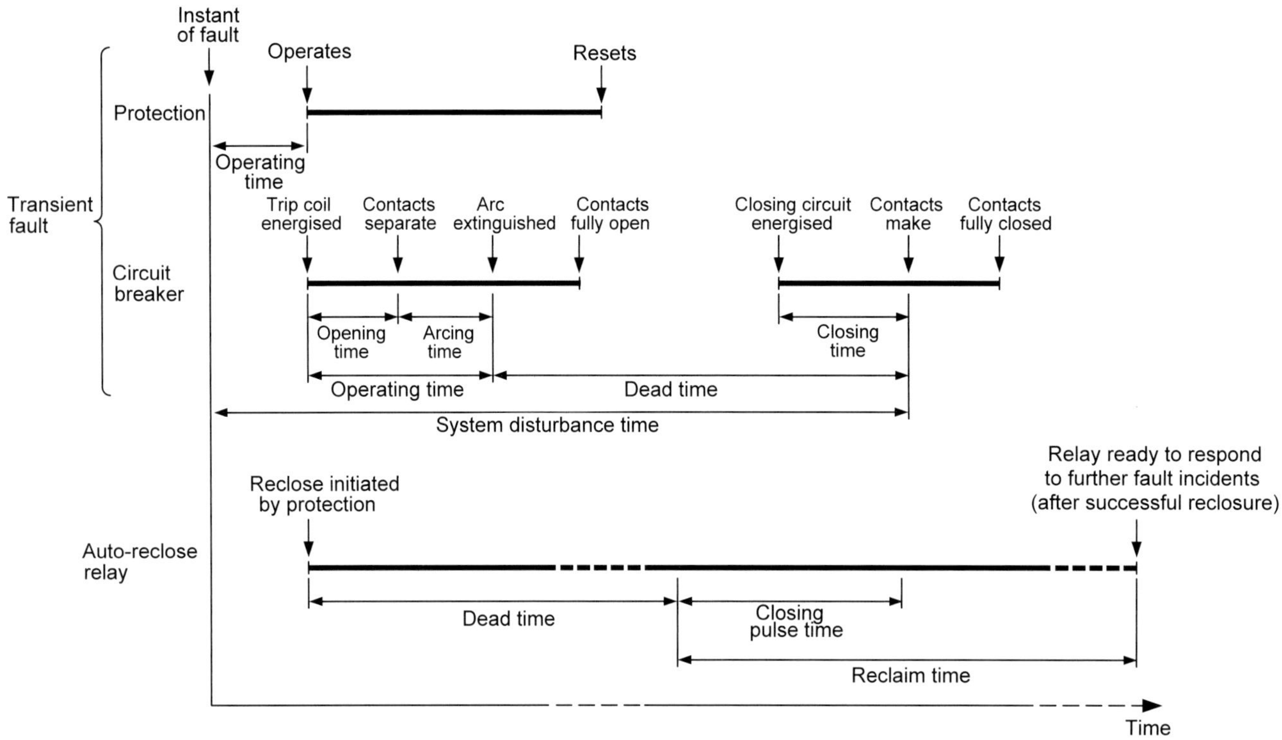

Figure 14.1: Single-shot auto-reclose scheme operation for a transient fault

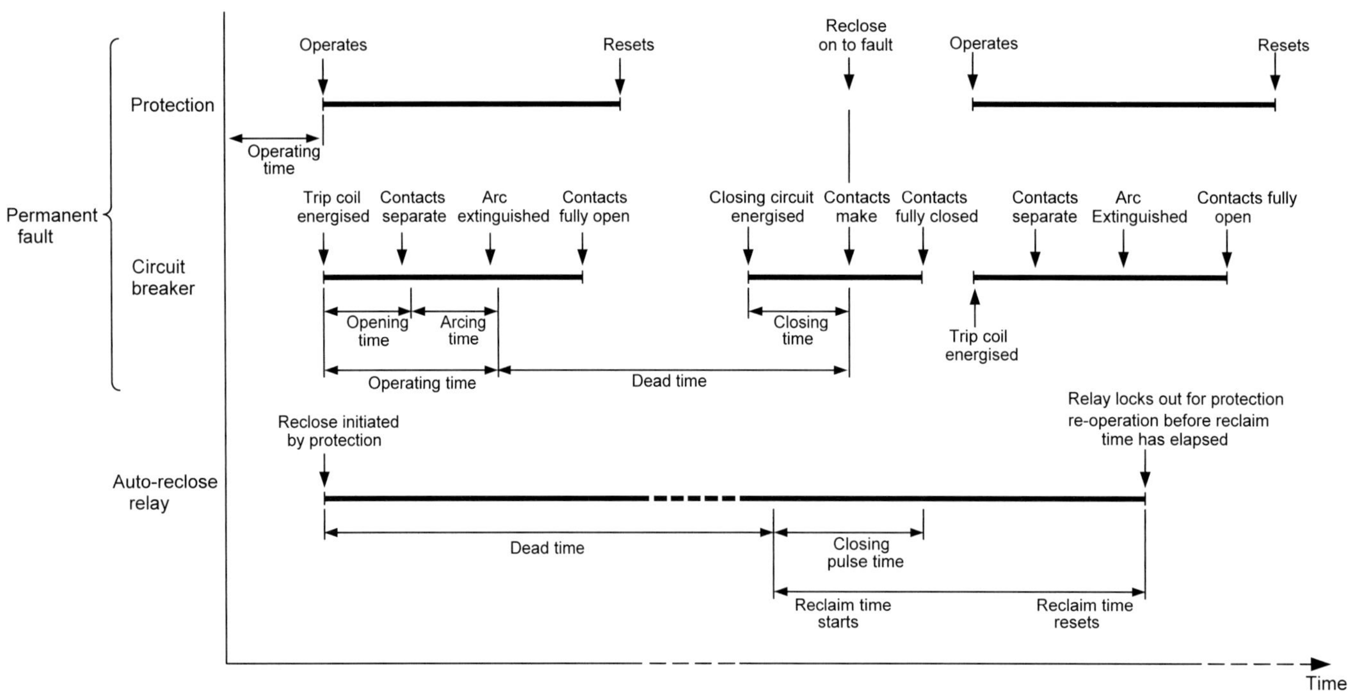

Figure 14.2: Single-shot auto-reclose scheme operation for a permanent fault

14.2 APPLICATION OF AUTO-RECLOSING

The most important parameters of an auto-reclose scheme are:

- dead time
- reclaim time
- single or multi-shot

These parameters are influenced by:

- type of protection
- type of switchgear
- possible stability problems
- effects on the various types of consumer loads

The weighting given to the above factors is different for HV distribution networks and EHV transmission systems and therefore it is convenient to discuss them under separate headings. Sections 14.3 and 14.4 cover the application of auto-reclosing to HV distribution networks while Sections 14.5 to 14.9 cover EHV schemes.

The rapid expansion in the use of auto-reclosing has led to the existence of a variety of different control schemes. The various features in common use are discussed in Section 14.10. The related subject of auto-closing, that is, the automatic closing of normally open circuit breakers, is dealt with in Section 14.11.

14.3 AUTO-RECLOSING ON HV DISTRIBUTION NETWORKS

On HV distribution networks, auto-reclosing is applied mainly to radial feeders where problems of system stability do not arise, and the main advantages to be derived from its use can be summarised as follows:

- reduction to a minimum of the interruptions of supply to the consumer
- instantaneous fault clearance can be introduced, with the accompanying benefits of shorter fault duration, less fault damage, and fewer permanent faults

As 80% of overhead line faults are transient, elimination of loss of supply from this cause by the introduction of auto-reclosing gives obvious benefits through:

- improved supply continuity
- reduction of substation visits

Instantaneous tripping reduces the duration of the power arc resulting from an overhead line fault to a minimum. The chance of permanent damage occurring to the line is reduced. The application of instantaneous protection may result in non-selective tripping of a number of circuit breakers and an ensuing loss of supply to a number of healthy sections. Auto-reclosing allows these circuit breakers to be reclosed within a few seconds. With transient faults, the overall effect would be loss of supply for a very short time but affecting a larger number of consumers. If only time-graded protection without auto-reclose were used, a smaller number of consumers might be affected, but for a longer time period.

When instantaneous protection is used with auto-reclosing, the scheme is normally arranged to inhibit the instantaneous protection after the first trip. For a permanent fault, the time-graded protection will give discriminative tripping after reclosure, resulting in the isolation of the faulted section. Some schemes allow a number of reclosures and time-graded trips after the first instantaneous trip, which may result in the burning out and clearance of semi-permanent faults. A further benefit of instantaneous tripping is a reduction in circuit breaker maintenance by reducing pre-arc heating when clearing transient faults.

When considering feeders that are partly overhead line and partly underground cable, any decision to install auto-reclosing would be influenced by any data known on the frequency of transient faults. Where a significant proportion of faults are permanent, the advantages of auto-reclosing are small, particularly since reclosing on to a faulty cable is likely to aggravate the damage.

14.4 FACTORS INFLUENCING HV AUTO-RECLOSE SCHEMES

The factors that influence the choice of dead time, reclaim time, and the number of shots are now discussed.

14.4.1 Dead Time

Several factors affect the selection of system dead time as follows:

- system stability and synchronism
- type of load
- CB characteristics
- fault path de-ionisation time
- protection reset time

These factors are discussed in the following sections.

14.4.1.1 System stability and synchronism

To reclose without loss of synchronism after a fault on the interconnecting feeder, the dead time must be kept to the minimum permissible consistent with de-ionisation of the fault arc. Other time delays that contribute to the total system disturbance time must also be kept as short as possible. The problem arises only on distribution networks with more than one power source, where power can be fed into both ends of an inter-connecting line. A typical example is embedded generation (see Chapter 17), or where a small centre of population with a local diesel generating plant may be connected to the rest of the supply system by a single tie-line.

The use of high-speed protection, such as unit protection or distance schemes, with operating times of less than 0.05s is essential. The circuit breakers must have very short operation times and then be able to reclose the circuit after a dead time of the order of 0.3s - 0.6s to allow for fault-arc de-ionisation.

It may be desirable in some cases to use synchronism check logic, so that auto-reclosing is prevented if the phase angle has moved outside specified limits. The matter is dealt with more fully in Section 14.9.

14.4.1.2 Type of load

On HV systems, the main problem to be considered in relation to dead time is the effect on various types of consumer load.

- Industrial consumers: Most industrial consumers operate mixed loads comprising induction motors, lighting, process control and static loads. Synchronous motors may also be used. The dead time has to be long enough to allow motor circuits to trip out on loss of supply. Once the supply is restored, restarting of drives can then occur under direction of the process control system in a safe and programmed manner, and can often be fast enough to ensure no significant loss of production or product quality
- Domestic consumers: It is improbable that expensive processes or dangerous conditions will be involved with domestic consumers and the main consideration is that of inconvenience and compensation for supply interruption. A dead time of seconds or a few minutes is of little importance compared with the loss of cooking facilities, central heating, light and audio/visual entertainment resulting from a longer supply failure that could occur without auto-reclosing

14.4.1.3 Circuit breaker characteristics

The time delays imposed by the circuit breaker during a tripping and reclosing operation must be taken into consideration, especially when assessing the possibility of applying high speed auto-reclosing.

- Mechanism resetting time: Most circuit breakers are 'trip free', which means that the breaker can be tripped during the closing stroke. After tripping, a time of the order of 0.2s must be allowed for the trip-free mechanism to reset before applying a closing impulse. Where high speed reclosing is required, a latch check interlock is desirable in the reclosing circuit
- Closing time: This is the time interval between the energisation of the closing mechanism and the making of the contacts. Owing to the time constant of the solenoid and the inertia of the plunger, a solenoid closing mechanism may take 0.3s to close. A spring-operated breaker, on the other hand, can close in less than 0.2s. Modern vacuum circuit breakers may have a closing time of less than 0.1s

The circuit breaker mechanism imposes a minimum dead time made up from the sum of (a) and (b) above. Figure 14.3 shows the performance of modern HV/EHV circuit breakers in this respect. Older circuit breakers may require longer times than those shown.

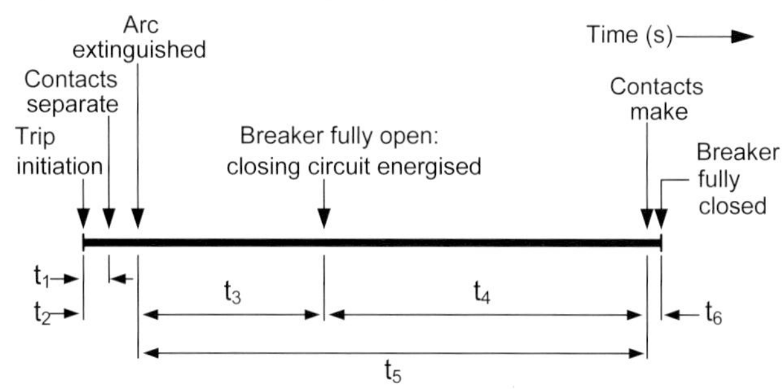

	Oil 11kV	Vacuum 15kV	Oil 132kV	Air 380kV	SF_6 132kV	SF_6 380kV
t_1	0.06	0.038	0.03	0.035	0.04	0.02
t_2	0.1	0.053	0.06	0.045	0.07	0.05
t_3	0.08	0.023	0.2	0.235	0.03	0.01
t_4	0.16	0.048	0.35	0.065	0.08	0.06
t_5	0.24	0.28	0.55	0.3	0.11	0.07
t_6	0.02	0.07	0.01	0.02	0.12	0.04

Note: 380kV data applicable to 400kV also. All times in seconds

Figure 14.3: Typical circuit breaker trip-close operation times

14.4.1.4 De-ionisation of the fault path

As mentioned above, successful high speed reclosure requires the interruption of the fault by the circuit breaker to be followed by a time delay long enough to allow the ionised air to disperse. This time is dependent on the system voltage, cause of fault, weather conditions and so on, but at voltages up to 66kV, 0.1s - 0.2s should be adequate. On HV systems, therefore, fault de-ionisation time is of less importance than circuit breaker time delays.

14.4.1.5 Protection reset time

If time delayed protection is used, it is essential that the timing device shall fully reset during the dead time, so that correct time discrimination will be maintained after reclosure on to a fault. The reset time of the electromechanical I.D.M.T. relay is

10 seconds or more when on maximum time setting, and dead times of at least this value may be required.

When short dead times are required, the protection relays must reset almost instantaneously, a requirement that is easily met by the use of static, digital and numerical I.D.M.T. relays.

14.4.2 Reclaim Time

Factors affecting the setting of the reclaim time are discussed in the following sections.

14.4.2.1 Type of protection

The reclaim time must be long enough to allow the protection relays to operate when the circuit breaker is reclosed on to a permanent fault. The most common forms of protection applied to HV lines are I.D.M.T. or definite time over-current and earth-fault relays. The maximum operating time for the former with very low fault levels could be up to 30 seconds, while for fault levels of several times rating the operating time may be 10 seconds or less.

In the case of definite time protection, settings of 3 seconds or less are common, with 10 seconds as an absolute maximum. It has been common practice to use reclaim times of 30 seconds on HV auto-reclose schemes. However, there is a danger with a setting of this length that during a thunderstorm, when the incidence of transient faults is high, the breaker may reclose successfully after one fault, and then trip and lock out for a second fault within this time. Use of a shorter reclaim time of, say, 15 seconds may enable the second fault to be treated as a separate incident, with a further successful reclosure.

Where fault levels are low, it may be difficult to select I.D.M.T. time settings to give satisfactory grading with an operating time limit of 15 seconds, and the matter becomes a question of selecting a reclaim time compatible with I.D.M.T. requirements.

It is common to fit sensitive earth-fault protection to supplement the normal protection to detect high resistance earth faults. This protection is usually set to have an operating time longer than that of the main protection. This longer time may have to be taken into consideration when deciding on a reclaim time. A broken overhead conductor in contact with dry ground or a wood fence may cause this type of fault. It is rarely if ever transient and may be a danger to the public. It is therefore common practice to use a contact on the sensitive earth fault relay to block auto-reclosing and lock out the circuit breaker.

Where high-speed protection is used, reclaim times of 1 second or less would be adequate. However, such short times are rarely used in practice, to relieve the duty on the circuit breaker.

14.4.2.2 Spring winding time

The reclaim time of motor-wound spring-closed breakers must be at least as long as the spring winding time, to ensure that the breaker is not subjected to a further reclosing operating with a partly wound spring.

14.4.3 Number of Shots

There are no definite rules for defining the number of shots for any particular auto-reclose application, but a number of factors must be taken into account.

14.4.3.1 Circuit breaker limitations

Important considerations are the ability of the circuit breaker to perform several trip and close operations in quick succession and the effect of these operations on the maintenance period. Maintenance periods vary according to the type of circuit breaker used and the fault current broken when clearing each fault. Use of modern numerical relays can assist, as they often have a CB condition-monitoring feature included that can be arranged to indicate to a Control Centre when maintenance is required. Auto-reclose may then be locked out until maintenance has been carried out.

14.4.3.2 System conditions

If statistical information on a particular system shows a moderate percentage of semi-permanent faults that could be burned out during 2 or 3 time-delayed trips, a multi-shot scheme may be justified. This is often the case in forest areas. Multi-shot schemes may also be applicable where fused 'tees' are used and the fault level is low, since the fusing time may not discriminate with the main I.D.M.T. relay. The use of several shots will heat the fuse to such an extent that it would eventually blow before the main protection operated.

14.5 AUTO-RECLOSING ON EHV TRANSMISSION LINES

The most important consideration in the application of auto-reclosing to EHV transmission lines is the maintenance of system stability and synchronism. The problems involved are dependent on whether the transmission system is weak or strong. With a weak system, loss of a transmission link may lead quickly to an excessive phase angle across the CB used for re-closure, thus preventing a successful re-closure. In a relatively strong system, the rate of change of phase angle will be slow, so that delayed auto-reclose can be successfully applied.

An illustration is the interconnector between two power systems as shown in Figure 14.4. Under healthy conditions, the amount of synchronising power transmitted, *P*, crosses the power/angle curve *OAB* at point *X*, showing that the phase displacement between the two systems is θ_0. Under fault conditions, the curve *OCB* is applicable, and the operating point changes to *Y*. Assuming constant power input to both ends of the line, there is now an accelerating power *XY*. As a result, the operating point moves to *Z*, with an increased phase displacement, θ_1, between the two systems. At this point the circuit breakers trip and break the connection. The phase displacement continues to increase at a rate dependent on the inertia of the two power sources. To maintain synchronism, the circuit breaker must be reclosed in a time short enough to prevent the phase angle exceeding θ_2. This angle is such that the area (2) stays greater than the area (1), which is the condition for maintenance of synchronism.

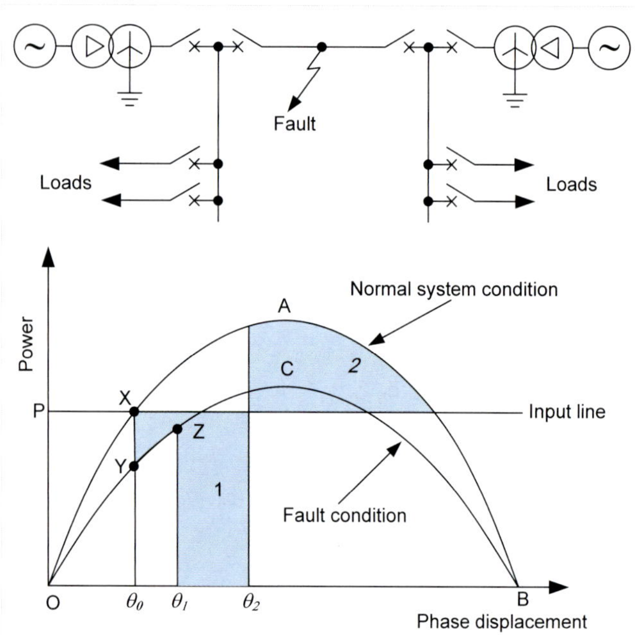

Figure 14.4: Effect of high-speed three-phase auto-reclosing on system stability for a weak system

This example, for a weak system, shows that the successful application of auto-reclosing in such a case needs high-speed protection and circuit breakers, and a short dead time. On strong systems, synchronism is unlikely to be lost by the tripping out of a single line. For such systems, an alternative policy of delayed auto-reclosing may be adopted. This enables the power swings on the system resulting from the fault to decay before reclosure is attempted.

The various factors to be considered when using EHV auto-reclose schemes are now dealt with. High-speed and delayed auto-reclose schemes are discussed separately.

14.6 HIGH SPEED AUTO-RECLOSING ON EHV SYSTEMS

The first requirement for the application of high-speed auto-reclosing is knowledge of the system disturbance time that can be tolerated without loss of system stability. This will normally require transient stability studies to be conducted for a defined set of power system configurations and fault conditions. With knowledge of protection and circuit breaker operating characteristics and fault arc de-ionisation times, the feasibility of high-speed auto-reclosing can then be assessed. These factors are now discussed.

14.6.1 Protection Characteristics

The use of high-speed protection equipment, such as distance or unit protection schemes, giving operating times of less than 40ms, is essential. In conjunction with fast operating circuit breakers, high-speed protection reduces the duration of the fault arc and thus the total system disturbance time.

It is important that the circuit breakers at both ends of a fault line should be tripped as rapidly as possible. The time that the line is still being fed from one end represents an effective reduction in the dead time, and may well jeopardise the chances of a successful reclosure. When distance protection is used, and the fault occurs near one end of the line, special measures have to be adopted to ensure simultaneous tripping at each end. These are described in Section 14.8.

14.6.2 De-Ionisation of the Fault Arc

It is important to know the time that must be allowed for complete de-ionisation of the arc, to prevent the arc restriking when the voltage is re-applied.

The de-ionisation time of an uncontrolled arc, in free air depends on the circuit voltage, conductor spacing, fault currents, fault duration, wind speed and capacitive coupling from adjacent conductors. Of these, the circuit voltage is the most important, and as a general rule, the higher the voltage the longer the time required for de-ionisation. Typical values for three-phase faults are given in Table 14.1.

Line Voltage (kV)	Minimum De-energisation Time (Seconds)
66	0.2
110	0.28
132	0.3
220	0.35
275	0.38
400	0.45
525	0.55

Table 14.1: Fault-arc de-ionisation times

If single-phase tripping and auto-reclosing is used, capacitive coupling between the healthy phases and the faulty phase tends to maintain the arc and hence extend the dead time required from the values given in the Table. This is a particular problem on long distance EHV transmission lines. However where shunt compensation is applied, a neutral reactor can often be used to balance system inter-phase capacitance and thus reduce the arcing time.

14.6.3 Circuit Breaker Characteristics

The high fault levels involved in EHV systems impose a very severe duty on the circuit breakers used in high-speed auto-reclose schemes. The accepted breaker cycle of break-make-break requires the circuit breaker to interrupt the fault current, reclose the circuit after a time delay of upwards of 0.2s and then break the fault current again if the fault persists. The types of circuit breaker commonly used on EHV systems are oil, air blast and SF_6 types.

14.6.3.1 Oil circuit breakers

Oil circuit breakers are used for transmission voltages up to 300kV, and can be subdivided into the two types: 'bulk oil' and 'small oil volume'. The latter is a design aimed at reducing the fire hazard associated with the large volume of oil contained in the bulk oil breaker.

The operating mechanisms of oil circuit breakers are of two types, 'fixed trip' and 'trip free', of which the latter is the most common. With trip-free types, the reclosing cycle must allow time for the mechanism to reset after tripping before applying the closing impulse.

Special means have to be adopted to obtain the short dead times required for high-speed auto-reclosing. Various types of tripping mechanism have been developed to meet this requirement.

The three types of closing mechanism fitted to oil circuit breakers are:

- solenoid
- spring
- pneumatic

CBs with solenoid closing are not suitable for high-speed auto-reclose due to the long time constant involved. Spring, hydraulic or pneumatic closing mechanisms are universal at the upper end of the EHV range and give the fastest closing time. Figure 14.3 shows the operation times for various types of EHV circuit breakers, including the dead time that can be attained.

14.6.3.2 Air blast circuit breakers

Air blast breakers have been developed for voltages up to the highest at present in use on transmission lines. They fall into two categories:

- pressurised head circuit breakers
- non-pressurised head circuit breakers

In pressurised head circuit breakers, compressed air is maintained in the chamber surrounding the main contacts. When a tripping signal is received, an auxiliary air system separates the main contacts and allows compressed air to blast through the gap to the atmosphere, extinguishing the arc. With the contacts fully open, compressed air is maintained in the chamber.

Loss of air pressure could result in the contacts reclosing, or, if a mechanical latch is employed, restriking of the arc in the de-pressurised chamber. For this reason, sequential series isolators, which isolate the main contacts after tripping, are commonly used with air blast breakers. Since these are comparatively slow in opening, their operation must be inhibited when auto-reclosing is required. A contact on the auto-reclose relay is made available for this purpose.

Non-pressurised head circuit breakers are slower in operation than the pressurised head type and are not usually applied in high-speed reclosing schemes.

14.6.3.3 SF_6 circuit breakers

Most EHV circuit breaker designs now manufactured use SF_6 gas as an insulating and arc-quenching medium. The basic design of such circuit breakers is in many ways similar to that of pressurised head air blast circuit breakers. Voltage withstand capability depends on a minimum gas pressure being available, and gas pressure monitors are fitted and arranged to block CB operation in the event of low gas pressure occurring. Sequential series isolators are normally used, to prevent damage to the circuit breaker in the event of voltage transients due to lightning strikes, etc. occurring when the CB is open. Provision should therefore be made to inhibit sequential series isolation during an auto-reclose cycle.

14.6.4 Choice of Dead Time

At voltages of 220kV and above, the de-ionisation time will probably dictate the minimum dead time, rather than any circuit breaker limitations. This can be deduced from Table 14.1. The dead time setting on a high-speed auto-reclose relay should be long enough to ensure complete de-ionisation of the arc. On EHV systems, an unsuccessful reclosure is more detrimental to the system than no reclosure at all.

14.6.5 Choice of Reclaim Time

Where EHV oil circuit breakers are concerned, the reclaim time should take account of the time needed for the closing mechanism to reset ready for the next reclosing operation.

14.6.6 Number of Shots

High-speed auto-reclosing on EHV systems is invariably single shot. Repeated reclosure attempts with high fault levels would have serious effects on system stability, so the circuit breakers are locked out after one unsuccessful attempt. Also, the incidence of semi-permanent faults that can be cleared by repeated reclosures is less likely than on HV systems. Multi-shot schemes have, however, occasionally been used on EHV systems, specifically to deal with bush fire faults prevalent in Africa.

14.7 SINGLE-PHASE AUTO-RECLOSING

Single phase to earth faults account for the majority of overhead line faults. When three-phase auto-reclosing is applied to single circuit interconnectors between two power systems, the tripping of all three phases may cause the two systems to drift apart in phase, as described in Section 14.5. No interchange of synchronising power can take place during the dead time. If only the faulty phase is tripped, synchronising power can still be interchanged through the healthy phases. Any difference in phase between the two systems will be correspondingly less, leading to a reduction in the disturbance on the system when the circuit breaker recloses.

For single-phase auto-reclosing each circuit breaker pole must be provided with its own closing and tripping mechanism; this is normal with EHV air blast and SF_6 breakers. The associated tripping and reclosing circuitry is therefore more complicated, as it must be inherently phase-selective.

On the occurrence of a phase-earth fault, single-phase auto-reclose schemes trip and reclose only the corresponding pole of the circuit breaker. The auto-reclose function in a relay therefore has three separate elements, one for each phase. Operation of any element energises the corresponding dead timer, which in turn initiates a closing pulse for the appropriate pole of the circuit breaker. A successful reclosure results in the auto-reclose logic resetting at the end of the reclaim time, ready to respond to a further fault incident. If the fault is persistent and reclosure is unsuccessful, it is usual to trip and lock out all three poles of the circuit breaker.

The above describes only one of many variants. Other possibilities are:

- three-phase trip and lockout for phase-phase or 3-phase faults, or if either of the remaining phases should develop a fault during the dead time
- use of a selector switch to give a choice of single or three-phase reclosing
- combined single and three-phase auto-reclosing; single phase to earth faults initiate single-phase tripping and reclosure, and phase-phase faults initiate three-phase tripping and reclosure

Modern numerical relays often incorporate the logic for all of the above schemes for the user to select as required. Use can be made of any user-definable logic feature in a numerical relay to implement other schemes that may be required.

The advantages of single-phase auto-reclosing are:

- the maintenance of system integrity
- on multiple earth systems, negligible interference with the transmission of load. This is because the current in the unfaulted phases can continue to flow until the fault is cleared and the faulty phase restored

The main disadvantage is the longer de-ionisation time resulting from capacitive coupling between the faulty and healthy lines. This leads to a longer dead time being required. Maloperation of earth fault relays on double circuit lines owing to the flow of zero sequence currents may also occur. These are induced by mutual induction between faulty and healthy lines (see Chapter 13 for details).

14.8 HIGH-SPEED AUTO-RECLOSING ON LINES EMPLOYING DISTANCE SCHEMES

The importance of rapid tripping of the circuit breakers at each end of a faulted line where high-speed auto-reclosing is employed has already been covered in Section 14.6. Simple distance protection presents some difficulties in this respect.

Owing to the errors involved in determining the ohmic setting of the distance relays, it is not possible to set Zone 1 of a distance relay to cover 100% of the protected line – see Chapter 11 for more details. Zone 1 is set to cover 80-85% of the line length, with the remainder of the line covered by time-delayed Zone 2 protection.

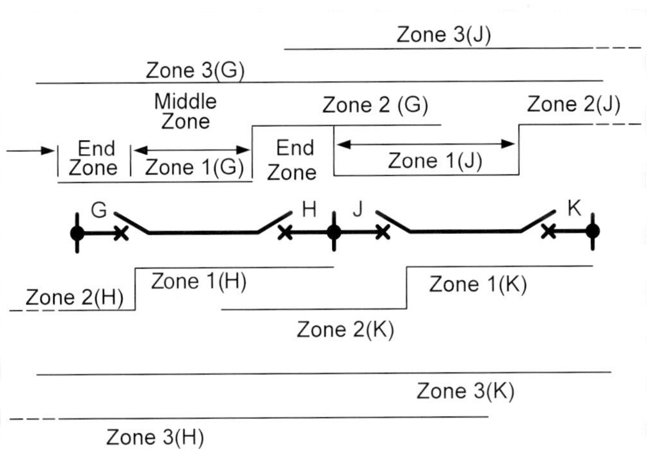

Figure 14.5: Typical three zone distance scheme

Figure 14.5 shows this for a typical three-zone distance scheme covering two transmission lines.

For this reason, a fault occurring in an end zone would be cleared instantaneously, by the protection at one end of the feeder. However, the CB at the other end opens in 0.3 - 0.4 seconds (Zone 2 time). High-speed auto-reclosing applied to the circuit breakers at each end of the feeder could result either in no dead time or in a dead time insufficient to allow de-ionisation of the fault arc. A transient fault could therefore be seen as a permanent one, resulting in the locking out of both circuit breakers.

Two methods are available for overcoming this difficulty. Firstly, one of the transfer-trip or blocking schemes that involves the use of an intertrip signal between the two ends of the line can be used. Alternatively, a Zone 1 extension scheme may be used to give instantaneous tripping over the whole line length. Further details of these schemes are given in Chapter 12, but a brief description of how they are used in conjunction with an auto-reclose scheme is given below.

14.8.1 Transfer-Trip or Blocking Schemes

This involves use of a signalling channel between the two ends of the line. Tripping occurs rapidly at both ends of the faulty line, enabling the use of high-speed auto-reclose. Some complication occurs if single-phase auto-reclose is used, as the signalling channel must identify which phase should be tripped, but this problem does not exist if a modern numerical relay is used.

Irrespective of the scheme used, it is customary to provide an auto-reclose blocking relay to prevent the circuit breakers auto-reclosing for faults seen by the distance relay in Zones 2 and 3.

14.8.2 Zone 1 Extension

In this scheme, the reach of Zone 1 is normally extended to 120% of the line length and is reset to 80% when a command from the auto-reclose logic is received. This auto-reclose logic signal should occur before a closing pulse is applied to the circuit breaker and remain operated until the end of the reclaim time. The logic signal should also be present when auto-reclose is out of service.

14.9 DELAYED AUTO-RECLOSING ON EHV SYSTEMS

On highly interconnected transmission systems, where the loss of a single line is unlikely to cause two sections of the system to drift apart significantly and lose synchronism, delayed auto reclosing can be employed. Dead times of the order of 5s - 60s are commonly used. No problems are presented by fault arc de-ionisation times and circuit breaker operating characteristics, and power swings on the system decay before reclosing. In addition, all tripping and reclose schemes can be three-phase only, simplifying control circuits in comparison with single-phase schemes. In systems on which delayed auto-reclosing is permissible, the chances of a reclosure being successful are somewhat greater with delayed reclosing than would be the case with high-speed reclosing.

14.9.1 Scheme Operation

The sequence of operations of a delayed auto-reclose scheme can be best understood by reference to Figure 14.6. This shows a transmission line connecting two substations A and B, with the circuit beakers at A and B tripping out in the event of a line fault. Synchronism is unlikely to be lost in a system that employs delayed auto-reclose. However, the transfer of power through the remaining tie-lines on the system could result in the development of an excessive phase difference between the voltages at points A and B. The result, if reclosure takes place, is an unacceptable shock to the system. It is therefore usual practice to incorporate a synchronism check relay into the reclosing system to determine whether auto-reclosing should take place.

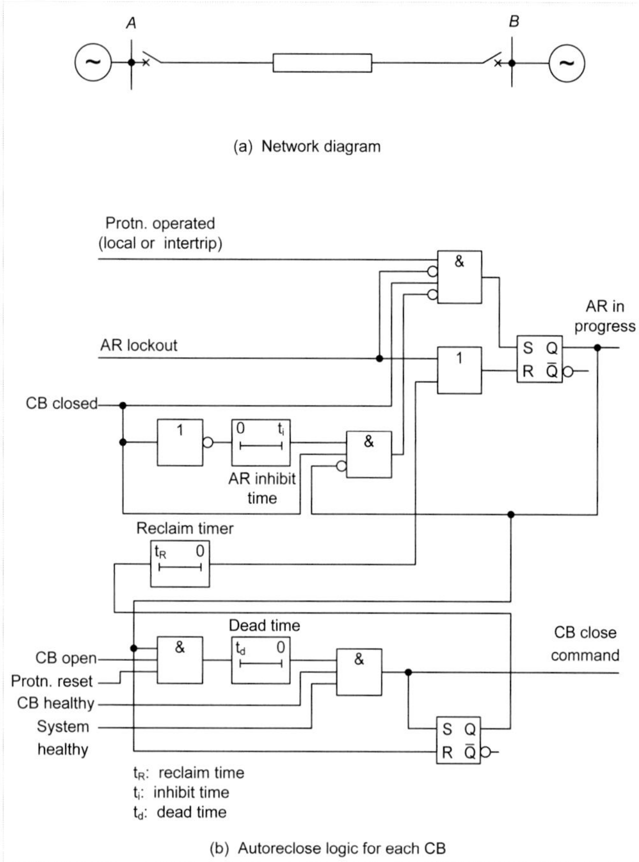

Figure 14.6: Delayed auto-reclose scheme logic

After tripping on a fault, it is normal procedure to reclose the breaker at one end first, a process known as 'live bus/dead line charging'. Reclosing at the other and is then under the control of a synchronism check relay element for what is known as 'live bus/live line reclosing'.

For example, if it were decided to charge the line initially from station *A*, the dead time in the auto-reclose relay at *A* would be set at, say, 5 seconds, while the corresponding timer in the auto-reclose relay at *B* would be set at, say, 15 seconds. The circuit beaker at *A* would then reclose after 5 seconds provided that voltage monitoring relays at *A* indicated that the busbars were alive and the line dead. With the line recharged, the circuit breaker at *B* would then reclose with a synchronism check, after a 2 second delay imposed by the synchronism check relay element.

If for any reason the line fails to 'dead line charge' from end *A*, reclosure from end *B* would take place after 15 seconds. The circuit breaker at *A* would then be given the opportunity to reclose with a synchronism check.

14.9.2 Synchronism Check Relays

The synchronism check relay element commonly provides a three-fold check:

- phase angle difference
- voltage
- rate of change of slip frequency

The phase angle setting is usually set to between 20° – 45°, and reclosure is inhibited if the phase difference exceeds this value. The scheme waits for a reclosing opportunity with the phase angle within the set value, but locks out if reclosure does not occur within a defined period, typically 5s.

A voltage check is incorporated to prevent reclosure under various circumstances. A number of different modes may be available. These are typically undervoltage on either of the two measured voltages, differential voltage, or both of these conditions.

The logic also incorporates a frequency difference check, either by direct measurement or by using a timer in conjunction with the phase angle check. In the latter case, if a 2 second timer is employed, the logic gives an output only if the phase difference does not exceed the phase angle setting over a period of 2 seconds. This limits the frequency difference (in the case of a phase angle setting of 20°) to a maximum of 0.11% of 50Hz, corresponding to a phase swing from +20° to -20° over the measured 2 seconds. While a significant frequency difference is unlikely to arise during a delayed auto-reclose sequence, the time available allows this check to be carried out as an additional safeguard.

As well as 'live bus/dead line' and 'live bus/live line' reclosing, sometimes 'live line/dead bus' reclosing may need to be implemented. A numerical relay will typically allow any combination of these modes to be implemented. The voltage settings for distinguishing between 'live' and 'dead' conditions must be carefully chosen. In addition, the locations of the VTs must be known and checked so that the correct voltage signals are connected to the 'line' and 'bus' inputs.

14.10 OPERATING FEATURES OF AUTO-RECLOSE SCHEMES

The extensive use of auto-reclosing has resulted in the existence of a wide variety of different control schemes. Some of the more important variations in the features provided are described below.

14.10.1 Initiation

Modern auto-reclosing schemes are invariably initiated by the tripping command of a protection relay function. Some older schemes may employ a contact on the circuit breaker. Modern digital or numerical relays often incorporate a comprehensive auto-reclose facility, thus eliminating the need for a separate auto-reclose relay.

14.10.2 Type of Protection

On HV distribution systems, advantage is often taken of auto-reclosing to use instantaneous protection for the first trip, followed by I.D.M.T. for subsequent trips in a single fault incident. In such cases, the auto-reclose relay must provide a means of isolating the instantaneous relay after the first trip. In older schemes, this may be done with a normally closed contact on the auto-reclose starting element wired into the connection between the instantaneous relay contact and the circuit breaker trip coil. With digital or numerical relays with in-built auto-reclose facilities, internal logic facilities will normally be used.

For certain utility companies, it is the rule to fit tripping relays to every circuit breaker. If auto-reclosing is required, self or electrically reset tripping relays must be used. If the later is used, a contact must be provided either in the auto-reclose logic or by separate trip relay resetting scheme to energise the reset coil before reclosing can take place.

14.10.3 Dead Timer

This will have a range of settings to cover the specified high-speed or delayed reclosing duty. Any interlocks that are needed to hold up reclosing until conditions are suitable can be connected into the dead timer circuit. Section 14.12.1 provides an example of this applied to transformer feeders.

14.10.4 Reclosing Impulse

The duration of the reclosing impulse must be related to the requirements of the circuit breaker closing mechanism. On auto-reclose schemes using spring-closed breakers, it is sufficient to operate a contact at the end of the dead time to energise the latch release coil on the spring-closing mechanism. A circuit breaker auxiliary switch can be used to cancel the closing pulse and reset the auto-reclose relay. With solenoid operated breakers, it is usual to provide a closing pulse of the order of 1 - 2 seconds, to hold the solenoid energised for a short time after the main contacts have closed. This ensures that the mechanism settles in the fully latched-in position. The pneumatic or hydraulic closing mechanisms fitted to oil, air blast and SF_6 circuit breakers use a circuit breaker auxiliary switch for terminating the closing pulse applied by the auto-reclose relay.

14.10.5 Anti-Pumping Devices

The function of an anti-pumping device is to prevent the circuit breaker closing and opening several times in quick succession. This might be caused by the application of a closing pulse while the circuit breaker is being tripped via the protection relays. Alternatively, it may occur if the circuit breaker is closed on to a fault and the closing pulse is longer than the sum of protection relay and circuit breaker operating times. Circuit breakers with trip free mechanisms do not require this feature.

14.10.6 Reclaim Timer

Electromechanical, static or software-based timers are used to provide the reclaim time, depending on the relay technology used. If electromechanical timers are used, it is convenient to employ two independently adjustable timed contacts to obtain both the dead time and the reclaim time on one timer. With static and software-based timers, separate timer elements are generally provided.

14.10.7 CB Lockout

If reclosure is unsuccessful the auto-reclose relay locks out the circuit breaker. Some schemes provide a lockout relay with a flag, with provision of a contact for remote alarm. The circuit breaker can then only be closed by hand; this action can be arranged to reset the auto-reclose relay element automatically. Alternatively, most modern relays can be configured such that a lockout condition can be reset only by operator action.

Circuit breaker manufacturers state the maximum number of operations allowed before maintenance is required. A number of schemes provide a fault trip counting function and give a warning when the total approaches the manufacturer's recommendation. These schemes will lock out when the total number of fault trips has reached the maximum value allowed.

14.10.8 Manual Closing

It is undesirable to permit auto-reclosing if circuit breaker closing is manually initiated. Auto-reclose schemes include the facility to inhibit auto-reclose initiation for a set time following manual CB closure. The time is typically in the range of 2 - 5 seconds.

14.10.9 Multi-Shot Schemes

Schemes providing up to three or four shots use timing circuits are often included in an auto-reclose relay to provide different, independently adjustable, dead times for each shot. Instantaneous protection can be used for the first trip, since each scheme provides a signal to inhibit instantaneous tripping after a set number of trips and selects I.D.M.T. protection for subsequent ones. The scheme resets if reclosure is successful within the chosen number of shots, ready to respond to further fault incidents.

14.11 AUTO-CLOSE SCHEMES

Auto-close schemes are employed to close automatically circuit breakers that are normally open when the supply network is healthy. This may occur for a variety of reasons, for instance the fault level may be excessive if the CBs were normally closed. The circuits involved are very similar to those used for auto-reclosing. Two typical applications are described in the following sections.

14.11.1 Standby Transformers

Figure 14.7 shows a busbar station fed by three transformers, *T1*, *T2* and *T3*. The loss of one transformer might cause serious overloading of the remaining two. However, connection of a further transformer to overcome this may increase the fault level to an unacceptable value.

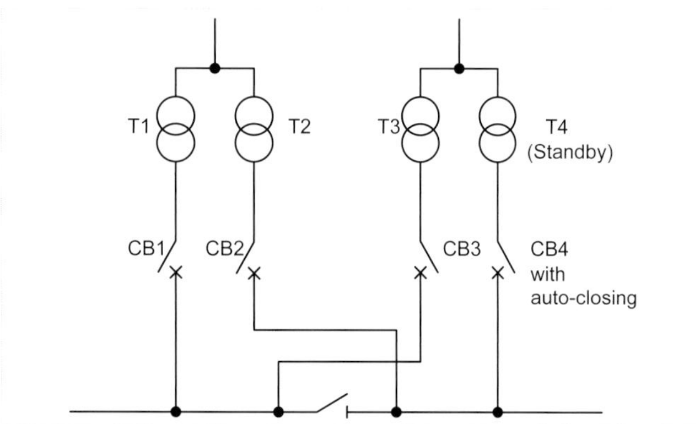

Figure 14.7: Standby transformer with auto-closing

The solution is to have a standby transformer *T4* permanently energised from the primary side and arranged to be switched into service if one of the others trips on fault.

The starting circuits for breaker *CB4* monitor the operation of transformer protection on any of the transformers *T1*, *T2* and *T3* together with the tripping of an associated circuit breaker *CB1* - *CB3*. In the event of a fault, the auto-close circuit is initiated and circuit breaker *CB4* closes, after a short time delay, to switch in the standby transformer. Some schemes employ an auto-tripping relay, so that when the faulty transformer is returned to service, the standby is automatically disconnected.

14.11.2 Bus Coupler or Bus Section Breaker

If all four power transformers are normally in service for the system of Figure 14.7, and the bus sections are interconnected by a normally-open bus section breaker instead of the isolator, the bus section breaker should be auto-closed in the event of the loss of one transformer, to spread the load over the remaining transformers. This, of course, is subject to the fault level being acceptable with the bus-section breaker closed.

Starting and auto-trip circuits are employed as in the stand-by scheme. The auto-close relay used in practice is a variant of one of the standard auto-reclose relays.

14.12 EXAMPLES OF AUTO-RECLOSE APPLICATIONS

The following sections describe auto-reclose facilities in common use for several standard substation configurations.

14.12.1 Double Busbar Substation

A typical double busbar station is shown in Figure 14.8. Each of the six EHV transmission lines brought into the station is under the control of a circuit breaker, CB1 to CB6 inclusive, and each transmission line can be connected either to the main or to the reserve busbars by manually operated isolators.

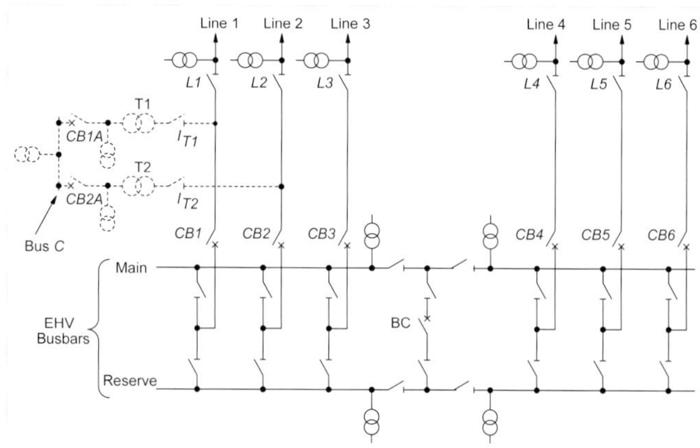

Figure 14.8: Double busbar substation

Bus section isolators enable sections of busbar to be isolated in the event of a fault and the bus coupler breaker *BC* permits sections of main and reserve bars to be interconnected.

14.12.1.1 Basic scheme – banked transformers omitted

Each line circuit breaker is provided with an auto-reclose relay that recloses the appropriate circuit breakers in the event of a line fault. For a fault on Line *1*, this would require opening of *CB1* and the corresponding CB at the remote end of the line. The operation of either the busbar protection or a VT Buchholz relay is arranged to lock out the auto-reclosing sequence. In the event of a persistent fault on Line 1, the line circuit breakers trip and lock out after one attempt at reclosure.

14.12.1.2 Scheme with banked transformers

Some utilities use a variation of the basic scheme in which Transformers *T1* and *T2* are banked off Lines *1* and *2*, as shown in Figure 14.8. This provides some economy in the number of circuit breakers required. The corresponding transformer circuits 1 and 2 are tee'd off Lines *1* and 2 respectively. The transformer secondaries are connected to a separate HV busbar system via circuit breakers *CB1A* and *CB2A*.

Auto-reclose facilities can be extended to cover the circuits for banked transformers where these are used. For example, a fault on line 1 would cause the tripping of circuit breakers *CB1*, *CB1A* and the remote line circuit breaker. When Line *1* is re-energised, either by auto-reclosure of *CB1* or by the remote circuit breaker, whichever is set to reclose first, transformer *T1* is also energised. *CB1A* will not reclose until the appearance of transformer secondary voltage, as monitored by the secondary VT; it then recloses on to the HV busbars after a short time delay, with a synchronism check if required.

In the event of a fault on transformer *T1*, the local and remote line circuit breakers and breaker *CB1A* trip to isolate the fault. Automatic opening of the motorised transformer isolator I_{T1} follows this. The line circuit breakers then reclose in the normal manner and circuit breaker *CB1A* locks out.

A shortcoming of this scheme is that this results in healthy transformer *T1* being isolated from the system; also, isolator *L1* must be opened manually before circuit breakers *CB1* and *CB1A*, can be closed to re-establish supply to the HV busbars via the transformer. A variant of this scheme is designed to instruct isolator *L1* to open automatically following a persistent fault on Line 1 and provide a second auto-reclosure of *CB1* and *CB1A*. The supply to Bus *C* is thereby restored without manual intervention.

14.12.2 Single Switch Substation

The arrangement shown in Figure 14.9 consists basically of two transformer feeders interconnected by a single circuit breaker *120*. Each transformer therefore has an alternative source of supply in the event of loss of one or other of the feeders.

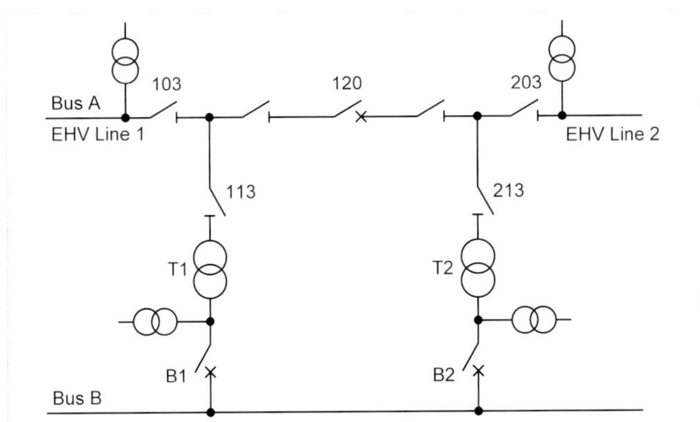

Figure 14.9: Single switch substation

For example, a transient fault on Line *1* causes tripping of circuit breakers *120* and *B1* followed by reclosure of CB *120*. If the reclosure is successful, Transformer *T1* is re-energised and circuit breaker *B1* recloses after a short time delay.

If the line fault is persistent, *120* trips again and the motorised line isolator *103* is automatically opened. Circuit breaker *120* recloses again, followed by *B1*, so that both transformers *T1* and *T2* are then supplied from Line *2*.

A transformer fault causes the automatic opening of the appropriate transformer isolator, lock-out of the transformer secondary circuit breaker and reclosure of circuit breaker *120*. Facilities for dead line charging or reclosure with synchronism check are provided for each circuit breaker.

14.12.3 Four-Switch Mesh Substation

The mesh substation shown in Figure 14.10 is extensively used by some utilities, either in full or part. The basic mesh has a feeder at each corner, as shown at mesh corners *MC2*, *MC3* and *MC4*. One or two transformers may also be banked at a mesh corner, as shown at *MC1*. Mesh corner protection is required if more than one circuit is fed from a mesh corner, irrespective of the CT locations – see Chapter 15 for more details.

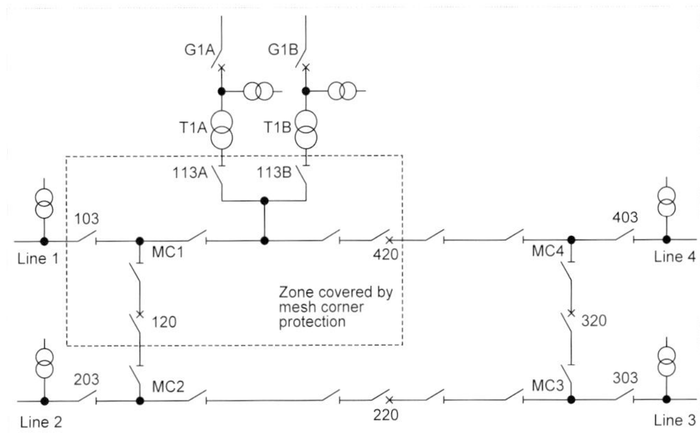

Figure 14.10: Four-switch mesh substation

Considerable problems can are encountered in the application of auto-reclosing to the mesh substation. For example, circuit breakers *120* and *420* in Figure 14.10 are tripped out for a variety of different types of fault associated with mesh corner 1 (*MC1*), and each requires different treatment as far as auto-reclosing is concerned. Further variations occur if the faults are persistent.

Following normal practice, circuit breakers must be reclosed in sequence, so sequencing circuits are necessary for the four mesh breakers. Closing priority may be in any order, but is normally *120*, *220*, *320*, and *420*.

A summary of facilities is now given, based on mesh corner *MC1* to show the inclusion of banked transformers; facilities at other corners are similar but omit the operation of equipment solely associated with the banked transformers.

14.12.3.1 Transient fault on Line 1

Tripping of circuit breakers *120*, *420*, *G1A* and *G1B* is followed by reclosure of *120* to give dead line charging of Line *1*.

Breaker *420* recloses in sequence, with a synchronism check. Breakers *G1A*, *G1B* reclose with a synchronism check if necessary.

14.12.3.2 Persistent fault on Line 1

Circuit breaker 120 trips again after the first reclosure and isolator 103 is automatically opened to isolate the faulted line. Breakers *120*, 420, *G1A*, and *G1B* then reclose in sequence as above.

14.12.3.3 Transformer fault (local transformer 1A)

Automatic opening of isolator *113A* to isolate the faulted transformer follows tripping of circuit breakers *120*, *420*, *G1A* and *G1B*. Breakers *120*, *420* and G1B then reclose in sequence, and breaker *G1A* is locked out.

14.12.3.4 Transformer fault (remote transformer)

For a remote transformer fault, an intertrip signal is received at the local station to trip breakers *120*, *420*, *G1A* and *G1B* and inhibit auto-reclosing until the faulted transformer has been isolated at the remote station. If the intertrip persists for 60 seconds it is assumed that the fault cannot be isolated at the remote station. Isolator *103* is then automatically opened and circuit breakers *120*, *420*, *G1A* and *G1B* are reclosed in sequence.

14.12.3.5 Transient mesh corner fault

Any fault covered by the mesh corner protection zone, shown in Figure 14.10, results in tripping of circuit breakers *120*, *420*, *G1A* and *G1B*. These are then reclosed in sequence.

There may be circumstances in which reclosure onto a persistent fault is not permitted – clearly it is not known in advance of reclosure if the fault is persistent or not. In these circumstances, scheme logic inhibits reclosure and locks out the circuit breakers.

14.12.3.6 Persistent mesh corner fault

The sequence described in Section 14.12.3.5 is followed initially. When CB *120* is reclosed, it will trip again due to the fault and lock out. At this point, the logic inhibits the reclosure of CB's *420*, *G1A* and *G1B* and locks out these CBs. Line isolator *103* is automatically opened to isolate the fault from the remote station.

14.12.4 Breaker and a Half Substations

A simplistic example of a breaker and a half substation is shown in Figure 14.11. The substation has two busbars, Bus 1 and Bus 2, with lines being energised via a *'diameter'* of three circuit breakers (CB1, CB2, CB3). Two line circuits can be energised from each diameter, shown as Line 1 and Line 2. It can therefore be seen that the ratio of circuit breakers to lines is one-and-a-half, or breaker and a half. The advantage of such a topology is that it reduces the number of costly circuit breakers required, compared to a double-bus installation, but also it means that for any line fault, the associated protection relay(s) must trip two circuit breakers to isolate it. Any autoreclose scheme will then need to manage the closure of two breakers (e.g. CB1 and CB2 for reclosing Line 1).

Utilities usually select from one of three typical scheme philosophies in such a scenario:

- Autoreclosure of an 'outer' (or 'diameter') breaker, leaving the closing of the centre breaker for manual remote control
- A leader-follower autoreclosing scheme
- Autoreclosure of both breakers simultaneously

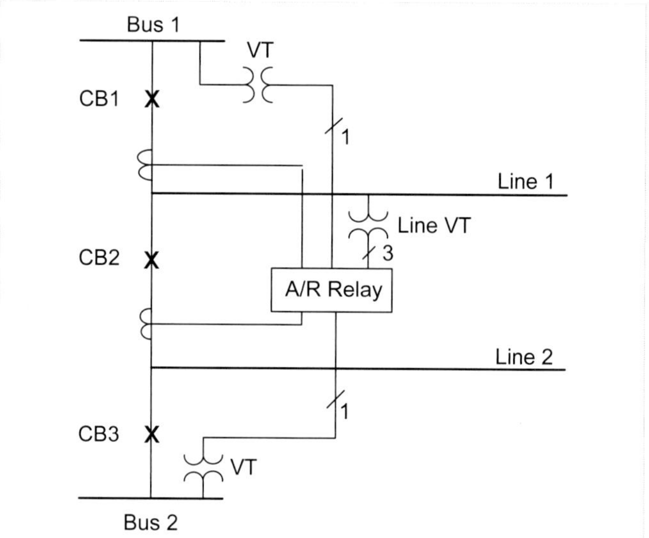

Figure 14.11: Breaker and a half example

The first option offers re-energisation of the line, but leaves the final topology restoration task of closing CB2 to the control operator.

A leader-follower scheme is one whereby just one circuit breaker is reclosed initially (CB1), and then only if this is successful, the second or *'follower'* breaker (CB2) is reclosed after a set follower time delay. The advantage here is that for a persistent fault there is only the increased interrupting duty of a switch-on-to-fault trip for a single circuit breaker, not two. Should a trip and lockout occur for CB1, then CB2 will also be driven to lockout. Figure 14.11 shows how a single autoreclose relay, associated with Line 1, can exert control upon both CB1 and CB2 (it shares control of CB2 with the autoreclose relay for Line 2). It is also important that before each circuit breaker is closed, the appropriate synchronism check is made. For this reason, the relay requires two bus synchronising voltages as inputs, in addition to the three-phase line VT input shown. CB1 can be permitted to close only if the voltage checks between Bus 1 and the Line VT are favourable, CB2 can be permitted to close only if the checks between Bus 2 and the Line VT are favourable

If a utility opts for a scheme which closes two circuit breakers simultaneously, the line voltage checks against both bus voltages need to be satisfied before the relay issues synchronised closing commands.

Chapter 15
Busbar Protection

15.1 INTRODUCTION

The protection scheme for a power system should cover the whole system against all probable types of fault. Unrestricted forms of line protection, such as overcurrent and distance systems, meet this requirement, although faults in the busbar zone are cleared only after some time delay. But if unit protection is applied to feeders and plant, the busbars are not inherently protected.

Busbars have often been left without specific protection, for one or more of the following reasons:

- the busbars and switchgear have a high degree of reliability, to the point of being regarded as intrinsically safe
- it was feared that accidental operation of busbar protection might cause widespread dislocation of the power system, which, if not quickly cleared, would cause more loss than would the very infrequent actual bus faults
- it was hoped that system protection or back-up protection would provide sufficient bus protection if needed

It is true that the risk of a fault occurring on modern metal-clad gear is very small, but it cannot be entirely ignored. However, the damage resulting from one uncleared fault, because of the concentration of fault MVA, may be very extensive indeed, up to the complete loss of the station by fire. Serious damage to or destruction of the installation would probably result in widespread and prolonged supply interruption.

Finally, system protection will frequently not provide the cover required. Such protection may be good enough for small distribution substations, but not for important stations. Even if distance protection is applied to all feeders, the busbar will lie in the second zone of all the distance protections, so a bus fault will be cleared relatively slowly, and the resultant duration of the voltage dip imposed on the rest of the system may not be tolerable.

With outdoor switchgear the case is less clear since, although the likelihood of a fault is higher, the risk of widespread damage resulting is much less. In general then, busbar protection is required when the system protection does not cover the busbars, or when, in order to maintain power system stability, high-speed fault clearance is necessary. Unit busbar protection provides this, with the further advantage that if the busbars are sectionalised, one section only need be isolated to clear a fault. The case for unit busbar protection is in fact strongest when there is sectionalisation.

15.2 BUSBAR FAULTS

The majority of bus faults involve one phase and earth, but faults arise from many causes and a significant number are interphase clear of earth. In fact, a large proportion of busbar faults result from human error rather than the failure of switchgear components.

With fully phase-segregated metalclad gear, only earth faults are possible, and a protection scheme need have earth fault sensitivity only. In other cases, an ability to respond to phase faults clear of earth is an advantage, although the phase fault sensitivity need not be very high.

15.3 PROTECTION REQUIREMENTS

Although not basically different from other circuit protection, the key position of the busbar intensifies the emphasis put on the essential requirements of speed and stability. The special features of busbar protection are discussed below.

15.3.1 Speed

Busbar protection is primarily concerned with:

- limitation of consequential damage
- removal of busbar faults in less time than could be achieved by back-up line protection, with the object of maintaining system stability

Some early busbar protection schemes used a low impedance differential system having a relatively long operation time, of up to 0.5 seconds. The basis of most modern schemes is a differential system using either low impedance biased or high impedance unbiased relays capable of operating in a time of the order of one cycle at a very moderate multiple of fault setting. To this must be added the operating time of any tripping relays, but an overall tripping time of less than two cycles can be achieved. With high-speed circuit breakers, complete fault clearance may be obtained in approximately 0.1 seconds. When a frame-earth system is used, the operating speed is comparable.

15.3.2 Stability

The stability of bus protection is of paramount importance. Bearing in mind the low rate of fault incidence, amounting to no more than an average of one fault per busbar in twenty years, it is clear that unless the stability of the protection is absolute, the degree of disturbance to which the power system is likely to be subjected may be increased by the installation of bus protection. The possibility of incorrect operation has, in the past, led to hesitation in applying bus protection and has also resulted in application of some very complex systems. Increased understanding of the response of differential systems to transient currents enables such systems to be applied with confidence in their fundamental stability. The theory of differential protection is given later in section 15.7.

Notwithstanding the complete stability of a correctly applied protection system, dangers exist in practice for a number of reasons. These are:

- interruption of the secondary circuit of a current transformer will produce an unbalance, which might cause tripping on load depending on the relative values of circuit load and effective setting. It would certainly do so during a through fault, producing substantial fault current in the circuit in question
- a mechanical shock of sufficient severity may cause operation, although the likelihood of this occurring with modern numerical schemes is reduced
- accidental interference with the relay, arising from a mistake during maintenance testing, may lead to operation

In order to maintain the high order of integrity needed for busbar protection, it is an almost invariable practice to make tripping depend on two independent measurements of fault quantities. Moreover, if the tripping of all the breakers within a zone is derived from common measuring relays, two separate elements must be operated at each stage to complete a tripping operation.

The two measurements may be made by two similar differential systems, or one differential system may be checked by a frame-earth system, by earth fault relays energised by current transformers in the transformer neutral-earth conductors or by voltage or overcurrent relays. Alternatively, a frame-earth system may be checked by earth fault relays.

If two systems of the unit or other similar type are used, they should be energised by separate current transformers in the case of high impedance unbiased differential schemes. The duplicate ring CT cores may be mounted on a common primary conductor but independence must be maintained throughout the secondary circuit.

In the case of low impedance, biased differential schemes that cater for unequal ratio CTs, the scheme can be energised from either one or two separate sets of main current transformers. The criteria of double feature operation before tripping can be maintained by the provision of two sets of ratio matching interposing CTs per circuit. When multi-contact tripping relays are used, these are also duplicated, one being energised from each discriminating relay; the contacts of the tripping relay are then series-connected in pairs to provide tripping outputs.

Separate tripping relays, each controlling one breaker only, are usually preferred. The importance of such relays is then no more than that of normal circuit protection, so no duplication is required at this stage. Not least among the advantages of using individual tripping relays is the simplification of trip circuit wiring, compared with taking all trip circuits associated with a given bus section through a common multi-contact tripping relay.

In double busbar installations, a separate protection system is applied to each section of each busbar. An overall check system is also provided, covering all sections of both busbars. The separate zones are arranged to overlap the busbar section switches, so that a fault on the section switch trips both the adjacent zones. This has sometimes been avoided in the past by giving the section switch a time advantage; the section switch is tripped first and the remaining breakers delayed by 0.5 seconds. Only the zone on the faulty side of the section switch will remain operated and trip, the other zone resetting and retaining that section in service. This gain, applicable only to very infrequent section switch faults, is obtained at the expense of seriously delaying the bus protection for all other faults. This practice is therefore not generally favoured. Some variations are dealt with later under the more detailed scheme descriptions. There are many combinations possible, but the essential principle is that no single accidental incident of a secondary nature shall be capable of causing an unnecessary trip of a bus section.

Security against maloperation is only achieved by increasing the amount of equipment that is required to function to complete an operation; and this inevitably increases the statistical risk that a tripping operation due to a fault may fail. Such a failure, leaving aside the question of consequential damage, may result in disruption of the power system to an extent as great, or greater, than would be caused by an unwanted trip. The relative risk of failure of this kind may be slight, but it has been thought worthwhile in some instances to provide a guard in this respect as well.

Security of both stability and operation is obtained by providing three independent channels (say X, Y and Z) whose outputs are arranged in a 'two-out-of three' voting arrangement, as shown in Figure 15.1.

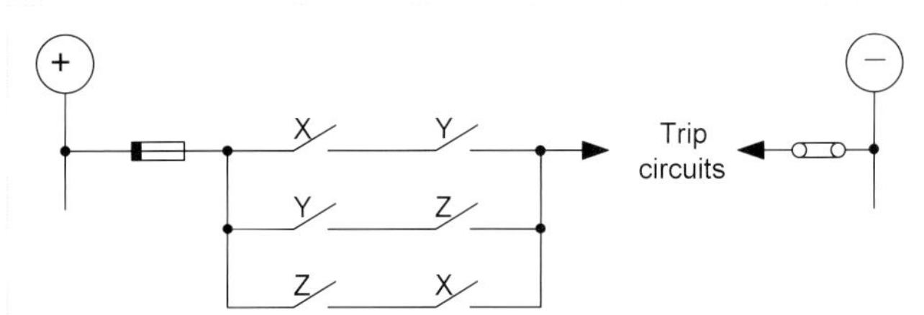

Figure 15.1: Two-out-of-three principle

15.4 TYPES OF PROTECTION SYSTEM

A number of busbar protection systems have been devised:

a. system protection used to cover busbars
b. frame-earth protection
c. differential protection
d. phase comparison protection
e. directional blocking protection

Of these, (a) is suitable for small substations only. Type (d) is often seen nowadays only as a supervision check element within biased differential numerical schemes. Type (e) is receiving greater acceptance nowadays when implemented as IEC 61850 GOOSE-based schemes using overcurrent relays, as described in section 15.11. Detailed discussion of types (b) and (c) occupies most of this chapter.

Early forms of biased differential protection for busbars, such as versions of 'Translay' protection and also a scheme using harmonic restraint, were superseded by unbiased high impedance differential protection.

The relative simplicity of the latter, and more importantly the relative ease with which its performance can be calculated, have ensured its success up to the present day.

But in the 1980's the advances in semiconductor technology, coupled with a more pressing need to be able to accommodate CTs of unequal ratio, led to the re-introduction of biased schemes, generally using static relay designs, particularly for the most extensive and onerous applications.

Frame-earth protection systems have been in use for many years, mainly associated with smaller busbar protection schemes at distribution voltages and for metalclad busbars (e.g. SF_6 insulated busbars). However, it has often been quite common for a unit protection scheme to be used in addition, to provide two separate means of fault detection.

The different types of protection are described in the following sections.

15.5 SYSTEM PROTECTION SCHEMES

System protection that includes overcurrent or distance systems will inherently give protection cover to the busbars.

Overcurrent protection will only be applied to relatively simple distribution systems, or as a back-up protection, set to give a considerable time delay. Distance protection will provide cover for busbar faults with its second and possibly subsequent zones. In both cases the busbar protection obtained is slow and suitable only for limiting the consequential damage.

The only exception is the case of a mesh-connected substation, in which the current transformers are located at the circuit breakers. Here, the busbars are included, in sections, in the individual zones of the main circuit protection, whether this is of unit type or not. In the special case when the current transformers are located on the line side of the mesh, the circuit protection will not cover the busbars in the instantaneous zone and separate busbar protection, known as mesh-corner protection, is generally used see section 15.7.2 for details.

15.6 FRAME-EARTH PROTECTION (HOWARD PROTECTION)

Frame leakage protection has been extensively used in the past in many different situations. There are several variations of frame leakage schemes available, providing busbar protection schemes with different capabilities. The following sections schemes have thus been retained for historical and general reference purposes. A considerable number of schemes are still in service and frame leakage may provide an acceptable solution in particular circumstances. However, the need to insulate the switchboard frame and provide cable gland insulation and the availability of alternative schemes using numerical relays, has contributed to a decline in use of frame leakage systems.

15.6.1 Single-Busbar Frame-Earth Protection

This is purely an earth fault system and, in principle, involves simply measuring the fault current flowing from the switchgear frame to earth. A current transformer is mounted on the earthing conductor and is used to energise a simple instantaneous relay as shown in Figure 15.2.

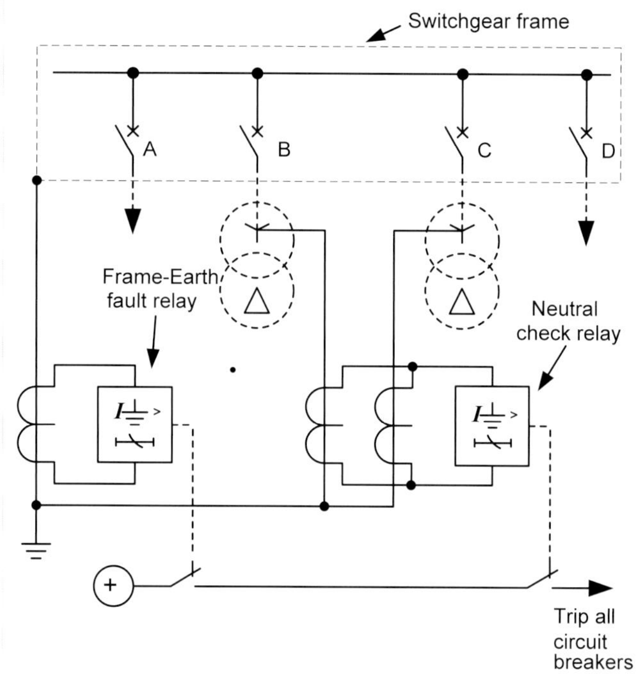

Figure 15.2: Single zone frame-earth protection

No other earth connections of any type, including incidental connections to structural steelwork are allowed. This requirement is so that:

- the principal earth connection and current transformer are not shunted, thereby raising the effective setting. An increased effective setting gives rise to the possibility of relay maloperation. This risk is small in practice
- earth current flowing to a fault elsewhere on the system cannot flow into or out of the switchgear frame via two earth connections, as this might lead to a spurious operation

The switchgear must be insulated as a whole, usually by standing it on concrete. Care must be taken that the foundation bolts do not touch the steel reinforcement; sufficient concrete must be cut away at each hole to permit grouting-in with no risk of touching metalwork. The insulation to earth finally achieved will not be high, a value of 10 ohms being satisfactory.

When planning the earthing arrangements of a frame-leakage scheme, the use of one common electrode for both the switchgear frame and the power system neutral point is preferred, because the fault path would otherwise include the two earthing electrodes in series. If either or both of these are of high resistance or have inadequate current carrying capacity, the fault current may be limited to such an extent that the protection equipment becomes inoperative. In addition, if the electrode earthing the switchgear frame is the offender, the potential of the frame may be raised to a dangerous value. The use of a common earthing electrode of adequate rating and low resistance ensures sufficient current for scheme operation and limits the rise in frame potential. When the system is resistance earthed, the earthing connection from the switchgear frame is made between the bottom of the earthing resistor and the earthing electrode.

Figure 15.3 illustrates why a lower limit of 10 ohms insulation resistance between frame and earth is necessary.

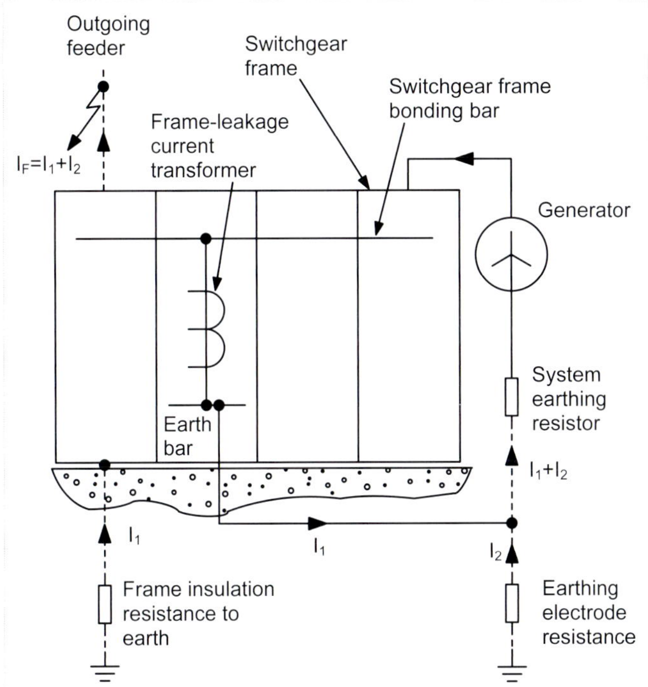

Figure 15.3: Current distribution for external fault

Under external fault conditions, the current I_F flows through the frame-leakage current transformer. If the insulation resistance is too low, sufficient current may flow to operate the frame-leakage relay, and, as the check feature is unrestricted, this will also operate to complete the trip circuit. The earth resistance between the earthing electrode and true earth is seldom greater than 1Ω, so with 10Ω insulation resistance the current I_1 is limited to *10%* of the total earth fault current I_1 and I_2. For this reason, the recommended minimum setting for the scheme is about 30% of the minimum earth fault current.

All cable glands must be insulated, to prevent the circulation of spurious current through the frame and earthing system by any voltages induced in the cable sheath. Preferably, the gland insulation should be provided in two layers or stages, with an interposing layer of metal, to facilitate the testing of the gland insulation. A test level of 5kV from each side is suitable.

15.6.2 Frame-Earth Protection - Sectioned Busbars

Section 15.6.1 covered the basic requirements for a system to protect switchgear as a whole. When the busbar is divided into sections, these can be protected separately, provided the frame is also sub-divided, the sections mutually insulated, and each provided with a separate earth conductor, current transformer and relay.

Ideally, the section switch should be treated as a separate zone, as shown in Figure 15.4, and provided with either a separate relay or two secondaries on the frame-leakage current transformer, with an arrangement to trip both adjacent zones. The individual zone relays trip their respective zone and the section switch.

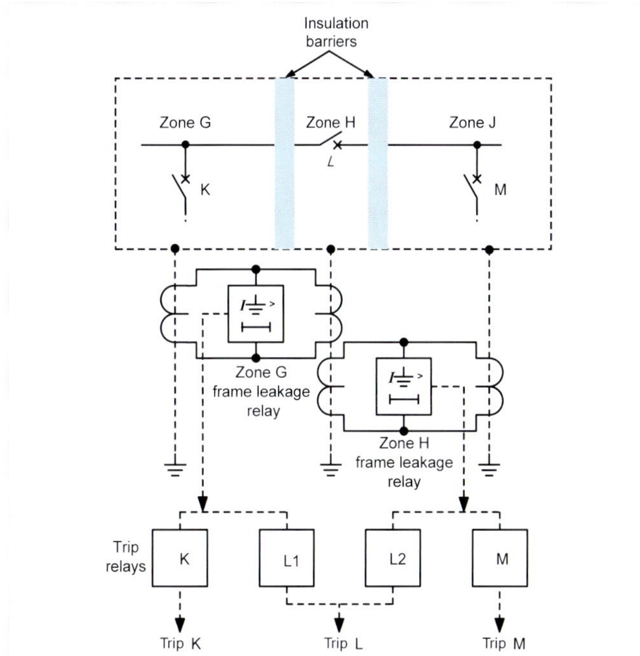

Figure 15.4: Three zone frame-earth scheme

If it is inconvenient to insulate the section switch frame on one side, this switch may be included in that zone. It is then necessary to intertrip the other zone after approximately 0.5 seconds if a fault persists after the zone including the section switch has been tripped. This is illustrated in Figure 15.5.

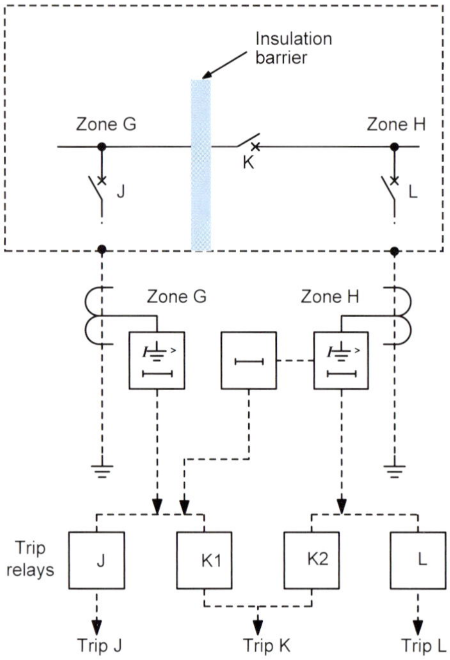

Figure 15.5: Frame-earth scheme: bus section breaker insulated on one side only

For the above schemes to function it is necessary to have a least one infeed or earthed source of supply, and in the latter case it is essential that this source of supply be connected to the side of the switchboard not containing the section switch. Further, if possible, it is preferable that an earthed source of supply be provided on both sides of the switchboard, in order to ensure that any faults that may develop between the insulating barrier and the section switch will continue to be fed with fault current after the isolation of the first half of the switchboard, and thus allow the fault to be removed. Of the two arrangements, the first is the one normally recommended, since it provides instantaneous clearance of busbar faults on all sections of the switchboard.

15.6.3 Frame-Earth Scheme - Double Bus Substation

It is not generally feasible to separately insulate the metal enclosures of the main and auxiliary busbars. Protection is therefore generally provided as for single bus installations, but with the additional feature that circuits connected to the auxiliary bus are tripped for all faults, as shown in Figure 15.6.

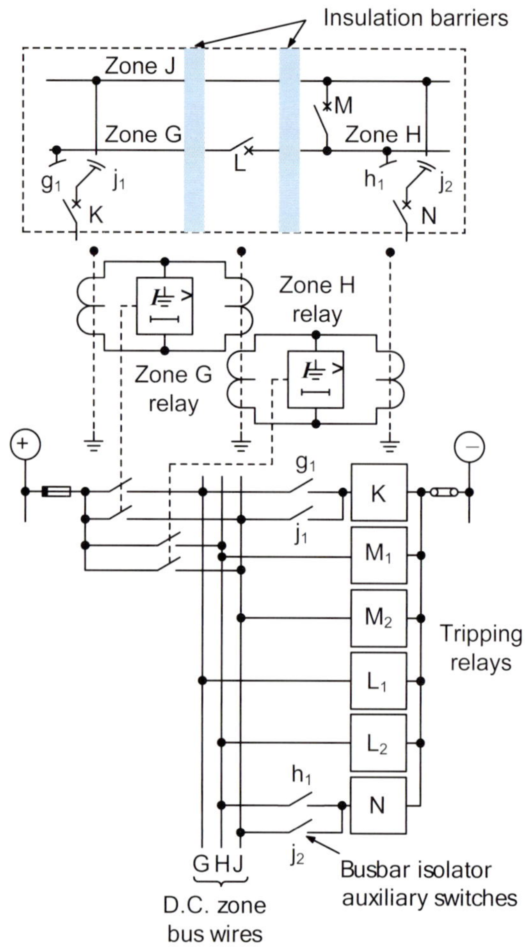

Figure 15.6: Frame-earth scheme for double busbar substation

15.6.4 Frame-Earth Protection - Check System

On all but the smallest equipments, a check system should be provided to guard against such contingencies as operation due to mechanical shock or mistakes made by personnel. Faults in the low voltage auxiliary wiring must also be prevented from causing operation by passing current to earth through the switchgear frame. A useful check is provided by a relay energised by the system neutral current, or residual current. If the neutral check cannot be provided, the frame-earth relays should have a short time delay.

When a check system is used, instantaneous relays can be used, with a setting of 30% of the minimum earth fault current and an operating time at five times setting of 15 milliseconds or less.

Figure 15.7 shows a frame-leakage scheme for a metalclad switchgear installation similar to that shown in Figure 15.4 and incorporating a neutral current check obtained from a suitable zero sequence current source, such as that shown in Figure 15.2.

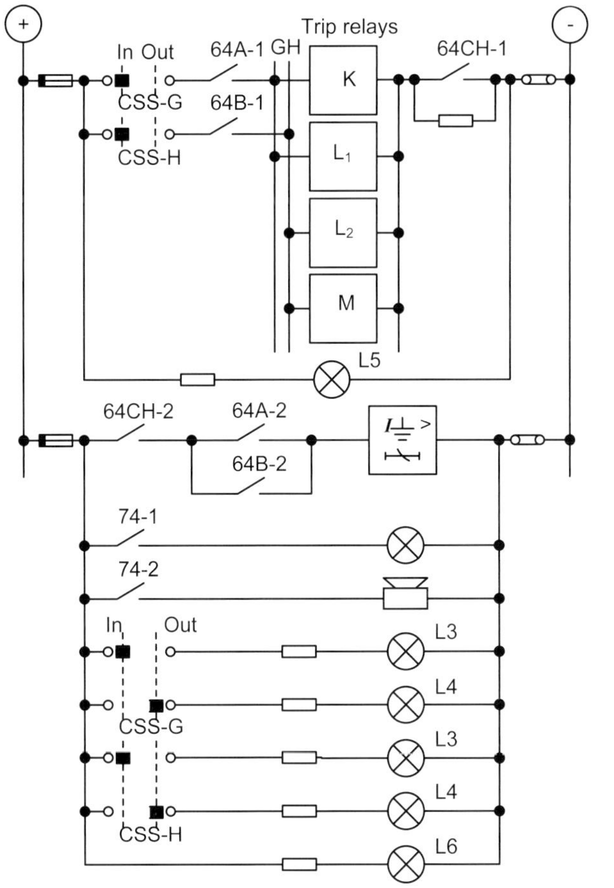

74 Alarm cancellation relay
CSS Control selector switch protection in/protection out
L3 Busbar protection in service lamp
L4 Busbar protection out of service lamp
L5 Tripping supply healthy lamp
L6 Alarm and indication supply healthy lamp

Figure 15.7: Typical tripping and alarm circuits for a frame-leakage scheme

The protection relays used for the discriminating and check functions are of the attracted armature type, with two normally open self reset contacts. The tripping circuits cannot be complete unless both the discriminating and check relays operate; this is because the discriminating and check relay contacts are connected in series. The tripping relays are of the attracted armature type.

It is usual to supervise the satisfactory operation of the protection scheme with audible and visual alarms and indications for the following:

- busbar faults
- busbar protection in service
- busbar protection out of service
- tripping supply healthy
- alarm supply healthy

To enable the protection equipment of each zone to be taken out of service independently during maintenance periods, isolating switches - one switch per zone - are provided in the trip supply circuits and an alarm cancellation relay is used.

15.7 DIFFERENTIAL PROTECTION PRINCIPLES

The Merz-Price principle is applicable to a multi-terminal zone such as a busbar. The principle is a direct application of Kirchhoff's first law. Usually, the circulating current arrangement is used, in which the current transformers and interconnections form an analogue of the busbar and circuit connections. A relay connected across the CT bus wires represents a fault path in the primary system in the analogue and hence is not energised until a fault occurs on the busbar; it then receives an input that, in principle at least, represents the fault current.

The scheme may consist of a single relay connected to the bus wires connecting all the current transformers in parallel, one set per circuit, associated with a particular zone, as shown in Figure 15.8(a). This will give earth fault protection for the busbar. This arrangement has often been thought to be adequate.

If the current transformers are connected as a balanced group for each phase together with a three-element relay, as shown in Figure 15.8(b), additional protection for phase faults can be obtained.

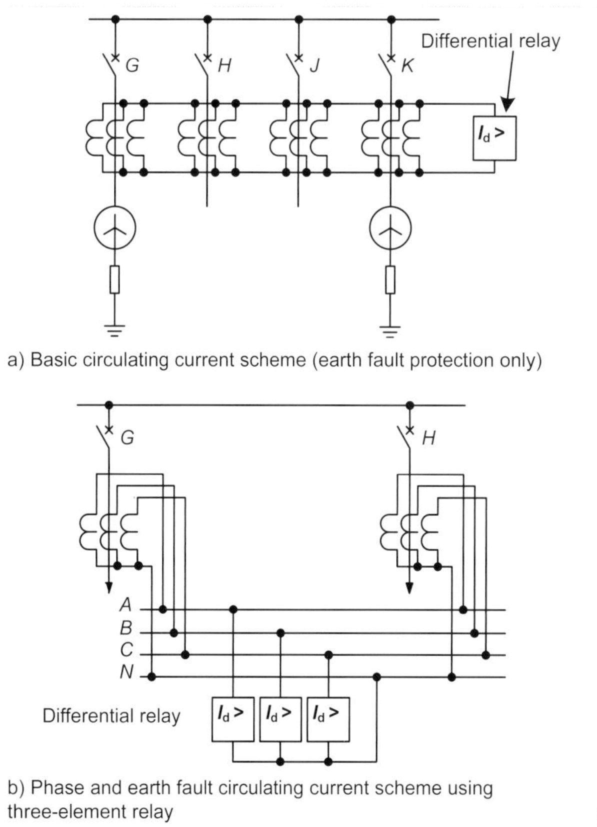

Figure 15.8: Circulating current scheme

The phase and earth fault settings are identical, and this

scheme is recommended for its ease of application and good performance.

15.7.1 Differential Protection for Sectionalised and Duplicate Busbars

Each section of a divided bus is provided with a separate circulating current system. The zones so formed are over-lapped across the section switches, so that a fault on the latter will trip the two adjacent zones. This is illustrated in Figure 15.9.

Tripping two zones for a section switch fault can be avoided by using the time-delayed technique of section 15.6.2. However instantaneous operation is the preferred choice.

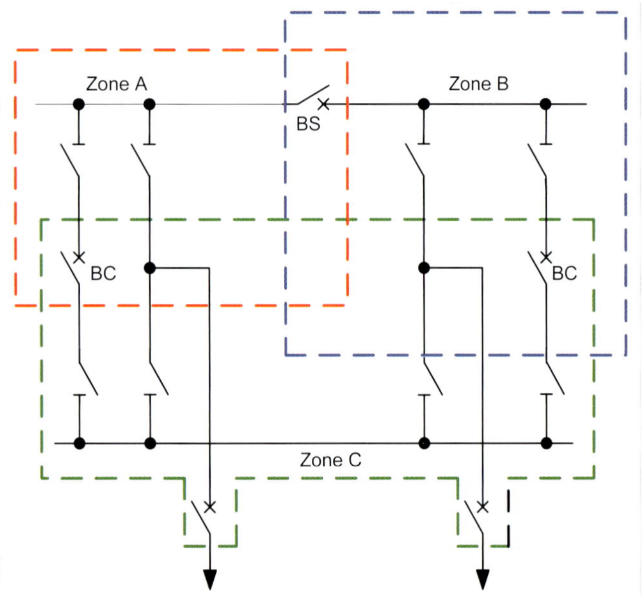

Figure 15.9: Zones of protection for double bus station

For double bus installation, the two busbars will be treated as separate zones. The auxiliary busbar zone will overlap the appropriate main busbar zone at the bus coupler.

Since any circuit may be transferred from one busbar to the other by isolator switches, these and the associated tripping circuit must also be switched to the appropriate zone by 'early make' and 'late break' auxiliary contacts. This is to ensure that when the isolators are closing, the auxiliary switches make before the main contacts of the isolator, and that when the isolators are opened, their main contacts part before the auxiliary switches open. The result is that the secondary circuits of the two zones concerned are briefly paralleled while the circuit is being transferred; these two zones have in any case been united through the circuit isolators during the transfer operation.

15.7.2 Location of Current Transformers

Ideally, the separate discriminating zones should overlap each other and also the individual circuit protections. The overlap should occur across a circuit breaker, so that the latter lies in both zones. For this arrangement it is necessary to install current transformers on both sides of the circuit breakers, which is economically possible with many but not all types of switchgear. With both the circuit and the bus protection current transformers on the same side of the circuit breakers, the zones may be overlapped at the current transformers, but a fault between the CT location and the circuit breaker will not be completely isolated. This matter is important in all switchgear to which these conditions apply, and is particularly important in the case of outdoor switchgear where separately mounted, multi-secondary current transformers are generally used. The conditions are shown in Figure 15.10.

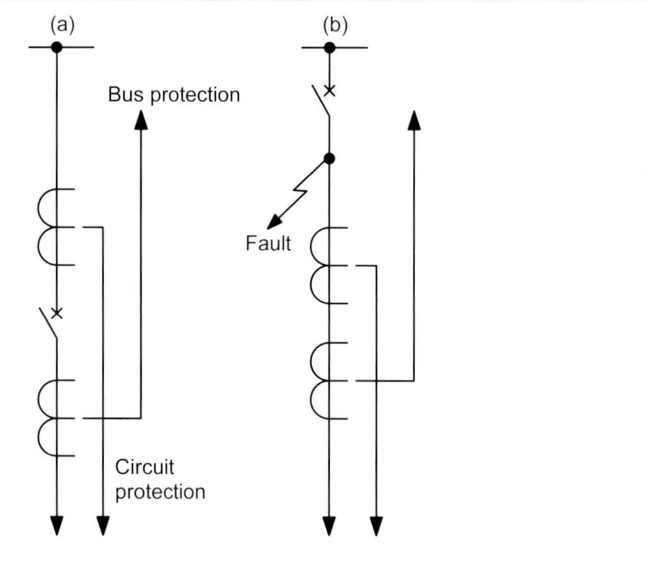

a. Current transformers mounted on both sides of breaker -no unprotected region

b. Current transformers mounted on circuit side only of breaker -fault shown not cleared by circuit protection

Figure 15.10: Unprotected zone with current transformers mounted on one side of the circuit breaker only

Figure 15.10(a) shows the ideal arrangement in which both the circuit and busbar zones are overlapped leaving no region of the primary circuit unprotected.

Figure 15.10(b) shows how mounting all current transformers on the circuit side of the breaker results in a small region of the primary circuit unprotected. This unprotected region is typically referred to as the 'short zone'. The fault shown will cause operation of the busbar protection, tripping the circuit breaker, but the fault will continue to be fed from the circuit, if a source of power is present. It is necessary for the bus protection to intertrip the far end of the circuit protection, if the latter is of the unit type.

With reference to Figure 15.10(b), special 'short zone' protection can be provided to detect that the circuit breaker has opened but that the fault current is still flowing. Under these conditions, the protection can initiate an intertrip to the remote end of the circuit. This technique may be used, particularly when the circuit includes a generator. In this case the intertrip proves that the fault is in the switchgear connections and not in the generator; the latter is therefore tripped electrically but not shut down on the mechanical side so as to be immediately ready for further service if the fault can be cleared.

15.7.2.1 CT locations for mesh-connected substations

The protection of busbars in mesh connected substations gives rise to additional considerations in respect of CT location. A single mesh corner is shown in Figure 15.11(a). Where only one connection to the mesh is made at a corner, CTs located as shown will provide protection not only to the line but the corner of the mesh included between them. However, this arrangement cannot be used where more than one connection is made to a mesh corner. This is because a fault on any of the connected circuits would result in disconnection of them all, without any means of determining the faulted connection. Protection CTs must therefore be located on each connection, as shown in Figure 15.11(b). This leaves the corner of the mesh unprotected, so additional CTs and a relay to provide mesh-corner protection are added, as also shown in Figure 15.11(b).

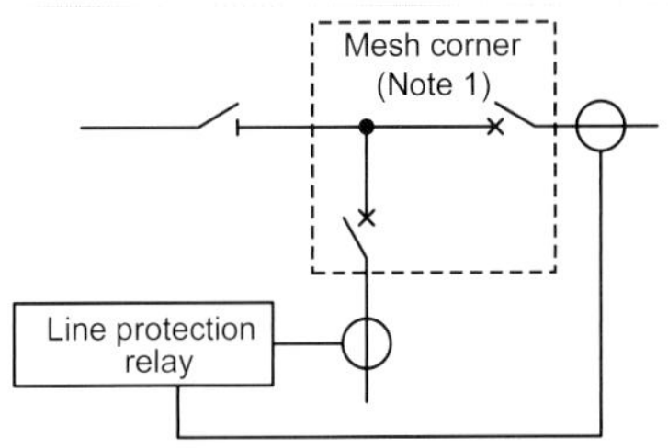

Note 1: Only 1 connection to the mesh corner permitted

(a) CT arrangements for protection including mesh corner

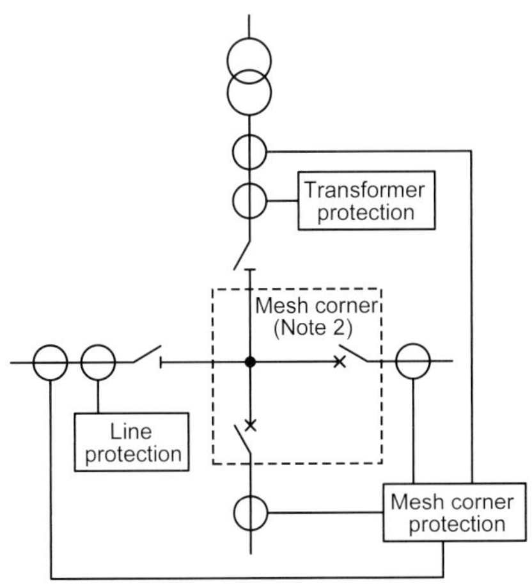

Note 2: Multiple circuits may be connected to the mesh corner

(b) CT arrangements for protection - additional mesh corner protection required

Figure 15.11: Mesh-corner protection

15.8 HIGH IMPEDANCE DIFFERENTIAL PROTECTION

This form of protection is still in common use. The considerations that have to be taken into account are detailed in the following sections.

15.8.1 Stability

The incidence of fault current with an initial unilateral transient component causes an abnormal build-up of flux in a current transformer, as described in section 15.8.2. When through-fault current traverses a zone protected by a differential system, the transient flux produced in the current transformers is not detrimental as long as it remains within the substantially linear range of the magnetising characteristic. With fault current of appreciable magnitude and long transient time constant, the flux density will pass into the saturated region of the characteristic; this will not in itself produce a spill output from a pair of balancing current transformers provided that these are identical and equally burdened. A group of current transformers, though they may be of the same design, will not be completely identical, but a more important factor is inequality of burden. In the case of a differential system for a busbar, an external fault may be fed through a single circuit, the current being supplied to the busbar through all other circuits. The faulted circuit is many times more heavily loaded than the others and the corresponding current transformers are likely to be heavily saturated, while those of the other circuits are not. Severe unbalance is therefore probable, which, with a relay of normal burden, could exceed any acceptable current setting. For this reason such systems were at one time always provided with a time delay. This practice is, however, no longer acceptable.

It is not feasible to calculate the spill current that may occur, but, fortunately, this is not necessary; an alternative approach provides both the necessary information and the technique required to obtain a high performance.

An equivalent circuit, as in Figure 15.12, can represent a circulating current system.

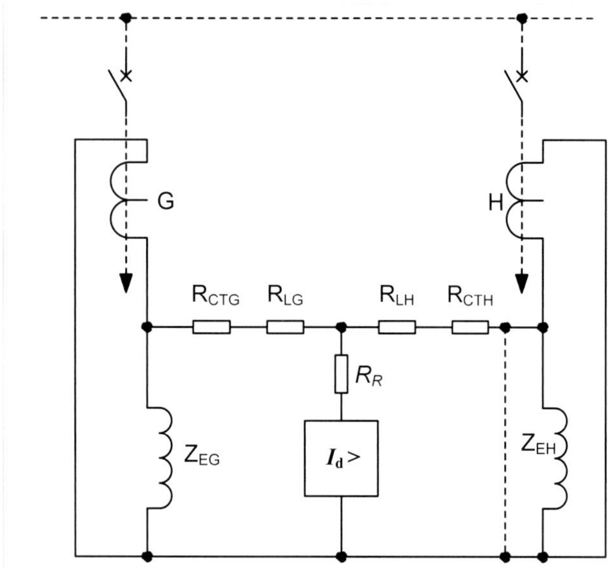

Figure 15.12: Equivalent circuit of circulating current system

The current transformers are replaced in the diagram by ideal current transformers feeding an equivalent circuit that represents the magnetising losses and secondary winding resistance, and also the resistance of the connecting leads. These circuits can then be interconnected as shown, with a relay connected to the junction points to form the complete equivalent circuit.

Saturation has the effect of lowering the exciting impedance, and is assumed to take place severely in current transformer *H* until, at the limit, the shunt impedance becomes zero and the CT can produce no output. This condition is represented by a short circuit, shown in broken line, across the exciting impedance. It should be noted that this is not the equivalent of a physical short circuit, since it is behind the winding resistance R_{CTH}.

Applying the Thévenin method of solution, the voltage developed across the relay will be given by:

$$V_f = I_f\left(R_{LH} + R_{CTH}\right)$$

Equation 15.1

The current through the relay is given by:

$$I_R = \frac{V_f}{R_R + R_{LH} + R_{CTH}} = \frac{I_f(R_{LH} + R_{CTH})}{R_R + R_{LH} + R_{CTH}}$$

Equation 15.2

If R_R is small, I_R will approximate to I_f, which is unacceptable. On the other hand, if R_R is large I_R is reduced. Equation 15.2 can be written, with little error, as follows:

$$I_R = \frac{V_f}{R_R} = \frac{I_f(R_{LH} + R_{CTH})}{R_R}$$

Equation 15.3

or alternatively:

$$I_R R_R = V_f = I_f\left(R_{LH} + R_{CTH}\right)$$

Equation 15.4

It is clear that, by increasing R_R, the spill current I_R can be reduced below any specified relay setting. R_R is frequently increased by the addition of a series-connected resistor which is known as the stabilising resistor.

It can also be seen from Equation 15.4 that it is only the voltage drop in the relay circuit at setting current that is important. The relay can be designed as a voltage measuring device consuming negligible current; and provided its setting voltage exceeds the value V_f of Equation 15.4, the system will be stable. In fact, the setting voltage need not exceed V_f, since the derivation of Equation 15.4 involves an extreme condition of unbalance between the G and H current transformers that is not completely realised. So a safety margin is built-in if the voltage setting is made equal to V_f.

It is necessary to realise that the value of I_f to be inserted in Equation 15.4 is the complete function of the fault current and the spill current I_R through the relay, in the limiting condition, will be of the same form. If the relay requires more time to operate than the effective duration of the d.c. transient component, or has been designed with special features to block the d.c. component, then this factor can be ignored and only the symmetrical value of the fault current need be entered in Equation 15.4. If the relay setting voltage, V_S, is made equal to V_f, that is, $I_f(R_L + R_{CT})$, an inherent safety factor of the order of two will exist.

In the case of a faster relay, capable of operating in one cycle and with no special features to block the d.c. component, it is the r.m.s. value of the first offset wave that is significant. This value, for a fully offset waveform with no d.c. decrement, is $\sqrt{3}I_f$. If settings are then chosen in terms of the symmetrical component of the fault current, the $\sqrt{3}$ factor which has been ignored will take up most of the basic safety factor, leaving only a very small margin.

Finally, if a truly instantaneous relay were used, the relevant value of I_f would be the maximum offset peak. In this case, the factor has become less than unity, possibly as low as 0.7. It is therefore possible to rewrite Equation 15.4 as:

$$I_{SL} = \frac{K \times V_S}{R_L + R_{CT}}$$

Equation 15.5

where:

I_{SL} = stability limit of scheme

V_S = relay circuit voltage setting

$R_L + R_{CT}$ = lead + CT winding resistance

K = factor depending on relay design (range 0.7 - 2.0)

It remains to be shown that the setting chosen is suitable in section 15.8.2.

The current transformers will have an excitation curve which has not so far been related to the relay setting voltage, the latter being equal to the maximum nominal voltage drop across the lead loop and the CT secondary winding resistance, with the maximum secondary fault current flowing through them. Under in-zone fault conditions it is necessary for the current transformers to produce sufficient output to operate the relay. This will be achieved provided the CT knee-point voltage exceeds the relay setting. In order to cater for errors, it is usual to specify that the current transformers should have a knee-point e.m.f. of at least twice the necessary setting voltage; a higher multiple is of advantage in ensuring a high speed of operation.

15.8.2 Effective Setting or Primary Operating Current

The minimum primary operating current is a further criterion of the design of a differential system. The secondary effective setting is the sum of the relay minimum operating current and the excitation losses in all parallel connected current transformers, whether carrying primary current or not. This summation should strictly speaking be vectorial, but is usually done arithmetically. It can be expressed as:

$$I_R = I_S + nI_{eS}$$

Equation 15.6

where:

I_R = effective setting

I_S = relay circuit setting current

I_{eS} = CT excitation current at relay voltage setting

n = number of parallel connected CTs

Having established the relay setting voltage from stability considerations, as shown in section 15.8.1, and knowing the excitation characteristic of the current transformers, the effective setting can be computed. The secondary setting is converted to the primary operating current by multiplying by the turns ratio of the current transformers. The operating current so determined should be considered in terms of the conditions of the application.

For a phase and earth fault scheme the setting can be based on the fault current to be expected for minimum plant and maximum system outage conditions. However, it should be remembered that:

- phase-phase faults give only 86% of the three-phase fault current
- fault arc resistance and earth path resistance reduce fault currents somewhat
- a reasonable margin should be allowed to ensure that relays operate quickly and decisively

It is desirable that the primary effective setting should not exceed 30% of the prospective minimum fault current.

In the case of a scheme exclusively for earth fault protection, the minimum earth fault current should be considered, taking into account any earthing impedance that might be present as well. Furthermore, in the event of a double phase to earth fault, regardless of the inter-phase currents, only 50% of the system e.m.f. is available in the earth path, causing a further reduction in the earth fault current. The primary operating current must therefore be not greater than 30% of the minimum single-phase earth fault current.

In order to achieve high-speed operation, it is desirable that settings should be still lower, particularly in the case of the solidly earthed power system. The transient component of the fault current in conjunction with unfavourable residual flux in the CT can cause a high degree of saturation and loss of output, possibly leading to a delay of several cycles additional to the natural operating time of the element. This will not happen to any large degree if the fault current is a larger multiple of setting; for example, if the fault current is five times the scheme primary operating current and the CT knee-point e.m.f. is three times the relay setting voltage, the additional delay is unlikely to exceed one cycle.

The primary operating current is sometimes designed to exceed the maximum expected circuit load in order to reduce the possibility of false operation under load current as a result of a broken CT lead. Desirable as this safeguard may be, it will be seen that it is better not to increase the effective current setting too much, as this will sacrifice some speed; the check feature in any case, maintains stability.

An overall earth fault scheme for a large distribution board may be difficult to design because of the large number of current transformers paralleled together, which may lead to an excessive setting. It may be advantageous in such a case to provide a three-element phase and earth fault scheme, mainly to reduce the number of current transformers paralleled into one group.

Extra-high-voltage substations usually present no such problem. Using the voltage-calibrated relay, the current consumption can be very small. A simplification can be achieved by providing one relay per circuit, all connected to the CT paralleling buswires. This enables the trip circuits to be confined to the least area and reduces the risk of accidental operation.

15.8.3 Check Feature

Schemes for earth faults only can be checked by a frame-earth system, applied to the switchboard as a whole, no subdivision being necessary.

For phase fault schemes, the check will usually be a similar type of scheme applied to the switchboard as a single overall zone. A set of current transformers separate from those used in the discriminating zones should be provided. No CT switching is required and no current transformers are needed for the check zone in bus-coupler and bus-section breakers.

15.8.4 Supervision of CT Secondary Circuits

Any interruption of a CT secondary circuit up to the paralleling interconnections will cause an unbalance in the system, equivalent to the load being carried by the relevant primary circuit. Even though this degree of spurious output is below the effective setting the condition cannot be ignored, since it is likely to lead to instability under any through fault condition.

Supervision can be carried out to detect such conditions by connecting a sensitive alarm relay across the bus wires of each zone. For a phase and earth fault scheme, an internal three-phase rectifier can be used to effect a summation of the bus wire voltages on to a single alarm element; see Figure 15.14 and Figure 15.13.

The alarm relay is set so that operation does not occur with the protection system healthy under normal load. Subject to this proviso, the alarm relay is made as sensitive as possible; the desired effective setting is 125 primary amperes or 10% of the lowest circuit rating, whichever is the greater.

Since a relay of this order of sensitivity is likely to operate during through faults, a time delay, typically of three seconds, is applied to avoid unnecessary alarm signals.

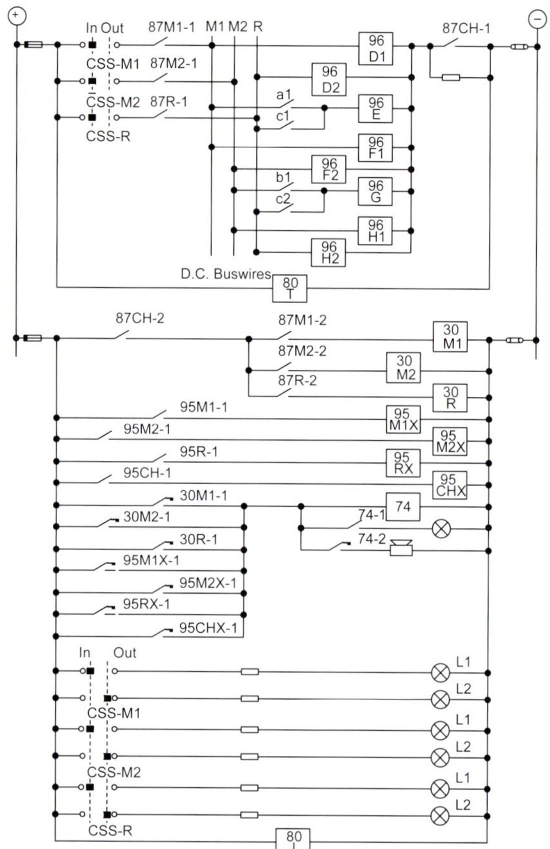

30 Zone indicating relay
74 Alarm cancellation relay
80 D.C. volts supervision relay
87 High impedance circulating current relay
95 Bus wires supervision relay
95X Zone bus wires shorting relay
CSS Control selector switch
L1 Indicating lamp protection in service
L2 Indicating lamp protection out of service

Figure 15.13: D.C. circuits for high impedance circulating current scheme

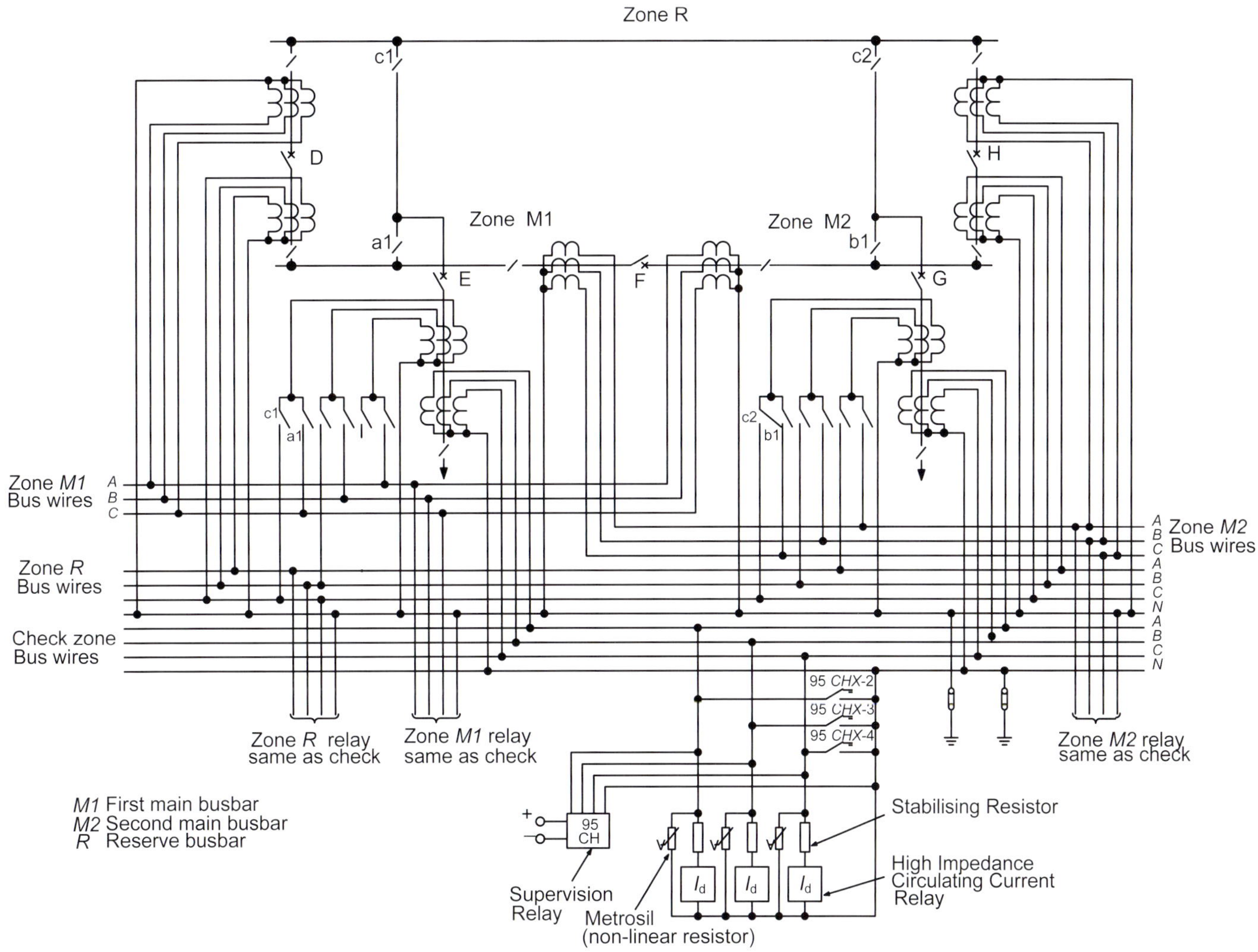

Figure 15.14: A.C. circuits for high impedance circulating current scheme for duplicate busbars

15.8.5 Arrangement of CT Connections

It is shown in Equation 15.4 how the setting voltage for a given stability level is directly related to the resistance of the CT secondary leads. This should therefore be kept to a practical minimum. Taking into account the practical physical laying of auxiliary cables, the CT bus wires are best arranged in the form of a ring around the switchgear site.

In a double bus installation, the CT leads should be taken directly to the isolator selection switches. The usual arrangement of cables on a double bus site is as follows:

- current transformers to marshalling kiosk
- marshalling kiosk to bus selection isolator auxiliary switches
- interconnections between marshalling kiosks to form a closed ring

The relay for each zone is connected to one point of the ring bus wire. For convenience of cabling, the main zone relays will be connected through a multicore cable between the relay panel and the bus section-switch marshalling cubicle. The reserve bar zone and the check zone relays will be connected together by a cable running to the bus coupler circuit breaker marshalling cubicle. It is possible that special circumstances involving onerous conditions may over-ride this convenience and make connection to some other part of the ring desirable.

Connecting leads will usually be not less than 7/0.67mm (2.5mm^2), but for large sites or in other difficult circumstances it may be necessary to use cables of, for example 7/1.04mm (6mm^2) for the bus wire ring and the CT connections to it. The cable from the ring to the relay need not be of the larger section.

When the reserve bar is split by bus section isolators and the two portions are protected as separate zones, it is necessary to common the bus wires by means of auxiliary contacts, thereby making these two zones into one when the section isolators are closed.

15.8.6 Summary of Practical Details

This section provides a summary of practical considerations when implementing a high-impedance busbar protection scheme.

15.8.6.1 Designed stability level

For normal circumstances, the stability level should be designed to correspond to the switchgear rating; even if the available short-circuit power in the system is much less than this figure, it can be expected that the system will be developed up to the limit of rating.

15.8.6.2 Current transformers

Current transformers must have identical turns ratios, but a turns error of one in 400 is recognised as a reasonable manufacturing tolerance. Also, they should preferably be of similar design; where this is not possible the magnetising characteristics should be reasonably matched.

Current transformers for use with high impedance protection schemes should meet the requirements of Class PX of IEC 60044-1.

15.8.6.3 Setting voltage

The setting voltage is given by the equation:

$$V_S \geq I_f\left(R_L + R_{CT}\right)$$

where:

I_f = Steady state through fault current

V_S = relay circuit voltage setting

R_L = CT lead loop resistance

R_{CT} = CT secondary winding resistance

15.8.6.4 Knee-point voltage of current transformers

This is given by the formula:

$$V_K \geq 2V_S$$

15.8.6.5 Effective setting (secondary)

The effective setting of the relay is given by:

$$I_R = I_S + nI_{eS}$$

where:

I_S = relay circuit setting current

I_{eS} = CT excitation current at relay voltage setting

n = number of parallel connected CTs

For the primary fault setting multiply I_R by the CT turns ratio.

15.8.6.6 Current transformer secondary rating

It is clear from Equation 15.4 and Equation 15.6 that it is advantageous to keep the secondary fault current low; this is done by making the CT turns ratio high. It is common practice to use current transformers with a secondary rating of 1A.

It can be shown that there is an optimum turns ratio for the current transformers; this value depends on all the application parameters but is generally about 2000/1. Although a lower ratio, for instance 400/1, is often employed, the use of the optimum ratio can result in a considerable reduction in the physical size of the current transformers.

15.8.6.7 Peak voltage developed by current transformers

Under in-zone fault conditions, a high impedance relay constitutes an excessive burden to the current transformers, leading to the development of a high voltage; the voltage waveform will be highly distorted but the peak value may be many times the nominal saturation voltage.

When the burden resistance is finite although high, an approximate formula for the peak voltage is:

$$V_p = 2\sqrt{2}\sqrt{V_K(V_F - V_K)}$$

Equation 15.7

where:

V_p = peak Voltage developed

V_K = saturation Voltage

V_F = prospective Voltage in absence of saturation

This formula does not hold for the open circuit condition and is inaccurate for very high burden resistances that approximate to an open circuit, because simplifying assumptions used in the derivation of the formula are not valid for the extreme condition.

Another approach applicable to the open circuit secondary condition is:

$$V_p = \sqrt{2}\frac{I_f}{I_{ek}}V_K$$

Equation 15.8

where:

I_f= fault current

I_{ek} = exciting current at knee-point voltage

V_k = knee-point voltage

Any burden connected across the secondary will reduce the voltage, but the value cannot be deduced from a simple combination of burden and exciting impedances.

These formulae are therefore to be regarded only as a guide to the possible peak voltage. With large current transformers, particularly those with a low secondary current rating, the voltage may be very high, above a suitable insulation voltage. The voltage can be limited without detriment to the scheme by connecting a ceramic non-linear resistor in parallel with the relay having a characteristic given by:

$$V = CI^{\beta}$$

where C is a constant depending on dimensions and β is a constant in the range 0.2 - 0.25.

The current passed by the non-linear resistor at the relay voltage setting depends on the value of *C*; in order to keep the shunting effect to a minimum it is recommended to use a non-linear resistor with a value of *C* of 450 for relay voltages up to 175V and one with a value of *C* of 900 for setting voltages up to 325V.

15.8.6.8 High impedance relay

Instantaneous attracted armature relays or numeric relays that mimic the high impedance function are used. Simple fast-operating relays would have a low safety factor constant in the stability equation, Equation 15.5, as discussed in section 15.8.1. The performance is improved by series-tuning the relay coil, thereby making the circuit resistive in effect. Inductive reactance would tend to reduce stability, whereas the action of capacitance is to block the unidirectional transient component of fault current and so raise the stability constant.

An alternative technique used in some relays is to apply the limited spill voltage principle shown in Equation 15.4. A tuned element is connected via a plug bridge to a chain of resistors; and the relay is calibrated in terms of voltage.

15.9 LOW IMPEDANCE BIASED DIFFERENTIAL PROTECTION

The principles of low impedance differential protection have been described in section 10.4.2, including the principle advantages to be gained by the use of a bias technique. Most modern busbar protection schemes use this technique.

The principles of a check zone, zone selection, and tripping arrangements can still be applied. Current transformer secondary circuits are not switched directly by isolator contacts but instead by isolator repeat relays after a secondary stage of current transformation. These switching relays form a replica of the busbar within the protection and provide the complete selection logic.

15.9.1 Stability

With some biased relays, the stability is not assured by the through current bias feature alone, but is enhanced by the addition of a stabilising resistor, having a value which may be calculated as follows.

The through current will increase the effective relay minimum operating current for a biased relay as follows:

$$I_R = I_S + BI_F$$

where:

I_R = effective minimum operating current

I_S = relay setting current

I_F = through fault current

B = percentage restraint

As I_F is generally much greater than I_S, the relay effective current, $I_R = BI_F$ approximately.

From Equation 15.4, the value of stabilising resistor is given by:

$$R_R = \frac{I_F(R_{LH} + R_{CTH})}{I_R} = \frac{(R_{LH} + R_{CTH})}{B}$$

It is interesting to note that the value of the stabilising resistance is independent of current level, and that there would appear to be no limit to the through fault stability level. This has been identified [15.1] as 'The Principle of Infinite Stability'.

The stabilising resistor still constitutes a significant burden on the current transformers during internal faults.

An alternative technique, used by the MBCZ system described in section 15.9.6, is to block the differential measurement during the portion of the cycle that a current transformer is saturated. If this is achieved by momentarily short-circuiting the differential path, a very low burden is placed on the current transformers. In this way the differential circuit of the relay is prevented from responding to the spill current.

It must be recognised though that the use of any technique for inhibiting operation, to improve stability performance for through faults, must not be allowed to diminish the ability of the relay to respond to internal faults.

15.9.2 Effective Setting or Primary Operating Current

For an internal fault, and with no through fault current flowing, the effective setting I_R is raised above the basic relay setting I_S by whatever biasing effect is produced by the sum of the CT magnetising currents flowing through the bias circuit. With low impedance biased differential schemes particularly where the busbar installation has relatively few circuits, these magnetising currents may be negligible, depending on the value of I_S.

The basic relay setting current was formerly defined as the minimum current required solely in the differential circuit to cause operation – Figure 15.15(a). This approach simplified analysis of performance, but was considered to be unrealistic, as in practice any current flowing in the differential circuit must flow in at least one half of the relay bias circuit causing the practical minimum operating current always to be higher than the nominal basic setting current. As a result, a later definition, as shown in Figure 15.15(b) was developed.

Conversely, it needs to be appreciated that applying the later definition of relay setting current, which flows through at least half the bias circuit, the notional minimum operation current in the differential circuit alone is somewhat less, as shown in Figure 15.15(b).

Using the definition presently applicable, the effective minimum primary operating current

$$= N\left[I_s + B\sum I_{es}\right]$$

Where N = CT ratio

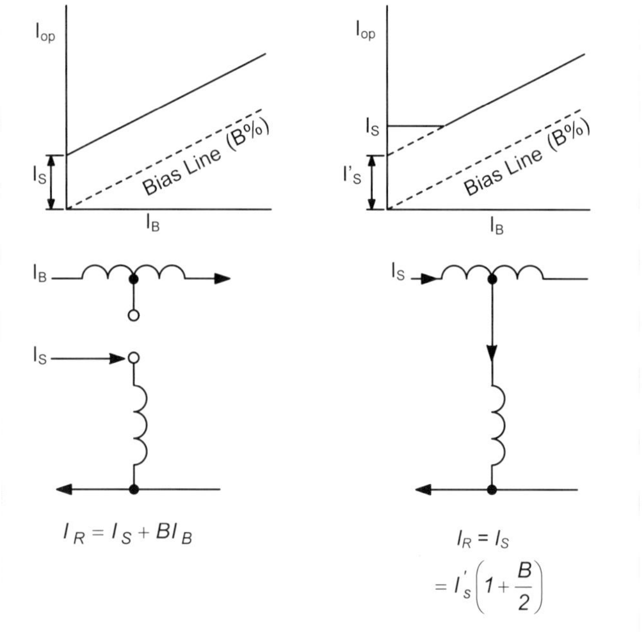

Figure 15.15: Definitions of relay setting current for biased relays

Unless the minimum effective operating current of a scheme has been raised deliberately to some preferred value, it will usually be determined by the check zone, when present, as the latter may be expected to involve the greatest number of current transformers in parallel. A slightly more onerous condition may arise when two discriminating zones are coupled, transiently or otherwise, by the closing of primary isolators.

It is generally desirable to attain an effective primary operating current that is just greater than the maximum load current, to prevent the busbar protection from operating spuriously from load current should a secondary circuit wiring fault develop. This consideration is particularly important where the check feature is either not used or is fed from common main CTs.

15.9.3 Check Feature

For some low impedance schemes, only one set of main CTs is required. This seems to contradict the general principle of all busbar protection systems with a check feature that complete duplication of all equipment is required, but it is claimed that the spirit of the checking principle is met by making operation of the protection dependent on two different criteria such as directional and differential measurements.

In the MBCZ scheme, described in section 15.9.6, the provision of auxiliary CTs as standard for ratio matching also provides a ready means for introducing the check feature duplication at the auxiliary CTs and onwards to the relays. This may be an attractive compromise when only one set of main CTs is available.

15.9.4 Supervision of CT Secondary Circuits

In low impedance schemes the integrity of the CT secondary circuits can also be monitored. A current operated auxiliary relay, or element of the main protection equipment, may be applied to detect any unbalanced secondary currents and give an alarm after a time delay. For optimum discrimination, the current setting of this supervision relay must be less than that of the main differential protection.

In modern busbar protection schemes, the supervision of the secondary circuits typically forms only a part of a comprehensive supervision facility.

15.9.5 Arrangement of CT connections

It is a common modern requirement of low impedance schemes that none of the main CT secondary circuits should be switched, in the previously conventional manner, to match the switching of primary circuit isolators.

The usual solution is to route all the CT secondary circuits back to the protection panel or cubicle to auxiliary CTs. It is then the secondary circuits of the auxiliary CTs that are switched as necessary. So auxiliary CTs may be included for this function even when the ratio matching is not in question.

In static protection equipment it is undesirable to use isolator auxiliary contacts directly for the switching without some form of insulation barrier. Position transducers that follow the opening and closing of the isolators may provide the latter.

Alternatively, a simpler arrangement may be provided on multiple busbar systems where the isolators switch the auxiliary current transformer secondary circuits via auxiliary relays within the protection. These relays form a replica of the busbar and perform the necessary logic. It is therefore necessary to route all the current transformer secondary circuits to the relay to enable them to be connected into this busbar replica.

Some installations have only one set of current transformers available per circuit. Where the facility of a check zone is still required, this can still be achieved with the low impedance biased protection by connecting the auxiliary current transformers at the input of the main and check zones in series, as shown in Figure 15.16.

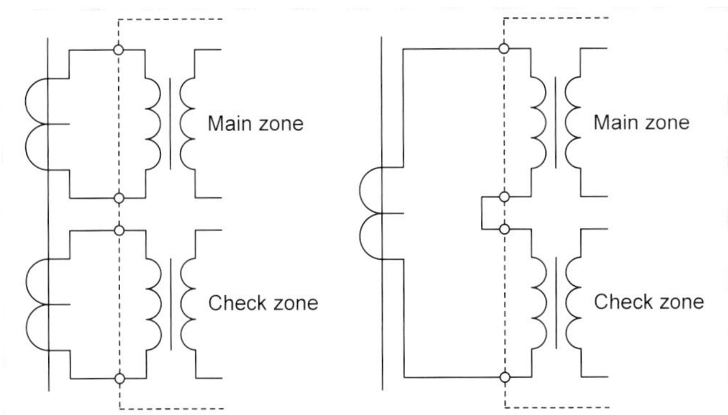

Figure 15.16: Alternative CT connections

15.9.6 Low Impedance Biased Differential Protection - Type MBCZ

Numerical schemes are now prevalent in the majority of new busbar protection installations. However, in order to appreciate the historical installed base, and due to the similarity of the basic operating principles, this section now considers a static scheme example – the MBCZ.

The Type MBCZ scheme conforms in general to the principles outlined earlier and comprises a system of standard modules that can be assembled to suit a particular busbar installation. Additional modules can be added at any time as the busbar is extended.

A separate module is used for each circuit breaker and also one for each zone of protection. In addition to these there is a common alarm module and a number of power supply units. Ratio correction facilities are provided within each differential module to accommodate a wide range of CT mismatch.

Figure 15.17 shows the correlation between the circuit breakers and the protection modules for a typical double busbar installation.

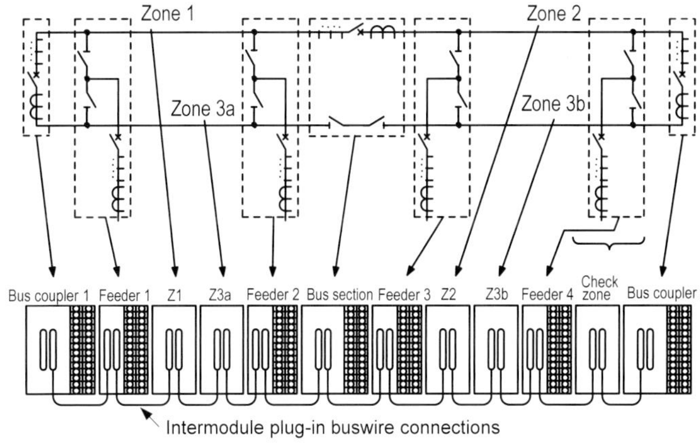

Figure 15.17: Type MBCZ busbar protection showing correlation between circuit breakers and protection modules

The modules are interconnected via a multicore cable that is plugged into the back of the modules. There are five main groups of buswires, allocated for:

- protection for main busbar
- protection for reserve busbar
- protection for the transfer busbar. When the reserve busbar is also used as a transfer bar then this group of buswires is used
- auxiliary connections used by the protection to combine modules for some of the more complex busbar configurations
- protection for the check zone

One extra module, not shown in this diagram, is plugged into the multicore bus. This is the alarm module, which contains the common alarm circuits and the bias resistors. The power supplies are also fed in through this module.

15.9.6.1 Bias

All zones of measurement are biased by the total current flowing to or from the busbar system via the feeders. This ensures that all zones of measurement will have similar fault sensitivity under all load conditions. The bias is derived from the check zone and fixed at 20% with a characteristic generally as shown in Figure 15.15(b). Thus some ratio mismatch is tolerable.

15.9.6.2 Stability with saturated current transformers

The traditional method for stabilising a differential relay is to add a resistor to the differential path. Whilst this improves stability it increases the burden on the current transformer for internal faults. The technique used in the MBCZ scheme overcomes this problem.

The MBCZ design detects when a CT is saturated and short-circuits the differential path for the portion of the cycle for which saturation occurs. The resultant spill current does not then flow through the measuring circuit and stability is assured.

This principle allows a very low impedance differential circuit to be developed that will operate successfully with relatively small CTs.

15.9.6.3 Operation for internal faults

If the CTs carrying fault current are not saturated there will be ample current in the differential circuit to operate the differential relay quickly for fault currents exceeding the minimum operating level, which is adjustable between 20% - 200% rated current.

When the only CT(s) carrying internal fault current become saturated, it might be supposed that the CT saturation detectors may completely inhibit operation by short-circuiting the differential circuit. However, the resulting inhibit pulses remove only an insignificant portion of the differential current, so operation of the relay is therefore virtually unaffected.

15.9.6.4 Discrepancy alarm feature

As shown in Figure 15.18, each measuring module contains duplicated biased differential elements and also a pair of supervision elements, which are a part of a comprehensive supervision facility.

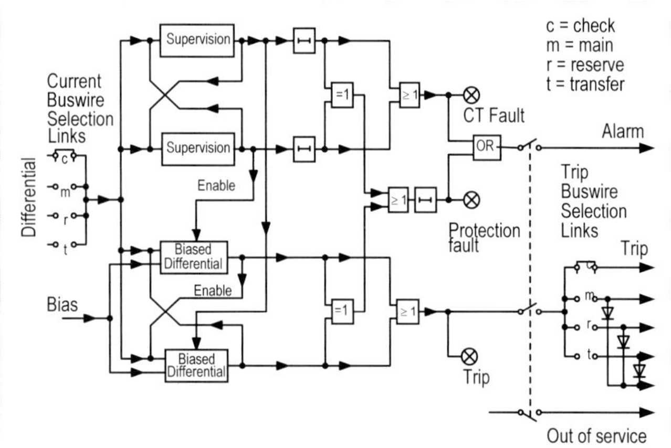

Figure 15.18: Block diagram of measuring unit

This arrangement provides supervision of CT secondary circuits for both open circuit conditions and any impairment of the element to operate for an internal fault, without waiting for an actual system fault condition to show this up. For a zone to operate it is necessary for both the differential supervision element and the biased differential element to operate. For a circuit breaker to be tripped it requires the associated main zone to be operated and also the overall check zone, as shown in Figure 15.19.

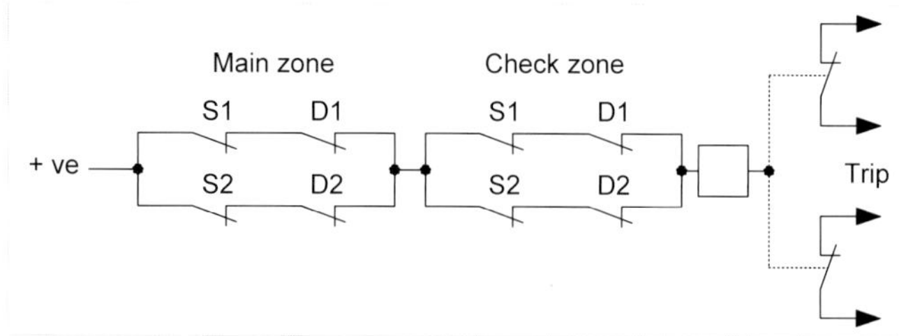

Figure 15.19: Busbar protection trip logic

15.9.6.5 Master/follower measuring units

When two sections of a busbar are connected together by isolators it will result in two measuring elements being connected in parallel when the isolators are closed to operate the two busbar sections as a single bar. The fault current will then divide between the two measuring elements in the ratio of their impedances. If both of the two measuring elements are of low and equal impedance the effective minimum operating current of the scheme will be doubled.

This is avoided by using a 'master/follower' arrangement. By making the impedance of one of the measuring elements very much higher than the other it is possible to ensure that one of the relays retains its original minimum operation current. Then to ensure that both the parallel-connected zones are tripped the trip circuits of the two zones are connected in parallel. Any measuring unit can have the role of 'master' or 'follower' as it is selectable by means of a switch on the front of the module.

15.9.6.6 Transfer tripping for breaker failure

Serious damage may result, and even danger to life, if a circuit breaker fails to open when called upon to do so. To reduce this risk breaker fail protection schemes were developed some years ago.

These schemes are generally based on the assumption that if current is still flowing through the circuit breaker a set time after the trip command has been issued, then it has failed to function. The circuit breakers in the next stage back in the system are then automatically tripped.

For a bus coupler or section breaker this would involve tripping all the infeeds to the adjacent zone, a facility that is included in the busbar protection scheme.

15.10 NUMERICAL BUSBAR PROTECTION SCHEMES

The application of numeric relay technology to busbar protection has become the preferred solution, overtaking the use of static. The very latest developments in the technology can be included, such as extensive use of a data bus to link the various units involved, and fault tolerance against loss of a particular link by providing multiple communications paths. The development process has been very rigorous, because the requirements for busbar protection in respect of immunity to maloperation are very high.

A philosophy that can be adopted is one of distributed processing of the measured values, as shown in Figure 15.20. Feeders each have their own processing unit, which collects together information on the state of the feeder (currents, voltages, CB and isolator status, etc.) and communicates it over high-speed fibre-optic data links to a central processing unit. For large substations, more than one central processing unit may be used, while in the case of small installations, all of the units can be co-located, leading to the appearance of a traditional centralised architecture.

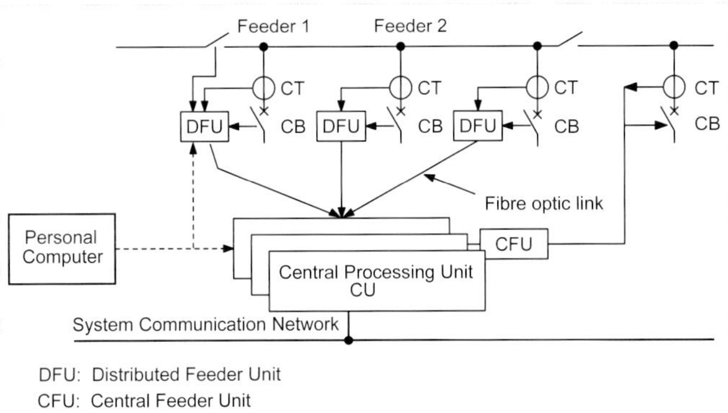

Figure 15.20: Architecture for numerical protection scheme

For simple feeders, interface units at a bay may be used with the data transmitted to a single centrally located feeder processing unit. The central processing unit performs the calculations required for the protection functions. Available protection functions are:

- protection
- backup overcurrent protection
- breaker failure
- dead zone protection (alternatively referred to as 'short zone' protection - see section 15.7.2)

In addition, monitoring functions such as CB and isolator monitoring, disturbance recording and transformer supervision are provided.

Because of the distributed topology used, synchronisation of the measurements taken by the Feeder Units is of vital importance. A high stability numerically-controlled oscillator is fitted in each of the central and feeder units, with time synchronisation between them. In the event of loss of the synchronisation signal, the high stability of the oscillator in the affected feeder unit(s) enables processing of the incoming data to continue without significant errors until synchronisation can be restored.

The feeder units have responsibility for collecting the required data, such as voltages and currents, and processing it into digital form for onwards transmission to the central processing unit. Modelling of the CT response is included, to eliminate errors caused by effects such as CT saturation. Disturbance recording for the monitored feeder is implemented, for later download as required. Because each feeder unit is concerned only with an individual feeder, the differential protection algorithms must reside in the central processing unit.

The differential protection algorithm can be much more sophisticated than with earlier technology, due to improvements in processing power. In addition to calculating the sum of the measured currents, the algorithm can also evaluate differences between successive current samples, since a large change above a threshold may indicate a fault – the threshold being chosen such that normal load changes, apart from inrush conditions do not exceed the threshold. The same considerations can also be applied to the phase angles of currents, and incremental changes in them.

One advantage gained from the use of numerical technology is the ability to easily re-configure the protection to cater for changes in configuration of the substation. For example, addition of an extra feeder involves the addition of an extra feeder unit, the fibre-optic connection to the central unit and entry via the HMI of the new configuration into the central processor unit. Figure 15.21 illustrates the latest numerical technology employed.

Figure 15.21: Busbar protection relay using the latest numerical technology (MiCOM P740 range)

15.10.1 Reliability Considerations

In considering the introduction of numerical busbar protection schemes, users have been concerned with reliability issues such as security and availability. Conventional high impedance schemes have been one of the main protection schemes used for busbar protection. The basic measuring element is simple in concept and has few components. Calculation of stability limits and other setting parameters is straightforward and scheme performance can be predicted without the need for costly testing. Practically high impedance schemes have proved to be a very reliable form of protection.

In contrast, modern numerical schemes are more complex with a much greater range of facilities and a higher component count. Based on low impedance bias techniques, and with a greater range of facilities to set, setting calculations can also be more complex.

However studies of the comparative reliability of conventional high impedance schemes and modern numerical schemes have shown that assessing relative reliability is not quite so simple as it might appear. The numerical scheme has two advantages over its older counterpart:

- there is a reduction in the number of external components such as switching and other auxiliary relays, many of the functions of which are performed internally within the software algorithms
- numerical schemes include sophisticated monitoring features which provide alarm facilities if the scheme is faulty. In certain cases, simulation of the scheme functions can be performed on line from the CT inputs through to the tripping outputs and thus scheme functions can be checked on a regular basis to ensure a full operational mode is available at all times

Reliability analyses using fault tree analysis methods have examined issues of dependability (i.e. the ability to operate when required) and security (i.e. the ability not to provide spurious/indiscriminate operation). These analyses have shown that:

- dependability of numerical schemes is better than conventional high impedance schemes
- security of numerical and conventional high impedance schemes are comparable

In addition, an important feature of numerical schemes is the in-built monitoring system. This considerably improves the potential availability of numerical schemes compared to conventional schemes as faults within the equipment and its operational state can be detected and alarmed. With the conventional scheme, failure to re-instate the scheme correctly after maintenance may not be detected until the scheme is required to operate. In this situation, its effective availability is zero until it is detected and repaired.

15.11 INTERLOCKED OVERCURRENT BUSBAR SCHEMES

Dedicated busbar protection, such as high impedance or biased differential, is commonplace for transmission and subtransmission systems. This ensures fast fault clearance, operating subcycle in some cases. The situation is a little different for distribution substations, where utilities take into account the large number of stations, the large number of feeder circuits, and may wish to consider an economical alternative solution.

Interlocked overcurrent schemes, also described as 'busbar blocking' or 'zone sequence interlocking' schemes can offer that economical alternative. The advantage is that the busbar protection requires no dedicated relay(s) to be installed, as it is configured to operate using logic facilities already available in the feeder manager overcurrent relays installed on the incoming and outgoing feeders. As feeder protection must in any case be installed for all circuits emanating from the busbar, the only additional cost to configure the busbar protection is to design and install the means for the individual relays to communicate peer-to-peer with each other.

Figure 15.22 shows a simple single incomer substation on a radial system, with four outgoing feeders. It can be appreciated that for any downstream fault, whether external to the substation, or in-zone on the busbar, the incoming feeder relay will detect the flow of fault current. The level of fault current can be similar, thus on the basis of overcurrent or earth fault detection alone, the incoming feeder relay cannot discriminate whether the fault is in-zone and requires a trip, or whether it is external (and tripping only by the correct outgoing feeder relay is required).

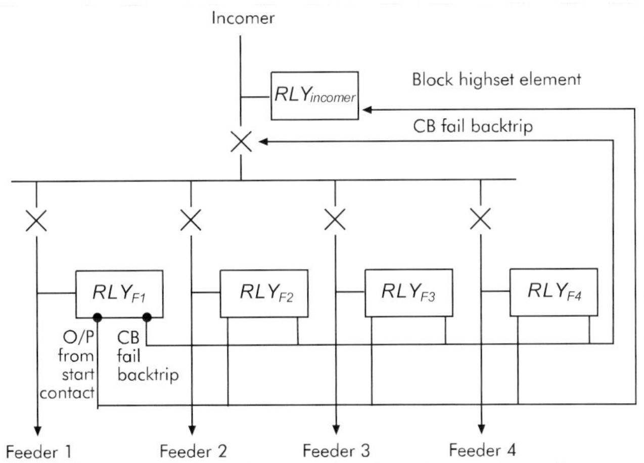

Figure 15.22: Simple busbar blocking scheme (single incomer)

However, most numerical feeder relays have the means to issue an instantaneous start output, to indicate that they are measuring an operating current above setting, and as such, their time-overcurrent function is in the process of timing out. Figure 15.22 shows how this facility can be enabled in the outgoing feeder relays, and communicated to the incomer relay as a 'block' signal. This block does not block the critical time-overcurrent function in the incomer relay, but merely blocks a definite-time high-set that has been specially-configured to offer busbar protection.

The high-set, measuring phase overcurrent, and often earth-fault too, is delayed by a margin marked as 'time to block' in Figure 15.23. This deliberate delay on operation is necessary to allow enough time for an external fault to be detected by an outgoing feeder relay, and for it to be communicated as a 'block' command, and acted upon by the incomer relay. Only if the incoming feeder relay measures fault current above the high-set setting, and no block arrives, it determines that the fault must lie on the protected busbar, and a bus zone trip command is issued.

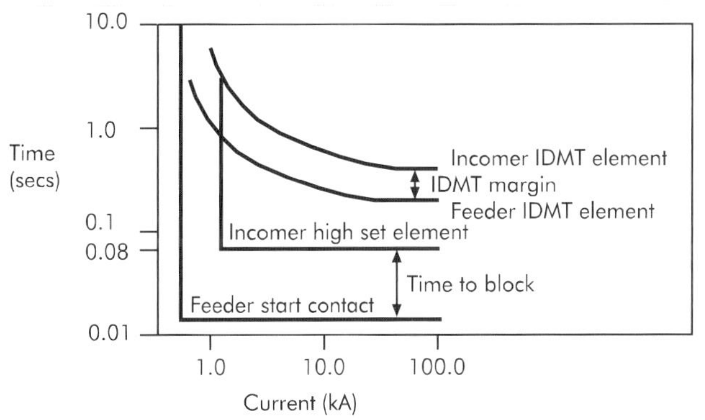

Figure 15.23: High-set bus-zone tripping: element co-ordination

The definite time delay - 'time to block' - will often be of the order of 50 to 100ms, meaning that the operating time for a genuine busbar fault will be of the order of 100ms. This delayed clearance may not be an issue, as system stability is typically not a primary concern for distribution systems. Indeed, 100ms operation would be much faster than waiting for the incoming feeder I.D.M.T. curve to time out, which would be the historical means by which the fault would be detected.

Figure 15.22 indicated that block signals would be exchanged between relays by hardwiring, and this is still a valid solution for simplicity. A more contemporary implementation would be to make use of IEC 61850 GOOSE logic to exchange such commands via Ethernet. GOOSE application is detailed in Chapter 24 - The Digital Substation. Whether deployed as hardwired, or Ethernet schemes, circuit breaker failure is commonly applied too.

Interlocked overcurrent busbar schemes may also be applied to sectionalised busbars, with more than one incomer, provided that directional relays are applied on any feeder which may act as an infeed, and at each bus section. However, such schemes are unlikely to be applicable in double busbar stations, where dedicated busbar protection is recommended.

15.12 REFERENCE

[15.1] The Behaviour of Current Transformers subjected to Transient Asymmetric Currents and the Effects on Associated Protective Relays. J.W. Hodgkiss. CIGRE Paper Number 329, Session 15-25 June 1960.

Chapter 16
Transformer and Transformer Feeder Protection

16.1 INTRODUCTION

The development of modern power systems has been reflected in the advances in transformer design. This has resulted in a wide range of transformers with sizes ranging from a few kVA to several hundred MVA being available for use in a wide variety of applications.

The considerations for a transformer protection package vary with the application and importance of the transformer. To reduce the effects of thermal stress and electrodynamic forces, it is advisable to ensure that the protection package used minimises the time for disconnection in the event of a fault occurring within the transformer. Small distribution transformers can be protected satisfactorily, from both technical and economic considerations, by the use of fuses or overcurrent relays. This results in time-delayed protection due to downstream co-ordination requirements. However, time-delayed fault clearance is unacceptable on larger power transformers used in distribution, transmission and generator applications, due to system operation/stability and cost of repair/length of outage considerations.

Transformer faults are generally classified into five categories:

- winding and terminal faults
- core faults
- tank and transformer accessory faults
- on–load tap changer faults
- abnormal operating conditions
- sustained or uncleared external faults

For faults originating in the transformer itself, the approximate proportion of faults due to each of the causes listed above is shown in Figure 16.1.

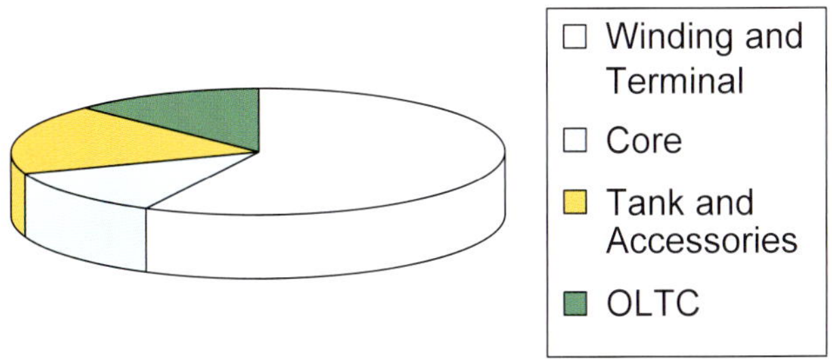

Figure 16.1: Transformer fault statistics

16.2 WINDING FAULTS

A fault on a transformer winding is controlled in magnitude by the following factors:

- source impedance
- neutral earthing impedance
- transformer leakage reactance
- fault voltage
- winding connection

Several distinct cases arise and are examined below.

16.2.1 Star-Connected Winding with Neutral Point Earthed Through an Impedance

The winding earth fault current depends on the earthing impedance value and is also proportional to the distance of the fault from the neutral point, since the fault voltage will be directly proportional to this distance.

For a fault on a transformer secondary winding, the corresponding primary current will depend on the transformation ratio between the primary winding and the short-circuited secondary turns. This also varies with the position of the fault, so that the fault current in the transformer primary winding is proportional to the square of the fraction of the winding that is short-circuited. The effect is shown in Figure 16.2. Faults in the lower third of the winding produce very little current in the primary winding, making fault detection by primary current measurement difficult.

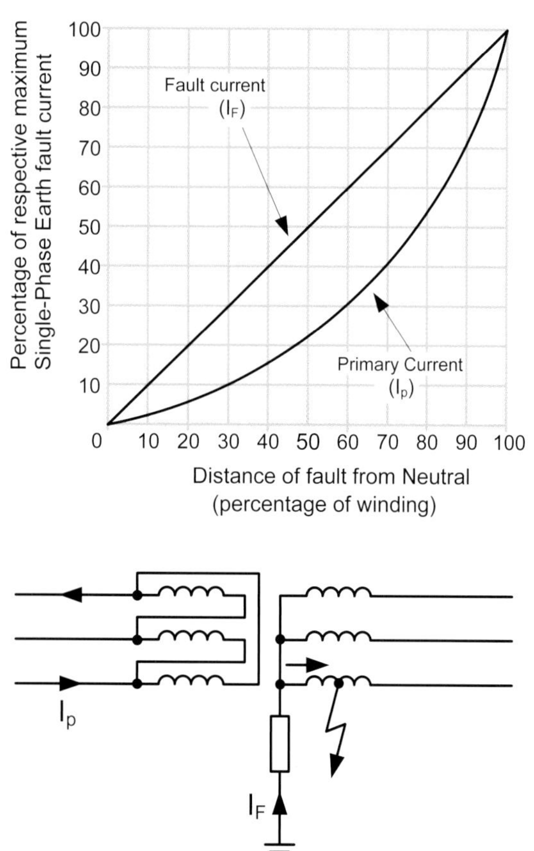

Figure 16.2: Earth fault current in resistance-earthed star winding

16.2.2 Star-Connected Winding with Neutral Point Solidly Earthed

The fault current is controlled mainly by the leakage reactance of the winding, which varies in a complex manner with the position of the fault. The variable fault point voltage is also an important factor, as in the case of impedance earthing. For faults close to the neutral end of the winding, the reactance is very low, and results in the highest fault currents. The variation of current with fault position is shown in Figure 16.3.

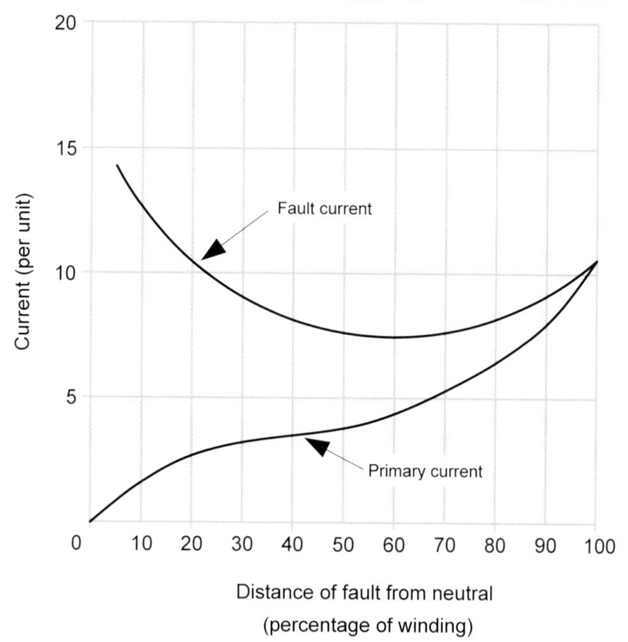

Figure 16.3: Earth fault current in solidly earthed star winding

For secondary winding faults, the primary winding fault current is determined by the variable transformation ratio; as the secondary fault current magnitude stays high throughout the winding, the primary fault current is large for most points along the winding.

16.2.3 Delta-Connected Winding

No part of a delta-connected winding operates with a voltage to earth of less than 50% of the phase voltage. The range of fault current magnitude is therefore less than for a star winding. The actual value of fault current will still depend on the method of system earthing; it should also be remembered that the impedance of a delta winding is particularly high to fault currents flowing to a centrally placed fault on one leg. The impedance can be expected to be between 25% and 50%, based on the transformer rating, regardless of the normal balanced through-current impedance. As the prefault voltage to earth at this point is half the normal phase voltage, the earth fault current may be no more than the rated current, or even less than this value if the source or system earthing impedance is appreciable. The current will flow to the fault from each side through the two half windings, and will be divided between two phases of the system. The individual phase currents may therefore be relatively low, resulting in difficulties in providing protection.

16.2.4 Phase to Phase Faults

Faults between phases within a transformer are relatively rare; if such a fault does occur it will give rise to a substantial current comparable to the earth fault currents discussed in Section 16.2.2.

16.2.5 Interturn Faults

In low voltage transformers, interturn insulation breakdown is unlikely to occur unless the mechanical force on the winding due to external short circuits has caused insulation degradation, or insulating oil (if used) has become contaminated by moisture.

A high voltage transformer connected to an overhead transmission system will be subjected to steep fronted impulse voltages, arising from lightning strikes, faults and switching operations. A line surge, which may be of several times the rated system voltage, will concentrate on the end turns of the winding because of the high equivalent frequency of the surge front. Part-winding resonance, involving voltages up to 20 times rated voltage may occur. The interturn insulation of the end turns is reinforced, but cannot be increased in proportion to the insulation to earth, which is relatively great. Partial winding flashover is therefore more likely. The subsequent progress of the fault, if not detected in the earliest stage, may well destroy the evidence of the true cause.

A short circuit of a few turns of the winding will give rise to a heavy fault current in the short-circuited loop, but the terminal currents will be very small, because of the high ratio of transformation between the whole winding and the short-circuited turns.

The graph in Figure 16.4 shows the corresponding data for a typical transformer of 3.25% impedance with the short-circuited turns symmetrically located in the centre of the winding.

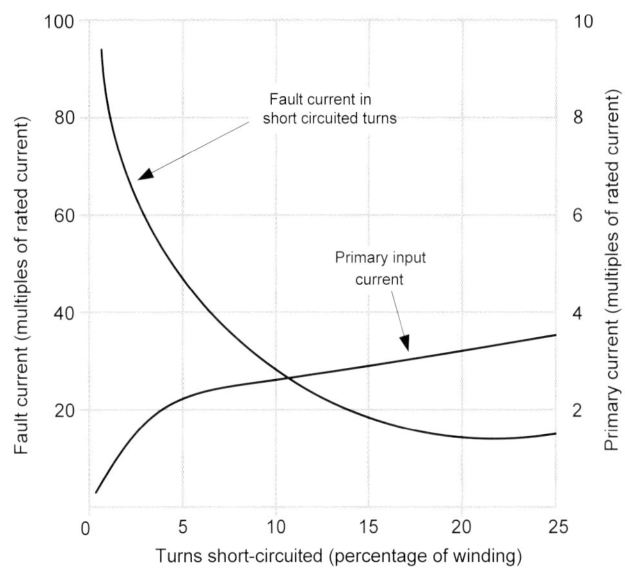

Figure 16.4: Interturn fault current/number of turns short-circuited

16.2.6 Core Faults

A conducting bridge across the laminated structures of the core can permit sufficient eddy-current to flow to cause serious overheating. The bolts that clamp the core together are always insulated to avoid this trouble. If any portion of the core insulation becomes defective, the resultant heating may reach a magnitude sufficient to damage the winding.

The additional core loss, although causing severe local heating, does not produce a noticeable change in input current and could not be detected by the normal electrical protection. However it is important that the condition is detected before a major fault has been created. In an oil-immersed transformer, core heating sufficient to cause winding insulation damage also causes breakdown of some of the oil with an accompanying evolution of gas. This gas escapes to the conservator and is used to operate a mechanical relay; see Section 16.15.3.

16.2.7 Tank Faults

Loss of oil through tank leaks ultimately produces a dangerous condition, either because of a reduction in winding insulation or because of overheating on load due to the loss of cooling.

Overheating may also occur due to prolonged overloading, blocked cooling ducts due to oil sludging or failure of the forced cooling system, if fitted.

16.2.8 Externally Applied Conditions

Sources of abnormal stress in a transformer are:

- overload
- system faults
- overvoltage
- reduced system frequency

16.2.8.1 Overload

Overload causes increased 'copper loss' and a consequent temperature rise. Overloads can be carried for limited periods and recommendations for oil-immersed transformers are given in IEC 60354.

The thermal time constant of naturally cooled transformers lies between 2.5-5 hours. Shorter time constants apply in the case of force-cooled transformers.

16.2.8.2 System faults

System short circuits produce a relatively intense rate of heating of the feeding transformers, the copper loss increasing in proportion to the square of the per unit fault current. The typical duration of external short circuits that a transformer can sustain without damage if the current is limited only by the self-reactance is shown in Table 16.1. IEC 60076 provides further guidance on short-circuit withstand levels.

Transformer Reactance (%)	Fault Current (Multiple of Rating)	Permitted Fault Duration (seconds)
4	25	2
5	20	2
6	16.6	2
7	14.2	2

Table 16.1: Transformer fault current withstand data

Maximum mechanical stress on windings occurs during the first cycle of the fault. Avoidance of damage is a matter of transformer design.

16.2.8.3 Overvoltages

Overvoltage conditions are of two kinds:

- transient surge voltages
- power frequency overvoltage

Transient overvoltages arise from faults, switching, and lightning disturbances and are liable to cause interturn faults, as described in Section 16.2.5. These overvoltages are usually limited by shunting the high voltage terminals to earth either with a plain rod gap or by surge diverters, which comprise a stack of short gaps in series with a non-linear resistor. The surge diverter, in contrast to the rod gap, has the advantage of extinguishing the flow of power current after discharging a surge, in this way avoiding subsequent isolation of the transformer.

Power frequency overvoltage causes both an increase in stress on the insulation and a proportionate increase in the working flux. The latter effect causes an increase in the iron loss and a disproportionately large increase in magnetising current. In addition, flux is diverted from the laminated core into structural steel parts. The core bolts, which normally carry little flux, may be subjected to a large flux diverted from the highly saturated region of core alongside. This leads to a rapid temperature rise in the bolts, destroying their insulation and damaging coil insulation if the condition continues.

16.2.8.4 Reduced system frequency

Reduction of system frequency has an effect with regard to flux density, similar to that of overvoltage.

It follows that a transformer can operate with some degree of overvoltage with a corresponding increase in frequency, but operation must not be continued with a high voltage input at a low frequency. Operation cannot be sustained when the ratio of voltage to frequency, with these quantities given values in per unit of their rated values, exceeds unity by more than a small amount, for instance if $V/f > 1.1$. If a substantial rise in system voltage has been catered for in the design, the base of 'unit voltage' should be taken as the highest voltage for which the transformer is designed.

16.3 MAGNETISING INRUSH

The phenomenon of magnetising inrush is a transient condition that occurs primarily when a transformer is energised. It is not a fault condition, and therefore transformer protection must remain stable during the inrush transient.

Figure 16.5(a) shows a transformer magnetising characteristic. To minimise material costs, weight and size, transformers are generally operated near to the 'knee point' of the magnetising characteristic. Consequently, only a small increase in core flux above normal operating levels will result in a high magnetising current.

Under normal steady-state conditions, the magnetising current associated with the operating flux level is relatively small (Figure 16.5(b)). However, if a transformer winding is energised at a voltage zero, with no remanent flux, the flux level during the first voltage cycle (2 x normal flux) will result in core saturation and a high non-sinusoidal magnetising current waveform – see Figure 16.5(c). This current is referred to as magnetising inrush current and may persist for several cycles.

Several factors affect the magnitude and duration of the magnetising current inrush:

- residual flux – worst-case conditions result in the flux peak value attaining 280% of normal value
- point on wave switching
- number of banked transformers
- transformer design and rating
- system fault level

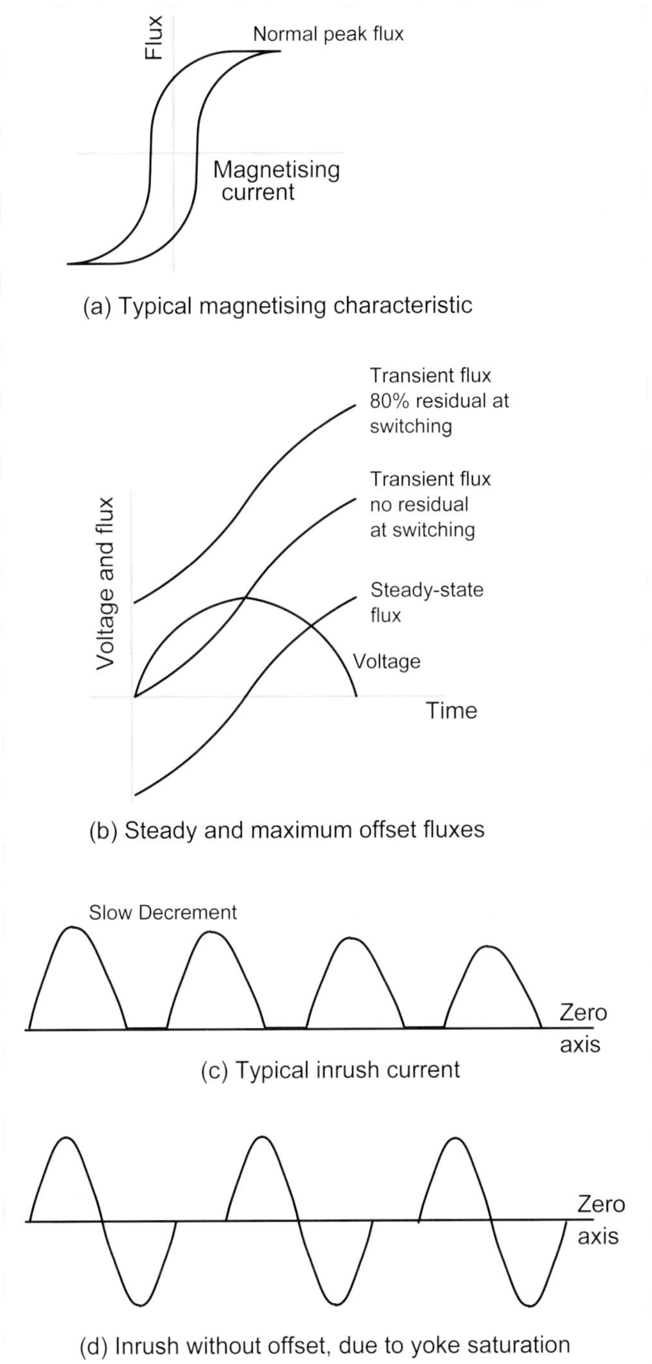

Figure 16.5: Transformer magnetising inrush

The very high flux densities quoted above are so far beyond the normal working range that the incremental relative permeability of the core approximates to unity and the inductance of the winding falls to a value near that of the 'air-cored' inductance. The current wave, starting from zero, increases slowly at first, the flux having a value just above the residual value and the permeability of the core being moderately high. As the flux passes the normal working value and enters the highly saturated portion of the magnetising characteristic, the inductance falls and the current rises rapidly to a peak that may be 500% of the steady state magnetising current. When the peak is passed at the next voltage zero, the following negative half cycle of the voltage wave reduces the flux to the starting value, the current falling symmetrically to zero. The current wave is therefore fully offset and is only restored to the steady state condition by the circuit losses. The time constant of the transient has a range between 0.1 second (for a 100kVA transformer) to 1.0 second (for a large unit). As the magnetising characteristic is non-linear, the envelope of the transient current is not strictly of exponential form; the magnetising current can be observed to be still changing up to 30 minutes after switching on.

Although correct choice of the point on the wave for a single–phase transformer will result in no transient inrush, mutual effects ensure that a transient inrush occurs in all phases for three-phase transformers.

16.3.1 Harmonic Content of Inrush Waveform

The waveform of transformer magnetising current contains a proportion of harmonics that increases as the peak flux density is raised to the saturating condition. The magnetising current of a transformer contains a third harmonic and progressively smaller amounts of fifth and higher harmonics. If the degree of saturation is progressively increased, not only will the harmonic content increase as a whole, but the relative proportion of fifth harmonic will increase and eventually exceed the third harmonic. At a still higher level the seventh would overtake the fifth harmonic but this involves a degree of saturation that will not be experienced with power transformers.

The energising conditions that result in an offset inrush current produce a waveform that is asymmetrical. Such a wave typically contains both even and odd harmonics. Typical inrush currents contain substantial amounts of second and third harmonics and diminishing amounts of higher orders. As with the steady state wave, the proportion of harmonics varies with the degree of saturation, so that as a severe inrush transient decays, the harmonic makeup of the current passes through a range of conditions.

16.4 TRANSFORMER OVERHEATING

The rating of a transformer is based on the temperature rise above an assumed maximum ambient temperature; under this condition no sustained overload is usually permissible. At a lower ambient temperature some degree of sustained overload can be safely applied. Short-term overloads are also permissible to an extent dependent on the previous loading conditions. IEC 60354 provides guidance in this respect.

The only certain statement is that the winding must not overheat; a temperature of about $95°C$ is considered to be the normal maximum working value beyond which a further rise of

$8°$- $10°C$, if sustained, will halve the insulation life of the unit.

Protection against overload is therefore based on winding temperature, which is usually measured by a thermal image technique. Protection is arranged to trip the transformer if excessive temperature is reached. The trip signal is usually routed via a digital input of a protection relay on one side of the transformer, with both alarm and trip facilities made available through programmable logic in the relay. Intertripping between the relays on the two sides of the transformer is usually applied to ensure total disconnection of the transformer.

Winding temperature protection may be included as a part of a complete monitoring package. See Section 16.18 for more details.

16.5 TRANSFORMER PROTECTION – OVERVIEW

The problems relating to transformers described in Section 16.4 require some means of protection. Table 16.2 summarises the problems and the possible forms of protection that may be used. The following sections provide more detail on the individual protection methods. It is normal for a modern relay to provide all of the required protection functions in a single package, in contrast to electromechanical types that would require several relays complete with interconnections and higher overall CT burdens.

Fault Type	Protection Used
Primary winding Phase-phase fault	Differential; Overcurrent
Primary winding Phase-earth fault	Differential; Overcurrent
Secondary winding Phase-phase fault	Differential
Secondary winding Phase-earth fault	Differential; Restricted Earth Fault
Interturn Fault	Differential, Buchholz
Core Fault	Differential, Buchholz
Tank Fault	Differential, Buchholz; Tank-Earth
Overfluxing	Overfluxing
Overheating	Thermal

Table 16.2: Transformer fault types/protection methods

16.6 TRANSFORMER OVERCURRENT PROTECTION

Fuses may adequately protect small transformers, but larger ones require overcurrent protection using a relay and CB, as fuses do not have the required fault breaking capacity.

16.6.1 Fuses

Fuses commonly protect small distribution transformers typically up to ratings of 1MVA at distribution voltages. In many cases no circuit breaker is provided, making fuse protection the only available means of automatic isolation. The fuse must have a rating well above the maximum transformer load current to withstand the short duration overloads that may occur. Also, the fuses must withstand the magnetising inrush currents drawn when power transformers are energised. High Rupturing Capacity (HRC) fuses, although very fast in operation with large fault currents, are extremely slow with currents of less than three times their rated value. It follows that such fuses will do little to protect the transformer, serving only to protect the system by disconnecting a faulty transformer after the fault has reached an advanced stage.

Table 16.3 shows typical ratings of fuses for use with 11kV transformers

Transformer Rating		Fuse	
kVA	Full Load Current (A)	Rated Current (A)	Operating Time at 3 x Rating (s)
100	5.25	16	3.0
200	10.5	25	3.0
315	15.8	36	10.0
500	26.2	50	20.0
1000	52.5	90	30.0

Table 16.3: Typical fuse ratings for use with distribution transformers

This table should be taken only as a typical example; considerable differences exist in the time characteristics of different types of HRC fuses. Furthermore grading with protection on the secondary side has not been considered.

16.6.2 Overcurrent Relays

With the advent of ring main units incorporating SF_6 circuit breakers and isolators, protection of distribution transformers can now be provided by overcurrent trips (e.g. tripping controlled by time limit fuses connected across the secondary windings of in-built current transformers) or by relays connected to current transformers located on the transformer primary side. Overcurrent relays are also used on larger transformers provided with standard circuit breaker control. Improvement in protection is obtained in two ways; the excessive delays of the HRC fuse for lower fault currents are avoided and an earth-fault tripping element is provided in addition to the overcurrent feature.

The time delay characteristic should be chosen to discriminate with circuit protection on the secondary side. A high-set instantaneous relay element is often provided, the current setting being chosen to avoid operation for a secondary short circuit. This enables high-speed clearance of primary terminal short circuits.

16.7 RESTRICTED EARTH FAULT PROTECTION

Conventional earth fault protection using overcurrent elements fails to provide adequate protection for transformer windings. This applies particularly to a star-connected winding with an impedance-earthed neutral, as discussed in Section 16.2.1.

The degree of protection is very much improved by the application of restricted earth fault protection (or REF protection). This is a unit protection scheme for one winding of the transformer. It can be a high impedance type as shown in Figure 16.6 or a biased low-impedance type. For the high-impedance type, the residual current of three line current transformers is balanced against the output of a current transformer in the neutral conductor. In the biased low-impedance version, the three phase currents and the neutral current become the bias inputs to a differential element.

The system is operative for faults within the region between current transformers, that is, for faults on the star winding in question. The system remains stable for all faults outside this zone.

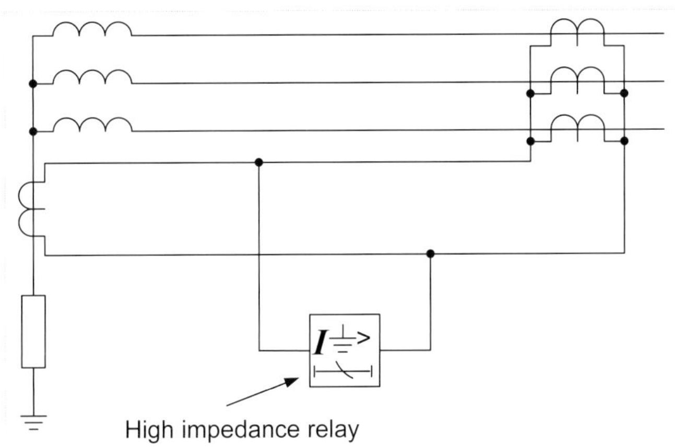

Figure 16.6: Restricted earth fault protection for a star winding

The gain in protection performance comes not only from using an instantaneous relay with a low setting, but also because the whole fault current is measured, not merely the transformed component in the HV primary winding (if the star winding is a secondary winding). Hence, although the prospective current level decreases as fault positions progressively nearer the neutral end of the winding are considered, the square law which controls the primary line current is not applicable, and with a low effective setting, a large percentage of the winding can be covered.

Restricted earth fault protection is often applied even when the neutral is solidly earthed. Since fault current then remains at a high value even to the last turn of the winding (Figure 16.2), virtually complete cover for earth faults is obtained. This is an improvement compared with the performance of systems that do not measure the neutral conductor current.

Earth fault protection applied to a delta-connected or unearthed star winding is inherently restricted, since no zero sequence components can be transmitted through the transformer to the other windings.

Both windings of a transformer can be protected separately with restricted earth fault protection, thereby providing high-speed protection against earth faults for the whole transformer with relatively simple equipment. A high impedance relay is used, giving fast operation and phase fault stability.

16.8 DIFFERENTIAL PROTECTION

The restricted earth fault schemes described above in Section 16.7 depend entirely on the Kirchhoff principle that the sum of the currents flowing into a conducting network is zero. A differential system can be arranged to cover the complete transformer; this is possible because of the high efficiency of transformer operation, and the close equivalence of ampere-turns developed on the primary and secondary windings. Figure 16.7 shows the principle. Current transformers on the primary and secondary sides are connected to form a circulating current system.

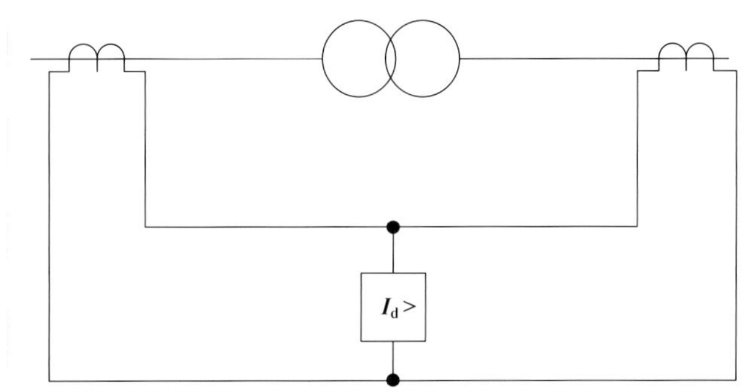

Figure 16.7: Principle of transformer differential protection

16.8.1 Basic Considerations for Transformer Differential Protection

In applying the principles of differential protection to transformers, a variety of considerations have to be taken into account. These include:

- correction for possible phase shift across the transformer windings (phase correction)
- the effects of the variety of earthing and winding arrangements (filtering of zero sequence currents)
- correction for possible unbalance of signals from current transformers on either side of the windings (ratio correction)
- the effect of magnetising inrush during initial energisation
- the possible occurrence of overfluxing

In traditional transformer differential schemes, the requirements for phase and ratio correction were met by the application of external interposing current transformers (ICTs), as a secondary replica of the main winding connections, or by a delta connection of the main CTs to provide phase correction only. Digital/numerical relays implement ratio and phase correction in the relay software instead, thus enabling most combinations of transformer winding arrangements to be catered for, irrespective of the winding connections of the primary CTs. This avoids the additional space and cost requirements of hardware interposing CTs.

16.8.2 Line Current Transformer Primary Ratings

Line current transformers have primary ratings selected to be approximately equal to the rated currents of the transformer windings to which they are applied. Primary ratings will usually be limited to those of available standard ratio CTs.

16.8.3 Phase Correction

Correct operation of transformer differential protection requires that the transformer primary and secondary currents, as measured by the relay, are in phase. If the transformer is connected delta/star, as shown in Figure 16.8, balanced three-phase through current suffers a phase change of 30°. If left uncorrected, this phase difference would lead to the relay seeing through current as an unbalanced fault current, and result in relay operation. Phase correction must be implemented.

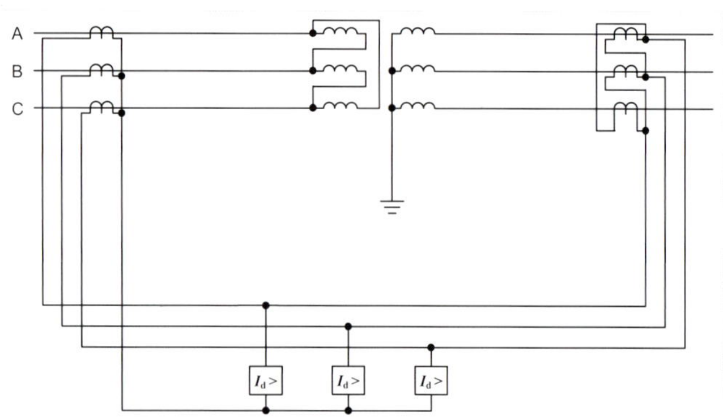

Figure 16.8: Differential protection for two-winding delta/star transformer

Electromechanical and static relays use appropriate CT/ICT connections to ensure that the primary and secondary currents applied to the relay are in phase.

For digital and numerical relays, it is common to use star-connected line CTs on all windings of the transformer and compensate for the winding phase shift in software. Depending on relay design, the only data required in such circumstances may be the transformer vector group designation. Phase compensation is then performed automatically. Caution is required if such a relay is used to replace an existing electromechanical or static relay, as the primary and secondary line CTs may not have the same winding configuration. Phase compensation and associated relay data entry requires more detailed consideration in such circumstances. Rarely, the available phase compensation facilities cannot accommodate the transformer winding connection, and in such cases interposing CTs must be used.

16.8.4 Filtering of Zero Sequence Currents

As described in Chapter 10.8, it is essential to provide some form of zero sequence filtering where a transformer winding can pass zero sequence current to an external earth fault. This is to ensure that out-of-zone earth faults are not seen by the transformer protection as an in-zone fault. This is achieved by use of delta-connected line CTs or interposing CTs for older relays, and hence the winding connection of the line and/or interposing CTs must take this into account, in addition to any phase compensation necessary. For digital/numerical relays, the required filtering is applied in the relay software. Table 16.4 summarises the phase compensation and zero sequence filtering requirements. An example of an incorrect choice of ICT connection is given in Section 16.19.1.

Transformer Connection		Transformer Phase Shift	Clock Face Vector	Phase Compensation Required	HV Zero Sequence Filtering	LV Zero Sequence Filtering
Yy0		0°	0	0°	Yes	Yes
Zd0					Yes	
Dz0						Yes
Dd0						
Yz1	Zy1	-30°	1	30°	Yes	Yes
Yd1					Yes	
Dy1						Yes
Yy6		-180°	1	180°	Yes	Yes
Zd6					Yes	
Dz6						Yes
Dd6						
Yz11	Zy11	30°	11	-30°	Yes	Yes
Yd11					Yes	
Dy11						Yes
YyH	YzH	(H / 12) x 360°	Hour 'H'	-(H / 12) x 360°	Yes	Yes
YdH	ZdH				Yes	
DzH	DyH					Yes
DdH						

Table 16.4: Current transformer connections for power transformers of various vector groups

16.8.5 Ratio Correction

Correct operation of the differential element requires that currents in the differential element balance under load and through fault conditions. As the primary and secondary line CT ratios may not exactly match the transformer rated winding currents, digital/numerical relays are provided with ratio correction factors for each of the CT inputs. The correction factors may be calculated automatically by the relay from knowledge of the line CT ratios and the transformer MVA rating. However, if interposing CTs are used, ratio correction may not be such an easy task and may need to take into account a factor of $\sqrt{3}$ if delta-connected CTs or ICTs are involved. If the transformer is fitted with a tap changer, line CT ratios and correction factors are normally chosen to achieve current balance at the mid tap of the transformer. It is necessary to ensure that current mismatch due to off-nominal tap operation will not cause spurious operation.

The example in Section 16.19.2 shows how ratio correction factors are used, and that of Section 16.19.3 shows how to set the ratio correction factors for a transformer with an unsymmetrical tap range.

16.8.6 Bias Setting

Bias is applied to transformer differential protection for the same reasons as any unit protection scheme – to ensure stability for external faults while allowing sensitive settings to pick up internal faults. The situation is slightly complicated if a tap changer is present. With line CT/ICT ratios and correction factors set to achieve current balance at nominal tap, an off-nominal tap may be seen by the differential protection as an internal fault. By selecting the minimum bias to be greater than sum of the maximum tap of the transformer and possible CT errors, maloperation due to this cause is avoided. Some relays use a bias characteristic with three sections, as shown in Figure 16.9. The first section is set higher than the transformer magnetising current. The second section is set to allow for off-nominal tap settings, while the third has a larger bias slope beginning well above rated current to cater for heavy through-fault conditions.

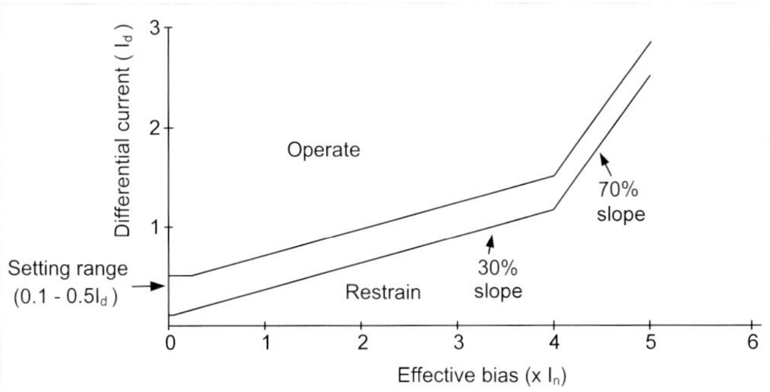

Figure 16.9: Typical bias characteristic

16.8.7 Transformers with Multiple Windings

The unit protection principle remains valid for a system having more than two connections, so a transformer with three or more windings can still be protected by the application of the above principles.

When the power transformer has only one of its three windings connected to a source of supply, with the other two windings feeding loads, a relay with only two sets of CT inputs can be used, connected as shown in Figure 16.10(a). The separate load currents are summated in the CT secondary circuits, and will balance with the infeed current on the supply side.

When more than one source of fault current infeed exists, there is a danger in the scheme of Figure 16.10(a) of current circulating between the two paralleled sets of current transformers without producing any bias. It is therefore important a relay is used with separate CT inputs for the two secondaries - Figure 16.10(b).

When the third winding consists of a delta-connected tertiary with no connections brought out, the transformer may be regarded as a two winding transformer for protection purposes and protected as shown in Figure 16.10(c).

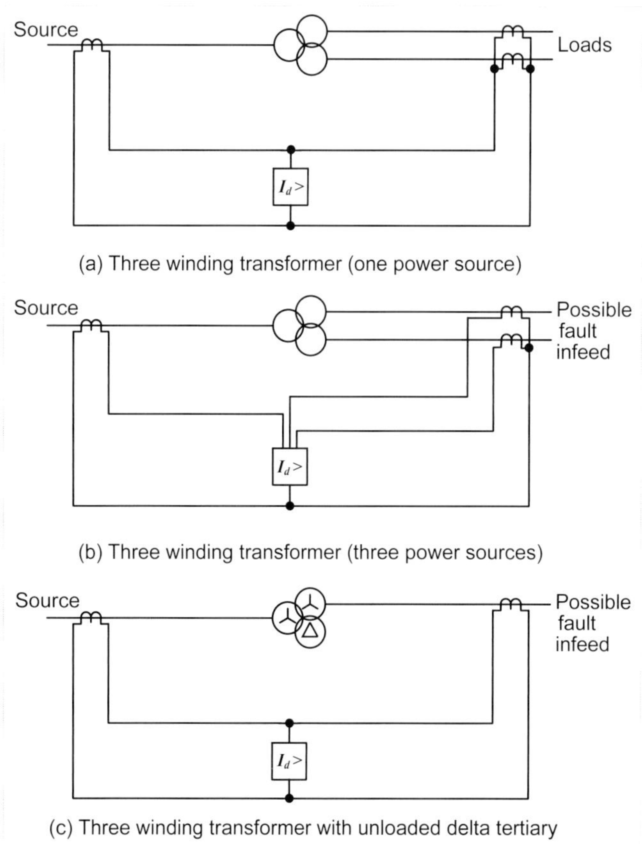

Figure 16.10: Differential protection arrangements for three-winding transformers (shown single phase for simplicity)

16.9 DIFFERENTIAL PROTECTION STABILISATION DURING MAGNETISING INRUSH CONDITIONS

The magnetising inrush phenomenon described in Section 16.3 produces current input to the energised winding which has no equivalent on the other windings. The whole of the inrush current appears, therefore, as unbalance and the differential protection is unable to distinguish it from current due to an internal fault. The bias setting is not effective and an increase in the protection setting to a value that would avoid operation would make the protection of little value. Methods of delaying, restraining or blocking of the differential element must therefore be used to prevent mal-operation of the protection.

16.9.1 Time Delay

Since the phenomenon is transient, stability can be maintained by providing a small time delay. However, because this time delay also delays operation of the relay in the event of a fault occurring at switch-on, the method is no longer used.

16.9.2 Harmonic Restraint

The inrush current, although generally resembling an in-zone fault current, differs greatly when the waveforms are compared. The difference in the waveforms can be used to distinguish between the conditions.

As stated before, the inrush current contains all harmonic orders, but these are not all equally suitable for providing bias. In practice, only the second harmonic is used.

This component is present in all inrush waveforms. It is typical of waveforms in which successive half period portions do not repeat with reversal of polarity but in which mirror-image symmetry can be found about certain ordinates.

The proportion of second harmonic varies somewhat with the degree of saturation of the core, but is always present as long as the uni-directional component of flux exists. The amount varies according to factors in the transformer design. Normal fault currents do not contain second or other even harmonics, nor do distorted currents flowing in saturated iron cored coils under steady state conditions.

The output current of a current transformer that is energised into steady state saturation will contain odd harmonics but not even harmonics. However, should the current transformer be saturated by the transient component of the fault current, the resulting saturation is not symmetrical and even harmonics are introduced into the output current. This can have the advantage of improving the through fault stability performance of a differential relay.

The second harmonic is therefore an attractive basis for a stabilising bias against inrush effects, but care must be taken to ensure that the current transformers are sufficiently large so that the harmonics produced by transient saturation do not delay normal operation of the relay. The differential current is passed through a filter that extracts the second harmonic; this component is then applied to produce a restraining quantity sufficient to overcome the operating tendency due to the whole of the inrush current that flows in the operating circuit. By this means a sensitive and high-speed system can be obtained.

16.9.3 Inrush Detection Blocking – Gap Detection Technique

Another feature that characterises an inrush current can be seen from Figure 16.5 where the two waveforms (c) and (d) have periods in the cycle where the current is zero. The minimum duration of this zero period is theoretically one quarter of the cycle and is easily detected by a simple timer T1 that is set to $\frac{1}{4f}$ seconds. Figure 16.11 shows the circuit in block diagram form. Timer T1 produces an output only if the current is zero for a time exceeding $\frac{1}{4f}$ seconds. It is reset when the instantaneous value of the differential current exceeds the setting reference.

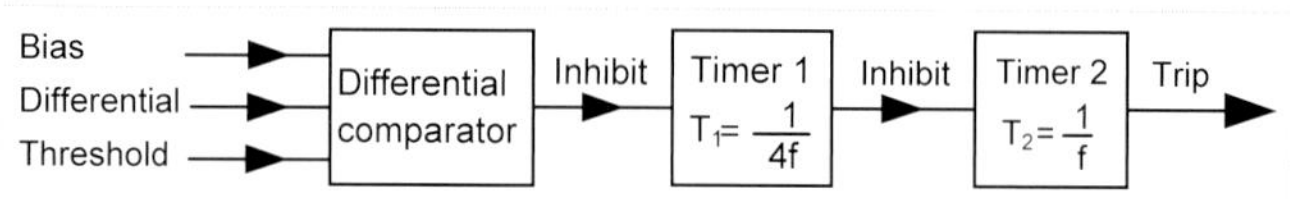

Figure 16.11: Block diagram to show waveform gap-detecting principle

As the zero in the inrush current occurs towards the end of the cycle, it is necessary to delay operation of the differential relay by $\frac{1}{f}$ seconds to ensure that the zero condition can be detected if present. This is achieved by using a second timer T2 that is held reset by an output from timer T1.

When no current is flowing for a time exceeding $\frac{1}{4f}$ seconds, timer T2 is held reset and the differential relay that may be controlled by these timers is blocked. When a differential current exceeding the setting of the relay flows, timer T1 is reset and timer T2 times out to give a trip signal in $\frac{1}{f}$ seconds. If the differential current is characteristic of transformer inrush then timer T2 will be reset on each cycle and the trip signal is blocked. Some numerical relays may use a combination of the harmonic restraint and gap detection techniques for magnetising inrush detection.

16.10 COMBINED DIFFERENTIAL AND RESTRICTED EARTH FAULT SCHEMES

The advantages to be obtained by the use of restricted earth fault protection, discussed in Section 16.7 lead to the system being frequently used in conjunction with an overall differential system. The importance of this is shown in Figure 16.12 from which it will be seen that if the neutral of a star-connected winding is earthed through a resistance of one per unit, an overall differential system having an effective setting of 20% will detect faults in only 42% of the winding from the line end.

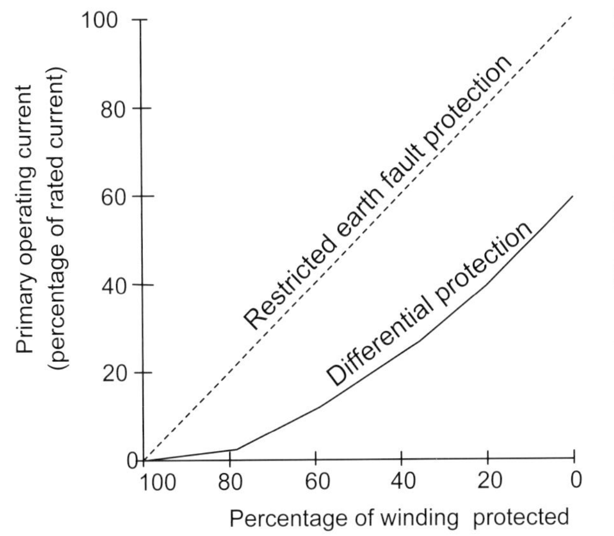

Figure 16.12: Amount of winding protected when transformer is resistance earthed and ratings of transformer and resistor are equal

Implementation of a combined differential/REF protection scheme is made easy if a numerical relay with software ratio/phase compensation is used. All compensation is made internally in the relay.

Where software ratio/phase correction is not available, either a summation transformer or auxiliary CTs can be used. The connections are shown in Figure 16.13 and Figure 16.14 respectively.

Care must be taken in calculating the settings, but the only significant disadvantage of the Combined Differential/REF scheme is that the REF element is likely to operate for heavy internal faults as well as the differential elements, thus making subsequent fault analysis somewhat confusing. However, the saving in CTs outweighs this disadvantage.

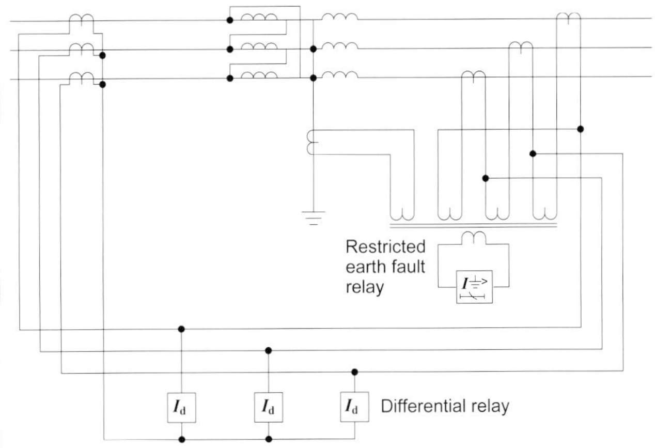

Figure 16.13: Combined differential and earth fault protection using summation current transformer

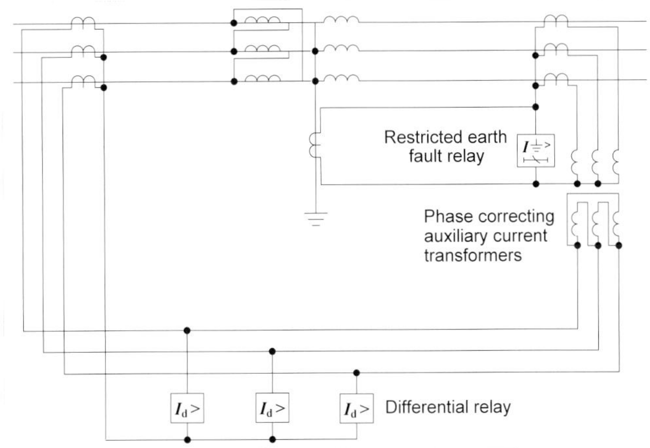

Figure 16.14: Combined differential and restricted earth fault protection using auxiliary CTs

16.10.1 Application When an Earthing Transformer is Connected Within the Protected Zone

A delta-connected winding cannot deliver any zero sequence current to an earth fault on the connected system, any current that does flow is in consequence of an earthed neutral elsewhere on the system and will have a 2-1-1 pattern of current distribution between phases. When the transformer in question represents a major power feed, the system may be earthed at that point by an earthing transformer or earthing reactor. They are frequently connected to the system, close to the main supply transformer and within the transformer protection zone. Zero sequence current that flows through the earthing transformer during system earth faults will flow through the line current transformers on this side, and, without an equivalent current in the balancing current transformers, will cause unwanted operation of the relays.

The problem can be overcome by subtracting the appropriate component of current from the main CT output. The earthing transformer neutral current is used for this purpose. As this represents three times the zero sequence current flowing, ratio correction is required. This can take the form of interposing

CT's of ratio 1/0.333, arranged to subtract their output from that of the line current transformers in each phase, as shown in Figure 16.15. The zero sequence component is cancelled, restoring balance to the differential system. Alternatively, numerical relays may use software to perform the subtraction, having calculated the zero sequence component internally.

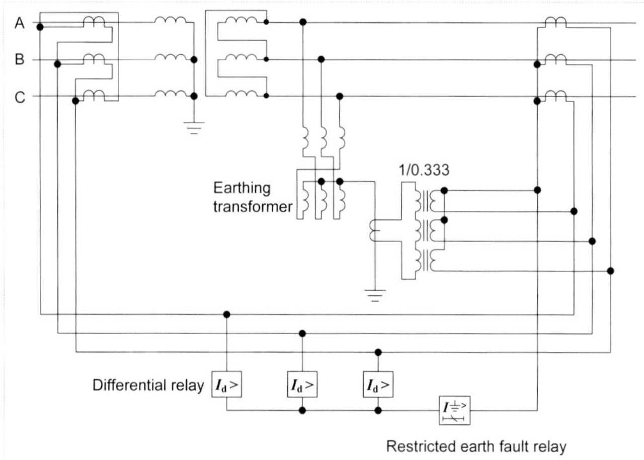

Figure 16.15: Differential protection with in-zone earthing transformer, with restricted earth fault relay

A high impedance relay element can be connected in the neutral lead between current transformers and differential relays to provide restricted earth fault protection to the winding.

As an alternative to the above scheme, the circulating current system can be completed via a three-phase group of interposing transformers that are provided with tertiary windings connected in delta. This winding effectively short-circuits the zero sequence component and thereby removes it from the balancing quantities in the relay circuit; see Figure 16.16.

Provided restricted earth fault protection is not required, the scheme shown in Figure 16.16 has the advantage of not requiring a current transformer, with its associated mounting and cabling requirements, in the neutral-earth conductor. The scheme can also be connected as shown in Figure 16.17 when restricted earth fault protection is needed.

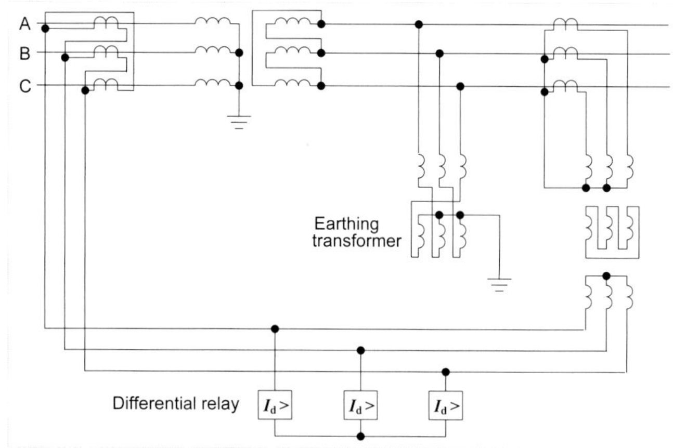

Figure 16.16: Differential protection with in-zone earthing transformer; no earth fault relay

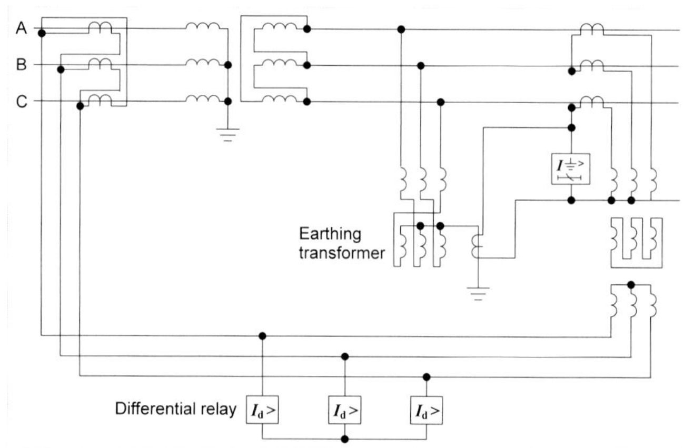

Figure 16.17: Differential protection with in-zone earthing transformer, with alternative arrangement of restricted earth fault relay

16.11 EARTHING TRANSFORMER PROTECTION

Earthing transformers not protected by other means can use the scheme shown in Figure 16.18. The delta-connected current transformers are connected to an overcurrent relay having three phase-fault elements. The normal action of the earthing transformer is to pass zero sequence current. The transformer equivalent current circulates in the delta formed by the CT secondaries without energising the relay. The latter may therefore be set to give fast and sensitive protection against faults in the earthing transformer itself.

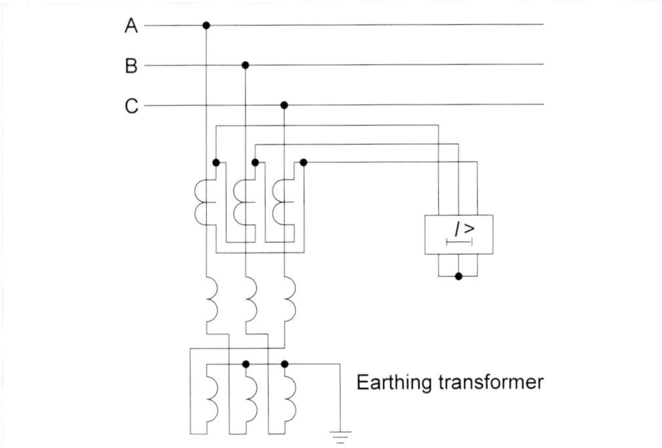

Figure 16.18: Earthing transformer protection

16.12 AUTOTRANSFORMER PROTECTION

Autotransformers are used to couple EHV power networks if the ratio of their voltages is moderate. An alternative to Differential Protection that can be applied to autotransformers is protection based on the application of Kirchhoff's law to a conducting network, namely that the sum of the currents flowing into all external connections to the network is zero.

A circulating current system is arranged between equal ratio current transformers in the two groups of line connections and the neutral end connections. If one neutral current transformer is used, this and all the line current transformers can be connected in parallel to a single element relay, providing a scheme responsive to earth faults only; see Figure 16.19(a).

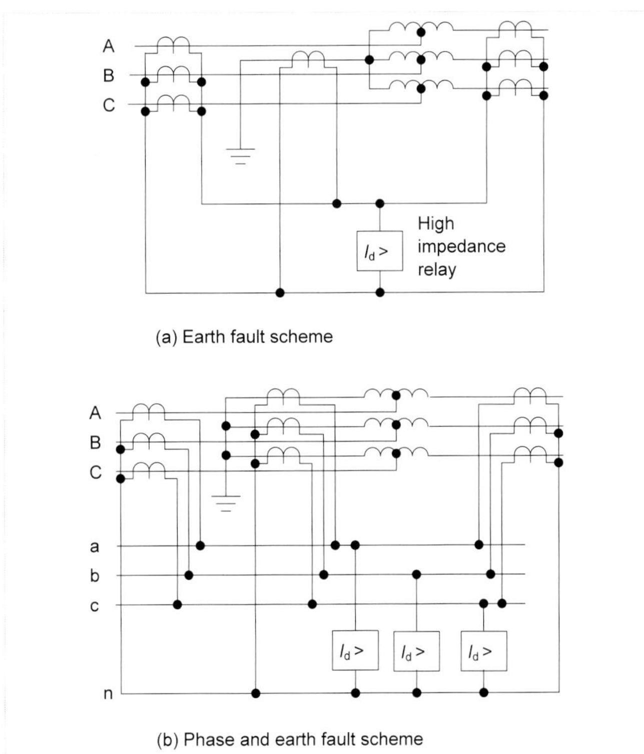

Figure 16.19: Protection of auto-transformer by high impedance differential relays

If current transformers are fitted in each phase at the neutral end of the windings and a three-element relay is used, a differential system can be provided, giving full protection against phase and earth faults; see Figure 16.19(b). This provides high-speed sensitive protection. It is unaffected by ratio changes on the transformer due to tap-changing and is immune to the effects of magnetising inrush current.

It does not respond to interturn faults, a deficiency that is serious in view of the high statistical risk quoted in Section 16.1. Such faults, unless otherwise cleared, will be left to develop into earth faults, by which time considerably more damage to the transformer will have occurred.

In addition, this scheme does not respond to any fault in a tertiary winding. Unloaded delta-connected tertiary windings are often not protected; alternatively, the delta winding can be earthed at one point through a current transformer that energises an instantaneous relay. This system should be separate from the main winding protection. If the tertiary winding earthing lead is connected to the main winding neutral above the neutral current transformer in an attempt to make a combined system, there may be 'blind spots' which the protection cannot cover.

16.13 OVERFLUXING PROTECTION

The effects of excessive flux density are described in Section 16.2.8. Overfluxing arises principally from the following system conditions:

- high system voltage
- low system frequency
- geomagnetic disturbances

The latter results in low frequency earth currents circulating through a transmission system.

Since momentary system disturbances can cause transient overfluxing that is not dangerous, time delayed tripping is required. The normal protection is an IDMT or definite time characteristic, initiated if a defined V/f threshold is exceeded. Often separate alarm and trip elements are provided. The alarm function would be definite time-delayed and the trip function would be an IDMT characteristic. A typical characteristic is shown in Figure 16.20.

Geomagnetic disturbances may result in overfluxing without the V/f threshold being exceeded. Some relays provide a 5th harmonic detection feature, which can be used to detect such a condition, as levels of this harmonic rise under overfluxing conditions.

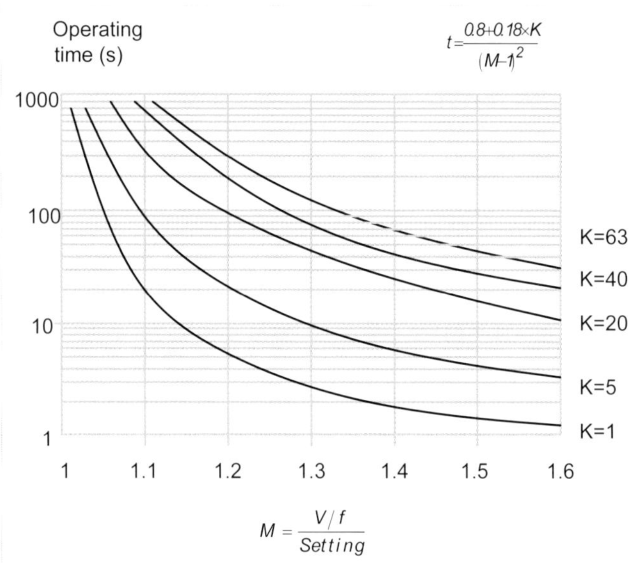

Figure 16.20: Typical IDMT characteristic for overfluxing protection

16.14 TANK-EARTH PROTECTION

This is also known as Howard protection. If the transformer tank is nominally insulated from earth (an insulation resistance of 10 ohms being sufficient) earth fault protection can be provided by connecting a relay to the secondary of a current transformer the primary of which is connected between the tank and earth. This scheme is similar to the frame-earth fault busbar protection described in Chapter 15.

16.15 OIL AND GAS DEVICES

All faults below oil in an oil-immersed transformer result in localised heating and breakdown of the oil; some degree of arcing will always take place in a winding fault and the resulting decomposition of the oil will release gases. When the fault is of a very minor type, such as a hot joint, gas is released slowly, but a major fault involving severe arcing causes a very rapid release of large volumes of gas as well as oil vapour. The action is so violent that the gas and vapour do not have time to escape but instead build up pressure and bodily displace the oil.

When such faults occur in transformers having oil conservators, the fault causes a blast of oil to pass up the relief pipe to the conservator. A Buchholz relay is used to protect against such conditions. Devices responding to abnormally high oil pressure or rate-of-rise of oil pressure are also available and may be used in conjunction with a Buchholz relay.

16.15.1 Oil Pressure Relief Devices

The simplest form of pressure relief device is the widely used 'frangible disc' that is normally located at the end of an oil relief pipe protruding from the top of the transformer tank.

The surge of oil caused by a serious fault bursts the disc, so allowing the oil to discharge rapidly. Relieving and limiting the pressure rise avoids explosive rupture of the tank and consequent fire risk. Outdoor oil-immersed transformers are usually mounted in a catchment pit to collect and contain spilt oil (from whatever cause), thereby minimising the possibility of pollution.

A drawback of the frangible disc is that the oil remaining in the tank is left exposed to the atmosphere after rupture. This is avoided in a more effective device, the sudden pressure relief valve, which opens to allow discharge of oil if the pressure exceeds a set level, but closes automatically as soon as the internal pressure falls below this level. If the abnormal pressure is relatively high, the valve can operate within a few milliseconds, and provide fast tripping when suitable contacts are fitted.

The device is commonly fitted to power transformers rated at 2MVA or higher, but may be applied to distribution transformers rated as low as 200kVA, particularly those in hazardous areas.

16.15.2 Sudden Pressure Rise Relay

This device detects rapid rise of pressure rather than absolute pressure and thereby can respond even quicker than the pressure relief valve to sudden abnormally high pressures. Sensitivities as low as 0.07bar/s are attainable, but when fitted to forced-cooled transformers the operating speed of the device may have to be slowed deliberately to avoid spurious tripping during circulation pump starts. Alternatively, sudden pressure rise relays may have their output supervised by instantaneous high-set overcurrent elements.

16.15.3 Buchholz Protection

Buchholz protection is normally provided on all transformers fitted with a conservator. The Buchholz relay is contained in a cast housing which is connected in the pipe to the conservator, as in Figure 16.21.

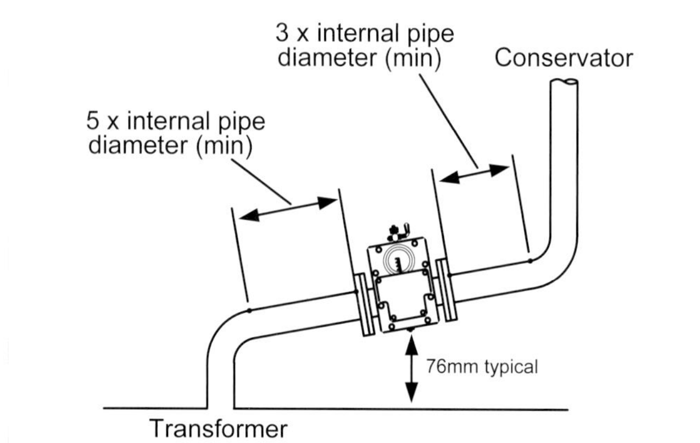

Figure 16.21: Buchholz relay mounting arrangement

A typical Buchholz relay will have two sets of contacts. One is arranged to operate for slow accumulations of gas, the other for bulk displacement of oil in the event of a heavy internal fault. An alarm is generated for the former, but the latter is usually direct-wired to the CB trip relay.

The device will therefore give an alarm for the following fault conditions, all of which are of a low order of urgency.

- hot spots on the core due to short circuit of lamination insulation
- core bolt insulation failure
- faulty joints
- interturn faults or other winding faults involving only lower power infeeds
- loss of oil due to leakage

When a major winding fault occurs, this causes a surge of oil, which displaces the lower float and thus causes isolation of the transformer. This action will take place for:

- all severe winding faults, either to earth or interphase
- loss of oil if allowed to continue to a dangerous degree

An inspection window is usually provided on either side of the gas collection space. Visible white or yellow gas indicates that insulation has been burnt, while black or grey gas indicates the presence of, dissociated oil. In these cases the gas will probably be inflammable, whereas released air will not. A vent valve is provided on the top of the housing for the gas to be released or collected for analysis. Transformers with forced oil circulation may experience oil flow to/from the conservator on starting/stopping of the pumps. The Buchholz relay must not operate in this circumstance.

Cleaning operations may cause aeration of the oil. Under such conditions, tripping of the transformer due to Buchholz operation should be inhibited for a suitable period.

Because of its universal response to faults within the transformer, some of which are difficult to detect by other means, the Buchholz relay is invaluable, whether regarded as a main protection or as a supplement to other protection schemes. Tests carried out by striking a high voltage arc in a transformer tank filled with oil, have shown that operation times of 0.05s-0.1s are possible. Electrical protection is generally used as well, either to obtain faster operation for heavy faults, or because Buchholz relays have to be prevented from tripping during oil maintenance periods. Conservators are fitted to oil-cooled transformers above 1000kVA rating, except those to North American design practice that use a different technique.

16.16 TRANSFORMER-FEEDER PROTECTION

A transformer-feeder comprises a transformer directly connected to a transmission circuit without the intervention of switchgear. Examples are shown in Figure 16.22.

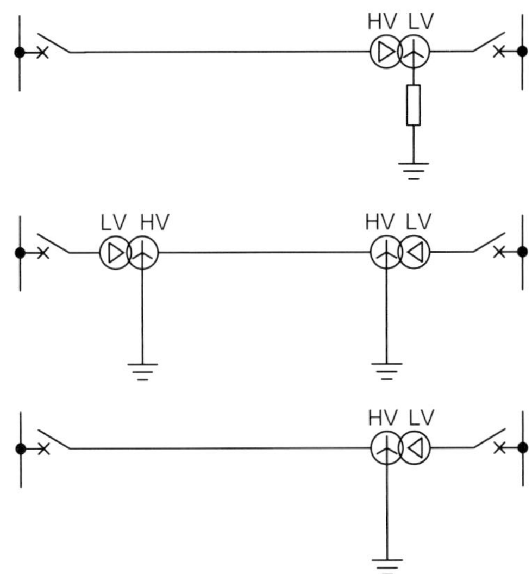

Figure *16.22*: Typical transformer-feeder circuits.

The saving in switchgear so achieved is offset by increased complication in the necessary protection. The primary requirement is intertripping, since the feeder protection remote from the transformer will not respond to the low current fault conditions that can be detected by restricted earth fault and Buchholz protections.

Either unrestricted or restricted protection can be applied; moreover, the transformer-feeder can be protected as a single zone or be provided with separate protections for the feeder and the transformer. In the latter case, the separate protections can both be unit type systems. An adequate alternative is the combination of unit transformer protection with an unrestricted system of feeder protection, plus an intertripping feature.

16.16.1 Non-Unit Schemes

The following sections describe how non-unit schemes are applied to protect transformer-feeders against various types of fault.

16.16.1.1 Feeder phase and earth faults

High-speed protection against phase and earth faults can be provided by distance relays located at the end of the feeder remote from the transformer. The transformer constitutes an appreciable lumped impedance. It is therefore possible to set a distance relay zone to cover the whole feeder and reach part way into the transformer impedance. With a normal tolerance on setting thus allowed for, it is possible for fast Zone 1 protection to cover the whole of the feeder with certainty without risk of over-reaching to a fault on the low voltage side.

Although the distance zone is described as being set 'half way into the transformer', it must not be thought that half the transformer winding will be protected. The effects of auto-transformer action and variations in the effective impedance of the winding with fault position prevent this, making the amount of winding beyond the terminals which is protected very small. The value of the system is confined to the feeder, which, as stated above, receives high-speed protection throughout.

16.16.1.2 Feeder phase faults

A distance scheme is not, for all practical purposes, affected by varying fault levels on the high voltage busbars and is therefore the best scheme to apply if the fault level may vary widely. In cases where the fault level is reasonably constant, similar protection can be obtained using high set instantaneous overcurrent relays. These should have a low transient over-reach (t), defined as:

$$\frac{I_S - I_F}{I_F} \times 100\%$$

where:

I_s = setting current

I_F = steady state r.m.s value of the fault current, which when fully offset, just operates the relay.

The instantaneous overcurrent relays must be set without risk of them operating for faults on the remote side of the transformer.

Referring to Figure 16.23, the required setting to ensure that the relay will not operate for a fully offset fault I_{F2} is given by:

$$I_s = 1.2\left(1+t\right) I_{F2}$$

Where I_{F2} is the fault current under maximum source conditions, that is, when Z_S is minimum, and the factor of 1.2 covers possible errors in the system impedance details used for calculation of I_{F2}, together with relay and CT errors.

As it is desirable for the instantaneous overcurrent protection to clear all phase faults anywhere within the feeder under varying system operating conditions, it is necessary to have a relay setting less than I_{F1} to ensure fast and reliable operation.

Let the setting ratio resulting from setting I_s be:

$$r = \frac{I_s}{I_{F1}}$$

Therefore,

$$r\, I_{F1} = 1.2\left(1+t\right) I_{F2}$$

Hence,

$$r = 1.2\left(1+t\right)\frac{Z_S + Z_L}{Z_S + Z_L + Z_T}$$

$$= 1.2\left(1+t\right)\frac{Z_S + Z_L}{\left(1+x\right)\left(Z_S + Z_L\right)}$$

$$= \frac{1.2\left(1+t\right)}{1+x}$$

where:

$$x = \frac{Z_T}{Z_S + Z_L}$$

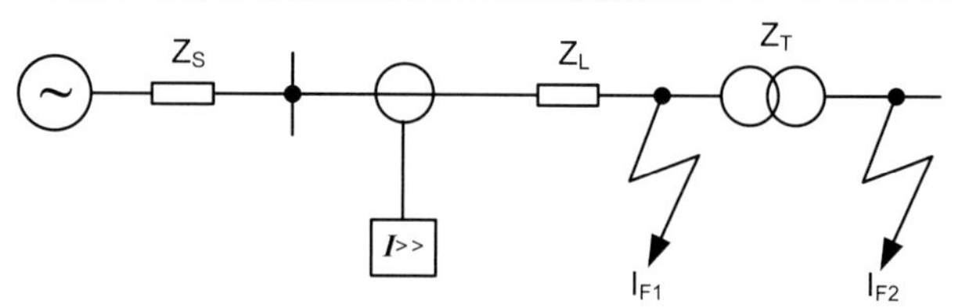

		Setting ratio $r = \frac{I_S}{I_{F1}}$			
	Transient over-reach (%) ►	5	25	50	100
$x = \frac{Z_T}{Z_S+Z_L}$	0.25	1.01	1.20	1.44	1.92
	0.5	0.84	1.00	1.20	1.60
	1.0	0.63	0.75	0.90	1.20
	2.0	0.42	0.50	0.60	0.80
	4.0	0.25	0.30	0.36	0.48
	8.0	0.14	0.17	0.20	0.27

I_S = Relay setting = $1.2(1+t)\ I_{F2}$

t = Transient over-reach (p.u.)

Figure 16.23: Over-reach considerations in the application of transformer-feeder protection

It can be seen that for a given transformer size, the most sensitive protection for the line will be obtained by using relays with the lowest transient overreach. It should be noted that where r is greater than 1, the protection will not cover the whole line. Also, any increase in source impedance above the minimum value will increase the effective setting ratios above those shown. The instantaneous protection is usually applied with a time delayed overcurrent element having a lower current setting. In this way, instantaneous protection is provided for the feeder, with the time-delayed element covering faults on the transformer.

When the power can flow in the transformer-feeder in either

direction, overcurrent relays will be required at both ends. In the case of parallel transformer-feeders, it is essential that the overcurrent relays on the low voltage side be directional, operating only for fault current fed into the transformer-feeder, as described in Section 9.14.3.

16.16.1.3 Earth faults

Instantaneous restricted earth fault protection is normally provided. When the high voltage winding is delta connected, a relay in the residual circuit of the line current transformers gives earth fault protection which is fundamentally limited to the feeder and the associated delta-connected transformer winding. The latter is unable to transmit any zero sequence current to a through earth fault.

When the feeder is associated with an earthed star-connected winding, normal restricted earth fault protection as described in Section 16.7 is not applicable because of the remoteness of the transformer neutral.

Restricted protection can be applied using a directional earth fault relay. A simple sensitive and high-speed directional element can be used, but attention must be paid to the transient stability of the element. Alternatively, a directional IDMT relay may be used, the time multiplier being set low. The slight inverse time delay in operation will ensure that unwanted transient operation is avoided.

When the supply source is on the high voltage star side, an alternative scheme that does not require a voltage transformer can be used. The scheme is shown in Figure 16.24. For the circuit breaker to trip, both relays *A* and *B* must operate, which will occur for earth faults on the feeder or transformer winding.

External earth faults cause the transformer to deliver zero sequence current only, which will circulate in the closed delta connection of the secondary windings of the three auxiliary current transformers. No output is available to relay *B*. Through phase faults will operate relay *B*, but not the residual relay *A*. Relay *B* must have a setting above the maximum load. As the earthing of the neutral at a receiving point is likely to be solid and the earth fault current will therefore be comparable with the phase fault current, high settings are not a serious limitation.

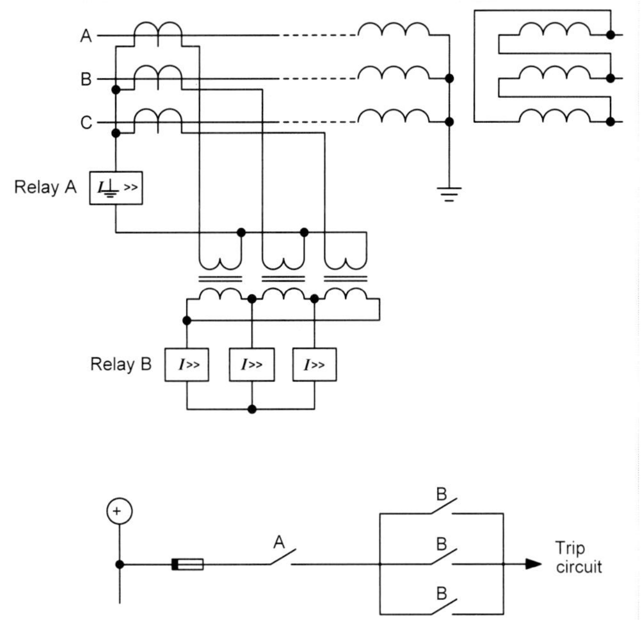

Figure 16.24: Instantaneous protection of transformer-feeder

Earth fault protection of the low voltage winding will be provided by a restricted earth fault system using either three or four current transformers, according to whether the winding is delta or star-connected, as described in Section 16.7.

16.16.1.4 In-zone capacitance

The feeder portion of the transformer-feeder will have an appreciable capacitance between each conductor and earth. During an external earth fault the neutral will be displaced, and the resulting zero sequence component of voltage will produce a corresponding component of zero sequence capacitance current. In the limiting case of full neutral displacement, this zero sequence current will be equal in value to the normal positive sequence current.

The resulting residual current is equal to three times the zero sequence current and hence to three times the normal line charging current. The value of this component of in-zone current should be considered when establishing the effective setting of earth fault relays.

16.16.2 Unit Schemes

The basic differences between the requirements of feeder and transformer protections lie in the limitation imposed on the transfer of earth fault current by the transformer and the need for high sensitivity in the transformer protection, suggesting that the two components of a transformer-feeder should be protected separately. This involves mounting current transformers adjacent to, or on, the high voltage terminals of the transformer. Separate current transformers are desirable for the feeder and transformer protections so that these can be arranged in two separate overlapping zones. The use of common current transformers is possible, but may involve the use of auxiliary current transformers, or special winding and connection arrangements of the relays. Intertripping of the remote circuit breaker from the transformer protection will be necessary, but this can be done using the communication facilities of the feeder protection relays.

Although technically superior, the use of separate protection systems is seldom justifiable when compared with an overall system or a combination of non-unit feeder protection and a unit transformer system.

An overall unit system must take into account the fact that zero sequence current on one side of a transformer may not be reproduced in any form on the other side. This represents little difficulty to a modern numerical relay using software phase/zero sequence compensation and digital communications to transmit full information on the phase and earth currents from one relay to the other. However, it does represent a more difficult problem for relays using older technology. The line current transformers can be connected to a summation transformer with unequal taps, as shown in Figure 16.25(a). This arrangement produces an output for phase faults and also some response for *A* and *B* phase-earth faults. However, the resulting settings will be similar to those for phase faults and no protection will be given for *C* phase-earth faults. An alternative technique is shown in Figure 16.25(b).

The *B* phase is taken through a separate winding on another transformer or relay electromagnet, to provide another balancing system. The two transformers are interconnected with their counterparts at the other end of the feeder-transformer by four pilot wires. Operation with three pilot cores is possible but four are preferable, involving little increase in pilot cost.

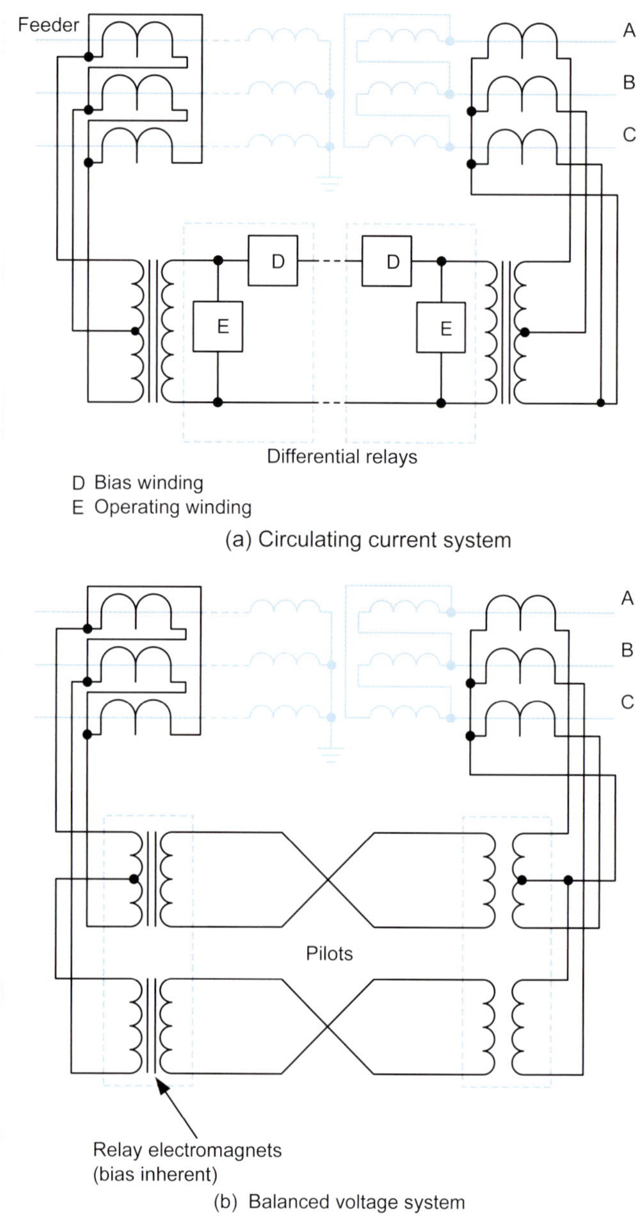

Figure 16.25: Methods of protection for transformer-feeders using electromechanical static technology

16.17 INTERTRIPPING

To ensure that both the high and low voltage circuit breakers operate for faults within the transformer and feeder, it is necessary to operate both circuit breakers from protection normally associated with one. The technique for doing this is known as intertripping.

The necessity for intertripping on transformer-feeders arises from the fact that certain types of fault produce insufficient current to operate the protection associated with one of the circuit breakers. These faults are:

- faults in the transformer that operate the Buchholz relay and trip the local low voltage circuit breaker, while failing to produce enough fault current to operate the protection associated with the remote high voltage circuit breaker
- earth faults on the star winding of the transformer, which, because of the position of the fault in the winding, again produce insufficient current for relay operation at the remote circuit breaker
- earth faults on the feeder or high voltage delta-connected winding which trip the high voltage circuit breaker only, leaving the transformer energised form the low voltage side and with two high voltage phases at near line-to-line voltage above earth. Intermittent arcing may follow and there is a possibility of transient overvoltage occurring and causing a further breakdown of insulation

Several methods are available for intertripping; these are discussed in Chapter 8.

16.17.1 Neutral Displacement

An alternative to intertripping is to detect the condition by measuring the residual voltage on the feeder. An earth fault occurring on the feeder connected to an unearthed transformer winding should be cleared by the feeder circuit, but if there is also a source of supply on the secondary side of the transformer, the feeder may be still live. The feeder will then be a local unearthed system, and, if the earth fault continues in an arcing condition, dangerous overvoltages may occur.

A voltage relay is energised from the broken-delta connected secondary winding of a voltage transformer on the high voltage line, and receives an input proportional to the zero sequence voltage of the line, that is, to any displacement of the neutral point; see Figure 16.26.

The relay normally receives zero voltage, but, in the presence of an earth fault, the broken-delta voltage will rise to three times the phase voltage. Earth faults elsewhere in the system may also result in displacement of the neutral and hence discrimination is achieved using definite or inverse time characteristics.

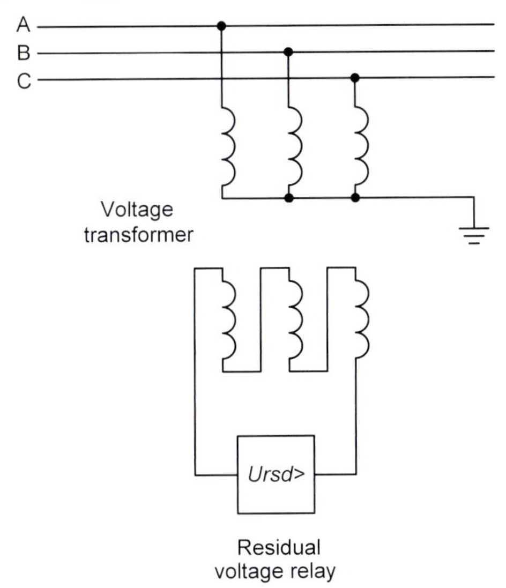

Figure 16.26: Neutral displacement detection using voltage transformer.

16.18 CONDITION MONITORING OF TRANSFORMERS

It is possible to provide transformers with measuring devices to detect early signs of degradation in various components and provide warning to the operator to avoid a lengthy and expensive outage due to failure. The technique, which can be applied to other plant as well as transformers, is called *condition monitoring*, as the intent is to provide the operator with regular information on the condition of the transformer. By reviewing the trends in the information provided, the operator can make a better judgement as to the frequency of maintenance, and detect early signs of deterioration that, if ignored, would lead to an internal fault occurring. Such techniques are an enhancement to, but are not a replacement for, the protection applied to a transformer.

The extent to which condition monitoring is applied to transformers on a system will depend on many factors, amongst which will be the policy of the asset owner, the suitability of the design (existing transformers may require modifications involving a period out of service – this may be costly and not justified), the importance of the asset to system operation, and the general record of reliability. Therefore, it should not be expected that all transformers would be, or need to be, so fitted.

A typical condition monitoring system for an oil-immersed transformer is capable of monitoring the condition of various transformer components as shown in Table 16.4. There can be some overlap with the measurements available from a digital/numerical relay. By the use of software to store and perform trend analysis of the measured data, the operator can be presented with information on the state of health of the transformer, and alarms raised when measured values exceed appropriate limits. This will normally provide the operator with early warning of degradation within one or more components of the transformer, enabling maintenance to be scheduled to correct the problem prior to failure occurring. The maintenance can obviously be planned to suit system conditions, provided the rate of degradation is not excessive.

Monitored Equipment	Measured Quantity	Health Information
Bushings	Voltage	Insulation quality
	Partial discharge measurement (wideband voltage)	
	Load current	Loading
		Permissible overload rating
		Hot-spot temperature
	Oil pressure	Insulation quality
Tank	Oil temperature	Hot-spot temperature
		Permissible overload rating
	Gas-in-oil content	Oil quality
		Winding insulation condition
	Moisture-in-oil content	Oil quality
	Buchholz gas content	Winding insulation condition
Tap changer	Position	Frequency of use of each tap position
	Drive power consumption	OLTC health
	Total switched load current	OLTC contact wear
	OLTC oil temperature	OLTC health

Monitored Equipment	Measured Quantity	Health Information
Coolers	Oil temperature difference	Cooler efficiency
	Cooling air temperature	
	Ambient temperature	
	Pump status	Cooling plant health
Conservator	Oil level	Tank integrity

Table 16.4: Condition monitoring for transformers

As asset owners become more conscious of the costs of an unplanned outage, and electric supply networks are utilised closer to capacity for long periods of time, the usefulness of this technique can be expected to grow. See Section 16.20 for further information on this topic.

16.19 EXAMPLES OF TRANSFORMER PROTECTION

This section provides three examples of the application of modern relays to transformer protection. The latest MiCOM P640 series relay provides advanced software to simplify the calculations, so an earlier Alstom type KBCH relay is used to show the complexity of the required calculations.

16.19.1 Provision of Vector Group Compensation and Zero-Sequence Filtering

Figure 16.27 shows a delta-star transformer to be protected using a unit protection scheme. With a main winding connection of Dyn11, suitable choices of primary and secondary CT winding arrangements, and software phase compensation are to be made. With the KBCH relay, phase compensation is selected by the user in the form of software-implemented ICTs.

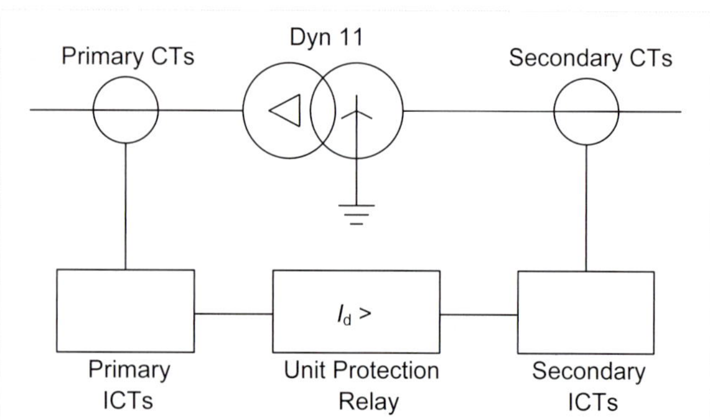

Figure 16.27: Transformer zero sequence filtering example

With the Dyn11 connection, the secondary voltages and currents are displaced by +30° from the primary. Therefore, the combination of primary, secondary and phase correction must provide a phase shift of –30° of the secondary quantities relative to the primary.

For simplicity, the CTs on the primary and secondary windings of the transformer are connected in star. The required phase shift can be achieved either by use of ICT connections on the primary side having a phase shift of +30° or on the secondary side having a phase shift of –30°. There is a wide combination of primary and secondary ICT winding arrangements that can provide this, such as *Yd10* (+60°) on the primary and *Yd3* (-90°) on the secondary. Another possibility is *Yd11* (+30°) on the primary and *Yy0* (0°) on the secondary. It is usual to choose the simplest arrangements possible, and therefore the latter of the above two possibilities might be selected.

However, the distribution of current in the primary and secondary windings of the transformer due to an external earth fault on the secondary side of the transformer must now be considered. The transformer has an earth connection on the secondary winding, so it can deliver zero sequence current to the fault. Use of star connected main CTs and *Yy0* connected ICTs provides a path for the zero sequence current to reach the protection relay. On the primary side of the transformer, the delta connected main primary winding causes zero-sequence current to circulate round the delta and hence will not be seen by the primary side main CTs. The protection relay will therefore not see any zero-sequence current on the primary side, and hence detects the secondary side zero sequence current incorrectly as an in-zone fault.

The solution is to provide the ICTs on the secondary side of the transformer with a delta winding, so that the zero-sequence current circulates round the delta and is not seen by the relay. Therefore, a rule can be developed that a transformer winding with a connection to earth must have a delta-connected main or ICT for unit protection to operate correctly.

Selection of *Yy0* connection for the primary side ICTs and *Yd1* (–30°) for the secondary side ICTs provides the required phase shift and the zero-sequence trap on the secondary side.

Modern numerical MiCOM relays employ a setting wizard, needing only vector group and zero sequence data to be entered. The relay then automatically adapts itself to suit the application.

16.19.2 Unit Protection of a Delta-Star Transformer

Figure 16.28 shows a delta-star transformer to which unit protection is to be applied, including restricted earth fault protection to the star winding.

Referring to the figure, the ICTs have already been correctly selected, and are conveniently applied in software. It therefore remains to calculate suitable ratio compensation (it is assumed that the transformer has no taps), transformer differential protection settings and restricted earth fault settings.

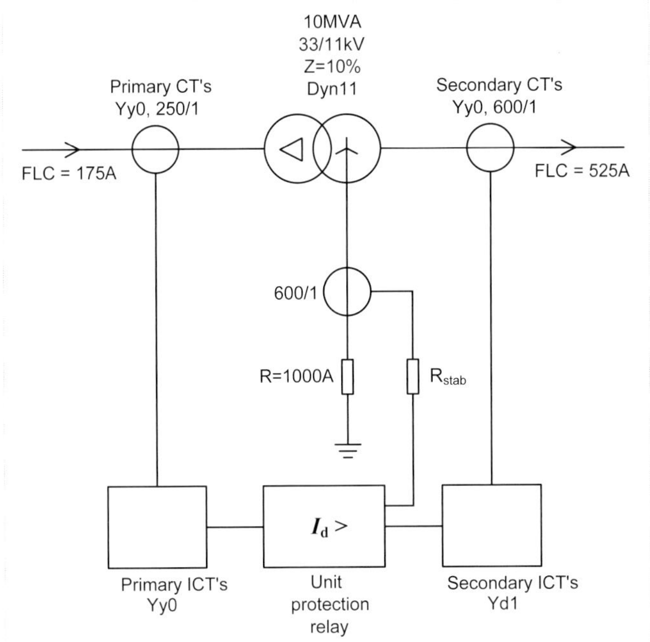

Figure 16.28: Transformer unit protection example

16.19.2.1 Ratio compensation

Transformer HV full load current on secondary of main CTs is:

$$\frac{175}{250} = 0.7$$

Ratio compensation $= \dfrac{1}{0.7} = 1.428$

Select nearest value = 1.43

LV secondary current $= \dfrac{525}{600} = 0.875$

Ratio compensation $= \dfrac{1}{0.875} = 1.14$

16.19.2.2 Transformer unit protection settings

A current setting of 20% of the rated relay current is recommended. This equates to 35A primary current. The KBCH relay has a dual slope bias characteristic with fixed bias slope settings of 20% up to rated current and 80% above that level. The corresponding characteristic is shown in Figure 16.29.

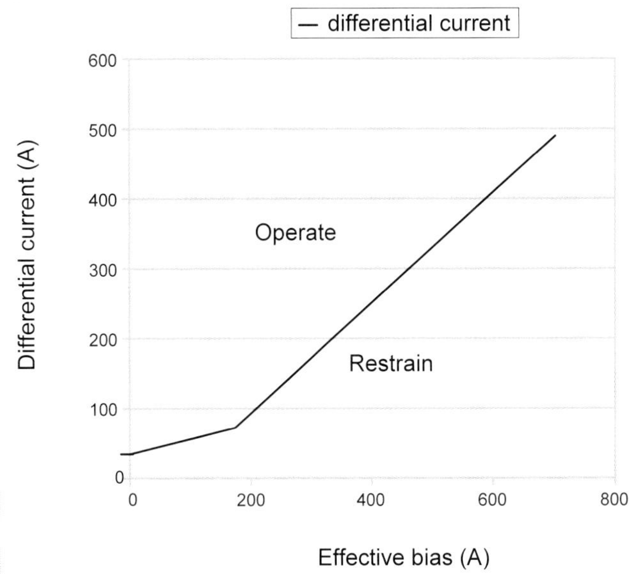

Figure 16.29: Transformer unit protection characteristic

16.19.2.3 Restricted earth fault protection

The KBCH relay implements high-impedance Restricted Earth Fault (REF) protection. Operation is required for a primary earth fault current of 25% rated earth fault current (i.e. 250A). The prime task in calculating settings is to calculate the value of the stabilising resistor R_{stab} and stability factor K.

A stabilising resistor is required to ensure through fault stability when one of the secondary CTs saturates while the others do not. The requirements can be expressed as:

$$V_S = I_S R_{stab}$$

And

$$V_S > K I_f \left(R_{ct} + 2R_l\right)$$

where:

V_S = stability voltage setting

V_K = CT knee point voltage

K = relay stability factor

I_S = relay current setting

R_{CT} = CT winding resistance

R_l = CT lead resistance

R_{stab} = stabilising resistor.

For this example:

V_K = 97V

R_{CT} = 3.7Ω

R_l = 0.057Ω

For the relay used, the various factors are related by the graph of Figure 16.30.

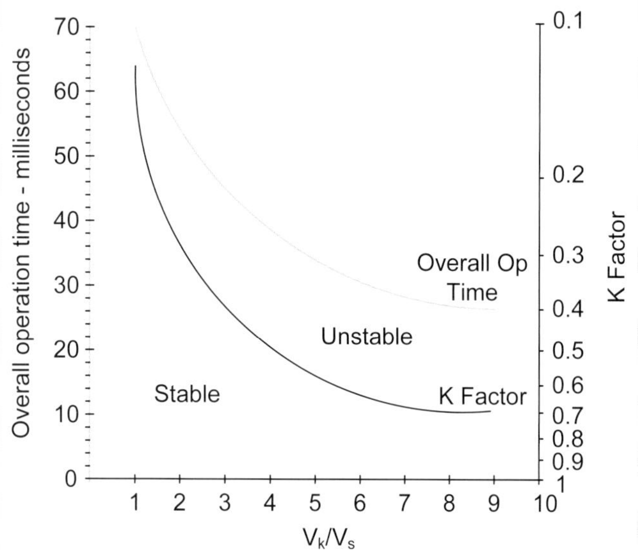

Figure 16.30: REF operating characteristic for KBCH relay

Starting with the desired operating time, the $\frac{V_K}{V_S}$ ratio and *K* factor can be found.

An operating of 40ms (2 cycles at 50Hz) is usually acceptable, and hence, from Figure 16.30,

$$\frac{V_K}{V_S} = 4, K = 0.5$$

The maximum earth fault current is limited by the earthing resistor to 1000A (primary). The maximum phase fault current can be estimated by assuming the source impedance to be zero, so it is limited only by transformer impedance to 5250A, or 10A secondary after taking account of the ratio compensation. Hence the stability voltage can be calculated as

$$V_S = 0.5 \times 10(3.7 + 2 \times 0.057) = 19.07V$$

Hence,

Calculated $V_K = 4 \times 19.07 = 76.28V$

However,

Actual $V_K = 91V$ and $V_K/V_S = 4.77$

Thus from Figure 16.30, with $K = 0.5$, the protection is unstable.

By adopting an iterative procedure for values of V_K/V_S and K, a final acceptable result of $\frac{V_K}{V_S} = 4.55, K = 0.6$ is obtained. This results in an operating time faster than 40ms.

The required earth fault setting current I_{op} is 250A. The chosen E/F CT has an exciting current I_e of 1%, and hence using the equation:

$$I_{op} = CTratio \times (I_S + nI_e)$$

where:

n = no of CTs in parallel (=4)

$I_s = 0.377$, use *0.38* nearest settable value.

The stabilising resistance R_{stab} can be calculated as 60.21Ω.

The relay can only withstand a maximum of 3kV peak under fault conditions. A check is required to see if this voltage is exceeded – if it is, a non-linear resistor, known as a Metrosil, must be connected across the relay and stabilising resistor. The peak voltage is estimated using the formula:

$$V_P = 2\sqrt{2V_K(V_f - V_K)}$$

where:

$$V_f = I_f(R_{CT} + 2R_l + R_{stab})$$

And

I_f = fault current in secondary of CT circuit and substituting values, $V_P = 544V$. Thus a Metrosil is not required.

16.19.3 Unit Protection for On-Load Tap Changing Transformer

The previous example deals with a transformer having no taps. In practice, most transformers have a range of taps to cater for different loading conditions. While most transformers have an off-load tap-changer, transformers used for voltage control in a network are fitted with an on-load tap-changer. The protection settings must then take the variation of tap-change position into account to avoid the possibility of spurious trips at extreme tap positions. For this example, the same transformer as in Section 16.19.2 will be used, but with an on-load tapping range of +5% to -15%. The tap-changer is located on the primary winding, while the tap-step usually does not matter.

16.19.3.1 Ratio correction

In accordance with Section 16.8.4 the mid-tap position is used to calculate the ratio correction factors. The mid tap position is –5%, and at this tap position:

Primary voltage to give rated secondary voltage:

$= 33 \times 0.95 = 31.35kV$ and Rated Primary Current = *184A*

Transformer HV full load current on secondary of main CTs is:

$$\frac{184}{250} = 0.737$$

$$\text{Ratio compensation} = \frac{1}{0.737} = 1.36$$

$$\text{LV secondary current} = \frac{525}{600} = 0.875$$

$$\text{Ratio compensation} = \frac{1}{0.875} = 1.14$$

Both of the above values can be set in the relay.

16.19.3.2 Bias slope setting

The on-load tapping range of +5% to -15% gives rise to a maximum excursion of ±10% from the -5% mid-tap position. As the differential scheme notionally balances at this mid-tap, this means that as an approximation, the maximum differential current that can flow when at top or bottom tap is 10% of the load (or fault current which may flow to an external fault). Those relays having an adjustable k1 bias slope setting should ensure that it is at least 10% higher than the percentage excursion.

16.20 TRANSFORMER ASSET MANAGEMENT

Due to the high capital cost of transformers, and the need for their in-service availability to be as high as possible to avoid constraining load flows demanded on the network, protection is no-longer the only concern. As cities expand, consumers' lifestyle expectations raise, and electric vehicle recharging loads become more prevalent, these combine to increase the demand on the network - and it is through the transformers, between the different voltage levels on the system, that the demand is traditionally supplied. This increases the focus on knowing the health of transformers, real-time, to be able to schedule condition-based maintenance. Maintenance or reconditioning at a time of the asset-owner's choosing is far more preferable than a forced unplanned outage due to failure.

This section provides an overview of techniques commonly available in modern numerical transformer protection relays, which can extend to asset management of the protected transformer.

16.20.1 Loss of life monitoring

Ageing of transformer insulation is a time-dependent function of temperature, moisture, and oxygen content. The moisture and oxygen contributions to insulation deterioration are minimised due to the preservation systems employed in the design of most modern transformers. Therefore, temperature is the key parameter in insulation ageing. Frequent excesses of overloading will shorten the life-expectancy of the transformer, due to the elevated winding temperatures.

Insulation deterioration is not uniform, and will be more pronounced at hot-spots within the transformer tank. Therefore, any asset management system intended to model the rate of deterioration and current estimated state of the insulation must do so based on simulated real-time hot spot temperature algorithms. These models may take ambient temperature, top-oil temperature, load current flowing, the status of oil pumps (pumping or not), and the status of radiator fans (forced cooling or not) as inputs.

The MiCOM P640 provides such a loss of life monitoring facility, according to the thermal model defined in IEEE Standard C57.91. The protection algorithm determines the current rate of losing life, and uses that to indicate the remaining years or hours until critical insulation health statuses are likely to be reached. Such criticalities will relate typically to known percentage degradations in the tensile strength of the insulation, degradation in the degree of polymerisation, and other life-loss factors. The asset owner can be alerted in advance that an outage will be required for reconditioning or rewinding, such that investment budgeting can be made years and months ahead of time.

16.20.2 Through-fault monitoring

Loss of life monitoring serves to track the deterioration caused by long term, repeated overloading. However, it is not the right technique to monitor short-term heavy fault currents which flow through the transformer, out to an external fault on the downstream power system (e.g. fault I_{F2} in Figure 16.23). Through faults are a major cause of transformer damage and failure, as they stress the insulation and mechanical integrity - such as the bracing of the windings.

A specific through-fault monitor is recommended to monitor currents which are due to external faults passing through, and so may range from 3.5 times up to tens of times the rated current of the transformer. The MiCOM P640 performs an I^2t calculation when the through current exceeds a user-set threshold, such that the heating effect of the square of the maximum phase current, and the duration of the fault event are calculated. Calculation results are added to cumulative values, and monitored so that utilities can schedule transformer maintenance or identify a need for system reinforcement.

Chapter 17
Generator and Generator Transformer Protection

17.1 INTRODUCTION

The core of an electric power system is the generation. With the exception of emerging fuel cell and solar-cell technology for power systems, the conversion of the fundamental energy into its electrical equivalent normally requires a 'prime mover' to develop mechanical power as an intermediate stage. The nature of this machine depends upon the source of energy and in turn this has some bearing on the design of the generator. Generators based on steam, gas, water or wind turbines, and reciprocating combustion engines are all in use. Electrical ratings extend from a few hundred kVA (or even less) for reciprocating engine and renewable energy sets, up to steam turbine sets exceeding 1200MVA.

Small and medium sized sets may be directly connected to a power distribution system. A larger set may be associated with an individual transformer, through which it is coupled to the EHV primary transmission system. Switchgear may or may not be provided between the generator and transformer. In some cases, operational and economic advantages can be attained by providing a generator circuit breaker in addition to a high voltage circuit breaker, but special demands will be placed on the generator circuit breaker for interruption of generator fault current waveforms that do not have an early zero crossing.

A unit transformer may be tapped off the interconnection between generator and transformer for the supply of power to auxiliary plant, as shown in Figure 17.1. The unit transformer could be of the order of 10% of the unit rating for a large fossil-fuelled steam set with additional flue-gas desulphurisation plant, but it may only be of the order of 1% of unit rating for a hydro set.

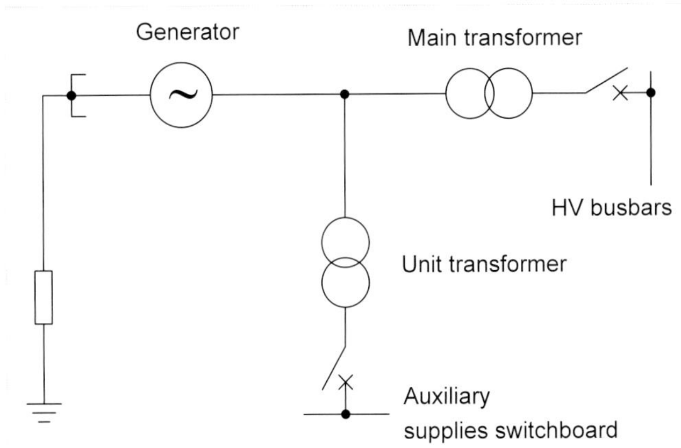

Figure 17.1: Generator-transformer unit

Industrial or commercial plants with a requirement for steam/hot water now often include generating plant utilising or producing steam to improve overall economics, as a Combined Heat and Power (CHP) scheme. The plant will typically have a connection to the public Utility distribution system, and such generation is referred to as 'embedded' generation. The generating plant may be capable of export of surplus power, or simply reduce the import of power from the supply Utility. This is shown in Figure 17.2.

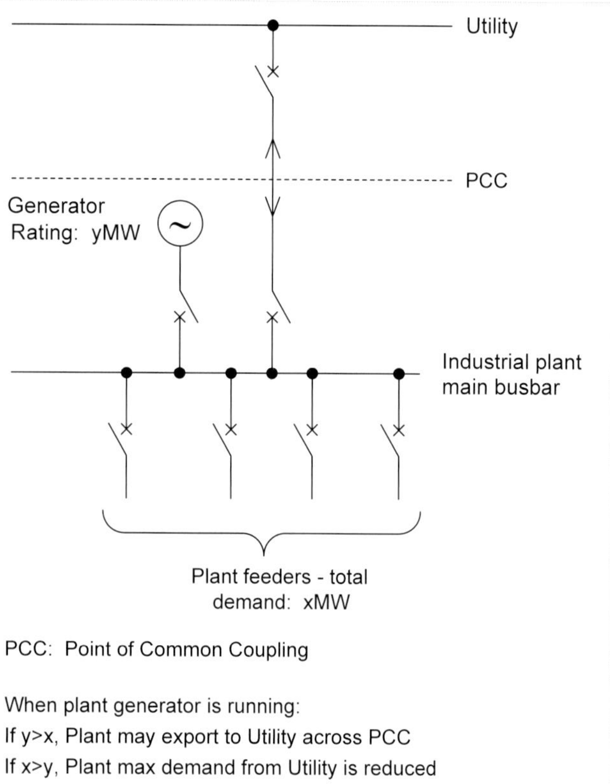

Figure 17.2: Embedded generation

A modern generating unit is a complex system comprising the generator stator winding, associated transformer and unit transformer (if present), the rotor with its field winding and excitation system, and the prime mover with its associated auxiliaries. Faults of many kinds can occur within this system for which diverse forms of electrical and mechanical protection are required. The amount of protection applied will be governed by economic considerations, taking into account the value of the machine, and the value of its output to the plant owner.

The following problems require consideration from the point of view of applying protection:

- stator electrical faults
- overload
- overvoltage
- unbalanced loading
- overfluxing
- inadvertent energisation
- rotor electrical faults
- loss of excitation
- loss of synchronism
- failure of prime mover
- lubrication oil failure
- overspeeding
- rotor distortion
- difference in expansion between rotating and stationary parts
- excessive vibration
- core lamination faults

17.2 GENERATOR EARTHING

The neutral point of a generator is usually earthed to facilitate protection of the stator winding and associated system. Earthing also prevents damaging transient overvoltages in the event of an arcing earth fault or ferroresonance.

For HV generators, impedance is usually inserted in the stator earthing connection to limit the magnitude of earth fault current. There is a wide variation in the earth fault current chosen, common values being:

- rated current
- 200A-400A (low impedance earthing)
- 10A-20A (high impedance earthing)

The main methods of impedance-earthing a generator are shown in Figure 17.3. Low values of earth fault current may limit the damage caused from a fault, but they simultaneously make detection of a fault towards the stator winding star point more difficult. Except for special applications, such as marine, LV generators are normally solidly earthed to comply with safety requirements. Where a step-up transformer is applied, the generator and the lower voltage winding of the transformer can be treated as an isolated system that is not influenced by the earthing requirements of the power system.

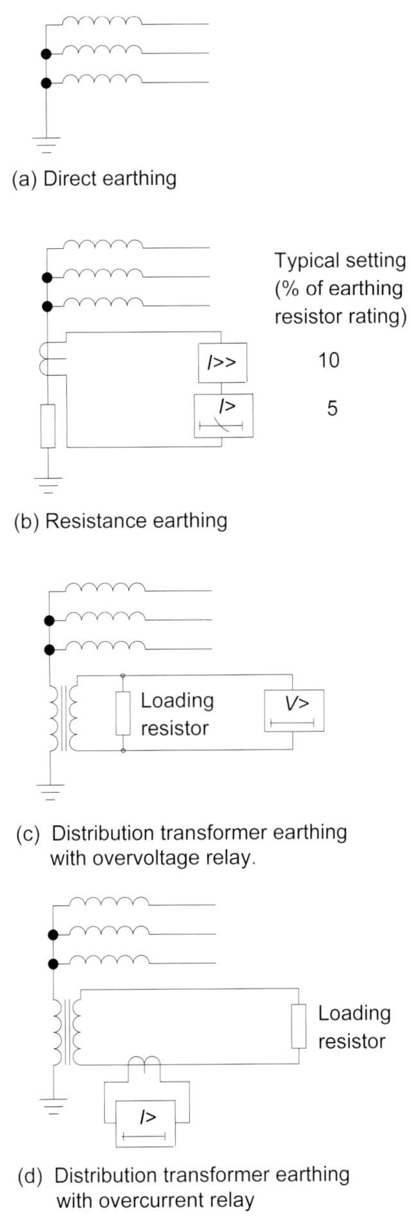

Figure 17.3: Methods of generator earthing

An earthing transformer or a series impedance can be used as the impedance. If an earthing transformer is used, the continuous rating is usually in the range 5-250kVA. The secondary winding is loaded with a resistor of a value which, when referred through the transformer ratio, will pass the chosen short-time earth-fault current. This is typically in the range of 5-20A. The resistor prevents the production of high transient overvoltages in the event of an arcing earth fault, which it does by discharging the bound charge in the circuit capacitance. For this reason, the resistive component of fault current should not be less than the residual capacitance current. This is the basis of the design, and in practice values of between 3-5 I_{co} are used.

It is important that the earthing transformer never becomes saturated; otherwise a very undesirable condition of ferroresonance may occur. The normal rise of the generated voltage above the rated value caused by a sudden loss of load or by field forcing must be considered, as well as flux doubling in the transformer due to the point-on-wave of voltage application. It is sufficient that the transformer be designed to have a primary winding knee-point e.m.f. equal to 1.3 times the generator rated line voltage.

17.3 STATOR WINDING FAULTS

Failure of the stator windings or connection insulation can result in severe damage to the windings and stator core. The extent of the damage will depend on the magnitude and duration of the fault current.

17.3.1 Earth Faults

The most probable mode of insulation failure is phase to earth. Use of an earthing impedance limits the earth fault current and hence stator damage.

An earth fault involving the stator core results in burning of the iron at the point of fault and welds laminations together. Replacement of the faulty conductor may not be a very serious matter (dependent on set rating/voltage/construction) but the damage to the core cannot be ignored, since the welding of laminations may result in local overheating. The damaged area can sometimes be repaired, but if severe damage has occurred, a partial core rebuild will be necessary. A flashover is more likely to occur in the end-winding region, where electrical stresses are highest. The resultant forces on the conductors would be very large and they may result in extensive damage, requiring the partial or total rewinding of the generator. Apart from burning the core, the greatest danger arising from failure to quickly deal with a fault is fire. A large portion of the insulating material is inflammable, and in the case of an air-cooled machine, the forced ventilation can quickly cause an arc flame to spread around the winding. Fire will not occur in a hydrogen-cooled machine, provided the stator system remains sealed. In any case, the length of an outage may be considerable, resulting in major financial impact from loss of generation revenue and/or import of additional energy.

17.3.2 Phase-Phase Faults

Phase-phase faults clear of earth are less common; they may occur on the end portion of stator coils or in the slots if the winding involves two coil sides in the same slot. In the latter case, the fault will involve earth in a very short time. Phase fault current is not limited by the method of earthing the neutral point.

17.3.3 Interturn Faults

Interturn faults are rare, but a significant fault-loop current can arise where such a fault does occur. Conventional generator protection systems would be blind to an interturn fault, but the extra cost and complication of providing detection of a purely interturn fault is not usually justified. In this case, an interturn fault must develop into an earth fault before it can be cleared. An exception may be where a machine has an abnormally complicated or multiple winding arrangement, where the probability of an interturn fault might be increased.

17.4 STATOR WINDING PROTECTION

To respond quickly to a phase fault with damaging heavy current, sensitive, high-speed differential protection is normally applied to generators rated in excess of 1MVA. For large generating units, fast fault clearance will also maintain stability of the main power system. The zone of differential protection can be extended to include an associated step-up transformer. For smaller generators, IDMT/instantaneous overcurrent protection is usually the only phase fault protection applied. Sections 17.5 to 17.8 detail the various methods that are available for stator winding protection.

17.5 DIFFERENTIAL PROTECTION OF DIRECT CONNECTED GENERATORS

The theory of circulating current differential protection is discussed fully in Section 10.4.

High-speed phase fault protection is provided, by use of the connections shown in Figure 17.4. This depicts the derivation of differential current through CT secondary circuit connections. This protection may also offer earth fault protection for some moderate impedance-earthed applications. Either biased differential or high impedance differential techniques can be applied. A subtle difference with modern, biased, numerical generator protection relays is that they usually derive the differential currents and biasing currents by algorithmic calculation, after measurement of the individual CT secondary currents. In such relay designs, there is full galvanic separation of the neutral-tail and terminal CT secondary circuits, as indicated in Figure 17.5(a). This is not the case for the application of high-impedance differential protection. This difference can impose some special relay design requirements to achieve stability for biased differential protection in some applications.

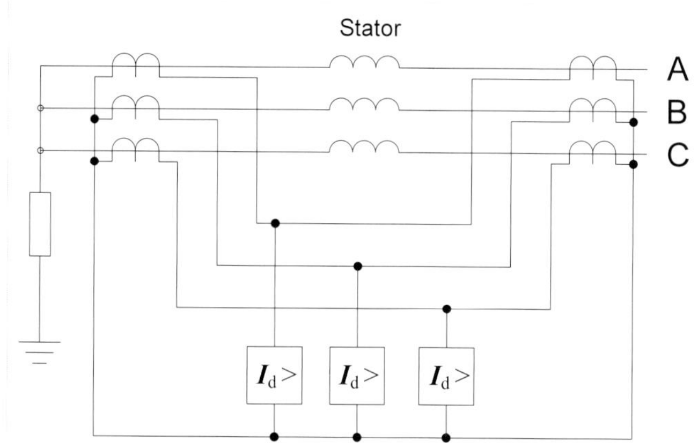

Figure 17.4: Stator differential protection

17.5.1 Biased Differential Protection

The relay connections for this form of protection are shown in Figure 17.5(a) and a typical bias characteristic is shown in Figure 17.5(b). The differential current threshold setting Is1 can be set as low as 5% of rated generator current, to provide protection for as much of the winding as possible. The bias slope break-point threshold setting Is2 would typically be set to a value above generator rated current, say 120%, to achieve external fault stability in the event of transient asymmetric CT saturation. Bias slope K2 setting would typically be set at 150%.

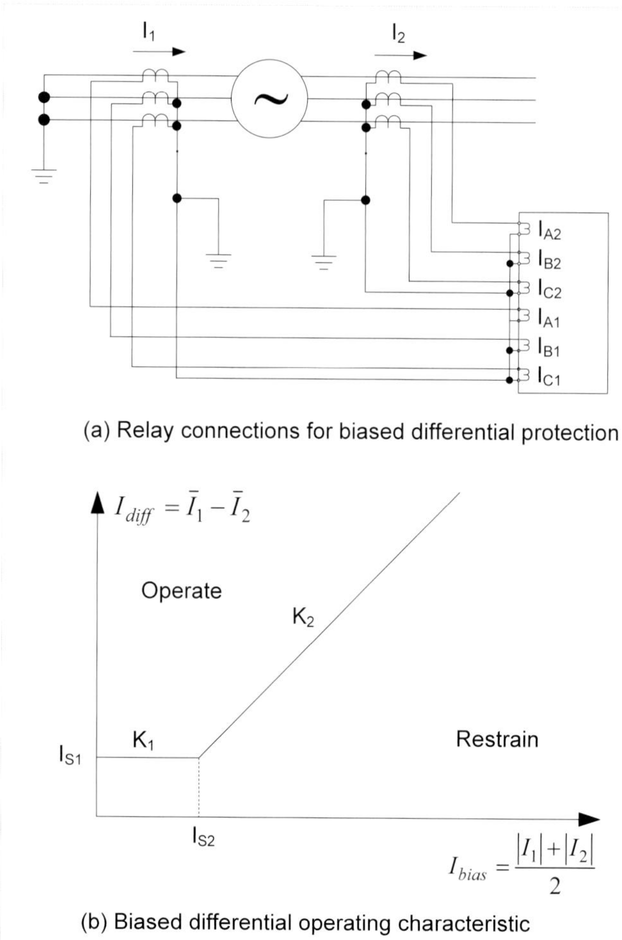

Figure 17.5: Typical generator biased differential protection

17.5.2 High Impedance Differential Protection

This differs from biased differential protection by the manner in which relay stability is achieved for external faults and by the fact that the differential current must be attained through the electrical connections of CT secondary circuits. If the impedance of each relay in Figure 17.4 is high, the event of one CT becoming saturated by the through fault current (leading to a relatively low CT impedance), will allow the current from the unsaturated CT to flow mainly through the saturated CT rather than through the relay. This provides the required protection stability where a tuned relay element is employed. In practice, external resistance is added to the relay circuit to provide the necessary high impedance. The principle of high-impedance protection application is illustrated in Figure 17.6, together with a summary of the calculations required to determine the value of external stabilising resistance.

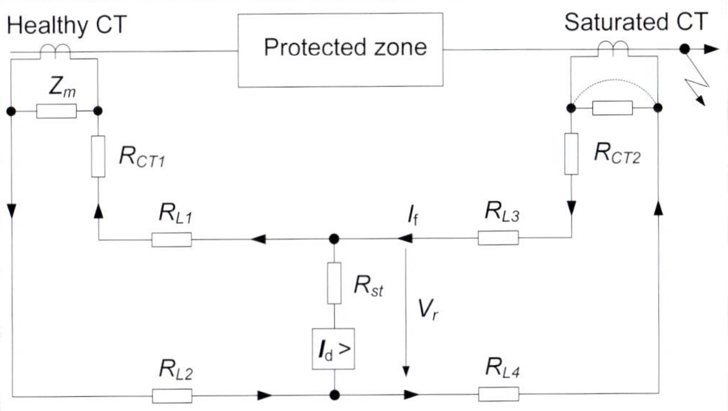

Voltage across relay circuit

$$V_r = I_f\left(R_{CT} + 2R_L\right)$$

and $V_s = KV_r$

where $1.0 < K \leq 1.5$

Stabilising resistor, R_{st}, limits spill current to $< I_s$ (relay setting)

$$R_{st} = \frac{V_s}{I_s} - R_R$$

where R_R= relay burden

Figure 17.6: Principle of high impedance differential protection

In some applications, protection may be required to limit voltages across the CT secondary circuits when the differential secondary current for an internal phase fault flows through the high impedance relay circuit(s), but this is not commonly a requirement for generator differential applications unless very high impedance relays are applied. Where necessary, shunt–connected, non-linear resistors, should be deployed, as shown in Figure 17.7.

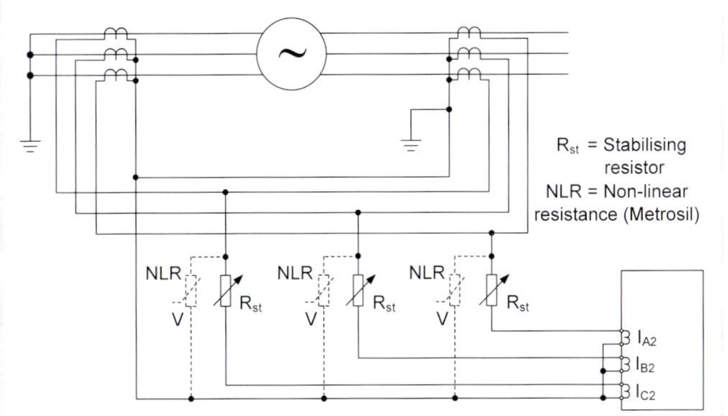

Figure 17.7: Relay connections for high impedance differential protection

To calculate the primary operating current, the following expression is used:

$$I_{op} = N \times \left(I_{s1} + nI_e\right)$$

where:

I_{op} = primary operating current

N = CT ratio

I_{s1} = relay setting

n = number of CTs in parallel with relay element

I_e = CT magnetising current at V_S

I_{s1} is typically set to 5% of generator rated secondary current.

It can be seen from the above that the calculations for the application of high impedance differential protection are more complex than for biased differential protection. However, the protection scheme is actually quite simple and it offers a high level of stability for through faults and external switching events. With the advent of multi-function numerical relays and with a desire to dispense with external components, high impedance differential protection is not as popular as biased differential protection in modern relaying practice.

17.5.3 CT Requirements

The CT requirements for differential protection will vary according to the relay used. Modern numerical relays may not require CTs specifically designed for differential protection to IEC 60044-1 class PX (or BS 3938 class X). However, requirements in respect of CT knee-point voltage will still have to be checked for the specific relays used. High impedance differential protection may be more onerous in this respect than biased differential protection. Many factors affect this, including the other protection functions fed by the CTs and the knee-point requirements of the particular relay concerned. Relay manufacturers are able to provide detailed guidance on this matter.

17.6 DIFFERENTIAL PROTECTION OF GENERATOR-TRANSFORMERS

A common connection arrangement for large generators is to operate the generator and associated step-up transformer as a unit without any intervening circuit breaker. The unit transformer supplying the generator auxiliaries is tapped off the connection between generator and step-up transformer. Differential protection can be arranged as follows.

17.6.1 Generator/Step-up Transformer Differential Protection

The generator stator and step-up transformer can be protected by a single zone of overall differential protection (Figure 17.8). This will be in addition to differential protection applied to the generator only. The current transformers should be located in the generator neutral connections and in the transformer HV connections. Alternatively, CTs within the HV switchyard may be employed if the distance is not technically prohibitive. Even where there is a generator circuit breaker, overall differential protection can still be provided if desired.

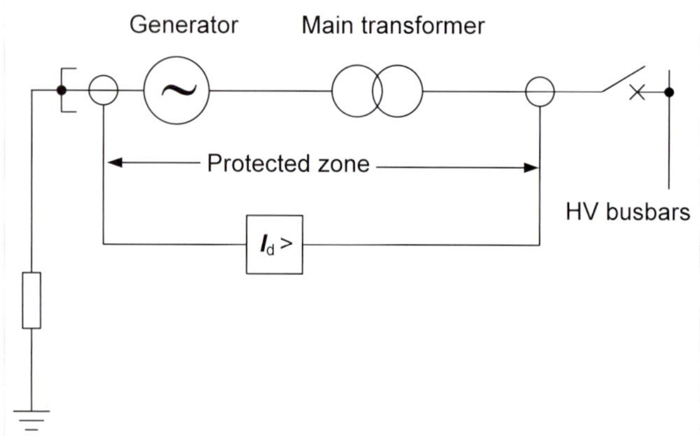

Figure 17.8: Overall generator-transformer differential protection

The current transformers should be rated according to Section 16.8.2. Since a power transformer is included within the zone of protection, biased transformer differential protection, with magnetising inrush restraint should be applied, as discussed in Section 16.8.5. Transient overfluxing of the generator transformer may arise due to overvoltage following generator load rejection. In some applications, this may threaten the stability of the differential protection. In such cases, consideration should be given to applying protection with transient overfluxing restraint/blocking (e.g. based on a 5th harmonic differential current threshold). Protection against sustained overfluxing is covered in Section 17.4.

17.6.2 Unit Transformer Differential Protection

The current taken by the unit transformer must be allowed for by arranging the generator differential protection as a three-ended scheme. Unit transformer current transformers are usually applied to balance the generator differential protection and prevent the unit transformer through current being seen as differential current. An exception might be where the unit transformer rating is extremely low in relation to the generator rating, e.g. for some hydro applications. The location of the third set of current transformers is normally on the primary side of the unit transformer. If located on secondary side of the unit transformer, they would have to be of an exceptionally high ratio, or exceptionally high ratio interposing CTs would have to be used. Thus, the use of secondary side CTs is not to be recommended. One advantage is that Unit Transformer faults would be within the zone of protection of the generator. However, the sensitivity of the generator protection to unit transformer phase faults would be considered inadequate, due to the relatively low rating of the transformer in relation to that of the generator. Thus, the unit transformer should have its own differential protection scheme. Protection for the Unit Transformer is covered in Chapter 16, including methods for stabilising the protection against magnetising inrush conditions.

17.7 OVERCURRENT PROTECTION

Overcurrent protection of generators may take two forms. Plain overcurrent protection may be used as the principal form of protection for small generators, and back-up protection for larger ones where differential protection is used as the primary method of generator stator winding protection. Voltage dependent overcurrent protection may be applied where differential protection is not justified on larger generators, or where problems are met in applying plain overcurrent protection.

17.7.1 Plain Overcurrent Protection

It is usual to apply time-delayed plain overcurrent protection to generators. For generators rated less than 1MVA, this will form the principal stator winding protection for phase faults. For larger generators, overcurrent protection can be applied as remote back-up protection, to disconnect the unit from any uncleared external fault. Where there is only one set of differential main protection, for a smaller generator, the overcurrent protection will also provide local back-up protection for the protected plant, in the event that the main protection fails to operate. The general principles of setting overcurrent relays are given in Chapter 9.

In the case of a single generator feeding an isolated system, current transformers at the neutral end of the machine should energise the overcurrent protection, to allow a response to winding fault conditions. Relay characteristics should be selected to take into account the fault current decrement behaviour of the generator, with allowance for the performance of the excitation system and its field-forcing capability. Without the provision of fault current compounding from generator CTs, an excitation system that is powered from an excitation transformer at the generator terminals will exhibit a pronounced fault current decrement for a terminal fault. With failure to consider this effect, the potential exists for the initial high fault current to decay to a value below the overcurrent protection pick-up setting before a relay element can operate, unless a low current setting and/or time setting is applied. The protection would then fail to trip the generator. The settings chosen must be the best compromise between assured operation in the foregoing circumstances and discrimination with the system protection and passage of normal load current, but this can be impossible with plain overcurrent protection.

In the more usual case of a generator that operates in parallel with others and which forms part of an extensive interconnected system, back-up phase fault protection for a generator and its transformer will be provided by HV overcurrent protection. This will respond to the higher-level backfeed from the power system to a unit fault. Other generators in parallel would supply this current and, being stabilised by the system impedance, it will not suffer a major decrement. This protection is usually a requirement of the power system operator. Settings must be chosen to prevent operation for external faults fed by the generator. It is common for the HV overcurrent protection relay to provide both time-delayed and instantaneous high-set elements. The time-delayed elements should be set to ensure that the protected items of plant cannot pass levels of through fault current in excess of their short-time withstand limits. The instantaneous elements should be set above the maximum possible fault current that the generator can supply, but less than the system-supplied fault current in the event of a generator winding fault. This back-up protection will minimise plant damage in the event of main protection failure for a generating plant fault and instantaneous tripping for an HV-side fault will aid the recovery of the power system and parallel generation.

17.7.2 Voltage-Dependent Overcurrent Protection

The plain overcurrent protection setting difficulty referred to in the previous section arises because allowance has to be made both for the decrement of the generator fault current with time and for the passage of full load current. To overcome the difficulty of discrimination, the generator terminal voltage can be measured and used to dynamically modify the basic relay current/time overcurrent characteristic for faults close to the generating plant. There are two basic alternatives for the application of voltage-dependent overcurrent protection, which are discussed in the following sections. The choice depends upon the power system characteristics and level of protection to be provided. Voltage-dependent overcurrent relays are often found applied to generators used on industrial systems, as an alternative to full differential protection.

17.7.2.1 Voltage Controlled Overcurrent Protection

Voltage controlled overcurrent protection has two time/current characteristics which are selected according to the status of a generator terminal voltage measuring element. The voltage threshold setting for the switching element is chosen according to the following criteria.

- during overloads, when the system voltage is sustained near normal, the overcurrent protection should have a current setting above full load current and an operating time characteristic that will prevent the generating plant from passing current to a remote external fault for a period in excess of the plant short-time withstand limits
- under close-up fault conditions, the busbar voltage must fall below the voltage threshold so that the second protection characteristic will be selected. This characteristic should be set to allow relay operation with fault current decrement for a close-up fault at the generator terminals or at the HV busbars. The protection should also time-grade with external circuit protection. There may be additional infeeds to an external circuit fault that will assist with grading

Typical characteristics are shown in Figure 17.9.

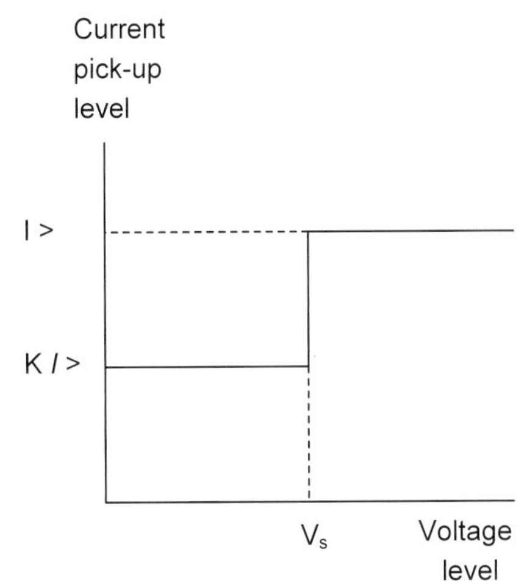

Figure 17.9: Voltage controlled relay characteristics

17.7.2.2 Voltage Restrained Overcurrent Protection

The alternative technique is to continuously vary the relay element pickup setting with generator voltage variation between upper and lower limits. The voltage is said to restrain the operation of the current element.

The effect is to provide a dynamic I.D.M.T. protection characteristic, according to the voltage at the machine terminals. Alternatively, the relay element may be regarded as an impedance type with a long dependent time delay. In consequence, for a given fault condition, the relay continues to operate more or less independently of current decrement in the machine. A typical characteristic is shown in Figure 17.10.

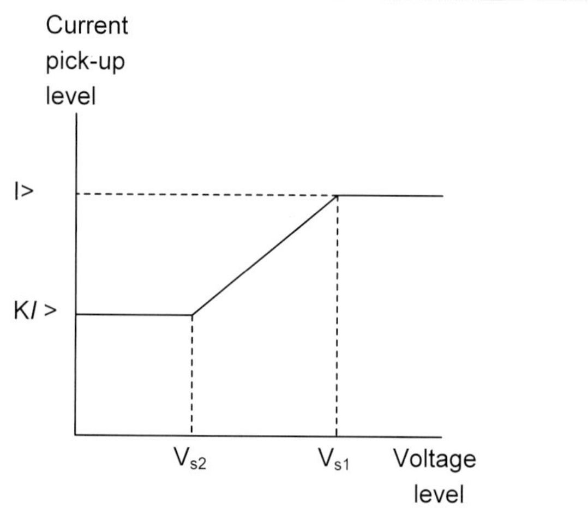

Figure 17.10: Voltage restrained relay characteristics

17.8 STATOR EARTH FAULT PROTECTION

Earth fault protection must be applied where impedance earthing is employed that limits the earth fault current to less than the pick-up threshold of the overcurrent and/or differential protection for a fault located down to the bottom 5% of the stator winding from the star-point. The type of protection required will depend on the method of earthing and connection of the generator to the power system.

17.8.1 Direct-Connected Generators

A single direct-connected generator operating on an isolated system will normally be directly earthed. However, if several direct-connected generators are operated in parallel, only one generator is normally earthed at a time. For the unearthed generators, a simple measurement of the neutral current is not possible, and other methods of protection must be used. The following sections describe the methods available.

17.8.1.1 Neutral Overcurrent Protection

With this form of protection, a current transformer in the neutral-earth connection energises an overcurrent relay element. This provides unrestricted earth-fault protection and so it must be graded with feeder protection. The relay element will therefore have a time-delayed operating characteristic. Grading must be carried out in accordance with the principles detailed in Chapter 9. The setting should not be more than 33% of the maximum earth fault current of the generator, and a lower setting would be preferable, depending on grading considerations.

17.8.1.2 Sensitive Earth Fault Protection

This method is used in the following situations:

a. direct-connected generators operating in parallel
b. generators with high-impedance neutral earthing, the earth fault current being limited to a few tens of amps
c. installations where the resistance of the ground fault path is very high, due to the nature of the ground

In these cases, conventional earth fault protection as described in Section 17.8.1.1 is of little use.

The principles of sensitive earth fault protection are described in Sections 9.17.1, 9.18 and 9.19. The earth fault (residual) current can be obtained from residual connection of line CTs, a line-connected CBCT, or a CT in the generator neutral. The latter is not possible if directional protection is used. The polarising voltage is usually the neutral voltage displacement input to the relay, or the residual of the three phase voltages, so a suitable VT must be used. For Petersen Coil earthing, a wattmetric technique (Section 9.19) can also be used.

For direct connected generators operating in parallel, directional sensitive earth fault protection may be necessary. This is to ensure that a faulted generator will be tripped before there is any possibility of the neutral overcurrent protection tripping a parallel healthy generator. When being driven by residually-connected phase CTs, the protection must be stabilised against incorrect tripping with transient spill current in the event of asymmetric CT saturation when phase fault or magnetising inrush current is being passed. Stabilising techniques include the addition of relay circuit impedance and/or the application of a time delay. Where the required setting of the protection is very low in comparison to the rated current of the phase CTs, it would be necessary to employ a single CBCT for the earth fault protection to ensure transient stability.

Since any generator in the paralleled group may be earthed, all generators will require to be fitted with both neutral overcurrent protection and sensitive directional earth fault protection. The setting of the sensitive directional earth fault protection is chosen to co-ordinate with generator differential protection and/or neutral voltage displacement protection to ensure that 95% of the stator winding is protected. Figure 17.11 illustrates the complete scheme, including optional blocking signals where difficulties in co-ordinating the generator and downstream feeder earth-fault protection occur.

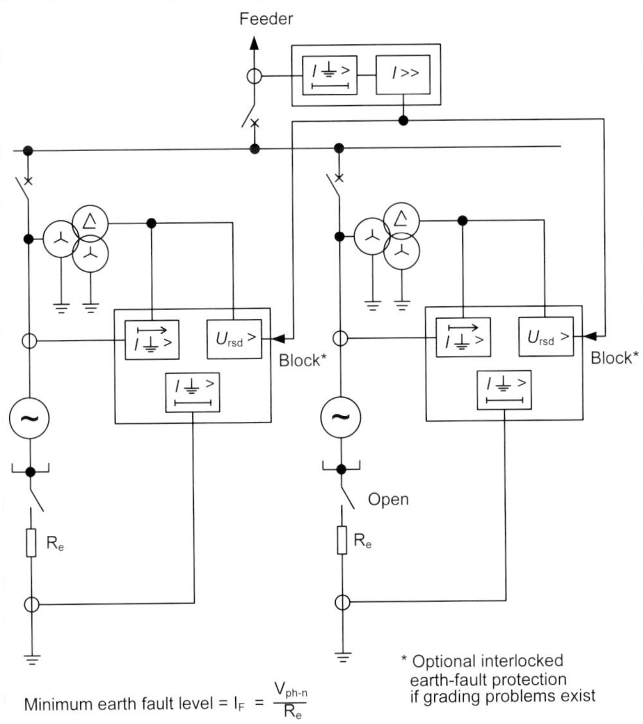

Figure 17.11: Comprehensive earth-fault protection scheme for direct-connected generators operating in parallel

For cases (b) and (c) above, it is not necessary to use a directional facility. Care must be taken to use the correct RCA setting – for instance if the earthing impedance is mainly resistive, this should be 0°. On insulated or very high impedance earthed systems, an RCA of –90° would be used, as the earth fault current is predominately capacitive.

Directional sensitive earth-fault protection can also be used for detecting winding earth faults. In this case, the relay element is applied to the terminals of the generator and is set to respond to faults only within the machine windings. Hence earth faults on the external system do not result in relay operation. However, current flowing from the system into a winding earth-fault causes relay operation. It will not operate on the earthed machine, so that other types of earth-fault protection must also be applied. All generators must be so fitted, since any can be operated as the earthed machine.

17.8.1.3 Neutral Voltage Displacement Protection

In a balanced network, the addition of the three phase-earth voltages produces a nominally zero residual voltage, since there would be little zero sequence voltage present. Any earth fault will set up a zero sequence system voltage, which will give rise to a non-zero residual voltage. This can be measured by a suitable relay element. The voltage signal must be derived from a VT that is suitable – i.e. it must be capable of transforming zero-sequence voltage, so 3-limb types and those without a primary earth connection are not suitable. This unbalance voltage provides a means of detecting earth faults. The relay element must be insensitive to third harmonic voltages that may be present in the system voltage waveforms, as these will sum residually.

As the protection is still unrestricted, the voltage setting of the relay must be greater than the effective setting of any downstream earth fault protection. It must also be time-delayed to co-ordinate with such protection. Sometimes, a second high-set element with short time delay is used to provide fast-acting protection against major winding earth faults. Figure 17.12 illustrates the possible connections that may be used.

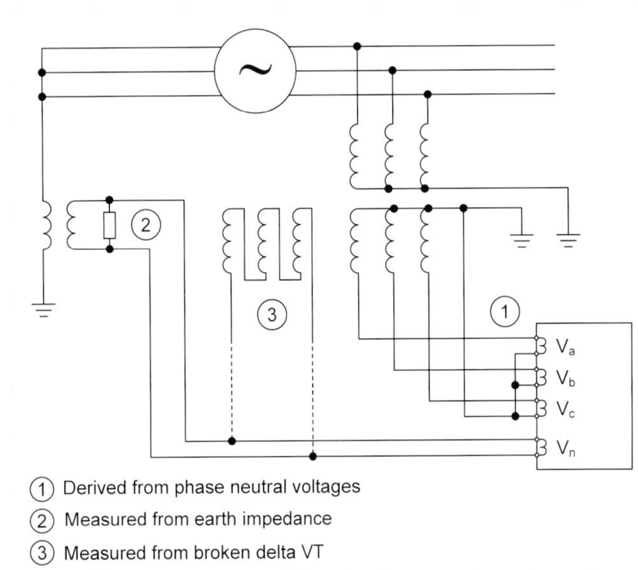

Figure 17.12: Neutral voltage displacement protection

17.8.2 Indirectly-Connected Generators

As noted in Section 17.2, a directly-earthed generator-transformer unit cannot interchange zero-sequence current with the remainder of the network, and hence an earth fault protection grading problem does not exist. The following sections detail the protection methods for the various forms of impedance earthing of generators.

17.8.2.1 High Resistance Earthing – Neutral Overcurrent Protection

A current transformer mounted on the neutral-earth conductor can drive an instantaneous and/or time delayed overcurrent relay element, as shown in Figure 17.13. It is impossible to provide protection for the whole of the winding, and Figure 17.13 also details how the percentage of winding covered can be calculated. For a relay element with an instantaneous setting, protection is typically limited to 90% of the winding. This is to ensure that the protection will not maloperate with zero sequence current during operation of a primary fuse for a VT earth fault or with any transient surge currents that could flow through the interwinding capacitance of the step-up transformer for an HV system earth fault.

A time-delayed relay is more secure in this respect, and it may have a setting to cover 95% of the stator winding. Since the generating units under consideration are usually large, instantaneous and time delayed relay elements are often applied, with settings of 10% and 5% of maximum earth fault current respectively; this is the optimum compromise in performance. The portion of the winding left unprotected for an earth fault is at the neutral end. Since the voltage to earth at this end of the winding is low, the probability of an earth fault occurring is also low. Hence additional protection is often not applied.

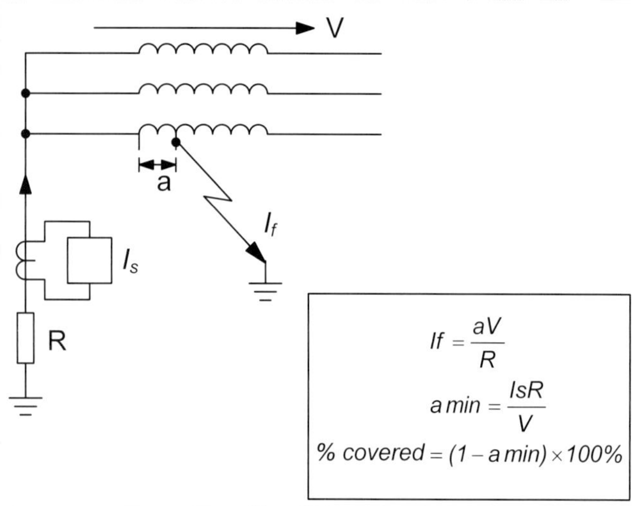

Figure 17.13: Earth fault protection of high-resistance earthed generator stator winding using a current element

17.8.2.2 Distribution Transformer Earthing Using a Current Element

In this arrangement, shown in Figure 17.14(a), the generator is earthed via the primary winding of a distribution transformer. The secondary winding is fitted with a loading resistor to limit the earth fault current. An overcurrent relay element energised from a current transformer connected in the resistor circuit is used to measure secondary earth fault current. The relay should have an effective setting equivalent to 5% of the maximum earth fault current at rated generator voltage, in order to protect 95% of the stator winding. The relay element response to third harmonic current should be limited to prevent incorrect operation when a sensitive setting is applied.

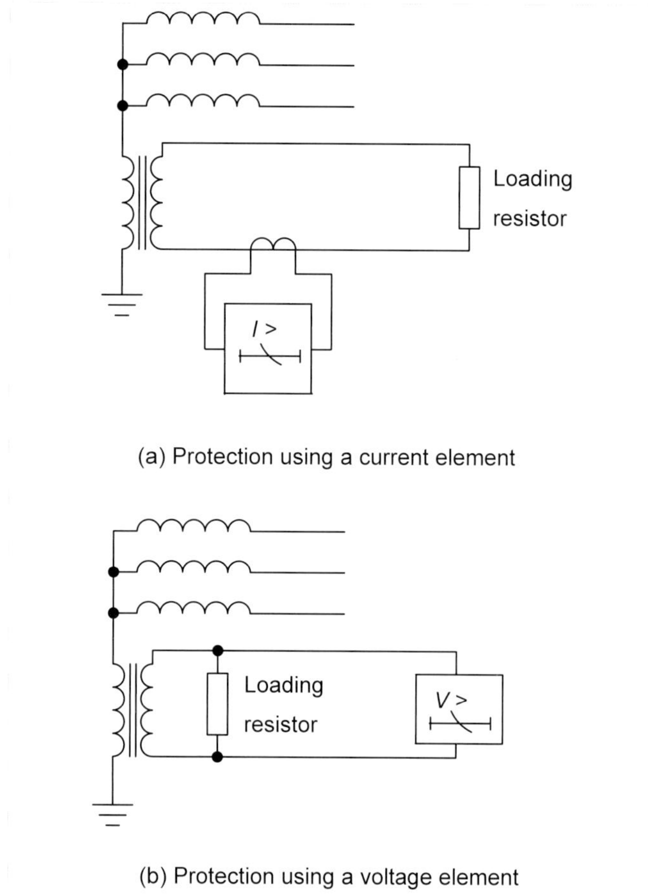

Figure 17.14: Generator winding earth-fault protection - distribution transformer earthing

As discussed in Section 17.8.2.1, the protection should be time delayed when a sensitive setting is applied, in order to prevent maloperation under transient conditions. It also must grade with generator VT primary protection (for a VT primary earth fault). An operation time in the range 0.5s-3s is usual. Less sensitive instantaneous protection can also be applied to provide fast tripping for a heavier earth fault condition.

17.8.2.3 Distribution Transformer Earthing Using a Voltage Element

Earth fault protection can also be provided using a voltage-measuring element in the secondary circuit instead. The setting considerations would be similar to those for the current operated protection, but transposed to voltage. The circuit diagram is shown in Figure 17.14(b).

Application of both voltage and current operated elements to a generator with distribution transformer earthing provides some advantages. The current operated function will continue to operate in the event of a short-circuited loading resistor and the voltage protection still functions in the event of an open-circuited resistor. However, neither scheme will operate in the event of a flashover on the primary terminals of the transformer or of the neutral cable between the generator and the transformer during an earth fault. A CT could be added in the neutral connection close to the generator, to energise a high-set overcurrent element to detect such a fault, but the fault current would probably be high enough to operate the phase differential protection.

17.8.2.4 Neutral Voltage Displacement Protection

This can be applied in the same manner as for direct-connected generators (Section 17.8.1.3). The only difference is that the are no grading problems as the protection is inherently restricted. A sensitive setting can therefore be used, enabling cover of up to 95% of the stator winding to be achieved.

17.8.3 Restricted Earth Fault Protection

This technique can be used on small generators not fitted with differential protection to provide fast acting earth fault protection within a defined zone that encompasses the generator. It is cheaper than full differential protection but only provides protection against earth faults. The principle is that used for transformer REF protection, as detailed in Section 16.7. However, in contrast to transformer REF protection, both biased low-impedance and high-impedance techniques can be used.

17.8.3.1 Low-Impedance Biased REF Protection

This is shown in Figure 17.15. The main advantage is that the neutral CT can also be used in a modern relay to provide conventional earth-fault protection and no external resistors are used. Relay bias is required, as described in Section 10.4.2, but the formula for calculating the bias is slightly different and also shown in Figure 17.15.

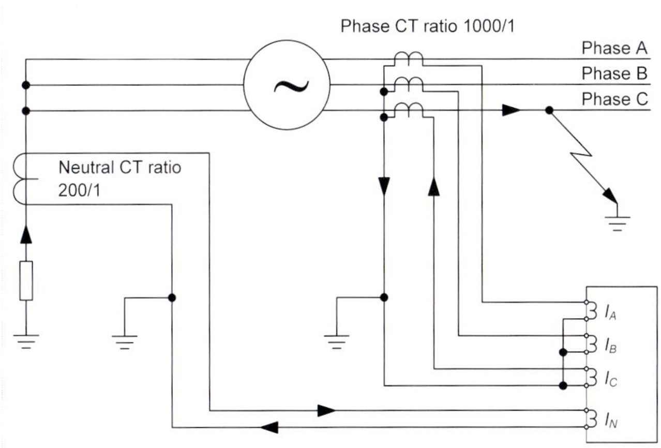

$$I_{BIAS} = \frac{(\text{highest of } I_A, I_B, I_C) + (I_N \times \text{scaling factor})}{2}$$

$$\text{where scaling factor} = \frac{\text{neutral CT ratio}}{\text{phase CT ratio}} = \frac{200}{1000} = 0.2$$

$$I_{DIFF} = \overline{I}_A + \overline{I}_B + \overline{I}_C + (\text{scaling factor} \times \overline{I}_N)$$

Figure 17.15: Low impedance biased REF protection of a generator

The initial bias slope is commonly set to 0% to provide maximum sensitivity, and applied up to the rated current of the generator. It may be increased to counter the effects of CT mismatch. The bias slope above generator rated current is typically set to 150% of rated value. The initial current setting is typically 5% of the minimum earth fault current for a fault at the machine terminals.

17.8.3.2 High Impedance REF Protection

The principle of high impedance differential protection is given in Chapter 10 and also described further in Section 17.5.2. The same technique can be used for earth-fault protection of a generator, using three residually connected phase CTs balanced against a similar single CT in the neutral connection. Settings of the order of 5% of maximum earth fault current at the generator terminals are typical. The usual requirements in respect of stabilising resistor and non-linear resistor to guard against excessive voltage across the relay must be taken, where necessary.

17.8.4 Earth Fault Protection for the Entire Stator Winding

All of the methods for earth fault protection detailed so far leave part of the winding unprotected. In most cases, this is of no consequence as the probability of a fault occurring in the 5% of the winding nearest the neutral connection is very low, due to the reduced phase to earth voltage. However, a fault can occur anywhere along the stator windings in the event of insulation failure due to localised heating from a core fault. In cases where protection for the entire winding is required, perhaps for alarm only, there are various methods available.

17.8.4.1 Measurement of Third Harmonic Voltage

One method is to measure the internally generated third harmonic voltage that appears across the earthing impedance due to the flow of third harmonic currents through the shunt capacitance of the stator windings etc. When a fault occurs in the part of the stator winding nearest the neutral end, the third harmonic voltage drops to near zero, and hence a relay element that responds to third harmonic voltage can be used to detect the condition. As the fault location moves progressively away from the neutral end, the drop in third harmonic voltage from healthy conditions becomes less, so that at around 20-30% of the winding distance, it no longer becomes possible to discriminate between a healthy and a faulty winding. Hence, a conventional earth-fault scheme should be used in conjunction with a third harmonic scheme, to provide overlapping cover of the entire stator winding. The measurement of third harmonic voltage can be taken either from a star-point VT or the generator line VT. In the latter case, the VT must be capable of carrying residual flux, and this prevents the use of 3-limb types. If the third harmonic voltage is measured at the generator star point, an undervoltage characteristic is used. An overvoltage characteristic is used if the measurement is taken from the generator line VT. For effective application of this form of protection, there should be at least 1% third harmonic voltage across the generator neutral earthing impedance under all operating conditions.

A problem encountered is that the level of third harmonic voltage generated is related to the output of the generator. The voltage is low when generator output is low. In order to avoid maloperation when operating at low power output, the relay element can be inhibited using an overcurrent or power elements (kW, kVAr or kVA) and internal programmable logic.

17.8.4.2 Use of Low-Frequency Voltage Injection

Another method for protecting the entire stator winding of a generator is to deploy signal injection equipment to inject a low frequency voltage between the stator star point and earth. An earth fault at any winding location will result in the flow of a measurable injection current to cause protection operation. This form of protection can provide earth fault protection when the generator is at standstill, prior to run-up. It is also an appropriate method to apply to variable speed synchronous machines. Such machines may be employed for variable speed motoring in pumped-storage generation schemes or for starting a large gas turbine prime mover.

17.9 OVERVOLTAGE PROTECTION

Overvoltages on a generator may occur due to transient surges on the network, or prolonged power frequency overvoltages may arise from a variety of conditions. Surge arrestors may be required to protect against transient overvoltages, but relay protection may be used to protect against power frequency overvoltages.

A sustained overvoltage condition should not occur for a machine with a healthy voltage regulator, but it may be caused by the following contingencies:

- defective operation of the automatic voltage regulator when the machine is in isolated operation
- operation under manual control with the voltage regulator out of service. A sudden variation of the load, in particular the reactive power component, will give rise to a substantial change in voltage because of the large voltage regulation inherent in a typical alternator
- sudden loss of load (due to tripping of outgoing feeders, leaving the set isolated or feeding a very small load) may cause a sudden rise in terminal voltage due to the trapped field flux and/or overspeed

Sudden loss of load should only cause a transient overvoltage while the voltage regulator and governor act to correct the situation. A maladjusted voltage regulator may trip to manual, maintaining excitation at the value prior to load loss while the generator supplies little or no load. The terminal voltage will increase substantially, and in severe cases it would be limited only by the saturation characteristic of the generator. A rise in speed simply compounds the problem. If load that is sensitive to overvoltages remains connected, the consequences in terms of equipment damage and lost revenue can be severe. Prolonged overvoltages may also occur on isolated networks, or ones with weak interconnections, due to the fault conditions listed earlier.

For these reasons, it is prudent to provide power frequency overvoltage protection, in the form of a time-delayed element, either IDMT or definite time. The time delay should be long enough to prevent operation during normal regulator action, and therefore should take account of the type of AVR fitted and its transient response. Sometimes a high-set element is provided as well, with a very short definite-time delay or instantaneous setting to provide a rapid trip in extreme circumstances. The usefulness of this is questionable for generators fitted with an excitation system other than a static type, because the excitation will decay in accordance with the open-circuit time constant of the field winding. This decay can last several seconds. The relay element is arranged to trip both the main circuit breaker (if not already open) and the excitation; tripping the main circuit breaker alone is not sufficient.

17.10 UNDERVOLTAGE PROTECTION

Undervoltage protection was historically rarely fitted to generators. It is sometimes used as an interlock element for another protection function or scheme, such as field failure protection or inadvertent energisation protection, where the abnormality to be detected leads directly or indirectly to an undervoltage condition. A transmission system undervoltage condition may arise when there is insufficient reactive power generation to maintain the system voltage profile and the condition must be addressed to avoid the possible phenomenon of system voltage collapse. However, it should be addressed by the deployment of 'system protection' schemes. The generation should not be tripped. The greatest case for undervoltage protection being required would be for a generator supplying an isolated power system or to meet the public Utility demands for connection of embedded generation (see Section 17.21).

In the case of generators feeding an isolated system, undervoltage may occur for several reasons, typically overloading or failure of the AVR. In some cases, the performance of generator auxiliary plant fed via a unit transformer from the generator terminals could be adversely affected by prolonged undervoltage.

Where undervoltage protection is required, it should comprise an undervoltage element and an associated time delay. Settings must be chosen to avoid maloperation during the inevitable voltage dips during power system fault clearance or associated with motor starting. Transient reductions in voltage down to 80% or less may be encountered during motor starting.

17.11 LOW FORWARD POWER/REVERSE POWER PROTECTION

Low forward power or reverse power protection may be required for some generators to protect the prime mover. Parts of the prime mover may not be designed to experience reverse torque or they may become damaged through continued rotation after the prime mover has suffered some form of failure.

17.11.1 Low Forward Power Protection

Low forward power protection is often used as an interlocking function to enable opening of the main circuit breaker for non-urgent trips – e.g. for a stator earth fault on a high-impedance earthed generator, or when a normal shutdown of a set is taking place. This is to minimise the risk of plant overspeeding when the electrical load is removed from a high-speed cylindrical rotor generator. The rotor of this type of generator is highly stressed mechanically and cannot tolerate much overspeed. While the governor should control overspeed conditions, it is not good practice to open the main circuit breaker simultaneously with tripping of the prime mover for non-urgent trips. For a steam turbine, for example, there is a risk of overspeeding due to energy storage in the trapped steam, after steam valve tripping, or in the event that the steam valve(s) do not fully close for some reason. For urgent trip conditions, such as stator differential protection operation, the risk involved in simultaneous prime mover and generator breaker tripping must be accepted.

17.11.2 Reverse Power Protection

Reverse power protection is applied to prevent damage to mechanical plant items in the event of failure of the prime mover. Table 17.1 gives details of the potential problems for various prime mover types and the typical settings for reverse power protection. For applications where a protection sensitivity of better than 3% is required, a metering class CT should be employed to avoid incorrect protection behaviour due to CT phase angle errors when the generator supplies a significant level of reactive power at close to zero power factor.

Table 17.1: Generator reverse power problems

Prime Mover	Motoring Power (% of rated)	Possible Damage	Protection Setting
Diesel Engine	5-25	Fire/explosion due to unburnt fuel	50% of motoring power
		Mechanical damage to gearbox/shafts	
Gas Turbine	10-15 (split shaft)	Gearbox damage	
	>50% (single shaft)		
Hydro	0.2-2 (blades out of water)	Blade and runner cavitation	
	>2 (blades in water)		
Steam Turbine	0.5-6	Turbine blade damage	
		Gearbox damage on geared sets	

The reverse power protection should be provided with a definite time delay on operation to prevent spurious operation with transient power swings that may arise following synchronisation or in the event of a power transmission system disturbance.

17.12 UNBALANCED LOADING

A three-phase balanced load produces a reaction field that, to a first approximation, is constant and rotates synchronously with the rotor field system. Any unbalanced condition can be resolved into positive, negative and zero sequence components. The positive sequence component is similar to the normal balanced load. The zero sequence component produces no main armature reaction.

17.12.1 Effect of Negative Sequence Current

The negative sequence component is similar to the positive sequence system, except that the resulting reaction field rotates in the opposite direction to the d.c. field system. Hence, a flux is produced which cuts the rotor at twice the rotational velocity, thereby inducing double frequency currents in the field system and in the rotor body. The resulting eddy-currents are very large and cause severe heating of the rotor. So severe is this effect that a single-phase load equal to the normal three-phase rated current can quickly heat the rotor slot wedges to the softening point. They may then be extruded under centrifugal force until they stand above the rotor surface, when it is possible that they may strike the stator core.

A generator is assigned a continuous negative sequence rating.

For turbo-generators this rating is low; standard values of 10% and 15% of the generator continuous rating have been adopted. The lower rating applies when the more intensive cooling techniques are applied, for example hydrogen-cooling with gas ducts in the rotor to facilitate direct cooling of the winding.

Short time heating is of interest during system fault conditions and it is usual in determining the generator negative sequence withstand capability to assume that the heat dissipation during such periods is negligible. Using this approximation it is possible to express the heating by the law:

$$I_2^2 t = K$$

where:

I_2 = negative sequence component (per unit of MCR)

t = time (seconds)

K = constant proportional to the thermal capacity of the generator rotor

For heating over a period of more than a few seconds, it is necessary to allow for the heat dissipated. From a combination of the continuous and short time ratings, the overall heating characteristic can be deduced to be:

$$M = \frac{I_2}{I_{2R}} = \sqrt{\frac{1}{1-e^{\frac{-\left(I_{2R}^{2}t\right)}{K}}}}$$

where:

I_{2R} = negative phase sequence continuous rating in per unit of MCR

The heating characteristics of various designs of generator are shown in Figure 17.16.

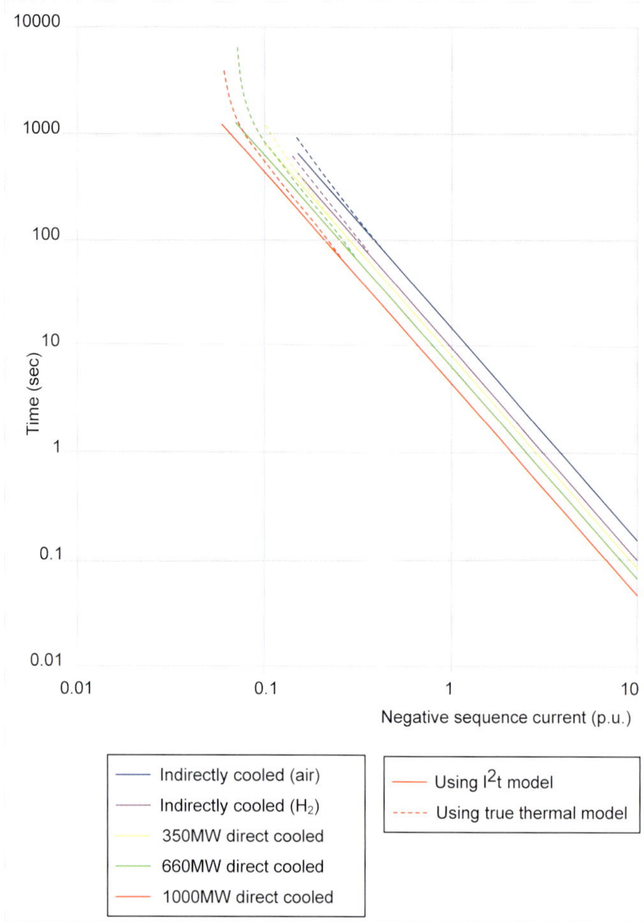

Figure 17.16: Typical negative phase sequence current withstand of cylindrical rotor generators

17.12.2 Negative Phase Sequence Protection

This protection is applied to prevent overheating due to negative sequence currents. Small salient-pole generators have a proportionately larger negative sequence capacity and may not require protection. Modern numerical relays derive the negative sequence current level by calculation, with no need for special circuits to extract the negative sequence component. A true thermal replica approach is often followed, to allow for:

- standing levels of negative sequence current below the continuous withstand capability. This has the effect of shortening the time to reach the critical temperature after an increase in negative sequence current above the continuous withstand capability
- cooling effects when negative sequence current levels are below the continuous withstand capability

The advantage of this approach is that cooling effects are modelled more accurately, but the disadvantage is that the tripping characteristic may not follow the withstand characteristic specified by the manufacturer accurately.

The typical relay element characteristic takes the form of

$$t = -\frac{K}{I_{2set}^2}\log_e\left[1-\left(\frac{I_{2set}}{I_2}\right)\right]$$

Equation 17.1

where:

t = time to trip

$$K = K_g \times \left(\frac{I_{flc}}{I_p}\right)^2$$

$$I_{2set} = I_{2cmr} \times \left(\frac{I_{flc}}{I_p}\right) \times I_n$$

K_g = negative sequence withstand coefficient (Figure 17.16)

I_{2cmr} = generator maximum continuous I_2 withstand

I_{flc} = generator rated primary current

I_p = CT primary current

I_n = Relay rated current

Figure 17.16 also shows the thermal replica time characteristic described by Equation 17.1, from which it will be seen that a significant gain in capability is achieved at low levels of negative sequence current. Such a protection element will also respond to phase-earth and phase-phase faults where sufficient negative sequence current arises. Grading with downstream power system protection relays is therefore required. A definite minimum time setting must be applied to the negative sequence relay element to ensure correct grading. A maximum trip time setting may also be used to ensure correct tripping when the negative sequence current level is only slightly in excess of the continuous withstand capability and hence the trip time from the thermal model may depart significantly from the rotor withstand limits.

17.13 PROTECTION AGAINST INADVERTENT ENERGISATION

Accidental energisation of a generator when it is not running may cause severe damage to it. With the generator at standstill, closing the circuit breaker results in the generator acting as an induction motor; the field winding (if closed) and the rotor solid iron/damper circuits acting as rotor circuits. Very high currents are induced in these rotor components, and also occur in the stator, with resultant rapid overheating and damage. Protection against this condition is therefore desirable.

A combination of stator undervoltage and overcurrent can be used to detect this condition. An instantaneous overcurrent element is used, and gated with a three-phase undervoltage element (fed from a VT on the generator side of the circuit breaker) to provide the protection. The overcurrent element can have a low setting, as operation is blocked when the generator is operating normally. The voltage setting should be low enough to ensure that operation cannot occur for transient faults. A setting of about 50% of rated voltage is typical. VT failure can cause maloperation of the protection, so the element should be inhibited under these conditions.

17.14 UNDER/OVERFREQUENCY/ OVERFLUXING PROTECTION

These conditions are grouped together because these problems often occur due to a departure from synchronous speed.

17.14.1 Overfluxing

Overfluxing occurs when the ratio of voltage to frequency is too high. The iron saturates owing to the high flux density and results in stray flux occurring in components not designed to carry it. Overheating can then occur, resulting in damage. The problem affects both direct-and indirectly-connected generators. Either excessive voltage, or low frequency, or a combination of both can result in overfluxing, a voltage to frequency ratio in excess of 1.05p.u. normally being indicative of this condition. Excessive flux can arise transiently, which is not a problem for the generator. For example, a generator can be subjected to a transiently high power frequency voltage, at nominal frequency, immediately after full load rejection. Since the condition would not be sustained, it only presents a problem for the stability of the transformer differential protection schemes applied at the power station (see Chapter 16 for transformer protection). Sustained overfluxing can arise during run up, if excitation is applied too early with the AVR in service, or if the generator is run down, with the excitation still applied. Other overfluxing instances have occurred from loss of the AVR voltage feedback signal, due to a reference VT problem. Such sustained conditions must be detected by a dedicated overfluxing protection function that will raise an alarm and possibly force an immediate reduction in excitation.

Most AVRs have an overfluxing protection facility included. This may only be operative when the generator is on open circuit, and hence fail to detect overfluxing conditions due to abnormally low system frequency. However, this facility is not engineered to protection relay standards, and should not be solely relied upon to provide overfluxing protection. A separate relay element is therefore desirable and provided in most modern relays.

It is usual to provide a definite time-delayed alarm setting and an instantaneous or inverse time-delayed trip setting, to match the withstand characteristics of the protected generator and transformer. It is very important that the VT reference for overfluxing protection is not the same as that used for the AVR.

17.14.2 Under/Overfrequency

The governor fitted to the prime mover normally provides protection against overfrequency. Underfrequency may occur as a result of overload of generators operating on an isolated system, or a serious fault on the power system that results in a deficit of generation compared to load. This may occur if a grid system suffers a major fault on transmission lines linking two parts of the system, and the system then splits into two. It is likely that one part will have an excess of generation over load, and the other will have a corresponding deficit. Frequency will fall fairly rapidly in the latter part, and the normal response is load shedding, either by load shedding relays or operator action. However, prime movers may have to be protected against excessively low frequency by tripping of the generators concerned.

With some prime movers, operation in narrow frequency bands that lie close to normal running speed (either above or below) may only be permitted for short periods, together with a cumulative lifetime duration of operation in such frequency bands. This typically occurs due to the presence of rotor torsional frequencies in such frequency bands. In such cases, monitoring of the period of time spent in these frequency bands is required. A special relay is fitted in such cases, arranged to provide alarm and trip facilities if either an individual or cumulative period exceeds a set time.

17.15 ROTOR FAULTS

The field circuit of a generator, comprising the field winding of the generator and the armature of the exciter, together with any associated field circuit breaker if it exists, is an isolated d.c. circuit which is not normally earthed. If an earth fault occurs, there will be no steady-state fault current and the need for action will not be evident.

Danger arises if a second earth fault occurs at a separate point in the field system, to cause the high field current to be diverted, in part at least, from the intervening turns. Serious damage to the conductors and possibly the rotor can occur very rapidly under these conditions.

More damage may be caused mechanically. If a large portion of the winding is short-circuited, the flux may adopt a pattern such as that shown in Figure 17.17. The attracting force at the surface of the rotor is given by:

$$F = \frac{B^2 A}{8\pi}$$

where:

A = area

B = flux density

It will be seen from Figure 17.17 that the flux is concentrated on one pole but widely dispersed over the other and intervening surfaces. The attracting force is in consequence large on one pole but very weak on the opposite one, while flux on the quadrature axis will produce a balancing force on this axis. The result is an unbalanced force that in a large machine may be of the order of 50-100 tons. A violent vibration is set up that may damage bearing surfaces or even displace the rotor by an amount sufficient to cause it to foul the stator.

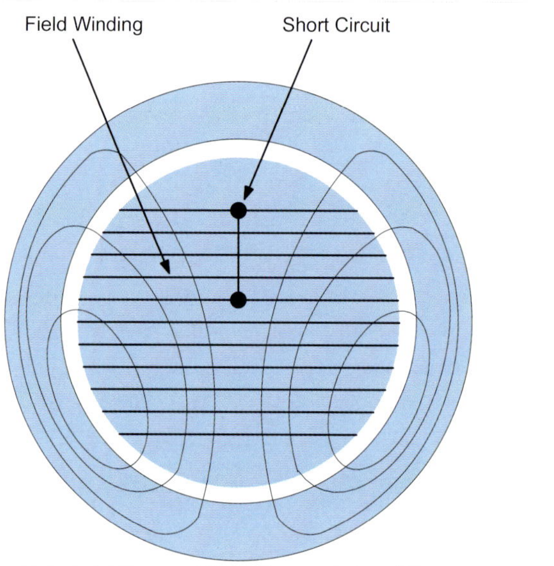

Figure 17.17: Flux distribution on rotor with partial winding short circuit

17.15.1 Rotor Earth-Fault Protection

Two methods are available to detect this type of fault. The first method is suitable for generators that incorporate brushes in the main generator field winding. The second method requires at least a slip-ring connection to the field circuit:

- Potentiometer method
- A.C. injection method

17.15.1.1 Potentiometer Method

This is a scheme that was fitted to older generators, and it is illustrated in Figure 17.18. An earth fault on the field winding would produce a voltage across the relay, the maximum voltage occurring for faults at the ends of the winding.

A 'blind spot' would exist at the centre of the field winding. To avoid a fault at this location remaining undetected, the tapping point on the potentiometer could be varied by a pushbutton or switch. The relay setting is typically about 5% of the exciter voltage.

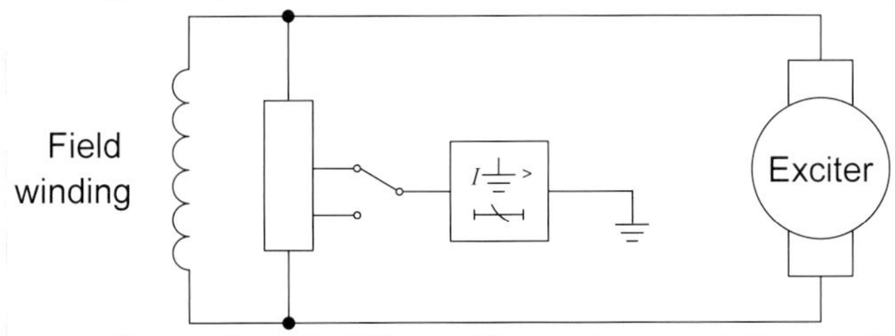

Figure 17.18: Earth fault protection of field circuit by potentiometer method

17.15.1.2 Injection Methods

Two methods are in common use. The first is based on low frequency signal injection, with series filtering, as shown in Figure 17.19(a). It comprises an injection source that is connected between earth and one side of the field circuit, through capacitive coupling and the measurement circuit. The field circuit is subjected to an alternating potential at substantially the same level throughout. An earth fault anywhere in the field system will give rise to a current that is detected as an equivalent voltage across the adjustable resistor by the relay. The capacitive coupling blocks the normal d.c. field voltage, preventing the discharge of a large direct current through the protection scheme. The combination of series capacitor and reactor forms a low-pass tuned circuit, the intention being to filter higher frequency rotor currents that may occur for a variety of reasons.

Other schemes are based on power frequency signal injection. An impedance relay element is used, a field winding earth fault reducing the impedance seen by the relay. These suffer the draw back of being susceptible to static excitation system harmonic currents when there is significant field winding and excitation system shunt capacitance. Greater immunity for such systems is offered by capacitively coupling the protection scheme to both ends of the field winding, where brush or slip ring access is possible (Figure 17.19(b)).

The low-frequency injection scheme is also advantageous in that the current flow through the field winding shunt capacitance will be lower than for a power frequency scheme. Such current would flow through the machine bearings to cause erosion of the bearing surface. For power frequency schemes, a solution is to insulate the bearings and provide an earthing brush for the shaft.

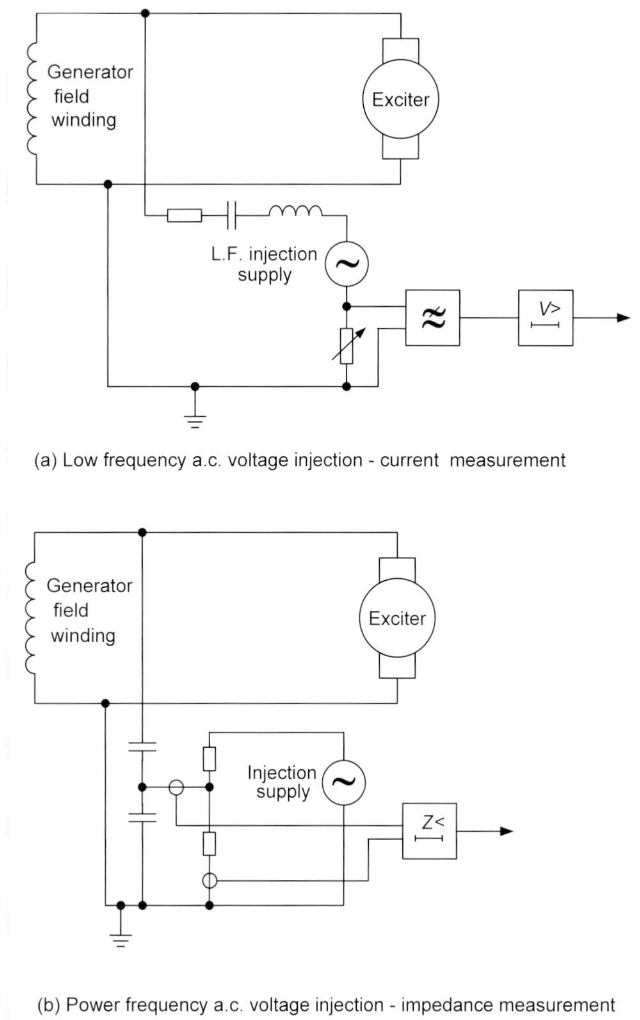

Figure 17.19: Earth fault protection of field circuit by a.c. injection

17.15.2 Rotor Earth Fault Protection for Brushless Generators

A brushless generator has an excitation system consisting of:

- a main exciter with rotating armature and stationary field windings
- a rotating rectifier assembly, carried on the main shaft line out
- a controlled rectifier producing the d.c. field voltage for the main exciter field from an a.c. source (often a small 'pilot' exciter)

Hence, no brushes are required in the generator field circuit. All control is carried out in the field circuit of the main exciter. Detection of a rotor circuit earth fault is still necessary, but this must be based on a dedicated rotor-mounted system that has a telemetry link to provide an alarm/data.

17.15.3 Rotor Shorted Turn Protection

As detailed in Section 17.15, a shorted section of field winding will result in an unsymmetrical rotor flux pattern and in potentially damaging rotor vibration. Detection of such an electrical fault is possible using a probe consisting of a coil placed in the airgap. The flux pattern of the positive and negative poles is measured and any significant difference in flux pattern between the poles is indicative of a shorted turn or turns. Automated waveform comparison techniques can be used to provide a protection scheme, or the waveform can be inspected visually at regular intervals. An immediate shutdown is not normally required unless the effects of the fault are severe. The fault can be kept under observation until a suitable shutdown for repair can be arranged. Repair will take some time, since it means unthreading the rotor and dismantling the winding.

Since short-circuited turns on the rotor may cause damaging vibration and the detection of field faults for all degrees of abnormality is difficult, the provision of a vibration a detection scheme is desirable – this forms part of the mechanical protection of the generator.

17.15.4 Protection Against Diode Failure

A short-circuited diode will produce an a.c. ripple in the exciter field circuit. This can be detected by a relay monitoring the current in the exciter field circuit, however such systems have proved to be unreliable. The relay would need to be time delayed to prevent an alarm being issued with normal field forcing during a power system fault. A delay of 5-10 seconds may be necessary.

Fuses to disconnect the faulty diode after failure may be fitted. The fuses are of the indicating type, and an inspection window can be fitted over the diode wheel to enable diode health to be monitored manually.

A diode that fails open-circuit occurs less often. If there is more than one diode in parallel for each arm of the diode bridge, the only impact is to restrict the maximum continuous excitation possible. If only a single diode per bridge arm is fitted, some ripple will be present on the main field supply but the inductance of the circuit will smooth this to a degree and again the main effect is to restrict the maximum continuous excitation. The set can be kept running until a convenient shutdown can be arranged.

17.15.5 Field Suppression

The need to rapidly suppress the field of a machine in which a fault has developed should be obvious, because as long as the excitation is maintained, the machine will feed its own fault even though isolated from the power system. Any delay in the decay of rotor flux will extend the fault damage. Braking the rotor is no solution, because of its large kinetic energy.

The field winding current cannot be interrupted instantaneously as it flows in a highly inductive circuit. Consequently, the flux energy must be dissipated to prevent an excessive inductive voltage rise in the field circuit. For machines of moderate size, it is satisfactory to open the field circuit with an air-break circuit breaker without arc blow-out coils. Such a breaker permits only a moderate arc voltage, which is nevertheless high enough to suppress the field current fairly rapidly. The inductive energy is dissipated partly in the arc and partly in eddy-currents in the rotor core and damper windings.

With generators above about 5MVA rating, it is better to provide a more definite means of absorbing the energy without incurring damage. Connecting a 'field discharge resistor' in parallel with the rotor winding before opening the field circuit breaker will achieve this objective. The resistor, which may have a resistance value of approximately five times the rotor winding resistance, is connected by an auxiliary contact on the field circuit breaker. The breaker duty is thereby reduced to that of opening a circuit with a low L/R ratio. After the breaker has opened, the field current flows through the discharge resistance and dies down harmlessly. The use of a fairly high value of discharge resistance reduces the field time constant to an acceptably low value, though it may still be more than one second. Alternatively, generators fitted with static excitation systems may temporarily invert the applied field voltage to reduce excitation current rapidly to zero before the excitation system is tripped.

17.16 LOSS OF EXCITATION PROTECTION

Loss of excitation may occur for a variety of reasons. If the generator was initially operating at only 20%-30% of rated power, it may settle to run super-synchronously as an induction generator, at a low level of slip. In doing so, it will draw reactive current from the power system for rotor excitation. This form of response is particularly true of salient pole generators. In these circumstances, the generator may be able to run for several minutes without requiring to be tripped. There may be sufficient time for remedial action to restore the excitation, but the reactive power demand of the machine during the failure may severely depress the power system voltage to an unacceptable level. For operation at high initial power output, the rotor speed may rise to approximately 105% of rated speed, where there would be low power output and where a high reactive current of up to 2.0p.u. may be drawn from the supply. Rapid automatic disconnection is then required to protect the stator windings from excessive current and to protect the rotor from damage caused by induced slip frequency currents.

17.16.1 Protection against Loss of Excitation

The protection used varies according to the size of generator being protected.

17.16.1.1 Small Generators (≤ 5MVA)

On the smaller machines, protection against asynchronous running has tended to be optional, but it may now be available by default, where the functionality is available within a modern numerical generator protection package. If fitted, it is arranged either to provide an alarm or to trip the generator. If the generator field current can be measured, a relay element can be arranged to operate when this drops below a preset value. However, depending on the generator design and size relative to the system, it may well be that the machine would be required to operate synchronously with little or no excitation under certain system conditions.

The field undercurrent relay must have a setting below the minimum exciting current, which may be 8% of that corresponding to the MCR of the machine. Time delay relays are used to stabilise the protection against maloperation in response to transient conditions and to ensure that field current fluctuations due to pole slipping do not cause the protection to reset.

If the generator field current is not measurable, then the technique detailed in the following section is utilised.

17.16.1.2 Large Generators (>5MVA)

For generators above about 5MVA rating, protection against loss of excitation and pole slipping conditions is normally applied.

Consider a generator connected to network, as shown in Figure 17.20. On loss of excitation, the terminal voltage will begin to decrease and the stator current will increase, resulting in a decrease of impedance viewed from the generator terminals and also a change in power factor.

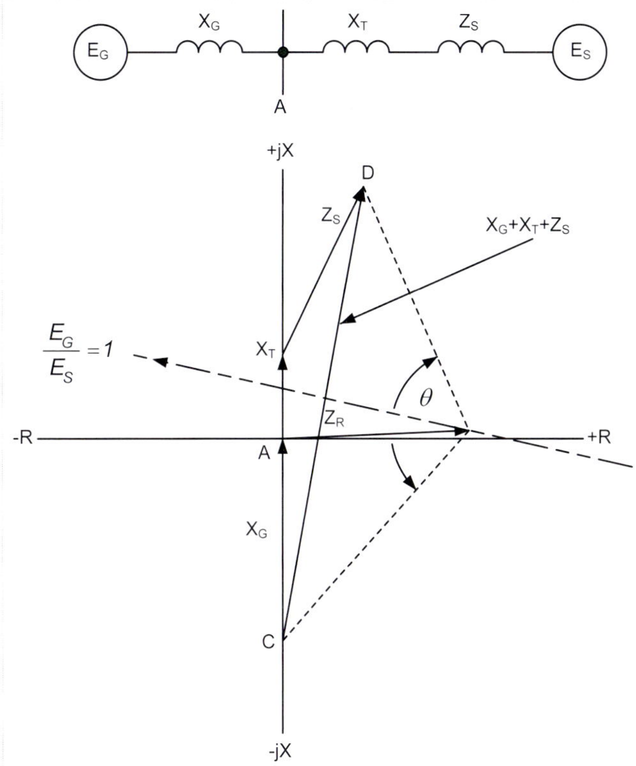

Figure 17.20: Basic interconnected system

A relay to detect loss of synchronism can be located at point A. It can be shown that the impedance presented to the relay under loss of synchronism conditions (phase swinging or pole slipping) is given by:

$$Z_R = \frac{(X_G + X_T + Z_S)n(n - \cos\theta - j\sin\theta)}{(n-\cos\theta)^2 + \sin^2\theta} - X_G$$

Equation 17.2

where:

$$n = \frac{E_G}{E_S} = \frac{generated}{system}\,voltage$$

θ = angle by which E_G leads E_S

If the generator and system voltages are equal the above expression becomes:

$$Z_R = \frac{(X_G + X_T + Z_S)\left(1 - \frac{j\cot\theta}{2}\right)}{2} - X_G$$

The general case can be represented by a system of circles with centres on the line CD; see Figure 17.21. Also shown is a typical machine terminal impedance locus during loss of excitation conditions.

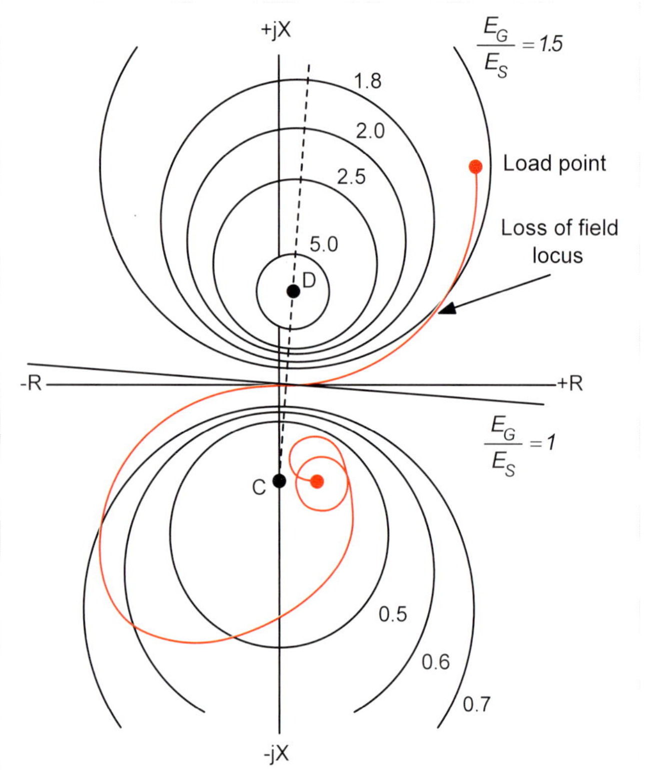

Figure 17.21: Swing curves and loss of synchronism locus

The special cases of E_G=E_S and E_G=0 result in a straight-line locus that is the right-angled bisector of CD and in a circular locus that is shrunk to point C, respectively.

When excitation is removed from a generator operating synchronously the flux dies away slowly, during which period the ratio of E_G/E_S is decreasing, and the rotor angle of the machine is increasing. The operating condition plotted on an impedance diagram therefore travels along a locus that crosses the power swing circles. At the same time, it progresses in the direction of increasing rotor angle. After passing the anti-phase position, the locus bends round as the internal e.m.f. collapses, condensing on an impedance value equal to the machine reactance. The locus is illustrated in Figure 17.21.

The relay location is displaced from point C by the generator reactance X_G. One problem in determining the position of these loci relative to the relay location is that the value of machine impedance varies with the rate of slip. At zero slip X_G is equal to X_d, the synchronous reactance, and at 100% slip X_G is equal to $X''d$, the sub-transient reactance. The impedance in a typical case has been shown to be equal to $X'd$, the transient reactance, at 50% slip, and to $2X'd$ with a slip of 0.33%. The slip likely to be experienced with asynchronous running is low, perhaps 1%, so that for the purpose of assessing the power swing locus it is sufficient to take the value X_G=$2X_d'$.

This consideration has assumed a single value for X_G. However, the reactance X_q on the quadrature axis differs from the direct-axis value, the ratio of X_d/X_g being known as the saliency factor. This factor varies with the slip speed. The effect of this factor during asynchronous operation is to cause X_G to vary at slip speed. In consequence, the loss of excitation impedance locus does not settle at a single point, but it continues to describe a small orbit about a mean point.

A protection scheme for loss of excitation must operate decisively for this condition, but its characteristic must not inhibit stable operation of the generator. One limit of operation corresponds to the maximum practicable rotor angle, taken to be 120°. The locus of operation can be represented as a circle on the impedance plane, as shown in Figure 17.22, stable operation conditions lying outside the circle.

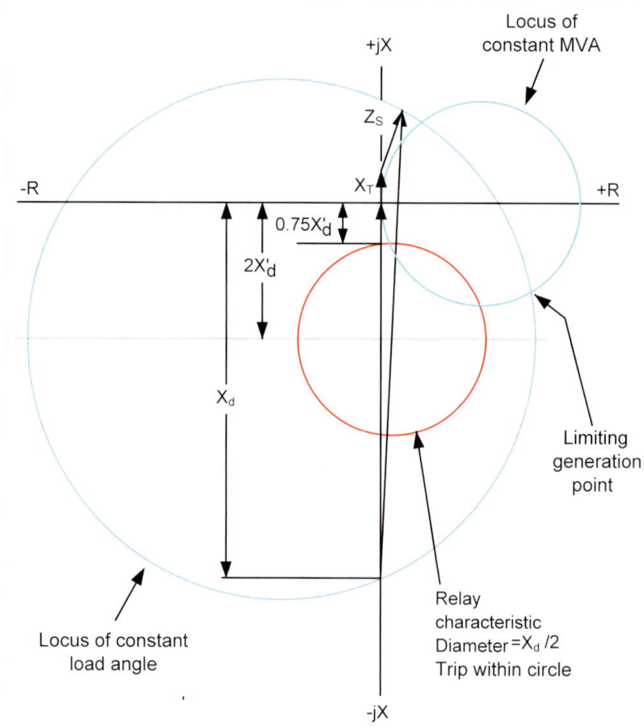

Figure 17.22: Locus of limiting operating conditions of synchronous machine

On the same diagram the full load impedance locus for one per unit power can be drawn. Part of this circle represents a condition that is not feasible, but the point of intersection with the maximum rotor angle curve can be taken as a limiting operating condition for setting impedance-based loss of excitation protection.

17.16.2 Impedance-Based Protection Characteristics

Figure 17.21 alludes to the possibility that a protection scheme for loss of excitation could be based on impedance measurement. The impedance characteristic must be appropriately set or shaped to ensure decisive operation for loss of excitation whilst permitting stable generator operation within allowable limits. One or two offset mho under impedance elements (see Chapter 11 for the principles of operation) are ideally suited for providing loss of excitation protection as long as a generator operating at low power output (20-30%P_n) does not settle down to operate as an induction generator. The characteristics of a typical two-stage loss of excitation protection scheme are illustrated in Figure 17.23. The first stage, consisting of settings X_{a1} and X_{b1} can be applied to provide detection of loss of excitation even where a generator initially operating at low power output (20-30%P_n) might settle down to operate as an induction generator.

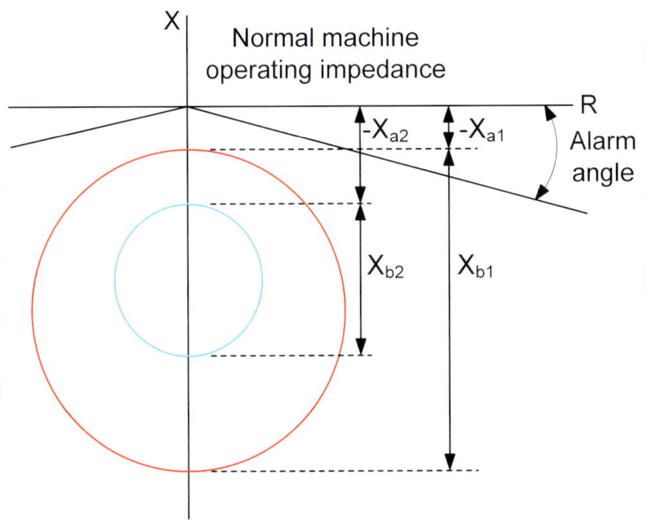

Figure 17.23: Loss of excitation protection characteristics

Pick-up and drop-off time delays t_{d1} and t_{do1} are associated with this impedance element. Timer t_{d1} is used to prevent operation during stable power swings that may cause the impedance locus of the generator to transiently enter the locus of operation set by X_{b1}. However, the value must short enough to prevent damage as a result of loss of excitation occurring. If pole-slipping protection is not required (see Section 17.17.2), timer t_{do1} can be set to give instantaneous reset. The second field failure element, comprising settings X_{b1}, X_{b2}, and associated timers t_{d2} and t_{do2} can be used to give instantaneous tripping following loss of excitation under full load conditions.

17.16.3 Protection Settings

The typical setting values for the two elements vary according to the excitation system and operating regime of the generator concerned, since these affect the generator impedance seen by the relay under normal and abnormal conditions. For a generator that is never operated at leading power factor, or at load angles in excess of 90° the typical settings are:

- impedance element diameter $X_{b1} = X_d$
- impedance element offset $X_{a1} = -0.5X'_d$
- time delay on pick-up, $t_{d1} = 0.5s - 10s$
- time delay on drop-off, $t_{do1} = 0s$

If a fast excitation system is employed, allowing load angles of up to 120° to be used, the impedance diameter must be reduced to take account of the reduced generator impedance seen under such conditions. The offset also needs revising. In these circumstances, typical settings would be:

- impedance element diameter $X_{b1} = 0.5X_d$
- impedance element offset $X_{a1} = -0.75X'_d$
- time delay on pick-up, $t_{d1} = 0.5\text{s} - 10\text{s}$
- time delay on drop-off, $t_{do1} = 0\text{s}$

The typical impedance settings for the second element, if used, are:

$$X_{b2} = \frac{kV^2}{MVA}$$

$$X_{a2} = -0.5X'_d \frac{kV^2}{MVA}$$

The time delay settings t_{d2} and t_{do2} are set to zero to give instantaneous operation and reset.

17.17 POLE SLIPPING PROTECTION

A generator may pole-slip, or fall out of synchronism with the power system for a number of reasons. The principal causes are prolonged clearance of a heavy fault on the power system, when the generator is operating at a high load angle close to the stability limit, or partial or complete loss of excitation. Weak transmission links between the generator and the bulk of the power system aggravate the situation. It can also occur with embedded generators running in parallel with a strong Utility network if the time for a fault clearance on the Utility network is slow, perhaps because only IDMT relays are provided. Pole slipping is characterised by large and rapid oscillations in active and reactive power. Rapid disconnection of the generator from the network is required to ensure that damage to the generator is avoided and that loads supplied by the network are not affected for very long.

Protection can be provided using several methods. The choice of method will depend on the probability of pole slipping occurring and on the consequences should it occur.

17.17.1 Protection Using Reverse Power Element

During pole-slipping, there will be periods where the direction of active power flow will be in the reverse direction, so a reverse power relay element can be used to detect this, if not used for other purposes. However, since the reverse power conditions are cyclical, the element will reset during the forward power part of the cycle unless either a very short pick-up time delay and/or a suitable drop-off time delay is used to eliminate resetting.

The main advantage of this method is that a reverse power element is often already present, so no additional relay elements are required. The main disadvantages are the time taken for tripping and the inability to control the system angle at which the generator breaker trip command would be issued, if it is a requirement to limit the breaker current interruption duty. There is also the difficulty of determining suitable settings. Determination of settings in the field, from a deliberate pole-slipping test is not possible and analytical studies may not discover all conditions under which pole-slipping will occur.

17.17.2 Protection Using an Under Impedance Element

With reference to Figure 17.21, a loss of excitation under impedance characteristic may also be capable of detecting loss of synchronism, in applications where the electrical centre of the power system and the generator lies 'behind' the relaying point. This would typically be the case for a relatively small generator that is connected to a power transmission system ($X_G >> (X_T + X_S)$). With reference to Figure 17.23; if pole-slipping protection response is required, the drop-off timer t_{do1} of the larger diameter impedance measuring element should be set to prevent its reset of in each slip cycle, until the t_{d1} trip time delay has expired. As with reverse power protection, this would be an elementary form of pole-slipping protection. It may not be suitable for large machines where rapid tripping is required during the first slip cycle and where some control is required for the system angle at which the generator circuit breaker trip command is given. Where protection against pole-slipping must be guaranteed, a more sophisticated method of protection should be used. A typical reset timer delay for pole-slipping protection might be 0.6s. For generator transformer units, the additional impedance in front of the relaying point may take the system impedance outside the under impedance relay characteristic required for loss of excitation protection. Therefore, the acceptability of this pole-slipping protection scheme will be dependent on the application.

17.17.3 Dedicated Pole-Slipping Protection

Large generator-transformer units directly connected to grid systems often require a dedicated pole-slipping protection scheme to ensure rapid tripping and with system angle control. Historically, dedicated protection schemes have usually been based on ohm-type impedance measurement characteristic.

17.17.3.1 Pole Slipping Protection by Impedance Measurement

Although a mho type element for detecting the change in impedance during pole-slipping can be used in some applications, but with performance limits, a straight line 'ohm' characteristic is more suitable. The protection principle is that of detecting the passage of the generator impedance through a zone defined by two such impedance characteristics, as shown in Figure 17.24. The characteristic is divided into three zones, A, B, and C. Normal operation of the generator lies in zone A. When a pole-slip occurs, the impedance traverses zones B and C, and tripping occurs when the impedance characteristic enters zone C.

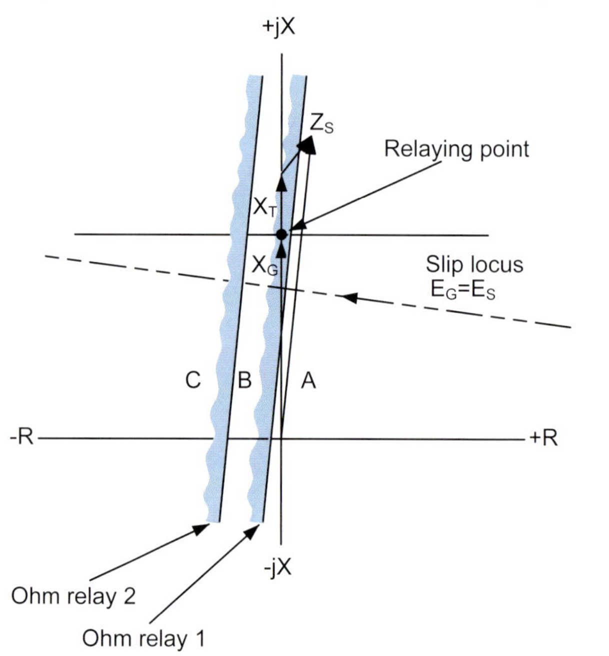

Figure 17.24: Pole slipping detection by ohm relays

Tripping only occurs if all zones are traversed sequentially. Power system faults should result in the zones not being fully traversed so that tripping will not be initiated. The security of this type of protection scheme is normally enhanced by the addition of a plain under impedance control element (circle about the origin of the impedance diagram) that is set to prevent tripping for impedance trajectories for remote power system faults. Setting of the ohm elements is such that they lie parallel to the total system impedance vector, and enclose it, as shown in Figure 17.24.

17.17.3.2 Use of Lenticular Characteristic

A more sophisticated approach is to measure the impedance of the generator and use a lenticular impedance characteristic to determine if a pole-slipping condition exists. The lenticular characteristic is shown in Figure 17.25. The characteristic is divided into two halves by a straight line, called the blinder.

The inclination, θ, of the lens and blinder is determined by the angle of the total system impedance. The impedance of the system and generator-transformer determines the forward reach of the lens, Z_A, and the transient reactance of the generator determines the reverse reach Z_B.

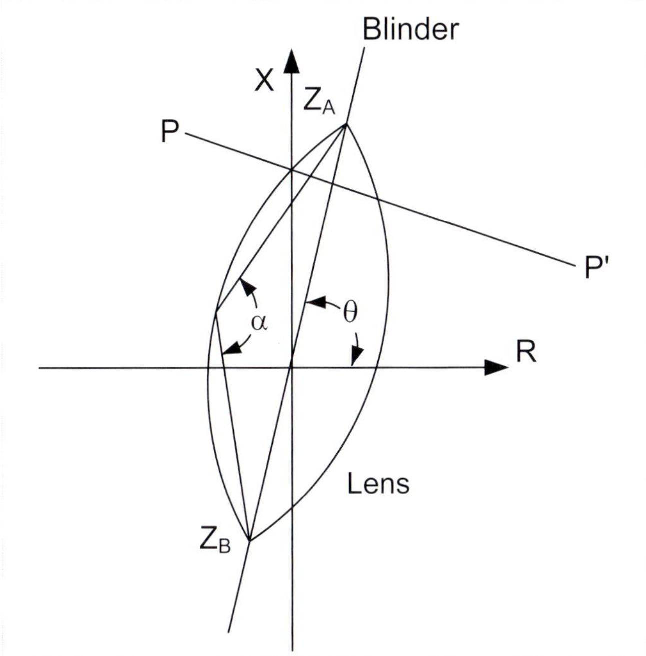

Figure 17.25: Pole-slipping protection using lenticular characteristic and blinder

The width of the lens is set by the angle α and the line PP', perpendicular to the axis of the lens, is used to determine if the centre of the impedance swing during a transient is located in the generator or power system.

Operation in the case of a generator is as follows. The characteristic is divided into 4 zones and 2 regions, as shown in Figure 17.26. Normal operation is with the measured impedance in zone R1. If a pole slip develops, the impedance locus will traverse though zones R2, R3, and R4. When entering zone R4, a trip signal is issued, provided the impedance lies below reactance line PP' and hence the locus of swing lies within or close to the generator – i.e. the generator is pole slipping with respect to the rest of the system.

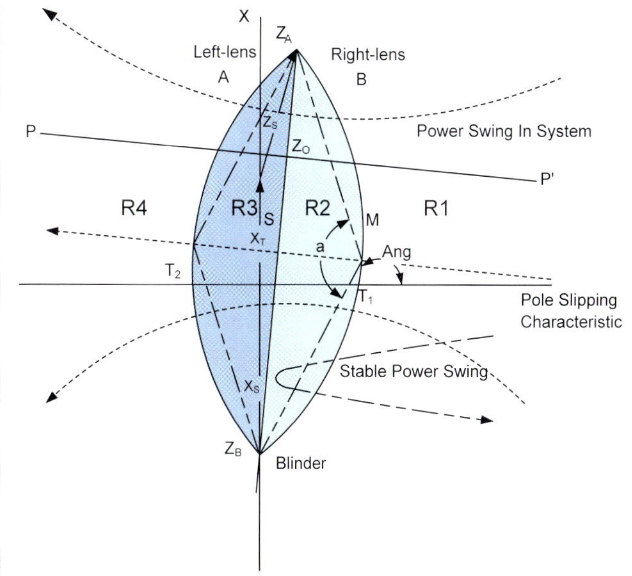

Figure 17.26: Definition of zones for lenticular characteristic

If the impedance locus lies above line PP', the swing lies far out in the power system – i.e. one part of the power system, including the protected generator, is swinging against the rest of the network. Tripping may still occur, but only if swinging is prolonged – meaning that the power system is in danger of complete break-up. Further confidence checks are introduced by requiring that the impedance locus spends a minimum time within each zone for the pole-slipping condition to be valid. The trip signal may also be delayed for a number of slip cycles even if a generator pole-slip occurs – this is to both provide confirmation of a pole-slipping condition and allow time for other relays to operate if the cause of the pole slip lies somewhere in the power system. Should the impedance locus traverse the zones in any other sequence, tripping is blocked.

17.18 STATOR OVERHEATING

Overheating of the stator may result from:

- overload
- failure of the cooling system
- overfluxing
- core faults

Accidental overloading might occur through the combination of full active load current component, governed by the prime mover output and an abnormally high reactive current component, governed by the level of rotor excitation and/or step-up transformer tap. With a modern protection relay, it is relatively simple to provide a current-operated thermal replica protection element to estimate the thermal state of the stator windings and to issue an alarm or trip to prevent damage.

Although current-operated thermal replica protection cannot take into account the effects of ambient temperature or uneven heat distribution, it is often applied as a back-up to direct stator temperature measuring devices to prevent overheating due to high stator current. With some relays, the thermal replica temperature estimate can be made more accurate through the integration of direct measuring resistance temperature devices.

Irrespective of whether current-operated thermal replica protection is applied or not, it is a requirement to monitor the stator temperature of a large generator in order to detect overheating from whatever cause.

Temperature sensitive elements, usually of the resistance type, are embedded in the stator winding at hot-spot locations envisaged by the manufacturer, the number used being sufficient to cover all variations. The elements are connected to a temperature sensing relay element arranged to provide alarm and trip outputs. The settings will depend on the type of stator winding insulation and on its permitted temperature rise.

17.19 MECHANICAL FAULTS

Various faults may occur on the mechanical side of a generating set. The following sections detail the more important ones from an electrical point of view.

17.19.1 Failure of the Prime Mover

When a generator operating in parallel with others loses its power input, it remains in synchronism with the system and continues to run as a synchronous motor, drawing sufficient power to drive the prime mover. This condition may not appear to be dangerous and in some circumstances will not be so. However, there is a danger of further damage being caused. Table 17.1 lists some typical problems that may occur.

Protection is provided by a low forward power/reverse power relay, as detailed in Section 17.11.

17.19.2 Overspeed

The speed of a turbo-generator set rises when the steam input is in excess of that required to drive the load at nominal frequency. The speed governor can normally control the speed, and, in any case, a set running in parallel with others in an interconnected system cannot accelerate much independently even if synchronism is lost. However, if load is suddenly lost when the HV circuit breaker is tripped, the set will begin to accelerate rapidly. The speed governor is designed to prevent a dangerous speed rise even with a 100% load rejection, but nevertheless an additional centrifugal overspeed trip device is provided to initiate an emergency mechanical shutdown if the overspeed exceeds 10%.

To minimise overspeed on load rejection and hence the mechanical stresses on the rotor, the following sequence is used whenever electrical tripping is not urgently required:

- trip prime mover or gradually reduce power input to zero
- allow generated power to decay towards zero
- trip generator circuit breaker only when generated power is close to zero or when the power flow starts to reverse, to drive the idle turbine

17.19.3 Loss of Vacuum

A failure of the condenser vacuum in a steam turbine driven generator results in heating of the tubes. This then produces strain in the tubes, and a rise in temperature of the low-pressure end of the turbine. Vacuum pressure devices initiate progressive unloading of the set and, if eventually necessary, tripping of the turbine valves followed by the high voltage circuit breaker. The set must not be allowed to motor in the event of loss of vacuum, as this would cause rapid overheating of the low-pressure turbine blades.

17.20 COMPLETE GENERATOR PROTECTION SCHEMES

From the preceding sections, it is obvious that the protection scheme for a generator has to take account of many possible faults and plant design variations. Determination of the types of protection used for a particular generator will depend on the nature of the plant and upon economic considerations, which in turn is affected by set size. Fortunately, modern, multi-function, numerical relays are sufficiently versatile to include all of the commonly required protection functions in a single package, thus simplifying the decisions to be made. The following sections provide illustrations of typical protection schemes for generators connected to a grid network, but not all possibilities are illustrated, due to the wide variation in generator sizes and types.

17.20.1 Direct-Connected Generator

A typical protection scheme for a direct-connected generator is shown in Figure 17.27. It comprises the following protection functions:

- stator differential protection
- overcurrent protection – conventional or voltage dependent
- stator earth fault protection
- overvoltage protection
- undervoltage protection

- overload/low forward power/reverse power protection (according to prime mover type)
- unbalanced loading
- overheating
- pole slipping
- loss of excitation
- underfrequency
- inadvertent energisation
- overfluxing
- mechanical faults

Figure 17.27 illustrates which trips require an instantaneous electrical trip and which can be time delayed until electrical power has been reduced to a low value. The faults that require tripping of the prime mover as well as the generator circuit breaker are also shown.

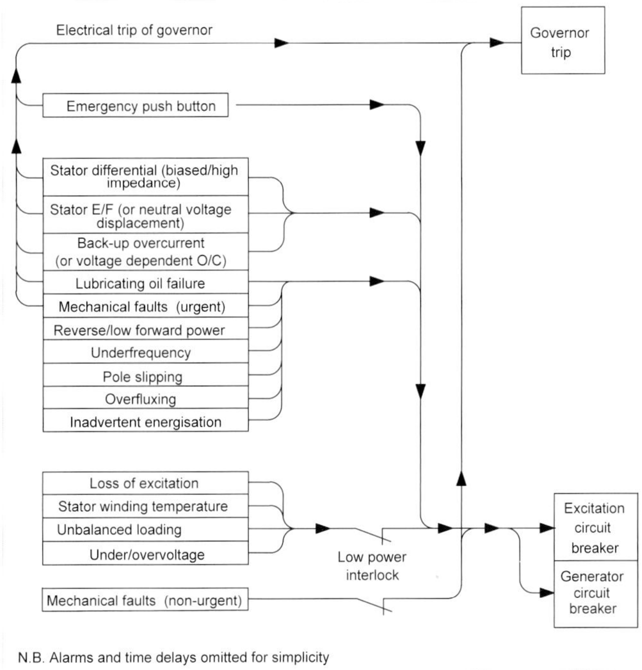

Figure 17.27: Typical protection arrangement for a direct-connected generator

17.20.2 Generator-Transformer Units

These units are generally of higher output than direct-connected generators, and hence more comprehensive protection is warranted. In addition, the generator transformer also requires protection, for which the protection detailed in Chapter 16 is appropriate. Overall biased generator/generator transformer differential protection is commonly applied in addition, or instead of, differential protection for the transformer alone. A single protection relay may incorporate all of the required functions, or the protection of the transformer (including overall generator/generator transformer differential protection) may utilise a separate relay. Figure 17.28 shows a typical overall scheme.

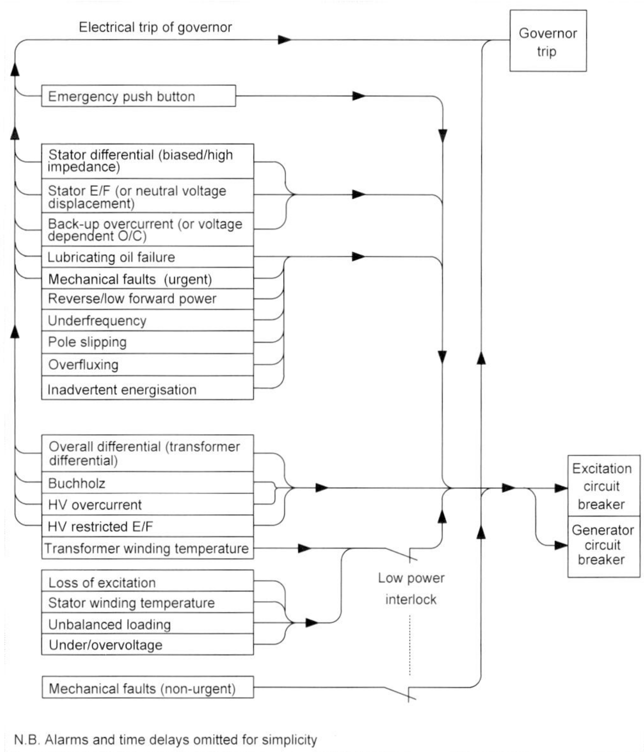

Figure 17.28: Typical tripping arrangements for generator-transformer unit

17.21 EMBEDDED GENERATION

In recent years, through de-regulation of the electricity supply industry and the ensuing commercial competition, many electricity users connected to MV power distribution systems have installed generating sets to operate in parallel with the public supply. The intention is either to utilise surplus energy from other sources, or to use waste heat or steam from the prime mover for other purposes. Parallel connection of generators to distribution systems did occur before de-regulation, but only where there was a net power import from the Utility. Power export to Utility distribution systems was a relatively new aspect. Since generation of this type can now be located within a Utility distribution system, as opposed to being centrally dispatched generation connected to a transmission system, the term 'Embedded Generation' is often applied. Figure 17.2 illustrates such an arrangement. Depending on size, the embedded generator(s) may be synchronous or asynchronous types, and they may be connected at any voltage appropriate to the size of plant being considered.

The impact of connecting generation to a Utility distribution system that was originally engineered only for downward power distribution must be considered, particularly in the area of protection requirements. In this respect, it is not important whether the embedded generator is normally capable of export to the Utility distribution system or not, since there may exist fault conditions when this occurs irrespective of the design intent.

If plant operation when disconnected from the Utility supply is required, underfrequency protection (Section 17.14.2) will become an important feature of the in-plant power system. During isolated operation, it may be relatively easy to overload the available generation, such that some form of load management system may be required. Similarly, when running in parallel with the Utility, consideration needs to be given to the mode of generator operation if reactive power import is to be controlled. The impact on the control scheme of a sudden break in the Utility connection to the plant main busbar also requires analysis. Where the in-plant generation is run using constant power factor or constant reactive power control, automatic reversion to voltage control when the Utility connection is lost is essential to prevent plant loads being subjected to a voltage outside acceptable limits.

Limits may be placed by the Utility on the amount of power/reactive power import/export. These may demand the use of an in-plant Power Management System to control the embedded generation and plant loads accordingly. Some Utilities may insist on automatic tripping of the interconnecting circuit breakers if there is a significant departure outside permissible levels of frequency and voltage, or for other reasons.

From a Utility standpoint, the connection of embedded generation may cause problems with voltage control and increased fault levels. The settings for protection relays in the vicinity of the plant may require adjustment with the emergence of embedded generation. It must also be ensured that the safety, security and quality of supply of the Utility distribution system is not compromised. The embedded generation must not be permitted to supply any Utility customers in isolation, since the Utility supply is normally the means of regulating the system voltage and frequency within the permitted limits. It also normally provides the only system earth connection(s), to ensure the correct performance of system protection in response to earth faults. If the Utility power infeed fails, it is also important to disconnect the embedded generation before there is any risk of the Utility power supply returning on to unsynchronised machines. In practice this generally requires the following protection functions to be applied at the 'Point of Common Coupling' (PCC) to trip the coupling circuit breaker:

- overvoltage
- undervoltage
- overfrequency
- underfrequency
- loss of Utility supply

In addition, particular circumstances may require additional protection functions:

- neutral voltage displacement
- reverse power
- directional overcurrent

In practice, it can be difficult to meet the protection settings or performance demanded by the Utility without a high risk of nuisance tripping caused by lack of co-ordination with normal power system faults and disturbances that do not necessitate tripping of the embedded generation. This is especially true when applying protection specifically to detect loss of the Utility supply (also called 'loss of mains') to cater for operating conditions where there would be no immediate excursion in voltage or frequency to cause operation of conventional protection functions.

17.21.1 Protection Against Loss of Utility Supply

If the normal power infeed to a distribution system, or to the part of it containing embedded generation is lost, the effects may be as follows:

a. embedded generation may be overloaded, leading to generator undervoltage/ underfrequency
b. embedded generation may be underloaded, leading to overvoltage/overfrequency
c. little change to the absolute levels of voltage or frequency if there is little resulting change to the load flow through the PCC

The first two effects are covered by conventional voltage and frequency protection. However, if condition (c) occurs, conventional protection may not detect the loss of Utility supply condition or it may be too slow to do so within the shortest possible auto-reclose dead-times that may be applied in association with Utility overhead line protection. Detection of condition (c) must be achieved if the requirements of the Utility are to be met. Many possible methods have been suggested, but the one most often used is the Rate of Change of Frequency (ROCOF) relay. Its application is based on the fact that the rate of change of small changes in absolute frequency, in response to inevitable small load changes, will be faster with the generation isolated than when the generation is in parallel with the public, interconnected power system. However, problems with nuisance tripping in response to national power system events, where the system is subject to significant frequency excursions following the loss of a large generator or a major power interconnector, have occurred.

This is particularly true for geographically islanded power systems, such as those of the British Isles. An alternative to ROCOF protection is a technique sometimes referred to as 'voltage vector shift' protection. In this technique the rate of phase change between the directly measured generator bus voltage is compared with a memorised a.c. bus voltage reference.

Sources of embedded generation are not normally earthed, which presents a potential safety hazard. In the event of an Utility system earth fault, the Utility protection should operate to remove the Utility power infeed. In theory, this should also result in removal of the embedded generation, through the action of the stipulated voltage/frequency protection and dependable 'loss of mains' protection. However, in view of safety considerations (e.g. fallen overhead line conductors in public areas), an additional form of earth fault protection may also be demanded to prevent the backfeed of an earth fault by embedded generation. The only way of detecting an earth fault under these conditions is to use neutral voltage displacement protection. The additional requirement is only likely to arise for embedded generation rated above 150kVA, since the risk of small embedded generators not being cleared by other means is negligible.

17.21.2 ROCOF Relay Description

A ROCOF relay detects the rate of change of frequency in excess of a defined setpoint. The signal is obtained from a voltage transformer connected close to the Point of Common Coupling (PCC). The principal method used is to measure the time period between successive zero-crossings to determine the average frequency for each half-cycle and hence the rate of change of frequency. The result is usually averaged over a number of cycles.

17.21.3 Voltage Vector Shift Relay Description

A voltage vector shift relay detects the drift in voltage phase angle beyond a defined setpoint as long as it takes place within a set period. Again, the voltage signal is obtained from a voltage transformer connected close to the Point of Common Coupling (PCC). The principal method used is to measure the time period between successive zero-crossings to determine the duration of each half-cycle, and then to compare the durations with the memorised average duration of earlier half-cycles in order to determine the phase angle drift.

17.21.4 Setting Guidelines

Should loss of the Utility supply occur, it is extremely unlikely that there will be an exact match between the output of the embedded generator(s) and the connected load. A small frequency change or voltage phase angle change will therefore occur, to which can be added any changes due to the small natural variations in loading of an isolated generator with time. Once the rate of change of frequency exceeds the setting of the ROCOF relay for a set time, or once the voltage phase angle drift exceeds the set angle, tripping occurs to open the connection between the in-plant and Utility networks.

While it is possible to estimate the rate of change of frequency from knowledge of the generator set inertia and MVA rating, this is not an accurate method for setting a ROCOF relay because the rotational inertia of the complete network being fed by the embedded generation is required. For example, there may be other embedded generators to consider. As a result, it is invariably the case that the relay settings are determined at site during commissioning. This is to ensure that the Utility requirements are met while reducing the possibility of a spurious trip under the various operating scenarios envisaged. However, it is very difficult to determine whether a given rate of change of frequency will be due to a 'loss of mains' incident or a load/frequency change on the public power network, and hence spurious trips are impossible to eliminate. Thus the provision of Loss of Utility Supply protection to meet power distribution Utility interface protection requirements, may actually conflict with the interests of the national power system operator. With the growing contribution of non-dispatched embedded generation to the aggregate national power demand, the loss of the embedded generation following a transmission system incident that may already challenge the security of the system can only aggravate the problem. There have been claims that voltage vector shift protection might offer better security, but it will have operation times that vary with the rate of change of frequency. As a result, depending on the settings used, operation times might not comply with Utility requirements under all circumstances. [Reference 17.1] provides further details of the operation of ROCOF relays and the problems that may be encountered.

Nevertheless, because such protection is a common requirement of some Utilities, the 'loss of mains' protection may have to be provided and the possibility of spurious trips will have to be accepted in those cases. Site measurements over a period of time of the typical rates of frequency change occurring may assist in negotiations of the settings with the Utility, and with the fine-tuning of the protection that may already be commissioned.

17.22 EXAMPLES OF GENERATOR PROTECTION SETTINGS

This section gives examples of the calculations required for generator protection. The first is for a typical small generator installed on an industrial system that runs in parallel with the

Utility supply. The second is for a larger generator-transformer unit connected to a grid system.

17.22.1 Protection Settings of a Small Industrial Generator

Generator Data							
KVA	KW	PF	Rated Voltage	Rated Current	Rated Frequency	Rated Speed	Prime Mover Type
6250	5000	0.8	11000	328	50	1500	Steam Turbine

Generator Parameters				
Generator Type	Xd p.u.	Xd' p.u.	CT Ratio	VT Ratio
Salient Pole	2.349	0.297	500/1	11000/110

Network Data			
Earthing Resistor	Maximum Earth Fault Current	Minimum Phase Fault Current	Maximum Downstream Phase Fault Current
31.7W	200A	145A	850A

Existing Protection						
CT Ratio	Overcurrent Settings			Earth Fault Settings		
	Characteristic	Setting	TMS	Characteristic	Setting	TMS
200/1	SI	144A	0.176	SI	48A	0.15

Salient details of the generator, network and protection required are given in Table 17.2. The example calculations are based on an Alstom MiCOM P343 relay in respect of setting ranges, etc.

Table 17.2: Data for small generator protection example

17.22.1.1 Differential Protection

Biased differential protection involves the determination of values for four setting values: I_{s1}, I_{s2}, K_1 and K_2 in Figure 17.5. I_{s1} can be set at 5% of the generator rating, in accordance with the recommendations for the relay, and similarly the values of I_{s2} (120%) and K_2 (150%) of generator rating. It remains for the value of K_1 to be determined. The recommended value is generally 0%, but this only applies where CTs that conform to IEC 60044-1 class PX (or the superseded BS 3938 Class X) are used – i.e. CTs specifically designed for use in differential protection schemes. In this application, the CTs are conventional class 5P CTs that meet the relay requirements in respect of knee-point voltage, etc. Where neutral tail and terminal CTs can saturate at different times due to transiently offset magnetising inrush or motor starting current waveforms with an r.m.s level close to rated current, and where there is a high L/R time constant for the offset, the use of a 0% bias slope may give rise to maloperation. Such waveforms can be encountered when plant of similar rating to the generator is being energised or started. Differences between CT designs or differing remanent flux levels can lead to asymmetric saturation and the production of a differential spill current. Therefore, it is appropriate to select a non-zero setting for K_1, and a value of 5% is usual in these circumstances.

17.22.1.2 Voltage Controlled Overcurrent Protection

This protection is applied as remote backup to the downstream overcurrent protection in the event of protection or breaker failure conditions. This ensures that the generator will not continue to supply the fault under these conditions.

At normal voltage, the current setting must be greater than the maximum generator load current of 328A. A margin must be allowed for resetting of the relay at this current (reset ratio =

95%) and for the measurement tolerances of the relay (5% of I_s under reference conditions), therefore the current setting is calculated as:

$$I_{vcset} > \frac{328}{0.95} \times 1.05$$
$$> 362.5A$$

The nearest settable value is 365A, or $0.73In$.

The minimum phase-phase voltage for a close-up single-phase to earth fault is 57%, so the voltage setting V_S must be less than this. A value of 30% is typically used, giving V_S = 33V. The current setting multiplying factor K must be chosen such that KI_S is less than 50% of the generator steady-state current contribution to an uncleared remote fault. This information is not available (missing data being common in protection studies). However, the maximum sustained close-up phase fault current (neglecting AVR action) is 145A, so that a setting chosen to be significantly below this value will suffice. A value of 87.5A (60% of the close-up sustained phase fault current) is therefore chosen, and hence K = 0.6. This is considered to be appropriate based on knowledge of the system circuit impedances. The TMS setting is chosen to co-ordinate with the downstream feeder protection such that:

- for a close-up feeder three-phase fault, that results in almost total voltage collapse as seen by the relay
- for a fault at the next downstream relay location, if the relay voltage is less than the switching voltage

It should also be chosen so that the generator cannot be subjected to fault or overload current in excess of the stator short-time current limits. A curve should be provided by the manufacturer, but IEC 60034-1 demands that an AC generator should be able to pass 1.5 times rated current for at least 30 seconds. The operating time of the downstream protection for a three-phase fault current of 850A is 0.682s, so the voltage controlled relay element should have a minimum operating time of 1.09s (0.4s grading margin used as the relay technology used for the downstream relay is not stated – see Table 9.2). With a current setting of 87.5A, the operating time of the voltage controlled relay element at a TMS of 1.0 is:

$$\frac{0.14}{\left(\frac{850}{87.5}\right)^{0.02} - 1} = 3.01s$$

Therefore a TMS of:

$$\frac{1.09}{3.01} = 0.362$$

is required. Use 0.375, nearest available setting.

17.22.1.3 Stator Earth Fault Protection

The maximum earth fault current, from Table 17.2, is 200A. Protection for 95% of the winding can be provided if the relay is set to detect a primary earth fault current of 16.4A, and this equates to a CT secondary current of 0.033A. The nearest relay setting is 0.04A, providing protection for 90% of the winding.

The protection must grade with the downstream earth fault protection, the settings of which are also given in Table 17.2. At an earth fault current of 200A, the downstream protection has an operation time of 0.73s. The generator earth fault protection must therefore have an operation time of not less than 1.13s. At a TMS of 1.0, the generator protection relay operating time will be:

$$\left[\frac{0.14}{\left(\frac{200}{20}\right)^{0.02} - 1}\right] s$$

= 2.97s, so the required TMS is $\frac{1.13}{2.97} = 0.38$.

Use a setting of 0.4, nearest available setting.

17.22.1.4 Neutral Voltage Displacement Protection

This protection is provided as back-up earth-fault protection for the generator and downstream system (direct-connected generator). It must therefore have a setting that grades with the downstream protection. The protection is driven from the generator star-connected VT, while the downstream protection is current operated.

It is therefore necessary to translate the current setting of the downstream setting of the current-operated earth-fault protection into the equivalent voltage for the NVD protection. The equivalent voltage is found from the formula:

$$V_{eff} = \frac{(I_{pe} \times Z_e) \times 3}{VTratio}$$
$$= \frac{48 \times 31.7 \times 3}{100}$$
$$= 45.6V$$

where:

V_{eff} = effective voltage setting

I_{pe} = downstream earth fault current setting

Z_e = earthing resistance

Hence a setting of 48V is acceptable. Time grading is

required, with a minimum operating time of the NVD protection of 1.13s at an earth fault current of 200A. Using the expression for the operation time of the NVD element:

$$t = \frac{K}{(M-1)} s$$

where:

$$M = \left(\frac{V}{V_{snvd}}\right)$$

and

V = voltage seen by relay

V_{snvd} = relay setting voltage

the value of K can be calculated as 3.34. The nearest settable value is 3.5, giving an operation time of 1.18s.

17.22.1.5 Loss of Excitation Protection

Loss of excitation is detected by a mho impedance relay element, as detailed in Section 17.16.2. The standard settings for the P340 series relay are:

$$X_a = 0.5'_d \times \left(\frac{CTratio}{VTratio}\right) \text{ secondary quantities}$$

$$= -0.5 \times 0.297 \times 19.36 \times \frac{500}{100}$$

$$= -14.5\Omega$$

$$X_b = X'_d \times \left(\frac{CTratio}{VTratio}\right)$$

$$= 2.349 \times 19.36 \times \frac{500}{100}$$

$$= 227\Omega$$

The nearest settings provided by the relay are

$$X_a = -14.5\Omega$$

$$X_b = 227\Omega$$

The time delay t_{d1} should be set to avoid relay element operation on power swings and a typical setting of 3s is used. This value may need to be modified in the light of operating experience. To prevent cyclical pick-up of the relay element without tripping, such as might occur during pole-slipping conditions, a drop-off time delay tdo1 is provided and set to 0.5s.

17.22.1.6 Negative Phase Sequence Current Protection

This protection is required to guard against excessive heating from negative phase sequence currents, whatever the cause. The generator is of salient pole design, so from IEC 60034-1, the continuous withstand is 8% of rating and the I^2t value is 20s. Using Equation 17.1, the required relay settings can found as I_2>> = 0.05 and K = 8.6s. The nearest available values are I_2>> = 0.05 and K = 8.6s. The relay also has a cooling time constant K_{reset} that is normally set equal to the value of K. To co-ordinate with clearance of heavy asymmetric system faults, that might otherwise cause unnecessary operation of this protection, a minimum operation time t_{min} should be applied. It is recommended to set this to a value of 1. Similarly, a maximum time can be applied to ensure that the thermal rating of the generator is not exceeded (as this is uncertain, data not available) and to take account of the fact that the P343 characteristic is not identical with that specified in IEC 60034. The recommended setting for t_{max} is 600s.

17.22.1.7 Overvoltage Protection

This is required to guard against various failure modes, e.g. AVR failure, resulting in excessive stator voltage. A two-stage protection is available, the first being a low-set time-delayed stage that should be set to grade with transient overvoltages that can be tolerated following load rejection. The second is a high-set stage used for instantaneous tripping in the event of an intolerable overvoltage condition arising.

Generators can normally withstand 105% of rated voltage continuously, so the low-set stage should be set higher than this value. A setting of 117.7V in secondary quantities (corresponding to 107% of rated stator voltage) is typically used, with a definite time delay of 10s to allow for transients due to load switch-off/rejection, overvoltages on recovery from faults or motor starting, etc.

The second element provides protection in the event of a large overvoltage, by tripping excitation and the generator circuit breaker (if closed). This must be set below the maximum stator voltage possible, taking into account saturation. As the open circuit characteristic of the generator is not available, typical values must be used. Saturation will normally limit the maximum overvoltage on this type of generator to 130%, so a setting of 120% (132V secondary) is typically used. Instantaneous operation is required. Generator manufacturers are normally able to provide recommendations for the relay settings. For embedded generators, the requirements of the local Utility may also have to be taken into account. For both elements, a variety of voltage measurement modes are available to take account of possible VT connections (single or three-phase, etc.), and conditions to be protected against. In this example, a three-phase VT connection is used, and overvoltages on any phase are to be detected, so a selection of 'Any' is used for this setting.

17.22.1.8 Underfrequency Protection

This is required to protect the generator from sustained overload conditions during periods of operation isolated from the Utility supply. The generating set manufacturer will normally provide the details of machine short-time capabilities. The example relay provides four stages of underfrequency protection. In this case, the first stage is used for alarm purposes and a second stage would be applied to trip the set.

The alarm stage might typically be set to 49Hz, with a time delay of 20s, to avoid an alarm being raised under transient conditions, e.g. during plant motor starting. The trip stage might be set to 48Hz, with a time delay of 0.5s, to avoid tripping for transient, but recoverable, dips in frequency below this value.

17.22.1.9 Reverse Power Protection

The relay setting is 5% of rated power.

$$= \left(\frac{0.05 \times 5 \times 10^6}{CTratio \times VTratio} \right)$$

$$= \left(\frac{0.05 \times 5 \times 10^6}{500 \times 100} \right)$$

$$= 5W$$

This value can be set in the relay. A time delay is required to guard against power swings while generating at low power levels, so use a time delay of 5s. No reset time delay is required.

A summary of the relay settings is given in Table 17.3.

Protection	Quantity	Value
Differential protection	I_{s1}	5%
	I_{s2}	120%
	K_1	5%
	K_2	150%
Stator earth fault	I_{se}	0.04
	TMS	0.4
Neutral Voltage Displacement	V_{snvd}	48V
	K	3.5
Loss of excitation	X_a	-14.5Ω
	X_b	227Ω
	t_{d1}	3s
	t_{do1}	0.5s
Voltage controlled overcurrent	I_{vcset}	0.73
	V_s	33
	K	0.6
	TMS	0.375
Negative phase sequence	I2>>	0.05
	K	8.6s
	K_{reset}	8.6s
	t_{min}	1.5s
	t_{max}	600s
Overvoltage	V> meas mode	three-phase
	V> operate mode	any
	V>1 setting	107%
	V>1 function	DT
	V>1 time delay	10s
	V>2 setting	120%
	V>2 function	DT
	V>2 time delay	0sec
Underfrequency	F<1 setting	49Hz
	F<1 time delay	20s
	F<2 setting	48Hz
	F<2 time delay	0.5s
Reverse Power	P1 function	reverse power
	P1 setting	5W
	P1 time delay	5s
	P1 DO time	0s

Table 17.3: Small generator protection example – relay settings

17.22.2 Large Generator Transformer Unit Protection

The data for this unit are given in Table 17.4. It is fitted with two main protection systems to ensure security of tripping in the event of a fault. To economise on space, the setting calculations for only one system, that using an Alstom MiCOM P343 relay are given. Settings are given in primary quantities throughout.

Parameter	Value	Unit
Generator MVA rating	187.65	MVA
Generator MW rating	160	MW
Generator voltage	18	kV
Synchronous reactance	1.93	pu
Direct-axis transient reactance	0.189	pu
Minimum operating voltage	0.8	pu
Generator negative sequence capability	0.08	pu
Generator negative sequence factor, K_g	10	
Generator third harmonic voltage under load	0.02	pu
Generator motoring power	0.02	pu
Generator overvoltage	alarm	1.1
	time delay	5
	trip	1.3
Generator undervoltage	not required	
Max pole slipping frequency	10	Hz
Generator transformer rating	360	MVA
Generator transformer leakage reactance	0.244	pu
Generator transformer overflux alarm	1.1	pu
Generator transformer overflux alarm	1.2	pu
Network resistance (referred to 18kV)	0.56	mΩ
Network reactance (referred to 18kV)	0.0199	Ω
System impedance angle (estimated)	80	deg
Minimum load resistance	0.8	Ω
Generator CT ratio	8000/1	
Generator VT ratio	18000/120	
Number of generators in parallel	2	

Table 17.4: System data for large generator protection example

17.22.2.1 Biased Differential Protection

The settings follow the guidelines previously stated. As 100% stator winding earth-fault protection is provided, high sensitivity is not required and hence I_{s1} can be set to 10% of generator rated current. This equates to 602A, and the nearest settable value on the relay is 640A (= 0.08 of rated CT current). The settings for K_1, I_{s2} and K_2 follow the guidelines in the relay manual.

17.22.2.2 Voltage Restrained Overcurrent Protection

The setting current I_{set} has to be greater than the full-load current of the generator (6019A). A suitable margin must be allowed for operation at reduced voltage, so use a multiplying factor of 1.2. The nearest settable value is 7200A. The factor K is calculated so that the operating current is less than the current for a remote end three phase fault. The steady-state current and voltage at the generator for a remote-end three-phase fault are given by the expressions:

$$I_{flt} = \frac{V_N}{\sqrt{(nR_f)^2 + (X_d + X_t + nX_f)^2}}$$

Where

I_{flt} = minimum generator primary current for a multi-phase feeder-end fault

V_N = no load phase-neutral generator voltage

X_d = generator d-axis synchronous reactance

X_t = generator transformer reactance

R_f = feeder resistance

X_f = feeder reactance

n = number of parallel generators

Hence,

$$I_{flt} = 2893A = 0.361 I_n$$

and

$$V_{flt} = \frac{V_N\sqrt{3\left((nR_f)^2 + (X_t + nX_f)^2\right)}}{\sqrt{(nR_f)^2 + (X_d + X_t + nX_f)^2}}$$

$$= 1304V$$

$$= 0.07V_N$$

A suitable value of K is therefore

$$\frac{0.361}{1.2} = 0.3.$$

A suitable value of V2set is 120% of Vflt, giving a value of 1565V. The nearest settable value is 3000V, minimum allowable relay setting. The value of $V1_{set}$ is required to be above the minimum voltage seen by the generator for a close-up phase-earth fault. A value of 80% of rated voltage is used for $V1_{set}$, 14400V.

17.2.2.3 Inadvertent Energisation Protection

This protection is a combination of overcurrent with undervoltage, the voltage signal being obtained from a VT on the generator side of the system. The current setting used is that of rated generator current of 6019A, in accordance with IEEE C37.102 as the generator is for installation in the USA. Use 6000A nearest settable value. The voltage setting cannot be more than 85% of the generator rated voltage to ensure operation does not occur under normal operation. For this application, a value of 50% of rated voltage is chosen.

17.22.2.4 Negative Phase Sequence Protection

The generator has a maximum steady-state capability of 8% of rating, and a value of K_g of 10. Settings of I_{2cmr} = 0.06 (=480A) and K_g = 10 are therefore used. Minimum and maximum time delays of 1s and 1300s are used to co-ordinate with external protection and ensure tripping at low levels of negative sequence current are used.

17.22.2.5 Overfluxing Protection

The generator-transformer manufacturer supplied the following characteristics:

Alarm $\frac{V}{f} > 1.1$

Trip $\frac{V}{f} > 1.2$ inverse time characteristic

Hence the alarm setting is

$$18000 \times \frac{1.05}{60} = 315\ {}^{V}\!/_{Hz}$$

A time delay of 5s is used to avoid alarms due to transient conditions. The trip setting is:

$$18000 \times \frac{1.2}{60} = 360\ {}^{V}\!/_{Hz}$$

A TMS value of 10 is selected, to match the withstand curve supplied by the manufacturer.

17.22.2.6 100% Stator Earth Fault Protection

This is provided by a combination of neutral voltage displacement and third harmonic undervoltage protection. For the neutral voltage displacement protection to cover 90% of the stator winding, the minimum voltage allowing for generator operation at a minimum of 92% of rated voltage is:

$$\frac{0.92 \times 18kV \times 0.1}{\sqrt{3}} = 956.1V$$

Use a value of 935.3V, nearest settable value that ensures 90% of the winding is covered. A 0.5s definite time delay is used to prevent spurious trips. The third harmonic voltage under normal conditions is 2% of rated voltage, giving a value of:

$$\frac{18kV \times 0.01}{\sqrt{3}} = 207.8V$$

The third harmonic undervoltage protection setting must be below this value, a factor of 80% is acceptable. Use a value of 166.3V and a time delay of 0.5s. Inhibition of the element at low generator output must be determined during commissioning.

17.22.2.7 Loss of Excitation Protection

The client requires a two-stage loss of excitation protection function. The first is alarm only, while the second provides tripping under high load conditions. To achieve this, the first impedance element of the P343 loss of excitation protection can be set in accordance with the guidelines of Section 17.16.3 for a generator operating at rotor angles up to 120°,as follows:

$$X_{b1} = 0.5X_d = 1.666\Omega$$
$$X_{a1} = -0.75X_d^{'} = -0.245\Omega$$

Use nearest settable values of 1.669Ω and 0.253Ω. A time delay of 5s is used to prevent alarms under transient conditions. For the trip stage, settings for high load are used, as given in Section 17.16.3:

$$X_{b2} = \frac{kV^2}{MVA} = \frac{18^2}{187.65} = 1.727\Omega$$
$$X_{a2} = -0.75X_d^{'} = -0.1406\Omega$$

The nearest settable value for X_{b2} is 1.725Ω. A time delay of 0.5s is used.

17.22.2.8 Reverse Power Protection

The manufacturer-supplied value for motoring power is 2% of rated power. The recommended setting is therefore 1.6MW. An instrumentation class CT is used in conjunction with the relay for this protection, to ensure accuracy of measurement. A time delay of 0.5s is used. The settings should be checked at the commissioning stage.

17.22.9 Over/Under-Frequency Protection

For under-frequency protection, the client has specified the following characteristics:

- Alarm: 59.3Hz, 0.5s time delay
- 1st stage trip: 58.7Hz, 100s time delay
- 2nd stage trip: 58.2Hz, 1s time delay

Similarly, the overfrequency is required to be set as follows:

- Alarm: 62Hz, 30s time delay
- Trip: 63.5Hz, 10s time delay

These characteristics can be set in the relay directly.

17.22.2.10 Overvoltage Protection

The generator manufacturers' recommendation is:

- Alarm: 110% voltage for 5s
- Trip: 130% voltage, instantaneous

This translates into the following relay settings:

- Alarm: 19800V, 5s time delay
- Trip: 23400V, 0.1s time delay

17.22.2.11 Pole Slipping Protection

This is provided by the method described in Section 17.7.2.2. Detection at a maximum slip frequency of 10Hz is required. The setting data, according to the relay manual, is as follows:

- Forward Reach, $Z_A = Z_n + Z_T = 0.02 + 0.22 = 0.24\Omega$
- Reverse Reach, $Z_B = Z_{GEN} = 2 \times X'_d = 0.652\Omega$
- Reactance Line,, $Z_C = 0.9 \times Z_t = 0.9 \times 0.22 = 0.198\Omega$

where:

Z_t = generator transformer leakage current

Z_n = network impedance

The nearest settable values are 0.243Ω, 0.656Ω, and 0.206Ω respectively.

The lens angle setting, α, is found from the equation:

$$\alpha_{\min} = 180^\circ - 2\tan^{-1}\left(\frac{1.54 - R_{t\min}}{Z_A + Z_B}\right)$$

and, substituting values,

$$\alpha_{\min} = 62.5^\circ$$

Use the minimum settable value of 90°. The blinder angle, θ, is estimated to be 80°, and requires checking during commissioning. Timers T_1 and T_2 are set to 15ms as experience has shown that these settings are satisfactory to detect pole slipping frequencies up to 10Hz.

This completes the settings required for the generator, and the relay settings are given in Table 17.5. Of course, additional protection is required for the generator transformer, according to the principles described in Chapter 16.

Protection	Quantity	Value	Protection	Quantity	Value
Differential protection	I_{s1}	8%	Reverse Power	P1 function	reverse power
	I_{s2}	100%		P1 setting	1.6MW
	K1	0%		P1 time delay	0.5s
	K_2	150%		P1 DO time	0s
100% Stator earth fault	$V_{n3H<}$	166.3V	Inadvertent energisation	Dead Mach I>	6000A
	V_{n3H} delay	0.5s		Dead Mach V<	9000V
Neutral Voltage Displacement	V_{snvd}	935.3V	Pole Slipping Protection	Z_a	0.243W
	Time Delay	0.5s		Z_b	0.656W
Loss of Excitation	X_{a1}	-0.245W		Z_c	0.206W
	X_{b1}	1.666W		a	90°
	t_{d1}	5s		q	80°
	X_{a2}	-0.1406W		T_1	15ms
	X_{b2}	1.725W		T_2	15ms
	t_{d2}	0.5s	Overfrequency	F>1 setting	62Hz
	t_{DO1}	0s		F>1 time delay	30s
Voltage restrained overcurrent	I_{set}	7200A		F>2 setting	63.5Hz
	K	3		F>2 time delay	10s
	V_{1set}	14400V	Reverse Power	P1 function	reverse power
	V_{2set}	3000V		P1 setting	1.6MW
Negative phase sequence	I2>>	0.06		P1 time delay	0.5s
	K_g	10		P1 DO time	0s
	K_{reset}	10	Underfrequency	F<1 setting	59.3Hz
	t_{min}	1s		F<1 time delay	0.5s
	t_{max}	1300s		F<2 setting	58.7Hz
Overvoltage	V> meas mode	three-phase		F<2 time delay	100s
	V> operate mode	any		F<3 setting	58.2Hz
	V>1 setting	19800V		F<3 time delay	1s
	V>1 function	DT			
	V>1 time delay	5s			
	V>2 setting	23400V			
	V>2 function	DT			
	V>2 time delay	0.1s			

Table 17.5: Relay settings for large generator protection example

17.23 REFERENCE

[17.1] Survey of Rate of Change of Frequency Relays and Voltage Phase Shift Relays for Loss of Mains Protection. ERA Report 95-0712R, 1995. ERA Technology Ltd.

Chapter 18
Industrial and Commercial Power System Protection

18.1 INTRODUCTION

As industrial and commercial operations processes and plants have become more complex and extensive (Figure 18.1), the requirement for improved reliability of electrical power supplies has also increased. The potential costs of outage time following a failure of the power supply to a plant have risen dramatically as well. The introduction of automation techniques into industry and commerce has naturally led to a demand for the deployment of more power system automation, to improve reliability and efficiency.

Figure 18.1: Large modern industrial plant

The protection and control of industrial power supply systems must be given careful attention. Many of the techniques that have been evolved for EHV power systems may be applied to lower voltage systems also, but typically on a reduced scale. However, industrial systems have many special problems that have warranted individual attention and the development of specific solutions.

Many industrial plants have their own generation installed. Sometimes it is for emergency use only, feeding a limited number of busbars and with limited capacity. This arrangement is often adopted to ensure safe shutdown of process plant and personnel safety. In other plants, the nature of the process allows production of a substantial quantity of electricity, perhaps allowing export of any surplus to the public supply system – at either at sub-transmission or distribution voltage levels. Plants that run generation in parallel with the public supply distribution network are often referred to as *co-generation* or *embedded* generation. Special protection arrangements may be demanded for the point of connection between the private and public utility plant (see Chapter 17 for further details).

Industrial systems typically comprise numerous cable feeders and transformers. Chapter 16 covers the protection of transformers and Chapters 9/10 the protection of feeders.

18.2 BUSBAR ARRANGEMENT

The arrangement of the busbar system is obviously very important, and it can be quite complex for some very large industrial systems. However, in most systems a single busbar divided into sections by a bus-section circuit breaker is common, as shown in Figure 18.2. Main and standby drives for a particular item of process equipment will be fed from different sections of the switchboard, or sometimes from different switchboards.

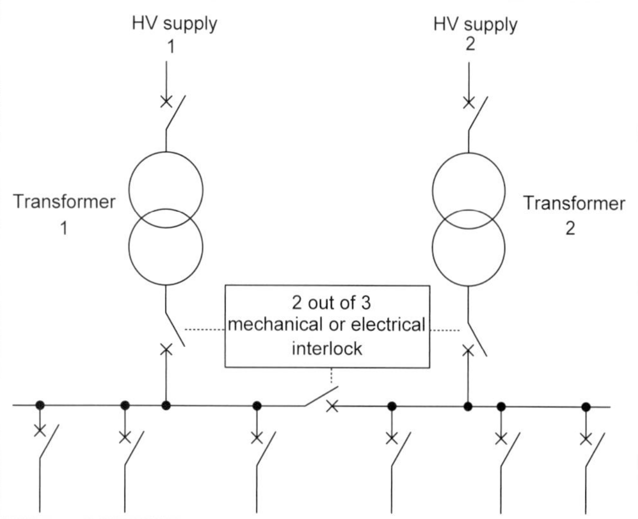

Figure 18.2: Typical switchboard configuration for an industrial plant

The main power system design criterion is that single outages on the electrical network within the plant will not cause loss of both the main and standby drives simultaneously. Considering a medium sized industrial supply system, shown in Figure 18.3, in more detail, it will be seen that not only are duplicate supplies and transformers used, but also certain important loads are segregated and fed from 'Essential Services Board(s)' (also known as 'Emergency' boards), distributed throughout the plant. This enables maximum utilisation of the standby generator facility. A standby generator is usually of the turbo-charged diesel-driven type. On detection of loss of incoming supply at any switchboard with an emergency section, the generator is automatically started. The appropriate circuit breakers will close once the generating set is up to speed and rated voltage to restore supply to the Essential Services sections of the switchboards affected, provided that the normal incoming supply is absent - for a typical diesel generator set, the emergency supply would be available within 10-20 seconds from the start sequence command being issued.

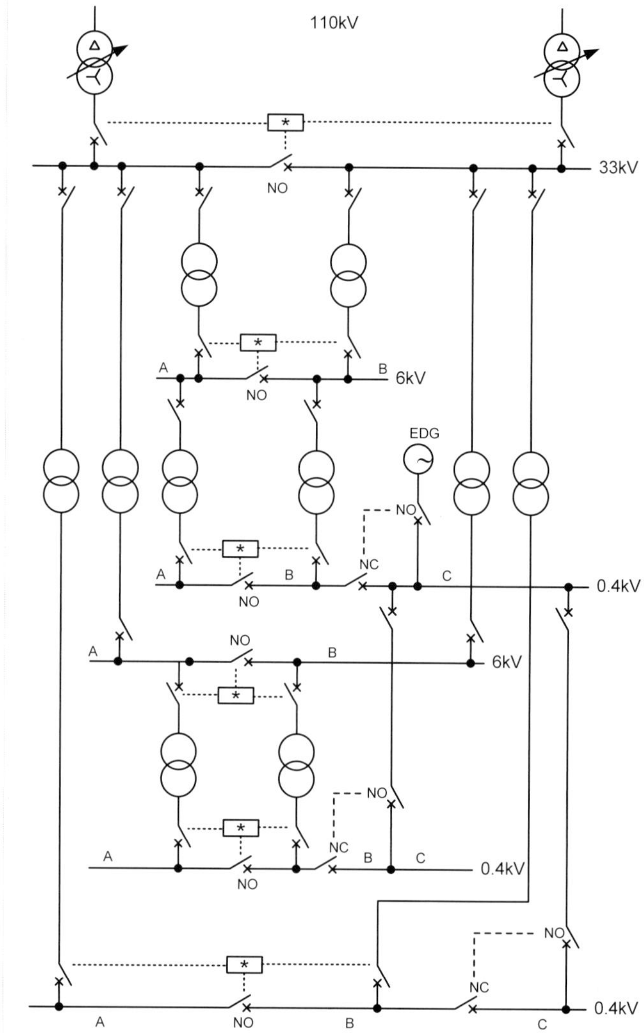

Figure 18.3: Typical industrial power system

The Essential Services Boards are used to feed equipment that is essential for the safe shut down, limited operation or preservation of the plant and for the safety of personnel. This will cover process drives essential for safe shutdown, venting systems, UPS loads feeding emergency lighting, process control computers, etc. The emergency generator may range in size from a single unit rated 20-30kW in a small plant up to several units of 2-10MW rating in a large oil refinery or similar plant. Large financial trading institutions may also have standby power requirements of several MW to maintain computer services.

18.3 DISCRIMINATION

Protection equipment works in conjunction with switchgear. For a typical industrial system, feeders and plant will be protected mainly by circuit breakers of various types and by fused contactors. Circuit breakers will have their associated overcurrent and earth fault relays. A contactor may also be equipped with a protection device (e.g. motor protection), but associated fuses are provided to break fault currents in excess of the contactor interrupting capability. The rating of fuses and selection of relay settings is carried out to ensure that discrimination is achieved – i.e. the ability to select and isolate only the faulty part of the system.

18.4 HRC FUSES

The protection device nearest to the actual point of power utilisation is most likely to be a fuse or a system of fuses and it is important that consideration is given to the correct application of this important device.

The HRC fuse is a key fault clearance device for protection in industrial and commercial installations, whether mounted in a distribution fuseboard or as part of a contactor or fuse-switch. The latter is regarded as a vital part of LV circuit protection, combining safe circuit making and breaking with an isolating capability achieved in conjunction with the reliable short-circuit protection of the HRC fuse. Fuses combine the characteristics of economy and reliability; factors that are most important in industrial applications.

HRC fuses remain consistent and stable in their breaking characteristics in service without calibration and maintenance. This is one of the most significant factors for maintaining fault clearance discrimination. Lack of discrimination through incorrect fuse grading will result in unnecessary disconnection of supplies, but if both the major and minor fuses are HRC devices of proper design and manufacture, this need not endanger personnel or cables associated with the plant.

18.4.1 Fuse Characteristics

The time required for melting the fusible element depends on the magnitude of current. This time is known as the 'pre-arcing' time of the fuse. Vaporisation of the element occurs on melting and there is fusion between the vapour and the filling powder leading to rapid arc extinction.

Fuses have a valuable characteristic known as 'cut-off', shown in Figure 18.4. When an unprotected circuit is subjected to a short circuit fault, the r.m.s. current rises towards a 'prospective' (or maximum) value. The fuse usually interrupts the short circuit current before it can reach the prospective value, in the first quarter to half cycle of the short circuit. The rising current is interrupted by the melting of the fusible element, subsequently dying away dying away to zero during the arcing period.

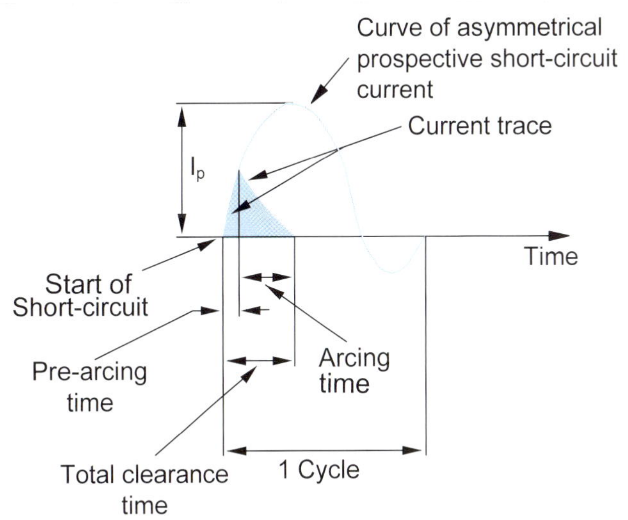

Figure 18.4: HRC fuse cut-off feature

Since the electromagnetic forces on busbars and connections carrying short circuit current are related to the square of the current, it will be appreciated that 'cut-off' significantly reduces the mechanical forces produced by the fault current and which may distort the busbars and connections if not correctly rated. A typical example of 'cut-off' current characteristics is shown in Figure 18.5. It is possible to use this characteristic during the design stage of a project to utilise equipment with a lower fault withstand rating downstream of the fuse, than would be the case if 'cut-off' was ignored. This may save on costs, but appropriate documentation and maintenance controls are required to ensure that only replacement fuses of very similar characteristics are used throughout the lifetime of the plant concerned – otherwise a safety hazard may arise.

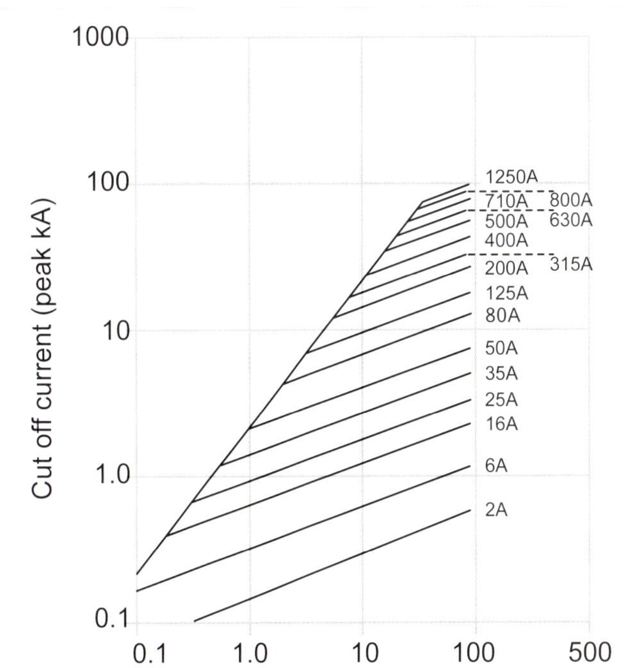

Figure 18.5: Typical fuse cut-off current characteristics

18.4.2 Discrimination Between Fuses

Fuses are often connected in series electrically and it is essential that they should be able to discriminate with each other at all current levels. Discrimination is obtained when the larger ('major') fuse remains unaffected by fault currents that are cleared by the smaller ('minor') fuse.

The fuse operating time can be considered in two parts:

- the time taken for fault current to melt the element, known as the 'pre-arcing time'
- the time taken by the arc produced inside the fuse to extinguish and isolate the circuit, known as the 'arcing time'

The total energy dissipated in a fuse during its operation consists of 'pre-arcing energy' and 'arc energy'. The values are usually expressed in terms of I^2t, where I is the current passing through the fuse and t is the time in seconds. Expressing the quantities in this manner provides an assessment of the heating effect that the fuse imposes on associated equipment during its operation under fault conditions.

To obtain positive discrimination between fuses, the total I^2t value of the minor fuse must not exceed the pre-arcing I^2t value of the major fuse. In practice, this means that the major fuse will have to have a rating significantly higher than that of the minor fuse, and this may give rise to problems of discrimination. Typically, the major fuse must have a rating of at least 160% of the minor fuse for discrimination to be obtained.

18.4.3 Protection of Cables by Fuses

PVC cable is allowed to be loaded to its full nominal rating only if it has 'close excess current protection'. This degree of protection can be given by means of a fuse link having a 'fusing factor' not exceeding 1.5, where:

- Fusing factor = Minimum Fusing Current/Current Rating

Cables constructed using other insulating materials (e.g. paper, XLPE) have no special requirements in this respect.

18.4.4 Effect of Ambient Temperature

High ambient temperatures can influence the capability of HRC fuses. Most fuses are suitable for use in ambient temperatures up to 35°C, but for some fuse ratings, derating may be necessary at higher ambient temperatures. Manufacturers' published literature should be consulted for the de-rating factor to be applied.

18.4.5 Protection of Motors

The manufacturers' literature should also be consulted when fuses are to be applied to motor circuits. In this application, the fuse provides short circuit protection but must be selected to withstand the starting current (possibly up to 8 times full load current), and also carry the normal full load current continuously without deterioration. Tables of recommended fuse sizes for both 'direct on line' and 'assisted start' motor applications are usually given. Examples of protection using fuses are given in Section 18.12.1.

18.5 INDUSTRIAL CIRCUIT BREAKERS

Some parts of an industrial power system are most effectively protected by HRC fuses, but the replacement of blown fuse links can be particularly inconvenient in others. In these locations, circuit breakers are used instead, the requirement being for the breaker to interrupt the maximum possible fault current successfully without damage to itself. In addition to fault current interruption, the breaker must quickly disperse the resulting ionised gas away from the breaker contacts, to prevent re-striking of the arc, and away from other live parts of equipment to prevent breakdown. The breaker, its cable or busbar connections, and the breaker housing, must all be constructed to withstand the mechanical forces resulting from the magnetic fields and internal arc gas pressure produced by the highest levels of fault current to be encountered.

The types of circuit breaker most frequently encountered in industrial system are described in the following sections.

18.5.1 Miniature Circuit Breakers (MCBs)

MCBs are small circuit breakers, both in physical size but more importantly, in ratings. The basic single pole unit is a small, manually closed, electrically or manually opened switch housed in a moulded plastic casing. They are suitable for use on 230V a.c. single-phase/400V a.c. three-phase systems and for d.c. auxiliary supply systems, with current ratings of up to 125A. Contained within each unit is a thermal element, in which a bimetal strip will trip the switch when excessive current passes through it. This element operates with a predetermined inverse-time/current characteristic. Higher currents, typically those exceeding 3-10 times rated current, trip the circuit breaker without intentional delay by actuating a magnetic trip overcurrent element. The operating time characteristics of MCBs are not adjustable. European Standard EN 60898-2 defines the instantaneous trip characteristics, while the manufacturer can define the inverse time thermal trip characteristic. Therefore, a typical tripping characteristic does not exist. The maximum a.c. breaking current permitted by the standard is 25kA.

Single-pole units may be coupled mechanically in groups to form 2, 3 or 4 pole units, when required, by assembly on to a rail in a distribution board. The available ratings make MCBs suitable for industrial, commercial or domestic applications, for protecting equipment such as cables, lighting and heating circuits, and also for the control and protection of low power motor circuits. They may be used instead of fuses on individual circuits, and they are usually 'backed-up' by a device of higher fault interrupting capacity.

Various accessory units, such as isolators, timers, and undervoltage or shunt trip release units may be combined with an MCB to suit the particular circuit to be controlled and protected. When personnel or fire protection is required, a residual current device (RCD) may be combined with the MCB. The RCD contains a miniature core balance current transformer that embraces all of the phase and neutral conductors to provide sensitivity to earth faults within a typical range of 0.05% to 1.5% of rated current, dependent on the RCD selected. The core balance CT energises a common magnetic trip actuator for the MCB assembly.

It is also possible to obtain current-limiting MCBs. These types open prior to the prospective fault current being reached, and therefore have similar properties to HRC fuses. It is claimed that the extra initial cost is outweighed by lifetime savings in replacement costs after a fault has occurred, plus the advantage of providing improved protection against electric shock if an RCD is used. As a result of the increased safety provided by MCBs fitted with an RCD device, they are tending to replace fuses, especially in new installations.

18.5.2 Moulded Case Circuit Breakers (MCCBs)

These circuit breakers are broadly similar to MCBs but have the following important differences:

- the maximum ratings are higher, with voltage ratings up to 1000V a.c./1200V d.c. Current ratings of 2.5kA continuous/180kA r.m.s break are possible, dependent upon power factor
- the breakers are larger, commensurate with the level of ratings. Although available as single, double or triple pole units, the multiple pole units have a common housing for all the poles. Where fitted, the switch for the neutral circuit is usually a separate device, coupled to the multi-pole MCCB
- the operating levels of the magnetic and thermal protection elements may be adjustable, particularly in the larger MCCBs
- because of their higher ratings, MCCBs are usually positioned in the power distribution system nearer to the power source than the MCBs
- the appropriate European specification is EN 60947-2

Care must be taken in the short-circuit ratings of MCCBs. MCCBs are given two breaking capacities, the higher of which is its ultimate breaking capacity. The significance of this is that after breaking such a current, the MCCB may not be fit for continued use. The lower, or service, short circuit breaking capacity permits continued use without further detailed examination of the device. The standard permits a service breaking capacity of as little as 25% of the ultimate breaking capacity. While there is no objection to use of MCCBs to break short-circuit currents between the service and ultimate values, the inspection required after such a trip reduces the usefulness of the device in such circumstances. It is also clearly difficult to determine if the magnitude of the fault current was in excess of the service rating.

The time-delay characteristics of the magnetic or thermal timed trip, together with the necessity for, or size of, a back-up device varies with make and size of breaker. Some MCCBs are fitted with microprocessor-controlled programmable trip characteristics offering a wide range of such characteristics. Time–delayed overcurrent characteristics may not be the same as the standard characteristics for dependent-time protection stated in IEC 60255-3. Hence, discrimination with other protection must be considered carefully. There can be problems where two or more MCBs or MCCBs are electrically in series, as obtaining selectivity between them may be difficult. There may be a requirement that the major device should have a rating of k times the minor device to allow discrimination, in a similar manner to fuses – the manufacturer should be consulted as to value of k. Careful examination of manufacturers' literature is always required at the design stage to determine any such limitations that may be imposed by particular makes and types of MCCBs. An example of co-ordination between MCCBs, fuses and relays is given in Section 18.12.2.

18.5.3 Air Circuit Breakers (ACBs)

Air circuit breakers are frequently encountered on industrial systems rated at 3.3kV and below. Modern LV ACBs are available in current ratings of up to 6.3kA with maximum breaking capacities in the range of 85kA-120kA r.m.s., depending on system voltage.

This type of breaker operates on the principle that the arc produced when the main contacts open is controlled by directing it into an arc chute. Here, the arc resistance is increased and hence the current reduced to the point where the circuit voltage cannot maintain the arc and the current reduces to zero. To assist in the quenching of low current arcs, an air cylinder may be fitted to each pole to direct a blast of air across the contact faces as the breaker opens, so reducing contact erosion.

Air circuit breakers for industrial use are usually withdrawable and are constructed with a flush front plate making them ideal for inclusion together with fuse switches and MCBs/MCCBs in modular multi-tier distribution switchboards, so maximising the number of circuits within a given floor area.

Older types using a manual or dependent manual closing mechanism are regarded as being a safety hazard. This arises under conditions of closing the CB when a fault exists on the circuit being controlled. During the close-trip operation, there is a danger of egress of the arc from the casing of the CB, with a consequent risk of injury to the operator. Such types may be required to be replaced with modern equivalents.

ACBs are normally fitted with integral overcurrent protection, thus avoiding the need for separate protection devices. However, the operating time characteristics of the integral protection are often designed to make discrimination with MCBs/MCCBs/fuses easier and so they may not be in accordance with the standard dependent time characteristics given in IEC 60255-3. Therefore, problems in co-ordination with discrete protection relays may still arise, but modern numerical relays have more flexible characteristics to alleviate such difficulties. ACBs will also have facilities for accepting an external trip signal, and this can be used in conjunction with an external relay if desired. Figure 18.6 shows the typical tripping characteristics available.

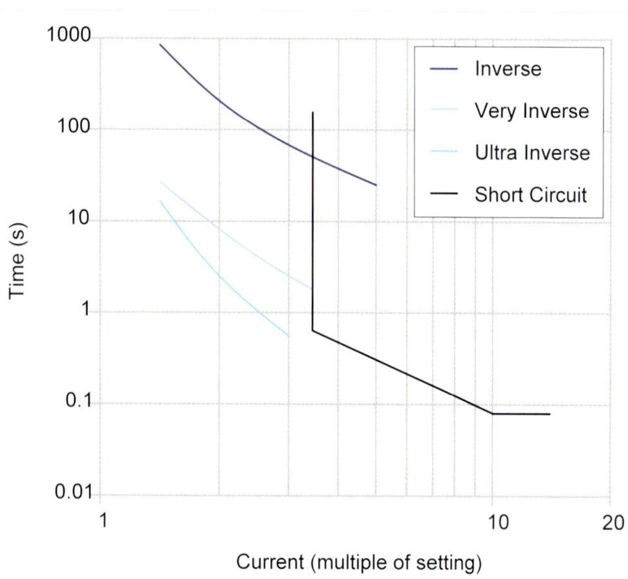

Figure 18.6: Typical tripping characteristics of an ACB

18.5.4 Oil Circuit Breakers (OCBs)

Oil circuit breakers have been very popular for many years for industrial supply systems at voltages of 3.3kV and above. They are found in both 'bulk oil' and 'minimum oil' types, the only significant difference being the volume of oil in the tank.

In this type of breaker, the main contacts are housed in an oil-filled tank, with the oil acting as the both the insulation and the arc-quenching medium. The arc produced during contact separation under fault conditions causes dissociation of the hydrocarbon insulating oil into hydrogen and carbon. The hydrogen extinguishes the arc. The carbon produced mixes with the oil. As the carbon is conductive, the oil must be changed after a prescribed number of fault clearances, when the degree of contamination reaches an unacceptable level.

Because of the fire risk involved with oil, precautions such as the construction of fire/blast walls may have to be taken when OCBs are installed.

18.5.5 Vacuum Circuit Breakers (VCBs)

In recent years, this type of circuit breaker, along with CBs using SF6, has replaced OCBs for new installations in industrial/commercial systems at voltages of 3.3kV and above.

Compared with oil circuit breakers, vacuum breakers have no fire risk and they have high reliability with long maintenance free periods. A variation is the vacuum contactor with HRC fuses, used in HV motor starter applications.

18.5.6 SF6 Circuit Breakers

In some countries, circuit breakers using SF6 gas as the arc-quenching medium are preferred to VCBs as the replacement for air- and oil-insulated CBs. Some modern types of switchgear cubicles enable the use of either VCBs or SF6-insulated CBs according to customer requirements. Ratings of up to 31.5kA r.m.s. fault break at 36kV and 40kA at 24kV are typical. SF6-insulated CBs also have advantages of reliability and maintenance intervals compared to air- or oil-insulated CBs and are of similar size to VCBs for the same rating.

18.6 PROTECTION RELAYS

When the circuit breaker itself does not have integral protection, then a suitable external relay will have to be provided. For an industrial system, the most common protection relays are time-delayed overcurrent and earth fault relays. Chapter 9 provides details of the application of overcurrent relays.

	CT connections	Phase elements	Residual current elements	System	Type of fault	Notes
(a)	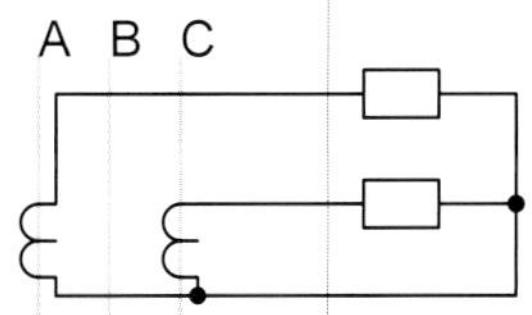			3Ph. 3w	Ph. - Ph.	Petersen coil and unearthed systems
(b)	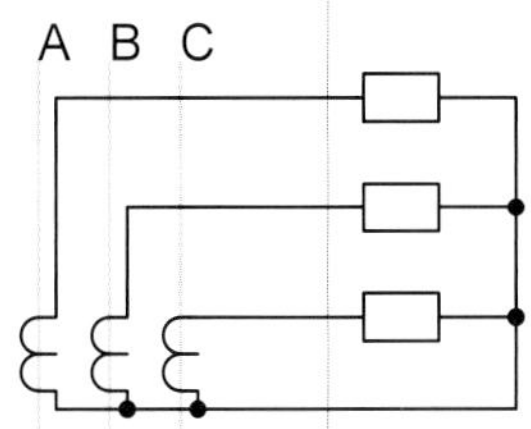			3Ph. 3w	(i) Ph.-Ph. (ii) Ph.–E*	* Earth fault protection only if earth fault current is not less than twice primary operating current
(c)				3Ph. 4w	(i) Ph. - Ph. (ii) Ph. – E (iii) Ph. - N	
(d)	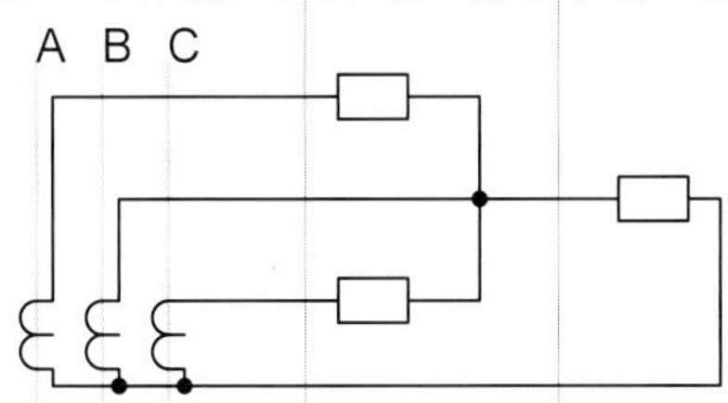			3Ph. 3w	(i) Ph. - Ph. (ii) Ph. – E	Phase elements must be in same phases at all stations. Earth fault settings may be less than full load
(e)	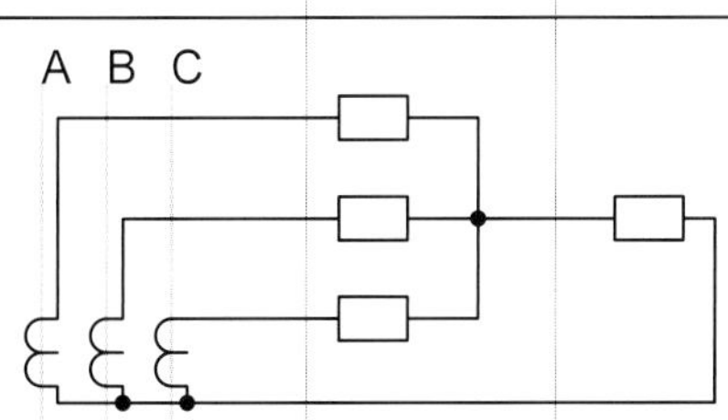			3Ph. 3w	(i) Ph. - Ph. (ii) Ph. – E	Earth fault settings may be less than full load
(f)				3Ph. 4w	(i) Ph. - Ph. (ii) Ph. – E (iii) Ph. - N	Earth fault settings may be less than full load but must be greater than largest Ph. – N load
(g)	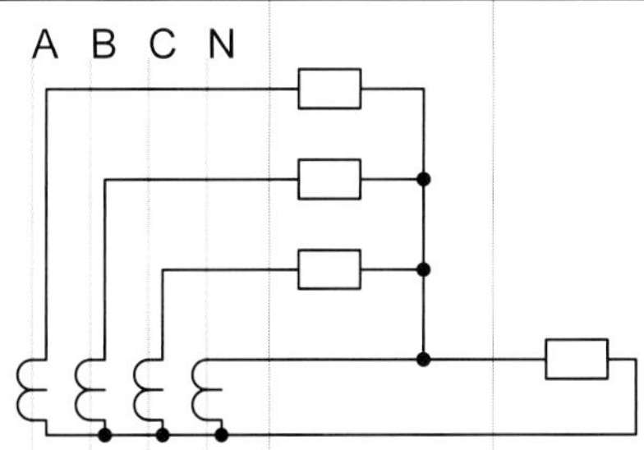			3Ph. 4w	(i) Ph. - Ph. (ii) Ph. – E (iii) Ph. - N	Earth-fault settings may be less than full load
(h)	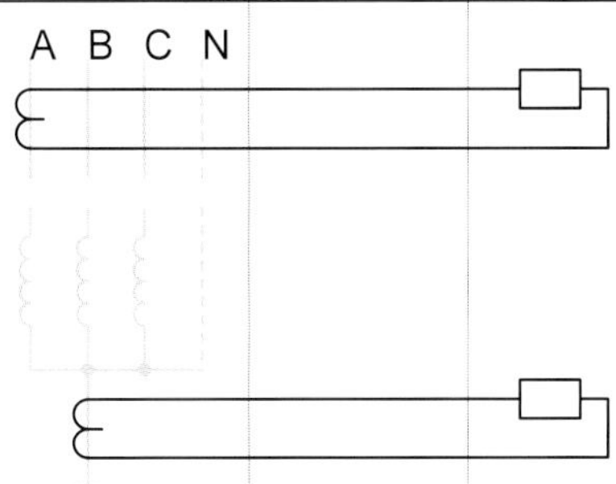			3Ph. 3w or 3Ph. 4w	Ph. - E	Earth-fault settings may be less than full load
Ph. = phase ; w = wire ; E = earth ; N = neutral						

Figure 18.7: Overcurrent and earth fault relay connections

Traditionally, for three wire systems, overcurrent relays have often been applied to two phases only for relay element economy. Even up until the last generation of static relays, economy was still a consideration in terms of the number of analogue current inputs that were provided. Two overcurrent elements can be used to detect any interphase fault, so it was conventional to apply two elements on the same phases at all relay locations. The phase CT residual current connections for an earth fault relay element are unaffected by such a convention. Figure 18.7 shows the possible relay connections and limitations on settings.

18.7 CO-ORDINATION PROBLEMS

There are a number of problems that commonly occur in industrial and commercial networks that are covered in the following sections.

18.7.1.1 Earth fault protection with residually-connected CTs

For four-wire systems, the residual connection of three phase CTs to an earth fault relay element will offer earth fault protection, but the earth fault relay element must be set above the highest single-phase load current to avoid nuisance tripping. Harmonic currents (which may sum in the neutral conductor) may also result in spurious tripping. The earth fault relay element will also respond to a phase-neutral fault for the phase that is not covered by an overcurrent element where only two overcurrent elements are applied. Where it is required that the earth fault protection should respond only to earth fault current, the protection element must be residually connected to three phase CTs and to a neutral CT or to a core-balance CT. In this case, overcurrent protection must be applied to all three phases to ensure that all phase-neutral faults will be detected by overcurrent protection. Placing a CT in the neutral earthing connection to drive an earth fault relay provides earth fault protection at the source of supply for a 4-wire system. If the neutral CT is omitted, neutral current is seen by the relay as earth fault current and the relay setting would have to be increased to prevent tripping under normal load conditions.

When an earth fault relay is driven from residually connected CTs, the relay current and time settings must be such that that the protection will be stable during the passage of transient CT spill current through the relay. Such spill current can flow in the event of transient, asymmetric CT saturation during the passage of offset fault current, inrush current or motor starting current. The risk of such nuisance tripping is greater with the deployment of low impedance electronic relays rather than electromechanical earth fault relays which presented significant relay circuit impedance. Energising a relay from a core balance type CT generally enables more sensitive settings to be obtained without the risk of nuisance tripping with residually connected phase CTs. When this method is applied to a four-wire system, it is essential that both the phase and neutral conductors are passed through the core balance CT aperture. For a 3-wire system, care must be taken with the arrangement of the cable sheath, otherwise cable faults involving the sheath may not result in relay operation (Figure 18.8).

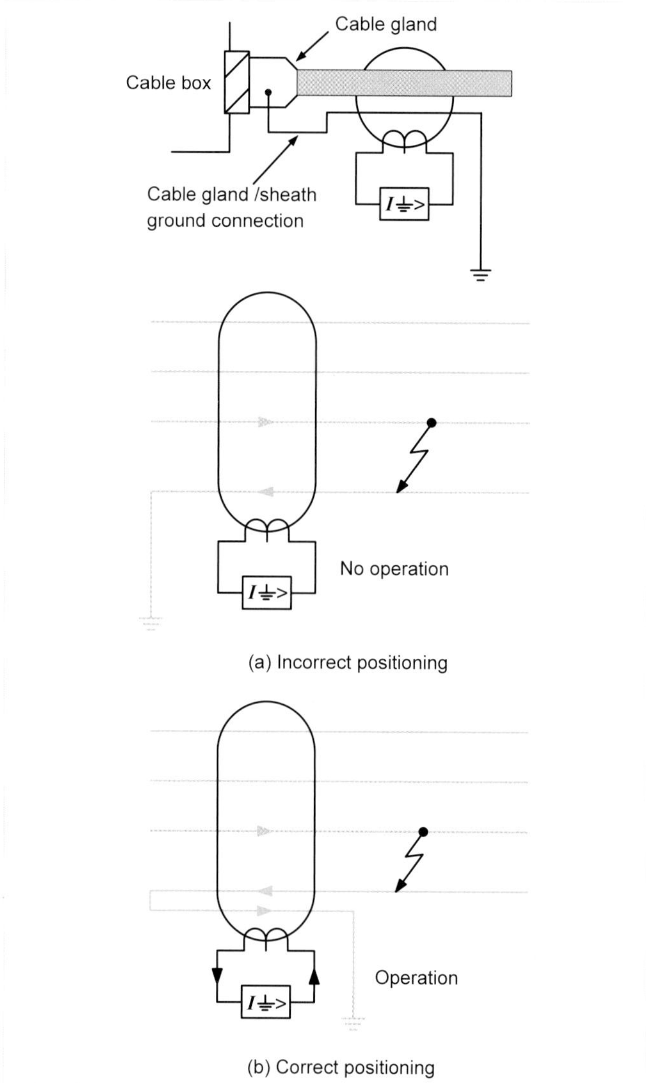

Figure 18.8: CBCT connection for four-wire system

18.7.2 Four-Wire Dual-Fed Substations

The co-ordination of earth fault relays protecting four-wire systems requires special consideration in the case of low voltage, dual-fed installations. Horcher [18.1] has suggested various methods of achieving optimum co-ordination. Problems in achieving optimum protection for common configurations are described below.

18.7.2.1 Use of 3-pole CBs

When both neutrals are earthed at the transformers and all circuit breakers are of the 3-pole type, the neutral busbar in the switchgear creates a double neutral to earth connection, as shown in Figure 18.9. In the event of an uncleared feeder earth fault or busbar earth fault, with both the incoming supply breakers closed and the bus section breaker open, the earth fault current will divide between the two earth connections. Earth fault relay R_{E2} may operate, tripping the supply to the healthy section of the switchboard as well as relay R_{E1} tripping the supply to the faulted section.

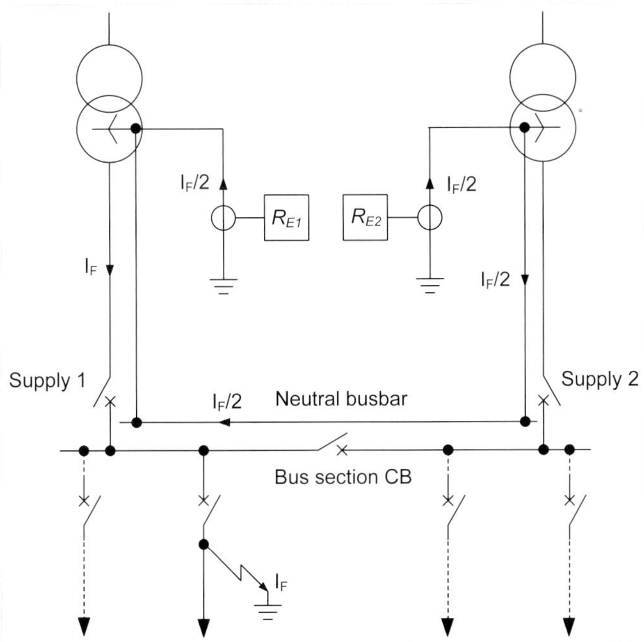

Figure 18.9: Dual fed four-wire systems: use of 3-pole CBs

If only one incoming supply breaker is closed, the earth fault relay on the energised side will see only a proportion of the fault current flowing in the neutral busbar. This not only significantly increases the relay operating time but also reduces its sensitivity to low-level earth faults.

The solution to this problem is to utilise 4-pole CBs that switch the neutral as well as the three phases. Then there is only a single earth fault path and relay operation is not compromised.

18.7.2.2 Use of single earth electrode

A configuration sometimes adopted with four-wire dual-fed substations where only a 3-pole bus section CB is used is to use a single earth electrode connected to the mid-point of the neutral busbar in the switchgear, as shown in Figure 18.10. When operating with both incoming main circuit breakers and the bus section breaker closed, the bus section breaker must be opened first should an earth fault occur, in order to achieve discrimination. The co-ordination time between the earth fault relays R_F and R_E should be established at fault level F_2 for a substation with both incoming supply breakers and bus section breaker closed.

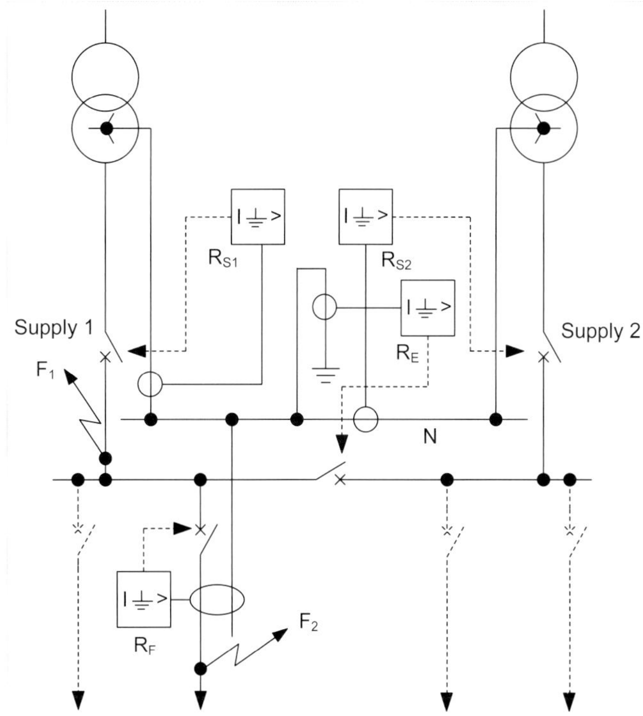

Figure 18.10: Dual fed four-wire systems: use of single point neutral earthing

When the substation is operated with the bus section switch closed and either one or both of the incoming supply breakers closed, it is possible for unbalanced neutral busbar load current caused by single phase loading to operate relay R_{S1} and/or R_{S2} and inadvertently trip the incoming breaker. Interlocking the trip circuit of each R_S relay with normally closed auxiliary contacts on the bus section breaker can prevent this.

However, should an earth fault occur on one side of the busbar when relays R_S are already operated, it is possible for a contact race to occur. When the bus section breaker opens, its break contact may close before the R_S relay trip contact on the healthy side can open (reset). Raising the pick-up level of relays R_{S1} and R_{S2} above the maximum unbalanced neutral current may prevent the tripping of both supply breakers in this case. However, the best solution is to use 4-pole circuit breakers, and independently earth both sides of the busbar.

If, during a busbar earth fault or uncleared feeder earth fault, the bus section breaker fails to open when required, the interlocking break auxiliary contact will also be inoperative. This will prevent relays R_{S1} and R_{S2} from operating and providing back-up protection, with the result that the fault must be cleared eventually by slower phase overcurrent relays. An alternative method of obtaining back-up protection could be to connect a second relay R_E, in series with relay R_E, having an operation time set longer than that of relays R_{S1} and R_{S2}. But since the additional relay must be arranged to trip both of the incoming supply breakers, back-up protection would be obtained but busbar selectivity would be lost.

An example of protection of a typical dual-fed switchboard is given in Section 18.12.3.

18.8 FAULT CURRENT CONTRIBUTION FROM INDUCTION MOTORS

When an industrial system contains motor loads, the motors will contribute fault current for a short time. They contribute to the total fault current via the following mechanism.

When an induction motor is running, a flux, generated by the stator winding, rotates at synchronous speed and interacts with the rotor. If a large reduction in the stator voltage occurs for any reason, the flux in the motor cannot change instantaneously and the mechanical inertia of the machine will tend to inhibit speed reduction over the first few cycles of fault duration. The trapped flux in the rotor generates a stator voltage equal initially to the back e.m.f. induced in the stator before the fault and decaying according to the X/R ratio of the associated flux and current paths. The induction motor therefore acts as a generator resulting in a contribution of current having both a.c. and d.c. components decaying exponentially. Typical 50Hz motor a.c. time constants lie in the range 10ms-60ms for LV motors and 60-200ms for HV motors. This motor contribution has often been neglected in the calculation of fault levels.

Industrial systems usually contain a large component of motor load, so this approach is incorrect. The contribution from motors to the total fault current may well be a significant fraction of the total in systems having a large component of motor load. Standards relating to fault level calculations, such as IEC 60909, require the effect of motor contribution to be included where appropriate. They detail the conditions under which this should be done, and the calculation method to be used. Guidance is provided on typical motor fault current contribution for both HV and LV motors if the required data is not known. Therefore, it is now relatively easy, using appropriate calculation software, to determine the magnitude and duration of the motor contribution, so enabling a more accurate assessment of the fault level for:

- discrimination in relay co-ordination
- determination of the required switchgear/busbar fault rating

For protection calculations, motor fault level contribution is not an issue that is generally important. In industrial networks, fault clearance time is often assumed to occur at 5 cycles after fault occurrence, and at this time, the motor fault level contribution is much less than just after fault occurrence. In rare cases, it may have to be taken into consideration for correct time grading for through-fault protection considerations, and in the calculation of peak voltage for high-impedance differential protection schemes.

It is more important to take motor contribution into account when considering the fault rating of equipment (busbars, cables, switchgear, etc.). In general, the initial a.c. component of current from a motor at the instant of fault is of similar magnitude to the direct-on-line starting current of the motor. For LV motors, 5xFLC is often assumed as the typical fault current contribution (after taking into account the effect of motor cable impedance), with 5.5xFLC for HV motors, unless it is known that low starting current HV motors are used. It is also accepted that similar motors connected to a busbar can be lumped together as one equivalent motor. In doing so, motor rated speed may need to be taken into consideration, as 2 or 4 pole motors have a longer fault current decay than motors with a greater number of poles. The kVA rating of the single equivalent motor is taken as the sum of the kVA ratings of the individual motors considered. It is still possible for motor contribution to be neglected in cases where the motor load on a busbar is small in comparison to the total load (again IEC 60909 provides guidance in this respect). However, large LV motor loads and all HV motors should be considered when calculating fault levels.

18.9 AUTOMATIC CHANGEOVER SYSTEMS

Induction motors are often used to drive critical loads. In some industrial applications, such as those involving the pumping of fluids and gases, this has led to the need for a power supply control scheme in which motor and other loads are transferred automatically on loss of the normal supply to an alternative supply. A quick changeover, enabling the motor load to be re-accelerated, reduces the possibility of a process trip occurring. Such schemes are commonly applied for large generating units to transfer unit loads from the unit transformer to the station supply/start-up transformer.

When the normal supply fails, induction motors that remain connected to the busbar slow down and the trapped rotor flux generates a residual voltage that decays exponentially. All motors connected to a busbar will tend to decelerate at the same rate when the supply is lost if they remain connected to the busbar. This is because the motors will exchange energy between themselves, so that they tend to stay 'synchronised' to each other. As a result, the residual voltages of all the motors decay at nearly the same rate. The magnitude of this voltage and its phase displacement with respect to the healthy alternative supply voltage is a function of time and the speed of the motors. The angular displacement between the residual motor voltage and the incoming voltage will be 180° at some instant. If the healthy alternative supply is switched on to motors which are running down under these conditions, very high inrush currents may result, producing stresses which could be of sufficient magnitude to cause mechanical damage, as well as a severe dip in the alternative supply voltage.

Two methods of automatic transfer are used:

- in-phase transfer system
- residual voltage system

The in-phase transfer method is shown in Figure 18.11(a). Normal and standby feeders from the same power source are used.

Phase angle measurement is used to sense the relative phase angle between the standby feeder voltage and the motor busbar voltage. When the voltages are approximately in phase, or just prior to this condition through prediction, a high-speed circuit breaker is used to complete the transfer. This method is restricted to large high inertia drives where the gradual run down characteristic upon loss of normal feeder supply can be predicted accurately.

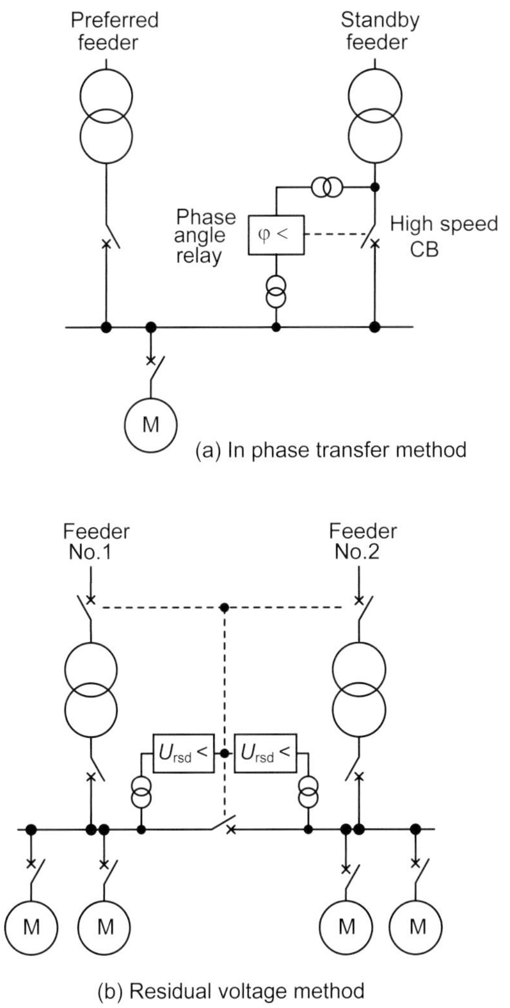

Figure 18.11: Auto-transfer systems

Figure 18.11(b) shows the residual voltage method, which is more common, especially in the petrochemical industry.

Two feeders are used, supplying two busbar sections connected by a normally open bus section breaker. Each feeder is capable of carrying the total busbar load. Each bus section voltage is monitored and loss of supply on either section causes the relevant incomer CB to open. Provided there are no protection operations to indicate the presence of a busbar fault, the bus section breaker is closed automatically to restore the supply to the unpowered section of busbar after the residual voltage generated by the motors running down on that section has fallen to a an acceptable level. This is between 25% and 40%, of nominal voltage, dependent on the characteristics of the power system. The choice of residual voltage setting will influence the re-acceleration current after the bus section breaker closes. For example, a setting of 25% may be expected to result in an inrush current of around 125% of the starting current at full voltage. Alternatively, a time delay could be used as a substitute for residual voltage measurement, which would be set with knowledge of the plant to ensure that the residual voltage would have decayed sufficiently before transfer is initiated.

The protection relay settings for the switchboard must take account of the total load current and the voltage dip during the re-acceleration period in order to avoid spurious tripping during this time. This time can be several seconds where large inertia HV drives are involved.

18.10 VOLTAGE AND PHASE REVERSAL PROTECTION

Voltage relays have been widely used in industrial power supply systems. The principle purposes are to detect undervoltage and/or overvoltage conditions at switchboards to disconnect supplies before damage can be caused from these conditions or to provide interlocking checks. Prolonged overvoltage may cause damage to voltage-sensitive equipment (e.g. electronics), while undervoltage may cause excessive current to be drawn by motor loads. Motors are provided with thermal overload protection to prevent damage with excessive current, but undervoltage protection is commonly applied to disconnect motors after a prolonged voltage dip. With a voltage dip caused by a source system fault, a group of motors could decelerate to such a degree that their aggregate re-acceleration currents might keep the recovery voltage depressed to a level where the machines might stall. Modern numerical motor protection relays typically incorporate voltage protection functions, thus removing the need for discrete undervoltage relays for this purpose (see Chapter 19). Older installations may still utilise discrete undervoltage relays, but the setting criteria remain the same.

Reverse phase sequence voltage protection should be applied where it may be dangerous for a motor to be started with rotation in the opposite direction to that intended. Incorrect rotation due to reverse phase sequence might be set up following some error after power system maintenance or repairs, e.g. to a supply cable. Older motor control boards might have been fitted with discrete relays to detect this condition. Modern motor protection relays may incorporate this function. If reverse phase sequence is detected, motor starting can be blocked. If reverse phase sequence voltage protection is not provided, the high-set negative phase sequence current protection in the relay would quickly detect the condition once the starting device is closed – but initial reverse rotation of the motor could not be prevented.

18.11 POWER FACTOR CORRECTION AND PROTECTION OF CAPACITORS

Loads such as induction motors draw significant reactive power from the supply system, and a poor overall power factor may result. The flow of reactive power increases the voltage-drops through series reactances such as transformers and reactors, it uses up some of the current carrying capacity of power system plant and it increases the resistive losses in the power system.

To offset the losses and restrictions in plant capacity they incur and to assist with voltage regulation, Utilities usually apply tariff penalties to large industrial or commercial customers for running their plant at excessively low power factor. The customer is thereby induced to improve the power factor of his system and it may be cost-effective to install fixed or variable power factor correction equipment to raise or regulate the plant power factor to an acceptable level.

Shunt capacitors are often used to improve power factor. The basis for compensation is shown in Figure 18.12, where $<\varphi_1$ represents the uncorrected power factor angle and $<\varphi_2$ the angle relating to the desired power factor, after correction.

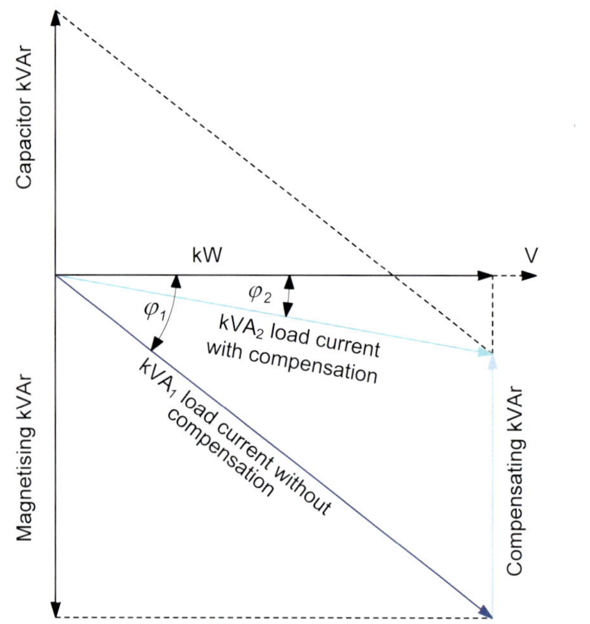

Figure 18.12: Power factor correction principle

The following may be deduced from this vector diagram:

Uncorrected power factor $= \dfrac{kW}{kVA_1} = \cos\angle\varphi_1$

Corrected power factor $= \dfrac{kW}{kVA_2} = \cos\angle\varphi_2$

Reduction in kVA = kVA_1 - kVA_2

If the kW load and uncorrected power factors are known, then the capacitor rating in kvar to achieve a given degree of correction may be calculated from:

Capacitor kVAr $= kW \times \left(\tan\angle\varphi_1 - \tan\angle\varphi_2\right)$

A spreadsheet can easily be constructed to calculate the required amount of compensation to achieve a desired power factor.

18.11.1 Capacitor Control

Where the plant load or the plant power factor varies considerably, it is necessary to control the power factor correction, since over-correction will result in excessive system voltage and unnecessary losses. In a few industrial systems, capacitors are switched in manually when required, but automatic controllers are standard practice. A controller provides automatic power factor correction, by comparing the running power factor with the target value. Based on the available groupings, an appropriate amount of capacitance is switched in or out to maintain an optimum average power factor. The controller is fitted with a 'loss of voltage' relay element to ensure that all selected capacitors are disconnected instantaneously if there is a supply voltage interruption. When the supply voltage is restored, the capacitors are reconnected progressively as the plant starts up. To ensure that capacitor groups degrade at roughly the same rate, the controller usually rotates selection or randomly selects groups of the same size in order to even out the connected time. The provision of overvoltage protection to trip the capacitor bank is also desirable in some applications. This would be to prevent a severe system overvoltage if the power factor correction (PFC) controller fails to take fast corrective action.

The design of PFC installations must recognise that many industrial loads generate harmonic voltages, with the result that the PFC capacitors may sink significant harmonic currents. A harmonic study may be necessary to determine the capacitor thermal ratings or whether series filters are required.

18.11.2 Motor P.F. Correction

When dealing with power factor correction of motor loads, group correction is not always the most economical method. Some industrial consumers apply capacitors to selected motor substations rather than applying all of the correction at the main incoming substation busbars. Sometimes, power factor correction may even be applied to individual motors, resulting in optimum power factor being obtained under all conditions of aggregate motor load. In some instances, better motor starting may also result, from the improvement in the voltage regulation due to the capacitor. Motor capacitors are often six-terminal units, and a capacitor may be conveniently connected directly across each motor phase winding.

Capacitor sizing is important, such that a leading power factor does not occur under any load condition. If excess capacitance is applied to a motor, it may be possible for self-excitation to occur when the motor is switched off or suffers a supply failure. This can result in the production of a high voltage or in mechanical damage if there is a sudden restoration of supply. Since most star/delta or auto-transformer starters other than the 'Korndorffer' types involve a transitional break in supply, it is generally recommended that the capacitor rating should not exceed 85% of the motor magnetising reactive power.

18.11.3 Capacitor Protection

When considering protection for capacitors, allowance should be made for the transient inrush current occurring on switch-on, since this can reach peak values of around 20 times normal current. Switchgear for use with capacitors is usually de-rated considerably to allow for this. Inrush currents may be limited by a resistor in series with each capacitor or bank of capacitors.

Protection equipment is required to prevent rupture of the capacitor due to an internal fault and also to protect the cables and associated equipment from damage in case of a capacitor failure. If fuse protection is contemplated for a three-phase capacitor, HRC fuses should be employed with a current rating of not less than 1.5 times the rated capacitor current.

Medium voltage capacitor banks can be protected by the scheme shown in Figure 18.14. Since harmonics increase capacitor current, the relay will respond more correctly if it does not have in-built tuning for harmonic rejection.

Double star capacitor banks are employed at medium voltage. As shown in Figure 18.13, a current transformer in the inter star-point connection can be used to drive a protection relay to detect the out-of-balance currents that will flow when capacitor elements become short-circuited or open-circuited. The relay will have adjustable current settings, and it might contain a bias circuit, fed from an external voltage transformer, that can be adjusted to compensate for steady-state spill current in the inter star-point connection.

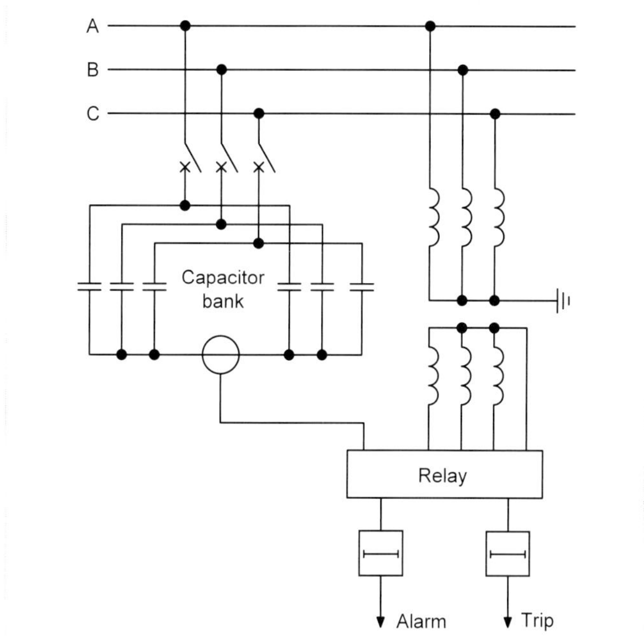

Figure 18.13: Protection of double star capacitor banks

Some industrial loads such as arc furnaces involve large inductive components and correction is often applied using very large high voltage capacitors in various configurations.

Another high voltage capacitor configuration is the 'split phase' arrangement where the elements making up each phase of the capacitor are split into two parallel paths. Figure 18.15 shows two possible connection methods for the relay. A differential relay can be applied with a current transformer for each parallel branch. The relay compares the current in the split phases, using sensitive current settings but also adjustable compensation for the unbalance currents arising from initial capacitor mismatch.

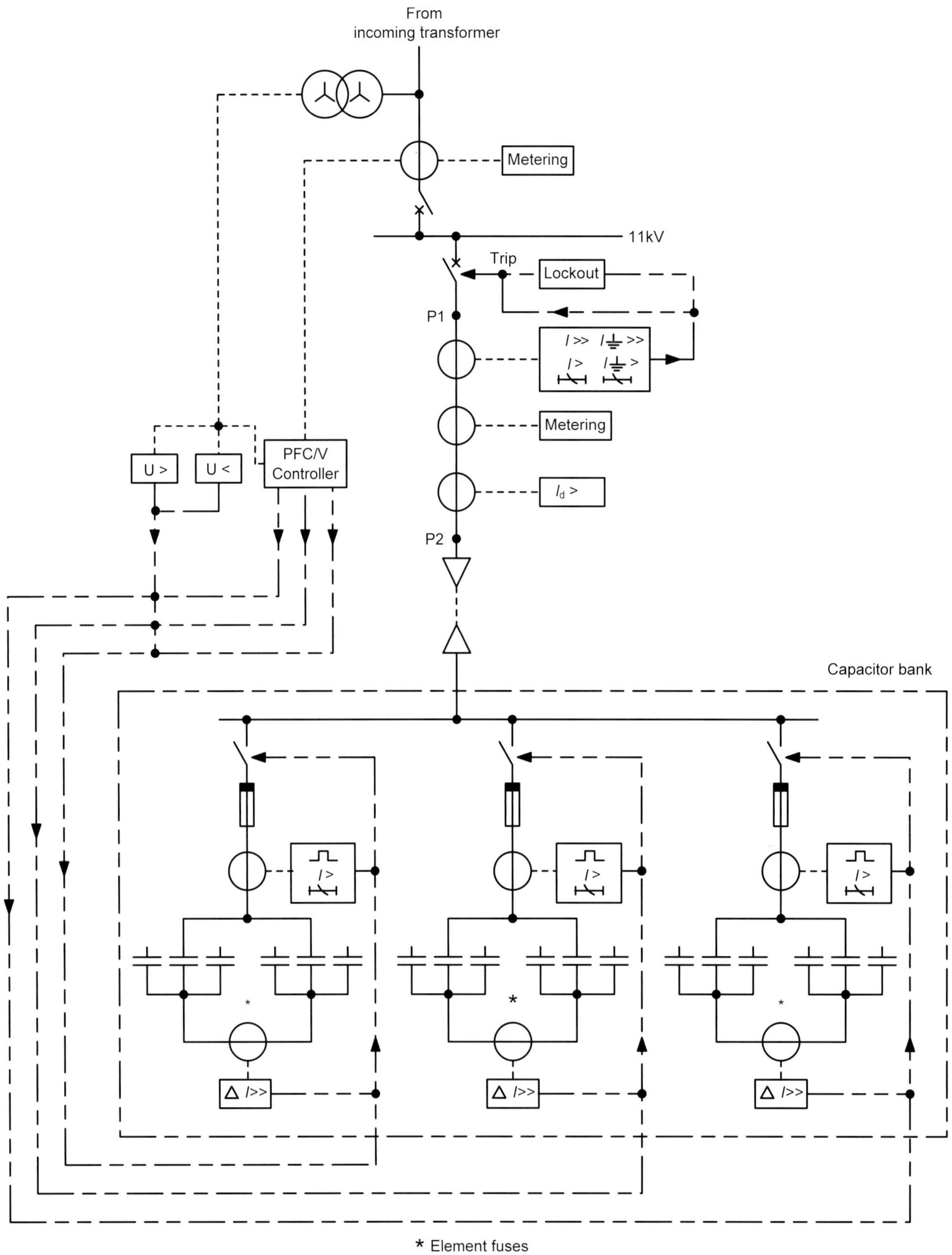

Figure 18.14: Protection of capacitor banks

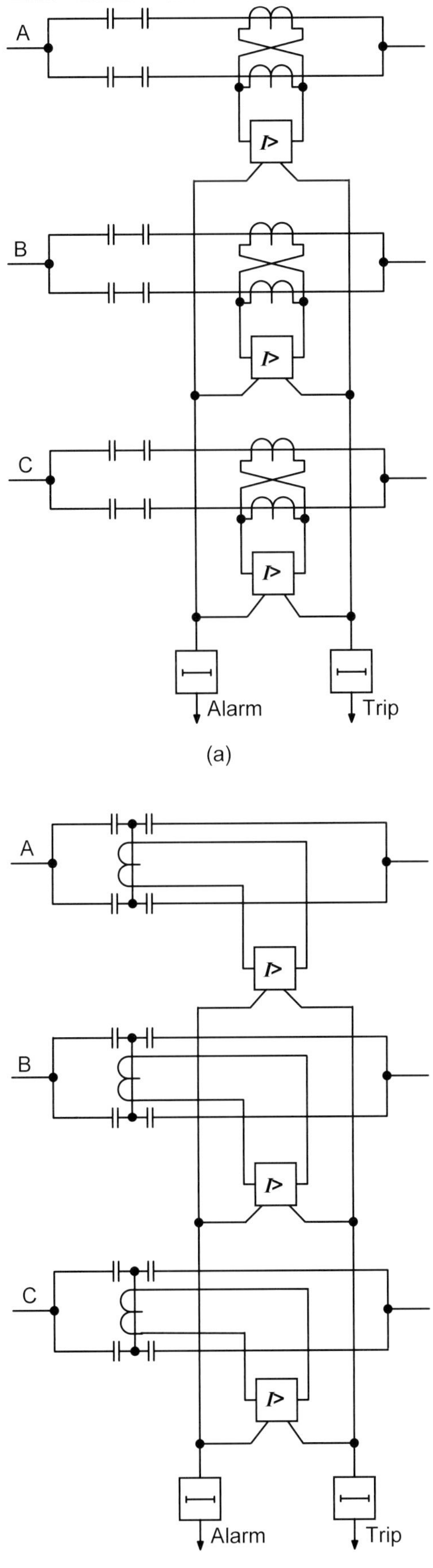

Figure 18.15: Differential protection of split phase capacitor banks

18.12 EXAMPLES

In this section, examples of the topics dealt with in the Chapter are considered.

18.12.1 Fuse Co-ordination

An example of the application of fuses is based on the arrangement in Figure 18.16(a). This shows an unsatisfactory scheme with commonly encountered shortcomings. It can be seen that fuses B, C and D will discriminate with fuse A, but the 400A sub-circuit fuse E may not discriminate, with the 500A sub-circuit fuse D at higher levels of fault current.

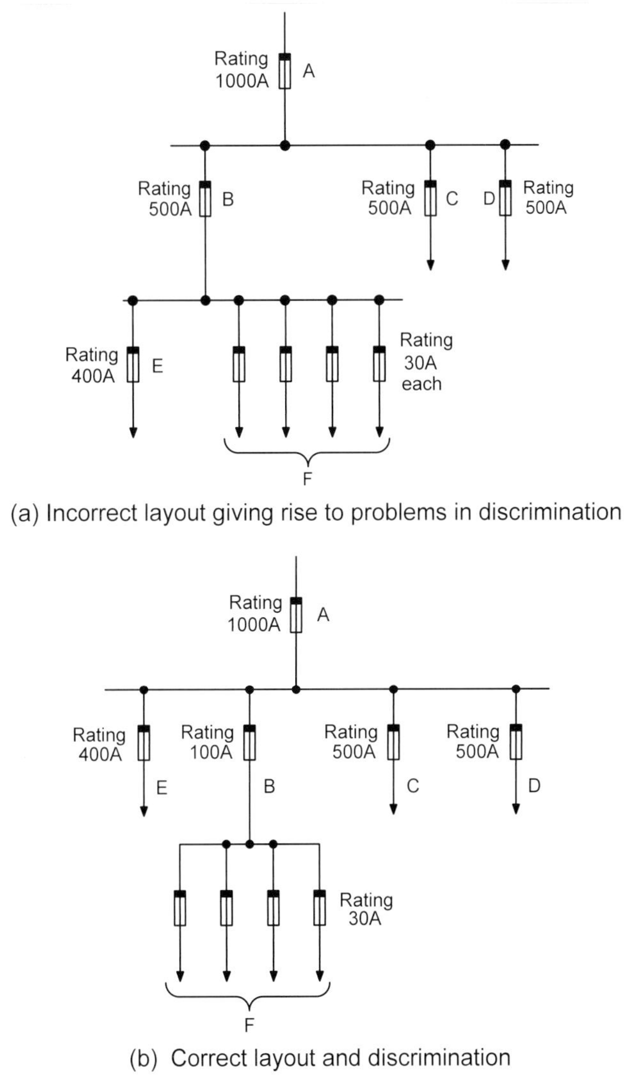

Figure 18.16: Fuse protection: effect of layout on discrimination

The solution, shown in Figure 18.16(b), is to feed the 400A circuit *E* direct from the busbars. The sub-circuit fuse *D* may now have its rating reduced from 500A to a value, of say 100A, appropriate to the remaining sub-circuit. This arrangement now provides a discriminating fuse distribution scheme satisfactory for an industrial system.

However, there are industrial applications where

discrimination is a secondary factor. In the application shown in Figure 18.17, a contactor having a fault rating of 20kA controls the load in one sub-circuit. A fuse rating of 630A is selected for the minor fuse in the contactor circuit to give protection within the through-fault capacity of the contactor.

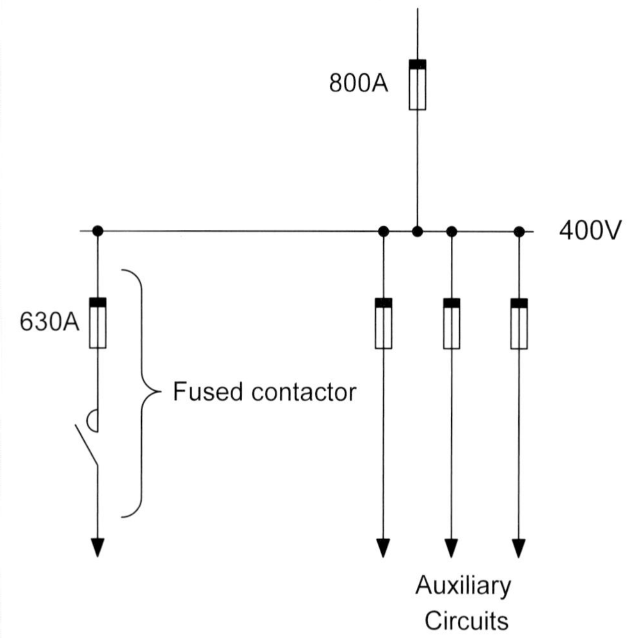

Figure 18.17: Example of back-up protection

The major fuse of 800A is chosen, as the minimum rating that is greater than the total load current on the switchboard. Discrimination between the two fuses is not obtained, as the pre-arcing I^2t of the 800A fuse is less than the total I^2t of the 630A fuse. Therefore, the major fuse will blow as well as the minor one, for most faults so that all other loads fed from the switchboard will be lost. This may be acceptable in some cases. In most cases, however, loss of the complete switchboard for a fault on a single outgoing circuit will not be acceptable, and the design will have to be revised.

18.12.2 Grading of Fuses/MCCBs/Overcurrent Relays

An example of an application involving a moulded case circuit breaker, fuse and a protection relay is shown in Figure 18.18. A 1MVA 3.3kV/400V transformer feeds the LV board via a circuit breaker, which is equipped with an Alstom MiCOM P14x numerical relay having a setting range of 8-400% of rated current and fed from 2000/1A CTs.

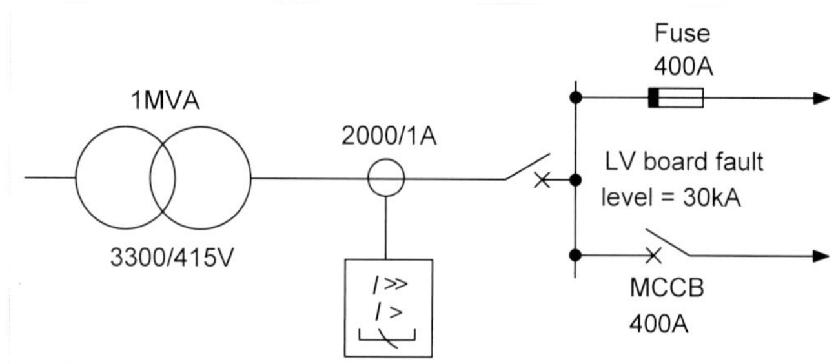

Figure 18.18: Network diagram for protection co-ordination example - fuse/MCCB/relay

Discrimination is required between the relay and both the fuse and MCCB up to the 40kA fault rating of the board. To begin with, the time/current characteristics of both the 400A fuse and the MCCB are plotted in Figure 18.19.

18.12.2.1 Determination of relay current setting

The relay current setting chosen must not be less than the full load current level and must have enough margin to allow the relay to reset with full load current flowing. The latter may be determined from the transformer rating:

$$\mathrm{FLC} = \frac{kVA}{kV \times \sqrt{3}}$$
$$= \frac{1000}{0.4 \times \sqrt{3}}$$
$$= 1443A$$

With the CT ratio of 2000/1A and a relay reset ratio of 95% of the nominal current setting, a current setting of at least 80% would be satisfactory, to avoid tripping and/or failure to reset with the transformer carrying full load current. However, choice of a value at the lower end of this current setting range would move the relay characteristic towards that of the MCCB and discrimination may be lost at low fault currents. It is therefore prudent to select initially a relay current setting of 100%.

18.12.2.2 Relay characteristic and time multiplier selection

An EI characteristic is selected for the relay to ensure discrimination with the fuse (see Chapter 9 for details). From Figure 18.19, it may be seen that at the fault level of 40kA the fuse will operate in less than 0.01s and the MCCB operates in approximately 0.014s. Using a fixed grading margin of 0.4s, the required relay operating time becomes 0.4 + 0.014 = 0.414s. With a CT ratio of 2000/1A, a relay current setting of 100%, and a relay TMS setting of 1.0, the extremely inverse curve gives a relay operating time of 0.2s at a fault current of 40kA. This is too fast to give adequate discrimination and indicates that the EI curve is too severe for this application. Turning to the VI relay characteristic, the relay operation time is found to be 0.71s at a TMS of 1.0. To obtain the required relay operating time of 0.414s:

$$\text{TMS setting} = \frac{0.414}{0.71} = 0.583$$

Use a TMS of 0.6, nearest available setting.

The use of a different form of inverse time characteristic makes it advisable to check discrimination at the lower current levels also at this stage. At a fault current of 4kA, the relay will operate in 8.1s, which does not give discrimination with the MCCB. A relay operation time of 8.3s is required. To overcome this, the relay characteristic needs to be moved away from the MCCB characteristic, a change that may be achieved by using a TMS of 0.625. The revised relay characteristic is also shown in Figure 18.19.

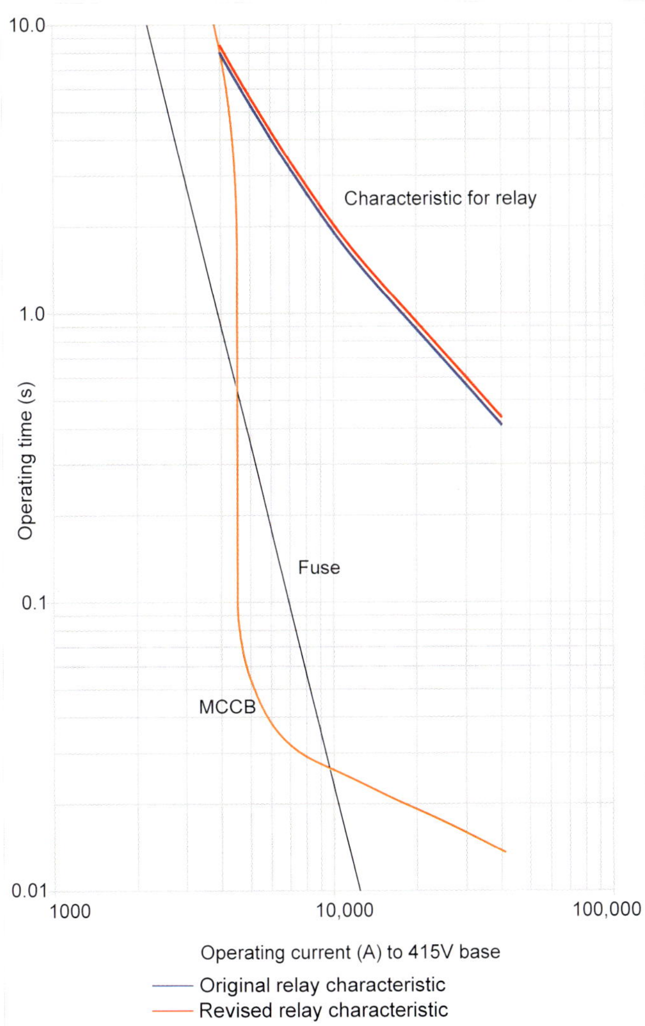

Figure 18.19: Grading curves for Fuse/MCCB/relay grading example

18.12.3 Protection of a Dual-Fed Substation

As an example of how numerical protection relays can be used in an industrial system, consider the typical large industrial substation of Figure 18.20. Two 1.6MVA, 11/0.4kV transformers feeding a busbar whose bus-section CB is normally open. The LV system is solidly earthed. The largest outgoing feeder is to a motor rated 160kW, 193kVA, and a starting current of 7 x FLC.

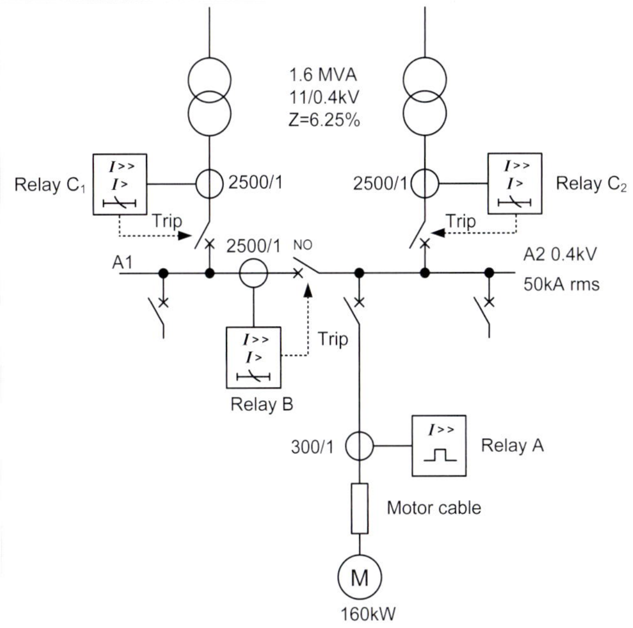

Figure 18.20: Relay grading example for dual-fed switchboard

The transformer impedance is to IEC standards. The LV switchgear and bus bars are fault rated at 50kA rms. To simplify the analysis, only the phase-fault LV protection is considered.

18.12.3.1 General considerations

Analysis of many substations configured as in Figure 18.20 shows that the maximum fault level and feeder load current is obtained with the bus-section circuit breaker closed and one of the infeeding CBs open. This applies so long as the switchboard has a significant amount of motor load. The contribution of motor load to the fault level at the switchboard is usually larger than that from a single infeeding transformer, as the transformer restricts the amount of fault current infeed from the primary side. The three-phase break fault level at the switchboard under these conditions is assumed to be 40kA rms.

Relays C are not required to have directional characteristics (see Section 9.14.3) as all three circuit breakers are only closed momentarily during transfer from a single infeeding transformer to two infeeding transformers configuration. This transfer is normally an automated sequence, and the chance of a fault occurring during the short period (of the order of 1s) when all three CBs are closed is taken to be negligibly small. Similarly, although this configuration gives the largest fault level at the switchboard, it is not considered from either a switchboard fault rating or protection viewpoint.

It is assumed that modern numerical relays are used. For

simplicity, a fixed grading margin of 0.3s is used.

18.12.3.2 Motor protection relay settings

From the motor characteristics given, the overcurrent relay settings (Relay A) can be found using the guidelines set out in Chapter 19 as:

Thermal element:

- current setting: 300A
- time constant: 20 mins

Instantaneous element:

- current setting: 2.32kA

These are the only settings relevant to the upstream relays.

18.12.3.3 Relay B settings

Relay B settings are derived from consideration of the loading and fault levels with the bus-section breaker between busbars A1 and A2 closed. No information is given about the load split between the two busbars, but it can be assumed in the absence of definitive information that each busbar is capable of supplying the total load of 1.6MVA. With fixed tap transformers, the bus voltage may fall to 95% of nominal under these conditions, leading to a load current of 2430A. The IDMT current setting must be greater than this, to avoid relay operation on normal load currents and (ideally) with aggregate starting/re-acceleration currents. If the entire load on the busbar was motor load, an aggregate starting current in excess of 13kA would occur, but a current setting of this order would be excessively high and lead to grading problems further upstream. It is unlikely that the entire load is motor load (though this does occur, especially where a supply voltage of 690V is chosen for motors – an increasingly common practice) or that all motors are started simultaneously (but simultaneous re-acceleration may well occur). What is essential is that relay B does not issue a trip command under these circumstances – i.e. the relay current/time characteristic is in excess of the current/time characteristic of the worst-case starting/re-acceleration condition. It is therefore assumed that 50% of the total bus load is motor load, with an average starting current of 600% of full load current (= 6930A), and that re-acceleration takes 3s. A current setting of 3000A is therefore initially used. The SI characteristic is used for grading the relay, as co-ordination with fuses is not required. The TMS is required to be set to grade with the thermal protection of relay A under 'cold' conditions, as this gives the longest operation time of Relay A, and the re-acceleration conditions. A TMS value of 0.41 is found to provide satisfactory grading, being dictated by the motor starting/re-acceleration transient. Adjustment of both current and TMS settings may be required depending on the exact re-acceleration conditions. Note that lower current and TMS settings could be used if motor starting/re-acceleration did not need to be considered.

The high-set setting needs to be above the full load current and motor starting/re-acceleration transient current, but less than the fault current by a suitable margin. A setting of 12.5kA is initially selected. A time delay of 0.3s has to used to ensure grading with relay A at high fault current levels; both relays A and B may see a current in excess of 25kA for faults on the cable side of the CB feeding the 160kW motor. The relay curves are shown in Figure 18.21.

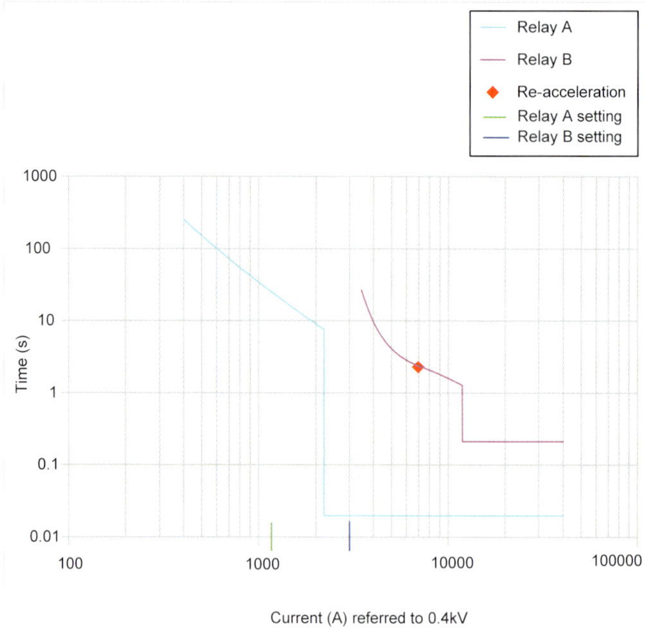

Figure 18.21: Grading of relays A and B

18.12.3.4 Relays C settings

The setting of the IDMT element of relays C_1 and C_2 has to be suitable for protecting the busbar while grading with relay B. The limiting condition is grading with relay B, as this gives the longest operation time for relays C.

The current setting has to be above that for relay B to achieve full co-ordination, and a value of 3250A is suitable. The TMS setting using the SI characteristic is chosen to grade with that of relay B at a current of 12.5kA (relay B instantaneous setting), and is found to be 0.45. The high-set element must grade with that of relay B, so a time delay of 0.62sec is required. The current setting must be higher than that of relay B, so use a value of 15kA. The final relay grading curves and settings are shown in Figure 18.22.

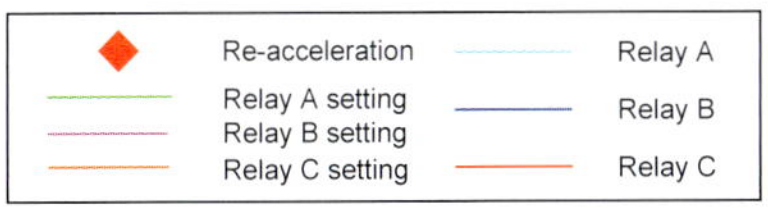

	Parameter	Value	Parameter	Value	Parameter	Value	Parameter	Value
Relay A	Ith	300A	Time Const	1200s	I>>	2320A	t_{dinst}	0
Relay B	I>	2500A	TMS	0.175	I>>	12500A	t_{dinst}	0.32s
Relay C	I>	2750A	TMS	0.25	I>>	15000	t_{dinst}	0.62s

(a) Relay settings

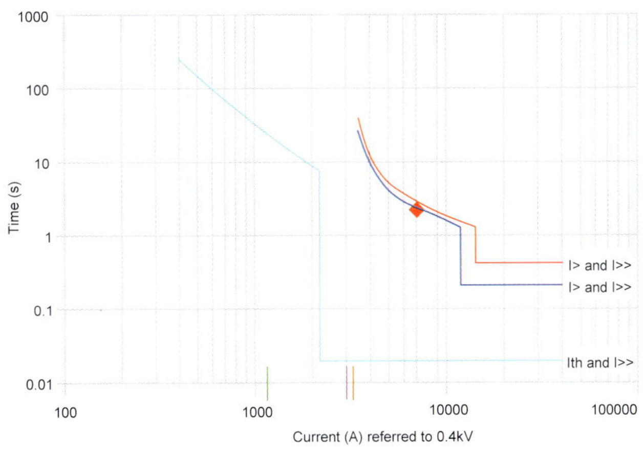

(b) Grading curves

Figure 18.22: Final relay grading curves

18.12.3.5 Comments on grading

While the above grading may appear satisfactory, the protection on the primary side of the transformer has not been considered. IDMT protection at this point will have to grade with relays C and with the through-fault short-time withstand curves of the transformer and cabling. This may result in excessively long operation times. Even if the operation time at the 11kV level is satisfactory, there is probably a Utility infeed to consider, which will involve a further set of relays and another stage of time grading, and the fault clearance time at the Utility infeed will almost certainly be excessive. One solution is to accept a total loss of supply to the 0.4kV bus under conditions of a single infeed and bus section CB closed. This is achieved by setting relays C such that grading with relay B does not occur at all current levels, or omitting relay B from the protection scheme. The argument for this is that network operation policy is to ensure loss of supply to both sections of the switchboard does not occur for single contingencies. As single infeed operation is not normal, a contingency (whether fault or maintenance) has already occurred, so that a further fault causing total loss of supply to the switchboard through tripping of one of relays B is a second contingency. Total loss of supply is therefore acceptable. The alternative is to accept a lack of discrimination at some point on the system, as already noted in Chapter 9. Another solution is to employ partial differential protection to remove the need for Relay A, but this is seldom used. The strategy adopted will depend on the individual circumstances.

18.13 REFERENCES

[18.1] Overcurrent Relay Co-ordination for Double Ended Substations. George R Horcher. IEEE. Vol. 1A-14 No. 6 1978.

Chapter 19
AC Motor Protection

19.1 INTRODUCTION

There are a wide range of a.c. motors and motor characteristics in existence, because of the numerous duties for which they are used. All motors need protection, but fortunately, the more fundamental problems affecting the choice of protection are independent of the type of motor and the type of load to which it is connected. There are some important differences between the protection of induction motors and synchronous motors, and these are fully dealt with in section 19.3.

Motor characteristics must be carefully considered when applying protection; while this may be regarded as stating the obvious, it is emphasised because it applies more to motors than to other items of power system plant. For example, the starting and stalling currents/times must be known when applying overload protection, and furthermore the thermal withstand of the machine under balanced and unbalanced loading must be clearly defined.

The conditions for which motor protection is required can be divided into two broad categories: imposed external conditions and internal faults. Table 19.1 provides details of most likely faults that require protection.

External Faults	Internal faults
Unbalanced supplies	Bearing failures
Undervoltages	Winding faults
Single phasing	Overloads
Reverse phase sequence	

Table 19.1: Causes of motor failures

19.2 MODERN RELAY DESIGN

The design of a modern numerical motor protection relay must be adequate to cater for the protection needs of any one of the vast range of motor designs in service, many of the designs having no permissible allowance for overloads. A relay offering comprehensive protection will have the following set of features:

Synchronous and asynchronous motors

- thermal protection
- extended start protection
- stalling protection
- number of starts limitation
- short circuit protection

- earth fault protection
- winding RTD measurement/trip
- negative sequence current detection
- undervoltage protection
- loss-of-load protection
- auxiliary supply supervision

Synchronous motors only

- out-of-step protection
- loss of supply protection

In addition, relays may offer options such as circuit breaker condition monitoring as an aid to maintenance. Manufacturers may also offer relays that implement a reduced functionality to that given above where less comprehensive protection is warranted (e.g. induction motors of low rating).

The following sections examine each of the possible failure modes of a motor and discuss how protection may be applied to detect that mode.

19.3 THERMAL (OVERLOAD) PROTECTION

The majority of winding failures are either indirectly or directly caused by overloading (either prolonged or cyclic), operation on unbalanced supply voltage, or single phasing, which all lead through excessive heating to the deterioration of the winding insulation until an electrical fault occurs. The generally accepted rule is that insulation life is halved for each 10°C rise in temperature above the rated value, modified by the length of time spent at the higher temperature. As an electrical machine has a relatively large heat storage capacity, it follows that infrequent overloads of short duration may not adversely affect the machine. However, sustained overloads of only a few percent may result in premature ageing and insulation failure.

Furthermore, the thermal withstand capability of the motor is affected by heating in the winding prior to a fault. It is therefore important that the relay characteristic takes account of the extremes of zero and full-load pre-fault current known respectively as the 'Cold' and 'Hot' conditions.

The variety of motor designs, diverse applications, variety of possible abnormal operating conditions and resulting modes of failure result in a complex thermal relationship. A generic mathematical model that is accurate is therefore impossible to create. However, it is possible to develop an approximate model if it is assumed that the motor is a homogeneous body, creating and dissipating heat at a rate proportional to temperature rise. This is the principle behind the 'thermal replica' model of a motor used for overload protection.

The temperature T at any instant is given by:

$$T = T_{max}\left(1 - e^{-\frac{t}{\tau}}\right)$$

where:

$T_{\max}$ = final steady state temperature

τ = heating time constant

Temperature rise is proportional to the current squared:

$$T = KI_R^2\left(1 - e^{-\frac{t}{\tau}}\right)$$

where:

I_R = current which, if flowing continuously, produces temperature $T_{\max}$ in the motor

Therefore, it can be shown that, for any overload current I, the permissible time t for this current to flow is:

$$t = \tau \log_e\left[\frac{1}{\left\{1 - \left(\frac{I_R}{I}\right)^2\right\}}\right]$$

In general, the supply to which a motor is connected may contain both positive and negative sequence components, and both components of current give rise to heating in the motor. Therefore, the thermal replica should take into account both of these components, a typical equation for the equivalent current being:

$$I_{eq} = \sqrt{\left(I_1^2 + KI_2^2\right)}$$

where

I_1 = positive sequence current

I_2 = negative sequence current

K = negative sequence rotor resistance / positive sequence rotor resistance at rated speed.

A typical value of K is 3.

Finally, the thermal replica model needs to take into account the fact that the motor will tend to cool down during periods of light load, and the initial state of the motor. The motor will have a cooling time constant, τ_r, that defines the rate of cooling. Hence, the final thermal model can be expressed as:

$$t = \tau \log_e \frac{\left(K^2 - A^2\right)}{\left(K^2 - 1\right)}$$

Equation 19.1

where:

τ = heating time constant

$$K = \frac{I_{eq}}{I_{th}}$$

A^2 = initial state of motor (cold or hot)

I_{th} = thermal setting current

Equation 19.1 takes into account the 'cold' and 'hot' characteristics defined in IEC 60255, part 8.

Some relays may use a dual curve characteristic for the heating time constant, and hence two values of the heating time constant are required. Switching between the two values takes place at a pre-defined motor current. This may be used to obtain better tripping performance during starting on motors that use a star-delta starter. During starting, the motor windings carry full line current, while in the 'run' condition, they carry only 57% of the current seen by the relay. Similarly, when the motor is disconnected from the supply, the heating time constant τ is set equal to the cooling time constant τ_r.

Since the relay should ideally be matched to the protected motor and be capable of close sustained overload protection, a wide range of relay adjustment is desirable together with good accuracy and low thermal overshoot.

Typical relay setting curves are shown in Figure 19.1.

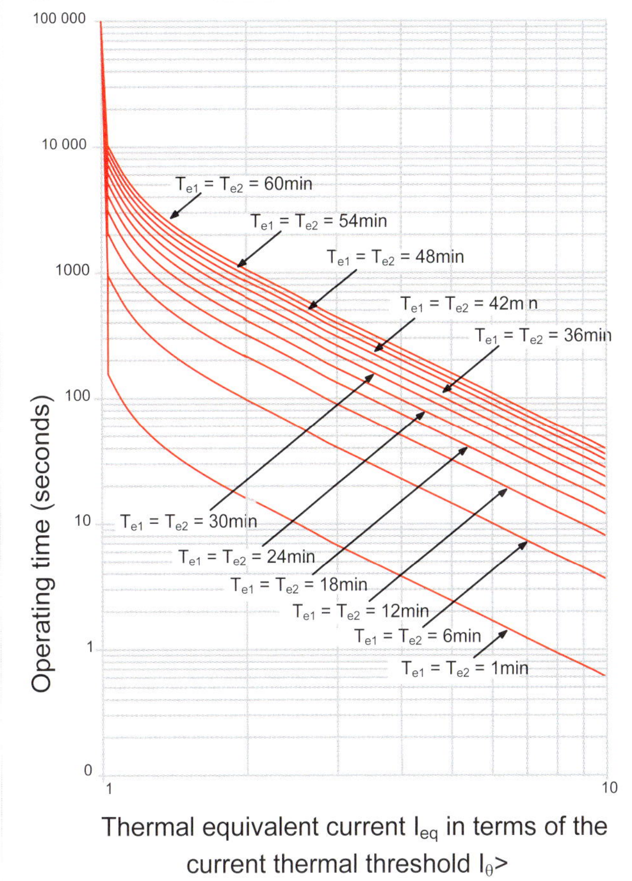

Figure 19.1: Thermal overload characteristic curves from cold – initial thermal state 0%

19.4 START/STALL PROTECTION

When a motor is started, it draws a current well in excess of full load rating throughout the period that the motor takes to run-up to speed. While the motor starting current reduces somewhat as motor speed increases, it is normal in protection practice to assume that the motor current remains constant throughout the starting period. The starting current will vary depending on the design of the motor and method of starting. For motors started DOL (direct-on-line), the nominal starting current can be 4-8 times full-load current. However, when a star-delta starter is used, the line current will only be $\frac{1}{\sqrt{3}}$ of the DOL starting current.

Should a motor stall whilst running, or fail to start, due to excessive loading, the motor will draw a current equal to its locked rotor current. It is not therefore possible to distinguish between a stall condition and a healthy start solely on the basis of the current drawn. Discrimination between the two conditions must be made based on the duration of the current drawn. For motors where the starting time is less than the safe stall time of the motor, protection is easy to arrange.

However, where motors are used to drive high inertia loads,

the stall withstand time can be less than the starting time. In these cases, an additional means must be provided to enable discrimination between the two conditions to be achieved.

19.4.1 Excessive Start Time/Locked Rotor Protection

A motor may fail to accelerate from rest for a number of reasons:

- loss of a supply phase
- mechanical problems
- low supply voltage
- excessive load torque
- etc.

A large current will be drawn from the supply, and cause extremely high temperatures to be generated within the motor. This is made worse by the fact that the motor is not rotating, and hence no cooling due to rotation is available. Winding damage will occur very quickly – either to the stator or rotor windings depending on the thermal limitations of the particular design (motors are said to be stator or rotor limited in this respect). The method of protection varies depending on whether the starting time is less than or greater than the safe stall time. In both cases, initiation of the start may be sensed by detection of the closure of the switch in the motor feeder (contactor or CB) and optionally current rising above a starting current threshold value – typically 200% of motor rated current. For the case of both conditions being sensed, they may have to occur within a narrow aperture of time for a start to be recognised.

Special requirements may exist for certain types of motors installed in hazardous areas (e.g. motors with type of protection EEx 'e') and the setting of the relay must take these into account. Sometimes a permissive interlock for machine pressurisation (on EEx 'p' machines) may be required, and this can be conveniently achieved by use of a relay digital input and the in-built logic capabilities.

19.4.1.1 Start time < safe stall time

Protection is achieved by use of a definite time overcurrent characteristic, the current setting being greater than full load current but less than the starting current of the machine. The time setting should be a little longer than the start time, but less than the permitted safe starting time of the motor. Figure 19.2 illustrates the principle of operation for a successful start.

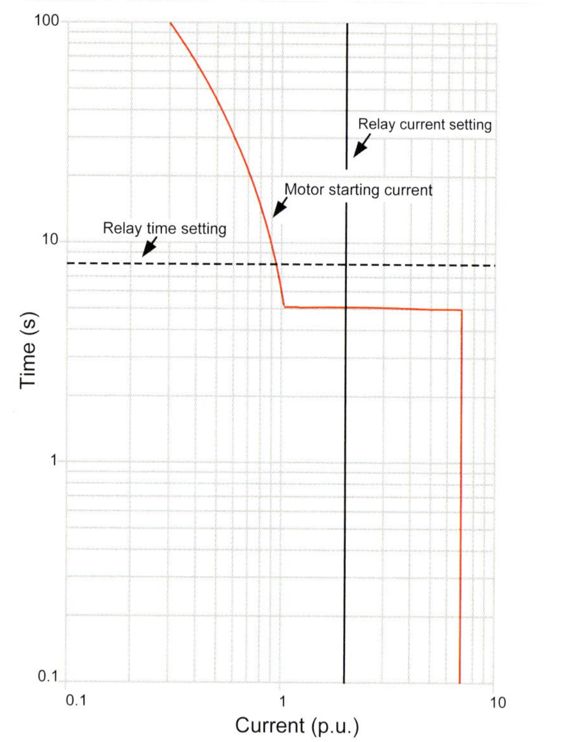

Figure 19.2: Relay setting for successful start: start time<stall time

19.4.1.2 Start time ≥ safe stall time

For this condition, a definite time overcurrent characteristic by itself is not sufficient, since the time delay required is longer than the maximum time that the motor can be allowed to carry starting current safely. An additional means of detection of rotor movement, indicating a safe start, is required. A speed-sensing switch usually provides this function. Detection of a successful start is used to select the relay timer used for the safe run-up time of the motor. This time can be longer than the safe stall time, as there is both a (small) decrease in current drawn by the motor during the start and the rotor fans begin to improve cooling of the machine as it accelerates. If a start is sensed by the relay through monitoring current and/or start device closure, but the speed switch does not operate, the relay element uses the safe stall time setting to trip the motor before damage can occur. Figure 19.3(a) illustrates the principle of operation for a successful start, and Figure 19.3(b) for an unsuccessful start.

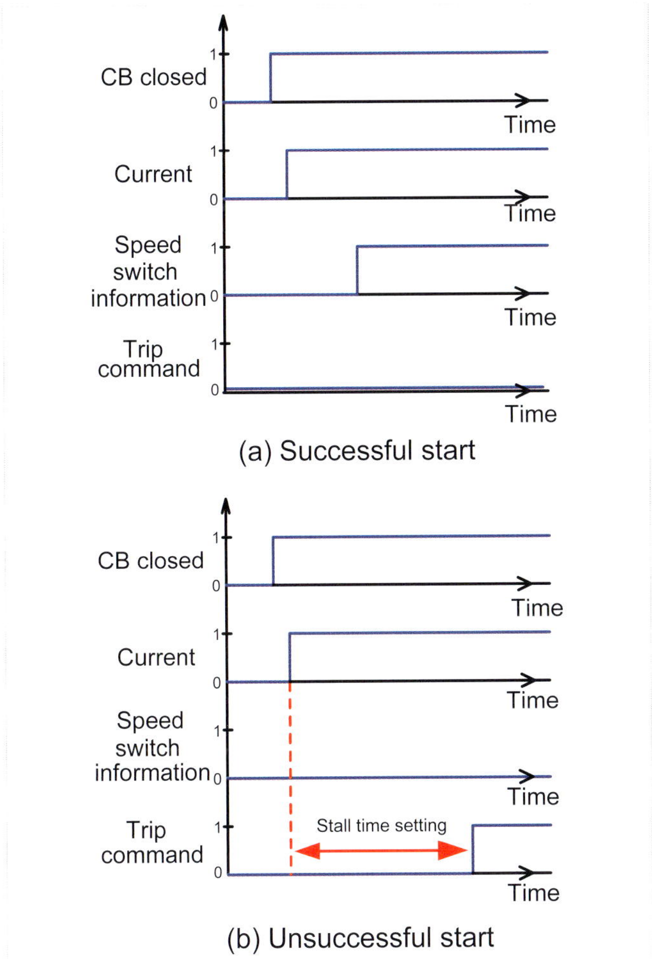

Figure 19.3: Relay settings for start time> stall time

19.4.2 Stall Protection

Should a motor stall when running or be unable to start because of excessive load, it will draw a current from the supply equivalent to the locked rotor current. It is obviously desirable to avoid damage by disconnecting the machine as quickly as possible if this condition arises.

Motor stalling can be recognised by the motor current exceeding the start current threshold after a successful start – i.e. a motor start has been detected and the motor current has dropped below the start current threshold within the motor safe start time. A subsequent rise in motor current above the motor starting current threshold is then indicative of a stall condition, and tripping will occur if this condition persists for greater than the setting of the stall timer. An instantaneous overcurrent relay element provides protection.

In many systems, transient supply voltage loss (typically up to 2 seconds) does not result in tripping of designated motors. They are allowed to re-accelerate upon restoration of the supply. During re-acceleration, they draw a current similar to the starting current for a period that may be several seconds. It is thus above the motor stall relay element current threshold.

The stall protection would be expected to operate and defeat the object of the re-acceleration scheme. A motor protection relay will therefore recognise the presence of a voltage dip and recovery, and inhibit stall protection for a defined period. The undervoltage protection element (section 19.11) can be used to detect the presence of the voltage dip and inhibit stall protection for a set period after voltage recovery. Protection against stalled motors in case of an unsuccessful re-acceleration is therefore maintained. The time delay setting is dependent on the re-acceleration scheme adopted and the characteristics of individual motors. It should be established after performing a transient stability study for the re-acceleration scheme proposed.

19.4.3 Number of Starts Limitation

Any motor has a restriction on the number of starts that are allowed in a defined period without the permitted winding, etc. temperatures being exceeded. Starting should be blocked if the permitted number of starts is exceeded. The situation is complicated by the fact the number of permitted 'hot' starts in a given period is less than the number of 'cold' starts, due to the differing initial temperatures of the motor. The relay must maintain a separate count of 'cold' and 'hot' starts. By making use of the data held in the motor thermal replica, 'hot' and 'cold' starts can be distinguished.

To allow the motor to cool down between starts, a time delay may be specified between consecutive starts (again distinguishing between 'hot' and 'cold' starts). The start inhibit is released after a time determined by the motor specification.

The overall protection function is illustrated in Figure 19.4.

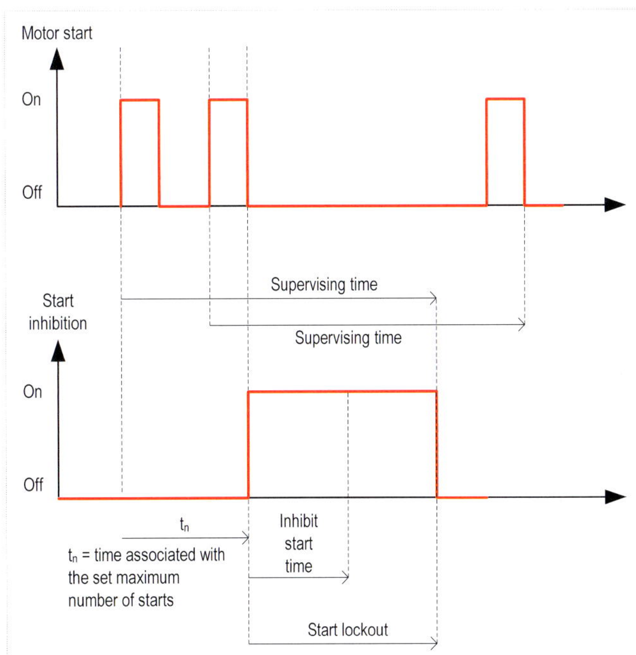

Figure 19.4: Number of starts limitation

In this example, the maximum number of starts within the Supervising Time has been reached, therefore the Inhibit Start Time is initiated. The remaining time is greater than the Inhibit Start Time, so the start inhibition remains for a duration equal to the supervising time minus the t_n.

19.5 SHORT-CIRCUIT PROTECTION

Motor short-circuit protection is often provided to cater for major stator winding faults and terminal flashovers. Because of the relatively greater amount of insulation between phase windings, faults between phases seldom occur. As the stator windings are completely enclosed in grounded metal the fault would very quickly involve earth, which would then operate the instantaneous earth fault protection. A single definite time overcurrent relay element is all that is required for this purpose, set to about 125% of motor starting current. The time delay is required to prevent spurious operation due to CT spill currents, and is typically set at 100ms. If the motor is fed from a fused contactor, co-ordination is required with the fuse, and this will probably involve use of a long time delay for the relay element. Since the object of the protection is to provide rapid fault clearance to minimise damage caused by the fault, the protection is effectively worthless in these circumstances. It is therefore only provided on motors fed via circuit breakers.

Differential (unit) protection may be provided on larger HV motors fed via circuit breakers to protect against phase-phase and phase-earth faults, particularly where the power system is resistance-earthed. The differential protection can be made quite sensitive allowing early detection of faults, thus damage to the motor can be minimised. The normal definite time overcurrent protection would not be sufficiently sensitive, and sensitive earth fault protection may not be provided. The user may wish to avoid the detailed calculations required of capacitance current in order to set sensitive non-directional earth fault overcurrent protection correctly on HV systems (Chapter 9) or there may be no provision for a VT to allow application of directional sensitive earth fault protection. There is still a lower limit to the setting that can be applied, due to spill currents from CT saturation during starting, while on some motors, neutral current has been found to flow during starting, even with balanced supply voltages, that would cause the differential protection to operate. For details on the application of differential protection, refer to Chapter 10. However, non-directional earth fault overcurrent protection will normally be cheaper in cases where adequate sensitivity can be provided.

19.6 EARTH FAULT PROTECTION

One of the most common faults to occur on a motor is a stator winding fault. Whatever the initial form of the fault (phase-phase, etc.) or the cause (cyclic overheating, etc.), the presence of the surrounding metallic frame and casing will ensure that it rapidly develops into a fault involving earth. Therefore, provision of earth fault protection is very important. The type and sensitivity of protection provided depends largely on the system earthing, so the various types will be dealt with in turn. It is common, however, to provide both instantaneous and time-delayed relay elements to cater for major and slowly developing faults.

19.6.1 Solidly-Earthed System

Most LV systems fall into this category, for reasons of personnel safety. Two types of earth fault protection are commonly found – depending on the sensitivity required.

For applications where a sensitivity of > 20% of motor continuous rated current is acceptable, conventional earth fault protection using the residual CT connection of Figure 19.5 can be used. A lower limit is imposed on the setting by possible load unbalance and/or (for HV systems) system capacitive currents.

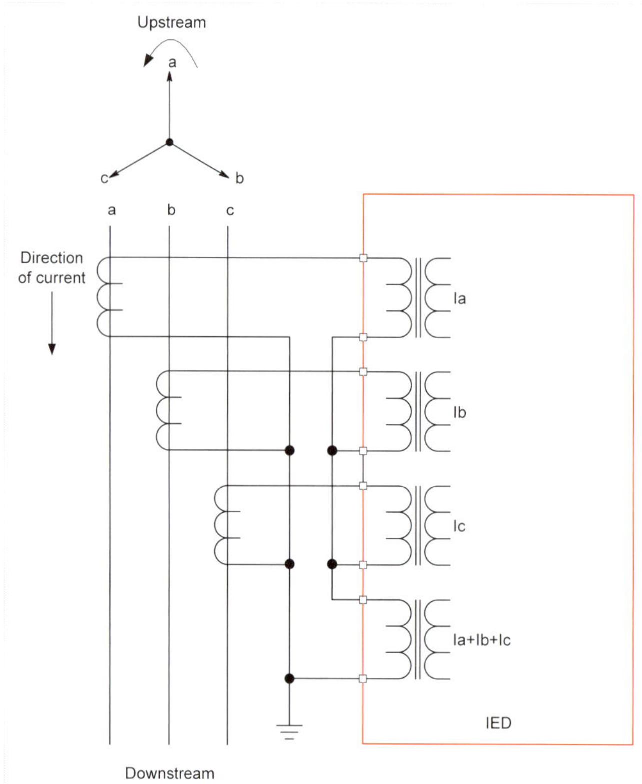

Figure 19.5: Residual CT connection for earth fault protection

Care must be taken to ensure that the relay does not operate from the spill current resulting from unequal CT saturation during motor starting, where the high currents involved will almost certainly saturate the motor CT's. It is common to use a stabilising resistor in series with the relay, with the value being calculated using the formula:

$$R_{stab} = \frac{I_{st}}{I_0}\left(R_{ct} + kR_l + R_r\right)$$

Equation 19.2

where:

I_{st} = starting current referred to CT secondary

I_0 = relay earth fault setting (A)

R_{stab} =stabilising resistor value (ohms)

R_{ct} =dc resistance of CT secondary (ohms)

R_l =CT single lead resistance (ohms)

R_r = relay resistance (ohms)

k = CT connection factor (1 for star point at CT, 2 for star point at relay).

The effect of the stabilising resistor is to increase the effective setting of the relay under these conditions, and hence delay tripping. When a stabilising resistor is used, the tripping characteristic should normally be instantaneous. An alternative technique, avoiding the use of a stabilising resistor is to use a definite time delay characteristic. The time delay used will normally have to be found by trial and error, as it must be long enough to prevent maloperation during a motor start, but short enough to provide effective protection in case of a fault.

Co-ordination with other devices must also be considered. A common means of supplying a motor is via a fused contactor. The contactor itself is not capable of breaking fault current beyond a certain value, which will normally be below the maximum system fault current – reliance is placed on the fuse in these circumstances. As a trip command from the relay instructs the contactor to open, care must be taken to ensure that this does not occur until the fuse has had time to operate. Figure 19.6(a) illustrates incorrect grading of the relay with the fuse, the relay operating first for a range of fault currents in excess of the contactor breaking capacity. Figure 19.6(b) illustrates correct grading. To achieve this, it may require the use of an intentional definite time delay in the relay.

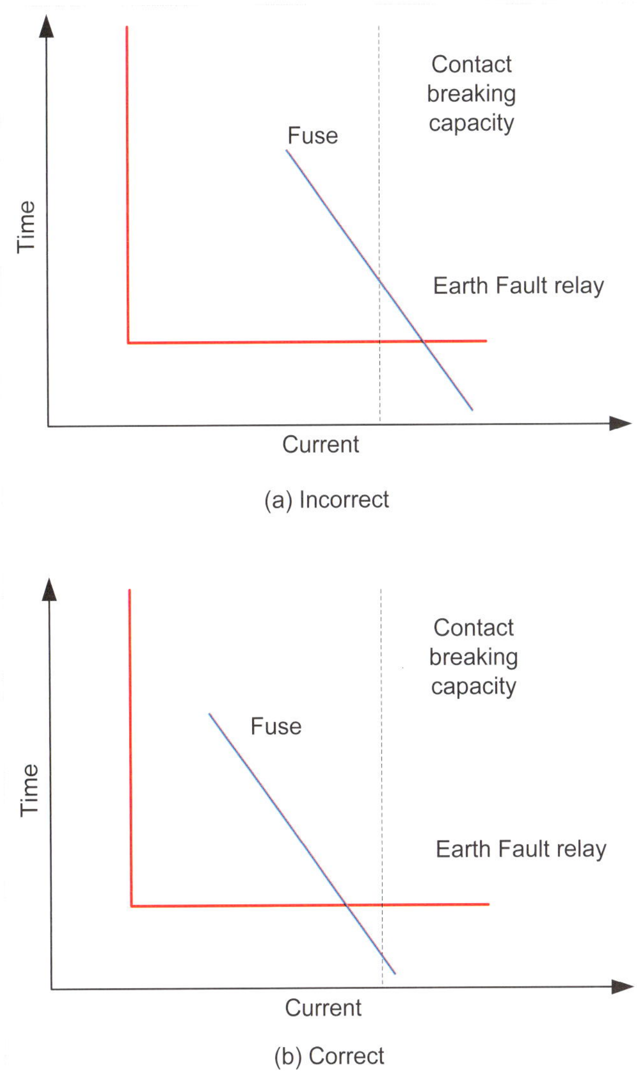

Figure 19.6: Grading of relay with fused contactor

If a more sensitive relay setting is required, it is necessary to use a core-balance CT (CBCT). This is a ring type CT, through which all phases of the supply to the motor are passed, plus the neutral on a four-wire system. The turns ratio of the CT is no longer related to the normal line current expected to flow, so can be chosen to optimise the pick-up current required. Magnetising current requirements are also reduced, with only a single CT core to be magnetised instead of three, thus enabling low settings to be used. Figure 19.7 illustrates the application of a core-balance CT, including the routing of the cable sheath to ensure correct operation in case of core-sheath cable faults.

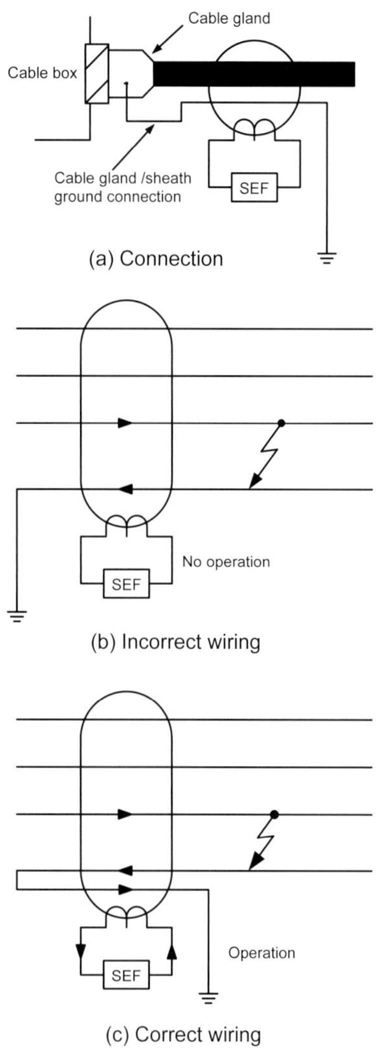

Figure 19.7: Application of core-balance CT

19.6.2 Resistance-Earthed Systems

These are commonly found on HV systems, where the intention is to limit damage caused by earth faults through limiting the earth-fault current that can flow. Two methods of resistance earthing are commonly used:

19.6.2.1 Low resistance earthing

In this method, the value of resistance is chosen to limit the fault current to a few hundred amps – values of 200A-400A being typical. With a residual connection of line CT's, the minimum sensitivity possible is about 10% of CT rated primary current, due to the possibility of CT saturation during starting. For a core-balance CT, the sensitivity that is possible using a simple non-directional earth fault relay element is limited to three times the steady-state charging current of the feeder. The setting should not be greater than about 30% of the minimum earth fault current expected. Other than this, the considerations in respect of settings and time delays are as for solidly earthed systems.

19.6.2.2 High resistance earthing

In some HV systems, high resistance earthing is used to limit the earth fault current to a few amps. In this case, the system capacitive charging current will normally prevent conventional sensitive earth fault protection being applied, as the magnitude of the charging current will be comparable with the earth fault current in the event of a fault. The solution is to use a sensitive directional earth fault relay. A core balance CT is used in conjunction with a VT measuring the residual voltage of the system, with a relay characteristic angle setting of +45°C (see Chapter 9 for details). The VT must be suitable for the relay and therefore the relay manufacturer should be consulted over suitable types – some relays require that the VT must be able to carry residual flux and this rules out use of a 3-limb, 3-phase VT. A setting of 125% of the single phase capacitive charging current for the whole system is possible using this method. The time delay used is not critical but must be fast enough to disconnect equipment rapidly in the event of a second earth fault occurring immediately after the first. Minimal damage is caused by the first fault, but the second effectively removes the current limiting resistance from the fault path leading to very large fault currents.

An alternative technique using residual voltage detection is also possible, and is described in the next section.

19.6.3 Insulated Earth System

Earth fault detection presents problems on these systems since no earth fault current flows for a single earth fault. However, detection is still essential as overvoltages occur on sound phases and it is necessary to locate and clear the fault before a second occurs. Two methods are possible:

- detection of the resulting unbalance in system charging currents
- residual overvoltage.

19.6.3.1 System charging current unbalance

Sensitive earth fault protection using a core-balance CT is required for this scheme. The principle is the same as already detailed, except that the voltage is phase shifted by +90°C instead of -90°C. To illustrate this, Figure 19.8 shows the current distribution in an Insulated system subjected to a C-phase to earth fault and Figure 19.9 the relay vector diagram for this condition. The residual current detected by the relay is the sum of the charging currents flowing in the healthy part of the system plus the healthy phase charging currents on the faulted feeder – i.e. three times the per phase charging current of the healthy part of the system. A relay setting of 30% of this value can be used to provide protection without the risk of a trip due to healthy system capacitive charging currents. As there is no earth fault current, it is also possible to set the relay at site after deliberately applying earth faults at various parts of the system and measuring the resulting residual currents.

If it is possible to set the relay to a value between the charging current on the feeder being protected and the charging current for the rest of the system, the directional facility is not required and the VT can be dispensed with.

The comments made in section 19.6.1 regarding grading with fused contactors also apply.

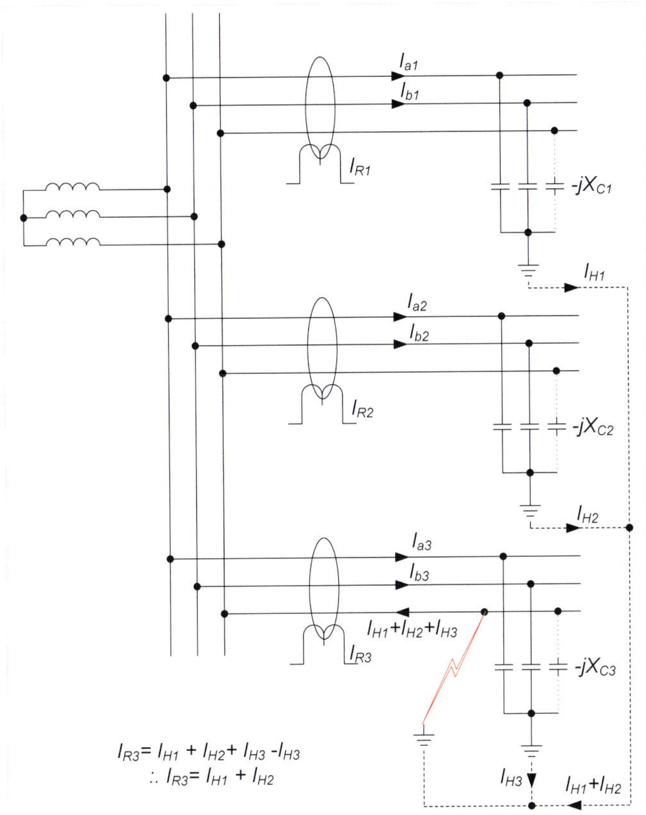

Figure 19.8: Current distribution in insulated-earth system for phase-earth fault

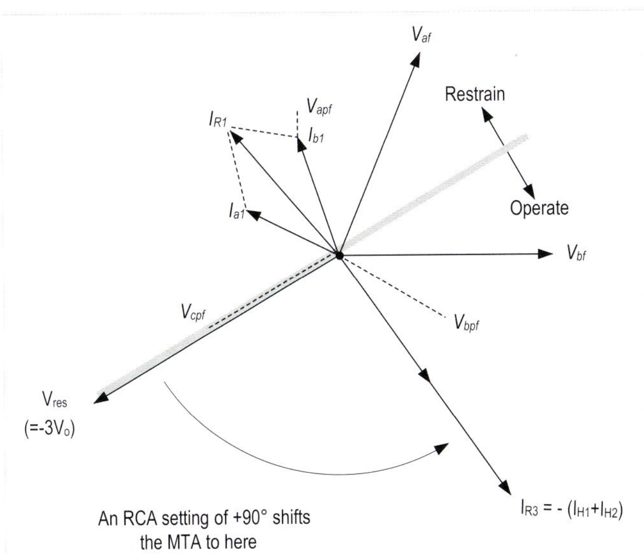

Figure 19.9: Relay vector diagram

19.6.3.2 Residual voltage method

A single earth fault results in a rise in the voltage between system neutral and earth, which may be detected by a relay measuring the residual voltage of the system (normally zero for a perfectly balanced, healthy system). Thus, no CTs are required, and the technique may be useful where provision of an extensive number of core-balance CTs is impossible or difficult, due to physical constraints or on cost grounds. The VTs used must be suitable for the duty, thus 3-limb, 3-phase VTs are not suitable, and the relay usually has alarm and trip settings, each with adjustable time delays. The setting voltage must be calculated from knowledge of system earthing and impedances, an example for a resistance-earthed system is shown in Figure 19.10.

Grading of the relays must be carried out with care, as the residual voltage will be detected by all relays in the affected section of the system. Grading has to be carried out with this in mind, and will generally be on a time basis for providing alarms (1st stage), with a high set definite time trip second stage to provide backup.

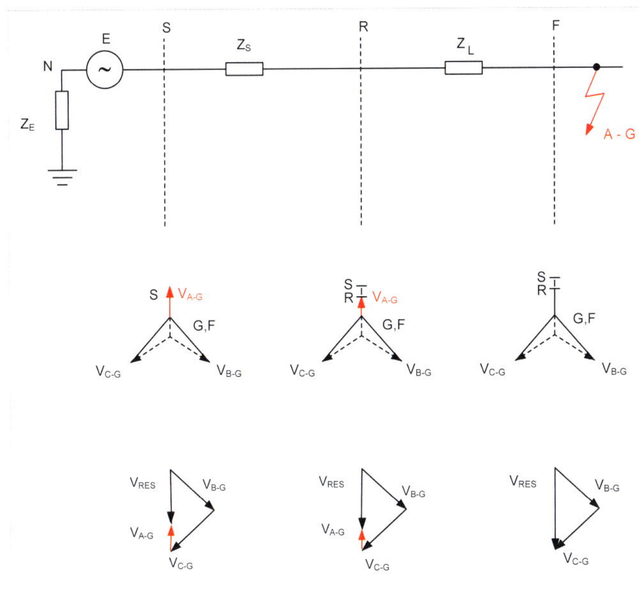

$$V_{RES} = \frac{Z_{SO} + 3Z_E}{2Z_{S1} + Z_{SO} + 2Z_{L1} + Z_{LO} + 3Z_E} \times 3E$$

Figure 19.10: Residual voltage earth-fault protection for resistance-earthed system

19.6.4 Petersen Coil Earthed System

Earthing of a HV power system using a reactor equal to the system shunt capacitance is known as Petersen Coil (or resonant coil) earthing. With this method, a single earth fault results in zero earth fault current flowing (for perfect balance between the earthing inductance and system shunt capacitance), and hence the system can be run in this state for a substantial period of time while the fault is located and corrected. The detailed theory and protection method is explained in Section 9.19.

19.7 NEGATIVE PHASE SEQUENCE PROTECTION

Negative phase sequence current is generated from any unbalanced voltage condition, such as unbalanced loading, loss of a single phase, or single-phase faults. The latter will normally be detected by earth-fault protection, however, a fault location in a motor winding may not result in the earth fault protection operating unless it is of the sensitive variety.

The actual value of the negative sequence current depends on the degree of unbalance in the supply voltage and the ratio of the negative to the positive sequence impedance of the machine. The degree of unbalance depends on many factors, but the negative sequence impedance is more easily determined. Considering the classical induction motor equivalent circuit with magnetising impedance neglected of Figure 19.11:

Motor positive sequence impedance at slip s:

$$=\left[\left(\frac{R_{1p}+R_{2p}}{(2-s)}\right)^{2}+\left(X_{1p}+X'_{2p}\right)^{2}\right]^{0.5}$$

Hence, at standstill (s=1.0), impedance

$$=\left[\left(R_{1p}+R'_{2p}\right)^{2}+\left(X_{1p}+X'_{2p}\right)^{2}\right]^{0.5}$$

The motor negative sequence impedance at slip s

$$=\left[\left(R_{1n}+\frac{R'_{2n}}{S}\right)^{2}+\left(X_{1n}+X'_{2n}\right)^{2}\right]^{0.5}$$

and, at normal running speed, the impedance

$$=\left[\left(R_{1n}+\frac{R'_{2n}}{2}\right)^{2}+\left(X_{1n}+X'_{2n}\right)^{2}\right]^{0.5}$$

where:

- suffix p indicates positive sequence quantities

and

- suffix n indicates negative sequence quantities

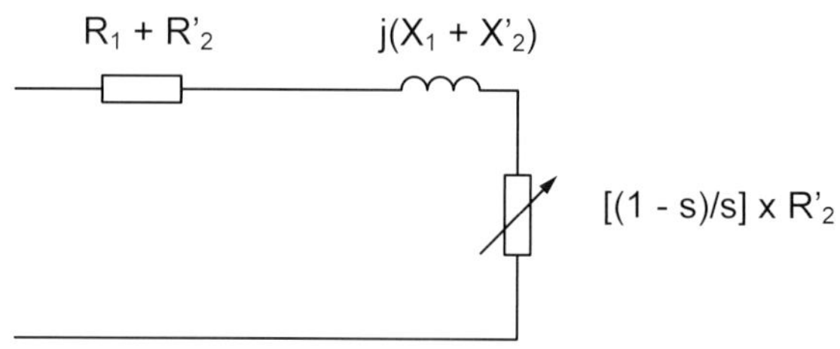

(a) Positive phase sequence equivalent circuit

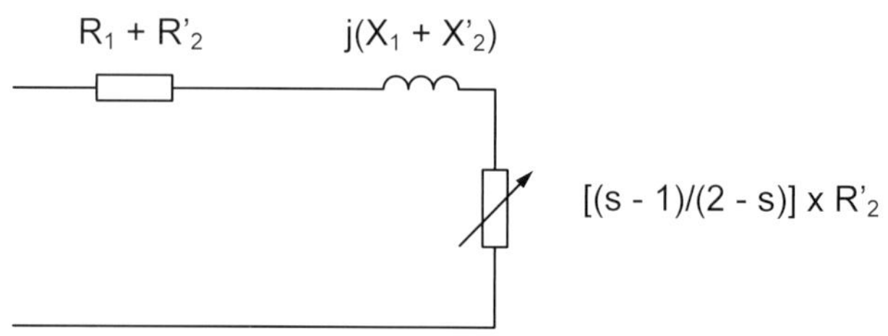

(b) Negative phase sequence equivalent circuit

Figure 19.11: Induction motor equivalent circuits

The resistance can be neglected as it is small compared with the reactance. Thus the negative sequence reactance at running speed is approximately equal to the positive sequence reactance at standstill. An alternative more meaningful way of expressing this is:

$$\frac{\textit{positive seq. impedance}}{\textit{negative seq. impedance}}=\frac{\textit{starting current}}{\textit{rated current}}$$

A typical LV motor starting current is 6 x full load current (FLC). Therefore, a 5% negative sequence voltage (due to, say, unbalanced loads on the system) would produce a 30% negative sequence current in the machine, leading to excessive heating. For the same motor, negative sequence voltages in excess of 17% will result in a negative sequence current larger than rated full load current.

Negative sequence current is at twice supply frequency. Skin effect in the rotor means that the heating effect in the rotor of a given negative sequence current is larger than the same positive sequence current. Thus, negative sequence current may result in rapid heating of the motor. Larger motors are more susceptible in this respect, as the rotor resistance of such machines tends to be higher. Protection against negative sequence currents is therefore essential.

Modern motor protection relays have a negative sequence current measurement capability, in order to provide such protection. The level of negative sequence unbalance depends largely upon the type of fault. For loss of a single phase at start, the negative sequence current will be 50% of the normal starting current. It is more difficult to provide an estimate of the negative sequence current if loss of a phase occurs while running. This is because the impact on the motor may vary widely, from increased heating to stalling due to the reduced torque available.

A typical setting for negative sequence current protection must take into account the fact that the motor circuit protected by the relay may not be the source of the negative sequence current. Time should be allowed for the appropriate protection to clear the source of the negative sequence current without introducing risk of overheating to the motor being considered. This indicates a two stage tripping characteristic, similar in principle to overcurrent protection. A low-set definite time-delay element can be used to provide an alarm, with an IDMT element used to trip the motor in the case of higher levels of negative sequence current, such as loss-of-phase conditions at start, occurring. Typical settings might be 20% of CT rated primary current for the definite time element and 50% for the IDMT element. The IDMT time delay has to be chosen to protect the motor while, if possible, grading with other negative sequence relays on the system. Some relays may not incorporate two elements, in which case the single element should be set to protect the motor, with grading being a secondary consideration.

19.8 FAULTS IN ROTOR WINDINGS

On wound rotor machines, some degree of protection against faults in the rotor winding can be given by an instantaneous stator current overcurrent relay element. As the starting current is normally limited by resistance to a maximum of twice full load, the instantaneous unit can safely be set to about three times full load if a slight time delay of approximately 30 milliseconds is incorporated. It should be noted that faults occurring in the rotor winding would not be detected by any differential protection applied to the stator.

19.9 RTD TEMPERATURE DETECTION

RTDs are used to measure temperatures of motor windings or shaft bearings. A rise in temperature may denote overloading of the machine, or the beginning of a fault in the affected part. A motor protection relay will therefore usually have the capability of accepting a number of RTD inputs and internal logic to initiate an alarm and/or trip when the temperature exceeds the appropriate setpoint(s). Occasionally, HV motors are fed via a unit transformer, and in these circumstances, some of the motor protection relay RTD inputs may be assigned to the transformer winding temperature RTDs, thus providing overtemperature protection for the transformer without the use of a separate relay.

19.10 BEARING FAILURES

There are two types of bearings to be considered: the anti-friction bearing (ball or roller), used mainly on small motors (up to around 350kW), and the sleeve bearing, used mainly on large motors.

The failure of ball or roller bearings usually occurs very quickly, causing the motor to come to a standstill as pieces of the damaged roller get entangled with the others. There is therefore very little chance that any relay operating from the input current can detect bearing failures of this type before the bearing is completely destroyed. Therefore, protection is limited to disconnecting the stalled motor rapidly to avoid consequential damage. Refer to Section 19.4 on stall protection for details of suitable protection.

Failure of a sleeve bearing can be detected by means of a rise in bearing temperature. The normal thermal overload relays cannot give protection to the bearing itself but will operate to protect the motor from excessive damage. Use of RTD temperature detection, as noted in Section 19.9, can provide suitable protection, allowing investigation into the cause of the bearing running hot prior to complete failure.

19.11 UNDERVOLTAGE PROTECTION

Motors may stall when subjected to prolonged undervoltage conditions. Transient undervoltages will generally allow a motor to recover when the voltage is restored, unless the supply is weak.

Motors fed by contactors have inherent undervoltage protection, unless a latched contactor is used. Where a specific undervoltage trip is required, a definite time undervoltage element is used. If two elements are provided, alarm and trip settings can be used. An interlock with the motor starter is required to block relay operation when the starting device is open, otherwise a start will never be permitted. The voltage and time delay settings will be system and motor dependent. They must allow for all voltage dips likely to occur on the system during transient faults, starting of motors, etc. to avoid spurious trips. As motor starting can result in a voltage depression to 80% of nominal, the voltage setting is likely to be below this value. Re-acceleration is normally possible for voltage dips lasting between 0.5-2 seconds, depending on system, motor and drive characteristics, and therefore the time delay will be set bearing these factors in mind.

19.12 LOSS-OF-LOAD PROTECTION

Loss-of-load protection has a number of possible functions. It can be used to protect a pump against becoming unprimed, or to stop a motor in case of a failure in a mechanical transmission (e.g. conveyor belt), or it can be used with synchronous motors to protect against loss-of-supply conditions. Implementation of the function is by a low forward power relay element, interlocked with the motor starting device to prevent operation when the motor is tripped and thus preventing a motor start. When starting against a very low load (e.g. a compressor), the function may also need to be inhibited for the duration of the start, to prevent maloperation.

The setting will be influenced by the function to be performed by the relay. A time delay may be required after pickup of the element to prevent operation during system transients. This is especially important for synchronous motor loss-of supply protection.

19.13 ADDITIONAL PROTECTION FOR SYNCHRONOUS MOTORS

The differences in construction and operational characteristics of synchronous motors mean that additional protection is required for these types of motor. This additional protection is discussed in the following sections.

19.13.1 Out-of-Step Protection

A synchronous motor may decelerate and lose synchronism (fall out-of-step) if a mechanical overload exceeding the peak motor torque occurs. Other conditions that may cause this condition are a fall in the applied voltage to stator or field windings. Such a fall may not need to be prolonged, a voltage dip of a few seconds may be all that is required. An out-of-step condition causes the motor to draw excessive current and generate a pulsating torque. Even if the cause is removed promptly, the motor will probably not recover synchronism, but eventually stall. Hence, it must be disconnected from the supply.

The current drawn during an out-of-step condition is at a very low power factor. Hence a relay element that responds to low power factor can be used to provide protection. The element must be inhibited during starting, when a similar low power factor condition occurs. This can conveniently be achieved by use of a definite time delay, set to a value slightly in excess of the motor start time.

The power factor setting will vary depending on the rated power factor of the motor. It would typically be 0.1 less than the motor rated power factor i.e. for a motor rated at 0.85 power factor, the setting would be 0.75.

19.13.2 Protection against Sudden Restoration of Supply

If the supply to a synchronous motor is interrupted, it is essential that the motor breaker be tripped as quickly as possible if there is any possibility of the supply being restored automatically or without the machine operator's knowledge. This is necessary in order to prevent the supply being restored out of phase with the motor generated voltage.

Two methods are generally used to detect this condition, in order to cover different operating modes of the motor.

19.13.2.1 Underfrequency protection

The underfrequency relay element will operate in the case of the supply failing when the motor is on load, which causes the motor to decelerate quickly. Typically, two elements are provided, for alarm and trip indications. The underfrequency setting value needs to consider the power system characteristics. In some power systems, lengthy periods of operation at frequencies substantially below normal occur, and should not result in a motor trip. The minimum safe operating frequency of the motor under load conditions must therefore be determined, along with minimum system frequency.

19.13.2.2 Low forward power protection

This can be applied in conjunction with a time delay to detect a loss-of-supply condition when the motor may share a busbar with other loads. The motor may attempt to supply the other loads with power from the stored kinetic energy of rotation. A low-forward-power relay can detect this condition. A time delay will be required to prevent operation during system transients leading to momentary reverse power flow in the motor.

19.14 MOTOR PROTECTION EXAMPLES

This section gives examples of the protection of HV and LV induction motors.

19.14.1 Protection of a HV Motor

Table 19.2 gives relevant parameters of a HV induction motor to be protected. Using an Alstom MiCOM P241 motor protection relay, the important protection settings are calculated in the following sections.

Quantity	Value
Rated output	1000kW CMR
Rated Voltage	3.3kV
Rated frequency	50Hz
Rated power factor/efficiency	0.9/0.92
Stall withstand time cold/hot	20/7 sec
Starting current	550% DOL
Permitted starts cold/hot	3/2
CT ratio	250/1
Start time at 100% voltage	4 sec
Start time at 80% voltage	5.5 sec
Heating/cooling time constant	25/75 mins

Quantity	Value
System earthing	Solid
Control device	Circuit Breaker

Table 19.2: Example motor data

19.14.1.1 Thermal protection

The current setting I_{TH} is set equal to the motor full load current, as it is a CMR rated (Continuous Maximum Rated) motor. Motor full load current can be calculated as 211A, therefore (in secondary quantities):

$$I_{TH} = \frac{211}{250} = 0.844$$

Use a value of 0.85, nearest available setting.

The relay has a parameter, *K*, to allow for the increased heating effect of negative sequence currents. In the absence of any specific information, use *K=3*.

Two thermal heating time constants are provided, τ_1 and τ_2. τ_2 is used for starting methods other than DOL, otherwise it is set equal to τ_1. τ_1 is set to the heating time constant, hence $\tau_1 = \tau_2$=25min. Cooling time constant τ_R is set as a multiple of τ_1. With a cooling time constant of 75min, $\tau_R = 3 \times \tau_1$.

19.14.1.2 Short circuit protection

Following the recommendations of Section 19.5, with a starting current of 550% of full load current, the short-circuit element is set to 1.25 x 5.5 x 211A = 1450A.
In terms of the relay nominal current I_n, the setting value is

$$\frac{1450}{250} = 5.8I_n$$

There is a minimum time delay of 100ms for currents up to 120% of setting to allow for transient CT saturation during starting and 40ms above this current value. These settings are satisfactory.

19.14.1.3 Earth-fault protection

It is assumed that no CBCT is fitted. A typical setting of 30% of motor rated current is used, leading to an earth fault relay setting of

$$0.3 \times \frac{211}{250} = 0.25I_n$$

A stabilising resistor is required, calculated in accordance with Equation 19.2 to prevent maloperation due to CT spill current during starting as the CTs may saturate. With the stabilising resistor present, instantaneous tripping is permitted. The alternative is to omit the stabilising resistor and use a definite time delay in association with the earth-fault element. However, the time delay must be found by trial and error during commissioning.

19.14.1.4 Locked rotor/Excessive start time protection

The current element must be set in excess of the rated current of the motor, but well below the starting current of the motor to ensure that a start condition is recognised (this could also be achieved by use of an auxiliary contact on the motor CB wired to the relay). A setting of 500A *($2I_n$)* is suitable. The associated time delay needs to be set to longer than the start time, but less than the cold stall time. Use a value of 15s.

19.14.1.5 Stall protection

The same current setting as for locked rotor protection can be used – 500A. The time delay has to be less than the hot stall time of 7s but greater than the start time by a sufficient margin to avoid a spurious trip if the start time happens to be a little longer than anticipated. Use a value of 6.5s.

The protection characteristics for sections 19.14.1.1-5 are shown in Figure 19.12.

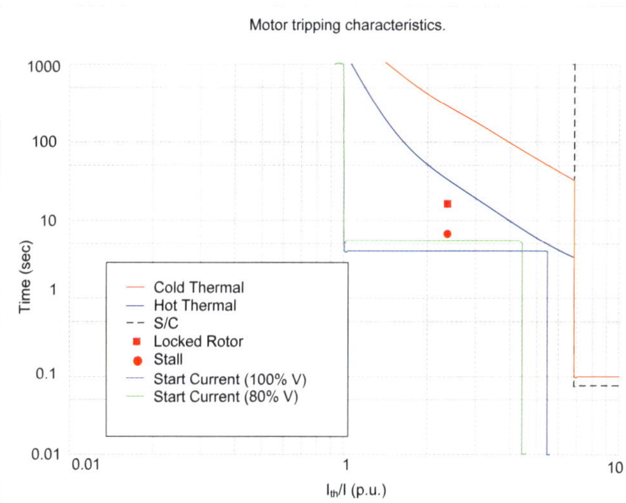

Figure 19.12: Protection characteristics for motor protection example

19.14.1.6 Negative phase sequence protection

Two protection elements are provided, the first is definite time-delayed to provide an alarm. The second is an IDMT element used to trip the motor on high levels of negative sequence current, such as would occur on a loss of phase condition at starting. In accordance with Section 19.7, use a setting of 20% with a time delay of 30s for the definite time element and 50% with a TMS of 1.0 for the IDMT element. The resulting characteristic is shown in Figure 19.13. The motor thermal protection, as it utilises a negative sequence component, is used for protection of the motor at low levels of negative sequence current.

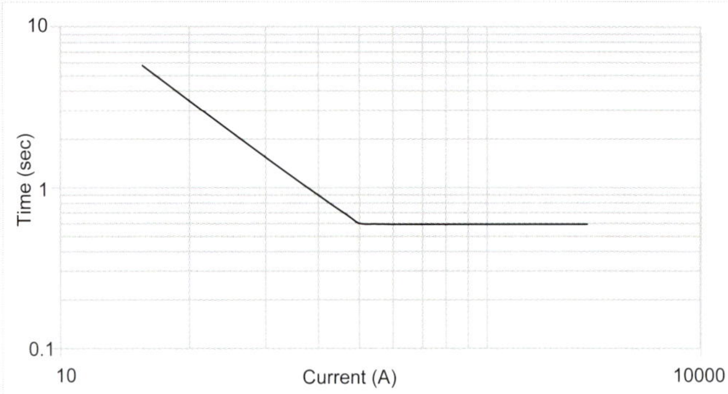

Figure 19.13: Motor protection example- negative sequence protection characteristic

19.14.1.7 Other protection considerations

If the relay can be supplied with a suitable voltage signal, stall protection can be inhibited during re-acceleration after a voltage dip using the undervoltage element (set to 80-85% of rated voltage). Undervoltage protection (set to approximately 80% voltage with a time delay of up to several seconds, dependent on system characteristics) and reverse phase protection can also be implemented to provide extra protection. Unless the drive is critical to the process, it is not justifiable to provide a VT specially to enable these features to be implemented.

19.14.2 Protection of an LV Motor

LV motors are commonly fed via fused contactors and therefore the tripping times of a protection relay for overcurrent must be carefully co-ordinated with the fuse to ensure that the contactor does not attempt to break a current in excess of its rating. Table 19.3(a) gives details of an LV motor and associated fused contactor.

Parameter	Symbol	Value	Unit
(a) LV motor example			
Standard		IEC 60034	
Motor Voltage		400	V
Motor kW		75	kW
Motor kVA		91.45	kVA
Motor FLC		132	A
Starting Current		670	%
Starting Time		4.5	sec
Contactor rating		300	A
Contactor breaking capacity		650	A
Fuse rating		250	A
(b) Relay settings			
Overcurrent		Disabled	-
Overload setting	Ib	4.4	A
Overload time delay	I>t	15	sec
Unbalance	I2	20	%
Unbalance time delay	I2>t	25	sec
Loss of phase time delay	<Ip	5	sec

Table 19.3: LV motor protection setting example

19.14.2.1 CT ratio

The relay is set in secondary quantities, and therefore a suitable CT ratio has to be calculated. From the relay manual, a CT with 5A secondary rating and a motor rated current in the range of 4-6A when referred to the secondary of CT is required. Use of a 150/5A CT gives a motor rated current of 4.4A when referred to the CT secondary, so use this CT ratio.

19.14.2.2 Overcurrent (short-circuit) protection

The fuse provides the motor overcurrent protection, as the protection relay cannot be allowed to trip the contactor on overcurrent in case the current to be broken exceeds the contactor breaking capacity. The facility for overcurrent protection within the relay is therefore disabled.

19.14.2.3 Thermal (overload) protection

The motor is an existing one, and no data exists for it except the standard data provided in the manufacturer's catalogue. This data does not include the thermal (heating) time constant of the motor.

In these circumstances, it is usual to set the thermal protection so that it lies just above the motor starting current.

The current setting of the relay, I_b, is found using the formula

$$I_b = 5 \times \frac{I_n}{I_p}$$

Where

I_n = motor rated primary current

I_p = CT primary current

Hence,

$$I_b = 5 \times \frac{132}{150} = 4.4A$$

With a motor starting current of 670% of nominal, a setting of the relay thermal time constant with motor initial thermal state of 50% of 15s is found satisfactory, as shown in Figure 19.14.

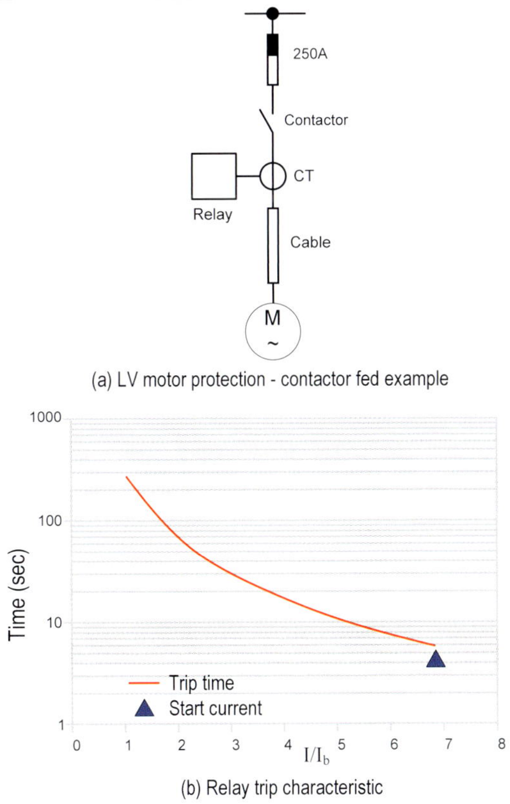

Figure 19.14: Motor protection example – contactor-fed motor

19.14.2.4 Negative sequence (phase unbalance) protection

The motor is built to IEC standards, which permit a negative sequence (unbalance) voltage of 1% on a continuous basis. This would lead to approximately 7% negative sequence current in the motor (Section 19.7). As the relay is fitted only with a definite time relay element, a setting of 20% (from Section 19.7) is appropriate, with a time delay of 25s to allow for short high-level negative sequence transients arising from other causes.

19.14.2.5 Loss of phase protection

The relay has a separate element for this protection. Loss of a phase gives rise to large negative sequence currents, and therefore a much shorter time delay is required. A definite time delay of 5s is considered appropriate.

The relay settings are summarised in Table 19.3(b).

Chapter 20
System Integrity Protection Schemes

20.1 INTRODUCTION

Operating electric power systems is continually becoming more complex. Increasing challenges are posed as the systems are operated closer to their stability limits. Electrical power systems continue to expand, integrating new components such as renewable energy sources, distributed generation and independent power producers, in an increasingly open energy market. Regulatory pressure has power system operators' attention focused to grow return on investment of their assets whilst power consumption is increasing and many power system infrastructures are ageing.

In the wake of deregulation, electricity markets load the grid in a less predictable and more dynamic way, while generation companies' decisions do not always account for transmission constraints.

Stability margins are narrowing since the system is being pushed to operate closer to its limits, and forecasting operational data is getting more difficult as grids connect over increasingly larger geographical footprints.

The result is that when things do go wrong, they can have a dramatic effect with cascade tripping and blackouts.

20.1.1 Blackouts

In 1965, 25 million people in parts of the United States and Canada lost their electricity supply for about 12 hours. The failure was attributed to a maintenance error and carries the reputable distinction of being the first large scale blackout. It is not an isolated event. Leaving aside 'planned' disruptions caused by, for example civil unrest such as the miners' strike in the UK in 1974, there are many such events that have caused disruption that is, at the least inconvenient, generally commercially damaging, and at worst, life-threatening.

In 1978, 80 percent of France was affected by a blackout caused by the breakdown of a transmission line.

In 1989, a geomagnetic storm caused an outage that left 6 million people in Québec, Canada without power for nine hours.

In 1999 a major blackout was triggered in Brazil following a lightning strike. Over 90 million people were affected.

In August 2003, a wide-area power collapse in the north of the United States of America and central Canada affected 50 million people. It is estimated to have cost billions of dollars, and at least eleven fatalities were attributed to the loss of power.

During the miners' strike of 1974, energy and electrical power systems had been at the forefront of the political agenda in the UK. The devastation caused by the North American blackout put electrical power systems onto the world political agenda. Supply of electrical energy had become, to many, no longer a luxury, but a basic human need.

Large-scale blackouts have continued since that time in countries including Australia, Japan, Peru, Greece, Russia and Italy, but technology is being developed and deployed to help prevent future occurrences.

Scenarios of major blackouts are hardly ever the same. The starting event may be a generation-load imbalance, a short circuit, human actions (or inaction), unexpected grid topology changes, lack of voltage support, or natural causes such as lightning storms or untrimmed trees. What they all share is a resulting failure process that cascades.

Occasionally, to prevent network components from overloading, parts of a power system automatically disconnect or shut down to avoid damage and isolate the problem component from the rest of the network. Under certain conditions, the shutdown of the component can cause significant current fluctuations or overloads in the remaining segments of the network. These fluctuations may propagate and amplify, leading to uncontrollable power system dynamic oscillations and cascading failure, which spreads to wider parts of the network and, in some cases, the entire grid.

20.1.2 Preserving Supply

Although rare, these widespread blackouts and their collection of critical consequences on business, public safety (lighting), public comfort (heating/air conditioning), telecommunications, etc., have driven the industry to develop automatic control systems to mitigate the occurrence of such large-scale failures caused by unexpected events. Modern power systems are designed and equipped to be resistant to this sort of cascading failure. Facilities and tools, which embed such resistance, include energy management system technology and system integrity protection schemes also known as special protection schemes, remedial action schemes, or wide area schemes.

Electrical power systems are designed and operated to withstand contingencies that are likely to occur as a result of a single event. Such events are sometimes referred to as class 1, or N-1 type contingencies. For these contingencies protection is likely to be afforded by relays applied to individual items of plant.

A combination of simultaneous events could be classified as a class 2, or N-2 type contingency, and a cascading event as a class 3, or N-3 contingency.

For N-2 and N-3 contingencies, system level action is required. This could involve generator tripping (possibly on a massive scale), load shedding or, ultimately, coordinated system separation, to preserve system stability and/or prevent serious damage to transmission and generation equipment.

As well as the N- contingencies, another way of looking at the system operation is by means of the state diagram introduced by Fink and Carlsen [20.1] as shown in Figure 20.1.

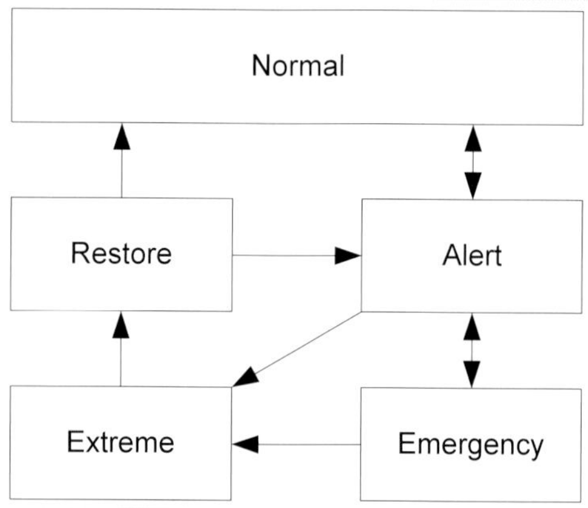

Figure 20.1: Fink Carlsen diagram

In this model, conventional protection and control is likely to be effective in the 'alert' and 'emergency' states where the load capacity and generating capacity remain matched. In the 'extreme' state, they are no longer matched and system integrity protection schemes are required.

Remedial actions performed by system integrity protection schemes are the second and third lines in the plan to protect the system.

20.1.3 Enabling Technologies

System integrity protection schemes have advanced thanks to technological developments, which have been widely adopted by the industry. Three technological developments in particular that have brought great benefit and advances in the field of measurement, protection, control and automation of electrical power systems are:

- advances in embedded computing
- wide-area communications and
- accurate time synchronisation

This chapter explores at how these enabling technologies can enhance the measurement of power system quantities, enabling wide-area monitoring and control, and 'bigger picture' protection solutions.

20.2 SUMMARY OF SYSTEM INTEGRITY PROTECTION SCHEMES

System integrity protection schemes (or 'special protection schemes', 'remedial action schemes', and in some cases 'wide-area schemes'), are tools to protect the system. Unlike conventional protection relays where the purpose is to isolate faulted plant, system integrity protection schemes initiate actions to correct the system. Typical system stress scenarios that a system integrity protection scheme might be required to act upon include:

- transient instability
- voltage degradation
- frequency degradation
- thermal overloading
- loss of synchronism
- large power swings
- cascade of overloads

To implement effective, intelligent system integrity protection schemes appropriate for the prevailing power system condition, it is essential that real time system data is made available.

Using this data, a system integrity protection scheme can initiate:

- generator tripping
- load rejection
- controlled system separation
- voltage clamping
- modification to the operational state of FACTS devices
- etc.

As with conventional protection the consequences of failing to operate when required, or operating when not required could be catastrophic and so system integrity protection schemes must be carefully designed. Four main design criteria applicable to system integrity protection schemes are [20.2]:

- Dependability:– The certainty that the system integrity protection scheme operates when required, that is, in all cases where emergency controls are required to avoid a collapse.
- Security:– The certainty that the system integrity protection scheme will not operate when not required, i.e. does not apply emergency controls unless they are necessary to avoid a collapse.
- Selectivity:– The ability to select the correct and minimum action to perform the intended function. That is, to avoid using disruptive controls such as load shedding if they are not necessary to avoid a collapse.
- Robustness:– The ability of the system integrity protection scheme to provide dependability, security and selectivity over the full range of dynamic and steady state operating conditions that it will encounter.

Conformance to these criteria is critical to the deployment of system integrity protection schemes.

Under normal system operation, the measures that a system integrity protection scheme may initiate will be inactive. They are activated when some system disturbance or stress condition occurs.

In terms of detection principles, system integrity protection schemes may be classified in two different categories: event-based and response based. Event-based systems act to detect the cause(s) of a disturbance; Response-based systems respond to the effects. Typically, event based systems will automatically activate remedial actions and will operate quickly. Response based systems may activate the remedial actions either automatically or manually, and longer operating times may apply.

System integrity protection schemes may exist in stand-alone devices, using local quantities only to provide fast acting decisions, or they may take quantities from a wider area of the system. The latter will have longer operating times as it relies on the communication of data between source of acquisition and decision-maker. Additionally, system integrity protection schemes may be made at the highest system level and be part of the energy management suite as shown in Figure 20.2.

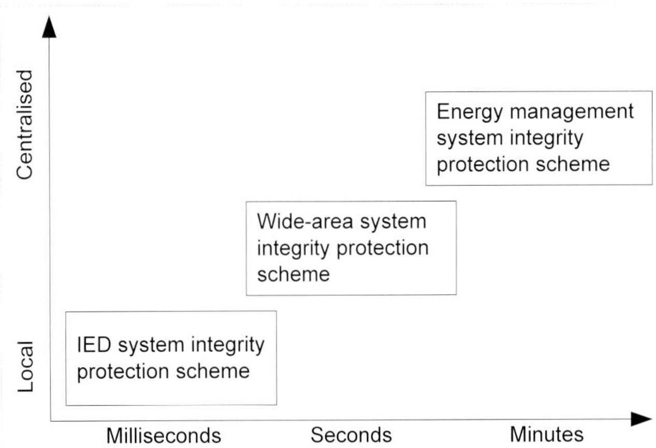

Figure 20.2: System integrity protection schemes residency and typical operating times.

Examples of IED (Intelligent Electronic Device) system integrity protection schemes include detecting changes to topology, and detecting loss of synchronism.

An example of a wide-area system integrity protection scheme is monitoring angular stability across the system and taking actions necessary to correct. Energy management system integrity protection schemes can be used to take a more 'static' view, looking at long term phenomena.

Implementing system integrity protection schemes requires real time data. When that data is coming from different sources it is crucial that the data is synchronised, and that it can be communicated effectively and efficiently. It is the possibility to accurately synchronise devices and to rapidly communicate the data that makes time-synchronised system integrity protection schemes possible.

20.3 TIME-SYNCHRONISED SYSTEM INTEGRITY PROTECTION SCHEMES

As discussed briefly in chapter 10, timing information from the global positioning satellite system (GPS) can be used to synchronise current differential unit protection and merging units (chapter 24). The GPS system consists of a number of satellites in space that are continually transmitting information.

The basic principle of using the GPS system to synchronise current differential protection is shown in Figure 20.3. This is an example of a communication-assisted scheme with a practical application.

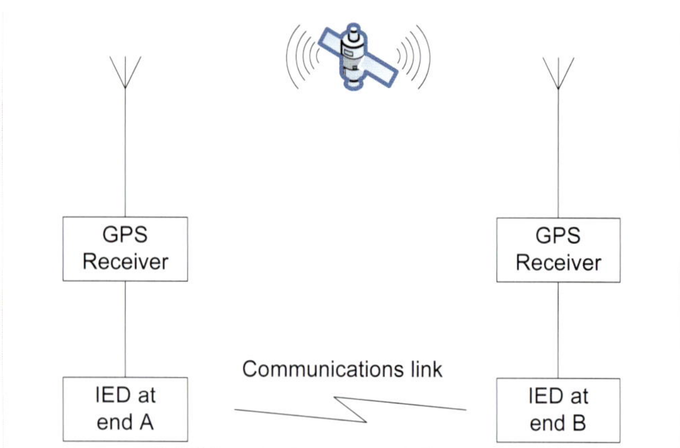

Figure 20.3: GPS timing synchronisation for current differential protection.

The basis of the scheme described by Figure 20.3 is that the GPS signals are picked up by antennae connected to synchronising units. These synchronising units decode the information to produce a highly accurate timing signal that synchronises the current signals used by the relays to perform the current differential algorithm. The timing signal presented to the relays is in the form of a pulse that is transmitted at the start of every second. This one pulse per second (1pps) is more than sufficiently accurate for multi-ended current differential protection and it is a logical progression to use the same accuracy of GPS timing information to synchronise the measurements of power system quantities across the whole grid.

The GPS 1pps signal can be used to synchronise the measurement of anything from anywhere in the world. In the electrical power system it can be used to synchronise the measurement of voltage and current from key nodes on the grid to enhance the monitoring and control of the grid.

So called phasor measurement units are synchronised using the GPS timing to produce time-synchronised phasor representations of the current and voltage signals that they take as input. These time-synchronised phasor representations, or synchrophasors, can be collected from different nodes on the grid and compared to get a wide-area system-level view.

Current and voltage synchrophasors from phasor measurement units can be used to observe the state of the system and improve the performance of different system level applications. Using synchrophasors to observe, measure, or monitor the system is sometimes termed 'wide-area monitoring' or WAM. Control actions taken on the basis of the interpretation of these WAMs is sometimes termed 'wide-area control', or WAC, and protection action taken on the WAMs is sometimes termed 'wide-area protection', or WAP. System integrity protection schemes based on wide-area technology are essentially control schemes. It is perhaps unwise to try to differentiate the application of wide-area technology to the separate fields of monitoring, control, and protection, rather to combine them into a single heading of 'wide-area monitoring, protection and control', or WAMPAC.

20.3.1 Wide-area Monitoring, Protection and Control (WAMPAC) System Integrity Protection Schemes

A wide-area scheme to accommodate system integrity protection could consist of phasor measurement units to generate the synchrophasor data signals. These signals need to be collected together before system level applications are exercised. Collection of synchrophasors is typically performed by a phasor data concentrator (PDC) which prepares the synchrophasor data for the application.

Wide-area technology based on synchrophasors can be used to enhance state estimation and provides opportunities for WAMPAC schemes as shown in Figure 20.4.

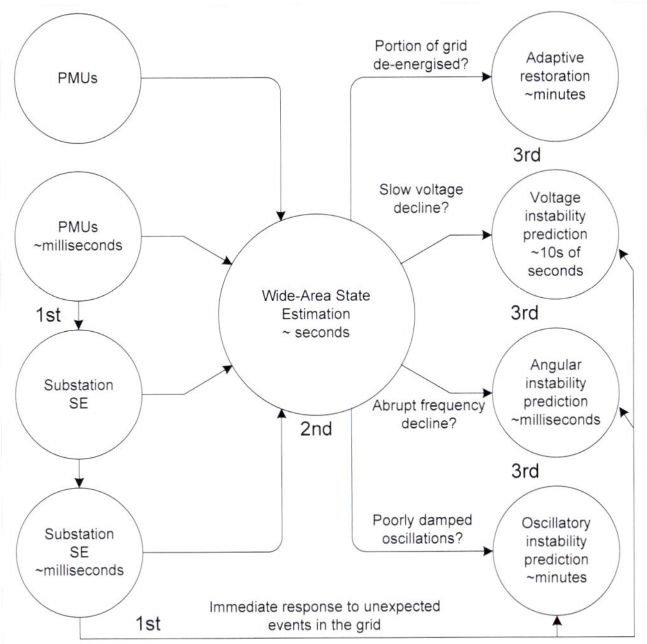

Figure 20.4: WAMPAC opportunities

Figure 20.4 shows applications based on the use of synchrophasors and state estimation (SE) for oscillatory stability, voltage stability, transient stability (or angular instability prediction), and system restoration.

The technology can also offer event triggering based on overcurrent, under/overvoltage, under/over-frequency and under/over-rate of change of frequency.

20.3.1.1 Synchrophasors for Oscillatory Stability

Synchrophasors can be used for oscillatory stability detection designed to detect slower inter-area system oscillations. These are characterised by power oscillations between two areas of generation, running at slightly different speeds and following small system perturbations such as load switching or tap changes. If there is insufficient system damping, these oscillations do not die away and can escalate leading to potential out-of-step conditions if no pre-emptive action is taken. If the phasor measurement units feature frequency tracking that can accurately measure and track small changes in signal frequencies, this can provide an accurate measurement for the fast detection of small signal oscillations to initiate damping actions.

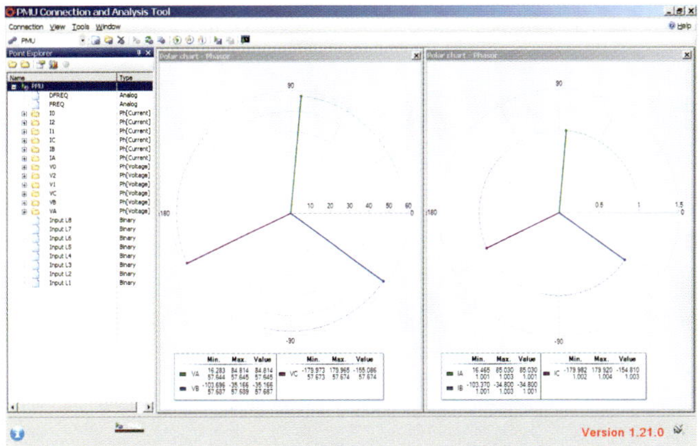

Figure 20.5: PMU data viewed in a simple visualisation tool

Synchronised measurements allow the comparison of the phase of the oscillatory modes to determine the parts of the system that are oscillating with respect to each other, and enable stabilising actions to be initiated as shown in Figure 20.6.

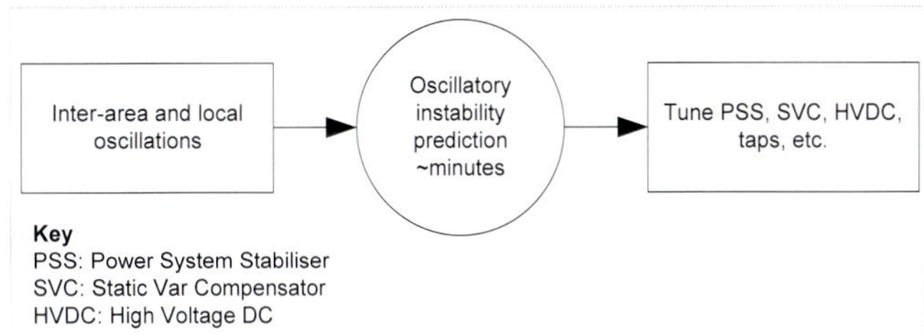

Figure 20.6: Oscillatory instability prediction

20.3.1.2 Synchrophasors for Voltage Stability

Synchrophasors can be used to predict voltage instability. By comparing voltage and power transfers on the line against specific stability margins, operators can be alerted to slow voltage declines and take remedial actions such as switching in reactors as shown in Figure 20.7.

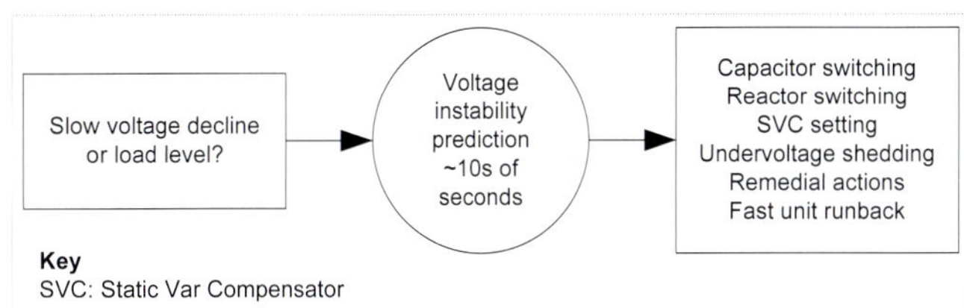

Figure 20.7: Voltage instability prediction

20.3.1.3 Synchrophasors for Transient Stability

Synchrophasors can be used to defend transient stability by predicting angular instability. By being measured in real-time rather than estimated, generator and bus-phase angles can be compared and stability threats can be presented to an operator for remedial action such as controlled islanding as shown in Figure 20.8.

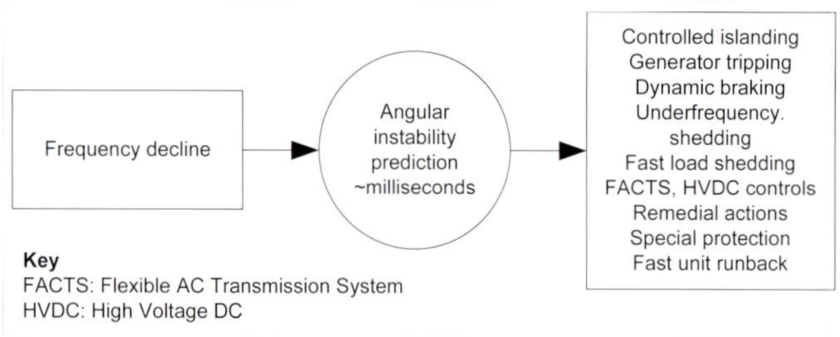

Figure 20.8: Angle instability prediction

Figure 20.9 shows the relative phase angles from two parts of the North American grid taken on August 14th 2003.

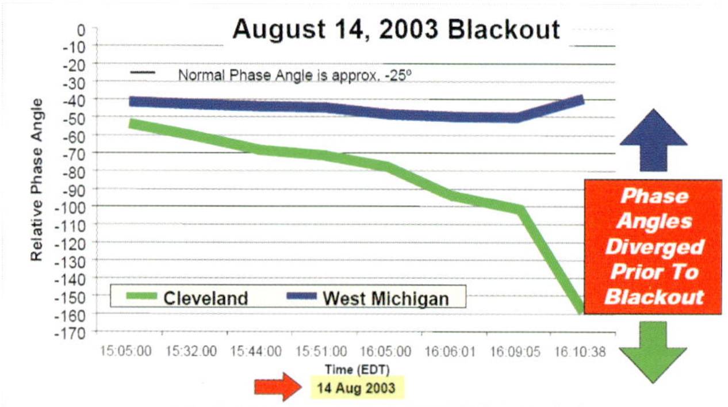

Figure 20.9: Phase angle divergence prior to blackout

The phase angles from the two parts of the system can be seen diverging. A major blackout followed. This is the type of event that the synchrophasor technology can help prevent. But what exactly is a synchrophasor?

20.3.2 Synchronised Phasor Measurements

A phasor is a representation of a sinusoidal quantity, representing the sinusoid as a vector rotating on the complex plane, defined in terms of amplitude and phase. As steady state power system quantities are sinusoids, it follows that they can be represented by phasors. When augmented by an accurate time synchronisation signal, time synchronised phasors, known as synchrophasors can be produced. The potential benefits of synchrophasor technology have long been recognised, but it is only since enabling technologies have developed that their effective application became reality.

In 1893, Charles Proteus Steinmetz presented a paper on a simplified mathematical description of the waveforms of alternating electricity. Steinmetz called his representation a phasor, and phasor notation became widely used in the field of electrical power systems (see section 3.4.1). The time synchronisation of phasors in what were to become known as phasor measurement units evolved and in 1988 the first phasor measurement units based on Steinmetz's technique, were developed. The first commercial phasor measurement unit was launched in 1992.

The deployment of phasor measurement units into electrical power systems, however, was limited initially as two main issues could not be addressed:

- Accuracy of time synchronisation
- Cost effective, high-speed, communications

With the global positioning satellite system, the ability to time synchronise anywhere in the world to an accuracy of better than a micro-second is achievable. Developments in the telecommunication markets have brought fast Ethernet to the substation, and with these developments, the hurdles are overcome, and the potential of synchrophasors can be exploited to the maximum.

20.3.3 Synchrophasor Definition

A phasor is a vector representation on the complex plane of a sinusoidal quantity such as an alternating current.

A synchrophasor is a phasor calculated from sampled values that are referenced to a common timing signal. The timing signal is global, meaning that a synchrophasor measured anywhere in the world will be referenced to the same time signal. This enables synchrophasors to be communicated for collation and comparison. This is the fundamental concept behind wide-area special protection schemes.

The IEEE 1344 standard for synchrophasors was published in 1995. It was reaffirmed in 2001, but in 2005 it was superseded by IEEE C37.118. This recognised the developments that would facilitate the effective application of synchrophasors to electrical power systems. IEEE C37.118 describes the standards for measurement of synchrophasors, the method of quantifying the measurements, as well as the testing and certification requirements for verifying accuracy. It also prescribes the data transmission format and protocol for synchrophasor communication.

The standard prescribes that all measurements are referenced to an accurate one pulse per second (1pps) signal. Generally it is anticipated that this will be derived from the global positioning satellite system, but it can be from any external source provided the accuracy is in accordance with the requirements.

Phasor representation defines the complex exponential as a point or vector on the complex plane according to the equation:

$$e^{j\theta} = \cos\theta + j\sin\theta$$

The IEEE C37.118 standard specifies that the angle θ is zero degrees when the maximum of the signal to be measured coincides with the leading edge of GPS 1pps pulse. It follows that it will be –π/2 radians (-90 degrees) if the positive zero crossing coincides with the 1pps pulse, +π/2 radians (+90 degrees) if the negative zero crossing coincides with the 1pps pulse, and π radians (180 degrees) when the reverse polarity peak of the signal to be measured coincides with the 1pps pulse.

The relationship of time domain and frequency domain signals relative to the 1pps signal is further described in Figure 20.10, where $X_r(n)$ and $X_i(n)$ are the RMS real and imaginary filtered components at a particular instance and θ is the phase angle as per the IEEE C37.118 standard definition.

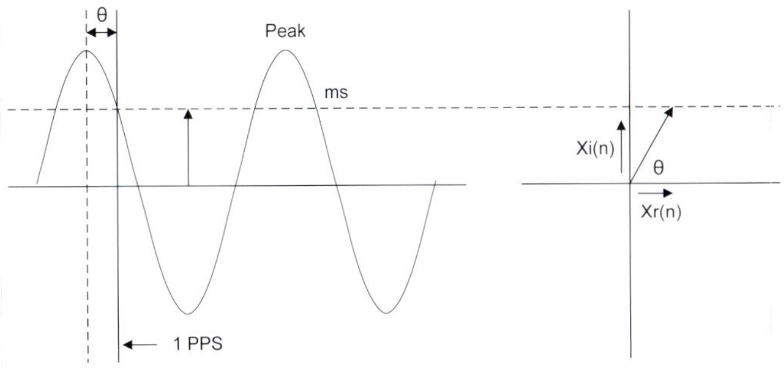

Figure 20.10: Time and frequency domain representation of synchrophasors

The measured angle θ is valid in the range of $-\pi$ to $+\pi$ radians.

A primary requirement for synchrophasors is the accuracy of the measured phasor. The IEEE C37.118 standard defines the total permitted vector error (TVE) for the static condition at nominal frequency as:

$$TVE = \sqrt{\frac{\left(X_r(n) - X_r\right)^2 + \left(X_i(n) - X_i\right)^2}{X_r^{\ 2} + X_i^{\ 2}}}$$

Where $X_r(n)$ and $X_i(n)$ are the measured real and imaginary components and X_r and X_i are the reference values.

Levels of accuracy are specified for the permissible total vector error as shown in Table 20.1 .

Influence quantity	Reference Condition	Range of influence quantity change with respect to reference and maximum allowable TVE (in %) for each compliance level			
		Level 0		Level 1	
		Range	TVE (%)	Range	TVE (%)
Signal frequency	$F_{nominal}$	+/- 0.5Hz	1	+/- 5Hz	1
Signal magnitude	100% rated	80% -120% rated	1	10% -120% rated	1
Phase angle	0 radians	+/-π radians	1	+/-π radians	1
Harmonic distortion	<0.2% (THD)	1%, any harmonic up to 50th	1	1%, any harmonic up to 50th	1
Out-of-band interfering signal at freq f_i *	<0.2% of input signal magnitude	1% of input signal magnitude	1	1% of input signal magnitude	1

* $|f_i - f_0| > F_s/2$, where F_s = phasor reporting rate and $f_0 = F_{nominal}$

Table 20.1: Influencing quantities and allowable error limits for compliance levels 0 – 1.

Devices offering compliance level 1 perform better in SIPS schemes, whereas level 0 devices are intended more for measurement purposes.

This measurement accuracy will vary with the magnitude and frequency of the input signal and may be depicted graphically as shown in Figure 20.11. The graph depicts a device with a total vector error typically less than 0.4% (y-axis), consistent for system frequency excursions off-nominal (z-axis), and maintained for all except very low current magnitudes (x-axis).

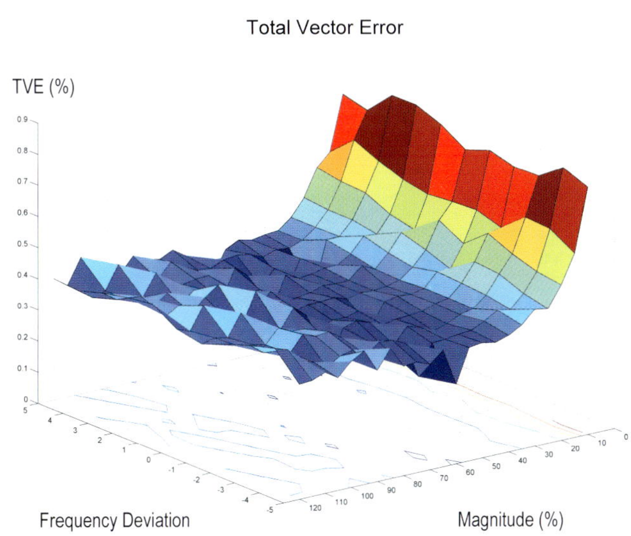

Figure 20.11: Graphical representation of synchrophasor total vector error

20.3.4 Synchrophasor Communication

As well as the generation of synchrophasors, the IEEE C37.118 standard also defines how the synchrophasor values are to be communicated.

Whilst the standard does not mandate that Ethernet communications are used for synchrophasor transmission, it is Ethernet that often enables wide-area phasor measurement technology, and that is reflected in this section. Some knowledge of Ethernet communications is assumed here, but if needed, a more detailed discussion on substation communications, including Ethernet, is presented in chapter 24: 'The Digital Substation'.

Four message types associated with synchrophasor communications are described in the standard:

- Data
- Configuration
- Header
- Command

All message types have a similar frame structure to package the information. The "Configuration", "Header" and "Command" types are used to set up the communications and are used infrequently. The "Data" type is used for the real-time transfer of synchrophasor values, and that will have the largest impact on the communications system.

According to the standard, synchrophasor signals are encapsulated into data packets, or frames, for transmission across, say, an Ethernet network. An example of a typical data frame is shown in Table 20.2.

Field function	Field size (bytes)
Message synchronisation	2
Frame size definition	2
PMU ID code*	2
Real time timing information	8
Status information	2
Phasor data values	Typically 48
System frequency measurement	2
Rate of change of system frequency	2
Digital data values	2
Frame checksum	2
Total	**Typically 72**

Table 20.2: Synchrophasor data frame format

* The PMU ID code is set to uniquely associate the synchrophasor values to the PMU responsible for producing them.

The rate at which the frames are sent is defined.

- For 50Hz operation the frame repetition rate can be set to 10, 25 or 50 frames per second
- For 60Hz operation the frame repetition rate can be set to 10, 12, 15, 20, 30, or 60 frames per second.

The faster the rate of frame repetition, the shorter the latency in responding to a developing system condition, allowing faster SIPS and state estimation reaction.

Clearly the more times per second that the data is sent, the greater the impact on the Ethernet communications.

20.3.4.1 Mapping Synchrophasors to Ethernet

When mapped to Ethernet the synchrophasor phasor data frames can use either transmission control protocol (TCP) or user datagram protocol (UDP) according to preference. TCP is a connection-oriented protocol. It manages message acknowledgement, re-transmission and time out. As such it can be considered to be reliable and ordered, but carrying overheads. UDP is a simpler protocol that broadcasts messages from the transmitter without checking the state of the receiver. As such it can be considered less reliable and not ordered, but lean. The choice will come down to the specific requirements of the application.

The TCP or UDP mapped messages are written to and read from using standard internet protocol (IP) input-output functions. These functions apply a numerical identifier for the data structures of the terminals of the communications. The terminal is referred to as a port and the numerical identifier is called the port number. The port number should be set to align with the recipient phasor data concentrator (see section 20.3.7).

Other settings that apply as the phasor message is mapped to the Ethernet layer are the IP address, subnet mask, and gateway address.

20.3.4.2 Communication Bandwidth Requirements

It can be useful to understand how much loading the addition of a phasor measurement unit can have on a communications system. A typical frame length of 72 bytes has already been mentioned. For a typical application with a frame length of 72 bytes and a repetition rate of 60 per second, the minimum bandwidth requirement for these raw frames is, therefore, 4320 bytes per second.

Ethernet conforms to the OSI (Open System Interconnect) 7 layer model (covered in chapter 24). Some layers require extra information to be added and so, in addition to the raw data frames, there is an overhead associated with the OSI 7-layer model.

For TCP/IP mapping the overhead is:

- TCP 24 bytes per frame
- IP 20 bytes per frame
- MAC 18 bytes per frame

So the TCP/IP overhead is 62 bytes of data per frame.

For UDP/IP mapping the overhead is:

- Source port 2 bytes
- Destination Port 2 bytes
- UDP length 2 bytes

- UDP checksum 2 bytes
- IP 20 bytes per frame

So the UDP/IP overhead is 28 bytes of data per frame.

As the C37.118 frame contains 72 bytes, the total number of bytes per frame is 134 for TCP and 100 for UDP.

With 8 bits per byte, the resulting bandwidth of the maximum frame rate is thus:

- 1072 * 50 = 53.6kbps for a maximum frame rate of 50 frames per second (TCP)
- 1072 * 60 = 64.32kbps for a maximum frame rate of 60 frames per second (TCP)
- 800 * 50 = 40kbps for a maximum frame rate of 50 frames per second (UDP)
- 800 * 60 = 48kbps for a maximum frame rate of 60 frames per second (UDP)

These figures are not absolute, but they can be of benefit in planning network requirements when installing and operating phasor measurement units.

20.3.5 Phasor Measurement Units

Synchrophasors are generated in phasor measurement units. A typical phasor measurement unit architecture is shown in Figure 20.12.

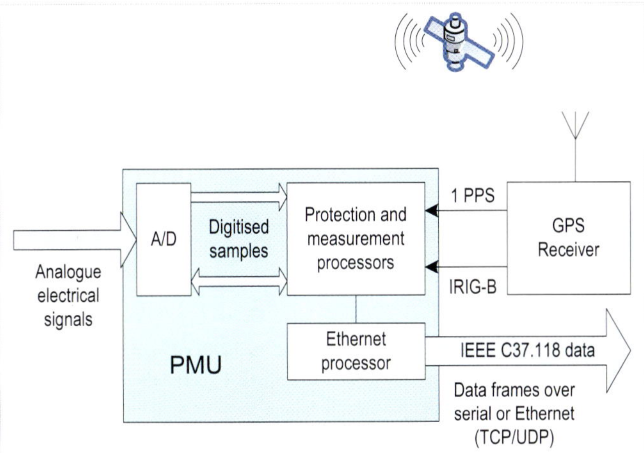

Figure 20.12: Typical Phasor measurement unit architecture

Electrical signals to be converted into synchrophasors are presented at the analogue inputs. They are sampled and converted into numerical values in an analogue to digital converter (A/D). The sampling is controlled by the measurement processor which is synchronised to the GPS system by a 1pps signal from a time synchronisation unit. The numerical values representing the digitised samples are passed to the measurement processor. The measurement processor determines the amplitude and phase quantities required for the synchrophasors according to their displacement from the 1pps signal. The measurement processor also calculates the derived quantities that are required to be presented as synchrophasors. These include sequence components, frequency, and rate-of-change of frequency, etc.

An IRIG-B signal from the time synchronisation unit allows the measurement processor to accurately time-stamp the synchrophasor values before passing them to the Ethernet processor. The Ethernet processor frames the synchrophasor values according to IEEE C37.118 before transmitting over either TCP/IP or UDP/IP.

20.3.5.1 Physical Deployment of Phasor Measurement Units

There is no need to install phasor measurement units at every bus on the system as this will not necessarily contribute to more efficient system operation. Indeed, too many phasor measurement units could cause congestion on the communications system, making data less manageable.

Phasor measurement units should be connected so that inter-area oscillations and other critical dynamics will be correctly monitored. They should be deployed on the network at critical buses to ensure that a sufficient picture of the system dynamics is available to the control centre to enable decisions to be made. System visualisation using synchrophasors from various parts of the system can show, for example, the angular, frequency and voltage differences between groups of generators. To do this, synchrophasors from various locations need to be brought together, and one method of achieving this is by means of phasor data concentrators, which is covered in section 20.3.7.

20.3.6 Time Synchronisation

Phasor measurement units require accurate time synchronisation.

IEEE-1588 describes a high-precision time protocol that can achieve sub-microsecond accuracy for time synchronisation to be used in measurement systems. It is an Ethernet protocol that can be connected to a local area network (LAN). Timing is derived from the global positioning satellite (GPS) system. Since the synchronism is delivered by the Ethernet LAN, only one GPS receiver is required to synchronise all devices on the LAN. This offers potential cost benefits when compared to using dedicated time synchronisation devices, especially as the PMUs are typically connected to the Ethernet for the transmission of synchrophasors. Given the criticality of delivering the time synchronisation signal, however, it may be more conservative to apply dedicated time synchronisation devices into the substation using connections that are separate to the LAN. A typical time synchronisation device might provide a one-pulse-per-second (1pps) signal, derived from the GPS signals, in order to provide precise alignment of the synchrophasors. To combat the effects of electrical interference in the substation environment, the 1pps signal is generally presented as a fibre-optic signal. The accuracy is typically better than 0.1μs. Although the 1pps signal is very accurate (Figure 20.13), the pulse on it only signifies that another second has passed (or is about to start). In order that the phasor measurement unit can provide an accurate time stamp for the synchrophasors, it is necessary to provide an additional 'real' time signal. This is achieved with an IRIG-B output, which delivers coordinated universal time (UTC) for very accurate time stamping. It is quite likely that the time synchronisation device will feature multiple outputs and be capable of synchronising multiple PMUs and merging units, providing substation-wide timing information.

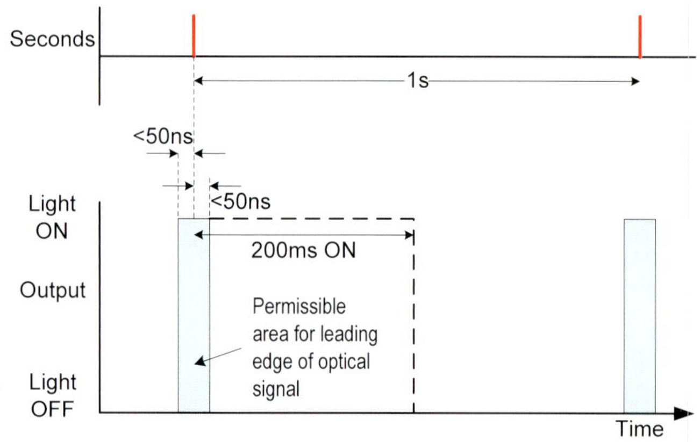

Figure 20.13: 1pps accuracy

The GPS system consists of more than 20 satellites rotating in orbit above the earth. The satellites are not in geo-stationary orbit, and so the constellation that can be 'seen' is constantly changing. In general, up to 7 satellites can be 'seen' at any time. When GPS time synchronising receivers are energised, they need to initialise their timing algorithms. Generally, this requires that at least 4 satellites are in view for a short while.

In order to receive the signals from the GPS satellites, an antenna is required. This needs to be mounted externally. A mounting pole is generally used as shown Figure 20.14.

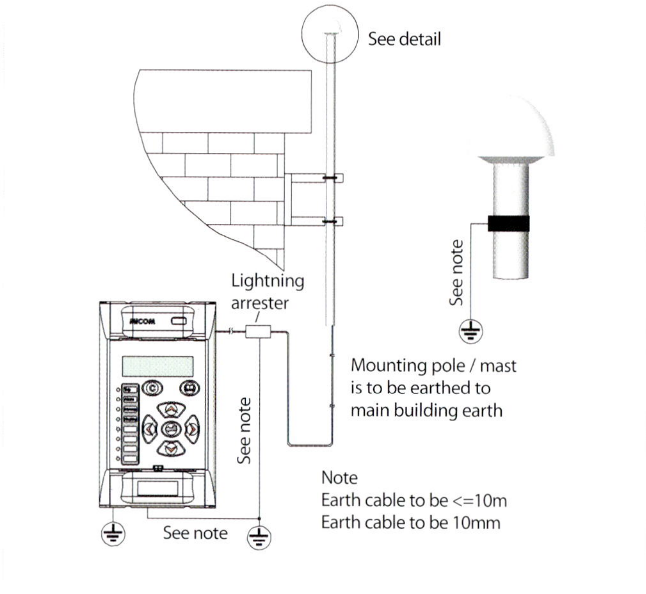

Figure 20.14: Antenna connection details

It is vital that the receiving antenna has the best possible mounting location to maintain reception with as many satellites as possible, thus the ideal mounting is shown in Figure 20.15.

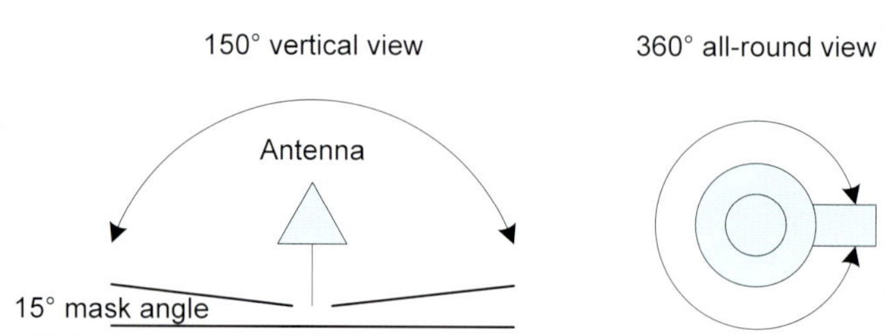

Figure 20.15: GPS antenna required line of sight

The antenna should be mounted so that it is above the roofline of the building to which it is attached, ensuring that there is a 360° horizontal view and a 150° vertical view of the sky (a masking angle of 15° with respect to the horizon is the maximum that should be permitted). Ideally there should be no obstructions in view such as metal structures or buildings. If the masking angle is greater than 15° with respect to the horizon (i.e. if is a large obstruction) the antenna must be re-sited or mounted on a longer antenna pole.

Due to the requirement for ensuring such a clear view of the GPS constellation, the antenna is likely to be the highest component in the vicinity. As such it is exposed to lightning, and suitable protection is required. A lightning arrestor is recommended for protection.

20.3.7 Phasor Data Concentrators

As shown in Figure 20.16, a phasor network consists of phasor measurement units (PMU) dispersed throughout the electricity system and connected to an energy management system via phasor data concentrators (PDC). Phasor data concentrators are devices which collect data from a number of phasor measurement units. They convert such data into a single data stream for onward transmission. Phasor measurement units are connected to a phasor data concentrator at either the substation or the control centre level.

Phasor data concentrators connect to supervisory control and data acquisition (SCADA) systems and/or energy management systems (EMS) at the central control facility to provide the synchrophasor information for the wide-area applications.

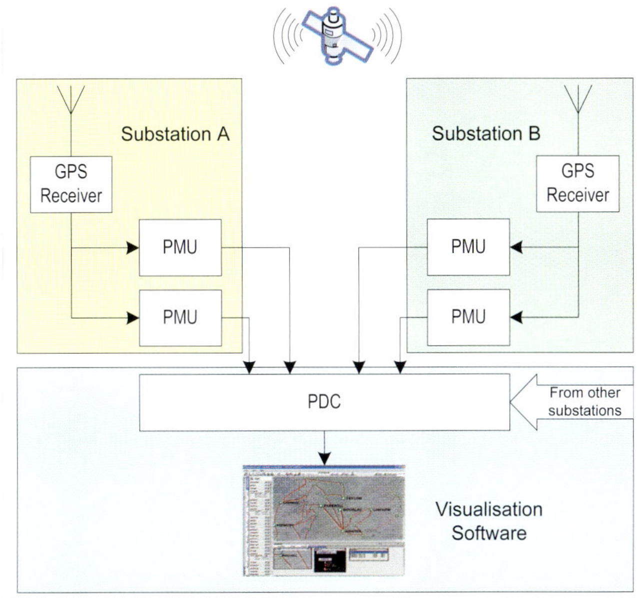

Figure 20.16: PMU and PDC architecture

The complete network requires rapid transfer of data within the frequency of sampling of the synchrophasor data.

The phasor data concentrator correlates the data, and controls and monitors the phasor measurement units. It also serves to archive the data, which is essential in the event that post-disturbance analysis might be undertaken in order to determine the root causes of cascade events. At the central control facility, the SCADA/EMS system presents system-wide data on all generators and substations in the system with frequent regular updates. These updates provide for a visualisation of the system which facilitates response-based schemes.

20.3.8 System Visualisation

System visualisation enhances wide-area schemes. Figure 20.17, Figure 20.18, and Figure 20.19, show visualisations of electrical networks as presented on an advanced energy management system.

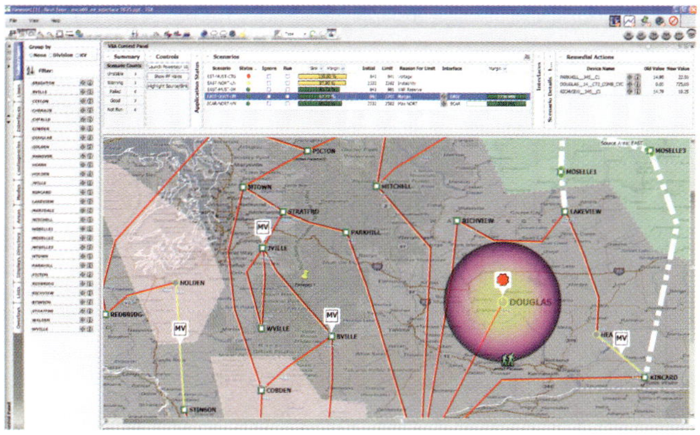

Figure 20.17: Stability state: annunciating voltage contours, weak elements and possible remedial actions

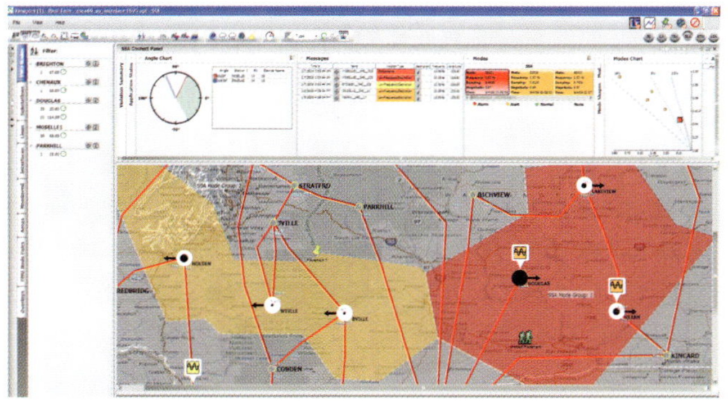

Figure 20.18: Visualisation of inter-area oscillations: frequency, severity and damping

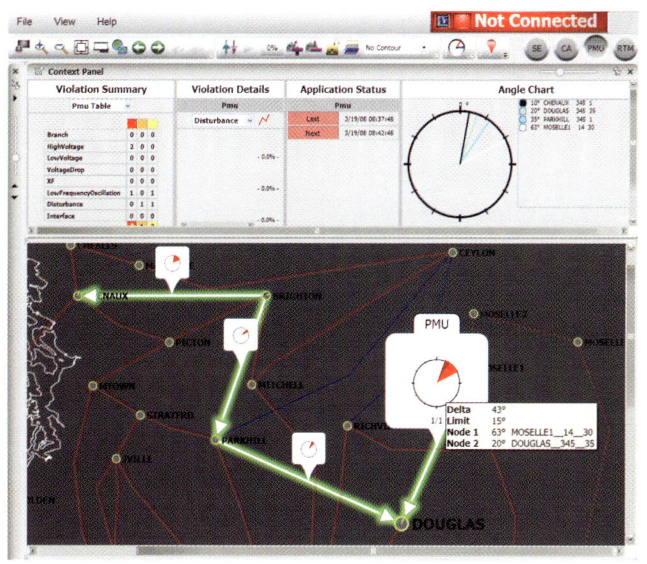

Figure 20.19: Angular separation across transmission lines: alerts to dispatchers

The figures show how phasor measurement units are providing EMS software with synchrophasor values and how virtual phase-meters connected to the lines give a clear picture of the real-time phase differences between buses. If the angular displacement exceeds a preset value, dispatchers are informed of the violation and the options available to them for remedial action.

System visualisation of real-time data realised thanks to wide-area technology provides power system operators greater awareness, and hence more precise control of the system. Underpinning this wide-area technology is the capability to accurately time synchronise the phasor measurement units.

20.4 NON-SYNCHRONISED SYSTEM INTEGRITY PROTECTION SCHEMES

The system integrity protection schemes described in the previous section require highly accurate time synchronisation and fast communications infrastructure. One might be tempted to consider that system integrity protection can only be afforded where these facilities are available. This is not, however, the case.

The numerical busbar protection outlined in chapter 15 can be considered as a distributed protection scheme as information is gathered from peripheral units at various points on the system for transmission to a central unit where protection monitoring and control is managed. This shows how distributed intelligence can work, and by increasing the geographical spread of other suitable devices, wide-area schemes are the next logical step.

Recognising however that sometimes information is available only from a single point in the system, system integrity protection schemes have been developed that require neither communications, nor accurate time synchronisation.

By taking signal inputs from single points in the power system, and by using special filters and techniques such as fuzzy logic, predictive out-of-step or loss of synchronism conditions can be detected and consequently acted upon to maintain system stability. Similar techniques can be used to detect and act on changes to system topology such as loss of power corridor.

By adding in busbar protection and relabeling the horizontal axis in Figure 20.2 , the communication requirement of system integrity protection can be highlighted, as shown in Figure 20.20.

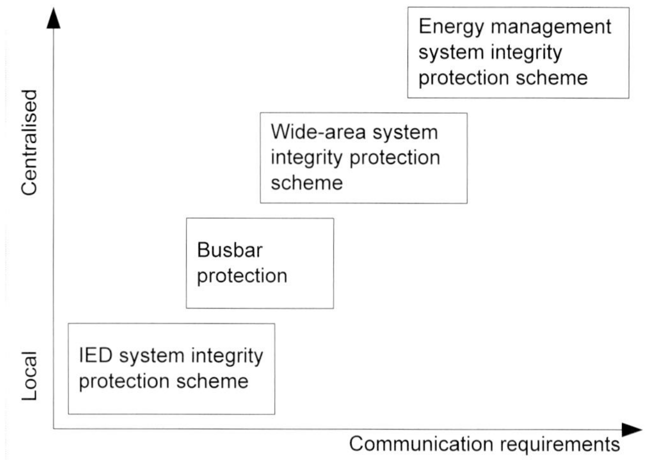

Figure 20.20: Communications requirements for system integrity protection schemes.

This shows 'local' IED system integrity protection schemes that do not require any communications, and these form the subject of the rest of this chapter.

20.4.1 Single-point special protection

In countries where large sources of power generation are geographically far from the principal loads, so-called transmission corridors of power provide the routes for the transmission of that power. One example of this is the Hydro-Québec network in Canada as shown in Figure 20.21.

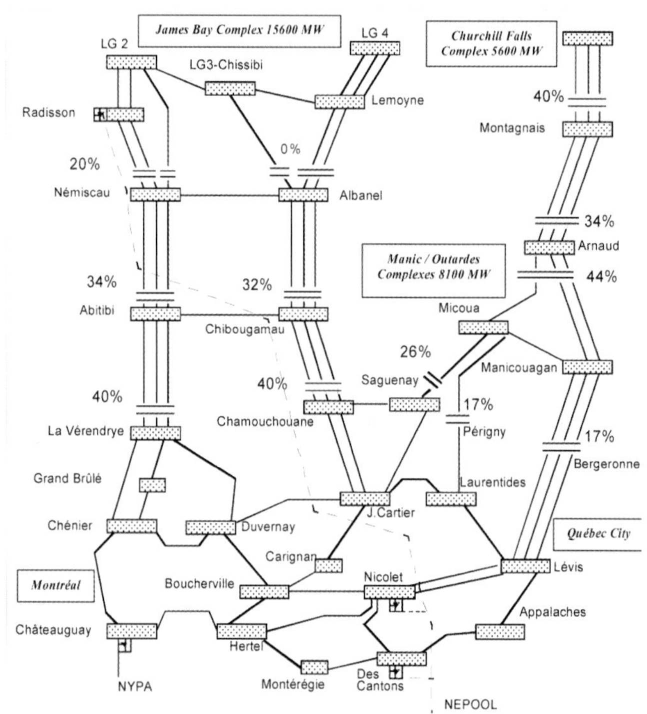

Figure 20.21: Hydro-Quebec 735kV series-compensated transmission system (reproduced courtesy of Hydro-Quebec)

Reliability of transmission of power on such systems can be achieved by the addition of series compensation and shunt reactors, but these bring added complexity. Protection against system disturbances may be provided to some extent by frequency relays or voltage relays set to initiate actions such as load-shedding on under-frequency conditions, but these may not be sufficient in themselves for all applications and sophisticated system integrity protection schemes have been developed to complement them. Two particular examples that do not require communications are described in sections entitled 20.4.2 System Topology Changes and 20.4.3 System Synchronism Stability. The former describes an example of an event-based system integrity protection scheme looking for changes in system topology. The latter describes an example of a response-based system integrity protection scheme looking for system instability.

Both of these system integrity protection schemes are able to anticipate system-threatening conditions using local variables only. They are based on Kalman filters and fuzzy-logic multi-criteria algorithms. These system integrity protection schemes are applicable to large power systems prone to transient and dynamic instabilities.

20.4.2 System Topology Changes

Detecting topology changes such as open line conditions allows timely initiation of system defence plans for severe contingencies such as a loss of transmission corridor. In the absence of communications, signalling an open line condition from a remote terminal is not possible and so, detecting a loss of corridor caused by a remote terminal action using only locally available signals can enable system stabilising actions to be initiated.

20.4.2.1 Open Line Detection

Open line detection, like any other protection, needs to be secure, reliable, and sufficiently fast.

To be secure, the detector should not operate in cases of low or zero power transfer on the monitored line, distortion on line voltages and currents during faults, fast variations in the power flow, reversal of the power flow, power swings, sub-synchronous resonances resulting from the parallel combination of shunt reactors and series capacitors, nor should it operate in the presence of harmonics.

To be reliable, the detector should detect line opening at the remote end, line closing at the local end when already open at the remote end, and line closing or opening with a shunt reactor connected.

The typical speed required for open line detection operation is two cycles or less.

As with many other aspects of electrical power systems, open line operation can be characterised by both transient and steady state conditions. Observation of the transient state aspects and steady state aspects of open line conditions reveals that transient events generally are associated with line opening, whilst steady state phenomena are more closely linked to closed lines.

An open line condition is very similar to a closed line which has no active power transfer. If the relationship of critical measured variables such as the voltage and the reactive power are observed when the active power of a connected line disappears, certain characteristics become apparent. Figure 20.22 shows that, for a given line length, an operating point (consisting of the angle shift and the voltage ratio between its ends) can be found so that, as seen from the sending end, both active and reactive powers disappear.

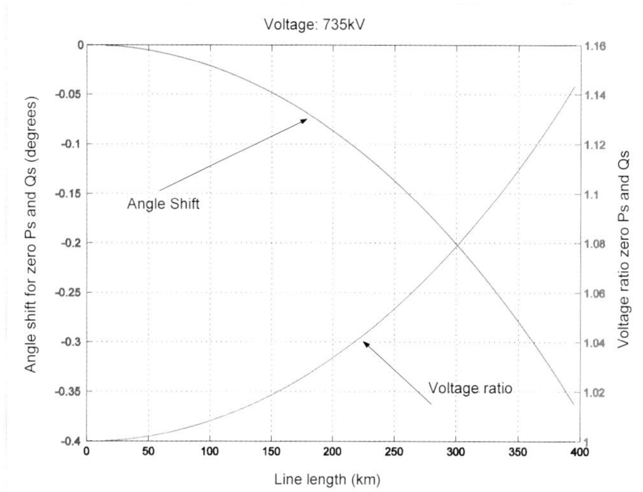

Figure 20.22: Voltage ratio and angle trajectories vary with line length for sending end active (Ps) and reactive (Qs) powers both at zero.

Although this critical operating point can be calculated, it is very sensitive: For short lines, say 30km, the typical sensitivities are 45% MW per degree of angle shift and 25% MVAr per unit of voltage ratio.

It is therefore difficult to accurately detect this condition based on power or current measurement alone, and thus it cannot form a reliable basis for detecting open lines.

Studying the transient state aspects of opening a line provides a better basis for open line detection. Observing the rate of change of active and reactive power for various system events produces interesting results. The rate of change of power (dP/dt, or DeltaP) can be studied for various system events and two in particular are illustrated in Figure 20.23 and Figure 20.24.

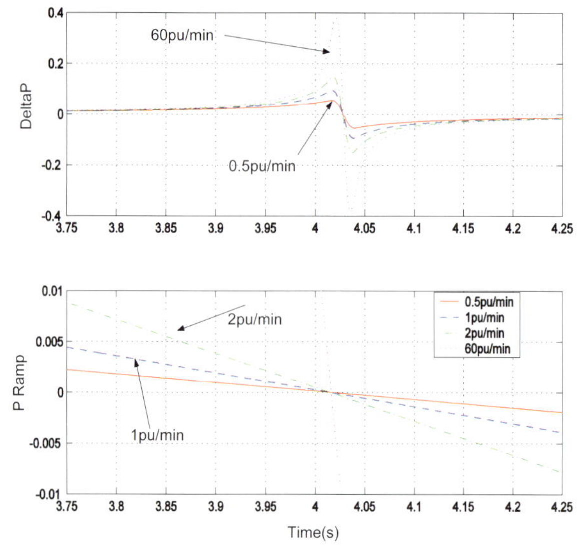

Figure 20.23: Rate of change of power (DeltaP) filter response to power flow reversal

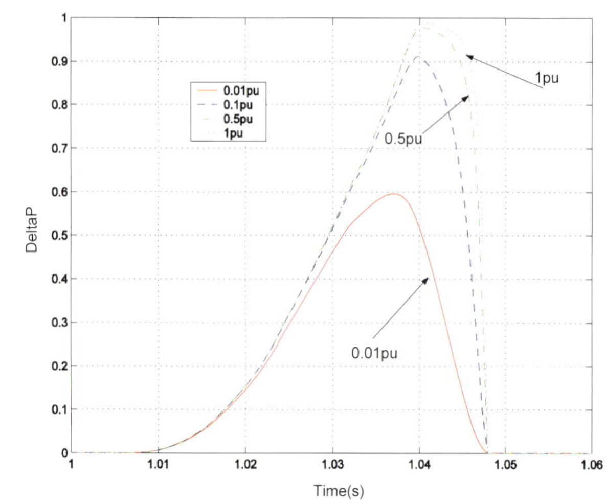

Figure 20.24: Rate of change of power (DeltaP) at breaker opening

Figure 20.23 shows power transfer ramping from a positive direction to a negative direction, thus passing through zero without apparently any significant effect. As the power ramp crosses zero, however, it can be seen that DeltaP reaches a maximum value a few cycles later. Although the maximum value of DeltaP varies according to the power slope, the maximum is reached at nearly the same time.

Figure 20.24 shows the effect on DeltaP of suddenly opening a line at various values of surge impedance loading. In all cases, when the line is opened, DeltaP increases sharply from its zero steady-state value to its maximum value.

It is apparent that observing DeltaP provides the capability for fast reliable detection of line opening.

20.4.2.2 Implementing Open Line Detection

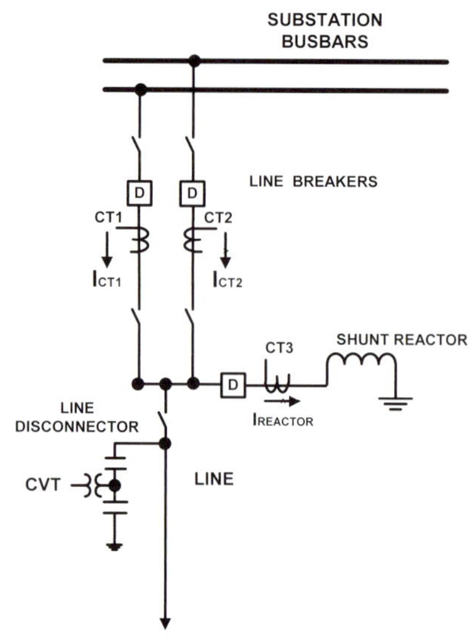

Figure 20.25: Connections for open line detection.

Figure 20.25 shows typical inputs used by an open line detection algorithm. The inputs are taken from a capacitive voltage transformer (CVT) and three sets of current transformers (CT1, CT2 associated with the circuit breakers and CT3 associated with the shunt reactor).

The functional diagram of the open line detection is shown in Figure 20.26. It consists of five principal elements.

- Anti-alias filtering
- Kalman filtering
- Pre-processing
- Fuzzy logic
- Output logic

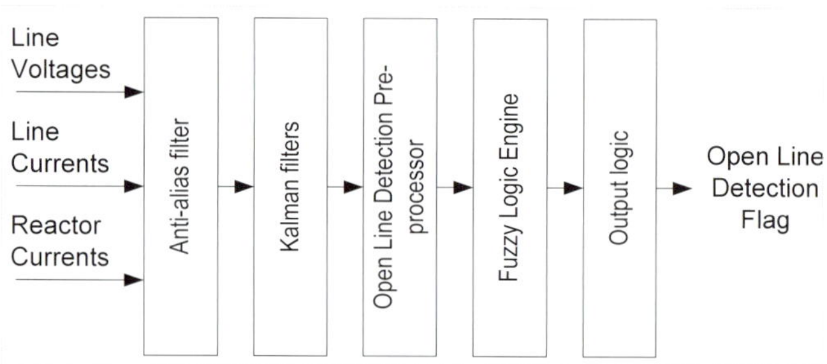

Figure 20.26: Open-line detection functional block diagram

Anti-aliasing filters limit the bandwidth of the input signals before high speed data acquisition converts the input signals into numeric values for provision to the principal filtering unit.

The signals are filtered to extract the fundamental phasor components.

These vectors are used as the inputs into a Kalman Filter. A Kalman Filter is used since it can extract the fundamental components and harmonic components. But unlike filters such as the Discrete Fourier Transform (DFT), a Kalman Filter can provide signal components at spectral frequencies other than the fundamental and its harmonics. Kalman Filters are well suited to the application of protecting compensated lines, since they can be used to detect, for example, the negative sequence component at fundamental as well as resonances that may transiently appear when a line is opened at both ends while its shunt reactors are connected. The outline of a Kalman filter is shown in Figure 20.27, and the state diagram is shown in Figure 20.28.

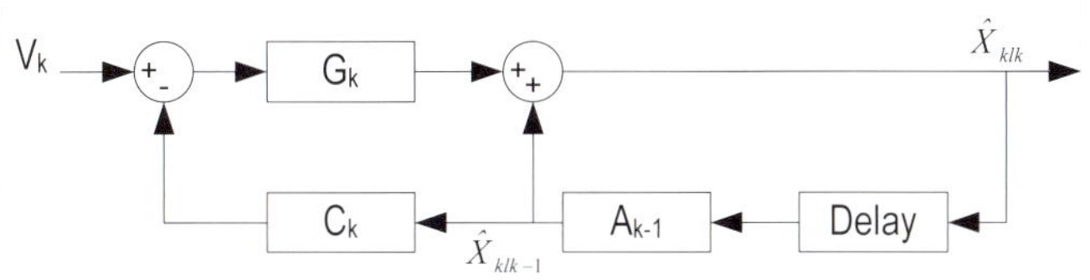

Figure 20.27: Kalman Filter block diagram

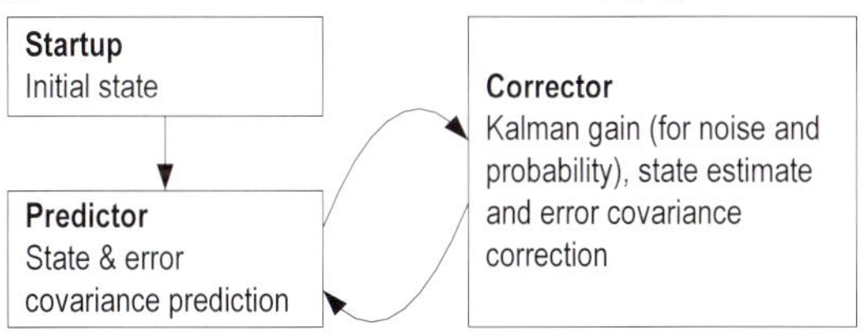

Figure 20.28: Kalman Filter state diagram.

The pre-processing unit takes the fundamental frequency phasor components of the inputs, the derived sequence and spectral components, and uses them to determine a number of decisions about the system based on quantities that reflect the 'openness' or 'closedness' of the system such as DeltaP, active power, reactive power, etc.. These decisions are used by the Fuzzy logic to produce a conclusion as to the state of the line (open or closed).

20.4.2.3 Fuzzy Logic

Conventional, or Boolean, logic is commonly used in power system protection. It is based on signals being in one state or the other (i.e. ON or OFF). Fuzzy logic is based on signals being in more than one state and where there may be overlap between states. Consider Figure 20.29, which shows a number of states for the active power.

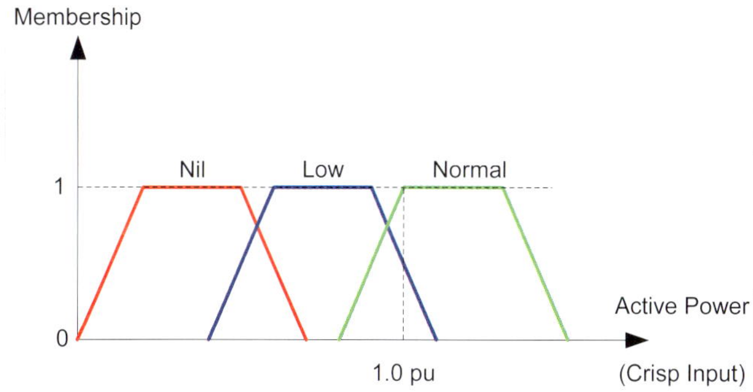

Figure 20.29: Fuzzy Logic Sets

The value 1.0pu is resident in both the "Low" state as well as the "Normal" state. It is not resident in the "Nil" state.

Its membership of the "Nil" state is 0, its membership of the "Normal" state is 1, and its membership of the "Low" state is somewhere between 0 and 1. This process is known as "fuzzification". Although called "fuzzy", it is, in fact, a repeatable process to which rules can be applied.

A fuzzy logic approach can resolve conflicts that may arise in doubtful cases such as during line energisation at no-load, or where capacitive voltage transformers induce electro-magnetic transients suggesting that the active power is not zero (line closed) while the line is actually open. Fuzzy logic is ideally suited to resolving such conflicts since the decision is based on several criteria with adaptable weighting factors.

The fuzzy decision system consists of three main steps:

- Fuzzification of the selected decisions features from the pre-processor
- Fuzzy logic inference on these features, using rules or criteria
- Crisp decision sent to the output

Fuzzification is a necessary step prior to a fuzzy-logic based reasoning system. It consists in transforming the crisp variables provided by the pre-processor into categories easily described by common language terms such as "Normal", "Small", "Large", "Very Large", etc.

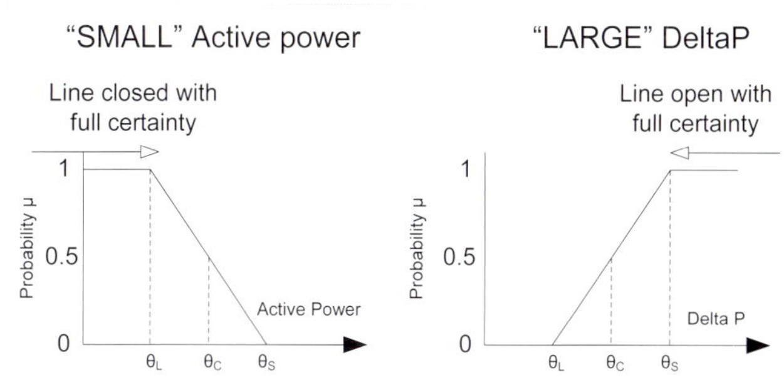

Figure 20.30: Fuzzification of typical decision features

Figure 20.30 illustrates this fuzzification process for two typical variables: the Active Power and the DeltaP. The first feature is defined as "SMALL" when its crisp value is below a given threshold (θ_L) while the DeltaP feature is defined as "LARGE" for crisp values higher than (θ_S).

A set of rules is applied to these fuzzy variables according to basic principles. "Reliability Rules" are used specifically to assess the open-line condition and "Security Rules" target closed-line situations.

An example for each type of rule is given below:

- Security: if (ActivePower is not SMALL) LINE is CLOSED
- Reliability: if (DeltaP is LARGE) LINE is OPEN

The fuzzy logic engine contains a number of these rules for detecting the open-line status under various circumstances (reliability rules), and a number of rules for detecting the closed-line status (security rules). In each category, the rules can be further classified in terms of transient and steady-state rules. For instance, the reliability rule above is a transient one while the security rule is a steady state.

The last stage of the process is called defuzzification. This consists of converting the evaluation results of the criteria, which are applied by the rules, into a single decision. Each criterion states that the line is open or closed with a given probability and defuzzification reconciles all partial decisions using an aggregation procedure into a single decision.

The output from the open line detector is the flag that is asserted when an open line is detected.

20.4.2.4 Open Line Detection Summary

In the context of interconnected and heavily loaded power systems, fast detection of topology change is considered an efficient means to instigate remedial actions for the defence of a power system against severe events. Open Line Detection, based on intelligent devices operating using locally-measured power system quantities can provide such detection for a typical topology change that could impact system stability.

20.4.3 System Synchronism Stability

In general, occurrence of extreme contingencies in a power system can lead to voltage, frequency or angular instabilities. Angular or frequency instabilities are related more to the incapacity of the system to provide appropriate real power flows between generation areas, or between generation and load areas to maintain system equilibrium. Such instabilities may lead to either under/overfrequency situations or loss-of-synchronism conditions.

In the first case, the frequency will be the same for the whole system but out of the normal range, whilst in the second case, area or sub-system frequencies will differ and induce loss-of-synchronism conditions characterised by periodic zero voltage, or virtual faults. Anticipation of such loss of synchronism and subsequent fast remedial action, may allow system stability to be maintained.

Prediction of an impending loss of synchronism, or 'out-of-step' condition, using signals from just a single point in the system can be achieved using conventional or novel algorithms.

Unlike conventional out-of-step relaying introduced in the earlier chapter on distance protection, the novel algorithms provide a system integrity protection scheme designed to **predict** out-of-step conditions on a transmission system. This prediction can be used to initiate actions to preserve power system stability such as engaging surge arrestors in order to protect primary equipment against over-voltages before they exceed the insulation rating, or islanding the system in advance into areas of equilibrium resulting in the minimum loss of load or generation.

20.4.3.1 Conventional Out-of-Step Protection

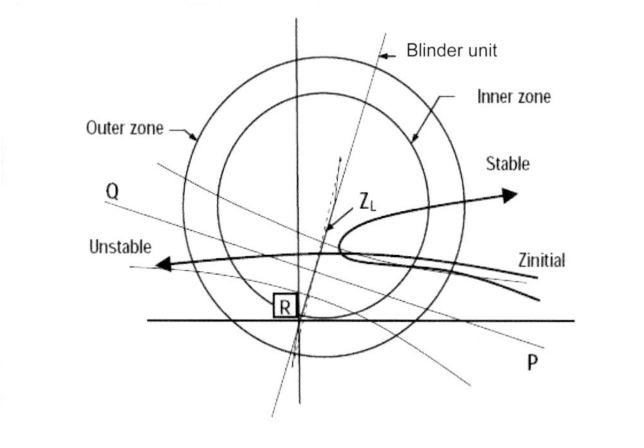

Figure 20.31: Impedance loci during power swings

Detection of power swings has been briefly introduced in chapters 11 and 17.

Classical approaches for transmission line applications make use of auxiliary elements in the R, X plane in combination with timers. Figure 20.31 illustrates this principle with two mho relays (outer and inner zones) and a blinder unit. $Z_{initial}$ is the pre-disturbance apparent impedance seen by a relay located at one end of a transmission line. Whilst a fault would cause an instantaneous movement of apparent impedance Z_{app} from the load area to the line impedance Z_L, a power swing will cause Z_{app} to move at a progressive rate and will be detected if the sequential pickup of the outer and inner mho units takes more than a predetermined time. It may be considered unstable if the blinder line is crossed, and stable if Z_{app} returns to the load area. Typically, this scheme will trip upon unstable swings and block for stable ones.

When tripping is required, excessive breaker stresses will be reduced if the two systems to be separated do not move close to anti-phase. It is therefore highly desirable either to anticipate the unstable swing or to delay the trip signal until after the pole slip. Anticipating the instability can rarely be done with such a scheme however, because the Z_{app} trajectory is hard to predict for all conditions prevailing in a real system and tripping on a stable swing is not desirable. On the other hand, with a delayed trip signal, the magnitude and duration of voltage excursions will be more severe and will decrease the chances of island survival.

These difficulties can be overcome with more sophisticated schemes for predicting the dynamic behaviour of power swings.

20.4.3.2 Predictive Out-of-Step

Rather than waiting for the out-of-step condition to occur, the novel predictive out-of-step protection provides early warning of impending system instability.

Predictive out-of-step protection can be achieved with a special algorithm called an RPS algorithm. The implementation of this algorithm is similar to that of the detection of open lines as already described. Kalman filters and fuzzy logic engines are used to apply multiple criteria that are indicative of impending loss of synchronism.

A specially designed Kalman filter extracts the voltage and current positive-sequence phasors used as the primary variables in these applications.

A transmission line model is used to compute values for the remote voltage and current phasors V_r and I_r, given the local phasors V_s and I_s. These sets are then used to derive decisional variables such as line angle and frequency shifts, local and remote $V\cos\varphi$ and derivative quantities.

The RPS variables are used to track and anticipate the stable or unstable evolution of power swings across the transmission line. Three angular variables, one concavity index, two voltage and six $V\cos\varphi$ variables are derived and processed to prepare them for the fuzzy-logic algorithm.

Any one of these variables individually can provide an indication of system stability, but given the wide range of operating conditions found in large power systems, the performance of single-variable-based algorithms is limited because of the margin for error in any one of them. Fuzzy logic enables multi-criteria algorithms to take advantage of the strength of all the variables by considering them all together to assess the stability condition. The construction of the fuzzy logic to implement the RPS algorithm is similar to that already described, with fuzzification giving linguistic categories to the signals and the fuzzy logic applying weighted rules to determine the RPS output, indicating whether the system is predicted to go out-of-step or not.

In practical applications, the out-of-step prediction is raised almost half a second before the instability, giving time to initiate remedial actions to prevent the event and protect the system.

20.4.4 Acknowledgements

Alstom Grid gratefully acknowledges the research, development and expert assistance provided by Hydro Québec during the joint development of the DLO (MiCOM P846 Open Line Detector and RPS (MiCOM P848 Predictive Out-Of-Step) products.

20.5 REFERENCES

[20.1] L. H. Fink and K. Carlsen, "Operating Under Stress and Strain," *IEEE Spectrum*, vol. 15, pp. 48–53, March 1978.

[20.2] CIGRE Task Force 38.02.19: "System Protection Schemes in Power Networks", Final Draft v 5.0, 2000.

[20.3] R Grondin, S Richards, et al, "Loss of Synchronism Detection. A Strategic Function for Power System Protection", CIGRE B5-205, 2006.

Chapter 21
Relay Testing and Commissioning

21.1 INTRODUCTION

The testing of protection equipment schemes presents a number of problems. This is because the main function of protection equipment is solely concerned with operation under system fault conditions, and cannot readily be tested under normal system operating conditions. This situation is aggravated by the increasing complexity of protection schemes, the use of relays with extensive functionality implemented in software, and the trend towards peer-to-peer logic implemented via Ethernet (as opposed to more traditional hardwired approaches).

The testing of protection equipment may be divided into four stages:

- type tests
- routine factory production tests
- commissioning tests
- periodic maintenance tests

21.1.1 Type Tests

Type tests are required to prove that a relay meets the published specification and complies with all relevant standards. Since the principal function of a protection relay is to operate correctly under abnormal power conditions, it is essential that the performance be assessed under such conditions. Comprehensive type tests simulating the operational conditions are therefore conducted at the manufacturer's works during the development and certification of the equipment.

The standards that cover most aspects of relay performance are IEC 60255 and IEEE C37.90. However compliance may also involve consideration of the requirements of IEC 61000, 60068 and 60529, while products intended for use in the EU also have to comply with the requirements of Directives (see Appendix C). Since type testing of a digital or numerical relay involves testing of software as well as hardware, the type testing process is very complicated and more involved than a static or electromechanical relay.

21.1.2 Routine Factory Production Tests

These are conducted to prove that relays are free from defects during manufacture. Testing will take place at several stages during manufacture, to ensure problems are discovered at the earliest possible time and hence minimise remedial work. The extent of testing will be determined by the complexity of the relay and past manufacturing experience.

21.1.3 Commissioning Tests

These tests are designed to prove that a particular protection scheme has been installed correctly prior to setting to work. All aspects of the scheme are thoroughly checked, from installation of the correct equipment through wiring checks and operation checks of the individual items of equipment, finishing with testing of the complete scheme.

21.1.4 Periodic Maintenance Checks

These are required to identify equipment failures and degradation in service, so that corrective action can be taken. Because a protection scheme only operates under fault conditions, defects may not be revealed for a significant period of time, until a fault occurs. Regular testing assists in detecting faults that would otherwise remain undetected until a fault occurs.

21.2 ELECTRICAL TYPE TESTS

Various electrical type tests must be performed, as follows:

21.2.1 Functional Tests

The functional tests consist of applying the appropriate inputs to the relay under test and measuring the performance to determine if it meets the specification. They are usually carried out under controlled environmental conditions. The testing may be extensive, even where only a simple relay function is being tested., as can be realised by considering the simple overcurrent relay element of Table 21.1.

To determine compliance with the specification, the tests listed in Table 21.2 are required to be carried out. This is a time consuming task, involving many engineers and technicians. Hence it is expensive.

Element	Range	Step Size
I>1	0.08 - 4.00In	0.01In
I>2	0.08 - 32In	0.01In
Directionality	Forward/Reverse/Non-directional	
RCA	-95° to +95°	1°
Characteristic	DT/IDMT	
Definite Time Delay	0 - 100s	0.01s
IEC IDMT Time Delay	IEC Standard Inverse IEC Very Inverse IEC Extremely Inverse UK Long Time Inverse	
Time Multiplier Setting (TMS)	0.025 - 1.2	0.005
IEEE IDMT Time Delay	IEEE Moderately Inverse IEEE Very Inverse IEEE Extremely Inverse US-C08 Inverse US-C02 Short Time Inverse	
Time Dial (TD)	0.5 - 15	0.1
IEC Reset Time (DT only)	0 - 100s	0.01s
IEEE Reset Time	IDMT/DT	
IEEE DT Reset Time	0 - 100s	0.01s
IEEE IDMT Reset Time	IEEE Moderately Inverse IEEE Very Inverse IEEE Extremely Inverse US-C08 Inverse US-C02 Short Time Inverse	

Table 21.1: Overcurrent relay element specification

Test no.	Description
Test 1	Three phase non-directional pick up and drop off accuracy over complete current setting range for both stages
Test 2	Three phase directional pick up and drop off accuracy over complete RCA setting range in the forward direction, current angle sweep
Test 3	Three phase directional pick up and drop off accuracy over complete RCA setting range in the reverse direction, current angle sweep
Test 4	Three phase directional pick up and drop off accuracy over complete RCA setting range in the forward direction, voltage angle sweep
Test 5	Three phase directional pick up and drop off accuracy over complete RCA setting range in the reverse direction, voltage angle sweep
Test 6	Three phase polarising voltage threshold test
Test 7	Accuracy of DT timer over complete setting range
Test 8	Accuracy of IDMT curves over claimed accuracy range
Test 9	Accuracy of IDMT TMS/TD
Test 10	Effect of changing fault current on IDMT operating times
Test 11	Minimum Pick-Up of Starts and Trips for IDMT curves
Test 12	Accuracy of reset timers
Test 13	Effect of any blocking signals, opto inputs, VTS, Autoreclose
Test 14	Voltage polarisation memory

Table 21.2: Overcurrent relay element functional type tests

When a modern numerical relay with many functions is considered, each of which has to be type-tested, the functional type-testing involved is a major issue. In the case of a recent relay development project, it was calculated that if one person had to do all the work, it would take 4 years to write the functional type-test specifications, 30 years to perform the tests and several years to write the test reports that result. Automated techniques/ equipment are clearly required, and are covered in Section 21.7.2.

21.2.2 Rating Tests

Rating type tests are conducted to ensure that components are used within their specified ratings and that there are no fire or electric shock hazards under a normal load or fault condition of the power system. This is in addition to checking that the product complies with its technical specification. The following are amongst the rating type tests conducted on protection relays, the specified parameters are normally to IEC 60255-1.

21.2.3 Thermal Withstand

The thermal withstand of VTs, CTs and output contact circuits is determined to ensure compliance with the specified continuous and short-term overload conditions. In addition to functional verification, the pass criterion is that there is no detrimental effect on the relay assembly, or circuit components, when the product is subjected to overload conditions that may be expected in service. Thermal withstand is assessed over a time period of 1s for CTs and 10s for VTs.

21.2.4 Relay Burden

The burdens of the auxiliary supply, optically isolated inputs, VTs and CTs are measured to check that the product complies with its specification. The burden of products with a high number of input/output circuits is application specific i.e. it increases according to the number of optically isolated input and output contact ports which are energised under normal power system load conditions. It is usually envisaged that not more than 50% of such ports will be energised concurrently in any application.

21.2.5 Relay Inputs

Relay inputs are tested over the specified ranges. Inputs include those for auxiliary voltage, VT, CT, frequency, optically isolated digital inputs and communication circuits.

21.2.6 Relay Output Contacts

Protection relay output contacts are type tested to ensure that they comply with the product specification. Particular withstand and endurance type tests have to be carried out using d.c., since the normal supply is via a station battery.

21.2.7 Insulation Resistance

The insulation resistance test is carried out according to IEC 60255-27, i.e. 500V d.c. $\pm$10%, for a minimum of 5 seconds. This is carried out between all circuits and case earth, between all independent circuits and across normally open contacts. The acceptance criterion for a product in new condition is a minimum of 100MΩ. After a damp heat test the pass criterion is a minimum of 10MΩ.

21.2.8 Auxiliary Supplies

Digital and numerical protection relays normally require an auxiliary supply to provide power to the on-board microprocessor circuitry and the interfacing opto-isolated input circuits and output relays. The auxiliary supply can be either a.c. or d.c., supplied from a number of sources or safe supplies - i.e. batteries, UPSs, etc., all of which may be subject to voltage dips, short interruptions and voltage variations. Relays are designed to ensure that operation is maintained and no damage occurs during a disturbance of the auxiliary supply.

Tests are carried out for both a.c. and d.c. auxiliary supplies and include mains variation both above and below the nominal rating, supply interruptions derived by open circuit and short circuit, supply dips as a percentage of the nominal supply, repetitive starts. The duration of the interruptions and supply dips range from 2ms to 60s intervals. A short supply interruption or dip up to 20ms, possibly longer, should not cause any malfunction of the relay. Malfunctions include the operation of output relays and watchdog contacts, the reset of microprocessors, alarm or trip indication, acceptance of corrupted data over the communication link and the corruption of stored data or settings. For a longer supply interruption, or dip in excess of 50ms, the relay self recovers without the loss of any function, data, settings or corruption of data. No operator intervention is required to restore operation after an interruption or dip in the supply.

In addition to the above, the relay is subjected a number of repetitive starts or a sequence of supply interruptions. Again the relay is tested to ensure that no damage or data corruption has occurred during the repetitive tests.

Specific tests carried out on d.c. auxiliary supplies include reverse polarity, a.c. waveform superimposed on the d.c. supply and the effect of a rising and decaying auxiliary voltage. All tests are carried out at various levels of loading of the relay auxiliary supply.

21.3 ELECTROMAGNETIC COMPATIBILITY TESTS

There are numerous tests that are carried out to determine the ability of relays to withstand the electrical environment in which they are installed. The substation environment is a very severe environment in terms of the electrical and electromagnetic interference that can arise. There are many sources of interference within a substation, some originating internally, others being conducted along the overhead lines or cables into the substation from external disturbances. The most common sources are:

- switching operations
- system faults
- lightning strikes

- conductor flashover
- telecommunication operations e.g. mobile phones

A whole suite of tests are performed to simulate these types of interference, and they fall under the broad umbrella of what is known as EMC, or Electromagnetic Compatibility tests.

Broadly speaking, EMC can be defined as:

'The ability of different equipment to co-exist in the same electromagnetic environment'

It is not a new subject and has been tested for by the military ever since the advent of electronic equipment. EMC can cause real and serious problems, and does need to be taken into account when designing electronic equipment.

EMC tests determine the impact on the relay under test of high-frequency electrical disturbances of various kinds. Relays manufactured or intended for use in the EU have to comply with Directive 2004/108/EC in this respect. To achieve this, in addition to designing for statutory compliance to this Directive, the following range of tests are carried out:

- d.c. interrupt test
- a.c. ripple on d.c. supply test
- d.c. ramp test
- high frequency disturbance test
- fast transient test
- surge immunity test
- power frequency interference test
- electrostatic discharge test
- conducted and radiated emissions tests
- conducted and radiated immunity tests
- magnetic field tests

21.3.1 D.C Interrupt Test

This is a test to determine the maximum length of time that the relay can withstand an interruption in the auxiliary supply without de-energising, e.g. switching off, and that when this time is exceeded and it does transiently switch off, that no maloperation occurs.

It simulates the effect of a loose fuse in the battery circuit, or a short circuit in the common d.c. supply, interrupted by a fuse. Another source of d.c. interruption is if there is a power system fault and the battery is supplying both the relay and the circuit breaker trip coils. When the battery energises the coils to initiate the circuit breaker trip, the voltage may fall below the required level for operation of the relay and hence a d.c. interrupt occurs. The test is specified in IEC 60255-11 and comprises interruptions of 10, 20, 30, 50, 100, 200, 300, 500, 1000, and 5000ms. For interruptions lasting up to and including 20ms, the relay must not de-energise of maloperate, while for longer interruptions it must not maloperate. Many modern products are capable of remaining energised for interruptions up to 50ms

The relay is powered from a battery supply, and both short circuit and open circuit interruptions are carried out. Each interruption is applied 10 times, and for auxiliary power supplies with a large operating range, the tests are performed at minimum, maximum, and other voltages across this range, to ensure compliance over the complete range.

21.3.2 A.C. Ripple on D.C. Supply

This test (IEC 60255-11) determines that the relay is able to operate correctly with a superimposed a.c. voltage on the d.c. supply. This is caused by the station battery being charged by the battery charger, and the relevant waveform is shown in Figure 21.1. It consists of a 15% peak-to-peak ripple superimposed on the d.c. supply voltage.

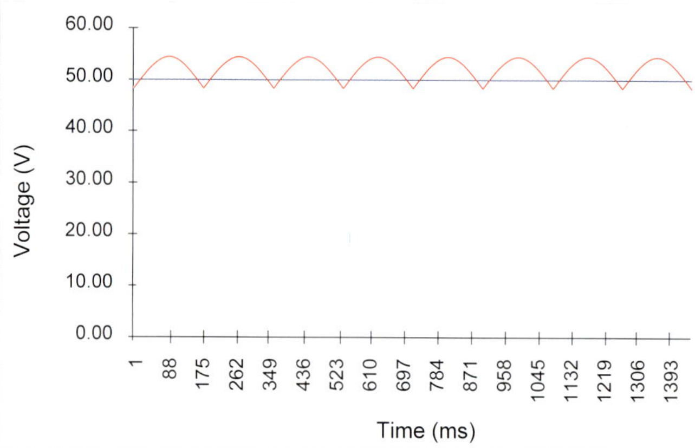

Figure 21.1: A.C. ripple superimposed on d.c. supply test

For auxiliary power supplies with a large operating range, the tests are performed at minimum, maximum, and other voltages across this range, to ensure compliance for the complete range. The interference is applied using a full wave rectifier network, connected in parallel with the battery supply. The relay must continue to operate without malfunction during the test.

21.3.3 D.C. Ramp Down/Ramp Up

This test simulates a failed station battery charger, which would result in the auxiliary voltage to the relay slowly ramping down. The ramp up part simulates the battery being recharged after discharging. The relay must power up cleanly when the voltage is applied and not maloperate.

There is no international standard for this test, so individual manufacturers can decide if they wish to conduct such a test and what the test specification shall be.

21.3.4 High Frequency Disturbance Test

The High Frequency Disturbance Test simulates high voltage transients that result from power system faults and plant switching operations. It consists of a 1MHz decaying sinusoidal waveform, as shown in Figure 21.2. The interference is applied across each independent circuit (differential mode) and between each independent circuit and earth (common mode) via an external coupling and switching network. The product is energised in both normal (quiescent) and tripped modes for this test, and must not maloperate when the interference is applied for a 2 second duration.

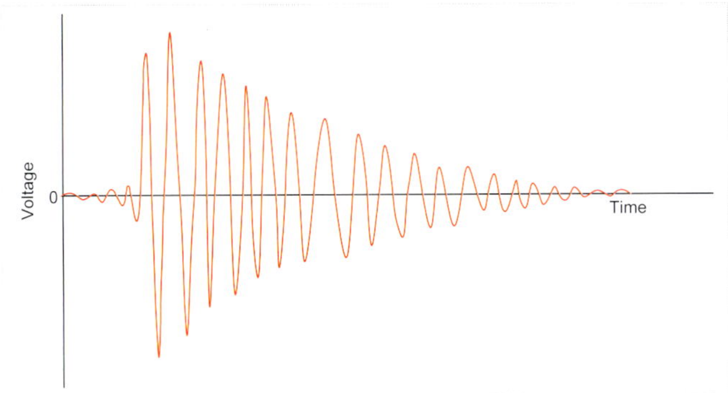

Figure 21.2: High Frequency Disturbance Test waveform

21.3.5 Fast Transient Test

The Fast Transient Test simulates the HV interference caused by disconnector operations in GIS substations or breakdown of the SF_6 insulation between conductors and the earthed enclosure. This interference can either be inductively coupled onto relay circuits or can be directly introduced via the CT or VT inputs. It consists of a series of 15ms duration bursts at 300ms intervals, each burst consisting of a train of 50ns wide pulses with very fast (5ns typical) rise times (Figure 21.3), with a peak voltage magnitude of 4kV.

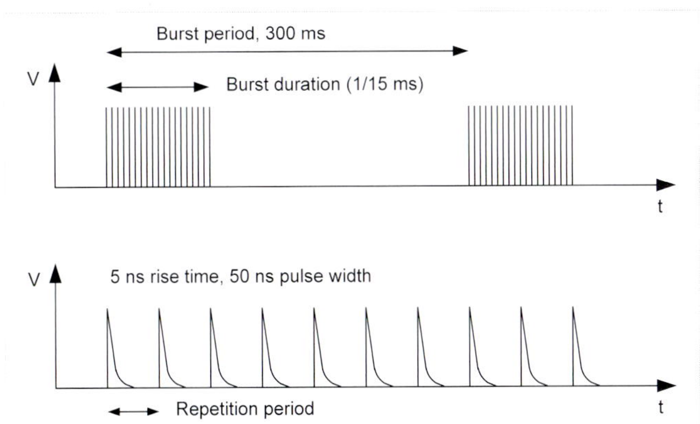

Figure 21.3 Fast Transient Test waveform

The product is energised in both normal (quiescent) and tripped modes for this test. It must not maloperate when the interference is applied in common mode via the integral coupling network to each circuit in turn, for 60 seconds. Interference is coupled onto communications circuits, if required, using an external capacitive coupling clamp.

21.3.6 Surge Immunity Test

The Surge Immunity Test simulates interference caused by major power system disturbances such as capacitor bank switching and lightning strikes on overhead lines within 5km of the substation. The test waveform has an open circuit voltage of 4kV for common mode surges and 2kV for differential mode surges. The waveshape consists on open circuit of a 1.2/50μs rise/fall time and a short circuit current of 8/20μs rise/fall time. The generator is capable of providing a short circuit test current of up to 2kA, making this test potentially destructive. The surges are applied sequentially under software control via dedicated coupling networks in both differential and common modes with the product energised in its normal (quiescent) state. The product shall not maloperate during the test, shall still operate within specification after the test sequence and shall not incur any permanent damage.

21.3.7 Power Frequency Interference

This test simulates the type of interference that is caused when there is a power system fault and very high levels of fault current flow in the primary conductors or the earth grid. This causes 50 or 60Hz interference to be induced onto control and communications circuits.

There is no international standard for this test, but one used by some Utilities is:

- 500V r.m.s., common mode
- 250V r.m.s., differential mode

applied to circuits for which power system inputs are not connected.

Tests are carried out on each circuit, with the relay in the following modes of operation:

- current and voltage applied at 90% of setting, (relay not tripped)
- current and voltage applied at 110% of setting, (relay tripped)
- main protection and communications functions are tested to determine the effect of the interference

The relay shall not maloperate during the test, and shall still perform its main functions within the claimed tolerance.

21.3.8 Electrostatic Discharge Test

This test simulates the type of high voltage interference that occurs when an operator touches the relay's front panel after being charged to a high potential. This is exactly the same phenomenon as getting an electric shock when stepping out of a car or after walking on a synthetic fibre carpet.

In this case the discharge is only ever applied to the front panel of the relay, with the cover both on and off. Two types of discharges are applied, air discharge and contact discharge. Air discharges are used on surfaces that are normally insulators, and contact discharges are used on surfaces that are normally conducting. IEC 60255-22-2 is the relevant standard this test, for which the test parameters are:

- cover on: Class 4, 8kV contact discharge, 15kV air discharge
- cover off: Class 3, 6kV contact discharge, 8kV air discharge

In both cases above, all the lower test levels are also tested.

The discharge current waveform is shown in Figure 21.4.

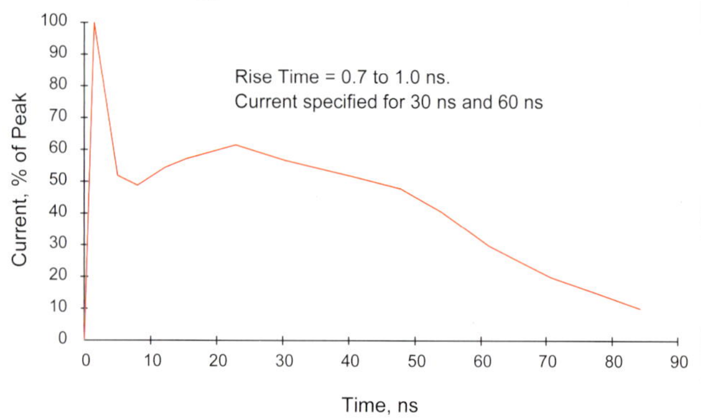

Figure 21.4: ESD Current Waveform

The test is performed with single discharges repeated on each test point 10 times with positive polarity and 10 times with negative polarity at each test level. The time interval between successive discharges is greater than 1 second. Tests are carried out at each level, with the relay in the following modes of operation:

- current and voltage applied at 90% of setting, (relay not tripped)
- current and voltage applied at 110% of setting, (relay tripped)
- main protection and communications functions are tested to determine the effect of the discharge

To pass, the relay shall not maloperate, and shall still perform its main functions within the claimed tolerance.

21.3.9 Conducted and Radiated Emissions Tests

These tests arise primarily from the essential protection requirements of the EU directive on EMC. These require manufacturers to ensure that any equipment to be sold in the countries comprising the European Union must not interfere with other equipment. To achieve this it is necessary to measure the emissions from the equipment and ensure that they are below the specified limits.

Conducted emissions are measured only from the equipment's power supply ports and are to ensure that when connected to a mains network, the equipment does not inject interference back into the network which could adversely affect the other equipment connected to the network.

Radiated emissions measurements are to ensure that the interference radiated from the equipment is not at a level that could cause interference to other equipment. This test is normally carried out on an Open Area Test Site (OATS) where there are no reflecting structures or sources of radiation, and therefore the measurements obtained are a true indication of the emission spectrum of the relay. An example of a plot obtained during conducted emissions tests is shown in Figure 21.5.

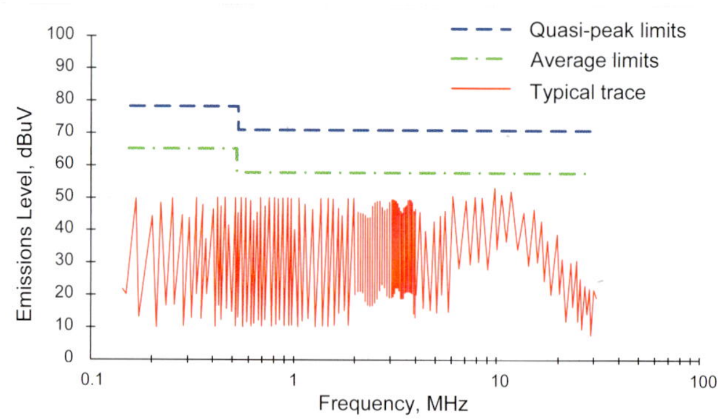

Figure 21.5: EMC Conducted Emissions test - example test plot

The test arrangements for the conducted and radiated emissions tests are shown in Figure 21.6.

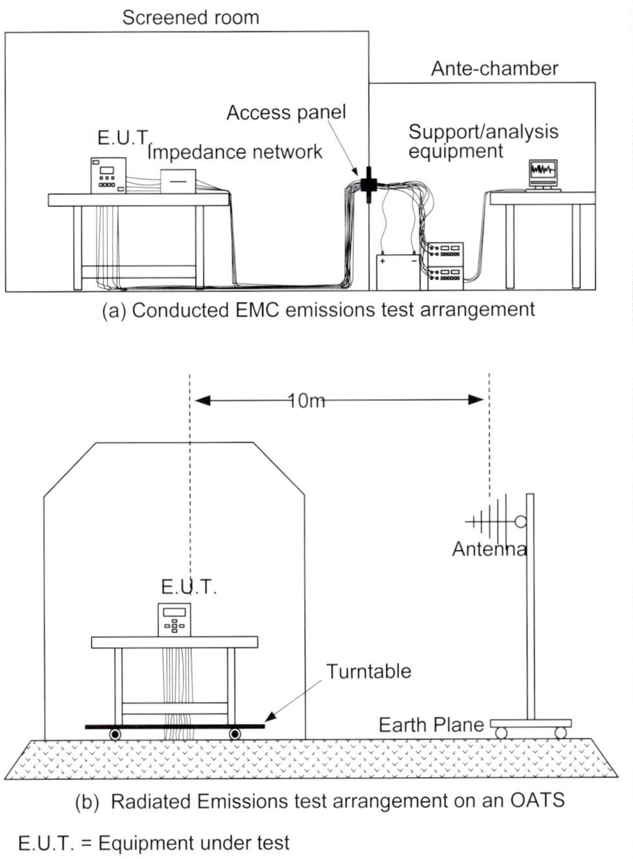

Figure 21.6: EMC test arrangements

When performing these two tests, the relay is in a quiescent condition, that is not tripped, with currents and voltages applied at 90% of the setting values. This is because for the majority of its life, the relay will be in the quiescent state and the emission of electromagnetic interference when the relay is tripped is considered to be of no significance. Tests are conducted in accordance with IEC 60255-25 and EN 55022, and are detailed in Table 21.3.

	Frequency Range	Specified Limits	Test Limits
Radiated	30 - 230MHz	30dB(μV/m) at 30m	40dB(μV/m) at 10m
Radiated	230 - 1000MHz	37dB(μV/m) at 30m	47dB(μV/m) at 10m
Conducted	0.15 - 0.5MHz	79dB(μV) quasi-peak 66dB(μV) average	79dB(μV) quasi-peak 66dB(μV) average
Conducted	0.5 - 30MHz	73dB(μV) quasi-peak 60dB(μV) average	73dB(μV) quasi-peak 60dB(μV) average

Table 21.3: Test criteria for Conducted and Radiated Emissions tests

21.3.10 Conducted and Radiated Immunity Tests

These tests are designed to ensure that the equipment is immune to levels of interference that it may be subjected to. The two tests, conducted and radiated, arise from the fact that for a conductor to be an efficient antenna, it must have a length of at least ¼ of the wavelength of the electromagnetic wave it is required to conduct.

If a relay were to be subjected to radiated interference at 150kHz, then a conductor length of at least

$$\lambda = \frac{300 \times 10^6}{\left(150 \times 10^3 \times 4\right)} = 500m$$

would be needed to conduct the interference. Even with all the cabling attached and with the longest PCB track length taken into account, it would be highly unlikely that the relay would be able to conduct radiation of this frequency, and the test therefore, would have no effect. The interference has to be physically introduced by conduction, hence the conducted immunity test. However, at the radiated immunity lower frequency limit of 80MHz, a conductor length of approximately 1.0m is required. At this frequency, radiated immunity tests can be performed with the confidence that the relay will conduct this interference, through a combination of the attached cabling and the PCB tracks.

Although the test standards state that all 6 faces of the equipment should be subjected to the interference, in practice this is not carried out. Applying interference to the sides and top and bottom of the relay would have little effect as the circuitry inside is effectively screened by the earthed metal case. However, the front and rear of the relay are not completely enclosed by metal and are therefore not at all well screened, and can be regarded as an EMC hole. Electromagnetic interference when directed at the front and back of the relay can enter freely onto the PCBs inside.

When performing these two tests, the relay is in a quiescent condition, that is not tripped, with currents and voltages applied at 90% of the setting values. This is because for the majority of its life, the relay will be in the quiescent state and the coincidence of an electromagnetic disturbance and a fault is considered to be unlikely.

However, spot checks are performed at selected frequencies when the main protection and control functions of the relay are exercised, to ensure that it will operate as expected, should it be required to do so.

The frequencies for the spot checks are in general selected to coincide with the radio frequency broadcast bands, and in particular, the frequencies of mobile communications equipment used by personnel working in the substation. This is to ensure that when working in the vicinity of a relay, the personnel should be able to operate their radios/mobile phones without fear of relay maloperation.

IEC 60255-22-3 specifies the radiated immunity tests to be conducted (ANSI/IEEE C37.90.2 is used for equipment compliant with North American standards), with signal levels of:

- IEC: Class III, 10V/m, 80MHz -2700MHz
- ANSI/IEEE: 35V/m 25MHz - 1000MHz with no modulation, and again with 100% pulse modulation

IEC 60255-22-6 is used for the conducted immunity test, with a test level of:

- Class III, 10V r.m.s., 150kHz - 80MHz.

21.3.11 Power Frequency Magnetic Field Tests

These tests are designed to ensure that the equipment is immune to magnetic interference. The three tests, steady state, pulsed and damped oscillatory magnetic field, arise from the fact that for different site conditions the level and waveshape is altered.

21.3.11.1 Steady state magnetic field tests

These tests simulate the magnetic field that would be experienced by a device located within close proximity of the power system. Testing is carried out by subjecting the relay to a magnetic field generated by two induction coils. The relay is rotated such that in each axis it is subjected to the full magnetic field strength. IEC 61000-4-8 is the relevant standard, using a signal level of:

- Level 5: 300A/m continuous and 1000A/m short duration

The test arrangement is shown in Figure 21.7.

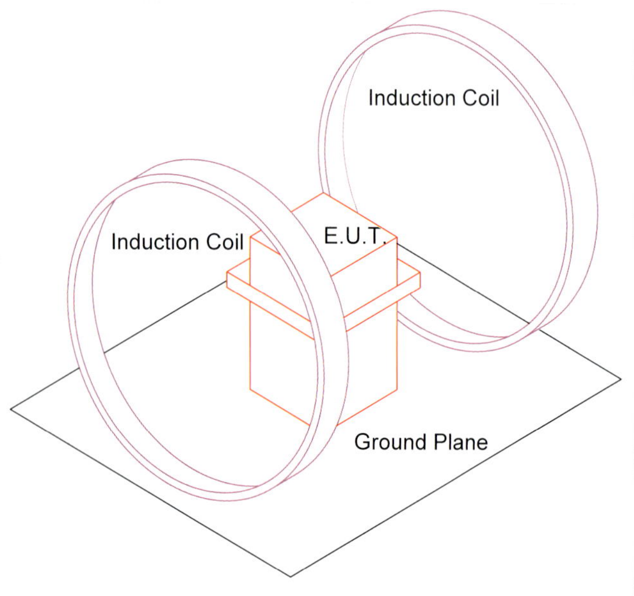

Figure 21.7: Power frequency magnetic field test arrangement

To pass the steady-state test, the relay shall not maloperate, and shall still perform its main functions within the claimed tolerance. During the application of the short duration test, the main protection function shall be exercised and verified that the operating characteristics of the relay are unaffected.

21.3.11.2 Pulsed magnetic field

These tests simulate the magnetic field that would be experienced by a device located within close proximity of the power system during a transient fault condition. According to IEC 61000-4-9, the generator for the induction coils shall produce a 6.4/16μs waveshape with test level 5, 100A/m with the equipment configured as for the steady state magnetic field test. The relay shall not maloperate, and shall still perform its main functions within the claimed tolerance during the test.

21.3.11.3 Damped oscillatory magnetic field

These tests simulate the magnetic field that would be experienced by a device located within close proximity of the power system during a transient fault condition. IEC 61000-4-10 specifies that the generator for the coil shall produce an oscillatory waveshape with a frequency of 0.1MHz and 1MHz, to give a signal level in accordance with Level 5 of 100A/m, and the equipment shall be configured as in Figure 21.7.

21.4 PRODUCT SAFETY TYPE TESTS

A number of tests are carried out to demonstrate that the product is safe when used for its intended application. The essential requirements are that the relay is safe and will not cause an electric shock or fire hazard under normal conditions and in the presence of a single fault. A number of specific tests to prove this may be carried out, as follows.

21.4.1 Dielectric Voltage Withstand

Dielectric Voltage Withstand testing is carried out as a routine test i.e. on every unit prior to despatch. The purpose of this test is to ensure that the product build is as intended by design. This is done by verifying the clearance in air, thus ensuring that the product is safe to operate under normal use conditions. The following tests are conducted unless otherwise specified in the product documentation:

- 2.0kV r.m.s., 50/60Hz for 1 minute between all terminals and case earth and also between independent circuits, in accordance with IEC 60255-27. Some communication circuits are excluded from this test, or have modified test requirements e.g. those using D-type connectors
- 1.5kV r.m.s., 50/60Hz for 1 minute across normally open contacts intended for connection to tripping circuits, in accordance with ANSI/IEEE C37.90

- 1.0kV r.m.s., 50/60Hz for 1 minute across the normally open contacts of watchdog or changeover output relays, in accordance with IEC 60255-27

The routine dielectric voltage withstand test time may be shorter than for the 1 minute type test time, to allow a reasonable production throughput, e.g. for a minimum of 1 second at 110% of the voltage specified for 1 minute.

21.4.2 Insulation Withstand for Overvoltages

The purpose of the High Voltage Impulse Withstand type test is to ensure that circuits and their components will withstand overvoltages on the power system caused by lightning. Three positive and three negative high voltage impulses, 5kV peak, are applied between all circuits and the case earth and also between the terminals of independent circuits (but not across normally open contacts). As before, different requirements apply in the case of circuits using D-type connectors.

The test generator characteristics are as specified in IEC 60255-27 and are shown in Figure 21.8. No disruptive discharge (i.e. flashover or puncture) is allowed.

If it is necessary to repeat either the Dielectric Voltage or High Voltage Impulse Withstand tests these should be carried out at 75% of the specified level, in accordance with IEC 60255-27, to avoid overstressing insulation and components.

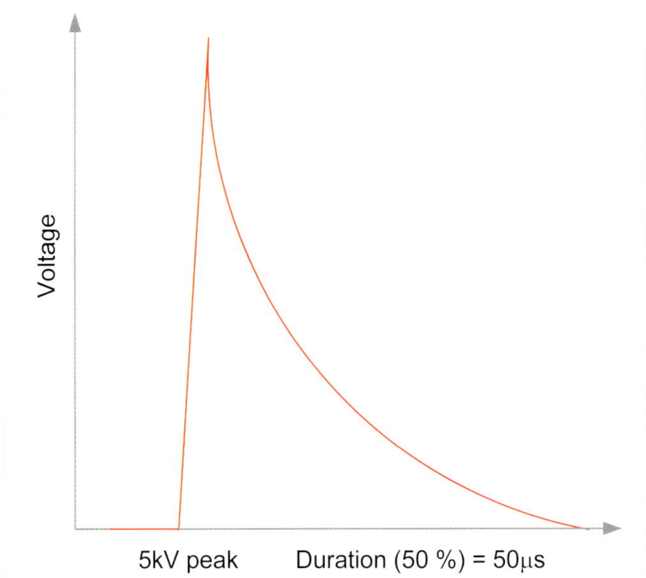

Figure 21.8: Test generator characteristics for insulation withstand test

21.4.3 Single Fault Condition Assessment

An assessment is made of whether a single fault condition such as an overload, or an open or short circuit, applied to the product may cause an electric shock or fire hazard. In the case of doubt, type testing is carried out to ensure that the product is safe.

21.4.4 Earth Bonding Impedance

Class 1 products that rely on a protective earth connection for safety are subjected to an earth bonding impedance (EBI) type test. This ensures that the earth path between the protective earth connection and any accessible earthed part is sufficiently low to avoid damage in the event of a single fault occurring. The test is conducted using a test voltage of 12V maximum and a test current of twice the recommended maximum protective fuse rating. After 1 minute with the current flowing in the circuit under test, the EBI shall not exceed 0.1Ω.

21.4.5 CE Marking

A CE mark on the product, or its packaging, shows that compliance is claimed against relevant EU directives e.g. Low Voltage Directive 2006/95/EC and Electromagnetic Compatibility (EMC) Directive 2004/108/EC.

21.4.6 Other Regional or Industry Norms

Many products intended for application in North America will additionally need certification to the requirements of UL (Underwriter's Laboratory), or CUL for Canada. Products for application in mining or explosive environments will often be required to demonstrate an ATEX claim

21.5 ENVIRONMENTAL TYPE TESTS

Various tests have to be conducted to prove that a relay can withstand the effects of the environment in which it is expected to work. They consist of: the following tests:

- temperature
- humidity
- enclosure protection
- mechanical

These tests are described in the following sections.

21.5.1 Temperature Test

Temperature tests are performed to ensure that a product can withstand extremes in temperatures, both hot and cold, during transit, storage and operating conditions. Storage and transit conditions are defined as a temperature range of -25°C to $+70^{\circ}$C and operating as -25°C to $+55^{\circ}$C. Many products now claim operating temperatures of +70°C or even higher.

Dry heat withstand tests are performed at 70°C for 96 hours with the relay de-energised. Cold withstand tests are performed at −40°C for 96 hours with the relay de-energised. Operating range tests are carried out with the product energised, checking all main functions operate within tolerance over the specified working temperature range −25°C to +55°C.

21.5.2 Humidity Test

The humidity test is performed to ensure that the product will withstand and operate correctly when subjected to 93% relative humidity at a constant temperature of 40°C for 56 days. Tests are performed to ensure that the product functions correctly within specification after 21 and 56 days. After the test, visual inspections are made for any signs of unacceptable corrosion and mould growth.

21.5.3 Cyclic Temperature/Humidity Test

This is a short-term test that stresses the relay by subjecting it to temperature cycling in conjunction with high humidity.

The test does not replace the 56 day humidity test, but is used for testing extension to ranges or minor modifications to prove that the design is unaffected.

The applicable standard is IEC60068-2-30 and test conditions of:

- +25°C ±3°C and 95% relative humidity/+55°C ±2°C
- 95% relative humidity

are used, over the 24 hour cycle shown in Figure 21.9.

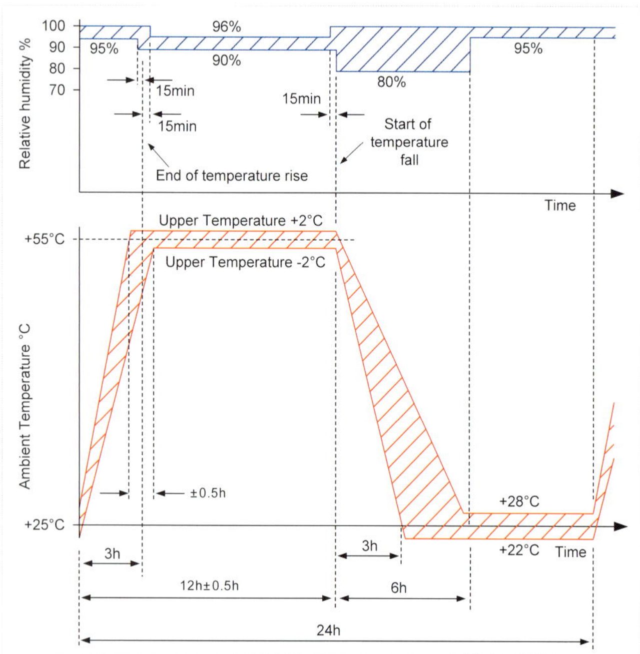

Figure 21.9: Cyclic temperature/humidity test profile

For these tests the relay is placed in a humidity cabinet, and energised with normal in-service quantities for the complete duration of the tests. In practical terms this usually means energising the relay with currents and voltages such that it is 10% from the threshold for operation. Throughout the duration of the test the relay is monitored to ensure that no unwanted operations occur.

Once the relay is removed from the humidity cabinet, its insulation resistance is measured to ensure that it has not deteriorated to below the claimed level. The relay is then functionally tested again, and finally dismantled to check for signs of component corrosion and growth.

The acceptance criterion is that no unwanted operations shall occur including transient operation of indicating devices. After the test the relay's insulation resistance should not have significantly reduced, and it should perform all of its main protection and communications functions within the claimed tolerance. The relay should also suffer no significant corrosion or growth, and photographs are usually taken of each PCB and the case as a record of this.

21.5.4 Enclosure Protection Test

Enclosure protection tests prove that the casing system and connectors on the product protect against the ingress of dust, moisture, water droplets (striking the case at pre-defined angles) and other pollutants. An 'acceptable' level of dust or water may penetrate the case during testing, but must not impair normal product operation, safety or cause tracking across insulated parts of connectors.

21.5.5 Mechanical Tests

Mechanical tests simulate a number of different mechanical conditions that the product may have to endure during its lifetime. These fall into two categories

- response to disturbances while energised
- response to disturbances during transportation (de-energised state)

Tests in the first category are concerned with the response to vibration, shock and seismic disturbance. The tests are designed to simulate normal in-service conditions for the product, for example earthquakes. These tests are performed in all three axes, with the product energised in its normal (quiescent) state. During the test, all output contacts are continually monitored for change using contact follower circuits. Vibration levels of 1g, over a 10Hz-150Hz frequency sweep are used. Seismic tests use excitation in a single axis, using a test frequency of 35Hz and peak displacements of 7.5mm and 3.5mm in the *x* and *y* axes respectively below the crossover frequency and peak accelerations of 2.0gn and 1.0gn in these axes above the crossover frequency.

The second category consists of vibration endurance, shock withstand and bump tests. They are designed to simulate the longer-term affects of shock and vibration that could occur during transportation. These tests are performed with the product de-energised. After these tests, the product must still operate within its specification and show no signs of permanent mechanical damage. Equipment undergoing a seismic type test is shown in Figure 21.10, while the waveform for the shock/bump test is shown in Figure 21.11

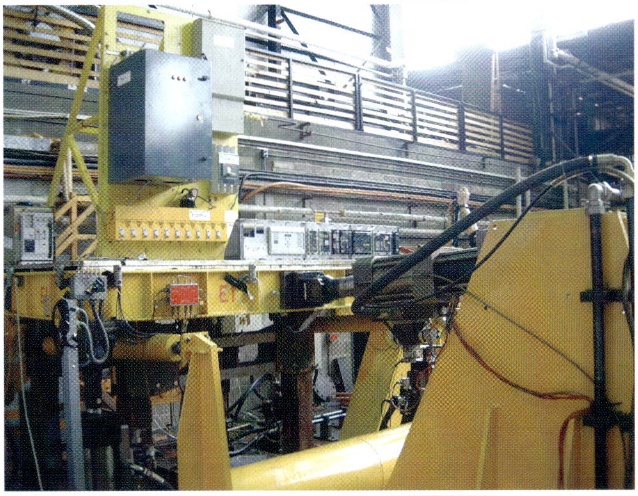

Figure 21.10: Relay undergoing seismic test

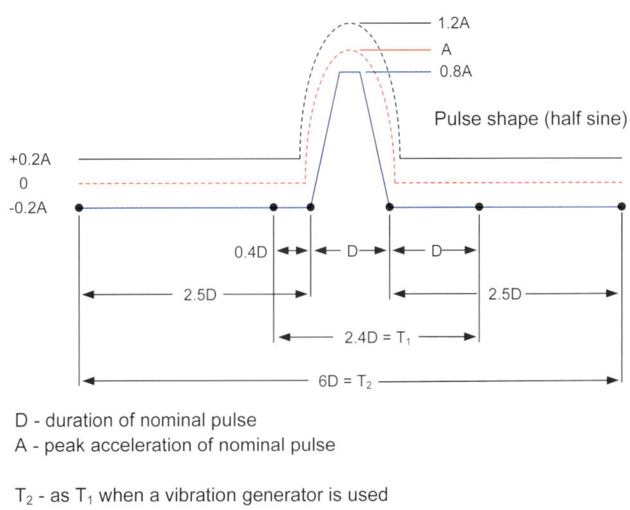

Figure 21.11: Shock/Bump Impulse waveform

The test levels for shock and bump tests are:

Shock response (energised):

- 3 pulses, each 10g, 11ms duration

Shock withstand (de-energised):

- 3 pulses, 15g, 11ms duration

Bump (de-energised):

- 1000 pulses, 10g, 16ms duration

21.6 SOFTWARE TYPE TESTS

Digital and numerical relays contain software to implement the protection and measurement functions of a relay. This software must be thoroughly tested, to ensure that the relay complies with all specifications and that disturbances of various kinds do not result in unexpected results. Software is tested in various stages:

- unit testing
- integration testing
- functional qualification testing

The purpose of unit testing is to determine if an individual function or procedure implemented using software, or small group of closely related functions, is free of data, logic, or standards errors. It is much easier to detect these types of errors in individual units or small groups of units than it is in an integrated software architecture and/or system. Unit testing is typically performed against the software detailed design and by the developer of the unit(s).

Integration testing typically focuses on these interfaces and also issues such as performance, timings and synchronisation that are not applicable in unit testing. Integration testing also focuses on 'stressing' the software and related interfaces.

Integration testing is 'black box' in nature, i.e. it does not take into account the structure of individual units. It is typically performed against the software architectural and detailed design. The unit developer normally performs the tests, unless high-integrity and/or safety critical software is involved, when the tester must be independent of the developer. The specified software requirements would typically also be used as a source for some of the test cases.

21.6.1 Static Unit Testing

Static Unit Testing (or static analysis as it is often called) analyses the unit(s) source code for complexity, precision tracking, initialisation checking, value tracking, strong type checking, macro analysis etc. While static unit testing can be performed manually, it is a laborious and error prone process and is best performed using a proprietary automated static unit analysis tool. It is important to ensure that any such tool is configured correctly and used consistently during development.

21.6.2 Dynamic Testing

Dynamic Testing is concerned with the runtime behaviour of the unit(s) being tested and so therefore, the unit(s) must be executed. Dynamic unit testing can be sub-divided into '*black box*' testing and '*white box*' testing. '*Black box*' testing verifies the implementation of the requirement(s) allocated to the unit(s). It takes no account of the internal structure of the unit(s) being tested. It is only concerned with providing known inputs and determining if the outputs from the unit(s) are correct for those inputs. '*White box*' testing is concerned with testing the internal structure of the unit(s) and measuring the test coverage, i.e. how much of the code within the unit(s) has been executed during the tests. The objective of the unit testing may, for example, be to achieve 100% statement coverage, in which every line of the code is executed at least once, or to execute every possible path through the unit(s) at least once.

21.6.3 Unit Testing Environment

Both Dynamic and Static Unit Testing are performed in the host environment rather than the target environment. Dynamic Unit Testing uses a *test harness* to execute the unit(s) concerned. The test harness is designed such that it simulates the interfaces of the unit(s) being tested - both software-software interfaces and software-hardware interfaces - using what are known as *stubs*. The test harness provides the test data to those units being tested and outputs the test results in a form understandable to a developer. There are many commercially available testing tools to automate test harness production and the execution of tests.

21.6.4 Software/Software Integration Testing

Software/Software Integration Testing is performed in the host environment. It uses a test harness to simulate inputs and outputs, hardware calls and system calls (e.g. the target environment operating system).

21.6.5 Software/Hardware Integration Testing

Software/Hardware Integration Testing is performed in the target environment, i.e. it uses the actual target hardware, operating system, drivers etc. It is usually performed after Software/Software Integration Testing. Testing the interfaces to the hardware is an important feature of Software/Hardware Integration Testing.

Test cases for Integration Testing are typically based on those defined for Validation Testing. However the emphasis should be on finding errors and problems. Performing a dry run of the validation testing often completes Integration Testing.

21.6.6 Validation Testing

The purpose of Validation Testing (also known as Software Acceptance Testing) is to verify that the software meets its specified functional requirements. Validation Testing is performed against the software requirements specification, using the target environment. In ideal circumstances, someone independent of the software development performs the tests. In the case of high-integrity and/or safety critical software, this independence is vital. Validation Testing is 'black box' in nature, i.e. it does not take into account the internal structure of the software. For relays, the non-protection functions included in the software are considered to be as important as the protection functions, and hence tested in the same manner.

Each validation test should have predefined evaluation criteria, to be used to decide if the test has passed or failed. The evaluation criteria should be explicit with no room for interpretation or ambiguity.

21.6.7 Traceability of Validation Tests

Traceability of validation tests to software requirements is vital. Each software requirement documented in the software requirements specification should have at least one validation test, and it is important to be able to prove this.

21.6.8 Software Modifications - Regression Testing

Regression Testing is not a type test in its' own right. It is the overall name given to the testing performed when an existing software product is changed. The purpose of Regression Testing is to show that unintended changes to the functionality (i.e. errors and defects) have not been introduced.

Each change to an existing software product must be considered in its' own right. It is impossible to specify a standard set of regression tests that can be applied as a 'catch-all' for introduced errors and defects. Each change to the software must be analysed to determine what risk there might be of unintentional changes to the functionality being introduced. Those areas of highest risk will need to be regression tested. The ultimate regression test is to perform the complete Validation Testing programme again, updated to take account of the changes made.

Regression Testing is extremely important. If it is not performed, there is a high risk of errors being found in the field. Performing it will not reduce to zero the chance of an error or defect remaining in the software, but it will reduce it. Determining the Regression Testing that is required is made much easier if there is traceability from properly documented software requirements through design (again properly documented and up to date), coding and testing.

21.7 DYNAMIC VALIDATION TYPE TESTING

There are two possible methods of dynamically proving the satisfactory performance of protection relays or schemes; the first method is by actually applying faults on the power system and the second is to carry out comprehensive testing on a power system simulator.

The former method is extremely unlikely to be used – lead times are lengthy and the risk of damage occurring makes the tests very expensive. It is therefore only used on a very limited basis and the faults applied are restricted in number and type. Because of this, a proving period for new protection equipment under service conditions has usually been required. As faults may occur on the power system at infrequent intervals, it can take a number of years before any possible shortcomings are discovered, during which time further installations may have occurred.

Power system simulators can be divided into two types:

- those which use analogue models of a power system
- those which model the power system mathematically using digital simulation techniques

21.7.1 Use of Power System Analogue Models

For many years, relays have been tested on analogue models of power systems such as artificial transmission lines, or test plant capable of supplying significant amounts of current. However, these approaches have significant limitations in the current and voltage waveforms that can be generated, and are not suitable for automated, unattended, testing programmes. While still used on a limited basis for testing electromechanical and static relays, a radically different approach is required for dynamic testing of numerical relays.

21.7.2 Use of Microprocessor Based Simulation Equipment

The complexity of numerical relays, reliant on software for implementation of the functions included, dictates some kind of automated test equipment. The functions of even a simple numerical overcurrent relay (including all auxiliary functions) can take several months of automated, 24 hours/day testing to test completely. If such test equipment was able to apply realistic current and voltage waveforms that closely match those found on power systems during fault conditions, the equipment can be used either for type testing of individual relay designs or of a complete protection scheme designed for a specific application. In recognition of this, a new generation of power system simulators has been developed, which is capable of providing a far more accurate simulation of power system conditions than has been possible in the past. The simulator enables relays to be tested under a wide range of system conditions, representing the equivalent of many years of site experience.

21.7.2.1 Simulation hardware

Equipment is now available to provide high-speed, highly accurate modelling of a power system. The equipment is based on distributed digital hardware under the control of real-time software models, and is shown in Figure 21.12. The modules have outputs linked to current and voltage sources that have a similar transient capability and have suitable output levels for direct connection to the inputs of relays –i.e. 110V for voltage and 1A/5A for current. Inputs are also provided to monitor the response of relays under test (contact closures for tripping, etc.) and these inputs can be used as part of the model of the power system. The software is also capable of modelling the dynamic response of CTs and VTs accurately.

The digital simulator can also be connected digitally to the relay(s) under test using IEC61850-8-1 for station bus, and IEC 61850-9-2 for process bus signals.

Figure 21.12: Digital power system simulator for relay/protection scheme testing

This equipment shows many advantages over traditional test equipment:

- the power system model is capable of reproducing high frequency transients such as travelling waves
- tests involving very long time constants can be carried out
- it is not affected by the harmonic content, noise and frequency variations in the a.c. supply
- it is capable of representing the variation in the current associated with generator faults and power swings
- saturation effects in CTs and VTs can be modelled
- a set of test routines can be specified in software and then left to run unattended (or with only occasional monitoring) to completion, with a detailed record of test results being available on completion
- the IEC61850 interface capabilities allow relays intended for applications in digital substations to be tested.

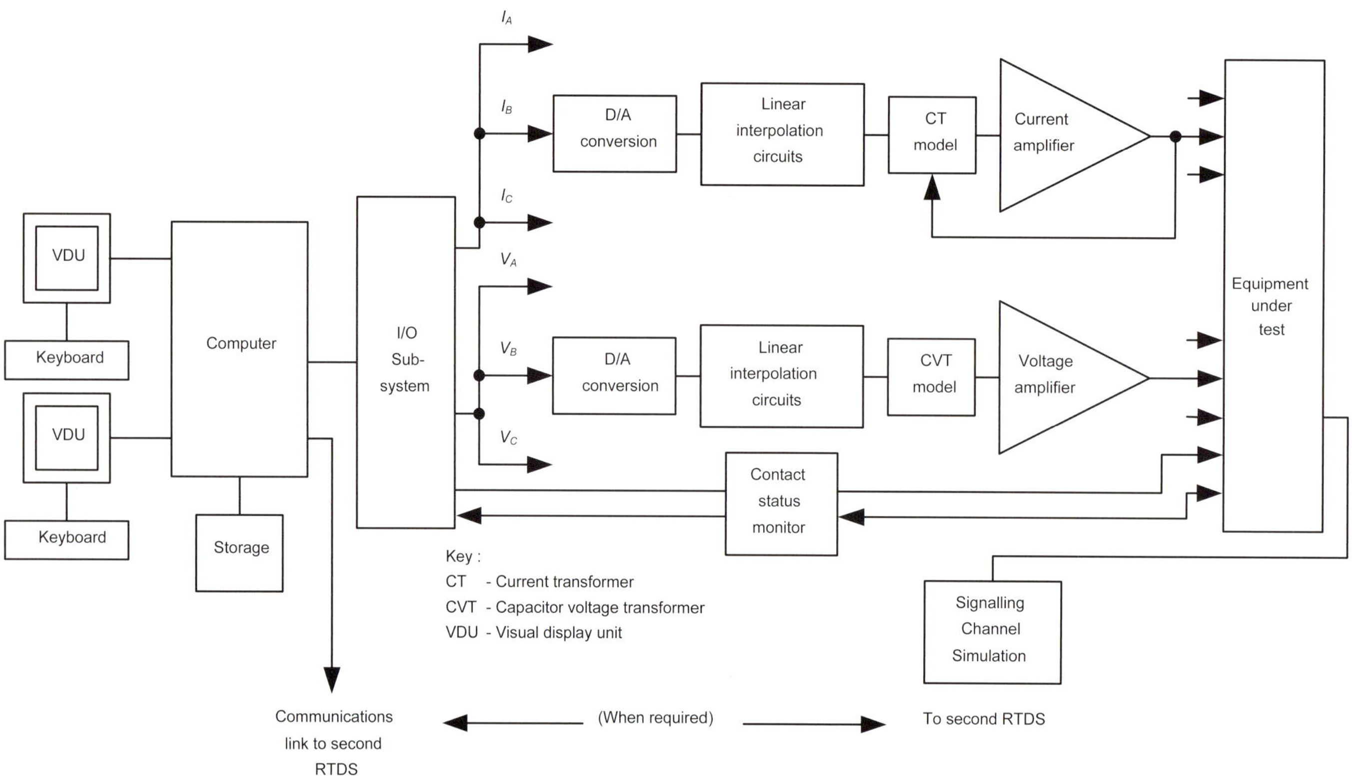

Figure 21.13: Block diagram of real-time digital relay test system

A block schematic of the equipment is shown in Figure 21.13. It is based around a computer which calculates and stores the digital data representing the system voltages and currents. The computer controls conversion of the digital data into analogue signals, and it monitors and controls the relays being tested.

21.7.2.2 Simulation software

Unlike most traditional software used for power systems analysis, the software used is suitable for the modelling the fast transients that occur in the first few milliseconds after fault inception. Two very accurate simulation programs are used, one based on time domain and the other on frequency domain techniques. In both programs, single and double circuit transmission lines are represented by fully distributed parameter models. The line parameters are calculated from the physical construction of the line (symmetrical, asymmetrical, transposed or non-transposed), taking into account the effect of conductor geometry, conductor internal impedance and the earth return path. It also includes, where appropriate, the frequency dependence of the line parameters in the frequency domain program. The frequency dependent variable effects are calculated using Fast Fourier Transforms and the results are converted to the time domain. Conventional current transformers and capacitor voltage transformers can be simulated.

The fault can be applied at any one point in the system and can be any combination of phase to phase or phase to earth, resistive, or non-linear phase to earth arcing faults. For series compensated lines, flashover across a series capacitor following a short circuit fault can be simulated.

The frequency domain model is not suitable for developing faults and switching sequences, therefore the widely used Electromagnetic Transient Program (EMTP), working in the time domain, is employed in such cases.

In addition to these two programs, a simulation program based on lumped resistance and inductance parameters is used. This simulation is used to represent systems with long time constants and slow system changes due, for example, to power swings.

21.7.2.3 Simulator applications

The simulator is used for checking the accuracy of calibration and performing type tests on a wide range of protection relays during their development. It has the following advantages over existing test methods:

- 'state of the art' power system modelling data can be used to test relays
- freedom from frequency variations and noise or harmonic content of the laboratory's own domestic supply
- the relay under test does not burden the power system simulation

- all tests are accurately repeatable
- wide bandwidth signals can be produced
- a wide range of frequencies can be reproduced
- selected harmonics may be superimposed on the power frequency
- the use of direct coupled current amplifiers allows time constants of any length
- capable of simulating slow system changes
- reproduces fault currents whose peak amplitude varies with time
- transducer models can be included
- automatic testing removes the likelihood of measurement and setting errors
- two such equipments can be linked together to simulate a system model with two relaying points

The simulator is also used for the production testing of relays, in which most of the advantages listed above apply. As the tests and measurements are made automatically, the quality of testing is also greatly enhanced. Further, in cases of suspected malfunction of a relay in the field under known fault conditions, the simulator can be used to replicate the power system and fault conditions, and conduct a detailed investigation into the performance of the relay. Finally, complex protection schemes can be modelled, using both the relays intended for use and software models of them as appropriate, to check the suitability of the proposed scheme under a wide variety of conditions.

Figure 21.14(a) shows a section of a particular power system modelled. The waveforms of Figure 21.14(b) show the three phase voltages and currents at the primaries of VT1 and CT1 for the fault condition indicated in Figure 21.14(a).

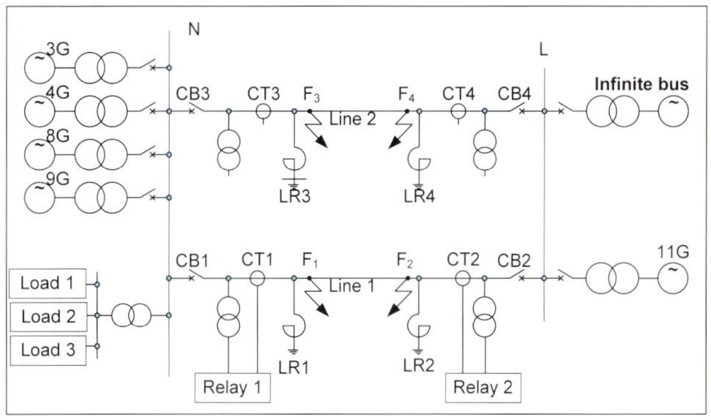

(a) Example power system

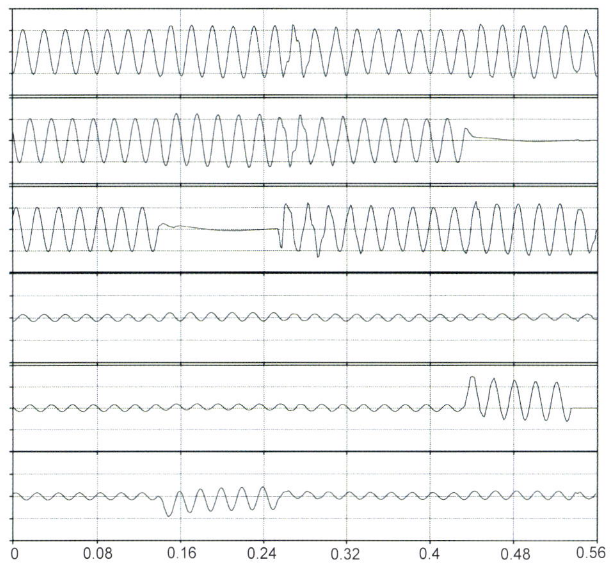

(b) Voltages and currents at VT1/CT1 power system

Figure 21.14: Example of application study

21.8 PRODUCTION TESTING

Production testing of protection relays is becoming far more demanding as the accuracy and complexity of the products increase. Electronic power amplifiers are used to supply accurate voltages and currents of high stability to the relay under test. The inclusion of a computer in the test system allows more complex testing to be performed at an economical cost, with the advantage of speed and repeatability of tests from one relay to another.

Figure 21.15 shows a modern computer-controlled test bench. The hardware is mounted in a special rack. Each unit of the test system is connected to the computer via an interface bus. Individual test programs for each type of relay are required, but the interface used is standard for all relay types. Control of input waveforms and analogue measurements, the monitoring of output signals and the analysis of test data are performed by the computer. A printout of the test results can also be produced if required.

Figure 21.15: Modern computer-controlled test bench

Because software is extensively tested at the type-testing stage, there is normally no need to check the correct functioning of the software. Checks are limited to determining that the analogue and digital I/O is functioning correctly. This is achieved for inputs by applying known voltage and current inputs to the relay under test and checking that the software has captured the correct values. Similarly, digital outputs are exercised by using test software to actuate each output and checking that the correct output is energised. Provided that appropriate procedures are in place to ensure that only type-tested software is downloaded, there is no need to check the correct functioning of the software in the relay. The final step is to download the software appropriate to the relay and store it in the relay's non-volatile memory.

21.9 COMMISSIONING TESTS

Installation of a protection scheme at site creates a number of possibilities for errors in the implementation of the scheme to occur. Even if the scheme has been thoroughly tested in the factory, wiring to the CTs and VTs on site may be incorrectly carried out, or the CTs/VTs may have been incorrectly installed. The impact of such errors may range from simply being a nuisance (tripping occurs repeatedly on energisation, requiring investigation to locate and correct the error(s)) through to failure to trip under fault conditions, leading to major equipment damage, disruption to supplies and potential hazards to personnel. The strategies available to remove these risks are many, but all involve some kind of testing at site.

Commissioning tests at site are therefore invariably performed before protection equipment is set to work. The aims of commissioning tests are:

- to ensure that the equipment has not been damaged during transit or installation
- to ensure that the installation work has been carried out correctly
- to prove the correct functioning of the protection scheme as a whole

The tests carried out will normally vary according to the protection scheme involved, the relay technology used, and the policy of the client. In many cases, the tests actually conducted are determined at the time of commissioning by mutual agreement between the client's representative and the commissioning team. Hence, it is not possible to provide a definitive list of tests that are required during commissioning. This section therefore describes the tests commonly carried out during commissioning.

The following tests are invariably carried out, since the protection scheme will not function correctly if faults exist.

- wiring diagram check, using circuit diagrams showing all the reference numbers of the interconnecting wiring
- general inspection of the equipment, checking all connections, wires on relays terminals, labels on terminal boards, etc.
- insulation resistance measurement of all circuits
- perform relay self-test procedure and external communications checks on digital/numerical relays
- test main current transformers
- test main voltage transformers
- check that protection relay alarm/trip settings have been entered correctly
- tripping and alarm circuit checks to prove correct functioning

In addition, the following checks may be carried out, depending on the factors noted earlier.

- secondary injection test on each relay to prove operation at one or more setting values
- primary injection tests on each relay to prove stability for external faults and to determine the effective current setting for internal faults (essential for some types of electromechanical relays)
- testing of protection scheme logic

This section details the tests required to cover the above items. Secondary injection test equipment is covered in Section 21.10. Section 21.11 details the secondary injection that may be carried out. Section 21.12 covers primary injection testing and Section 21.13 details the checks required on any logic involved in the protection scheme. Finally section 21.14 details the tests required on alarm/tripping circuits tripping/alarm circuits.

21.9.1 Insulation Tests

All the deliberate earth connections on the wiring to be tested should first be removed, for example earthing links on current transformers, voltage transformers and d.c. supplies. Some insulation testers generate impulses with peak voltages exceeding 5kV. In these instances any electronic equipment should be disconnected while the external wiring insulation is checked.

The insulation resistance should be measured to earth and between electrically separate circuits. The readings are recorded and compared with subsequent routine tests to check for any deterioration of the insulation.

The insulation resistance measured depends on the amount of wiring involved, its grade, and the site humidity. Generally, if the test is restricted to one cubicle, a reading of several hundred megohms should be obtained. If long lengths of site wiring are involved, the reading could be only a few megohms.

21.9.2 Relay Self-Test Procedure

Digital and numerical relays will have a self-test procedure that is detailed in the appropriate relay manual. These tests should be followed to determine if the relay is operating correctly. This will normally involve checking of the relay watchdog circuit, exercising all digital inputs and outputs and checking that the relay analogue inputs are within calibration by applying a test current or voltage. For these tests, the relay outputs are normally disconnected from the remainder of the protection scheme, as it is a test carried out to prove correct relay, rather than scheme, operation.

Unit protection schemes involve relays that need to communicate with each other. This leads to additional testing requirements. The communications path between the relays is tested using suitable equipment to ensure that the path is complete and that the received signal strength is within specification. Numerical relays may be fitted with loopback test facilities that enable either part of or the entire communications link to be tested from one end.

After completion of these tests, it is usual to enter the relay settings required. This can be done manually via the relay front panel controls, or using a portable PC and suitable software. Whichever, method is used, a check by a second person that the correct settings have been used is desirable, and the settings recorded. Programmable scheme logic that is required is also entered at this stage.

21.9.3 Current Transformer Tests

The following tests are normally carried out prior to energisation of the main circuits.

21.9.3.1 Polarity check

Each current transformer should be individually tested to verify that the primary and secondary polarity markings are correct; see Figure 21.16. The ammeter connected to the secondary of the current transformer should be a robust moving coil, permanent magnet, centre-zero type. A low voltage battery is used, via a single-pole push-button switch, to energise the primary winding. On closing the push-button, the d.c. ammeter, *A*, should give a positive flick and on opening, a negative flick.

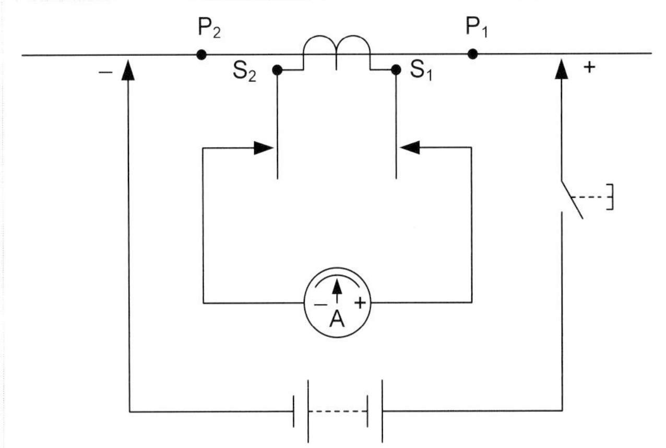

Figure 21.16: Current transformer polarity check

21.9.3.2 Magnetisation Curve

Several points should be checked on each current transformer magnetisation curve. This can be done by energising the secondary winding from the local mains supply through a variable auto-transformer while the primary circuit remains open; see Figure 21.17. The characteristic is measured at suitable intervals of applied voltage, until the magnetising current is seen to rise very rapidly for a small increase in voltage. This indicates the approximate knee-point or saturation flux level of the current transformer. The magnetising current should then be recorded at similar voltage intervals as it is reduced to zero.

Care must be taken that the test equipment is suitably rated. The short-time current rating must be in excess of the CT secondary current rating, to allow for measurement of the saturation current. This will be in excess of the CT secondary current rating. As the magnetising current will not be sinusoidal, a moving iron or dynamometer type ammeter should be used.

It is often found that current transformers with secondary ratings of 1A or less have a knee-point voltage higher than the local mains supply. In these cases, a step-up interposing transformer must be used to obtain the necessary voltage to check the magnetisation curve.

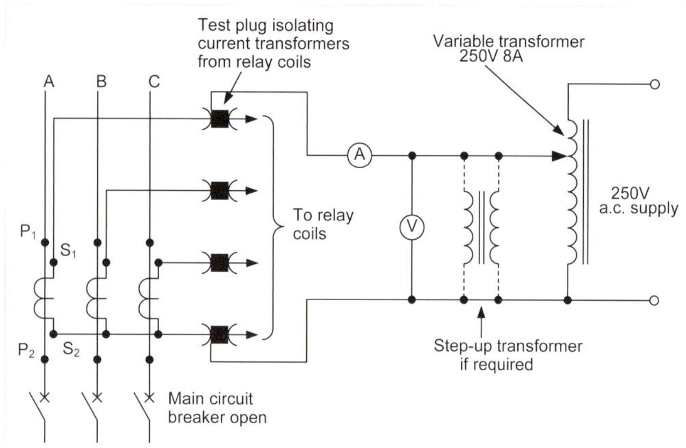

Figure 21.17: Testing current transformer magnetising curve

21.9.4 Voltage Transformer Tests

Voltage transformers require testing for polarity and phasing.

21.9.4.1 Polarity check

The voltage transformer polarity can be checked using the method for CT polarity tests. Care must be taken to connect the battery supply to the primary winding, with the polarity ammeter connected to the secondary winding. If the voltage transformer is of the capacitor type, then the polarity of the transformer at the bottom of the capacitor stack should be checked.

21.9.4.2 Ratio check

This check can be carried out when the main circuit is first made live. The voltage transformer secondary voltage is compared with the secondary voltage shown on the nameplate.

21.9.4.3 Phasing check

The secondary connections for a three-phase voltage transformer or a bank of three single-phase voltage transformers must be carefully checked for phasing. With the main circuit alive, the phase rotation is checked using a phase rotation meter connected across the three phases, as shown in Figure 21.18. Provided an existing proven VT is available on the same primary system, and that secondary earthing is employed, all that is now necessary to prove correct phasing is a voltage check between, say, both 'A' phase secondary outputs. There should be nominally little or no voltage if the phasing is correct. However, this test does not detect if the phase sequence is correct, but the phases are displaced by 120° from their correct position, i.e. phase A occupies the position of phase C or phase B in Figure 21.18. This can be checked by removing the fuses from phases B and C (say) and measuring the phase-earth voltages on the secondary of the VT. If the phasing is correct, only phase A should be healthy, phases B and C should have only a small residual voltage.

Correct phasing should be further substantiated when carrying out 'on load' tests on any phase-angle sensitive relays, at the relay terminals. Load current in a known phase CT secondary should be compared with the associated phase to neutral VT secondary voltage. The phase angle between them should be measured, and should relate to the power factor of the system load.

If the three-phase voltage transformer has a broken-delta tertiary winding, then a check should be made of the voltage across the two connections from the broken delta V_N and V_L, as shown in Figure 21.18. With the rated balanced three-phase supply voltage applied to the voltage transformer primary windings, the broken-delta voltage should be below 5V with the rated burden connected.

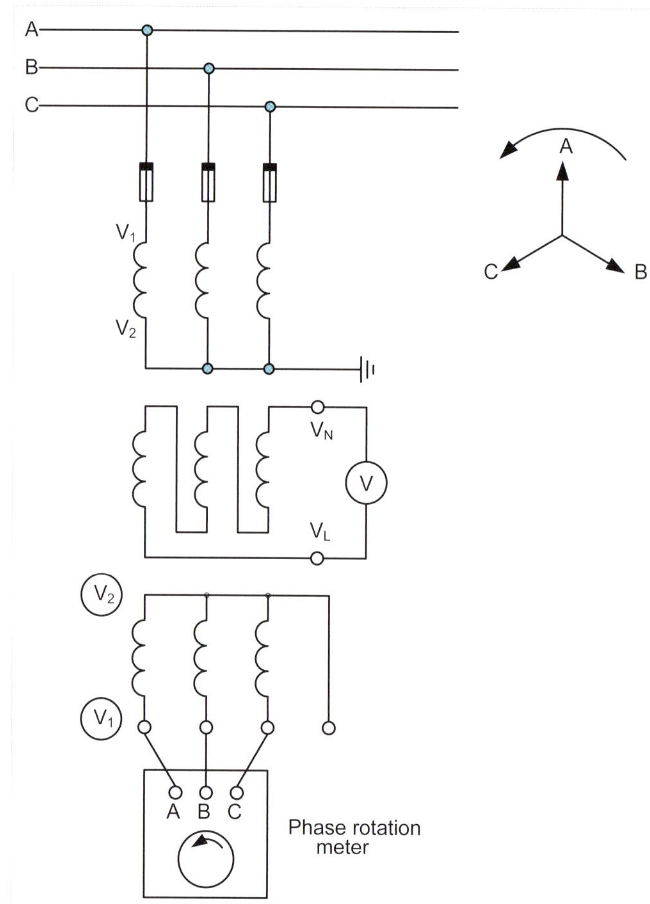

Figure 21.18: Voltage transformer phasing check

21.9.5 Protection Relay Setting Checks

At some point during commissioning, the alarm and trip settings of the relay elements involved will require to be entered and/or checked. Where the complete scheme is engineered and supplied by a single contractor, the settings may already have been entered prior to despatch from the factory, and hence this need not be repeated. The method of entering settings varies according to the relay technology used. For electromechanical and static relays, manual entry of the settings for each relay element is required. This method can also be used for digital/numerical relays. However, the amount of data to be entered is much greater, and therefore it is usual to use appropriate software, normally supplied by the manufacturer, for this purpose. The software also makes the essential task of making a record of the data entered much easier.

Once the data has been entered, it should be checked for compliance with the recommended settings as calculated from the protection setting study. Where appropriate software is used for data entry, the checks can be considered complete if the data is checked prior to download of the settings to the relay. Otherwise, a check may required subsequent to data entry by inspection and recording of the relay settings, or it may be considered adequate to do this at the time of data entry. The recorded settings form an essential part of the commissioning documentation provided to the client.

21.10 SECONDARY INJECTION TEST EQUIPMENT

Secondary injection tests are always done prior to primary injection tests. The purpose of secondary injection testing is to prove the correct operation of the protection scheme that is downstream from the inputs to the protection relay(s). Secondary injection tests are always done prior to primary injection tests. This is because the risks during initial testing to the LV side of the equipment under test are minimised. The primary (HV) side of the equipment is disconnected, so that no damage can occur. These tests and the equipment necessary to perform them are generally described in the manufacturer's manuals for the relays, but brief details are given below for the main types of protection relays.

21.10.1 Test Blocks/Plugs for Secondary Injection Equipment

It is common practice to provide test blocks or test sockets in the relay circuits so that connections can readily be made to the test equipment without disturbing wiring. Test plugs of either multi-finger or single-finger design (for monitoring the current in one CT secondary circuit) are used to connect test equipment to the relay under test.

The top and bottom contact of each test plug finger is separated by an insulating strip, so that the relay circuits can be completely isolated from the switchgear wiring when the test plug is inserted. To avoid open-circuiting CT secondary terminals, it is therefore essential that CT shorting jumper links are fitted across all appropriate 'live side' terminals of the test plug BEFORE it is inserted. With the test plug inserted in position, all the test circuitry can now be connected to the isolated 'relay side' test plug terminals. Some modern test blocks incorporate the live-side jumper links within the block and these can be set to the 'closed' or 'open' position as appropriate, either manually prior to removing the cover and inserting the test plug, or automatically upon removal of the cover. Removal of the cover also exposes the colour-coded face-plate of the block, clearly indicating that the protection scheme is not in service, and may also disconnect any d.c. auxiliary supplies used for powering relay tripping outputs.

Withdrawing the test plug immediately restores the connections to the main current transformers and voltage transformers and removes the test connections. Replacement of the test block cover then removes the short circuits that had been applied to the main CT secondary circuits. Where several relays are used in a protection scheme, one or more test blocks may be fitted on the relay panel enabling the whole scheme to be tested, rather than just one relay at a time.

Test blocks usually offer facilities for the monitoring and secondary injection testing of any power system protection scheme. The test block may be used either with a multi-fingered test plug to allow isolation and monitoring of all the selected conductor paths, or with a single finger test plug that allows the currents on individual conductors to be monitored. A modern test block and test plugs are illustrated in Figure 21.19.

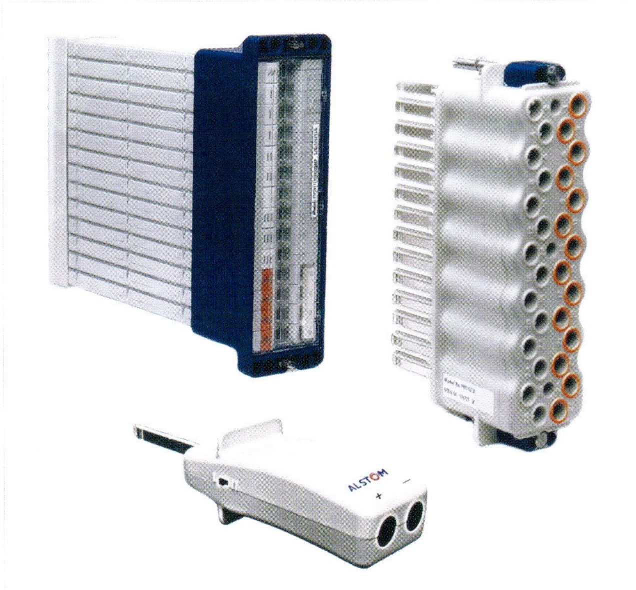

Figure 21.19: Modern test block/plugs

21.10.2 Secondary Injection Test Sets

The type of the relay to be tested determines the type of equipment used to provide the secondary injection currents and voltages. Many electromechanical relays have a non-linear current coil impedance when the relay operates and this can cause the test current waveform to be distorted if the injection supply voltage is fed directly to the coil. The presence of harmonics in the current waveform may affect the torque of electromechanical relays and give unreliable test results, so some injection test sets use an adjustable series reactance to control the current. This keeps the power dissipation small and the equipment light and compact.

Many test sets are portable and include precision ammeters and voltmeters and timing equipment. Test sets may have both voltage and current outputs. The former are high-voltage, low current outputs for use with relay elements that require signal inputs from a VT as well as a CT. The current outputs are high-current, low voltage to connect to relay CT inputs. It is important, however, to ensure that the test set current outputs are true current sources, and hence are not affected by the load impedance of a relay element current coil. Use of a test set with a current output that is essentially a voltage source can give rise to serious problems when testing electromechanical relays. Any significant impedance mismatch between the output of the test set and the relay current coil during relay operation will give rise to a variation in current from that desired and possible error in the test results. The relay operation time may be greater than expected (never less than expected) or relay 'chatter' may occur. It is quite common for such errors to only be found much later, after a fault has caused major damage to equipment through failure of the primary protection to operate. Failure investigation then shows that the reason for the primary protection to operate is an incorrectly set relay, due in turn to use of a test set with a current output consisting of a voltage-source when the relay was last tested. Figure 21.20 shows typical waveforms resulting from use of test set current output that is a voltage source – the distorted relay coil current waveform gives rise to an extended operation time compared to the expected value.

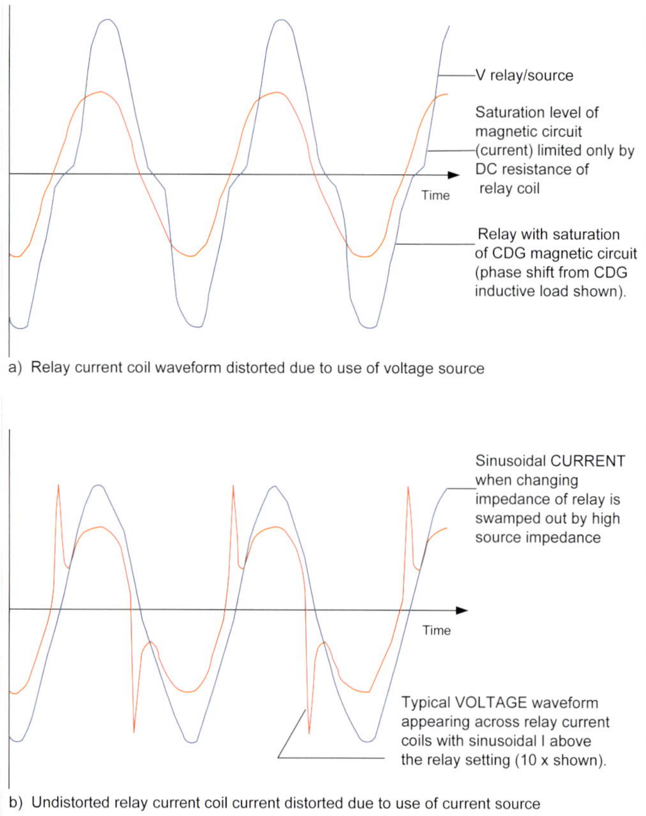

Figure 21.20: Relay current coil waveforms

Modern test sets are computer based. They comprise a PC (usually a standard laptop PC with suitable software) and a power amplifier that takes the low-level outputs from the PC and amplifies them into voltage and current signals suitable for application to the VT and CT inputs of the relay. The phase angle between voltage and current outputs will be adjustable, as also will the phase angles between the individual voltages or currents making up a 3-phase output set. Much greater precision in the setting of the magnitudes and phase angles is possible, compared to traditional test sets. Digital signals to exercise the internal logic elements of the relays may also be provided. The alarm and trip outputs of the relay are connected to digital inputs on the PC so that correct operation of the relay, including accuracy of the relay tripping characteristic can be monitored and displayed on-screen, saved for inclusion in reports generated later, or printed for an immediate record to present to the client. Optional features may include GPS time synchronising equipment and remote-located amplifiers to facilitate testing of unit protection schemes, and digital I/O for exercising the programmable scheme logic of modern relays. Some test sets offer a digital interface for IEC61850 GOOSE I/O monitoring, and for virtual "injection" of IEC 61850-9-2 sampled values.

The software for modern test sets is capable of testing the functionality of a wide variety of relays, and conducting a set of tests automatically. Such sets ease the task of the commissioning engineer. The software will normally offer options for testing, ranging from a test carried out at a particular point on the characteristic to complete determination of the tripping characteristic automatically. This feature can be helpful if there is any reason to doubt that the relay is operating correctly with the tripping characteristic specified. Figure 21.21 illustrates a modern PC-based test set.

Figure 21.21: Modern PC-based secondary injection test set

Traditional test sets use an arrangement of adjustable transformers and reactors to provide control of current and voltage without incurring high power dissipation. Some relays require adjustment of the phase between the injected voltages and currents, and so phase shifting transformers may be used.

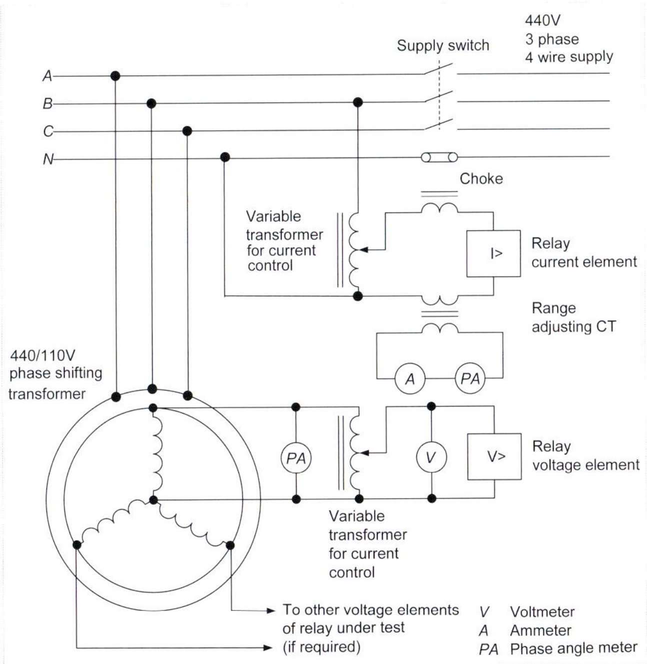

Figure 21.22: Circuit diagram for traditional test set for directional/distance relays

21.11 SECONDARY INJECTION TESTING

The purpose of secondary injection testing is to check that the protection scheme from the relay input terminals onwards is functioning correctly with the settings specified. This is achieved by applying suitable inputs from a test set to the inputs of the relays and checking if the appropriate alarm/trip signals occur at the relay/control room/CB locations. The extent of testing will be largely determined by the client specification and relay technology used, and may range from a simple check of the relay characteristic at a single point to a complete verification of the tripping characteristics of the scheme, including the response to transient waveforms and harmonics and checking of relay bias characteristics. This may be important when the protection scheme includes transformers and/or generators.

The testing should include any scheme logic. If the logic is implemented using the programmable scheme logic facilities available with most digital or numerical relays, appropriate digital inputs may need to be applied and outputs monitored (see Section 21.13). It is clear that a modern test set can facilitate such tests, leading to a reduced time required for testing.

21.11.1 Schemes using Digital or Numerical Relay Technology

The policy for secondary injection testing varies widely. In some cases, manufacturers recommend, and clients accept, that if a digital or numerical relay passes its' self-test, it can be relied upon to operate at the settings used and that testing can therefore be confined to those parts of the scheme external to the relay. In such cases, secondary injection testing is not required at all. More often, it is required that one element of each relay (usually the simplest) is exercised, using a secondary injection test set, to check that relay operation occurs at the conditions expected, based on the setting of the relay element concerned. Another alternative is for the complete functionality of each relay to be exercised. This is rarely required with a digital or numerical relay, probably only being carried out in the event of a suspected relay malfunction.

To illustrate the results that can be obtained, Figure 21.23 shows the results obtained by a modern test set when determining the reach settings of a distance relay using a search technique.

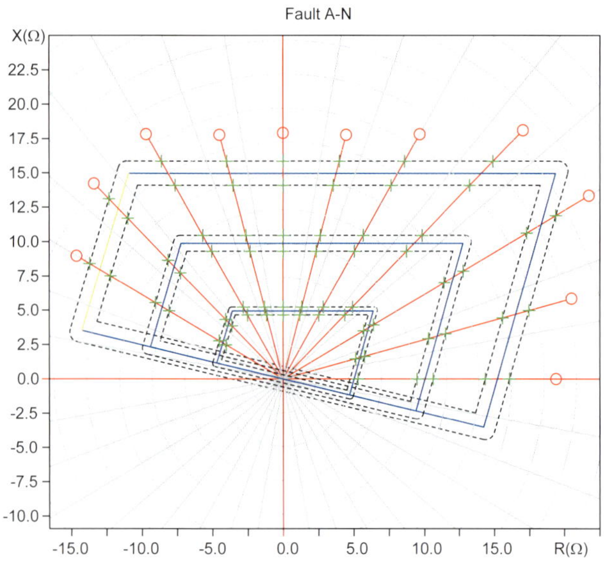

Figure 21.23: Distance relay zone checking using search technique and tolerance bands

21.11.2 Schemes using Electromechanical/Static Relay Technology

Schemes using single function electromechanical or static relays will usually require each relay to be exercised. Thus a scheme with distance and back-up overcurrent elements will require a test on each of these functions, thereby taking up more time than if a digital or numerical relay is used. Similarly, it may be important to check the relay characteristic over a range of input currents to confirm parameters for an overcurrent relay such as:

- the minimum current that gives operation at each current setting
- the maximum current at which resetting takes place
- the operating time at suitable values of current
- the time/current curve at two or three points with the time multiplier setting TMS at 1
- the resetting time at zero current with the TMS at 1

Similar considerations apply to distance and unit protection relays of these technologies.

21.11.3 Test Circuits for Secondary Injection Testing

The test circuits used will depend on the type of relay and test set being used. Unless the test circuits are simple and obvious, the relay commissioning manual will give details of the circuits to be used. When using the circuits in this reference, suitable simplifications can easily be made if digital or numerical relays are being tested, to allow for their built-in measurement capabilities – external ammeters and voltmeters may not be required.

All results should be carefully noted and filed for record purposes. Departures from the expected results must be thoroughly investigated and the cause determined. After rectification of errors, all tests whose results may have been affected (even those that may have given correct results) should be repeated to ensure that the protection scheme has been implemented according to specification.

21.12 PRIMARY INJECTION TESTS

This type of test involves the entire circuit; current transformer primary and secondary windings, relay coils, trip and alarm circuits, and all intervening wiring are checked. There is no need to disturb wiring, which obviates the hazard of open-circuiting current transformers, and there is generally no need for any switching in the current transformer or relay circuits. The drawback of such tests is that they are time consuming and expensive to organise. Increasingly, reliance is placed on all wiring and installation diagrams being correct and the installation being carried out as per drawings, and secondary injection testing being completed satisfactorily. Under these circumstances, the primary injection tests may be omitted. However, wiring errors between VTs/CTs and relays, or incorrect polarity of VTs/CTs may not then be discovered until either spurious tripping occurs in service, or more seriously, failure to trip on a fault. This hazard is much reduced where digital/numerical relays are used, since the current and voltage measurement/display facilities that exist in such relays enable checking of relay input values against those from other proven sources. Many connection/wiring errors can be found in this way, and by isolating temporarily the relay trip outputs, unwanted trips can be avoided.

Primary injection testing is, however, the only way to prove correct installation and operation of the whole of a protection scheme. As noted in the previous section, primary injection tests are always carried out after secondary injection tests, to ensure that problems are limited to the VTs and CTs involved, plus associated wiring, all other equipment in the protection scheme having been proven satisfactory from the secondary injection tests.

21.12.1 Test Facilities

An alternator is the most useful source of power for providing the heavy current necessary for primary injection. Unfortunately, it is rarely available, since it requires not only a spare alternator, but also spare busbars capable of being connected to the alternator and circuit under test. Therefore, primary injection is usually carried out by means of a portable injection transformer (Figure 21.24), arranged to operate from the local mains supply and having several low voltage, heavy current windings. These can be connected in series or parallel according to the current required and the resistance of the primary circuit. Outputs of 10V and 1000A can be obtained. Alternatively, modern PC-controlled test sets have power amplifiers capable of injecting currents up to about 200A for a single unit, with higher current ratings being possible by using multiple units in parallel.

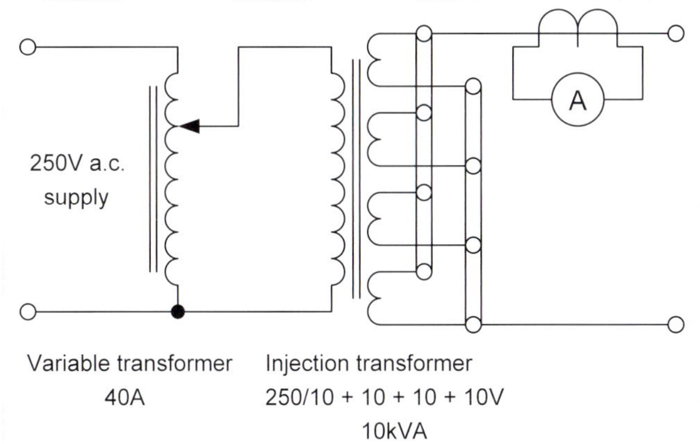

Figure 21.24: Traditional primary injection test set

If the main current transformers are fitted with test windings, these can be used for primary injection instead of the primary winding. The current required for primary injection is then greatly reduced and can usually be obtained using secondary injection test equipment. Unfortunately, test windings are not often provided, because of space limitations in the main current transformer housings or the cost of the windings.

21.12.2 CT Ratio Check

Current is passed through the primary conductors and measured on the test set ammeter, A_1 in Figure 21.25. The secondary current is measured on the ammeter A_2 or relay display, and the ratio of the value on A_1 to that on A_2 should closely approximate to the ratio marked on the current transformer nameplate.

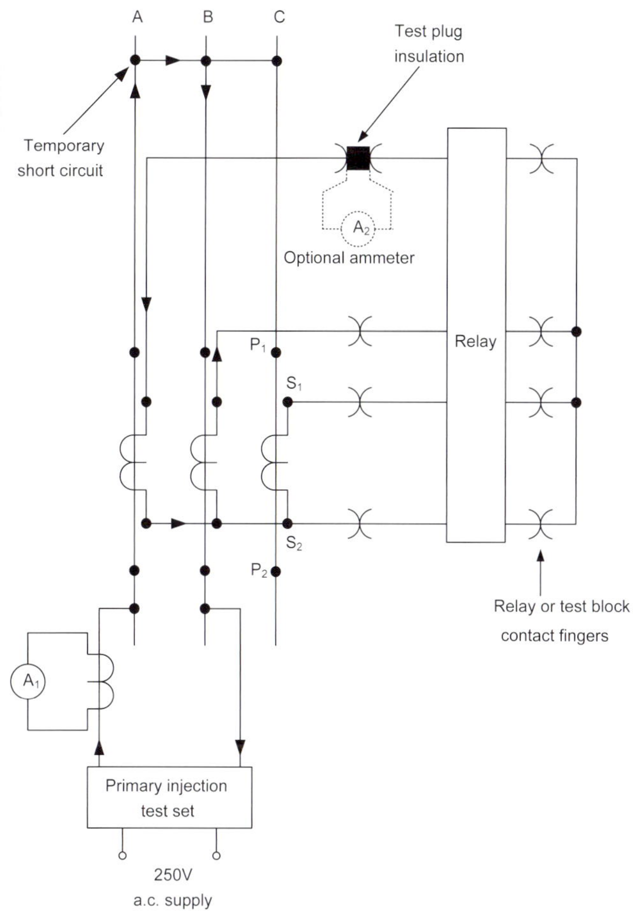

Figure 21.25: Current transformer ratio check

21.12.3 CT Polarity Check

If the equipment includes directional, differential or earth fault relays, the polarity of the main current transformers must be checked. It is not necessary to conduct the test if only overcurrent relays are used.

The circuit for checking the polarity with a single-phase test set is shown in Figure 21.26. A short circuit is placed across the phases of the primary circuit on one side of the current transformers while single-phase injection is carried out on the other side. The ammeter connected in the residual circuit, or relay display, will give a reading of a few milliamperes with rated current injected if the current transformers are of correct polarity. A reading proportional to twice the primary current will be obtained if they are of wrong polarity. Because of this, a high-range ammeter should be used initially, for example one giving full-scale deflection for twice the rated secondary current. If an electromechanical earth-fault relay with a low setting is also connected in the residual circuit, it is advisable to temporarily short-circuit its operating coil during the test, to prevent possible overheating. The single-phase injection should be carried out for each pair of phases.

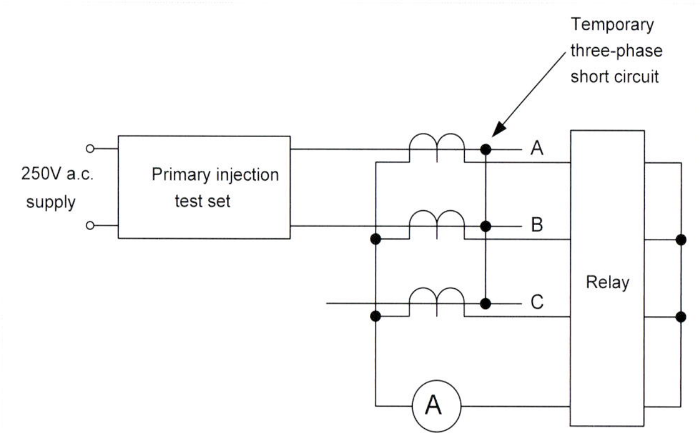

Figure 21.26: Polarity check on main current transformers

21.12.4 Primary Injection Testing of Relay Elements

As with secondary injection testing, the tests to be carried out will be those specified by the client, and/or those detailed in the relay commissioning manual. Digital and numerical relays usually require far fewer tests to prove correct operation, and these may be restricted to observations of current and voltage on the relay display under normal load conditions.

21.13 TESTING OF PROTECTION SCHEME LOGIC

Protection schemes often involve the use of logic to determine the conditions under which designated circuit breakers should be tripped. Simple examples of such logic can be found in Chapters 9-14. Traditionally, this logic was implemented by means of discrete relays, separate from the relays used for protection. Such implementations would occur where electromechanical or static relay technology is used. However, digital and numerical relays normally include programmable logic as part of the software within the relay, together with associated digital I/O. This facility (commonly referred to as Programmable Scheme Logic, or PSL) offers important advantages to the user, by saving space and permitting modifications to the protection scheme logic through software if the protection scheme requirements change with time. Changes to the logic are carried out using software hosted on a PC (or similar computer) and downloaded to the relay. Use of languages defined in IEC 61131, such as ladder logic or Boolean algebra is common for such software, and is readily understood by Protection Engineers. Further, there are several commonly encountered protection functions that manufacturers may supply with relays as one or more 'default' logic schemes.

Because software is used, it is essential to carefully test the logic during commissioning to ensure correct operation. The only exception to this may be if the relevant 'default' scheme is used. Such logic schemes will have been proven during relay type testing, and so there is no need for proving tests during commissioning. However, where a customer generates the scheme logic, it is necessary to ensure that the commissioning tests conducted are adequate to prove the functionality of the scheme in all respects. A specific test procedure should be prepared, and this procedure should include:

- checking of the scheme logic specification and diagrams to ensure that the objectives of the logic are achieved
- testing of the logic to ensure that the functionality of the scheme is proven
- testing of the logic, as required, to ensure that no output occurs for the relevant input signal combinations

The degree of testing of the logic will largely depend on the criticality of the application and complexity of the logic. The responsibility for ensuring that a suitable test procedure is produced for logic schemes other than the 'default' one(s) supplied lies with the specifier of the logic. Relay manufacturers cannot be expected to take responsibility for the correct operation of logic schemes that they have not designed and supplied.

21.14 TRIPPING AND ALARM ANNUNCIATION TESTS

If primary and/or secondary injection tests are not carried out, the tripping and alarm circuits will not have been checked. Even where such checks have been carried out, CB trip coils and/or Control Room alarm circuits may have been isolated. In such cases, it is essential that all of the tripping and alarm circuits are checked.

This is done by closing the protection relay contacts manually and checking that:

- the correct circuit breakers are tripped
- the alarm circuits are energised
- the correct flag indications are given
- there is no maloperation of other apparatus that may be connected to the same master trip relay or circuit breaker

Many designs of withdrawable circuit breaker can be operated while in the maintenance position, so that substation operation can continue unaffected except for the circuit controlled by the circuit breaker involved. In other cases, isolators can be used to avoid the need for busbar de-energisation if the circuit involved is not ready for energisation.

21.15 PERIODIC MAINTENANCE TESTS

Periodic testing is necessary to ensure that a protection scheme continues to provide satisfactory performance for many years after installation. All equipment is subject to gradual degradation with time, and regular testing is intended to identify the equipment concerned so that remedial action can be taken before scheme maloperation occurs. However, due care should be taken in this task, otherwise faults may be introduced as a direct result of the remedial work.

The clearance of a fault on the system is correct only if the number of circuit breakers opened is the minimum necessary to remove the fault. A small proportion of faults are incorrectly cleared, the main reasons being:

- limitations in protection scheme design
- faulty relays
- defects in the secondary wiring
- incorrect connections
- incorrect settings
- known application shortcomings accepted as improbable occurrences
- pilot wire faults due to previous unrevealed damage to a pilot cable
- various other causes, such as switching errors, testing errors, and relay operation due to mechanical shock

The self-checking facilities of numerical relays assist in minimising failures due to faulty relays. Defects in secondary wiring and incorrect connections are virtually eliminated if proper commissioning after scheme installation/alteration is carried out. The possibility of incorrect settings is minimised by regular reviews of relay settings. Network fault levels change over time, and hence setting calculations may need to be revised. Switching and testing errors are minimised by adequate training of personnel, use of proven software, and well-designed systematic working procedures. All of these can be said to be within the control of the user.

The remaining three causes are not controllable, while two of these three are unavoidable – engineering is not science and there will always be situations that a protection relay cannot reasonably be expected to cover at an affordable cost.

21.15.1 Frequency of Inspection and Testing

Although protection equipment should be in sound condition when first put into service, problems can develop unchecked and unrevealed because of its infrequent operation. With digital and numerical relays, the in-built self-testing routines can be expected to reveal and annunciate most faults, but this does not cover any other components that, together, comprise the protection scheme. Regular inspection and testing of a protection scheme is therefore required. In practice, the frequency of testing may be limited by lack of staff or by the operating conditions on the power system.

It is desirable to carry out maintenance on protection equipment at times when the associated power apparatus is out of service. This is facilitated by co-operation between the maintenance staff concerned and the network operations control centre. Maintenance tests may sometimes have to be made when the protected circuit is on load. The particular equipment to be tested should be taken out of commission and adequate back-up protection provided for the duration of the tests. Such back-up protection may not be fully discriminative, but should be sufficient to clear any fault on the apparatus whose main protection is temporarily out of service.

Maintenance is assisted by the displays of measured quantities provided on digital and numerical relays. Incorrect display of a quantity is a clear indication that something is wrong, either in the relay itself or the input circuits.

21.15.2 Maintenance Tests

Primary injection tests are normally only conducted out during initial commissioning. If scheme maloperation has occurred and the protection relays involved are suspect, or alterations have been made involving the wiring to the relays from the VTs/CTs, the primary injection tests may have to be repeated.

Secondary injection tests may be carried out at suitable intervals to check relay performance, and, if possible, the relay should be allowed to trip the circuit breakers involved. The interval between tests will depend upon the criticality of the circuit involved, the availability of the circuit for testing and the technology of the relays used. Secondary injection testing is only necessary on the selected relay setting and the results should be checked against those obtained during the initial commissioning of the equipment.

It is better not to interfere with relay contacts at all unless they are obviously corroded. The performance of the contacts is fully checked when the relay is actuated.

Insulation tests should also be carried out on the relay wiring to earth and between circuits, using a 1000V tester. These tests are necessary to detect any deterioration in the insulation resistance.

21.16 PROTECTION SCHEME DESIGN FOR MAINTENANCE

If the following principles are adhered to as far as possible, the danger of back-feeds is lessened and fault investigation is made easier:

- test blocks should be used, to enable a test plug to be used, and a defective unit to be replaced quickly without interrupting service
- circuits should be kept as electrically separate as possible, and the use of common wires should be avoided, except where these are essential to the correct functioning of the circuits
- each group of circuits which is electrically separate from other circuits should be earthed through an independent earth link
- where a common voltage transformer or d.c. supply is used for feeding several circuits, each circuit should be fed through separate links or fuses. Withdrawal of these should completely isolate the circuit concerned
- power supplies to protection schemes should be segregated from those supplying other equipment and provided with fully discriminative circuit protection
- a single auxiliary switch should not be used for interrupting or closing more than one circuit
- terminations in relay panels require good access, as these may have to be altered if extensions are made. Modern panels are provided with special test facilities, so that no connections need be disturbed during routine testing
- junction boxes should be of adequate size and, if outdoors, must be made waterproof
- all wiring should be ferruled for identification
- electromechanical relays should have high operating and restraint torques and high contact pressures; jewel bearings should be shrouded to exclude dust and the use of very thin wire for coils and connections should be avoided. Dust-tight cases with an efficient breather are essential on these types of electromechanical element
- static, digital and numerical relays should have test facilities to assist in fault finding. The relay manual should clearly detail the expected results at each test point when healthy

Chapter 22
Power System Measurements

22.1 INTRODUCTION

The accurate measurement of the voltage, current or other parameter of a power system is a prerequisite to any form of control, ranging from automatic closed-loop control to the recording of data for statistical purposes. Measurement of these parameters can be accomplished in a variety of ways, including the use of direct-reading instruments as well as electrical measuring transducers.

There are a wide range of measurement devices used to collect data and convert it into useful information for the operator. Generally, the devices are classified as to whether the measurements are to be used locally or remotely.

Local Information

- Instruments or meters (may be directly connected, or via a transducer)
- Measurement centres

Transmitted Data

- Transducers

Remote Systems

- Measurement centres
- Disturbance recorders
- Power quality recorders (see chapter 23)

22.2 GENERAL TRANSDUCER CHARACTERISTICS

Transducers produce an accurate d.c. analogue output, usually a current, which corresponds to the parameter being measured (the measurand). They provide electrical isolation by transformers, sometimes referred to as 'Galvanic Isolation', between the input and the output. This is primarily a safety feature, but also means that the cabling from the output terminals to any receiving equipment can be lightweight and have a lower insulation specification. The advantages over discrete measuring instruments are as follows:

- mounted close to the source of the measurement, reducing instrument transformer burdens and increasing safety through elimination of long wiring runs
- ability to mount display equipment remote from the transducer

- ability to use multiple display elements per transducer
- the burden on CTs/VTs is considerably less

Outputs from transducers may be used in many ways – from simple presentation of measured values for an operator, to being utilised by a network automation scheme to determine the control strategy.

Transducers may have single or multiple inputs and/or outputs. The inputs, outputs and any auxiliary circuits will all be isolated from each other. There may be more than one input quantity and the measurand may be a function of one or more of them.

Whatever measurement transducer is being used, there will usually be a choice between discrete and modular types, the latter being plug-in units to a standard rack. The location and user-preferences will dictate the choice of transducer type.

22.2.1 Transducer Inputs

The input of a transducer is often taken from transformers and these may be of many different types. Ideally, to obtain the best overall accuracy, metering-class instrument transformers should be used since the transformer errors will be added, albeit algebraically, to the transducer errors. However, it is common to apply transducers to protection-class instrument transformers and that is why transducers are usually characterised to be able to withstand significant short-term overloads on their current inputs. A typical specification for the current input circuits of a transducer suitable for connection to protection-class instrument transformers is to withstand:

- 300% of full-load current continuously
- 2500% for three seconds
- 5000% for one second

The input impedance of any current input circuit will be kept as low as possible, and that for voltage inputs will be kept as high as possible. This reduces errors due to impedance mismatch.

22.2.2 Transducer Outputs

The output of a transducer is usually a current source. This means that, within the output voltage range (compliance voltage) of the transducer, additional display devices can be added without limit and without any need for adjustment of the transducer. The value of the compliance voltage determines the maximum loop impedance of the output circuit, so a high value of compliance voltage facilitates remote location of an indicating instrument.

Where the output loop is used for control purposes, appropriately rated Zener diodes are sometimes fitted across the terminals of each of the devices in the series loop to guard against the possibility of their internal circuitry becoming open circuit. This ensures that a faulty device in the loop does not cause complete failure of the output loop. The constant-current nature of the transducer output simply raises the voltage and continues to force the correct output signal round the loop.

22.2.3 Transducer Accuracy

Accuracy is usually of prime importance, but in making comparisons, it should be noted that accuracy can be defined in several ways and may only apply under very closely defined conditions of use. The following attempts to clarify some of the more common terms and relate them to practical situations, using the terminology given in IEC 60688.

The accuracy of a transducer will be affected, to a greater or lesser extent, by many factors, known as *influence quantities*, over which the user has little, or no, control.

Table 22.1 provides a complete list of influence quantities. The accuracy is checked under an agreed set of conditions known as *reference conditions*. The reference conditions for each of the influence quantities can be quoted as a single value (e.g. 20°C) or a range (e.g. 10-40°C).

Input current	Input voltage
Input quantity distortion	Input quantity frequency
Power factor	Unbalanced currents
Continuous operation	Output load
Interaction between measuring elements	Ambient temperature
Auxiliary supply voltage	Auxiliary supply frequency
External magnetic fields	Self heating
Series mode interference	Common mode interference
External heat	

Table 22.1: Transducer influence quantities

The error determined under reference conditions is referred to as the *intrinsic error*. All transducers having the same intrinsic error are grouped into a particular accuracy class, denoted by the *class index*. The class index is the same as the intrinsic error expressed as a percentage (e.g. a transducer with an intrinsic accuracy of 0.1% of full scale has a class index of 0.1). The class index system used in IEC 60688 requires that the variation for each of the influence quantities be strictly related to the *intrinsic error*. This means that the higher the accuracy claimed by the manufacturer, the lower must be all of the variations.

Because there are many influence quantities, the variations are assessed individually, whilst maintaining all the other influence quantities at reference conditions.

The *nominal range of use* of a transducer is the normal

operating range of the transducer as specified by the manufacturer. The nominal range of use will naturally be wider than the reference value or range. Within the nominal range of use of a transducer, additional errors accumulate resulting in an additional error. This additional error is limited for any individual influence quantity to, at most, the value of the class index.

Table 22.2 gives performance details of a typical range of transducers (accuracy class 0.5) according to the standard.

Influence Quantity	Reference Range	Max. Error-Reference Range %	Nominal Working Range	Max. Error-Nominal Range
Input current,In	In=1A, 5A 20...120%	0.5%	0-120%	0.5%
Input voltage,Vn	Vn=50...500V 80...120%	0.25%	0-120%	0.5%
Input frequency	45...65Hz	0.5%	-	-
Power factor	Cos φ=0.5...1	0.25%	Cos φ=0...1	0.5%
Unbalanced current	0...100%	0.5%	-	-
Interaction between measuring elements	Current input 0...360°	0.25%°	-	-
Continuous operation	Continuous >6h	0.5%	-	-
Self Heating	1...30min	0.5%	-	-
Output load	10...100%	0.25%	-	-
Waveform crest factor	1.41 (sine wave)	-	1.2...1.8	0.5%
Ambient temperature	0°-50° C	0.5%	-10°-60° C	1.0%
Aux. supply d.c. voltage	24...250V DC	0.25%	19V-300V	0.25%
A.C. Aux. Supply frequency, fn	90...110% fn	0.25%	-	-
External magnetic fields	0...0.4kA/m	0.5%	-	-
Output series mode interference	1V 50Hz r.m.s. in series with output	0.5%	-	-
Output common mode interference	100V 50Hz r.m.s. output to earth	0.5%	-	-

Table 22.2: Typical transducer performance

Confusion also arises in specifying the performance under real operating conditions. The output signal is often a d.c. analogue of the measurand, but is obtained from alternating input quantities and will, inevitably, contain a certain amount of alternating component, or ripple. *Ripple* is defined as the peak-to-peak value of the alternating component of the output signal although some manufacturers quote 'mean-to-peak' or 'r.m.s.' values. To be meaningful, the conditions under which the value of the ripple has been measured must be stated, e.g. 0.35% r.m.s. = 1.0% peak-to-peak ripple.

Under changing conditions of the measurand, the output signal does not follow the changes instantaneously but is time-delayed. This is due to the filtering required to reduce ripple or, in transducers using numerical technology, prevent aliasing. The amount of such a delay is called the *response time*. To a certain extent, ripple and response time are interrelated. The response time can usually be shortened at the expense of increased ripple, and vice-versa. Transducers having shorter response times than normal can be supplied for those instances where the power system suffers swings, dips, and low frequency oscillations that must be monitored.

Transducers having a current output have a maximum output voltage, known as the *compliance voltage*. If the load resistance is too high and hence the compliance voltage is exceeded, the output of the transducer is no longer accurate.

Certain transducers are characterised by the manufacturer for use on systems where the waveform is not a pure sinusoid. They are commonly referred to as '*true r.m.s. sensing*' types. For these types, the distortion factor of the waveform is an influence quantity. Other transducers are referred to as '*mean-sensing*' and are adjusted to respond to the r.m.s. value of a pure sine wave. If the input waveform becomes distorted, errors will result. For example, the error due to third harmonic distortion can amount to 1% for every 3% of harmonic.

Once installed, the user expects the accuracy of a transducer to remain stable over time. The use of high quality components and conservative power ratings will help to ensure long-term stability, but adverse site conditions can cause performance changes which may need to be compensated for during the lifetime of the equipment.

22.3 DIGITAL TRANSDUCER TECHNOLOGY

Digital power system transducers make use of the same technology as that described for digital and numerical relays in Chapter 7. The analogue signals acquired from VTs and CTs are filtered to avoid aliasing, converted to digital form using A/D conversion, and then signal processing is carried out to extract the information required. Basic details are given in Chapter 7. Sample rates of 64 samples/cycle or greater may be used, and the accuracy class is normally 0.2 or 0.5.

Outputs may be both digital and analogue. The analogue outputs will be affected by the factors influencing accuracy as described above. Digital outputs are typically in the form of a communications link, with RS232 or RS485 serial, and RJ45 Ethernet connections commonly available. The response time may suffer compared to analogue transducers, depending on the rate at which values are transferred to the communications link and the delay in processing data at the receiving end. In fact, all of the influence quantities that affect a traditional analogue transducer also are present in a digital transducer in some form, but the errors resulting may be much less than in an analogue transducer and it may be more stable over a long period of time.

The advantages of a transducer using numerical technology are:

- improved long-term stability
- more accurate r.m.s measurements
- improved communications facilities
- programmability of scaling
- wider range of functions
- reduced size

The improved long term stability reduces costs by extending the intervals between re-calibration. More accurate r.m.s measurements provide the user with data of improved accuracy, especially on supplies with significant harmonic content. The improved communications facilities permit many transducers to share the same communications link, and each transducer to provide several measurements. This leads to economy in interconnecting wiring and number of transducers used. Remote or local programmable scaling of the transducer permits scaling of the transducer in the field. The scaling can be changed to reflect changes in the network, or to be re-used elsewhere. Changes can be downloaded via the communications link, thus removing the need for a site visit. It also minimises the risk of the user specifying an incorrect scaling factor and having to return the transducer to the manufacturer for adjustment. Suppliers can keep a wider range of transducers suitable for a wide range of applications and inputs in stock, thus reducing delivery times. Transducers are available with a much wider range of functions in one package, thus reducing space requirements in a switchboard. Functions available include harmonics up to the 31st, energy, and maximum demand information. The latter are useful for tariff negotiations.

22.4 ANALOGUE TRANSDUCER TECHNOLOGY

All analogue transducers have the following essential features:

- an input circuit having impedance Z_{in}
- isolation (no electrical connection) between input and output
- an ideal current source generating an output current, I_1, which is an accurate and linear function of Q_{in}, the input quantity
- a parallel output impedance, Z_o. This represents the actual output impedance of the current source and shunts a small fraction, I_2, of the ideal output
- an output current, I_o, equal to $(I_1 - I_2)$

These features are shown diagrammatically in Figure 22.1.

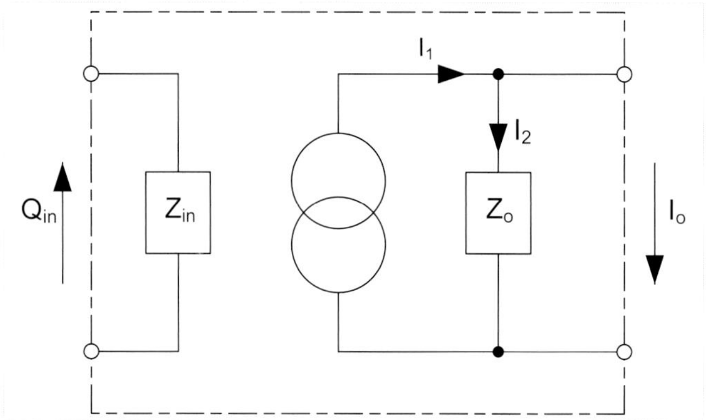

Figure 22.1: Schematic of an analogue transducer

Output ranges of 0-10mA, 0-20mA, and 4-20mA are common. Live zero (e.g. 4-20mA), suppressed zero (e.g. 0-10mA for 300-500kV) and linear inverse range (e.g. 10-0mA for 0-15kV) transducers normally require an auxiliary supply. The dual-slope type has two linear sections to its output characteristic, for example, an output of 0-2mA for the first part of the input range, 0-8kV, and 2-10mA for the second part, 8-15kV.

22.5 TRANSDUCER SELECTION

The selection of the correct transducer to perform a measurement function depends on many factors. These are detailed below.

22.5.1 Current Transducers

Current transducers are usually connected to the secondary of an instrument current transformer with a rated output of 1 or 5 amps. Mean-sensing and true r.m.s. types are available. If the waveform contains significant amounts of harmonics, a true r.m.s sensing type must be used for accurate measurement of the input. They can be self-powered, except for the true r.m.s. types, or when a live zero output (for example 4-20mA) is required. They are not directional and, therefore, are unable to distinguish between 'export' and 'import' current. To obtain a directional signal, a voltage input is also required.

22.5.2 Voltage Transducers

Connection is usually to the secondary of an instrument voltage transformer but may be direct if the measured quantity is of sufficiently low voltage. The suppressed zero type is commonly used to provide an output for a specific range of input voltage where measurement of zero on the input quantity is not required. The linear inverse type is often used as an aid to synchronising.

22.5.3 Frequency

Accurate measurement of frequency is of vital importance to transmission system operators but not quite so important, perhaps, for the operator of a diesel generator set. Accuracy specifications of 0.1% and 0.01% are available, based on percent of centre scale frequency. This means, for example that a device quoted as 0.1% and having a centre scale value of 50Hz will have a maximum error of +/- 50mHz under reference conditions.

22.5.4 Phase Angle

Transducers for the measurement of phase angle are frequently used for the display of power factor. This is achieved by scaling the indicating instrument in a non-linear fashion, following the cosine law. For digital indicators and SCADA equipment, it is necessary for the receiving equipment to provide appropriate conversions to achieve the correct display of power factor. Phase angle transducers are available with various input ranges. When the scaling is $-180^{\circ}...0^{\circ}...180^{\circ}$, there is an ambiguous region, of about +/-2° at the extremes of the range. In this region, where the output is expected to be, for example, −10mA or +10mA, the output may jump sporadically from one of the full-scale values to the other. Transducers are also available for the measurement of the angle between two input voltages. Some types of phase angle transducer use the zero crossing point of the input waveform to obtain the phase information and are thus prone to error if the input contains significant amounts of harmonics.

Calculating the power factor from the values of the outputs of a watt and a var transducer will give a true measurement in the presence of harmonics.

22.5.5 Power Quantities

The measurement of active power (watts) and reactive power (vars) is generally not quite as simple as for the other quantities. More care needs to be taken with the selection of these types because of the variety of configurations. It is essential to select the appropriate type for the system to be measured by taking into account factors such as system operating conditions (balanced or unbalanced load), the number of current and voltage connections available and whether the power flow is likely to be 'import', 'export', or both. The range of the measurand will need to encompass all required possibilities of over-range under normal conditions so that the transducer and its indicating instrument, or other receiving equipment, is not used above the upper limit of its effective range. Figure 22.2 illustrates the connections to be used for the various types of measurement.

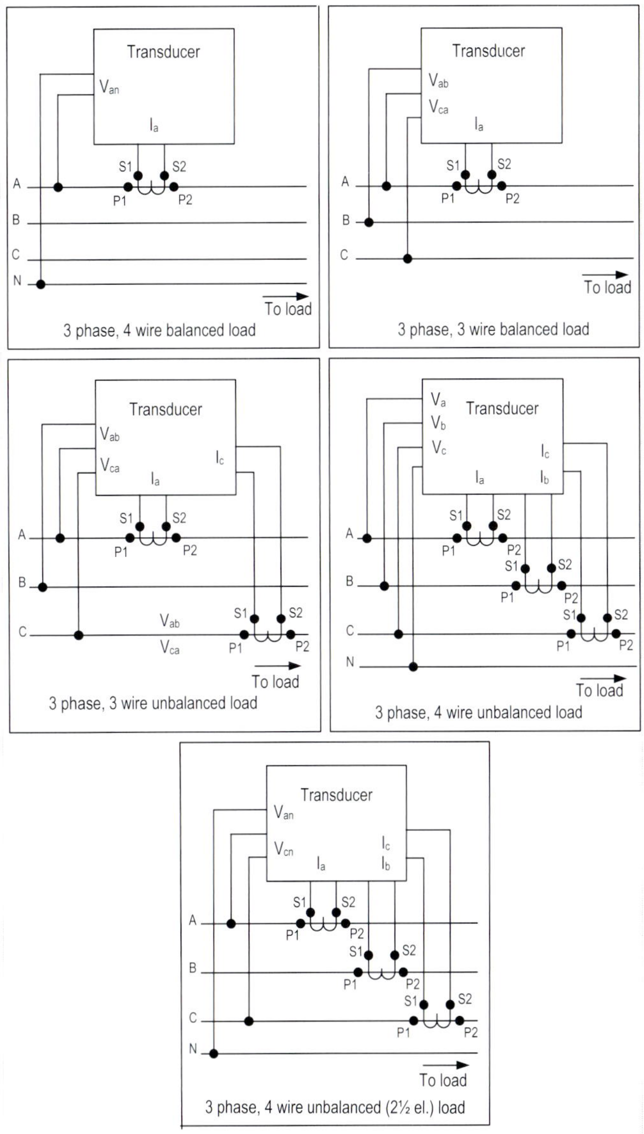

Figure 22.2: Connections for 3-phase watt/VAr transducers

22.5.6 Scaling

The relationship of the output current to the value of the measurand is of vital importance and needs careful consideration. Any receiving equipment must, of course, be used within its rating but, if possible, some kind of standard should be established.

As an example, examine the measurement of a.c. voltage. The primary system has a nominal value of 11kV and the transformer has a ratio of 11kV/110V. To specify the conversion coefficient for the voltage transducer to be 110V/10mA would not necessarily be the optimum. One of the objectives must be to have the capability of monitoring the voltage over a range of values so an upper limit must be selected – for instance +20%, or 132V. Using the original conversion coefficient, the maximum output of the transducer is required to be 12mA. This is within the capability of most 0-10mA transducers, the majority of which can accommodate an over-range of 25%, but it does mean any associated analogue indicating instrument must have a sensitivity of 12mA. However, the scale required on this instrument is now 0-13.2kV, which may lead to difficulty in drawing the scale in such a way as to make it readable (and conforms to the relevant standard). In this example, it would be more straightforward to establish the full-scale indication as 15kV and to make this equivalent to 10mA, thus making the specification of the display instrument much easier. The transducer will have to be specified such that an input of 0-150V gives an output of 0-10mA. In the case of transducers with a 4-20mA output, great care is required in the output scaling, as there is no over-range capability. The 20mA output limit is a fixed one from a measurement point of view. Such outputs are typically used as inputs to SCADA systems, and the SCADA system is normally programmed to assume that a current magnitude in excess of 20mA represents a transducer failure. In addition, a reading below 4mA also indicates a failure, usually an open circuit in the input connection. Thus, using the above example, the output might be scaled so that 20mA represents 132V and hence the nominal 110V input results in an output of 16.67mA. A more convenient scaling might be to use 16mA as representing 110V, with 20mA output being equal to 137.5V (i.e. 25% over-range instead of the 20% required). It would be incorrect to scale the transducer so that 110V input was represented by 20mA output, as the over-range capability required would not be available.

Similar considerations apply to current transducers and, with added complexity, to watt transducers, where the ratios of both the voltage and the current transformers must be taken into account. In this instance, the output will be related to the primary power of the system. It should be noted that the input current corresponding to full-scale output may not be exactly equal to the secondary rating of the current transformer but this does not matter - the manufacturer will take this into account.

Some of these difficulties do not need to be considered if the transducer is only feeding, for example, a SCADA outstation. Any receiving equipment that can be programmed to apply a scaling factor to each individual input can accommodate most input signal ranges. The main consideration will be to ensure that the transducer is capable of providing a signal right up to the full-scale value of the input, that is, it does not saturate at the highest expected value of the measurand.

22.5.7 Auxiliary Supplies

Some transducers do not require any auxiliary supply. These are termed 'self-powered' transducers. Of those that do need a separate supply, the majority have a biased, or live zero output, such as 4-20mA. This is because a non-zero output cannot be obtained for zero input unless a separate supply is available. Transducers that require an auxiliary supply are generally provided with a separate pair of terminals for the auxiliary circuit so that the user has the flexibility of connecting the auxiliary supply input to the measured voltage, or to a separate supply. However, some manufacturers have standardised their designs such that they appear to be of the self-powered type, but the auxiliary supply connection is actually internal. For a.c. measuring transducers, the use of a d.c. auxiliary supply enables the transducer to be operated over a wider range of input.

The range of auxiliary supply voltage over which a transducer can be operated is specified by the manufacturer. If the auxiliary voltage is derived from an input quantity, the range of measurement will be restricted to about +/-20% of the nominal auxiliary supply voltage. This can give rise to problems when attempting to measure low values of the input quantity.

22.6 MEASUREMENT CENTRES

A Measurement Centre is different from a transducer in three ways.

- It can measure a large number of instantaneous parameters, and with an internal clock, calculate time-based parameters such as maximum demand
- It has many different forms of communication to transmit the data ranging from simple pulsed contacts to multiple digital communication ports
- It has a local display so that information, system status and alarms can be displayed to the operator.

This is largely impractical if analogue technology for signal processing is used, but no such limitation exists if digital or numerical technology is adopted. Therefore, Measurement Centres are generally only found implemented using these technologies. As has been already noted in Chapter 7, a numerical relay can provide many measurements of power system quantities. Therefore, an alternative way of looking at a Measurement Centre that uses numerical technology is that it is a numerical relay, stripped of its protection functions and incorporating a wide range of power system parameter measurements.

This is rather an oversimplification of the true situation, as there are some important differences. A protection relay has to provide the primary function of protection over a very large range of input values; from perhaps 5% to 500% or greater of rated values. The accuracy of measurement, whilst important, is not required to be as accurate as, for instance, metering for tariff purposes. Metering does not have to cover quite such a wide range of input values, and therefore the accuracy of measurement is often required to be higher than for a protection relay. Additional functionality over that provided by the measurement functions of a protection relay is often required – for a typical set of functions provided by a measurement centre, see Table 22.3.

On the other hand, the fundamental measurement process in a measurement centre based on numerical technology is identical to that of a numerical relay, so need not be repeated here. The only differences are the ranges of the input quantities and the functionality. The former is dealt with by suitable design of the input signal conditioning and A/D conversion, the latter is dealt with by the software provided.

R.M.S. line currents	R.M.S. line voltages
Neutral current	R.M.S. phase voltages
Average current	Average voltage
Negative sequence voltage	Negative sequence current
Power (each phase and total)	Reactive Power (each phase and total)
Apparent Power (each phase and total)	Power factor (each phase and total)
Phase angle (voltage/current) – each phase	Demand time period
Demand current in period	Demand power in period
Demand reactive power in period	Demand VA in period
Demand power factor in period	Maximum demand current (each phase and total) since reset
Maximum demand (W and var) since reset	Energy, Wh
Energy, varh	Frequency
Individual harmonics (to 31st)	%THD (voltage) – each phase and total
%THD (current) – each phase and total	Programmable multiple analogue outputs

Table 22.3: Typical function set provided by a Measurement Centre

The advantages of a Measurement Centre are that a comprehensive set of functions are provided in a single item of equipment, taking up very little extra space compared to a discrete transducer for a much smaller number of parameters. Therefore, when the requisite CTs and VTs are available, it may well make sense to use a Measurement Centre even if not all of the functionality is immediately required. History shows that more and more data is required as time passes, and incorporation of full functionality at the outset may make sense. Figure 22.3 illustrates a typical transducer.

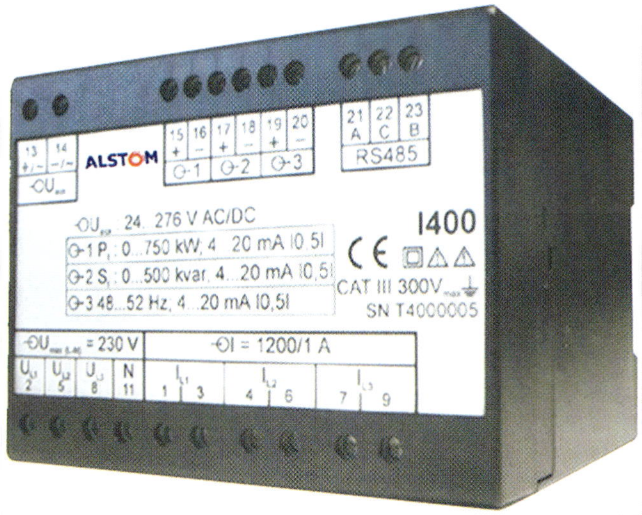

Figure 22.3: Typical transducer

22.7 TARIFF METERING

Tariff metering is a specialised form of measurement, being concerned with the measurement of electrical power, reactive power or energy for the purposes of charging the consumer. As such, it must conform to the appropriate national standards for such matters. Primary tariff metering is used for customer billing purposes, and may involve a measurement accuracy of 0.2% of reading, even for readings that are 5% or less of the nominal rated value. Secondary tariff metering is applied where the user wishes to include his own metering as a check on the primary tariff metering installed by the supplier, or within a large plant or building to gain an accurate picture of the consumption of energy in different areas, perhaps for the purpose of energy audits or internal cost allocation. The accuracy of such metering is rather less, an overall accuracy of 0.5% over a wide measurement range being typically required. As this is the overall accuracy required, each element in the metering chain (starting with the CTs/VTs) must have an accuracy rather better than this. Careful attention to wiring and mounting of the transducer is required to avoid errors due to interference, and the accuracy may need to be maintained over a fairly wide frequency range. Thus a tariff metering scheme requires careful design of all of the equipment included in the scheme. Facilities are normally included to provide measurements over a large number of defined time periods (e.g. 24 half-hour periods for generator energy tariff metering) so that the exporter of the energy can produce an overall invoice for the user according to the correct rates for each tariff period. The time intervals that these periods cover may change according to the time of year (winter, spring, etc.) and therefore flexibility of programming of the energy metering is required. Remote communications are invariably required, so that the data is transferred to the relevant department on a regular basis for invoicing purposes.

For primary tariff metering, security of information is a major design factor. Meters will also have tamper-proof physical indication.

22.8 SYNCHRONISERS

Synchronisers are required at points on a power system where two supplies (either generator and grid, or two grid supplies) may need to be paralleled. They are more than just a measuring device, as they will provide contact closures to permit circuit breaker closing when conditions for paralleling (synchronising) are within limits. However, they are not regarded as protection relays, and so are included in this Chapter for convenience. There are two types of synchronisers - auto-synchronisers and check synchronisers.

22.8.1 Check Synchronisers

The function of a check synchroniser is to determine if two voltages are in synchronism, or nearly so, and provide outputs under these conditions. The outputs are normally in the form of volt-free contacts, so that they may be used in CB control circuits to permit or block CB closing. When applied to a power system, the check synchroniser is used to check that it is safe to close a CB to connect two independent networks together, or a generator to a network, as in Figure 22.4. In this way, the check synchroniser performs a vital function in blocking CB closure when required.

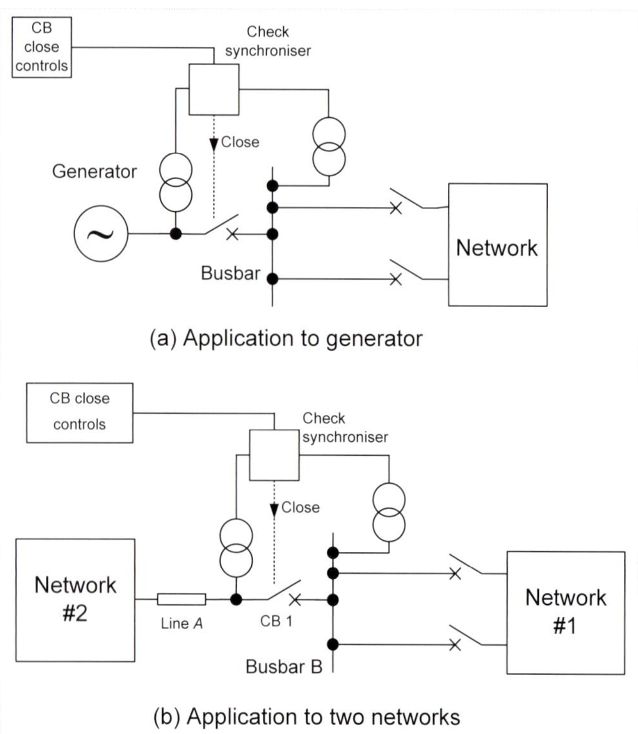

Figure 22.4: Check synchroniser applications

Synchronism occurs when two a.c. voltages are of the same frequency and magnitude, and have zero phase difference. The check synchroniser, when active, monitors these quantities and enables CB close circuits when the differences are within pre-set limits. While CB closure at the instant of perfect synchronism is the ideal, this is very difficult to obtain in practice and some mismatch in one or more of the monitored quantities can be tolerated without leading to excessive current/voltage transients on CB closure. The check synchroniser has programmable error limits to define the limits of acceptability when making the comparison.

The conditions under which a check synchroniser is required to provide an output are varied. Consider the situation of a check synchroniser being used as a permissive device in the closing control circuit of a CB that couples two networks together at a substation. It is not sufficient to assume that both networks will be live, situations where either Line *A* or Busbar *B* may be dead may have to be considered, leading to the functionality shown in Table 22.4(a).

(a) Check synchroniser functionality	
Live bus/live line synchronising	Live bus/dead line synchronising
Dead bus/live line synchronising	Network supply voltage #1 deviation from nominal
Network supply voltage #2 deviation from nominal	Voltage difference within limits
Frequency difference within limits	Phase angle difference within limits
CB closing advance time	CB closing pulse time
Maximum number of synchronising attempts	
(b) Additional functions for auto-synchroniser	
Incoming supply frequency deviation from nominal	Incoming supply voltage raise/lower signal
Incoming supply voltage raise/lower mode (pulse/continuous)	Incoming supply frequency raise/lower mode (pulse/continuous)
Incoming supply voltage setpoint	Incoming supply frequency setpoint
Voltage raise/lower pulse time	Frequency raise/lower pulse time

Table 22.4: Synchroniser function set

When the close signal is permitted, it may be given only for a limited period of time, to minimise the chances of a CB close signal remaining after the conditions have moved outside of limits. Similarly, circuits may also be provided to block closure if the CB close signal from the CB close controls is present prior to satisfactory conditions being present – this ensures that an operator must be monitoring the synchronising displays and only initiating closure when synchronising conditions are correct, and also detects synchronising switch contacts that have become welded together.

A check synchroniser does not initiate any adjustments if synchronising conditions are not correct, and therefore acts only as a permissive control in the overall CB closing circuit to provide a check that conditions are satisfactory. In a substation, check-synchronisers may be applied individually to all required CBs. Alternatively, a reduced number may be installed, together with suitable switching arrangements in the signal input/output circuits so that a single device may be selected to cover several CBs.

22.8.2 Auto-synchroniser

An auto-synchroniser contains additional functionality compared to a check synchroniser. When an auto-synchroniser is placed in service, it measures the frequency and magnitude of the voltages on both sides of the circuit breaker, and automatically adjusts one of the voltages if conditions are not correct. Application of auto-synchronisers is normally restricted to generators – i.e. the situation shown in Figure 22.4(a), replacing the check synchroniser with an auto-synchroniser. This is because it is generally not possible to adjust either of the network voltages by changing the settings of one or a very few equipments in a network. When applied to a generator, it is relatively easy to adjust the frequency and magnitude of the generated voltage by transmitting signals to the Governor and AVR respectively.

The auto-synchroniser will check the voltage of the incoming generator against the network voltage for compliance with the following:

a. slip frequency within limits (i.e. difference in frequency between the generator and network)
b. phase difference between the voltages within limits
c. voltage magnitude difference within limits

The CB close command is issued automatically when all three conditions are satisfied. Checks may also be made that the network frequency and voltage is within pre-set limits, and if not the synchronising sequence is locked out. This prevents synchronising under unusual network conditions, when it may not be desirable. This facility should be used with caution, since under some emergency conditions, it could block the synchronising of a generator that was urgently required in service to help assist in overcoming the condition.

If (a) above is not within limits, signals are sent automatically to the governor of the generating set to adjust the speed setpoint appropriately. In the case of (c) not in limits, similar signals are sent to the Automatic Voltage Regulator to raise or lower the setpoint. The signals are commonly in the form of pulses to raise or lower the setpoint, but could be continuous signals if that is what the particular equipment requires. It is normal for the speed and voltage of the generator to be slightly higher than that of the network, and this can be accommodated either by initial settings on the Governor/AVR or by providing setpoint values in the synchroniser. This ensures stable synchronising and export of power at lagging power factor to the network by the generator after CB closure. The possibility of tripping due to reverse/low forward power conditions and/or field failure/under-excitation is avoided. Use of an auto-synchroniser also helps avoid human error if manual synchronising were employed – there is potential for damage to equipment, primarily the generator, if synchronising outside of permitted limits occurs.

To ensure that the CB is closed at the correct instant, the CB close time is normally a required data item. The auto-synchroniser calculates from a knowledge of this and the slip frequency the correct time in advance of phase co-incidence to issue the CB close command. This ensures that the CB closes as close to the instant of phase co-incidence as possible. Upon receipt of the signal indicating 'CB closed' a further signal to raise frequency may be sent to the governor to ensure stable export of power is achieved. Conversely, failure of the CB to close within a set time period will reset the auto-synchroniser, ready for another attempt, and if further attempts are still unsuccessful, the auto-synchroniser will lock out and raise an alarm.

Practice in respect of fitting of auto-synchronisers varies widely between Utilities. Where policy is flexible, it is most common when the time to synchronise is important – i.e. emergency standby and peak lopping sets. Many Utilities still relay on manual synchronising procedures. It is also possible for both an auto-synchroniser and check-synchroniser to be fitted in series. This provides protection against internal failure of the auto-synchroniser leading to a CB close command being given incorrectly.

22.9 DISTURBANCE RECORDERS

Power systems suffer from various types of disturbances. In post-fault analysis, it is beneficial to have a detailed record of a disturbance to enable the initiating event to be distinguished from the subsequent effects. Especially where the disturbance causes further problems (e.g. single-phase fault develops into 3-phase), a detailed recording of the fault may be required to distinguish between cause and effect. If the effects of a fault are spread over a wide area, records of the disturbance from a number of locations can assist in determining the location of the disturbance. The equipment used for this purpose is known as a disturbance, or fault, recorder.

22.9.1 Disturbance Recorder Features

A disturbance recorder will normally have the following capabilities:

- multi-channel analogue input waveform recording
- multi-channel digital input recording

- storage of several fault records, ready for download/analysis
- recording time of several seconds per disturbance
- triggering from any analogue or digital input channel, or quantity derived from a combination of inputs, or manually
- distance to fault location for one or more feeders
- variable pre/post trigger recording length
- time synchronisation (IRIG-B, GPS, etc.)
- programmable sampling rates
- standard data transfer formats (IEEE COMTRADE (IEC 60255-24), etc.
- communication links to control centre (Ethernet, modem, etc.)
- self-monitoring/diagnostics

Analogue channels are provided to record the important currents and voltages at the fault recorder location. High resolution is required to ensure accurate capture of the waveforms, with 16 bit A/D conversion being usual. Digital inputs are provided to capture signals such as CB opening, protection relay operation, intertrip signals, etc. so that a complete picture of the sequence of events can be built up. The information can then be used to check that the sequence of operations post-fault is correct, or assist in determining the cause of an unexpected sequence of operations. To avoid loss of the disturbance data, sufficient memory is provided to capture and store the data from several faults prior to transfer of the data for analysis. Flexibility in the triggering arrangements is extremely important, as it is pointless to install a disturbance recorder, only for it to miss recording events due to lack of appropriate triggering facilities. It is normal for triggering to be available if the relevant threshold is crossed on any analogue or digital channel, or a quantity derived from a combination of the inputs.

Power system disturbances may last from periods of a few seconds to several minutes. To ensure that maximum benefit is obtained from the investment, a disturbance recorder must be able to capture events over a wide range of timescales. This leads to the provision of programmable sampling rates, to ensure that short-term transients are captured with sufficient resolution while also ensuring that longer-term ones have sufficient of the transient captured to enable a meaningful analysis to be undertaken. The record for each disturbance is divided into sections covering pre-fault, fault, and post–fault periods, and each of these periods may have different sampling rates. Time synchronisation is also a vital feature, to enable a recording from one recorder to be aligned with another of the same event from a different recorder to obtain a complete picture of events.

Since most disturbance recorders are fitted in substations that are normally unmanned, the provision to download captured information is essential. Each fault recording will contain a large amount of data, and it is vital that the data is uniquely identified in respect of recorder, fault event, channel, etc. Standards exist in field to facilitate the interchange of data, of which perhaps the best known is the IEEE COMTRADE format, now also an international standard (IEC 60255-24). Once downloaded, the data from a disturbance recorder can be analysed by various PC software packages. The software will often have the ability to perform harmonic and other analysis.

Chapter 23
Power Quality

23.1 INTRODUCTION

Over the last forty years or so, the amount of equipment containing electronics has increased dramatically. Such equipment can both cause and be affected by electromagnetic disturbances. A disturbance that affects a process control computer in a large industrial complex could easily result in shutdown of the process. The lost production and product recycling during start-up represents a large cost to the business. Similarly, a protection relay affected by a disturbance through conduction or radiation from nearby conductors could trip a feeder or substation, causing loss of supply to a large number of consumers. At the other end of the scale, a domestic user of a PC has to re-boot the PC due to a transient voltage dip, causing annoyance to that and other similarly affected users. Therefore, transporters and users of electrical energy have become much more interested in the nature and frequency of disturbances in the power supply. The topic has become known by the title of *Power Quality*.

23.2 CLASSIFICATION OF POWER SYSTEM DISTURBANCES

To make the study of Power Quality problems useful, the various types of disturbances need to be classified by magnitude and duration. This is especially important for manufacturers and users of equipment that may be at risk. Manufacturers need to know what is expected of their equipment, and users, through monitoring, can determine if an equipment malfunction is due to a disturbance or problems within the equipment itself. Not surprisingly, standards have been introduced to cover this field. They define the types and sizes of disturbance, and the tolerance of various types of equipment to the possible disturbances that may be encountered. The principal standards in this field are IEC 61000, EN 50160, and IEEE 1159. Standards are essential for manufacturers and users alike, to define what is reasonable in terms of disturbances that might occur and what equipment should withstand.

Table 23.1 provides a broad classification of the disturbances that may occur on a power system, some typical causes of them and the potential impact on equipment. From this Table, it will be evident that the electricity supply waveform, often thought of as composed of pure sinusoidal quantities, can suffer a wide variety of disturbances. The following sections of this Chapter describe the causes in more detail, along with methods of measurement and possible remedial measures.

Category	Causes	Impacts
Voltage dips	Local and remote faults Inductive loading Switch on of large loads	Tripping of sensitive equipment Resetting of control systems Motor stalling/tripping
Voltage surges	Capacitor switching Switch off of large loads Phase faults	Tripping of sensitive equipment Damage to insulation and windings Damage to power supplies for electronic equipment
Overvoltage	Load switching Capacitor switching System voltage regulation	Problems with equipment that requires constant steady-state voltage
Harmonics	Industrial furnaces Non-linear loads Transformers/generators Rectifier equipment	Mal-operation of sensitive equipment and relays Capacitor fuse or capacitor failures Telephone interference
Power frequency variation	Loss of generation Extreme loading conditions	Negligible most of time Motors run slower De-tuning of harmonic filters
Voltage fluctuation	AC motor drives Inter-harmonic current components Welding and arc furnaces	Flicker in: Fluorescent lamps Incandescent lamps
Rapid voltage change	Motor starting Transformer tap changing	Light flicker Tripping of equipment
Voltage imbalance	Unbalanced loads Unbalanced impedances	Overheating in motors/generators Interruption of 3-phase operation
Short and long voltage interruptions	Power system faults Equipment failures Control malfunctions CB tripping	Loss of supply to customer equipment Computer shutdowns Motor tripping
Undervoltage	Heavy network loading Loss of generation Poor power factor Lack of var support	All equipment without backup supply facilities
Transients	Lightning Capacitive switching Non –linear switching loads System voltage regulation	Control system resetting Damage to sensitive electronic components Damage to insulation

Table 23.1: Power quality issues

Table 23.2 lists the limits given in Standard EN 50160 and notes where other standards have similar limits.

Type of disturbance	Voltage Level	Limits from EN50160	Measurement period	Typical duration	Other applicable standards
Voltage Variation	230V	+/- 10%	95% of 1 week	-	
Voltage Dips	230V	<10%<90%	10-1000/year	10ms –1sec	IEEE 1159
Rapid voltage changes	230V	+/-5% to +/-10%	Several per day	Short duration	
Rapid voltage changes	1kV-35kV	<6%	Per day	Short duration	IEEE 1159
Short Interruptions	230V	>99%	20-200 per year	Up to 3 mins	EN61000-4-11
Long Interruptions	230V	>99%	10-50 per year	>3 mins	IEEE 1159
Transient Overvoltage	230V	Generally <6kV	Not specified	<1ms	IEEE 1159
Voltage unbalance	230V				
Undervoltage	230V	<-10%	Not specified	>1 min	IEEE 1159
Voltage surge	230V	<150% of nominal voltage	Not specified	>200ms	IEEE 1159
Voltage fluctuations	230V	3%	10 min	<200ms	IEC 60827
Frequency variation		+/- 1%	95% of 1 week	Not specified	Measured over 10s
Frequency variation		+4%, -6%	100% of 1 week	Not specified	Measured over 10s
Harmonics		THD<8% up to 40th	95% of 1 week	Not specified	

Table 23.2: Power system disturbance classification to EN 50160

For computer equipment, a common standard that manufacturers use is the ITI (Information Technology Industry) curve, illustrated in Figure 23.1. Voltage disturbances that lie in the area indicated as 'safe' should not cause a malfunction in any way. However, some disturbances at LV levels that lie within the boundaries defined by EN50160 might cause a malfunction because they do not lie in the safe area of the ITI curve. It may be necessary to check carefully which standards are applicable when considering equipment susceptibility.

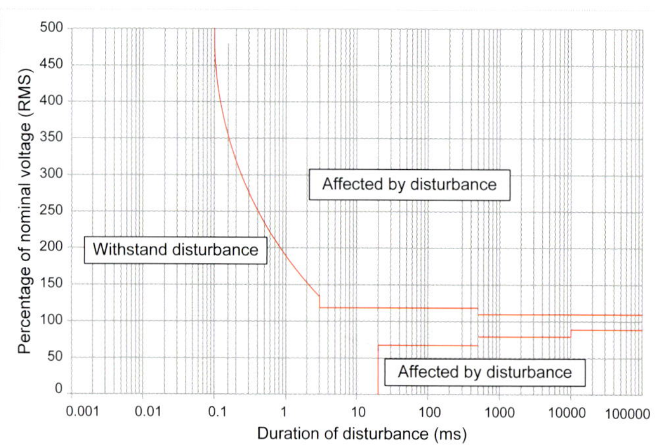

Figure 23.1: ITI curve for equipment susceptibility

23.3 CAUSES AND IMPACT OF POWER QUALITY PROBLEMS

Each of the Power Quality disturbance categories detailed in Table 23.1 is now examined in more detail as to the possible causes and the impact on consumers.

23.3.1 Voltage Dips

Figure 23.2 shows the profile of a voltage dip, together with the associated definitions. The major cause of voltage dips on a supply system is a fault on the system, that is sufficiently remote electrically that a voltage interruption does not occur. Other sources are the starting of large loads (especially common in industrial systems), and, occasionally, the supply of large inductive loads.

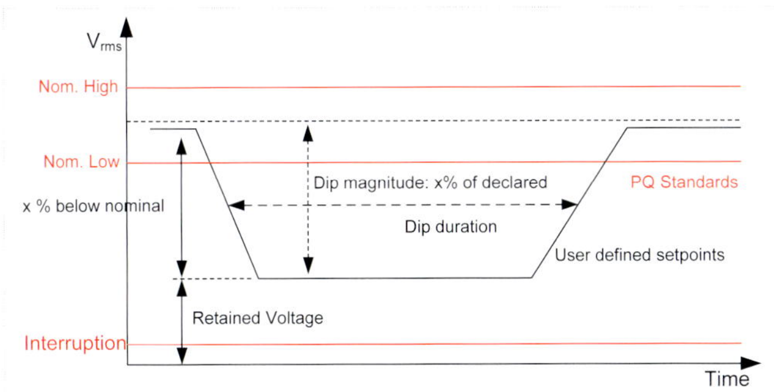

Figure 23.2: Voltage dip profile

Voltage dips due to the latter are usually due to poor design of the network feeding the consumer. A voltage dip is the most common supply disturbance causing interruption of production in an industrial plant. Faults on a supply network will always occur, and in industrial systems, it is often practice to specify equipment to ride-through voltage dips of up to 0.2s. The most common exception is contactors, which may well drop out if the voltage dips below 80% of rated voltage for more than 50-100ms. Motor protection relays that have an undervoltage element setting that is too sensitive is another cause. Since contactors are commonly used in circuits supplying motors, the impact of voltage dips on motor drives, and hence the process concerned, requires consideration. Other network-related fault causes are weather–related (such as snow, ice, wind, salt spray, dust) causing insulator flashover, collisions due to birds, and excavations damaging cables. Multiple voltage dips, as illustrated in Figure 23.3, cause more problems for equipment than a single isolated dip.

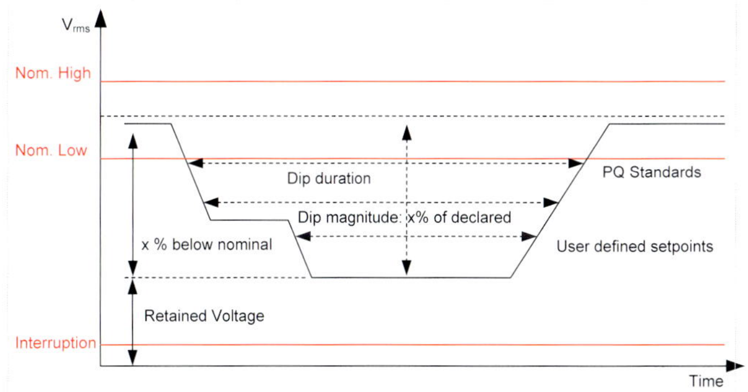

Figure 23.3: Multiple voltage dip

The impact on consumers may range from the annoying (non-periodic light flicker) to the serious (tripping of sensitive loads and stalling of motors). Where repeated dips occur over a period of several hours, the repeated shutdowns of equipment can give rise to serious production problems. Figure 23.4 shows an actual voltage dip, as captured by a Power Quality recorder.

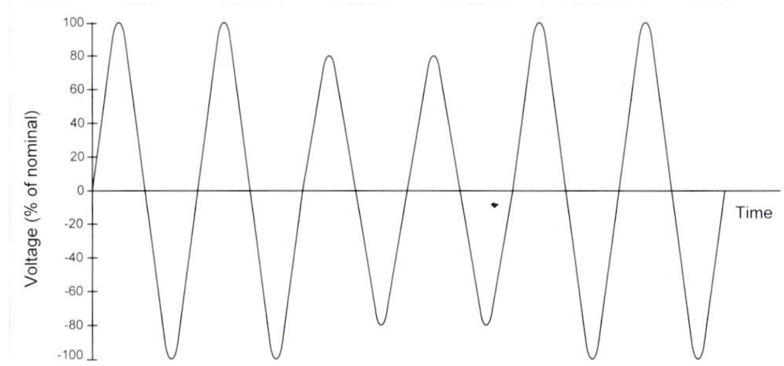

Figure 23.4: Recording of a voltage dip

Typical data for undervoltage disturbances on power systems during evolving faults are shown in Figure 23.5. Disturbances that lie in the front right-hand portion of the histogram are the ones that cause most problems, but fortunately these are quite rare.

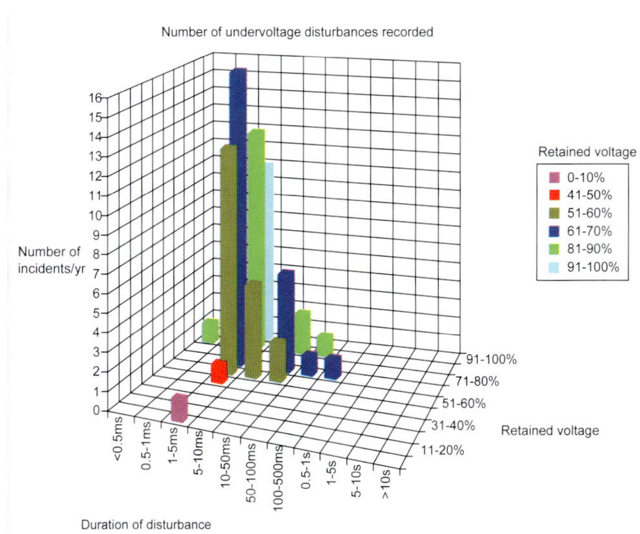

Figure 23.5: Undervoltage disturbance histogram

23.3.2 Voltage Surges/Spikes

Voltage surges/spikes are the opposite of dips – a rise that may be nearly instantaneous (spike) or takes place over a longer duration (surge). These are most often caused by lightning strikes and arcing during switching operations on circuit breakers/contactors (fault clearance, circuit switching, especially switch-off of inductive loads). Figure 23.6 shows the profile of a voltage surge.

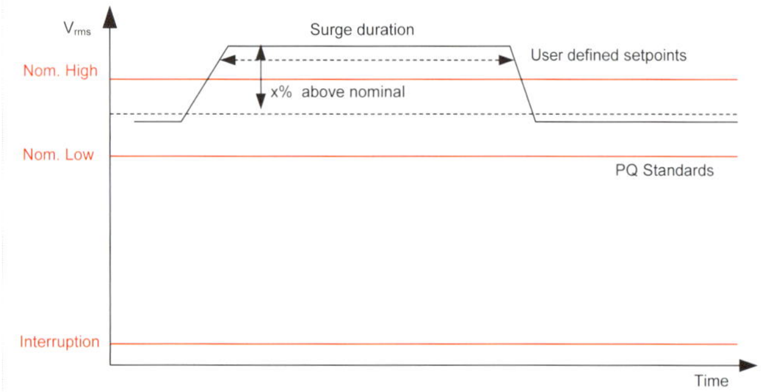

Figure 23.6: Voltage surge profile

Equipment may suffer serious damage from these causes, ranging from insulation damage to destruction of sensitive electronic devices. The damage may be immediate and obvious by the fact that equipment stops working, through to failure at a much later date from deterioration initiated from a surge or spike of voltage. These latter failures are very difficult to distinguish from random failures due to age, minor manufacturing defects, etc.

23.3.3 Overvoltages

Sustained overvoltages are not common. The most likely causes are maladjusted voltage regulators on generators or on-load tap changers, or incorrectly set taps on fixed-tap transformers. Equipment failures may immediately result in the case of severe overvoltages, but more likely is accelerated degradation leading to premature failure without obvious cause. Some equipment that is particularly sensitive to overvoltages may have to be shut down by protective devices.

23.3.4 Harmonics

This is a very common problem in the field of Power Quality. The main causes are Power Electronic Devices, such as rectifiers, inverters, UPS systems, static var compensators, etc. Other sources are electric discharge lamps, arc furnaces and arc welders. In fact, any non-linear load will be a source of harmonics. Figure 23.7 illustrates a supply waveform that is distorted due to the presence of harmonics.

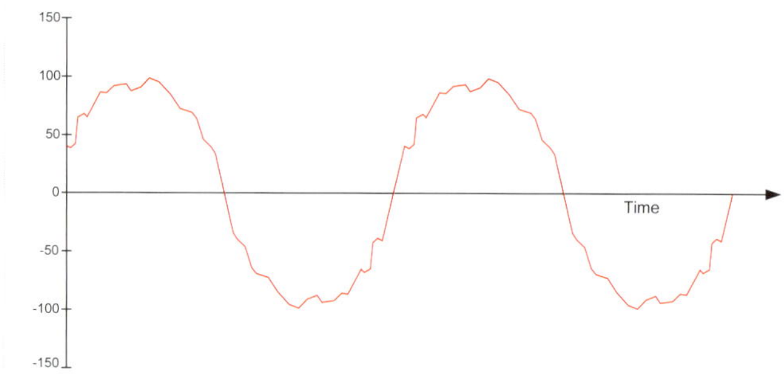

Figure 23.7: Supply waveform distorted due to the presence of harmonics

Harmonics usually lead to heating in rotating equipment (generators and motors), and transformers, leading to possible shutdown. Capacitors may be similarly affected. If harmonic levels are sufficiently high enough, protective devices may shut the equipment down to avoid damage. Some equipment, such as certain protection devices, may maloperate and cause unnecessary shutdowns. Special provision may have to be made to filter harmonics from the measured signals in these circumstances. Interference may be caused to communication systems. Overloading of neutral conductors in LV systems has also occurred (the harmonics in each phase summing in the neutral conductor, not cancelling) leading to failure due to overheating. This is a particular risk in buildings that have a large number of PCs etc. In such cases a neutral conductor rated at up to 150% of the phase conductors has been known to be required. Busbar risers in buildings are also at risk, due to harmonic-induced vibration causing joint securing bolts, etc. to work loose.

23.3.5 Frequency Variations

Frequency variations that are large enough to cause problems are most often encountered in small isolated networks, due to faulty or maladjusted governors. Other causes are serious overloads on a network, or governor failures, though on an interconnected network, a single governor failure will not cause widespread disturbances of this nature. Network overloads are most common in areas with a developing electrical infrastructure, where a reduction in frequency may be a deliberate policy to alleviate overloading. Serious network faults leading to islanding of part of an interconnected network can also lead to frequency problems.

Few problems are normally caused by this problem. Processes where product quality depends on motor speed control may be at risk but such processes will normally have closed-loop speed controllers. Motor drives will suffer output changes, but process control mechanisms will normally take care of this. Extreme under- or overfrequency may require the tripping of generators, leading to the possibility of progressive network collapse through network overloading/underfrequency causes.

23.3.6 Voltage Fluctuations

These are mainly caused by load variations, especially large rapid ones such as are likely to occur in arc and induction heating furnaces, rolling mills, mine winders, and resistance welders.

Flicker in incandescent lamps is the most usual effect of voltage fluctuations. It is a serious problem, with the human eye being particularly sensitive to light flicker in the frequency range of 5-15Hz. Because of the wide use of such lamps, the effects are widespread and inevitably give rise to a large number of complaints. Fluorescent lamps are also affected, though to a lesser extent.

23.3.7 Voltage Unbalance

Unbalanced loading of the network normally causes voltage unbalance. However, parts of the supply network with unbalanced impedances (such as untransposed overhead transmission lines) will also cause voltage unbalance, though the effect of this is normally small.

Overheating of rotating equipment results from voltage imbalance. In serious cases, tripping of the equipment occurs to protect it from damage, leading to generation/load imbalance or loss of production.

23.3.8 Supply Interruptions

Faults on the power system are the most common cause, irrespective of duration. Other causes are failures in equipment, and control and protection malfunctions.

Electrical equipment ceases to function under such conditions, with undervoltage protection devices leading to tripping of some loads. Short interruptions may be no more than an inconvenience to some consumers (e.g. domestic consumers), but for commercial and industrial consumers (e.g. semiconductor manufacture) may lead to lengthy serious production losses with large financial impact. Longer interruptions will cause production loss in most industries, as induction and synchronous motors cannot tolerate more than 1-2 seconds interruption without having to be tripped, if only to prevent excessive current surges and resulting large voltage dips on supply restoration. On the other hand, vital computer systems are often fed via a UPS supply that may be capable of supplying power from batteries for several hours in the event of a mains supply failure. More modern devices such as Dynamic Voltage Restorers can also be used to provide continuity of supply due to a supply interruption. For interruptions lasting some time, a standby generator can be provide a limited supply to essential loads, but cannot be started in time to prevent an interruption occurring.

23.3.9 Undervoltage

Excessive network loading, loss of generation, incorrectly set transformer taps and voltage regulator malfunctions cause undervoltage. Loads with a poor power factor (see Chapter 18 for Power Factor Correction) or a general lack of reactive power support on a network also contribute. The location of power factor correction devices is often important, incorrect location resulting in little or no improvement.

The symptoms of undervoltage problems are tripping of equipment through undervoltage trips. Lighting will run at reduced output. Undervoltage can also indirectly lead to overloading problems as equipment takes an increased current to maintain power output (e.g. motor loads). Such loads may then trip on overcurrent or thermal protection.

23.3.10 Transients

Transients on the supply network are due to faults, control and protection malfunctions, lightning strikes, etc.

Voltage-sensitive devices and insulation of electrical equipment may be damaged, as noted above for voltage surges/spikes. Control systems may reset. Semiconductor manufacture can be seriously affected unless the supplies to critical process plant are suitably protected.

Chapter 24
The Digital Substation

24.1 INTRODUCTION

The Digital Substation is a term applied to electrical substations where operation is managed between distributed intelligent electronic devices (IEDs) interconnected by communications networks. It became possible by using computing technology in the substation environment. In the third edition of the Protective Relays Application Guide, first printed in June 1987, a new chapter was introduced: "The application of microprocessors to substation control". Microprocessors were introduced into substation products such as protection devices to improve the performance of the main product functions. Features such as improved accuracy and stability were delivered. Communications ports were also incorporated, but these were more like a developer's debug tool than the sophisticated ones we might recognise today.

Communication facilities developed, however. They were refined to provide connection to SCADA equipment to reduce hardwiring. Communications could provide operators with informative interfaces using software packages running on personal computers. Significantly they developed to enable the transmission of data between the different substation devices.

As computing power became greater and cheaper, it became possible to integrate multiple functionalities into single devices, resulting in fewer devices being required to implement the same traditional functionality.

As a result of functional integration and communications, the traditional divisions between protection, monitoring and control became blurred. The term 'substation automation' was becoming used by protection engineers. This is reflected in a later revision to PRAG: In July 2002 PRAG was replaced by the first edition of NPAG and "The application of microprocessors to substation control" was removed in favour of new chapters associated with "Automation".

Automation offered exciting possibilities, and improvement in communications and computing resulted in more sophisticated IEDs for substation automation becoming available.

Along a similar time line, innovative approaches for instrumentation transformers were being developed, with so-called merging units converting indicative signals of power system quantities into data streams for communication to IEDs.

The technology was now available to communicate current

and voltage information to the different IEDs within the substation on a so called 'process bus', whilst the IEDs communicated information between themselves and the control centre on a so called 'station bus'. The digital substation was close to becoming a reality, but despite attempts at standardisation, proprietary solutions hampered the end goal of "plug-and-play".

The breakthrough came with the introduction and adoption of IEC61850, an open standard for electrical substation automation. This standard caters for data modelling, station buses, process buses, etc., facilitating seamless interoperability.

The digital substation brings major benefits in terms of design and engineering, installation, and operation. Off-the-shelf solutions can be offered, modifications can be easily accommodated, cabling (and hence costs), are reduced, and embedded diagnostics assure system integrity.

This chapter takes a tour through the enablers of the digital substation, looking at communications (and how to ensure that they are sufficiently secure, reliable, and dependable), current and voltage transformation coupled with merging units, and concluding with an introduction to the standard IEC61850.

24.2 COMMUNICATIONS

Communication is the conveyance of information from a source to a destination. In 1949 C E Shannon and W Weaver [24.1] produced a mathematical model for communications as shown in Figure 24.1.

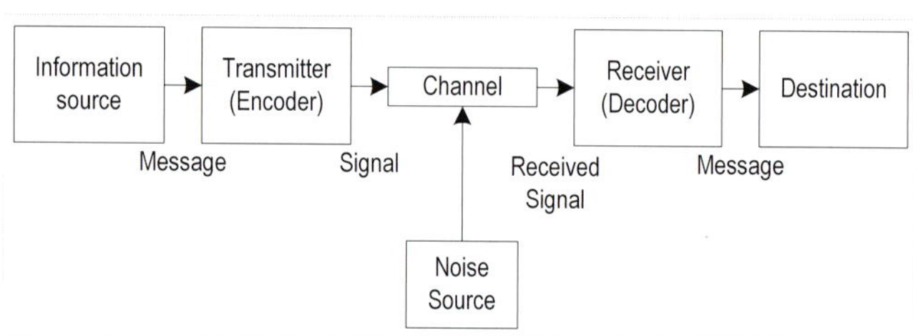

Figure 24.1: The Shannon Weaver model

The model shows that for the information to transfer from the source to the destination, a transmitter (encoder) and a receiver (decoder), both connected by a channel are required. It also demonstrates that sources of noise can interfere with the process. Considering communication between two people, the source is the brain of the person who wants to get the message across, the transmitter could be that person's voice and the channel is the atmosphere. At the receiving end, the receiver would be the other person's ears and the destination that person's brain. Audible noise could drown out the message causing it not to be received and this is a possible cause of communication breakdown, but it is not the only one.

If the source is speaking but the destination is not listening (i.e. they are not synchronised), communication won't be successful. Similarly, if the two people talking don't share a common language, and/or if certain protocols are not observed (e.g. not both speaking at the same time), failure is inevitable. Successful communication requires common understanding between source and destination.

24.2.1 The OSI Model for Computer Communications

In the context of substation automation, communication is the transfer of information from one computing device to another but, as with the example above, problems with synchronisation, language, and protocol can all cause communication failure.

The substation communication might be in the form of a dedicated link between two devices, or it may be over some form of communications network.

The International Standards Organisation (ISO) recognised the need for a framework for inter-device data communications, and in 1984 introduced the Open Systems Interconnection (OSI) model.

The OSI model divides the data communication process into seven distinct layers. Each of the seven layers defines how the data is handled during the different stages of transmission. Each layer provides a service for the layer immediately above it.

A model representing the seven layers is shown in Figure 24.2 with the purpose of the different layers being described below.

Layer Type	Data Unit	Layer	Function
Host Layers	Data	7 Application layer	Communication application
		6 Presentation layer	Encryption and data representation
		5 Session layer	Inter-host communication
	Segment	4 Transport layer	End-to-end connections and reliability
Media Layers	Packet	3 Network layer	Logical addressing
	Frame	2 Link layer	Physical addressing
	Bit	1 Physical layer	Media, signal and binary transmission

Figure 24.2: The OSI seven layer model

24.2.1.1 OSI Layer 1 – The Physical Layer

Every data message is transmitted on some medium. This medium usually takes the form of cables, wires, or optical fibres, but it could just as well be wireless, with the data being carried on electromagnetic waves. The physical layer specifies the type of medium to be used between one end of the data exchange and the other, the type of connectors to be used and the voltage and current levels, and if applicable, optical characteristics defining the state of the data bits. A very common medium, particularly in Ethernet systems started with the CAT 5 UTP cable (Category 5 Unshielded Twisted Pair), which consists of four colour-coded twisted pairs of wires terminated with RJ45 connectors. As the name implies, this type of cable is not shielded against electromagnetic interference. There is also a version of this cable called CAT 5 STP (Category 5 Shielded Twisted Pair) that has a metallic outer sheath, providing electromagnetic shielding. This type of cable is often used in noisy environments like the substation. CAT 5, has more recently evolved into new standards such as CAT 6 and CAT 7, which offer even more robust shielding, fire protection etc.

24.2.1.2 OSI Layer 2 – The Data Link Layer

The physical layer provides the Data Link layer with bits. The Data Link layer now provides some intelligence to this sequence of bits by defining Data Frames. These Data Frames are packets of data containing the data to be transmitted and some control information governing the transmission. The control information comprises flags to indicate the start and end of the message. The standards used at this layer must ensure that the control flags are not mistaken for data and that the data frames are checked for errors. An example is given in Figure 24.3.

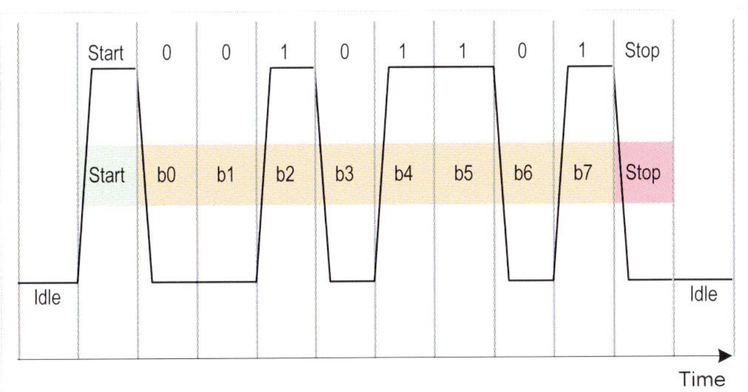

Figure 24.3: Example of a data frame

A commonly used Data Link protocol is Ethernet. Examples of equipment that work at this layer are Ethernet switches and bridges.

24.2.1.3 OSI Layer 3 – The Network Layer

The network layer is concerned with packet delivery. Logical paths are established between the sending and receiving equipment by adding information onto the data frame which defines where the packet has come from, and where the packet is going to. This takes the form of a logical source and destination address for each packet of information. A commonly used Network protocol is IP (Internet Protocol).

An example of equipment that operates at the network layer is a router.

24.2.1.4 OSI Layer 4 – The Transport Layer

The transport layer is the first layer that is not concerned with the mechanics of the data transfer. It is concerned with managing and sequencing the packets once they have arrived. There is an array of transport layer protocols ranging from the very simple, which simply accepts the data as it comes in, not caring about whether the packets have errors nor even if they are in the right order, through to quite complex protocols which check the data for errors, send out an acknowledgement to the sending equipment, order the packets into the correct sequence and present the data to the next layer guaranteed error-free and correctly sequenced.

An example of a simple protocol is UDP (User Datagram Protocol). A more sophisticated widely used transport layer protocol used is TCP (Transmission Control Protocol).

Layer 4 functionality is achieved by the devices at either end of the transmission path, which is usually a computer. In the context of substation protection, this would be the IED.

24.2.1.5 OSI Layer 5 – The Session Layer

The first four layers establish a means of reliable communication between two IEDs (computers), but they do not deal with intelligent management of the communication. This is performed at the session layer. It is the first layer that has user interaction. Each communication session is governed by criteria pertinent to the session. One communication session may be downloading a file from a web site, whilst another may be working on a file situated on a remote server. Session layer software can implement password control, monitor system usage, and allow a user interaction with the communication.

24.2.1.6 OSI Layer 6 – The Presentation Layer

As the name implies, the presentation layer concerns itself with how the data is presented. A good example of this is ASCII (American Standard Code for Information Interchange). An ASCII code is an 8-bit binary code, which defines the familiar character set used to produce text. For example the character 'A' is defined by the binary code 01000001, which is 41 hexadecimal or 65 decimal. The presentation layer, if compliant with the ASCII format, can interpret this.

24.2.1.7 OSI Layer 7 – The Application Layer

The application layer is the layer that interacts with the user. This is in the main software that allows the user to define the communication. Examples of application level protocols would be file transfer protocol (FTP), hypertext transfer protocol (HTTP), post-office protocol 3 (POP3), or simple mail transfer protocol (SMTP).

24.2.2 Communications Between IEDs

The OSI model outlines the need for compatibility between all communication layers between computers (which are generally in the form of IEDs in the substation). This section explores the lowest two layers; the physical and data link layers.

24.2.2.1 Physical Connection to IEDs for Substation Control and Automation

In substation control and automation systems, connection to IED communications ports at the physical layer (OSI layer 1) is generally to one of three standards:

- EIA 232
- EIA 485
- Ethernet

EIA (Electronic Industries Association) is a set of standards defined for connecting electronic devices together. They were formerly known as the recommended standards (RS) and often people will still refer to them as RS 232 and RS 485.

Note: Although, traditionally the standards were designed with metallic connections between devices, more recent times have seen an increase in optical fibre as the medium connecting networks and devices. Many of the physical layer and data link layer protocols have now been modified to work over optical fibre as well.

EIA 232

EIA 232 is an electrical connection allowing full duplex communication between two devices. Prior to the development of the USB port, it was the typical serial port found on most computers and is characterised by Table 24.1

Item	Details
Max. number of transmitters	1
Max. number of receivers	1
Connection type	25 core shielded
Mode of operation	DC coupling
Maximum distance of transmission	15m
Maximum data rate	20kbits/s
Transmitter voltage	5V min, 15V max
Receiver intensity	3V
Driver slew rate	30V/µS

Table 24.1: EIA(RS)232 characteristics

EIA 232 was designed to allow computers to connect using Modems and is suitable only for point-to-point connection. It uses switched single-sided 12V (nominal) signals for data transmission as well as handshake signals to control the communication. Due to Modem limitations, it is not generally used at speeds in excess of 9,600 bits per second. No isolation is specified and so it is only suitable for connection over very short distances. For permanent connection in a substation environment, the use of optical isolating units should be considered to avoid damage caused by induced transients. If transmission over longer distances is required, some form of EIA 232 to fibre-optic converter and a fibre-optic communications link should be used. Since EIA 232 provides only point to point connectivity it is not used in automation systems, rather it is limited to relaying information to a SCADA system.

EIA 485

EIA 485 is an electrical connection characterised by Table 24.2.

Item	Details
Max. number of transmitters	32
Max. number of receivers	32
Connection type	Shielded Twisted Pair
Mode of operation	Differential
Maximum distance of transmission	1200m
Maximum data rate	10Mbits/s
Transmitter voltage	1.5V min
Receiver intensity	300mV

Table 24.2: EIA(RS)485 characteristics

EIA 485 provides a two-wire half-duplex connection designed for multi-drop connections as shown in Figure 24.4 which makes it more suitable than EIA 232 for use in automation schemes.

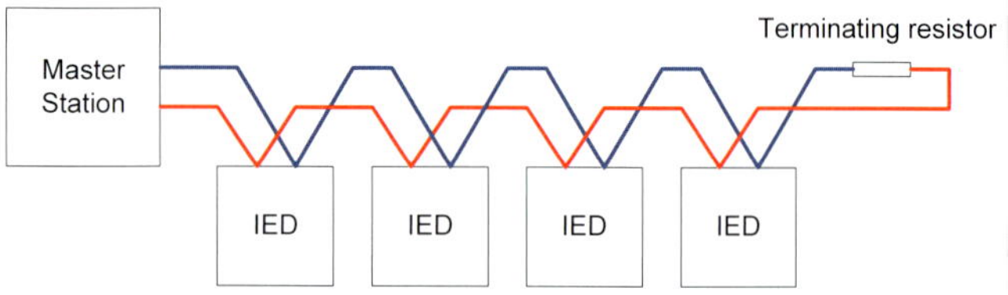

Figure 24.4: Multi-drop connection of RS485 devices

EIA 485 uses differential signalling on twisted pairs and can be isolated. The multi-drop connection (sometimes called daisy-chaining) generally has a limit of 1km, and although in theory the number of devices connected (nodes) is not limited, a practical limit is normally 32 per bus, with repeaters being used if further expansion is required. Communication speeds of 64kbits per second (kbps) can be reliably achieved. EIA 485 allows relatively simple networks to be constructed very efficiently and cost effectively.

Ethernet

Ethernet is a standard, which defines the connection of computing devices to local area networks (LANs). As per

IEEE 802.3, standard signalling speeds are 10Mbps, 100Mbps or 1Gbps. It is the most widespread LAN technology. The specification allows connection to be made either electrically using an RJ45 connector, or by direct fibre-optic connection. Ethernet over fibre-optic cables provide a mechanism for extremely high speed and noise resistant communication, making it the ideal communication medium for the substation. Ethernet is discussed in detail, later in the chapter.

24.2.2.2 Network Topologies

When linking multiple products together to form a computer network, in addition to the physical connection of the device, it is also necessary to consider the network topology as well as the protocol or language by which information is exchanged.

For automation systems to be effective, information must be communicated reliably between different devices in the system. In early centralised systems, a hierarchical series of connections is required for devices at the acquisition and process levels to communicate data upwards and to receive commands in return. This tree-type structure typifies the kind of connection that the point-to-point EIA 232 communications can bring.

The multi-drop capabilities of EIA 485 do more to encourage distribution of control since the EIA 485 network will have devices connected on the same multi-drop bus, with a master connected to many slaves.

Systems based on Ethernet technology provide more scope for different topologies by virtue of equipment such as switches, hubs, bridges and routers.

24.2.3 Serial Communications

This is a form of communication whereby bits of data are exchanged serially through the signalling channel. With parallel communications multiple bits of data are exchanged in parallel. An example of parallel communication is the LPT printer port that used to feature on computers, printers, and office equipment before the widespread deployment of Ethernet in the working environment.

EIA 232, EIA 485, and Ethernet are all forms of serial communication. The much higher speed of Ethernet sets it apart from the others. As a consequence, in the domain of substation automation, communications based on EIA 232 and EIA 485 tend to be referred to as "serial" communications, with Ethernet being singled out for separate attention. This chapter adheres to that convention and continues with a discussion on "serial" communications based on EIA 232 and EIA 485, later followed by Ethernet communication.

24.2.3.1 Serial Communications Protocols

A communications protocol is a set of standard rules for data representation, signalling, authentication and error detection, which defines the transfer of information over a communication channel. Put another way, for devices to be able to speak to each other, they need to share a common language and rules of engagement.

When digital communication facilities were first added to computer based protection devices, no standard protocol existed for this communication; manufacturers developed their own proprietary solutions to exploit the benefits of the communication interfaces. An example of a proprietary protocol is the Courier protocol that was developed by the former GEC Measurements (now Alstom).

A substation control system is required to communicate with all the distributed functionality in the system. If devices from different vendors are to be included in the system, then the different protocols will need to be supported. This increases the system engineering work and consequently the cost.

Consider a relatively simple system, with a bay controller connected to IEDs from different manufacturers A, B, and C as depicted in Figure 24.5.

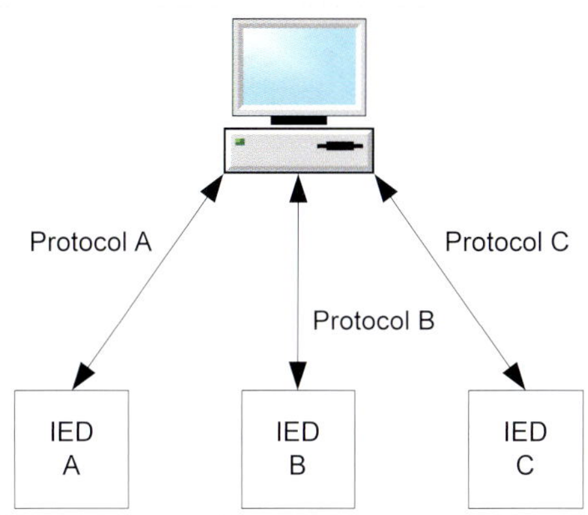

Figure 24.5: Multi-protocol system

If the three products all contain key elements to implement the control system, then either the bay controller would need to support three communications protocols (as shown in the figure), or some form of protocol converter devices would need to be included. Either way, it can be seen that the use of different protocols increases the engineering time, and the component costs associated with implementing the system.

From the systems' engineering perspective, as well as the utilities' perspective, the adoption of a standard protocol by the manufacturers brings clear benefits.

Efforts were made to develop a standard solution and from the protocols available, three open standards for IED serial communications emerged:

- MODBUS
- IEC 60870-5-103
- DNP3

Modbus

Modbus was published in 1979 by Modicon. Originally designed for programmable logic controllers of the time, it is a master-slave protocol that allows a master to read or write bits in registers of the slave devices. It can be implemented on simple serial connections (EIA 485), but has also been migrated to Ethernet.

IEC 60870-5-103

IEC 60870-5-103 was developed by Technical Committee 57 of the International Electro-technical Committee (IEC TC57). It is based on, and is a superset of, the German VDEW communications standard. Primarily serving European clients, it defines the standards for communication between protection equipment and the devices of the control system. As well as simple electrical connection, the standard also defines a direct communication interface implemented over optical fibre.

DNP3

DNP3 was developed by Harris in the 1990s. It drew on early work of TC57 on the IEC 60870-5 (before it was standardised) to develop something with a specific focus on the North American market that could be quickly implemented. Like Modbus, it is also supported on Ethernet as well as simple serial connection. Although originally targeted at the North American market, it has been deployed elsewhere.

The use of these open communications standards simplified the system engineering, making future upgrading and expansion easier, hence reducing overall life-cycle costs.

In general, the open standards of Modbus, IEC60870-5-103 and DNP3 cover layers 1-4 of the OSI model. They allow automation systems engineers to mix and match IEDs from different manufacturers with the same protocol, satisfying the clients' needs for protection and control better than if a mix of proprietary protocols were used. Each of the protocols has particular advantages, but they all suffer disadvantages too:

- Although they are standard protocols, all have ambiguities that result in different implementations in different IEDs.
- The IED data model (i.e. the way the data representing the IED is structured and presented over the communications link) varies between vendors, and can sometimes vary between different versions of the same IED from the same vendor.
- Different functionalities are supported by the different protocols, and different IED vendors may support different levels of those functionalities.
- The provision of private codes in the IEC 60870-5-103 protocol permits much greater functionality, but at the same time hinders interoperability of equipment from different vendors because there is no need for the format of such messages to be made public. In effect, the use of 'private' messages by vendors of devices essentially turns the standard into several proprietary ones.

Due to continuing developments in the fields of computing and communications, there is a need for a truly open communications architecture allowing 'plug-and-play' connectivity between devices from different manufacturers for effective, efficient substation automation. The IEC61850, standard, based on Ethernet satisfies that need.

24.2.4 Ethernet Communications

Ethernet defines the connection of computing devices to local area networks (LANs). It is the most widespread LAN technology. Standardised in IEEE 802.3, it describes the requirements for network connection according to layer 1 (the physical layer) and layer 2 (the data link layer) of the OSI model.

At the physical layer the specification allows connection to be made either electrically using an RJ45 connector, or by direct fibre-optic connection using LC or ST (otherwise known as BFOC2.5) type connectors.

Standard signalling speeds for Ethernet communications are 10Mbps, 100Mbps, 1Gbps and 10Gbps.

At the data link layer (layer 2 of the OSI model), MAC (Media Access Control) addresses take care of the physical addressing of devices on a network. These physical addresses are set in the manufacturing process. MAC uses a 48 bit addressing scheme allowing a unique address for every piece of equipment manufactured.

The most commonly used network protocol (layer 3 of the OSI model) is the Internet Protocol (IP), and the most commonly used transport layer mechanism (OSI layer 4) is the Transmission Control Protocol (TCP). Often these are put together as 'TCP/IP' and are synonymous with Ethernet and the internet.

At the network layer, IP is used to transfer packets of data on the network and across network boundaries from the source to the destination. To do this, devices are assigned IP addresses and these addresses have two purposes: The first is to identify the host network (which may be subdivided into sub-networks), whilst the second is to identify the device on the network or sub-network (subnet). The IP address is a 32 bit binary number, which can be written in the form nnn.nnn.nnn.nnn, where nnn represents a binary octet, and is an integer value between 0 and 255. A typical IP address is, therefore, 192.168.0.2. Each device on the network will have an IP address that is unique and may be configured by a network server, a system administrator, or by an address auto-configuration. The most significant bits of the address are used for network identification and are known as the routing prefix. The least significant bits are known as the rest field and provide unique identification of the device on the local network to which it is connected. Typically the two most significant octets identify the network. The two least significant octets are used for device identification, but can also be used to create subnets. A so called subnet mask is then used to identify the subnet address. For example if the third octet were used as a subnet identifier, then performing a logical AND between the IP address and 255.255.255.0 (the subnet mask) would identify that, in the case above, the subnet address is 192.168.0.0 and the device is at address 2 on that subnet.

When a packet of data is placed on the IP network, it is sent to its destination network by means of routers, switches and hubs. These devices are discussed in the next section.

24.2.4.1 Ethernet Hubs, Switches and Routers

An Ethernet hub is, as the name suggests, a hub for linking multiple Ethernet devices together. Hubs work at the physical layer. Devices on the network transmit data packages to the hub and the hub passes them to all connected devices. There is little arbitration and data collision is common. As such, they are not appropriate for use in mission critical applications such as substation automation.

Ethernet switches operate at the data link layer. Ethernet switches are intelligent devices that provide a common connection point for devices on a network. They store incoming packages and forward them to the appropriate destination on the LAN when the channel is free. Full duplex operation provides for consistent, reliable operation. Figure 24.6 shows a typical Ethernet switch outline.

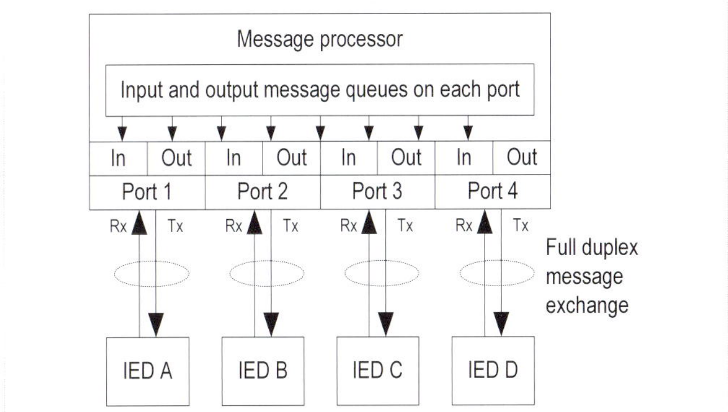

Figure 24.6: Ethernet Switch outline (4-port)

Ethernet routers are used as gateways between the LAN and wide-area networks (WANs). They are similar to switches, but they have fewer ports and act at the network layer. The information is presented to the router in the form of a sequence of IP packets. Each packet has a logical source address and destination address associated with it, and the router passes the packet on to the next step of the destination. Packets with the same destination address do not necessarily follow the same route. If one route gets congested, for example, the router may decide to send the packet onwards via an alternative route (rather like a motorist may choose to avoid road works). For this reason, packets making up a message may arrive at their destination out of sequence. In many cases, it is critical that the message arrives error-free and in the correct sequence. Higher level protocols in the computer, such as TCP, ensure that this happens.

24.2.4.2 Ethernet Topologies

The topology of a network is the way in which devices are connected together. Figure 24.7 shows the two simplest topologies which are the star and the ring so-called because they resemble stars and rings when drawn.

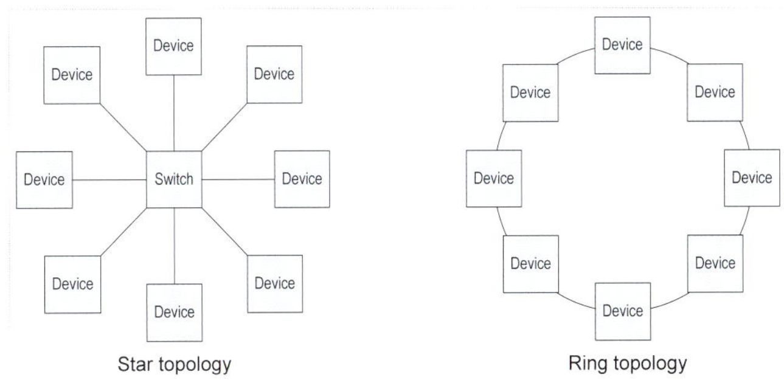

Figure 24.7: Simple star and ring network topologies

In the star topology, a physical connection runs from each device on the network to a central location, usually an Ethernet switch.

In the ring topology, a physical connection is daisy-chained around the devices in the form of a ring.

Ethernet is inherently a star-based network topology. Initially it did not allow ring topology connection due to the risk of traffic getting stuck in an endless loop around a ring and the consequent paralysis of the network as the amount of stuck traffic increased. IEEE 802.1D introduced the Spanning Tree Protocol (STP) which was developed to overcome this limitation, and allow Ethernet to be used in rings as shown in Figure 24.8.

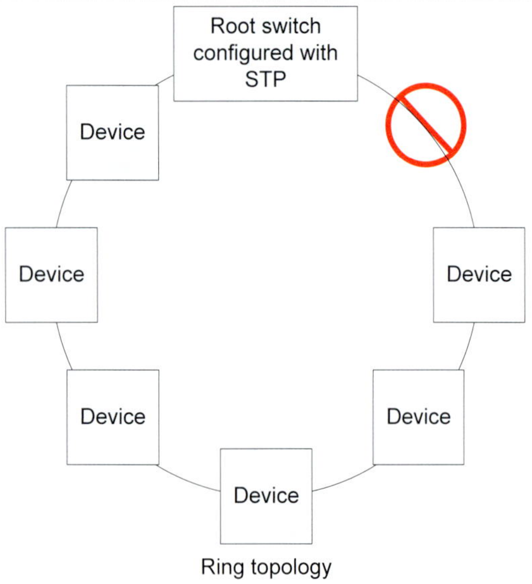

Figure 24.8: Ethernet topology with STP

STP was limited in terms of speed, so IEEE 802.1W Rapid Spanning Tree Protocol (RSTP) was developed from STP to provide faster performance and hence more effective Ethernet ring topologies.

24.2.4.3 Principles of Redundancy in Communications networks

The term "redundancy" can be open to misinterpretation as it wrongly implies that something may not be needed. In the context of communication networks, what it actually means is: "Redundancy is any resource that would not be needed if there were no failures". Redundancy is therefore a provision of transparent backup. It is required where failure cannot be tolerated, such as critical applications like transmission substation automation.

Figure 24.9 shows how redundancy can be incorporated into a star network. A connection from each node goes to a different switch, providing an alternative path. This topology is called Dual Homing Star Topology.

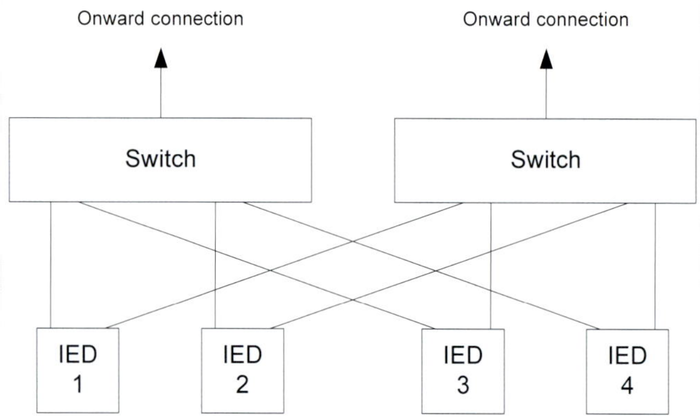

Figure 24.9: Redundant connections in star topology

Figure 24.10 shows simple redundancy with a ring topology. The cable is daisy-chained from device to device in a ring, the idea being that if the link from one direction fails, the link in the other direction can be used to effect transactions. In the event that there is a break at one point of the ring, appropriate redundancy protocols can automatically readjust the ring such that the data is sent back in the opposite direction, ensuring it will get to its destination.

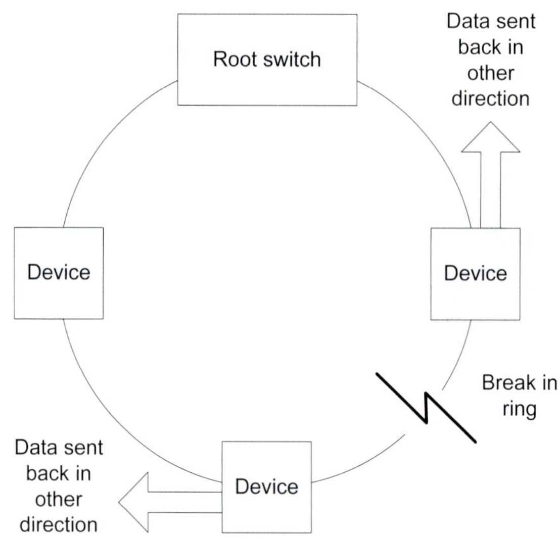

Figure 24.10: Redundancy in a ring topology

This inherent redundancy is one of the features that RSTP offers.

In this example, an alternative path is provided in a 'standby' arrangement. When the primary path fails, this is detected and the alternative path is switched into action. This type of redundancy is known as dynamic redundancy, and it needs a certain amount of time to operate.

Where an alternative path is provided in a 'workby' arrangement such as that shown in Figure 24.9, it is called static redundancy. Should one path fail, the alternative paths are continually working and, should one path fail, no switching time is lost because the other path is already operating.

24.2.4.4 Redundancy requirements for substation automation networks

Dynamic redundancy is cost-effective, but requires switchover time. Static redundancy provides seamless switchover but generally costs more.

The Dual Homing Star, and Rapid Spanning Tree protocols both offer redundancy, but neither presents an ideal solution for the needs of substation automation. To overcome the limitations, the standard IEC 62439-3 (Industrial communication networks - High availability automation networks) was developed for substation automation.

Of particular interest is IEC 62439-3 clause 4 which describes a Parallel Redundancy Protocol (PRP), and IEC 62439-3 clause 5 which describes High-availability Seamless Redundancy (HSR).

Switchover delays and grace time are critical to the substation automation system. The switchover delay is dictated by the grace time and is the most constraining factor in fault-tolerant systems. The grace time is the time that the plant allows for recovery before taking emergency actions.

Some examples of grace times for automation systems are given in IEC62439-3 and are shown in Table 24.3.

Applications	Typical grace time
Uncritical automation applications, e.g. enterprise systems	20s
Automation management, e.g. manufacturing	2s
General automation, e.g. process automation, power plants	0.2s
Time-critical automation, e.g. synchronised drives	0.02s

Table 24.3: Examples of grace time

Table 24.4 provides examples of recovery times for various protocols used for implementing redundancy.

Protocol	Description	Frame Loss	Typical recovery time
IP	IP routing	Yes	30 seconds
STP	Spanning Tree Protocol	Yes	20 seconds
RSTP	Rapid Spanning Tree Protocol	Yes	2 seconds
CRP	Cross-network Redundancy Protocol	Yes	1 second
MRP	Media Redundancy Protocol	Yes	200 milliseconds
BRP	Beacon Redundancy Protocol	Yes	8 milliseconds
PRP	Parallel Redundancy Protocol	No	0 seconds
HSR	High-availability Seamless Ring	No	0 seconds

Table 24.4: Typical recovery times for common redundancy protocols

From Table 24.3 and Table 24.4 it can be seen that systems based on PRP and HSR are appropriately suited to the needs of substation automation networks, providing true static redundancy. This is also known as 'bumpless' redundancy.

24.2.4.5 IEC 62439-3 PRP

PRP is standardised in IEC62439-3 (clause 4) for use in Dual Homing Star Topology networks. PRP is capable of providing bumpless redundancy for real-time systems, and hence becomes the reference standard for star-topology networks in the substation environment.

A PRP compatible device has two ports operating in parallel, each port being connected to a separate LAN segment. IEC62439-3 assigns the term DANP (Doubly Attached Node running PRP) to such devices.

Figure 24.11 shows an example of a PRP network. The doubly attached nodes DANP 1 and DANP 2 have full node redundancy, whilst the singly attached nodes SAN 1 and SAN 4 do not have any redundancy. Singly attached nodes can, however, be connected to both LANs, via a device that converts a singly attached node into a doubly attached node. These devices are referred to as redundancy boxes or RedBoxes. Devices with single network cards such as PCs, printers, etc., are singly attached nodes that may be connected into the PRP network via a RedBox as shown in figure Figure 24.11.

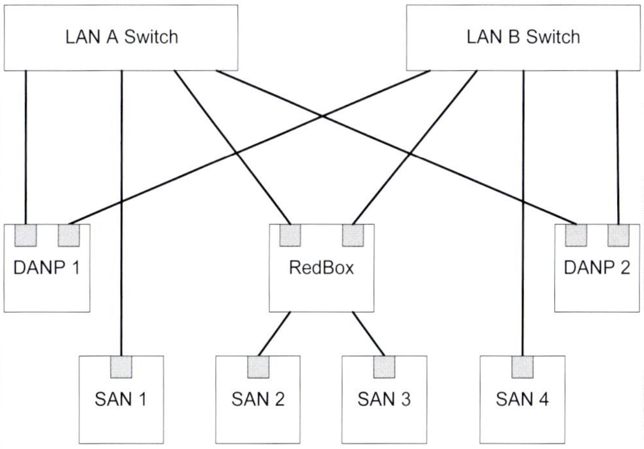

Figure 24.11: Example PRP redundant network

Both ports share the same MAC address, so there is no impact on the functioning of the address resolution protocol. Every data frame is seen by both ports.

When a node sends a frame of data, the frame is duplicated on both ports and thus on both LAN segments, providing a redundant path for the data frame in the event of failure of one of the segments. When both LAN segments are operational, as is the normal case, each port receives identical frames and these identical frames need to be carefully handled. The handling brings overheads, but bumpless redundancy is assured.

24.2.4.6 IEC 62439-3 HSR

HSR is standardised in IEC62439-3 (clause 5) for use in Ring

Topology networks. HSR is capable of providing bumpless redundancy for real-time systems and becomes the reference standard for ring-topology networks in the substation environment.

HSR works on the premise that each device connected in the ring is a doubly attached node running HSR. IEC62439 assigns the term DANH (Doubly Attached Node running HSR) to such devices. As per PRP, singly attached nodes are connected via a so-called Redbox.

Figure 24.12 shows a simple HSR network, where a doubly attached node is sending a multicast frame (that is a frame that is intended for multiple recipients on the network). The frame (C frame) is duplicated, and each duplicate frame is tagged with the destination MAC address and the sequence number. The frames differ only in their sequence number, which is used to identify one copy from another. For convenience, the duplicate frames are labelled the A frame and B frame. Each frame is sent to the network via a separate port. The destination DANH receives two identical frames from each port, removes the HSR tag of the first frame received and passes this to its upper layers. This now becomes the D frame. The duplicate frame is discarded.

The nodes forward frames from one port to another unless the particular node is the node that originally injected it into the ring.

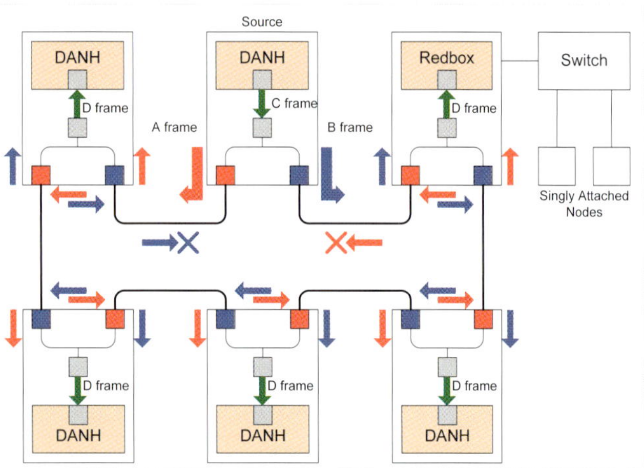

Figure 24.12: HSR for multicast traffic

With unicast frames (frames that are intended for a single destination), there is just one destination and the frames are sent to that destination alone. All non-recipient devices simply pass the frames on. They do not process them in any way. In other words, D frames are produced only for the receiving DANH. This is illustrated in Figure 24.13

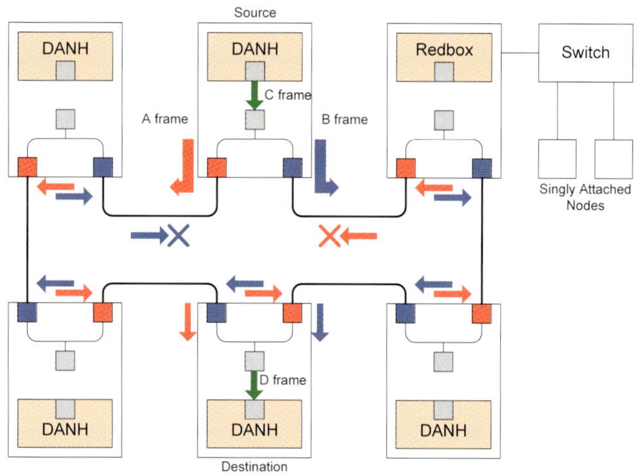

Figure 24.13: HSR for unicast traffic

The use of PRP and HSR delivers network reliability, but the networks also need to be secure. Security within network communications comes under the banner 'Cyber Security', which will be discussed in the next section.

24.3 CYBER SECURITY

When communications were first introduced into the substation environment for control and automation, they were fully contained within the substation and based on proprietary protocols. As a consequence they enjoyed inherent security. They were "secure by isolation" (if the substation network is not connected to the outside world, it can't be accessed from the outside world), and "secure by obscurity" (if the formats and protocols are proprietary, it can be very difficult, to interpret and hack into them).

The increasing sophistication of substation protection, control and automation schemes coupled with the advancement of technology and the desire for vendor interoperability has resulted in standardisation of networks and data interchange within substations. In wide-area applications, substations can be interconnected with open networks such as corporate networks or the internet, which use open protocols for communication. Open protocols mean that the security that obscurity brought cannot be assumed, and interconnection via open networks means that the security that isolation brought them cannot be assumed either. This leaves the networks vulnerable to so-called cyber-attacks.

Cyber-security provides protection against unauthorised disclosure, transfer, modification, or destruction of information and/or information systems, whether accidental or intentional. Factors to be considered in the context of cyber-security include:

- Confidentiality (preventing unauthorised access to information).

- Integrity (preventing unauthorised modification).
- Availability / Authentication (preventing denial of service and assuring authorised access to information).
- Non-Repudiation (preventing denial of an action taking place).
- Traceability/Detection (monitoring and logging of activity to detect intrusion and analyse incidents).

The threats to cyber-security may be unintentional (e.g. natural disasters, human error), or intentional (e.g. cyber-attacks by hackers).

Cyber-security is attainable with a range of measures such as closing down vulnerability loopholes, assuring availability, implementing adequate security processes and procedures, and providing appropriate technology such as firewalls.

Examples of vulnerabilities are:

- Indiscretions by personnel (e.g. users keep passwords in locations visible to others).
- Bypassing of controls (e.g. users turn off security measures).
- Bad practice (users do not change default passwords, or everyone uses the same password to access all substation equipment).
- Inadequate technology (e.g. substation is not firewalled).

Examples of availability issues are:

- Equipment overload, resulting in reduced or no performance.
- Expiry of a certificate prevents access to equipment.

Processes and procedures are required to assure secure exchange of the following categories of information:

- Security Context: This defines information that allows users to have access to devices. It includes passwords, permissions and user credentials.
- Log and Event Management: This includes security logs, which are stored in different IEDs.
- Settings: This includes information about the IED, such as the number of used and unused ports and performance statistics.
- Datagrams: Packets of information exchanged between IEDs.

The OSI 7-layer model enables security measures to be applied at the Transport layer (layer 4) and the Application layer (layer 7).

At the Transport layer the dialog between two devices is controlled. The connections between the two devices in question are established, managed, and terminated at this layer. Applying security at this layer is known as Transport Layer Security, and the most common protocol to achieve this is the Transport Layer Security protocol (TLS), otherwise known as Secure Socket Layer (SSL).

Security measures applied at the Transport Layer (layer 4) guarantee confidentiality, integrity and authenticity, but because they rely on security provided by the transport layer, this type of secure data exchange is limited to point-to-point communication.

A more flexible solution can be implemented by applying security at the Application Layer (layer 7), whereby messages themselves are secured independently of the transport on which they are exchanged. Application of security measures at layer 7 produces a Service Oriented Architecture (SOA).

SOA is based on web services that are message oriented. Secure messages can be sent between any of the devices in the network and are not limited to point-to-point communication. SOA solutions are not prescribed to any specific profile, but the Device Profile for Web Services (DPWS) protocol provides an appropriate level of security for substation automation. Figure 24.14 depicts layer 4 (TLS) and layer 7 (DPWS) security application.

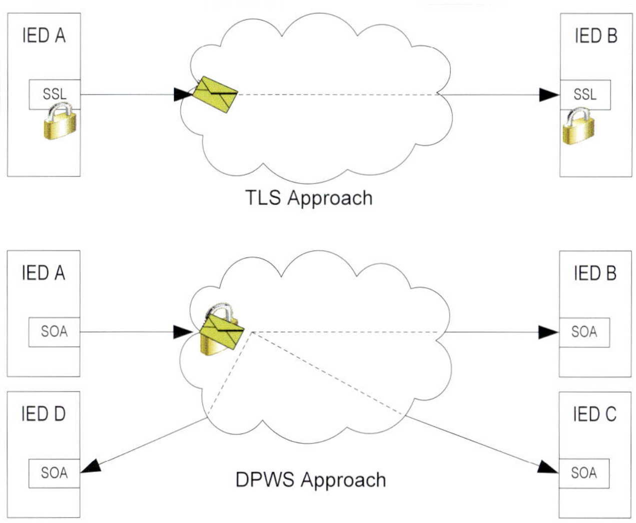

Figure 24.14: TLS vs DPWS

DPWS enables secure web services on devices with limited resources. This means that DPWS is very applicable to IEDs. DPWS is widely deployed having been embedded in the Windows 7 operating system and the DotNet framework for previous Windows versions.

24.3.1 Role Based Access Control

In order to have a complete cyber-security solution, other criteria must be taken into consideration, such as

Authentication, Authorisation and Accounting (AAA).

In Role Based Access Control (RBAC), different classes of user can be granted different rights of access to different information in devices on the network.

A typical RBAC model is shown in Figure 24.15.

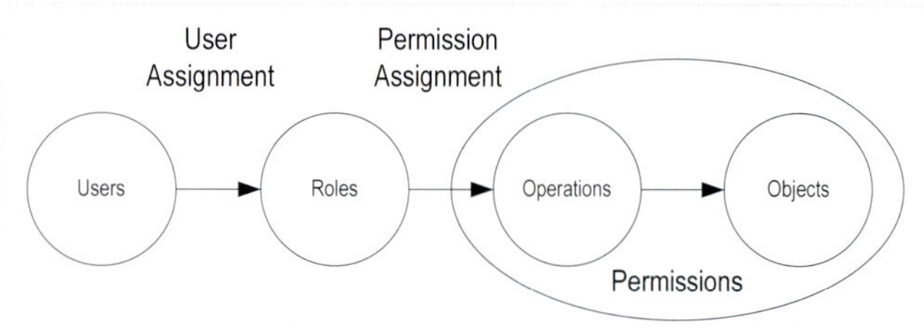

Figure 24.15: RBAC Model

The RBAC model has five basic elements: users, roles, permissions, objects and operations.

- Objects are entities that contain or receive information (e.g. files, directories, printers, etc.).
- Operations are executable functions (read, write, insert, delete, etc.)
- Permissions can contain objects and operations.
- Users acquire permissions through roles.

Transactions are controlled through RBAC, so that access is granted only to what, where, and when, and to whom it should be granted.

A typical transaction is shown in Figure 24.16, in which an AAA server stores security-related data such as user credentials and permissions.

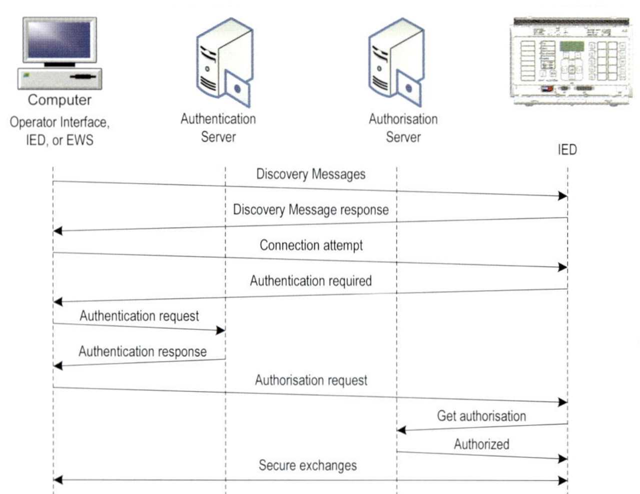

Figure 24.16: Message Flow

Note: Authorisation and Authentication functionality is usually contained in one server. This figure shows two servers to illustrate the detailed interactions.

Once a device is attached to the network, an automatic device discovery process may be performed. Following that, the user selects the device he/she would like to interact with. The chosen device requests a token from the user to prove that he/she has been authenticated by the security server. The user transfers this token to the end device (in this case, the IED). The IED then requests the user roles and credentials from the authorisation server. If the roles and credentials are certified, a secure exchange of data can occur between the devices' applications. This secure data exchange includes for example log events, messages stored at the IED level, etc., for which the user has been granted access.

24.3.2 Cyber Security Standards

There are several standards which apply to substation cyber-security some of which are outlined in Table 24.5.

Standard	Country	Outline
NERC CIP (North American Electric Reliability Corporation)	USA	Framework for the protection of the grid critical Cyber Assets
BDEW (German Association of Energy and Water Industries)	Germany	Requirements for Secure Control and Telecommunication Systems
ANSI ISA 99	USA	Relevant for electric power utility completing existing standard and identifying new topics such as patch management
IEEE 1686	International	International Standard for substation IED cyber security capabilities
IEC 62351	International	Power system data and communication. protocol
ISO/IEC 27002	International	Framework for the protection of the grid critical Cyber Assets
NIST SP800-53 (National Institute of Standards and Technology)	USA	Complete framework for SCADA SP800-82 and Industrial control system cyber security
CPNI Guidelines (Centre for the Protection of National Infrastructure)	UK	Clear and valuable good practices for Process Control and SCADA security

Table 24.5: Standards applicable to cyber-security

The NERC CIP standards, in conjunction with IEEE 1686 are widely applied since they are compulsory in the USA, with compliance auditing having started in 2007.

24.3.2.1 NERC Compliance

The North American Electric Reliability Corporation (NERC) created a set of standards for the protection of critical infrastructure. These are known as the CIP standards (Critical Infrastructure Protection). These were introduced to ensure the protection of critical cyber assets, which control or have an influence on the reliability of North America's bulk electric systems.

The group of CIP standards are listed below:

- CIP-002-1 Critical Cyber Assets
- CIP-003-1 Security Management Controls

- CIP-004-1 Personnel and Training
- CIP-005-1 Electronic Security
- CIP-006-1 Physical Security
- CIP-007-1 Systems Security Management
- CIP-008-1 Incident Reporting and Response Planning
- CIP-009-1 Recovery Plans

CIP-002-1 Critical Cyber Assets

CIP 002 is concerned with the identification of critical assets, such as overhead lines and transformers, as well as critical cyber-assets, such as IEDs that use routable protocols to communicate outside or inside the Electronic Security Perimeter, or are accessible by dial-up. Utilities are required to create and maintain a register of these assets.

CIP-003-1 Security Management Controls

CIP 003 requires the implementation of a cyber security policy, with associated documentation, that demonstrates Management commitment and ability to secure its critical cyber-assets. The standard also requires change control practices whereby all entity or vendor-related changes to hardware and software components are documented and maintained.

CIP-004-1 Personnel and Training

CIP 004 requires that personnel having authorised cyber-access or authorised physical access to critical cyber-assets, (including contractors and service engineers), have an appropriate level of training to assure security.

CIP-005-1 Electronic Security

CIP 005 requires the establishment of an Electronic Security Perimeter (ESP), which provides:

- The disabling of ports and services that are not required
- Permanent monitoring and access to logs
- Vulnerability assessments (yearly at a minimum)
- Documentation of network changes

CIP-006-1 Physical Security

CIP 006 requires that physical security controls, providing perimeter monitoring and logging along with robust access controls, are implemented and documented. All cyber-assets used for physical security are considered critical and should be treated as such.

CIP-007-1 Systems Security Management

CIP 007 addresses system security management and covers the following points:

- Test procedures
- Ports and services
- Security patch management
- Antivirus
- Account management
- Monitoring
- Annual vulnerability assessment

These should be defined and documented for all critical cyber assets.

CIP-008-1 Incident Reporting and Response Planning

CIP 008 requires that an incident response plan be developed, including the definition of an incident response team, their responsibilities and associated procedures.

CIP-009-1 Recovery Plans

CIP 009 requires that a disaster recovery plan should be created and implemented. It should be tested with annual drills.

24.3.2.2 IEEE 1686

IEEE 1686 is an IEEE Standard for substation IEDs' cyber-security capabilities. Used in conjunction with the NERC CIP standards, it proposes practical and achievable mechanisms to achieve secure operations as outlined below:

- Passwords are 8 characters long and can contain upper-case, lower-case, numeric and special characters.
- Passwords are never displayed or transmitted to a user.
- IED functions and features are assigned to different password levels. The assignment is fixed.
- Record of an audit trail listing events in the order in which they occur, held in a circular buffer.
- Records contain all defined fields from the standard and record all defined function event types where the function is supported.
- No password defeat mechanism exists. Instead a secure recovery password scheme is implemented.
- Unused ports (physical and logical) may be disabled.

Whilst the potential threats from cyber-attack are grave, methods for assuring security in the real and cyber worlds are constantly evolving and, correctly applied, should help keep the lights on.

24.4 CURRENT AND VOLTAGE – DIGITAL TRANSFORMATION

Developments in communication have done much to realise the digital substation, but to realise a full digital substation it is necessary to have everything in digital form. Whilst much substation protection, control, and automation technology has always been digital (trip signals, interlocking signals, etc.), the principal power system inputs of voltage and current have traditionally been presented in analogue form.

It has been traditional to present scaled versions of power system currents and voltages to measuring devices, protective relays etc. Scaled versions can easily be produced using conventional electrical transformers, although capacitor dividers may be additionally employed for transforming very high voltages.

As is shown in chapter 6, conventional transformers based on iron cores introduce measurement errors. Due to the wide dynamic range of current signals on power systems, current transformers for protection need large cores to avoid saturation under fault conditions. Due to the nature of the magnetic core material, however, these large cores produce significant errors at nominal current, which renders them impractical for metering purposes. Therefore metering-class transformers need to be introduced resulting in increasing costs.

The iron core is a source of inaccuracy due to the need to magentise the core, as well as the effect of flux remanence, flux leakage, eddy current heating, etc. Conventional wired 1A/5A current transformers (CT) circuits have thermal overload constraints, and pose increasing burdens on the core as cross-site wire run lengths increase. This can degrade protection performance, potentially leading to the need to duplicate CTs. Conventional voltage transformer (VT) circuits may experience ferro-resonance phenomena, with thermal overstress resulting. Capacitor voltage transformers (CVT) can produce high frequency interference signals.

As introduced in chapter 6, techniques that do not require the iron core of conventional transducers can overcome the limitations. The solutions use different sensor technologies such as optical and Rogowski coils. In practical implementations, the techniques require sophisticated solutions employing digital signal processors and micro-processors in numerical products. Since such solutions are eminently able to support digital communications, it is a logical progression to present numerical representations of the measured quantities to other substation devices via communication links, rather than reproducing scaled analogue waveforms. This presentation of analogue power system quantities in the form of standardised digital communication signals is the final element in realising the digital substation.

Solutions providing signal transformation based on technology other than wound transformers are often referred to as non-conventional instrument transformers (NCIT), and the devices that provide the standardised digital communication equivalents of the power system signals are referred to as merging units.

As described in the earlier chapter on transformers, NCIT technologies may be based on optical techniques, or Rogowski coils, and overcome the limitation of iron-cored transformers by delivering:

- Single devices providing measurement class accuracy with dynamic ranges also capable of faithfully re-producing fault currents.
- Reliable, repeatable accuracy.
- High measurement bandwidth for rated frequency, harmonics, and sub-harmonics.
- Low electrical stress insulation – no premature ageing, moisture ingress, or risk of explosion.

As well as having a lower risk of explosion, NCIT devices are also inherently safer since they cannot be 'open-circuited'.

An NCIT device based on Rogowski coil technology is shown in Figure 24.17.

Figure 24.17: Rogowski coil installation in GIS

Merging units present signals such as power system voltages and currents to IEDs within the substation, in the form of numerical values adhering to standardised definitions. The use of NCIT sensors has made it possible for raw measurement information to be fed into so-called Merging Units for further distribution. It is these merging units that are one of the main contributors to the digital substation, and these will be discussed in the next section.

24.4.1 Merging Unit

The merging unit (MU) is the interface device which takes as its inputs connections from the instrument transformer sensors, and performs signal processing to generate and distribute output sampled value streams. The merging unit is thus the gateway to the data from the instrument transformer, as it has the intelligence to interpret the effects of the specialist physical characteristics of the instrument transformer sensor, and convert the output to a numerical value adhering to standardised definitions for communication with substation IEDs.

As well as connecting with NCIT technology such as optical sensors and Rogowski coils, merging units can also be provided in conjunction with conventional transformer technology such that numerical equivalents of system current and voltage can be provided over communication buses.

Using merging units with conventional current and voltage transformers allows the different life-cycle expectations of primary and secondary plant to de de-coupled as per Figure 24.18.

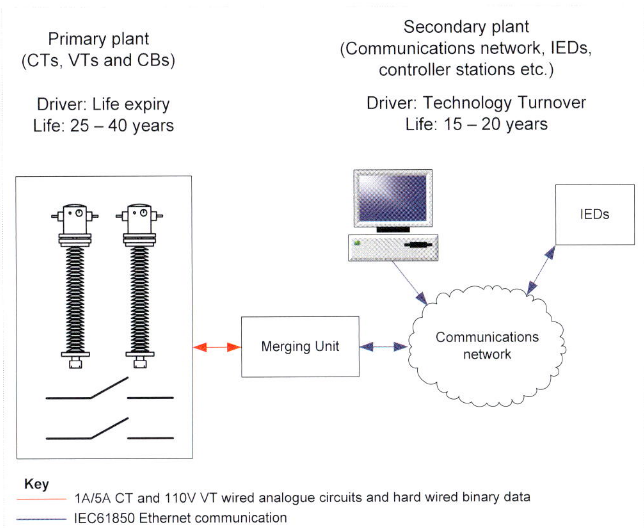

Figure 24.18: De-coupling primary and secondary plant with merging units

Key features of merging units are:

- They can support signal processing for all transformer types including conventional and NCIT.
- They provide accurately time-stamped sampled analogue values.
- They deliver Ethernet multicast transmission of sampled analogue values via a process bus.
- They can support Ethernet connection to the station bus.
- They feature watchdog self monitoring of the NCIT sensors as well as the MU itself.

The connections to the process bus and station bus are in accordance with the IEC61850 standard that is introduced later in this chapter and is shown in Figure 24.19.

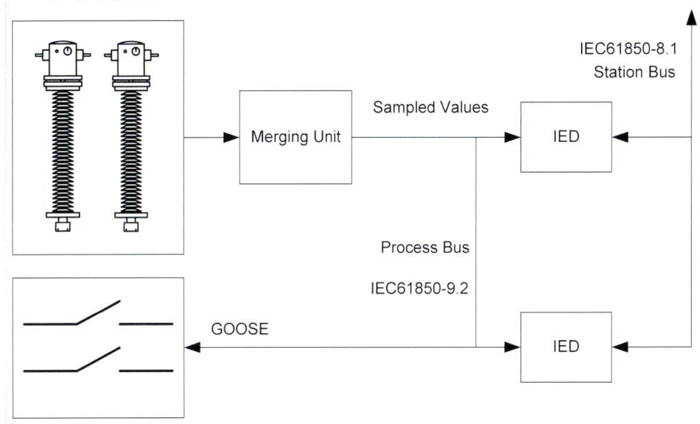

Figure 24.19: Location of MU in the substation

Each merging unit module offers signal processing to provide sampled values of phase currents (Ia, Ib, Ic), phase voltages (Va, Vb, Vc), plus residual current and residual voltage. The sampled value frames are multicast via Ethernet, using a fibre optic, or copper Ethernet connection.

The outputs of the merging units need to be accurately time stamped. This requires the merging units to be accurately time-synchronised, and this is discussed in the next section.

24.4.1.1 Merging Unit Time Synchronisation

The accurate time synchronisation needed by the merging units is realised in the same way as for synchronising phasor measurement units (PMUs) described in chapter 20.

Synchronisation can be achieved thanks to the global positioning satellite system. Synchronising signals may either be delivered over fibre-optic links in the form of one-pulse-per-second (1pps) signals or over Ethernet according to IEEE1588.

In 1956 the American Inter Range Instrumentation Group standardised the different time code formats. These were published in the IRIG Document 104-60. This was revised in 1970 to IRIG Document 104-70, and published later as IRIG Standard 200-70.

The name of an IRIG code format consists of a single letter plus 3 subsequent digits. Each letter or digit reflects an attribute of the corresponding IRIG code. The following tables contain the meanings of the suffixes and descriptions of the abbreviations used.

Suffix position	Suffix description	Suffix	Specification
First letter	Format	A	1000 PPS
		B	100 PPS
		D	1 PPM
		E	10 PPS
		G	10000 PPS
		H	1 PPS

Suffix position	Suffix description	Suffix	Specification
First digit	Modulation Frequency	0	DC Level Shift, width coded, no carrier
		1	Sine wave carrier, amplitude modulated
		2	Manchester Modulated Code
Second digit	Frequency/Resolution	0	No carrier/index count interval
		1	100 Hz / 10 milliseconds
		2	1 kHz / 1 milliseconds
		3	10 kHz / 100 microseconds
		4	100 kHz / 10 microseconds
		5	1 MHz / 1 microsecond
Third digit	Coded expressions	0	BCD_{TOY}, CF, SBS
		1	BCD_{TOY}, CF
		2	BCD_{TOY}
		3	BCD_{TOY}, SBS
		4	BCD_{TOY}, BCD_{YEAR}, CF, SBS
		5	BCD_{TOY}, BCD_{YEAR}, CF
		6	BCD_{TOY}, BCD_{YEAR},
		7	BCD_{TOY}, BCD_{YEAR}, SBS

Table 24.6: Serial time code formats

Abbreviation	Meaning
PPS	Pulses Per Second
PPM	Pulses Per Minute
DCLS	DC Level Shift
BCD	Binary Coded Decimal, coding of time (HH,MM,SS,DDD)
CF	Control Functions depending on the user application
SBS	Straight Binary Second of day (0....86400)
TOY	Time Of Year

Table 24.7: Suffix Descriptions

There are many subsets of the IRIG-B format. IRIG-B is the standard for time synchronisation using 100 PPS. It was this flavour that was embraced by the utility industry to provide real-time information exchange between substations. For IED time synchronisation, IRIG-B12x is typically used for modulated signals and IRIG-B00x for demodulated signals.

The IRIG-B time code signal is a sequence of one second time frames. Each frame is split up into ten 100ms slots as follows:

- Time-slot 1: Seconds
- Time-slot 2: Minutes
- Time-slot 3: Hours
- Time-slot 4: Days
- Time-slot 5 and 6: Control functions
- Time-slots 7 to 10: Straight binary time of day

The first four time-slots define the time in binary-coded decimal (BCD). Time-slots 5 and 6 are used for control functions, which control deletion commands and allow different data groupings within the synchronisation strings. Time-slots 7-10 define the time in SBS (Straight Binary Second of day).

Each frame starts with a position reference and a position identifier. Each time-slot is further separated by an 8mS position identifier.

A typical 1 second time frame is illustrated in Figure 24.20. If the control function or SBS time-slots are not used, the bits defined within those fields are set as a string of zeroes.

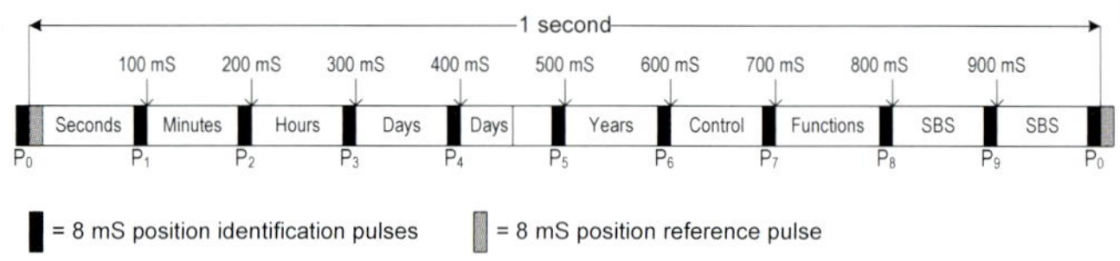

Figure 24.20: IRIG-B frame construction

The 74-bit time code contains 30 bits of BCD time-of-year information in days, hours, minutes and seconds, 17 bits of SBS, 9 bits for year information and 18 bits for control functions.

The BCD code (seconds sub-word) begins at index count 1 with binary coded bits occurring between position identifier bits P0 and P6: 7 for seconds, 7 for minutes, 6 for hours, 10 for days and 9 for year information between position identifiers P5 and P6 to complete the BCD word. An index marker occurs between the decimal digits in each sub-word to provide separation for visual resolution.

The SBS word begins at index count 80 and is between position identifiers P8 and P0 with a position identifier bit, P9 between the 9th and 10th SBS coded bits. The SBS time code recycles each 24-hour period.

The eighteen control bits occur between position identifiers P6 and P8 with a position identifier every 10 bits.

The frame rate is 1.0 second with resolutions of 10ms (dc level shift) and 1ms (modulated 1 kHz carrier).

With IEDs interconnected via Ethernet on the station bus, and with time synchronised merging units multicasting time stamped analogue sampled values via the process bus, the building blocks for the digital substation are all in place. The need for a common understanding between source and destination has already been highlighted. The introduction of standard protocols such as Modbus, DNP3 and IEC60870-5-103 go some way to opening the doors to common understanding, but it was the introduction of IEC61850 that brought true possibilities for interoperability and 'plug-and-play' opportunities for substation automation.

24.5 IEC61850

In the early 1990s the Electric Power Research Institute (EPRI) in the US started work on a Utility Communications Architecture (UCA). The goal was to produce industry consensus regarding substation integrated control, protection, and data acquisition, to allow interoperability of substation devices from different manufacturers. The work developed to produce UCA2 which showed that true interoperability was possible. UCA2 was taken forward by IEC TC57 to produce the standard IEC61850 that revolutionises substation automation.

24.5.1 Benefits of IEC61850

IEC 61850 is the international standard for Ethernet-based communication in substations. It enables integration of all protection, control, measurement and monitoring functions within a substation, and additionally provides the means for interlocking and inter-tripping.

IEC 61850 is more than a protocol, or even a collection of protocols. It is a comprehensive standard, which was designed from the ground up to operate over modern networking technologies. It delivers functionality that is not available from legacy communications protocols. These unique characteristics of IEC 61850 can significantly reduce costs associated with designing, installing, commissioning and operating power systems.

Some of the key features and capabilities of IEC61850 are as follows:

- Use of a virtualised model: In addition to the protocols that define how the data is transmitted over the network, the virtualised model also allows definition of data, services, and device behaviour.
- Use of names for data: Every element of IEC 61850 data is named using descriptive strings.
- Object names are standardised: Names are not dictated by the device vendor nor configured by the user. They are defined in the standard and provided in a power system context, which allows the engineer to immediately identify the meaning of the data.
- Devices are self-describing: Client applications can download the description of all the data, without any manual configuration of data objects or names.
- High-Level Services: IEC61850's abstract communications service interface supports a wide variety of services, such as generic object-oriented substation events (GOOSE), sampled measured values (SMV), logs, etc.
- Standardised substation configuration language (SCL): SCL enables the configuration of a device and its role in the power system to be precisely defined using extensible mark-up language (XML) files.

The major benefits of the standard are as follows:

- Eliminates procurement ambiguity: As well as for configuration purposes, SCL can be used to precisely define user requirements for substations and devices.
- Lower installation cost: IEC 61850 enables devices to exchange data using GOOSE over the station LAN without having to wire separate links for each IED. By using the station LAN to exchange these signals, this reduces infrastructure costs associated with wiring, trenching and ducting.
- Lower transducer costs: A single merging unit can deliver measurement signals to many devices using a single transducer. This reduces transducer, wiring, calibration, and maintenance costs.
- Lower commissioning costs: IEC 61850-compatible devices do not require much manual configuration. Also, client applications do not need to be manually configured for each point they need to access, because they can retrieve this information directly from the device or import it via an SCL file. Many applications require nothing more than the setting up of a network address. Most manual configuration is therefore eliminated, drastically reducing errors, rework, and therefore costs.
- Lower equipment migration costs: The cost associated with equipment migrations is reduced due to the standardised naming conventions and device behaviour.
- Lower extension costs: Adding devices and applications into an existing IEC 61850-based network can be done with little impact on existing equipment.
- Lower integration costs. IEC 61850 networks are capable of delivering data without separate communication front-ends or reconfiguring devices. This means that the cost associated with integrating substation data is substantially reduced.
- Implement new capabilities: IEC61850 enables new and innovative applications that would be too costly to produce otherwise. This is because all data associated with a substation is available on its LAN in a standard format and accessible using standard protocols.

24.5.2 Structure of the IEC 61850 standard

The IEC 61850 standard consists of ten parts, as summarised in Table 24.8.

Part	Title
Part 1	Introduction and overview
Part 2	Glossary
Part 3	General requirements

Part	Title
Part 4	System and project management
Part 5	Communication requirements for functions and device models
Part 6	Configuration description language for communication in electrical substations related to IEDs
Part 7	Basic communication structure for substation and feeder equipment
Part 7.1	- Principles and models
Part 7.2	- Abstract communication service interface (ACSI)
Part 7.3	- Common data classes
Part 7.4	- Compatible logical node classes and data classes
Part 8	Specific Communication Service Mapping (SCSM)
Part 8.1	- Mappings to MMS (ISO 9506-1 and ISO 9506-2) and to ISO/IEC 8802-3
Part 9	Specific Communication Service Mapping (SCSM)
Part 9.1	- Sampled values over serial unidirectional multidrop point to point link
Part 9.2	- Sampled values over ISO/IEC 8802-3
Part 10	Part 10: Conformance testing

Table 24.8: Structure of IEC 61850 standard

Parts 1 and 2 introduce the standard, provide a summary and a glossary of all terms used throughout the standard. Parts 3, 4, and 5 of the standard start by identifying the general and specific functional requirements for communication in a substation (key requirements stated above). These requirements are then used as forcing functions to aid in the identification of the services and data models needed, application protocol required, and the underlying transport, network, data link, and physical layers that will meet the overall requirements.

Part 6 of the standard defines an XML-based Substation Configuration Language (SCL). SCL allows formal description of the relations between the substation automation system and the substation switchyard. Each device must provide an SCL file that describes its own configuration.

Part 7 specifies the communication structure for substations. It consists of 4 sub-sections. IEC 61850 abstracts the definition of the data items and the services from the underlying protocols. These abstract definitions allow data objects and services to be mapped to any protocol that can meet the data and service requirements. The definition of the abstract services is found in part 7.2 of the standard and the abstraction of the data objects (referred to as Logical Nodes) is found in part 7.4. Many of the data objects belong to common categories such as Status, Control, Measurement, and Substitution. These are known as Common Data Classes (CDCs). These CDCs, which define common building blocks for creating the larger data objects, are defined in part 7.3.

The abstract definitions of data and services can be mapped onto various suitable protocols. Section 8.1 of the standard defines the mapping of these onto the Manufacturing Messaging Specification (MMS). Section 9.2 defines the mapping of Sample Measured Values onto an Ethernet data frame. The 9.2 document defines what has become known as the process bus.

24.5.3 The IEC 61850 data model

A typical communications protocol defines how data is transmitted over a medium, but does not specify how the data should be organised in terms of the application. This approach requires power system engineers to manually configure objects and map them to power system variables at a low level (register numbers, index numbers, I/O modules, etc.).

IEC 61850 is different in this respect. In addition to the protocol elements, it specifies a comprehensive model for how power system devices should organise data in a manner that is consistent across all types and brands of devices. This eliminates much of the tedious non-power system configuration effort because the devices can configure themselves.

Some devices use an SCL file to configure the objects. If this is the case, the engineer needs only to import this SCL file into the device. The client application will then extract the object definitions from the device over the network. This significantly reduces the effort needed to configure the device.

The data model of any IEC 61850 IED can be viewed as a hierarchy of information, whose nomenclature and categorisation is defined and standardised in the IEC 61850 specification. The IEC61850 data model is represented conveniently by Figure 24.21.

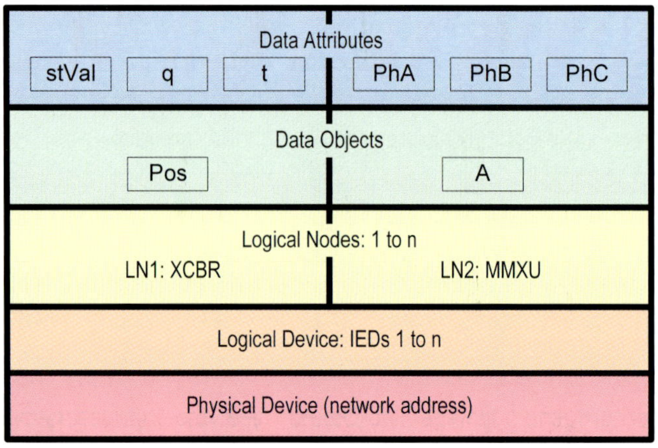

Figure 24.21: IEC 61850 device model

The device model starts with a physical device that connects to the network – typically, the IED. The physical device is defined by its network IP address. Within each physical device, there may be one or more logical devices. The IEC 61850 logical device model allows a single physical device to act as a proxy or gateway for multiple logical devices.

Each logical device contains one or more logical nodes. A logical node is a named grouping of data and associated services that is logically related to some power system function. These logical nodes are categorised into 13 different groups and 86 different classes. Table 24.9 lists these groups and specifies the group designators of each.

Logical node groups	Group designator
System logical nodes	L
Protection functions	P
Protection related functions	R
Supervisory control	C
Generic function references	G
Interfacing and archiving	I
Automatic control	A
Metering and measurement	M
Switchgear	X
Instrument transformer	T
Power transformer and related functions	Y
Further power system equipment	Z
Sensors	S

Table 24.9: Logical node categorisation

The names of these logical nodes have been standardised in IEC61850 and cannot be changed. The naming convention is as follows:

<Single letter group designator><three-letter mnemonic abbreviation of function><instance ID>

For example: PDIR1 belongs to the group "Protection functions" and is the "Directional element" for the first stage (of overcurrent protection, typically). A complete list of logical node names is provided in the standard, and a selection of those most pertinent to readers is given in appendix C.

Each logical node contains one or more elements of data. There are several hundred data elements, which can be broadly categorized into seven groups:

- System information
- Physical device information
- Measurands
- Metered values
- Controllable data
- Status information
- Settings

Each data element has a unique purposeful name determined by the standard, along with a set of attributes. These data names provide a mnemonic description of its function. For example, a circuit breaker is modelled as an XCBR logical node. This contains a variety of data including:

- **Loc** for determining if operation is remote or local
- **OpCnt** for an operations count
- **Pos** for the position
- **BlkOpn** block breaker open commands
- **BlkCls** block breaker close commands
- **CBOpCap** for the circuit breaker operating capability

Each element of data conforms to the specification of a common data class (CDC), which describes the type and structure of the data within the logical node. There are CDCs for:

- Status information
- Measured information
- Controllable status information
- Controllable analogue set point information
- Status settings
- Analogue settings

Each CDC has a set of attributes each with a defined name, defined type, and specific purpose. A set of functional constraints (FC) groups these attributes into categories. For example, there is a CDC called SPS (Single Point Status).

In the Single Point Status (SPS) CDC there are functional constraints for status (ST) attributes, substituted value (SV) attributes, description (DC) attributes, and extended definition (EX) attributes. In this example the status attributes of the SPS class consist of a status value (stVal), a quality flag (q), and a time stamp (t).

Example:
Suppose that a logical device called **Breaker1** consists of a single circuit breaker logical node **XCBR1**. To determine if the breaker is in the remote or local mode of operation, the following expression would be used:

- Breaker1/XCBR1STLoc$stVal.

The returned value would indicate the mode of operation.

24.5.4 Mapping IEC 61850 to a protocol stack

The IEC 61850 abstract model needs to work over a real set of protocols, which are convenient and practical to implement, and which can operate within the computing environments commonly found in the power industry. The IEC61850 standard takes advantage of existing protocols to achieve each necessary layer of communication.

Figure 24.22 shows a simplified version of the IEC61850 protocol stack, and how it fits in with the OSI model.

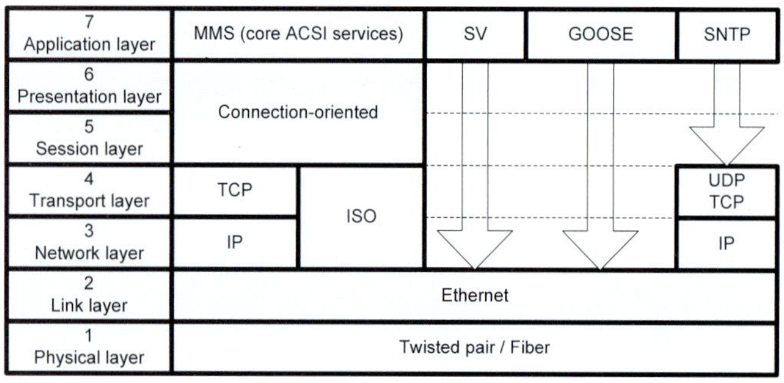

Figure 24.22: IEC 61850 protocol stack

IEC 61850 part 8.1 maps the abstract objects and services to the Manufacturing Message Specification (MMS) protocols specified in ISO9506. MMS is the only public ISO-compliant protocol that can easily support the complex naming and service models specified by IEC 61850. MMS is used because it supports complex named objects and a rich set of flexible services that supports the mapping to IEC 61850 in a straightforward manner.

Part 8.1 also defines profiles for the lower layers of the communication stack. MMS operates over connection-oriented ISO, or TCP/IP. SNTP operates over TCP/IP or UDP/IP. Sampled Values and GOOSE data map directly into the Ethernet data frame thereby eliminating processing of any middle layers and providing very fast response.

24.5.5 Substation Configuration Language

IEC 61850-6-1 specifies an XML-based Substation Configuration Language (SCL) to describe the configuration of IEC 61850 based systems. SCL specifies a hierarchy of configuration files, which enable the various levels of the system to be described. These are:

- System specification description (SSD)
- IED capability description (ICD)
- Substation configuration description (SCD)
- Configured IED description (CID) files

These files are constructed in the same method and format, but have different scopes depending on the need. Even though an IEC 61850 client can extract the configuration from an IED when it is connected over a network, there are several benefits of having a formal off-line description language for reasons other than configuring IEC 61850 client applications. These benefits are as follows:

- SCL enables off-line system development tools to generate the files needed for IED configuration automatically from the power system design. This significantly reduces the cost and effort associated with IED configuration by eliminating most of the manual configuration tasks.
- SCL enables the sharing of IED configuration details among users and suppliers. This helps to reduce inconsistencies and misunderstandings in system configuration requirements. Users can specify and provide their own SCL files to ensure that IEDs are configured according to their requirements.
- SCL allows IEC 61850 applications to be configured off-line without requiring a network connection to the IED.

Figure 24.23 illustrates the position and workflow context of the configuration files in an IEC61850 project

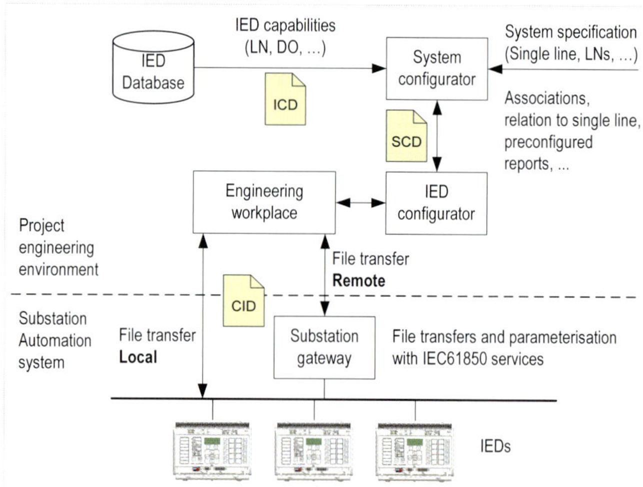

Figure 24.23: IEC 61850 project configuration

24.5.6 IEC61850-9.2

In IEC61850-9.2, the wiring that traditionally connected transformers and switchgear to protection control and monitoring devices is replaced by a process bus, where the process bus is a high-speed Ethernet communications network.

Merging units form the data acquisition layer in the network and present sampled measured values to IEDs via the process bus.

The interface of the merging unit with the primary plant may either be in the form of conventional transformers, or with non-conventional instrument transformers (NCIT), but the output is the same: time-stamped sampled measured values.

The use of the IEC 61850-9.2 process bus introduces a number of opportunities for lifecycle cost reduction. This lifecycle cost comprises configuration, installation and commissioning costs, plus costs beyond commissioning. These later-life costs include any need for extension, upgrade or retrofitting, and also any periodic maintenance.

Some key points where the process bus offers distinct advantage are described in the following paragraphs:

Mid-life secondary equipment renewal:

Referring again to Figure 24.18, the primary equipment is of high capital cost, and cannot be de-energised easily – hence it needs to be left in-service as long as is economically viable. This period of time can be of the order of 40 years. The secondary system's life expectancy is often dependent on advances in technology and increased software-related functionality being requested, and not due to physical failure or end-of-life.

Merging units in a process-bus solution can become a convenient boundary between the two disparate renewal cycles, such that secondary side retrofits/renewals can be undertaken without open-circuiting CTs. This mid-life refit can be done quicker, and more cheaply.

Substation extensions and upgrades can be undertaken as a software modification:
There is minimal panel hardwiring, hence upgrading is easier, as a controlled software procedure.

Deploying standard bay solutions capitalises on copycat scheme roll-out:
As the fundamental scheme performance is defined in software, if this is standardised, it can be rolled-out in multiple sites and multiple bays. Standardisation allows the investment cost in developing the solutions to be amortised over multiple instances.

Current transformer savings:
A protection scheme is entirely dependent on its input transducers (typically CTs and VTs) to deliver its performance. Thus, any lowering of the requirement (size, dimensioning, accuracy class, number of cores) will mean total scheme cost savings.

In a process bus scheme, there is minimal or zero physical burden for the transformers to power – hence sizes and material costs can be reduced. Merging units can also digitise the outputs of conventional CTs, close to the CT itself. This reduces the burden, and hence the CT knee-point voltage requirement, especially in the case of 5A CT installations.

Configurable (digitised) analogue signal sources for IEDs:
In a process bus solution, there is not the need to switch VT or CT sources with auxiliary contacts – this can be done as a "soft" function, rather than requiring make-before-break or other special auxiliary contact arrangements.

Multipurpose IEDs, not dedicated devices:
The lack of an internal CT in the IEDs means they can perform both protection and measurement applications. In protection applications the dynamic range to measure fault current magnitudes is required, and in measurement applications accuracy is the key.

24.6 CONCLUSION

IEC61850-9.2, and IEC61850 in general are at the heart of the digital substation. Together they have revolutionised substation engineering. This chapter has aimed to equip engineers with an understanding of the terminology and concepts in order to be able to function in this new world focused on Ethernet communication technology.

24.7 REFERENCES

[24.1] Shannon, C. E., & Weaver, W. (1949). The mathematical theory of communication. Urbana, Illinois: University of Illinois Press.

[24.2] Reece, I. and Walker, S. (1997) Teaching, Training and Learning. Sunderland: Business Education Publishers Ltd.

Chapter 25
Substation Control and Automation

25.1 INTRODUCTION

The complex interlocking and sequence control requirements that prevail in most substations of any significant size lend themselves naturally to the application of substation automation. These requirements can be readily expressed in mathematical logic (such as truth tables and Boolean algebra) and thus ease the application of computers and associated software. Hence, computers have been applied to the control of electrical networks for many years, and examples of them being applied to substation control/automation were in use in the early 1970's. The first applications were naturally in the bulk power transmission field, as a natural extension of a trend to centralised control rooms for such systems. The large capital investment in such systems and the consequences of major system disruption made the cost of such schemes justifiable. In the last twenty years or so, continuing cost pressures on utilities and advances in digital hardware have led to the application of computers to substation control and automation as a first-choice solution.

25.2 TOPOLOGY AND FUNCTIONALITY

The topology of a substation control system is the architecture of the digital system used. The functionality of such a system is the complete set of functions that can be implemented in the control system – noting that a particular substation may only utilise a subset of the functionality possible.

All digital control systems utilise one of two basic topologies, the basic concepts of which are illustrated in Figure 25.1:

- centralised
- distributed

Early examples of substation automation used the centralised concept, due to limitations in technology, both of processor power and communication techniques. Latest examples use a distributed architecture, in that a number of Intelligent Electronic Devices (IEDs), such as numerical relays, may be linked to a local processor. The local processor may control one or more bays in a substation. All of the local processors are, in turn, connected to a Human Machine Interface (or HMI), and possibly also to a local or remote SCADA system for overall network monitoring/control.

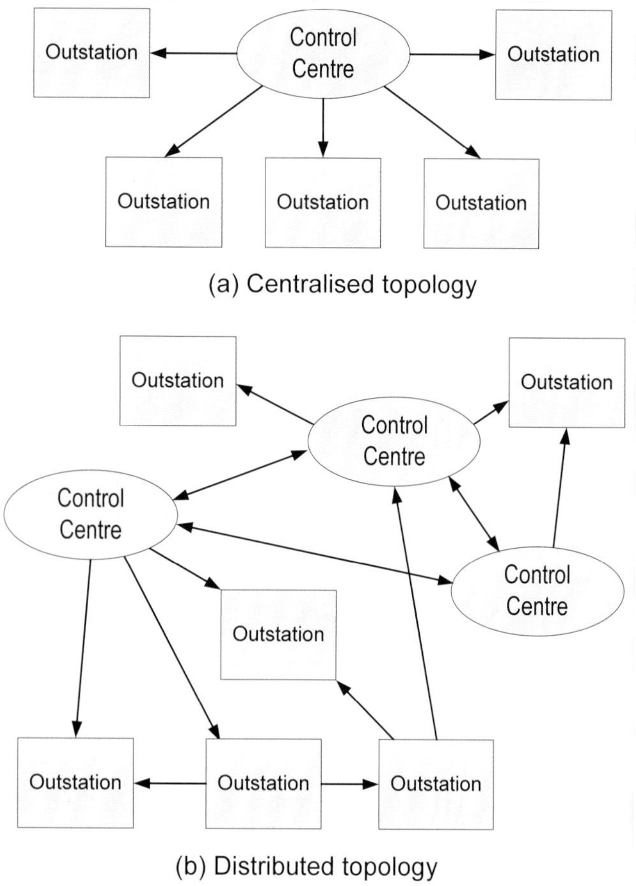

Figure 25.1: Basic substation automation system topologies

25.2.1 System Elements

The main system elements in a substation control system are:

- IEDs, implementing a specific function or functions on a circuit or busbar in a substation. The most common example of an IED is a numerical protection relay, but it could also be a measurement device, interface unit to older relays, etc.
- Bay Module (or "Bay Controller"). This device will normally contain all of the software required for the control and interlocking of a single bay in the substation, and sufficient I/O to interface to all of the devices required for measurement/protection/control of the bay. The I/O may include digital and analogue I/O (for interfacing to discrete devices such as breaker tripping and closing circuits, isolator motors, non-digital legacy relays) and communications links to IEDs.
- RTU (Remote Terminal Unit). This is a device installed at a substation, which collects analogue, digital and status data and transmits this in a suitable format to the remote SCADA master.
- Human Machine Interface (HMI). This is the principal user interface and would normally take the form of a computer. A desktop PC is commonly used, but specialised computers are also possible, while normally unmanned substations may dispense with a permanently installed HMI and rely on operations/maintenance staff bringing a laptop computer equipped with the appropriate software with them when attendance is required. Sometimes, one or more printers are linked to the HMI in order to provide hard-copy event records or other reports.
- A communications bus or busses, linking the various devices. Where a substation automation system is being retrofitted to an existing substation, it may be necessary to use existing communications busses to communicate with some legacy devices. This can lead to a multiplicity of communications busses and protocols within the automation system
- A link to a remote SCADA system. This may be provided by a dedicated interface unit, be part of the HMI computer or part of an IED. A dedicated gateway is used to provide a secure means of communicating outside the substation, encrypting the data and making it immune from cyber attacks.

25.2.2 System Requirements

A substation control/automation scheme will normally be required to possess the following features:

- control of all substation electrical equipment from a central point
- monitoring of all substation electrical equipment from a central point
- interface to remote SCADA system
- control of electrical equipment in a bay locally
- monitoring of electrical equipment in a bay locally
- status monitoring of all connected substation automation equipment
- system database management
- energy management
- condition monitoring of substation electrical equipment (switchgear, transformers, relays, etc.)

The system may be required to be fault-tolerant, implying that redundancy in devices and communications paths is provided. The extent of fault-tolerance provided will depend on the size and criticality of the substation to the operator, and the normal manning status (ie. whether manned or unmanned). Many of the functions may be executed from a remote location, such as a Regional or National Control Centre, in addition to the substation itself.

Certain of the above functions will be required even in the

most elementary application. However, the selection of the complete set of functions required for a particular application is essentially the responsibility of the end-user Utility. Due to a modular, 'building block' approach to software design, it is normally relatively easy to add functionality at a later stage. This often occurs through changing operators' needs and/or electrical network development. Compatibility of the underlying database of network data must be addressed to ensure historical data can still be accessed.

25.3 HARDWARE IMPLEMENTATION

To form a digital substation control system (DCS), the various elements described above must be assembled into some form of topology. Three major hardware topologies can be identified as being commonly used, as follows:

25.3.1 HMI-based Topology

This takes the form of Figure 25.2. The software to implement the control/automation functions resides in the HMI computer and this has direct links to IEDs using one or more communications protocols. The link to a remote SCADA system is normally also provided in the HMI computer, though a separate interface unit may be provided to offload some of the processor requirements from the HMI computer, especially if a proprietary communications protocol to the SCADA system is used.

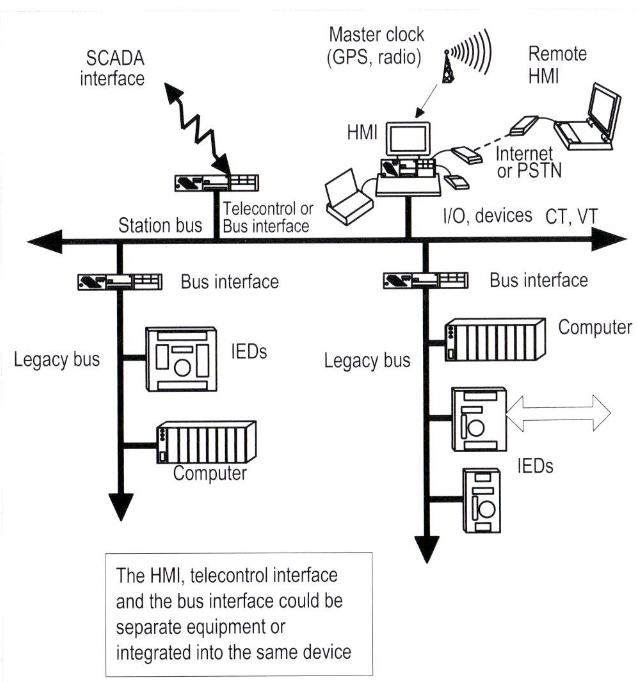

Figure 25.2: HMI-based hardware topology

For this topology, a powerful HMI computer is clearly required if large numbers of IEDs are to be accommodated. In practice, costs usually dictate the use of a standard PC, and hence there will be limitations on substation size that it can be applied to because of a resulting limit to the number of IEDs that can be connected. The other important issue is one of reliability and availability – there is only one computer that can control the substation and therefore only local manual control will be possible if the computer fails for any reason. Such a topology is therefore only suited to small MV substations where the consequences of computer failure (requiring a visit from a repair crew to remedy) are acceptable. Bay Modules are not used, the software for control and interlocking of each substation bay runs as part of the HMI computer software.

25.3.2 RTU-based Topology

This topology is an enhancement of the HMI topology and is shown in Figure 25.3. A digital RTU is used to host the automation software, freeing the HMI computer for operator interface duties only. The HMI computer can therefore be less powerful and usually takes the form of a standard PC, or for unmanned substations, visiting personnel can use a laptop PC.

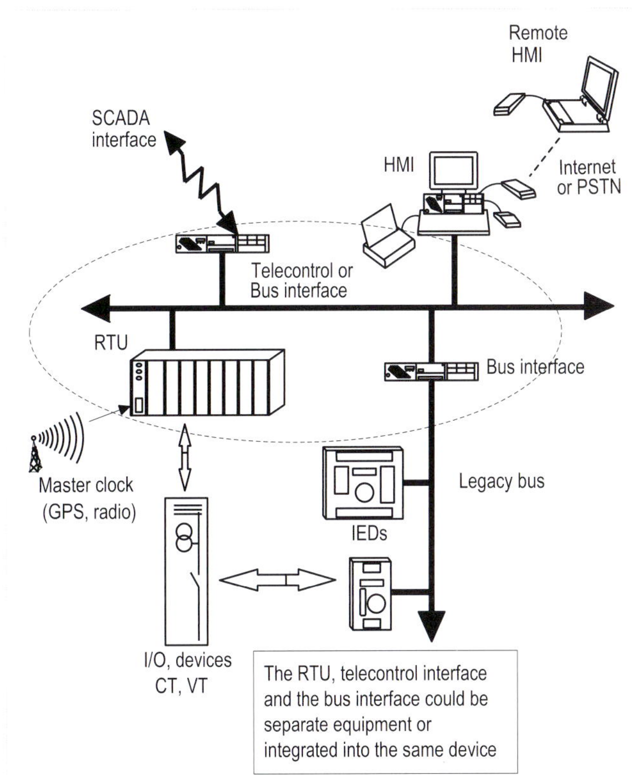

Figure 25.3: RTU-based topology

A greater number of I/O points can be accommodated than in the HMI topology, while the possibility exists of hosting a wider variety of communications protocols for IEDs and the remote SCADA connection. Bay Modules are not required, the associated software for interlocking and control sequences is part of the RTU software.

25.3.3 Decentralised Topology

This topology is illustrated in Figure 25.4. In it, each bay of the substation is controlled by a Bay Module (alternatively commonly described as a Bay Controller). This device houses the control and interlocking software, interfaces to the various IEDs required as part of the control and protection for the bay, and interfaces to the HMI. It is possible to use an HMI computer to take local control of an individual bay for commissioning/testing and fault finding purposes. The amount of data from the various substation I/O points dictates that a separate SCADA interface unit is provided (often called an RTU or Gateway), while it is possible to have more than one HMI computer, the primary one being dedicated to operations and others for engineering use. Optionally, a remote HMI computer may be made available via a separate link. It is always desirable in such schemes to separate the real-time operations function from engineering tasks, which do not have the same time-critical importance.

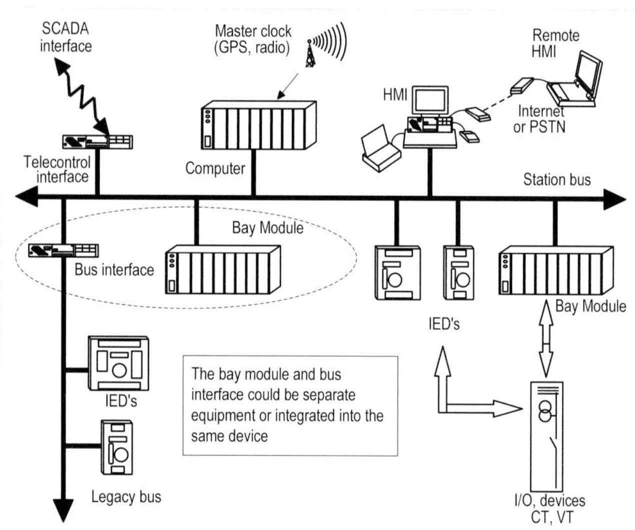

Figure 25.4: Decentralised topology

The connection between the various Bay Modules and the HMI computer is of some interest. Simplest is the star arrangement of Figure 25.5(a). This is the least-cost solution but suffers from two disadvantages. Firstly, a break in the link will result in loss of remote control of the bay affected; only local control via a local HMI computer connected to the bay is then possible. Secondly, the number of communications ports available on the HMI computer will limit the number of Bay Modules.

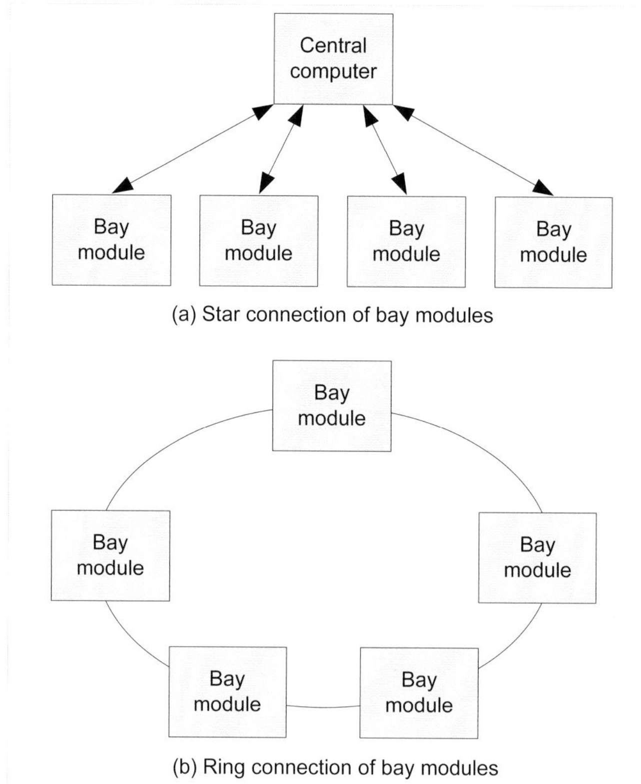

Figure 25.5: Methods of hardware interconnection

Of course, it is possible to overcome the first problem by duplicating links and running the links in physically separate routes.

An alternative is to connect the Bay Modules, HMI computer and SCADA gateway in a ring, as shown in Figure 25.5(b). By using a communications architecture such as found in a LAN network, each device is able to talk to any other device on the ring without any message conflicts. A single break in the ring does not result in loss of any facilities. The detection of ring breakage and re-configuration required can be made automatically. Thus, the availability and fault tolerance of the network is improved. Multiple rings emanating from the HMI computer can be used if the number of devices exceeds the limit for a single ring. It can be easier to install on a step-by-step basis for retrofit applications, but of course, all these advantages have a downside. The cost of such a topology is higher than that of the other solutions, so this topology is reserved for situations where the highest reliability and availability are required.

It is usual to have more than one operators' HMI, either for operational reasons or for fault-tolerance. The system computer may be duplicated on a 'hot-standby' or 'dual-redundant' basis, or tasks may be normally shared between two or more system computers with each of them having the capability of taking over the functions of one of the others in the event of a failure.

The total I/O count in a major substation will become large

Functional area	Functionality			
Interlocking	CB's	Isolators	Contactors	
Tripping sequences	CB failure	Intertripping	Simultaneous trips	
Switching sequences	Automatic transformer changeover	Automatic busbar changeover	Restoration of supply following fault	Network re-configuration
Load management	Load shedding	Load restoration	Generator despatch	
Transformer supervision	OLTC control	Load management		
Energy monitoring	Import/export control	Energy management	Power factor control	
Switchgear monitoring	AIS monitoring	GIS monitoring		
Equipment status	Relay status	CB status	Isolator status	
Parameter setting	Relays	Transformers	Switching sequences	IED configuration
HMI functionality	Access control	One-line views	System views	Event logging
HMI functionality	Trend curves	Harmonic analysis	Remote access	Disturbance analysis
HMI functionality	Interface to SCADA	Alarm processing		

Table 25.1: Typical substation automation functionality

and it must be ensured that the computer hardware and communications links have sufficient performance to ensure prompt processing of incoming data. Overload in this area can lead to one or more of the following:

- undue delay in updating the system status diagrams/events log/alarm log in response to an incident
- corruption of system database, so that the information presented to the operator is not an accurate representation of the state of the actual electrical system
- system lockup

As I/O at the bay level, both digital and analogue will typically be handled by intelligent relays or specialised IEDs, it is therefore important to ensure that these devices have sufficient I/O capacity. If additional IEDs have to be provided solely for ensuring adequate I/O capacity, cost and space requirements will increase. There will also be an increase in the number of communications links required.

25.4 SUBSTATION AUTOMATION FUNCTIONALITY

The hardware implementation provides the physical means to implement the functionality of the substation automation scheme. The software provided in the various devices is used to implement the functionality required. The software may be quite simple or extremely complex – Table 25.1 illustrates the functionality that may be provided in a large scheme.

The description of the electrical network and the characteristics of the various devices associated with the network are held within the computer as a database or set of databases. Within each database, data is organised into tables, usually on a 'per device' basis that reflects the important characteristics of the device and its interrelationship with other devices on the network. Electrical system configuration changes require modification of the database using an appropriate software tool. The tool is normally a high level, user-friendly interface, so that modifications to the one-line can be drawn directly on-screen, with 'pick-and-place' facilities. This work would normally be done offline on the Engineers' workstation, if available, or as a background task on the control computer if not. Careful and extensive checking of the data is required, both before and after entry into the database, to ensure that no errors have been made. Full testing on the new configuration using a simulator is recommended prior to use of the new database on the main control computer to ensure that there is minimal possibility of errors.

The software is written as a set of well-proven, standard modules, so there is little or no need for new modules to be written and tested for a particular substation. The required data for the calculations performed by the software is held in the network database. This means that adding functionality later is not difficult, so long as the database design has considered this possibility. There may be problems if the electrical system configuration is altered or additional functionality added, in reading historical data prior to the change. Training of operations personnel will inevitably be required in operation of the system, configuration management and automation system maintenance. Automation system suppliers will be able to provide configuration management and system maintenance services under contract if required, often with defined cost schedules and response times so financial management of the automation scheme once installed is well-defined.

The issuing of commands to switching devices in the system has to be carefully structured, in order to prevent commands that would cause a hazard being issued. A hierarchical structure is commonly used as shown in Figure 25.6, beginning with the requirement for an operator wishing to issue a command to switching devices to log-in to the system using a password, or other means of authentication.

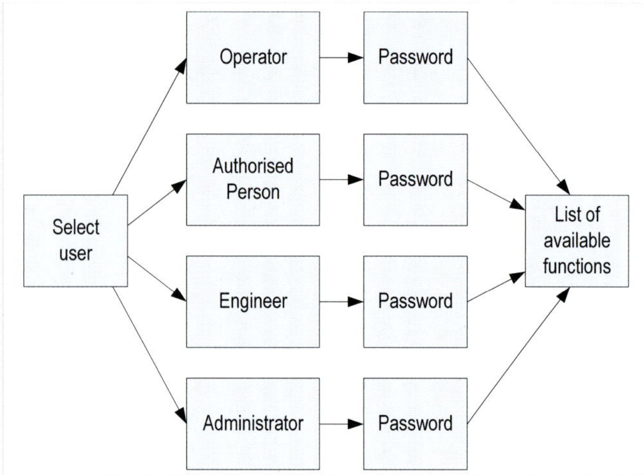

Figure 25.6 Hierarchical command structure

Different levels of authority, allowing for restrictions on the type and/or location of switching commands capable of being issued by a particular operator may be implemented at this stage – this is termed as 'role-based access'. The next level in the hierarchy is to structure the issuing of commands on an 'issue/confirm/execute' basis (Figure 25.7), so that the operator is given an opportunity to check that the command entered is correct prior to its execution.

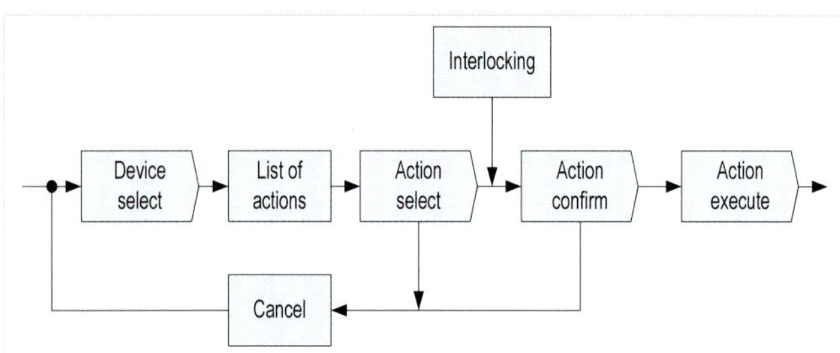

Figure 25.7: Device selection/operation

The final level in the hierarchy is implemented in software at the bay level and is actioned after the operator confirms that the switching action is to be executed. At this stage, prior to execution, the operation is checked against:

- devices locked out - i.e. prevented from operation
- interlocking of devices/switching sequences

to ensure that the command issued is safe to carry out. The action is cancelled and the operator informed if it is not safe to proceed, otherwise the action is executed and the operator informed when it is complete.

In a number of systems, some routine switching operations (e.g. transfer of a feeder from one busbar to the other in a double-bus substation) are automated in software. The operator need only request the 'bus-transfer' action to be carried out on a particular feeder, and the software is able to work out the correct switching sequence required. This minimises the possibility of operator error, but at the expense of some extra complexity in the software and more extensive checking at the factory test stage. However, since software is modular in nature, substation electrical topology is restricted to a number of standard configurations and such sequences are common, the software development is essentially a one-off activity for any particular substation control system. The development cost can be spread over the sale of a number of such systems, and hence the cost to any individual user is small compared with the potential benefits.

25.5 SYSTEM CONFIGURATION AND TESTING

These tasks, along with project management, are the most time consuming tasks in the process of realising a control and monitoring system for an electrical network. The strategies available for dealing with these problems vary between manufacturers, but typical approaches are as follows.

25.5.1 System Configuration

Software tools exist that assist in configuring a modern substation or network automation system. The extent to which the task is automated will vary, but all require as a minimum the details of the network to be controlled, extending to the individual device level (circuit breaker, isolator, disconnector, etc.). Where communication to an existing SCADA system is required, data on the logical addresses expected by the SCADA system and devices controlled remotely from the SCADA system will also be part of the data input. Use can also be made of existing databases that cover pre-defined network configurations – for example the interlocking equations for a substation bay.

Software tools will check the data for consistency, prior to creation of:

- the required equipment that forms the automation scheme, together with the required interconnections
- the databases for each individual device
- The data will divided into domains, according to the use made of the data:
- process – CB/isolator position, interlocking equation, values of current/voltage
- system – number of bay computers, hardware configuration of each bay computer, automated sequences
- graphical – the links between each mimic display and the data to be displayed
- operator – security access levels, alarm texts, etc.
- external constraints – data addresses for external database access

Once all the data has been defined, the configurator tools can define the hardware configuration to provide the required functions at least cost, and the data required for implementation of the automation scheme.

25.5.2 System Testing

The degree of testing to be carried will be defined by the customer and encapsulated in a specification for system testing. It is normal for testing of the complete functionality of the scheme to be required prior to despatch from the manufacturer – termed a 'factory acceptance test' (FAT). The larger and more complex the automation scheme, the more important for all parties that such FAT testing is carried out. It is accepted wisdom that the earlier problems are discovered, the cheaper and quicker it is to fix them. Remediation of problems on-site during commissioning is the most expensive and time-consuming activity. Manual testing of a network automation scheme is only practical for small networks, due to the cost of testing. Simulation tools are necessary for all other automation schemes. These tools fall into two categories:

- simulator tools that re-create the network to be controlled by the automation system.
- test management tools

25.5.2.1 Simulator tools

Simulator tools are dedicated to the network being tested. They will normally be provided with a simulation language that the test team can use to play scenarios, and hence determine how the automation system will react to various stimuli.

Process simulator tools may be hardware and/or software based and emulate the response of the various devices to be controlled or measured (breakers, isolators, instrument transformers or protection relays). They must be capable of closely following the dynamic response of such devices under multiple and cascade simulation scenarios. Specific tools and libraries are developed as required, which simulate the response of equipment within the control span of the automation equipment, or that of equipment outside of the span of control, in order that the response of the automation system can be tested.

Communications simulator tools are used both to load the internal communications network within the automation system to ensure that all devices are communicating correctly and that performance of the overall automation system is within specification during periods of high communications traffic loading. These simulators are standardised and a single simulator may be able to emulate several items of equipment. External communications simulators test the communications with an external system, such as a remote control centre.

25.5.3 Test Strategy

The strategy adopted for the testing of the automation system must naturally satisfy client requirements, and generally follows on of two approaches:

- a single test is carried out when all equipment for the scheme has been assembled
- incremental tests are carried out as the automation system is built up, with simulators used to represent missing equipment

The former solution is quickest and cheapest, but can give rise to problems where it is not easy to locate problems down to the device level. It is therefore used principally when an upgrade to an existing system is being carried out.

It is usual for all of the functionality to be tested, including that specified for normal conditions and specified levels of degradation within the automation system. This leads to a large number of tests being required.

25.5.4 Management of System Tests

The large number of tests required to demonstrate the compliance of an automation system with specification makes manual techniques for management of the tests cumbersome and time consuming. The end result is increased cost and timescale. Moreover, each test may result in a large amount of data to be analysed. The results of the analysis need to be presented in an easily understood form and stored as quality records, for traceability. If changes are made to software for any reason over the lifetime of the equipment, the different versions must be stored, together with a record of what the changes between versions were, and why they were made. The management of this becomes very complex, and software tools are normally used to address the issues of test schedules, test result presentation, software version control, and configuration management.

Appendix A
Terminology

The Protection Engineer must be familiar with a range of technical terms in all areas of protection engineering. Below is a list of terms and their meanings that are now commonly encountered in the Protection and Control field.

AC

Alternating Current

ACB

Air Circuit Breaker

ACCURACY

The accuracy of a transducer is defined by the limits of intrinsic error and by the limits of variations.

ACCURACY CLASS

A number used to indicate the accuracy range of a measurement transducer, according to a defined standard.

ACTIVE POWER (WATT) TRANSDUCER

A transducer used for the measurement of active electrical power

ADC

Analogue to Digital Converter

A/D CONVERSION

The process of converting an analogue signal into an equivalent digital one, involving the use of an analogue to digital converter

ADJUSTMENT

The operation intended to bring a transducer into a state of performance suitable for its use

AI

Analogue Input

AIS

Air Insulated Switchgear

ALARM

An alarm is any event (see below) tagged as an alarm during the configuration phase

ALF

Accuracy Limitation Factor of a current transformer

ALL-OR-NOTHING RELAY
An electrical relay which is intended to be energised by a quantity, whose value is either higher than that at which it picks up or lower than that at which it drops out

ANTI-PUMPING DEVICE
A feature incorporated in a Circuit Breaker or reclosing scheme to prevent repeated operation where the closing impulse lasts longer than the sum of the relay and CB operating times

AO
Analogue Output

AR (OR A/R)
Auto Reclose:

A function associated with CB, implemented to carry out reclosure automatically to try to clear a transient fault

ARCING TIME
The time between instant of separation of the CB contacts and the instant of arc extinction

ARIP
Auto-Reclose In Progress

AUTO-TRANSFORMER
A power transformer that does not provide galvanic isolation between primary and secondary windings

AUX
Auxiliary

AUXILIARY CIRCUIT
A circuit which is usually energised by the auxiliary supply but is sometimes energised by the measured quantity

AUXILIARY RELAY
An all-or-nothing relay energised via another relay, for example a measuring relay, for the purpose of providing higher rated contacts, or introducing a time delay, or providing multiple outputs from a single input.

AUXILIARY SUPPLY
An a.c. or d.c. electrical supply other than the measured quantity which is necessary for the correct operation of the transducer

AVR
Automatic Voltage Regulator

AWG
American Wire Gauge

B
Susceptance

BACK-UP PROTECTION
A protection system intended to supplement the main protection in case the latter should be ineffective, or to deal with faults in those parts of the power system that are not readily included in the operating zones of the main protection

BAR
Block Auto-Reclose signal

BAY
The switchgear, isolators and other equipment typically associated with a single circuit, bus section, or bus coupler.

BC
Bay Computer (or Bay Controller)

Computer dedicated to the control of one or several bays within a substation

BCD
Binary Coded Decimal

BCP
Bay Control Point

A local keypad at bay level to control the elements of a single bay

BIASED RELAY
A relay in which the characteristics are modified by the introduction of some quantity other than the actuating quantity, and which is usually in opposition to the actuating quantity

BIAS CURRENT
The current used as a bias quantity in a biased relay

BIOS
Basic Input/Output System (of a computer or microprocessor)

BURDEN

The loading imposed by the circuits of the relay on the energising power source or sources, expressed as the product of voltage and current (volt-amperes, or watts if d.c.) for a given condition, which may be either at 'setting' or at rated current or voltage.

The rated output of measuring transformers, expressed in VA, is always at rated current or voltage and it is important, in assessing the burden imposed by a relay, to ensure that the value of burden at rated current is used

C

Capacitance

C/O

Changeover Contacts ('form c')

CAD

Computer Aided Design

CALIBRATION

The set of operations, which establish under specified conditions, the relationship between values indicated by a transducer and the corresponding values of a quantity realised by a reference standard.

(This should not be confused with 'adjustment')

CB

Circuit Breaker

CBCT

Core Balance Current Transformer

CH

Channel. Usually a communication or signalling channel

CHARACTERISTIC ANGLE

The angle between the vectors representing two of the energising quantities applied to a relay and used for the declaration of the performance of the relay

CHARACTERISTIC CURVE

The curve showing the operating value of the characteristic quantity corresponding to various values or combinations of the energising quantities

CHARACTERISTIC IMPEDANCE RATIO (C.I.R.)

The maximum value of the System Impedance Ratio up to which the relay performance remains within the prescribed limits of accuracy

CHARACTERISTIC QUANTITY

A quantity, the value of which characterizes the operation of the relay, for example, current for an overcurrent relay, voltage for a voltage relay, phase angle for a directional relay, time for an independent time delay relay, impedance for an impedance relay

CHECK PROTECTION SYSTEM

An auxiliary protection system intended to prevent tripping due to inadvertent operation of the main protection system

CHP

Combined Heat and Power

CIP

Critical Infrastructure Protection standards, for example covering cyber security.

CIRCUIT INSULATION VOLTAGE

The highest circuit voltage to earth on which a circuit of a transducer may be used and which determines its voltage test

CLASS INDEX

The number which designates the accuracy class

CLOSING IMPULSE TIME

The time during which a closing impulse is given to the CB

CLOSING TIME

The time for a CB to close, from the time of energisation of the closing circuit to making of the CB contacts

COMPLIANCE VOLTAGE (ACCURACY LIMITING OUTPUT VOLTAGE)

For current output signals only, the output voltage up to which the transducer meets its accuracy specification

CONJUNCTIVE TEST

A test of a protection system including all relevant components and ancillary equipment appropriately interconnected. The test may be parametric or specific

CONVERSION COEFFICIENT

The relationship of the value of the measurand to the corresponding value of the output

CORE BALANCE CURRENT TRANSFORMER

A ring-type Current Transformer in which all primary conductors are passed through the aperture of the CBCT. Hence the secondary current is proportional only to any imbalance in current. Used for sensitive earth-fault protection

COUNTING RELAY

A relay that counts the number of times it is energised and actuates an output after a desired count has been reached.

COURIER

A proprietary data protocol for exchanging information associated with substation automation.

CYCLIC REDUNDANCY CHECK

CSV

Character (or Comma) Separated Values format

A widely used format for the exchange of data between different software, in which the individual data items a separated by a known character – usually a comma

CT

Current Transformer

CURRENT TRANSDUCER

A transducer used for the measurement of a.c. current

CCVT (OR CVT)

Capacitor Coupled Voltage Transformer

A voltage transformer that uses capacitors to obtain voltage division. These are used at EHV voltages instead of electromagnetic VTs, for size and cost reasons

DAR

Delayed auto-reclose

DBMS

Data Base Management system

DC

Direct current

DCP

Device Control Point:

local keypad on device level to control the switchgear, often combined with local/remote switch

DCS

Digital Control System

DEAD TIME (AUTO-RECLOSE)

The time between the fault arc being extinguished and the CB contacts re-making

DEF

Directional Earth Fault protection

DE-IONISATION TIME (AUTO-RECLOSE)

The time required for dispersion of ionised air after a fault is cleared so that the arc will not re-strike on re-energisation

DELAYED AUTO-RECLOSE

An auto-reclosing scheme which has a time delay in excess of the minimum required for successful operation

DEPENDABILITY

A measure of a protection scheme's ability to operate correctly when it is called upon

DEPENDENT TIME MEASURING RELAY

A measuring relay for which times depend, in a specified manner, on the value of the characteristic quantity

DFT

Discrete Fourier Transform

DG

Distributed Generation

DI

Digital Input

DIGITAL SIGNAL PROCESSOR

A microprocessor optimised in both hardware architecture and software instruction set for the processing of analogue signals digitally, through use of the DFT and similar techniques

DIGITAL SIGNAL PROCESSING

A technique for the processing of digital signals by various filter algorithms to obtain some desired characteristics in the output. The input signal to the processing algorithm is usually the digital representation of an analogue signal, obtained by A/D conversion

DIRECT-ON-LINE (DOL)

A method of motor starting, in which full line voltage is applied to a stationary motor

DIRECTIONAL RELAY

A protection relay in which the tripping decision is dependent in part upon the direction in which the measured quantity is flowing

DISCRIMINATION

The ability of a protection system to distinguish between power system conditions for which it is intended to operate and those for which it is not intended to operate

DISTORTION FACTOR

The ratio of the r.m.s. value of the harmonic content to the r.m.s. value of the non-sinusoidal quantity

DLR

Dynamic Line Rating. This is a technique to adapt thermal protection to include additional line heating and cooling factors, in addition to the traditional inputs of current and ambient temperature. Wind speed, wind direction and solar radiation sensors may be added to the scheme in order to more accurately simulate the thermal state and allow greater line loadability.

DNP

Distributed Network Protocol. A communication protocol used on secondary networks between HMI, substation computers or Bay Computers and protective devices

DO

Digital Output

DROP-OUT (OR DROP-OFF)

A relay drops out when it moves from the energised position to the un-energised position

DROP-OUT/PICK-UP RATIO

The ratio of the limiting values of the characteristic quantity at which the relay resets and operates. This value is sometimes called the differential of the relay

DSP

Digital Signal Processing

DT

Definite time

EARTH FAULT PROTECTION SYSTEM

A protection system which is designed to respond only to faults to earth

EARTHING TRANSFORMER

A three-phase transformer intended essentially to provide a neutral point to a power system for the purpose of earthing

EFFECTIVE RANGE

The range of values of the characteristic quantity or quantities, or of the energising quantities to which the relay will respond and satisfy the requirements concerning it, in particular those concerning precision

EFFECTIVE SETTING

The 'setting' of a protection system including the effects of current transformers. The effective setting can be expressed in terms of primary current or secondary current from the current transformers and is so designated as appropriate

ELECTRICAL RELAY

A device designed to produce sudden predetermined changes in one or more electrical circuits after the appearance of certain conditions in the electrical circuit or circuits controlling it

NOTE: The term 'relay' includes all the ancillary equipment calibrated with the device

ELECTROMECHANICAL RELAY

An electrical relay in which the designed response is developed by the relative movement of mechanical elements under the action of a current in the input circuit

EMC

Electro-Magnetic Compatibility

EMBEDDED GENERATION

Generation that is connected to a distribution system (possibly at LV instead of HV) and hence poses particular problems in respect of electrical protection

E.M.F.

Electro-motive Force (or voltage)

ENERGISING QUANTITY

The electrical quantity, either current or voltage, which along or in combination with other energising quantities, must be applied to the relay to cause it to function

EPROM

Electrically Programmable Read Only Memory

ERROR (OF A TRANSDUCER)

The actual value of the output minus the intended value of the output, expressed algebraically

EUT

Equipment Under Test in certification activities

EVENT

An event is any information acquired or produced by the digital control system

FAT

Factory Acceptance Test. Validation procedures witnessed by the customer at the factory

FAULT PASSAGE INDICATOR

A sensor that detects the passage of current in excess of a set value (i.e. current due to a fault) at the location of the sensor. Hence, it indicates that the fault lies downstream of the sensor

FBD

Functional Block Diagram: One of the IEC 61131-3 programming languages

FFT

Fast Fourier Transform

FIDUCIAL VALUE

A clearly specified value to which reference is made in order to specify the accuracy of a transducer

(For transducers, the fiducial value is the span, except for transducers having a reversible and symmetrical output when the fiducial value may be either the span or half the span as specified by the manufacturer. It is still common practice, however, for statements of accuracy for frequency transducers to refer to 'percent of centre-scale frequency' and, for phase angle transducers, to an error in electrical degrees.)

FLC

Full Load Current (full nominal rated load current for the electrical equipment)

FPI

Fault Passage Indicator

FREQUENCY TRANSDUCER

A transducer used for the measurement of the frequency of an a.c. electrical quantity

FULL DUPLEX COMMUNICATIONS

A communications system in which data can travel simultaneously in both directions

FUZZY LOGIC

A probability-based extension to straight binary logic, whereby intermediate fractional values between limits of 0 and 1 are used

G

Conductance

GATEWAY

The Gateway is a computer which provides interfaces between the local computer system and one or several SCADA (or RCC) systems

GIS

Gas Insulated Switchgear (usually SF_6)

GLOBAL POSITIONING SYSTEM

A system used for locating objects on Earth precisely, using a system of satellites in geostationary orbit in Space. Used by some numerical relays to obtain accurate UTC time information

GMT

Greenwich Mean Time

GOOSE

Generic Object Oriented Substation Event. An IEC61850 Ethernet control mechanism in which status and value data is grouped into a data set and transmitted within a set time period. Typically applied for peer-peer communication between devices such as relays and other IEDs

GPS
Global Positioning System

HALF- DUPLEX COMMUNICATIONS
A communications system in which data can travel in both directions, but only in one direction at a time

HI-Z
Sometimes used to indicate high impedance unit protection, but sometimes used to indicate downed conductor protection.

HIGH-SPEED RECLOSING
A reclosing scheme where re-closure is carried out without any time delay other than that required for de-ionisation, etc.

HMI
Human Machine Interface. The means by which a human inputs data to and receives data from a computer-based system. This usually takes the form of a Personal Computer (PC) (desktop or portable) with keyboard, screen and pointing device. This term is also used to describe the front panel interface on numerical relays.

HRC
High Rupturing Capacity (applicable to fuses)

HSR
High Speed Auto-Reclose. Also denotes High-availability Seamless ring for redundant Ethernet topology as per IEC62439

HV
High Voltage

HVDC
High Voltage Direct Current

I
Current

I/O
Input/Output

ICCP
Term used for IEC 60870-6-603 protocol

ICT
Interposing Current Transformer

IDMT
Inverse Definite Minimum Time

IEC
International Electrotechnical Commission

IED
Intelligent Electronic Device. Equipment containing a microprocessor and software used to implement one or more functions in relation to an item of electrical equipment (e.g. a bay controller, remote SCADA interface/protocol converter). A microprocessor-based numerical relay is also an IED. IED is a generic term used to describe any microprocessor-based equipment, apart from a computer

IEEE
Institute of Electrical and Electronics Engineers

IGBT
Insulated Gate Bipolar Transistor. These are typically used in high-speed high-break static output contacts

INDEPENDENT TIME MEASURING RELAY
A measuring relay, the specified time for which can be considered as being independent, within specified limits, of the value of the characteristic quantity

INFLUENCE QUANTITY
A quantity which is not the subject of the measurement but which influences the value of the output signal for a constant value of the measurand

INPUT QUANTITY
The quantity, or one of the quantities, which constitute the signals received by the transducer from the measured system

INSTANTANEOUS RELAY
A relay that operates and resets with no intentional time delay.

NOTE: All relays require some time to operate; it is possible, within the above definition, to discuss the operating time characteristics of an instantaneous relay

INSULATED GATE BIPOLAR TRANSISTOR
A special design of transistor that is suitable for handling high voltages and currents (relative to an ordinary transistor). Frequently used in static power control equipment (inverters, controlled rectifiers, etc) due to the flexibility of control of the output

INTRINSIC ERROR

An error determined when the transducer is under reference conditions

INVERSE TIME DELAY RELAY

A dependent time delay relay having an operating time which is an inverse function of the electrical characteristic quantity

INVERSE TIME RELAY WITH DEFINITE MINIMUM TIME (IDMT)

An inverse time relay having an operating time that tends towards a minimum value with increasing values of the electrical characteristic quantity

IP

Internet Protocol

IRIG-B

An international standard for time synchronisation

ISO

International Standards Organisation

K-BUS

A proprietary serial communications link used by Alstom's range of IEDs.

KNEE-POINT E.M.F.

That sinusoidal e.m.f. applied to the secondary terminals of a current transformer, which, when increased by 10%, causes the exciting current to increase by 50%

L

Inductance

LAN

Local Area Network

LCD

Liquid Crystal Display

LD

Ladder Diagram. One of the IEC 61131-3 programming languages. This can also mean an IEC61850 Logical Device.

LDC

Line drop compensation

LED

Light Emitting Diode

LIMITING VALUE OF THE OUTPUT CURRENT

The upper limit of output current which cannot, by design, be exceeded under any conditions

LN

Logical Node. Standardised IEC61850 data model describing the logical attributes of a protection or control function

LOCAL CONTROL MODE

When set for a given control point it means that the commands can be issued from this local point within the substation

LOCK-OUT (AUTO-RECLOSE)

Prevention of a CB reclosing after tripping

LOL (LOSS OF LIFE)

Loss of life in the context of transformer asset monitoring

LOW-SPEED AUTO-RECLOSE

See Delayed auto-reclose

LV

Low Voltage

MAIN PROTECTION

The protection system which is normally expected to operate in response to a fault in the protected zone

MAXIMUM PERMISSIBLE VALUES OF THE INPUT CURRENT AND VOLTAGE

Values of current and voltage assigned by the manufacturer which the transducer will withstand indefinitely without damage

MCB

Miniature Circuit Breaker

MCCB

Moulded Case Circuit Breaker

MEAN-SENSING TRANSDUCER

A transducer which actually measures the mean (average) value of the input waveform but which is adjusted to give an output corresponding to the r.m.s. value of the input when that input is sinusoidal

MEASURAND

A quantity subjected to measurement

MEASUREMENT CENTRE

A non-protection IED capable of measuring a large number of electrical system parameters

MEASURING ELEMENT

A unit or module of a transducer which converts the measurand, or part of the measurand, into a corresponding signal

MEASURING RANGE

That part of the span where the performance complies with the accuracy requirements

MEASURING RELAY

An electrical relay intended to switch when its characteristic quantity, under specified conditions and with a specified accuracy attains its operating value

METERING (NON-TARIFF)

Values computed depending on the values of digital or analogue inputs during variable periods

METERING (TARIFF)

Energy values computed from digital and/or analogue inputs during variable periods and dedicated to energy measurement for billing (tariff) purposes

MODBUS

Communication protocol used on secondary networks between HMI, substation computers or Bay Computers and protective devices

MU

Merging Unit. A merging unit connects voltage and current measured quantities to the protection and control devices, typically using IEC61850-9-2 Ethernet sampled values. The main task of the merging unit is to merge current and voltage data from the three phases, and in some applications it must translate the input received from non-conventional sensors.

MULTI-ELEMENT TRANSDUCER

A transducer having two or more measuring elements. The signals from the individual elements are combined to produce an output signal corresponding to the measurand

MULTI-SECTION TRANSDUCER

A transducer having two or more independent measuring circuits for one or more functions

MULTI-SHOT RECLOSING

A reclosing scheme that permits more than one reclosing operation of a CB after a fault occurs before lock-out occurs

MV

Medium Voltage

N

Neutral

N/C

Normally Closed

N/O

Normally Open

NCIT

Non-Conventional Instrument Transformer, or transducer. Typically used to describe optical, Rogowski, or other advanced solutions.

NERC

North American Electricity Reliability Corporation

NOMINAL RANGE OF USE

A specified range of values which it is intended that an influence quantity can assume without the output signal of the transducer changing by amounts in excess of those specified

NOTCHING RELAY

A relay which switches in response to a specific number of applied impulses

NPS

Negative Phase Sequence

NUMERICAL RELAY

A protection relay which utilises Digital Signal Processing to execute the protection algorithms in software

NV
Non-volatile. Typically used to describe a secure memory facility in numerical relays.

NVD
Neutral Voltage Displacement

OCB
Oil Circuit Breaker

OFF-LOAD TAP CHANGER
A tap changer that is not designed for operation while the transformer is supplying load

OHL
Overhead line

OLTC
On Load Tap Changer.

ON LOAD TAP CHANGER
A tap changer that can be operated while the transformer is supplying load.

OPENING TIME
The time between energisation of a CB trip coil and the instant of contact parting

OPERATING CURRENT (OF A RELAY)
The current at which a relay will pick up

OPERATING TIME (CB)
The time between energisation of a CB trip coil and arc extinction

OPERATING TIME (RELAY)
With a relay de-energised and in its initial condition, the time which elapses between the application of a characteristic quantity and the instant when the relay operates

OPERATING TIME CHARACTERISTIC
The curve depicting the relationship between different values of the characteristic quantity applied to a relay and the corresponding values of operating time

OPERATING VALUE
The limiting value of the characteristic quantity at which the relay actually operates

OPGW
Optical Ground Wire – a ground wire that includes optical fibres to provide a communications link

OPTO
An optically coupled logic input (also referred to as a binary input)

OSI 7-LAYER MODEL
The Open Systems Interconnection 7-layer model is a model developed by ISO for modelling of a communications network.

OUTPUT COMMON MODE INTERFERENCE VOLTAGE
An unwanted alternating voltage which exists between each of the output terminals and a reference point

OUTPUT CURRENT (OF A TRANSDUCER)
The current produced by the transducer which is an analogue function of the measurand

OUTPUT LOAD
The total effective resistance of the circuits and apparatus connected externally across the output terminals

OUTPUT POWER (OF A TRANSDUCER)
The power available at the transducer output terminals

OUTPUT SERIES MODE INTERFERENCE VOLTAGE
An unwanted alternating voltage appearing in series between the output terminals and the load

OUTPUT SIGNAL
An analogue or digital representation of the measurand

OUTPUT SPAN (SPAN)
The algebraic difference between the lower and upper nominal values of the output signal

OVERCURRENT RELAY
A protection relay whose tripping decision is related to the degree by which the measured current exceeds a set value.

OVERSHOOT TIME
The overshoot time is the difference between the operating time of the relay at a specified value of the input energising

quantity and the maximum duration of the value of input energising quantity which, when suddenly reduced to a specific value below the operating level, is insufficient to cause operation

PARAMETRIC CONJUNCTIVE TEST

A conjunctive test that ascertains the range of values of each parameter for which the test meets specific performance requirements

PCB

Printed Circuit Board

PCC

Point of Common Coupling

PDC

See Phasor Data Concentrator

PHASE ANGLE TRANSDUCER

A transducer used for the measurement of the phase angle between two a.c. electrical quantities having the same frequency

PHASOR DATA CONCENTRATOR

An IED, or software function, able to receive Synchrophasor frames from a number of PMUs, archive them in non-volatile memory, and/or retransmit them to upstream equipment with the data content merged, filtered or at a different resolution than the incoming data

PHASOR MEASUREMENT UNIT (PMU)

An IED whose function is to accurately measure time-synchronised current, voltage or other phasor vectors, and communicate these to other devices in a standard format.

PICK-UP

A relay is said to 'pick-up' when it changes from the de-energised position to the energised position

PILOT CHANNEL

A means of interconnection between relaying points for the purpose of protection

PLC

Programmable Logic Controller

A specialised computer for implementing control sequences using software

PLCC

Power Line Carrier Communications

PMU

See Phasor Measurement Unit

POINT OF COMMON COUPLING

The interface between an in-plant network containing embedded generation and the utility distribution network to which the in-plant network is connected

POW

Point-on- Wave

Point-on-wave switching is the process to control moment of switching to minimise the effects (inrush currents, overvoltages)

POWER FACTOR

The factor by which it is necessary to multiply the product of the voltage and current to obtain the active power

POWER QUALITY

Classification of power system disturbances according to standards such as EN50160

POWER LINE CARRIER COMMUNICATIONS

A means of transmitting information over a power transmission line by using a carrier frequency superimposed on the normal power frequency

PPS

Positive Phase Sequence. Also Pulse Per Second in time synchronising applications

PROTECTED ZONE

The portion of a power system protected by a given protection system or a part of that protection system

PROTECTION EQUIPMENT

The apparatus, including protection relays, transformers and ancillary equipment, for use in a protection system

PROTECTION RELAY

A relay designed to initiate disconnection of a part of an electrical installation or to operate a warning signal, in the case of a fault or other abnormal condition in the installation. A protection relay may include more than one electrical element and accessories

PROTECTION SCHEME

The co-ordinated arrangements for the protection of one or more elements of a power system. A protection scheme may comprise several protection systems

PROTECTION SYSTEM

A combination of protection equipment designed to secure, under predetermined conditions, usually abnormal, the disconnection of an element of a power system, or to give an alarm signal, or both

PROTOCOL

A set of rules that define the method in which a function is carried out – commonly used in respect of communications links, where it defines the hardware and software features necessary for successful communication between devices.

PRP

Parallel Redundancy Protocol. A redundant star Ethernet topology as per IEC62439

PSB

Power Swing Blocking

PSM

Plug Setting Multiple – a term used in conjunction with electromechanical relays, denoting the ratio of the fault current to the current setting of the relay

PSTN

Public Switched Telephone Network

PT100

Platinum resistance temperature probe

PTP

Precision time Protocol, typically as per IEEE1588

R

Resistance

RATIO CORRECTION

A feature of digital/numerical relays that enables compensation to be carried out for a CT or VT ratio that is not ideal

RATING

The nominal value of an energising quantity that appears in the designation of a relay. The nominal value usually corresponds to the CT and VT secondary ratings

RBAC

Role-Based Access Control

RCA

Relay Characteristic Angle. The centre of the directional characteristic

RCD

Residual Current Device. A protection device which is actuated by the residual current

RCP

Remote Control Point

The Remote Control Point is a SCADA interface. Several RCP's may be managed with different communication protocols. Physical connections are done at a Gateway or at substation computers or at a substation HMI

REACTIVE POWER (VAR) TRANSDUCER

A transducer used for the measurement of reactive electrical power

RECLAIM TIME (AUTO-RECLOSE)

The time between a successful closing operation, measured from the time the auto-reclose relay closing contact makes until a further reclosing sequence is permitted in the event of a further fault occurring

REF

Restricted Earth Fault

REFERENCE CONDITIONS

Conditions of use for a transducer prescribed for performance testing, or to ensure valid comparison of results of measurement

REFERENCE RANGE

A specified range of values of an influence quantity within which the transducer complies with the requirements concerning intrinsic errors

REFERENCE VALUE

A specified single value of an influence quantity at which the transducer complies with the requirements concerning intrinsic errors

RELAY
See Protection relay

RESETTING VALUE
The limiting value of the characteristic quantity at which the relay returns to its initial position

RESIDUAL CURRENT
The algebraic sum, in a multi-phase system, of all the line currents

RESIDUAL VOLTAGE
The algebraic sum, in a multi-phase system, of all the line-to-earth voltages

RESPONSE TIME
The time from the instant of application of a specified change of the measurand until the output signal reaches and remains at its final steady value or within a specified band centred on this value

REVERSIBLE OUTPUT CURRENT
An output current which reverses polarity in response to a change of sign or direction of the measurand

RIPPLE CONTENT OF THE OUTPUT
With steady-state input conditions, the peak-to-peak value of the fluctuating component of the output

RMS
Root Mean Square

RMS SENSING TRANSDUCER
A transducer specifically designed to respond to the true r.m.s. value of the input and which is characterised by the manufacturer for use on a specified range of waveforms

RMU
Ring Main Unit

ROCOF
Rate Of Change Of Frequency (protection relay)

RS
Recommended Standard. Terminology still commonly used when referring to EIA serial communication

RSTP
Rapid Spanning Tree Protocol

RTD
Resistance Temperature Detector

RTOS
Real Time Operating System

RTU
Remote Terminal Unit. An IED used specifically for data gathering and control within a substation. Sometimes may include control/ monitoring/storage functions

SAT
Site Acceptance Test

Validation procedures for equipment executed with the customer on site

SCADA
Supervisory Control and Data Acquisition

SCL
Substation Configuration Language

Normalised configuration language for substation modelling (as expected by IEC 61850-6)

SCP
Substation Control Point

HMI computers at substation level allowing the operators to control the substation

SCS
Substation Control System

SECURITY
A measure of a protection scheme's ability to restrain and prevent spurious operation, when no operation is required

SEF
Sensitive Earth Fault

SETTING
The limiting value of a 'characteristic' or 'energising' quantity at which the relay is designed to operate under specified conditions. Such values are usually marked on the relay and

may be expressed as direct values, percentages of rated values, or multiples

SFC

Sequential Function Chart: One of the IEC 61131-3 programming languages

SIMPLEX COMMUNICATIONS SYSTEM

A communications system in which data can only travel in one direction

SINGLE-SHOT RECLOSING

An auto-reclose sequence that provides only one reclosing operation, lock-out of the CB occurring if it subsequently trips

S.I.R.

System Impedance Ratio

SINGLE ELEMENT TRANSDUCER

A transducer having one measuring element

SIPS

System Integrity Protection Schemes

SNTP

Simple Network Time Protocol

SOE

Sequence Of Events

SOTF

Switch On To Fault (protection)

SPECIFIC CONJUNCTIVE TEST

A conjunctive test using specific values of each of the parameters

SPRING WINDING TIME

For spring-closed CB's, the time for the spring to be fully charged after a closing operation

ST

Structured Text: One of the IEC 61131-3 programming languages

STABILITY (OF A TRANSDUCER)

The ability of a transducer to keep its performance characteristics unchanged during a specified time, all conditions remaining constant

STABILITY (OF A PROTECTION SYSTEM)

The quantity whereby a protection system remains inoperative under all conditions other than those for which it is specifically designed to operate

STABILITY LIMITS (OF A PROTECTION SYSTEM)

The r.m.s. value of the symmetrical component of the through fault current up to which the protection system remains stable

STARTING RELAY

A unit relay which responds to abnormal conditions and initiates the operation of other elements of the protection system

STATCOM

A particular type of Static Var Compensator, in which Power Electronic Devices are used to generate the reactive power required, rather than capacitors and inductors

STATIC RELAY

An electrical relay in which the designed response is developed by electronic, magnetic, optical or other components without mechanical motion. Excludes relays using digital/numeric technology

STC

Short Time Current (rating of a CT)

STORAGE CONDITIONS

The conditions, defined by means of ranges of the influence quantities, such as temperature, or any special conditions, within which the transducer may be stored (non-operating) without damage

SVC

Static Var Compensator

SYNCHROPHASOR

A vector synchronised to real time, accurately defining the magnitude and phase of a power system quantity. Typically IEEE C37.118 applies

SYSTEM DISTURBANCE TIME (AUTO-RECLOSE)

The time between fault inception and CB contacts making on successful re-closure

SYSTEM IMPEDANCE RATIO
The ratio of the power system source impedance to the impedance of the protected zone

T101
Term used for IEC 60870-5-101 protocol

TAP CHANGER
A mechanism, usually fitted to the primary winding of a transformer, to alter the turns ratio of the transformer by small discrete amounts over a defined range

TCP/IP
Transmission Control Protocol/Internet Protocol. A common protocol for the transmission of messages over the Internet

TCS
Trip Circuit Supervision

TC57
Technical Committee 57 working for the IEC and responsible for producing standards in the field of Protection (e.g. IEC 61850)

TF
Transfer Function of a device (usually an element of a control system. Also Transient Factor (of a CT)

THD
Total Harmonic Distortion

THROUGH FAULT CURRENT
The current flowing through a protected zone to a fault beyond that zone

TIME DELAY
A delay intentionally introduced into the operation of a relay system

TIME DELAY RELAY
A relay having an intentional delaying device

TIME DIAL (TD)
An alternative to the time multiplier setting (TMS), applied to IEEE inverse-time curves

TIME MULTIPLIER SETTING (TMS)
Used to Adapt the speed of operation of inverse-time curves

TPI
Tap Position Indicator (for transformers)

TRANSDUCER (ELECTRICAL MEASURING TRANSDUCER)
A device that provides a d.c. output quantity having a definite relationship to the a.c. measurand

TRANSDUCER WITH OFFSET ZERO (LIVE ZERO)
A transducer which gives a predetermined output other than zero when the measurand is zero

TRANSDUCER WITH SUPPRESSED ZERO
A transducer whose output is zero when the measurand is less than a certain value

TVE
Total Vector Error. A composite percentage error measurement, used to describe the accuracy requirement for phasor measurement units.

UDP
User Datagram Protocol.

UNIT ELECTRICAL RELAY
A single relay that can be used alone or in combinations with others

UNIT PROTECTION
A protection system that is designed to operate only for abnormal conditions within a clearly defined zone of the power system

UNRESTRICTED PROTECTION
A protection system which has no clearly defined zone of operation and which achieves selective operation only by time grading

UCA
Utility Communications Architecture. UCA2 was a forerunner of IEC61850

UPS
Uninterruptible Power Supply

USB
Universal Serial Bus

UTC
Universal Time Coordinated. The precise internationally recognised time reference, equivalent to GMT.

V
Voltage

VCB
Vacuum Circuit Breaker

VDEW
Term used for IEC 60870-5-103 protocol. The VDEW protocol is a subset of the IEC 60870-5-103 protocol

VECTOR GROUP COMPENSATION
A feature of digital and numerical relays that compensates for the phase angle shift that occurs in transformers (including VTs) due to use of dissimilar winding connections – e.g. transformers connected delta/star

VOLTAGE TRANSDUCER
A transducer used for the measurement of a.c. voltage

VT
Voltage Transformer

WAMPAC
Wide-Area Monitoring/Measurement, Protection and Control

WAN
Wide Area Network

X
Reactance

Y
Admittance

Z
Impedance

ZONE
The boundary limits, typically for a protected unit in the power system, within which protection is expected to operate

Appendix B
ANSI/IEC Symbols

The most common methods for indicating protection device functions are defined by the IEEE (ANSI) C37-2, which uses an alphanumeric code and IEC61850, which uses a numeric code. There are also other common designations, some of which originate from IEC 60617. A list of common ANSI device numbers and their equivalents is given in this appendix.

FUNCTIONS	ANSI	IEC 61850	Other Common Designations
Arc flash Detector	AFD	SARC	
Automatic Tap Change Control		ATCC	AVR
Autoreclose	79	RREC	DAR
Blocking (eg. Power Swing Blocking) or "out-of-step"	68	RPSB	PSB, OST, OOS
Breaker Failure	50BF	RBRF	BFP
Broken Conductor	46BC	PTOC	I2/I1>
Busbar Differential	87BB	PDIF	Idiff>
Circuit Breaker	52	XCBR	CB
Circuit Breaker Closed Auxiliary Contacts	52a	XCBR	
Circuit Breaker Open Auxiliary Contacts	52b	XCBR	
Circuit Switch	89	XSWI	
Clock or Timing Source	CLK		
Closing Coil	CC		
Cold Load Pick-Up	51CLP		CLP
Current Transformer Supervision		RVCS	CTS
Data Communications Device	16		
Digital Fault Recorder	DFR	RDRE, RADR, RBDR	
Delta Directional Comparison			ΔI / ΔV
Directional Earth Fault Overcurrent	67N	PTOC	IN>, t>, DEF
Directional Over Power	32O	PDOP	P>
Directional Overcurrent	67P	PTOC	I>
Directional Under Power	32U	PDUP	P<
Distance	21	PDIS	Z
Distance Aided Schemes	21/85	PSCH	
Disturbance Recorder	DDR	RDRE, RADR, RBDR	DR
Dynamic Line Rating	49DLR	PTTR	DLR
Earth Fault Overcurrent	51N	PTOC	IN>
Ethernet Switch	16ES		
Fault Locator	21FL	RFLO	DTF
Fuse Fail Overcurrent	51FF	PTOC	
Generator Differential	87G	PDIF	Idiff>
High Impedance Earth Fault Detection	HIZ	PHIZ	Hi-Z
Human Machine Interface	HMI	IHMI	
Instantaneous Overcurrent	50	PTOC	I>
Interlocked Overcurrent Busbar Protection Scheme	51BB	PTOC	
Interlocking	3	CILO	
Interturn Fault	50	PDIF	
Line Differential	87L	PDIF	Idiff>
Load Restoration	81R	PTOF	

FUNCTIONS	ANSI	IEC 61850	Other Common Designations
Lockout Relay	86		
Loss of Field / Under Excitation	40	PDUP	
Loss of Life	LoL	MMTR	
Motor Anti-Backspin			ABS
Motor Differential	87M	PDIF	Idiff>
Motor Emergency Restart	66/86		
Motor Locked Rotor	51LR	PMRI/PMSS	
Motor Number of Starts	66	PMRI/PMSS	
Motor Reacceleration Authorisation	27LV		
Motor Restart Inhibition	49, 66	PMRI	
Motor Starting Time Supervision	48, 51LR, 49R	PMRI/PMSS	
Negative Sequence Overvoltage	47	PTOV	V2>
Negative Phase Sequence Thermal	46T	PTTR	
Neutral Admittance			YN
Neutral Displacement Voltage (Residual Overvoltage)	59N	PTOV	VN>
Out of Step Trip (Pole Slip)	78	PPAM	OST, OOS
Overfrequency	81O	PTOF	f>
Overspeed	12		ω>
Overtemperature	26	PTTR	θ>
Overvoltage	59	PTOV	V>
Phase Angle	78		φ>
Phase Comparison	87P	PDIF	
Phase-Balance (eg. Negative Sequence) Current	46	PTOC	I2>
Phasor Data Concentrator	PDC		
Phasor Measurement Unit	PMU		
Pilot-Wire or Carrier Communications	85	PSCH	
Point-on-Wave Switching		CPOW	POW
Pole Dead (Circuit Breaker Open)		PTUV	
Positive Sequence Undervoltage	47	PTUV	V1<
Power Factor (Over)	55O	POPF	PF>, cosφ>
Power Factor (Under)	55U	PUPF	PF<, cosφ<
Power Quality Monitor	PQM	QFVR, QITR, QIUB, QVTR, QVUB, QVVR	PQ
Push Button	PB		
Rate of Change of Frequency	81R	PFRC	df/dt
Remote Terminal Unit / Data Concentrator	RTU		
Restricted Earth Fault (Biased)	64N	PDIF	Idiff>, REF
Restricted Earth Fault (High Imp.)	64N	PDIF	Idiff>, REF
Reverse Power	32R	PDOP	
Rotor Earth Fault	64R	PEFI	

FUNCTIONS	ANSI	IEC 61850	Other Common Designations
Rotor Thermal Overload	49R	PTTR	
Router	16ER		
Sensitive Directional Earth Fault Overcurrent	67SEF	PTOC	Isef>
Sequence of Events Recorder	SER		SOE
Switch on to Fault	SOTF	PSOF	
Stator Earth Fault	64S	PTOC	
100% Stator Earth Fault - 3rd harmonic undervoltage, 3rd harmonic overvoltage, low frequency injection	27TN, 59TN, 64S	PTUV/PTOV/PEFI	
Stator Thermal Overload	49S	PTTR	
Stub Bus Protection	50ST	PTOC	
Substation Metering	MET	MMTR, MMXU	
Synchronism Check	25	RSYN	CS
Thermal Device (eg. RTD, thermistor)	26	PTTR	RTD
Thermal Overload	49	PTTR	
Through Fault Monitoring	Thru	MMXU	TF
Time Delay	2		t>, TD
Time Overcurrent (eg. IDMT)	51	PTOC	I>, t>
Transformer Differential	87T	PDIF	Idiff>
Transformer Inrush Detection	68	PHAR	
Transient Earth Fault		PTEF	TGF
Trip Circuit Monitor / Supervision	TCM		TCS
Trip Coil	TC		
Tripping Relay	94	PTRC	
Turbine Abnormal Frequency	81AB	PTAF	
Undercurrent	37	PTUC	I<
Underfrequency	81U	PTUF	f<
Underspeed or Zero Speed Device	14	PZSU	ω<
Undervoltage	27	PTUV	V<
Unintentional Energisation (Dead Machine Protection)	50/27	PDMP	
Voltage Balance	60	PTOV	
Voltage Dependent Overcurrent	51V	PVOC	
Volts per Hertz Overfluxing	24	PVPH	V/Hz
Voltage Transformer Supervision	VTS	RVCS	
Voltage Vector Shift		PVSP	ΔV
Wattmetric Earth Fault	64W	PSDE	PN>

Appendix C
Typical Standards Applicable to Protection and Control Numerical Devices

There are many different institutions, both international and national, which issue standards used by manufacturers of protection and control devices. The standards shown here can be supplemented by the electrical power system operators' own particular standards. The most used standards bodies in the field of power systems engineering are:

- EU: European Union directives
- IEC: International Electrotechnical Commission
- IEEE: Institute of Electrical and Electronic Engineers

The list of standards shown here is not exhaustive, but serves to demonstrate the diverse nature of this industry, and its associated compliancy requirements.

C.1 DIELECTRIC, IMPULSE AND INSULATION

Standard	Typical Details
Impulse, EN 60255-27:2005	5kV 1.2/50μs impulse, common and differential mode – CT, VT, opto inputs, relays, power supply, IRIG-B & terminal block communications connections.
Dielectric withstand, EN 60255-27:2005	2kV rms. for 1 minute between all terminals connected together and case earth. 2kV rms. for 1 minute between all terminals of independent circuits with terminals on each independent circuit connected together. 1kV rms. for 1 minute across watchdog contacts.
ANSI dielectric withstand, ANSI/IEEE C37.90. (1989) (Reaffirmed 1994)	1kV rms. for 1 minute across open contacts of the watchdog contacts. 1kV rms. for 1 minute across open contacts of changeover output contacts. 1.5kV rms. for 1 minute across normally open output contacts.
Insulation resistance, EN 60255-27:2005	100 MΩ minimum.

C.2 DC AUXILIARY SUPPLY TESTS

Standard	Typical Details
DC voltage interruptions, IEC 60255-11:2008	50ms with no loss of protection. 100ms, 200ms, 300ms, 0.5s, 1s, 5s with temporary loss of protection.
DC voltage dip, IEC 60255-11:2008	40% @ 200ms, 70% @ 500ms with no loss of protection.
Alternating component (ripple) in DC supply voltage, IEC 60255-11:2008	AC 100Hz ripple superimposed on DC max. and min. auxiliary supply at 12% of highest rated DC.
Gradual shut down / start up, IEC 60255-11:2008	
Reverse polarity, IEC 60255-11:2008	

C.3 AC VOLTAGE DIPS & SHORT INTERRUPTIONS

Standard	Typical Details
AC voltage interruptions, EN 61000-4-11:2004 and EN60255-11 : 2008	10ms, 20ms with no loss of protection. 50ms, 100ms, 200ms, 500ms, 5s with temporary loss of protection.
AC voltage dip, EN 61000-4-11:2004 and EN60255-11 : 2008	60% @ 200ms, 30% @ 500ms, 20% @ 5s with no loss of protection.

C.4 HIGH FREQUENCY DISTURBANCE TESTS

Standard	Typical Details
1MHz burst Immunity test, EN 60255-22-1:2008 Class III	2.5kV common mode between independent circuits and between independent circuits and case earth. 1kV differential mode across terminals of the same circuit (except metallic contacts).

C.5 FAST TRANSIENT

Standard	Typical Details
Fast Transient, EN 61000-4-4:2004 Level 4	4kV 5kHz, power supply and earth ports. 2kV 5kHz, all other circuits (excluding power supply and earth ports).
Fast Transient, EN 60255-22-4:2008 Class A.	4kV 5kHz and 100kHz, applied to all circuits excluding communication ports. 2kV 5kHz and 100kHz, applied to communication ports.

C.6 CONDUCTED EMISSIONS

Standard	Typical Details
Power supply, EN55022: 2006 and EN60255-25: 2000	0.15 - 0.5MHz, 79dBμV (quasi peak) 66dBμV (average). 0.5 - 30MHz, 73dBμV (quasi peak) 60dBμV (average).
Permanently connected communications ports, EN55022: 2006 and EN60255-25: 2000	0.15 - 0.5MHz, 97dBμV (quasi peak) 84dBμV (average). 0.5 - 30MHz, 87dBμV (quasi peak) 74dBμV (average).

C.7 RADIATED EMISSIONS

Standard	Typical Details
Radiated emissions, EN55022:2006 (EN60255-25:2000)	30 - 230MHz, 40dBμV/m at 10m measurement distance. 230 - 1000MHz, 47dBμV/m at 10m measurement distance.

C.8 CONDUCTED IMMUNITY

Standard	Typical Details
Conducted immunity, EN 61000-4-6: 2009 Level 3 (EN60255-22-6: 2001)	10V emf @ 1kHz 80% am, 150kHz to 80MHz. Spot tests at 27MHz, 68MHz.

C.9 RADIATED IMMUNITY

Standard	Typical Details
Radiated immunity, EN 61000-4-3: 2006 Level 3	10 V/m 80MHz - 1GHz @ 1kHz 80% am.
Radiated immunity from digital radio telephones, EN 61000-4-3: 2006 Level 4	30 V/m 800MHz – 960MHz and 1.4GHz – 2GHz @ 1kHz 80% am.
Radiated immunity, EN 60255-22-3: 2007 Class III)	10 V/m 80MHz - 1GHz and 1.4GHz – 2.7GHz @ 1kHz 80% am. Spot tests at 80MHz, 160MHz, 380MHz, 450MHz, 900MHz, 1850MHz and 2150MHz.

C.10 ANSI/IEEE RADIATED IMMUNITY

Standard	Typical Details
ANSI Radiated Immunity/IEEE C37.90.2 2004	35 V/m 80MHz - 1GHz, @ 1kHz 80% am. 35 V/m 80MHz - 1GHz, 100% pulse modulated. Spot tests at 80MHz, 160MHz, 450MHz, 900MHz and 900MHz (200Hz rep. freq., pulse modulated).

C.11 ELECTROSTATIC DISCHARGE

Feature	Details
ESD, EN61000-4-2:2009 Level 3 and Level 4 (EN60255-22-2: 2008)	Level 4: 15kV air discharge. Tests carried out both with and without cover fitted. Level 4: 8kV contact discharge. Tests carried out both with and without cover fitted. Level 4: 15kV indirect discharge. Tests carried out both with and without cover fitted.

C.12 IEC SURGE IMMUNITY

Feature	Details
Power supply, EN61000-4-5:2006 Level 4 (EN60255-22-5:2002)	4kV common mode 12Ω source impedance, 2kV differential mode 2Ω source impedance, level 4.
CT and VT inputs, opto inputs, output relays, EN61000-4-5:2006 Level 4 (EN60255-22-5:2002)	4kV common mode 42Ω source impedance, 2kV differential mode 42Ω source impedance, Level 4.
Cable screen, EN61000-4-5:2006 Level 4 (EN60255-22-5:2002)	4kV common mode 2Ω source impedance

C.13 POWER FREQUENCY MAGNETIC FIELD IMMUNITY

Standard	Typical Details
Power Frequency Magnetic Field Immunity, EN61000-4-8:1993+A1:2001 Level 5.	100A/m field applied continuously in all planes for the EUT in a quiescent and tripping state (EUT – 'Equipment Under Test'). 1000A/m field applied for 3s in all planes for the EUT in a quiescent and tripping state.

C.14 PULSED MAGNETIC FIELD IMMUNITY

Standard	Typical Details
Pulse Magnetic Field Immunity, EN 61000-4-9:1993+A1:2001 Level 5	1000A/m field applied in all planes for the EUT in a quiescent state

C.15 DAMPED OSCILLATORY MAGNETIC FIELD IMMUNITY

Standard	Typical Details
Damped oscillatory Magnetic Field Immunity, EN61000-4-10:1993+A1:2001 Level 5	100A/m field applied in all planes at 100kHz / 1MHz with burst duration of 2 seconds.

C.16 DAMPED OSCILLATORY TESTS

Standard	Typical Details
Damped oscillatory tests, 100 kHz and 1 MHz, EN 61000-4-18: 2007	2.5kV common mode. Power supply, relay contacts, CT, VT, opto input, communications, IRIG-B. 1kV differential mode. Power supply, relay contacts, CT, VT, opto input.
Damped oscillatory tests, 3MHz, 10MHz, 30MHz, EN 61000-4-18: 2007	Power supply, relay contacts, CT, VT, opto input, communications, IRIG-B

C.17 POWER FREQUENCY TEST

Standard	Typical Details
Power frequency test, EN 60255-22-7: 2003.	300V rms. common mode. Voltage applied to all non-mains frequency inputs. 150V rms. differential mode. Voltage applied to all non-mains frequency inputs.

C.18 ANSI/IEEE SURGE WITHSTAND CAPABILITY

Standard	Typical Details
Oscillatory SWC Test, ANSI/IEEE C37.90.1 2002	2.5kV, 1MHz - common and differential mode – applied to all circuits except for terminal block communications, which are tested common mode only.
Fast Transient SWC Tests, ANSI/IEEE C37.90.1 2002	4kV crest voltage - common and differential mode - applied to all circuits except for terminal block communications, which are tested common mode only.

C.19 OPERATING AND STORAGE TEMPERATURE

C.19.1 Cold Tests

	Typical Details			
Standard	Temp. °C	Type of Test	Duration hr	EUT energised
Start – up, EN 60068-2-1:2007	+20	Functional verification tests	2	Yes
Intermediate, EN 60068-2-1:2007	0	Functional verification tests	2	Yes
Intermediate, EN 60068-2-1:2007	-10	Functional verification tests	2	Yes
Low temp claim, EN 60068-2-1:2007	-25	Functional verification tests	96	Yes
Low temp claim, EN 60068-2-1:2007	-25	Storage	96	No
Cold start, EN 60068-2-1:2007	-25 (typical)	Operate	2	Yes
Last test, EN 60068-2-1:2007	+20	Functional verification tests	2	Yes

C.19.2 Dry Heat Tests

	Typical Details			
Standard	Temp. °C	Type of Test	Duration hr	EUT energised
Start – up, EN 60068-2-2:2007	+20	Functional verification tests	2	Yes
Intermediate, EN 60068-2-2:2007	+40	Functional verification tests	2	Yes
High temp claim, EN 60068-2-2:2007	+70 (typical)	Operate	96	Yes
High temp claim, EN 60068-2-2:2007	+70 (typical)	Storage	96	No
Hot Start, EN 60068-2-2:2007	+55	Operate	2	Yes

C.19.3 Change of Temperature Tests

Standard	Typical Details				
	Ambient Temperature	Lower Temperature	High Temperature	Rate of change of temperature	Exposure Time
Change of Temperature, EN 60068-2-14:2000	+20 °C	-25 °C	+55 °C	1°C/min	3 hours

C.20 SIX-DAY CYCLIC TEMPERATURE WITH HUMIDITY

Standard	Typical Details
Damp heat cyclic, EN 60068-2-30: 2005	six (12 + 12 hour cycles) of 55°C ±2°C 93% ±3% RH and 25°C ±3°C 93% ±3% RH. (RH – 'Relative Humidity')

C.21 56-DAY HUMIDITY (IEC 60068-2-3)

Standard	Typical Details
Damp heat, steady state, EN 60068-2-78: 2002	40° C ± 2° C and 93% relative humidity (RH) +2% -3%, duration 56 days.

C.22 MECHANICAL TESTS

Standard	Typical Details
Vibration test per IEC 60255-21-1:1996	Response: class 2, Endurance: class 2
Shock and bump immunity per IEC 60255-21-2:1995	Shock response: class 2, Shock withstand: class 1, Bump withstand: class 2
Seismic test per IEC 60255-21-3: 1995	Class 2

C.23 ENCLOSURE PROTECTION TESTS

C.23.1 Water, Dust and Foreign Object Ingress

Standard	Typical Details
IP52, EN 60529:1989	Front face Protected against vertically falling drops of water with the product in 4 fixed positions of 15° tilt with a flow rate of 3mm/minute for 2.5 minutes. Protected against dust, limited ingress permitted.
IP50, EN 60529:1989	Sides of case Protected against solid foreign objects of 25mm diameter and greater.
IP10, EN 60529:1989	Connectors of case Protected against solid foreign objects of 50mm diameter and greater

C.23.2 Corrosive Environment

Standard	Typical Details
IEC60068-2-60: 1995 part 2, test Ke, method class 3	Industrial environment / poor environment control / mixed gas flow test. 21 days exposure to elevated concentrations of H_2S, NO_2 and SO_2 at 75% relative humidity and +30°C

C.23.3 Creepage and Clearances

Standard	Typical Details
IEC60255-27:2005	Pollution Degree: 3 Overvoltage Category: III Impulse Test Voltage: 5kV

C.24 EUROPEAN UNION DIRECTIVES

Standard	Typical Details
EMC Compliance Per 89/336/EEC:	Compliance to the European Commission Directive on EMC. EN50263: 2000
Product Safety Per 73/23/EEC:	Compliance with European Commission Low Voltage Directive. Compliance is demonstrated by reference to generic safety standards: EN61010-1: 2001 EN60950-1: 2002
2002/95/EC (Restriction on Use of Hazardous Substances Directive – 'RoHS')	All products to be free of lead, mercury, cadmium, hexavent chromium, polybrominated biphenyls (PBB) and polybrominated diphenyl ether (PBDE).

Appendix D
Company Data and Nomenclature

D.1 ALSTOM GROUP

Alstom is a global leader in major infrastructure projects. It has three main activities, championed by sectors which are expert in their respective domains. Alstom draws on its cutting edge technologies to ensure performance, environmental preservation and competitiveness.

D.1.1 Alstom Power

Alstom Power is the Number 1 in turnkey power plants, power plant service and air quality control systems. The portfolio includes coal, gas, oil, nuclear, hydro and wind, as well as CO_2 capture technologies.

D.1.2 Alstom Transport

Alstom Transport is a world-leader in rail equipment and service provision. It specialises in high-speed and very-high-speed trains, and also urban transportation.

D.1.3 Alstom Grid

The third sector, Alstom Grid, is the newest to be born in Alstom. Although a new name, Alstom Grid is more than that. The name embodies a strong heritage in electrical transmission, from pioneering the earliest AC and DC power grids to developing the new technologies of today's and tomorrow's 'smarter' grids. Today, Alstom carries with it the pioneering spirit, technologies and expertise of all of its ancestors, GEC, Delle, English Electric, Sprecher & Shuh, AREVA, Cegelec, ESCA, FIR – to name just a few.

As part of Alstom, "we are shaping the future": delivering customer-valued solutions to build smarter, more stable, more efficient and environmentally-friendly electrical grids worldwide.

www.alstom.com/grid

D.2 ALSTOM GRID'S AUTOMATION ACTIVITIES

Alstom Grid provides utilities and operators with the key software, hardware and expertise required to keep energy networks operating at maximum efficiency. Mission-critical solutions are engineered to protect, control & manage electrical grids and systems.

The portfolio includes:

- Smarter Network Management Solutions
- Substation Automation Solutions
- Generation and load management
- Grid stability solutions
- Protection and control

The strong leadership and customer support ethos from our 2500 employees continually builds upon our strong brands such as MiCOM, e-terra, MODULEX and iSTAT – to name just a few. In rounded terms, such solutions contribute to an installed base of 200 000 numerical relays, 1 000 digital control systems, and over a million electromechanical relays.

D.3 TERMINOLOGY REFERENCE – PROTECTION RELAYS

Many readers of this book may find it convenient to have a guide to the catalogue of Alstom and its predecessor's products, for cross-referencing. Further details for current products such as MiCOM can be found on the website:

www.alstom.com/protectionrelays

Details for all products, including the legacy installed base, and the access for general support can be found using the link at:

www.alstom.com/contactcentre

The Alstom Grid Contact Centre is a unique facility - an access point for the wealth of technical expertise in our Centres of Excellence, in addition to our local Application specialists.

D.3.1 MiCOM

MiCOM protection relays have a model numbering system which prefixes the device type with a letter 'P'. In the same way, a control device would begin with 'C', a measuring device 'M', a transducer instrument 'I', and a software product 'S'.

Within the MiCOM protection family, the letter P is followed by a numerical digit which is used to define the application of an IED:

- P1** Overcurrent and feeder management devices
- P2** Motor management and protection
- P3** Generator protection, and PCC mains decoupling
- P4** Distance protection
- P5** Line unit protection, including current differential and phase comparison
- P6** Transformer protection
- P7** Busbar protection
- P8** Breaker fail, reclosing, system integrity protection or phasor measurement
- P9** Voltage and frequency

** is a wildcard to denote the remaining digits, used to define the platform upon which a device is built, and the unique model within the application.

The most common and powerful platform in a utility environment is the MiCOM P*40 series, with an installed base of over 100 000 IEDs.

D.3.2 Non-MiCOM Relays

D.3.2.1 MIDOS M & MIDOS K

MIDOS is the family name for the Modular Integrated Draw-Out System for protective relays from Alstom Grid. Launched in the 1980's, this family offers the convenience of withdrawability, and standard 4U case sizing for ease of scheme integration. The signature products in this family were the MCGG22-MCGG82 overcurrent relays, launched in 1982, which became the world's first volume-seller microprocessor digital relays. All such products have a type name beginning with 'M'.

Specialist products for traditional applications such as pilot-wire differential (MBCI) and buswire supervision of high-impedance schemes (MVTP) continue, however sales of MIDOS relays today are more prevalent in the hinged armature market. Tripping and auxiliary relays are still extremely useful to perform contact multiplication, and to create a visible, mechanical, lockout.

The type designations of the most common hinged armature relays are:

- MCAA Current operated auxiliary relays, typically for series flagging
- MCAG Current-calibrated high impedance busbar or REF protection
- MFAC Voltage-calibrated high impedance busbar or REF protection
- MVAA Voltage operated auxiliary relays
- MVAJ Tripping relays
- MVAW Interposing relays
- MVAX Discrete trip circuit supervision relays

In addition to the MIDOS M-series of relays, the MIDOS K-series was launched in 1992, bringing numerical applications into a compact platform. Unique applications on the platform still include the KVGC202 for voltage regulation (transformer tap change control).

Associated with all 4U modular products, the MIDOS MMLG test block can be used as an access point for test switching and periodic secondary injection.

D.3.2.2 Modular Relays

Modular Relays were first manufactured in the late 1970s and marked the appearance of analogue to digital integrated circuits and customer-designed application schemes in their software. Examples are:

Full scheme static distance relays such as the Micromho SHNB, followed by the Quadramho SHPM101 in the 1980's.

Modular relays housed in an 80TE – 19" rack, based on a common hardware architecture, for generator and transmission applications such as the LGPG, Optimho LFZP and the LFCB digital / numerical current differential which became available from the mid 1980s onwards.

D.4 NOMENCLATURE – PROTECTION RELAYS

Many of the products from Alstom Grid and it's predecessors were described by a three letter, or four letter system. Those described with three letters (for example the CDG induction disc overcurrent relays, still in manufacture today) have no leading letter to denote the family of the product, meaning that the device form factor is not the 4U MIDOS standard.

For non-MiCOM products, the following may act as a guide to the nomenclature:

D.4.1 1st Technology Digit

Where the device type is four alpha (four letters), the first letter denotes the platform of the device technology, for example:

- K: K-Series MIDOS numerical relays (1992+)
- L: Modular relays occupying a full 19 inch (80TE) rack, typically for transmission applications (eg. LFZP Optimho, LFCB digital/numerical current differential - 1986)
- M: MIDOS series
- S: Static protection (for example the subcycle Micromho from 1980)

D.4.2 2nd Application Digit

In the three alpha system, the first letter denotes the device primary application, often the energising quantity of the device. The application letter is the 2nd letter under the four-alpha designation, examples include:

- A: Auxiliary functions
- B: Biased differential (eg. MBCI, KBCH)
- C: Current-operated
- E: Directional
- F: Frequency, or transmission Feeder (eg. LFZP)
- G: Generator (eg. LGPG)
- H: Pilot-wire differential (eg. MHOR)
- I: Communications Interface (eg. KITZ)
- V: Voltage-operated (eg. KVGC)
- W: Power measuring

D.4.3 3rd Application Digit

The following digit is used to define the application in more detail, examples include:

- A: Attracted-armature electromechanical, or auxiliary / re-closing
- C: Current-operated
- D: Directional-comparison transmission scheme (LFDC)
- G: General (eg. the KCGG is a general current-operated feeder relay)
- N: Negative phase sequence
- V: Voltage-operated
- T: Time-delayed
- Z: Distance impedance relay

D.4.4 4th Application Digit

The last letter is used to uniquely define the family, examples include:

- B: Biased differential
- D: Definite-time
- R: With recording, or reclosing
- S: Check synchroniser
- H: Harmonic restraint
- I: Instantaneous
- U: Inverse-time
- X: Trip circuit supervision

Laboratory validation of a numerical IED

Dynamic simulator scheme approval for a numerical IED

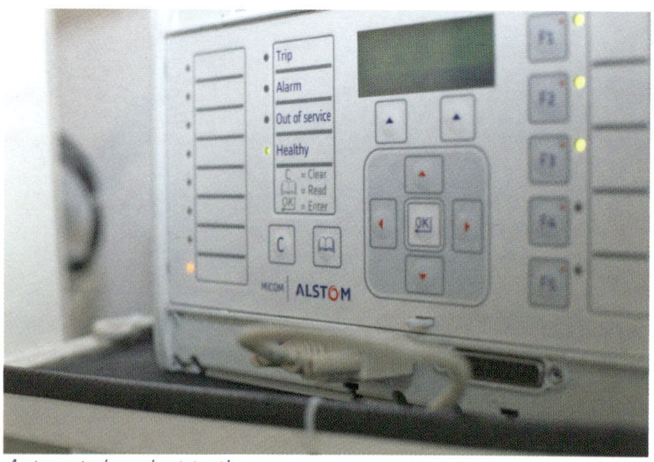

Automated product testing

Engineering of a turnkey scheme

Factory acceptance testing

On-site post delivery services

we are shaping the future | ALSTOM

INDEX

E

H

N

S

T

DISCLAIMER OF WARRANTIES AND LIMITATION OF LIABILITIES